Cylindrical Antennas and Arrays

Cylindrical antennas are the radiating elements in most major radio communication systems, including broadcasting networks, cellular telephone systems and radar. In this book, the authors present practical theoretical methods for determining current distributions, input admittances and impedances, and field patterns of a wide variety of cylindrical antennas and arrays, including the isolated antenna, the two-element array, the circular array, curtain arrays, Yagi and log-periodic arrays, planar arrays and three-dimensional arrays. Coverage also includes the analysis of horizontal and vertical antennas over, on and in the earth and sea, large resonant arrays of electrically short dipoles and a chapter on the theory and techniques of experimental measurement. Written by three of the leading engineers in the field, and based on world-class research carried out at Harvard over the last 40 years, *Cylindrical Antennas and Arrays* is destined to become the basic reference for practicing engineers and advanced students for many years to come.

Ronold W. P. King is Gordon McKay Professor of Applied Physics, Emeritus at Harvard University. Since joining the Harvard faculty in 1938, he has supervised the research of over 100 Ph.D. students. He is a Life Fellow of the IEEE; a Fellow of the American Physical Society; and a member of the Bavarian Academy of Sciences and the American Academy of Arts and Sciences. He holds the Distinguished Service Award from the University of Wisconsin (1973), the Centennial Medal of the IEEE (1985), the Harold Pender Award from the University of Pennsylvania (1986), the Distinguished Achievement Award of the IEEE Antennas & Propagation Society (1991), the IEEE Graduate Teaching Award (1997), and the Chen-To Tai Distinguished Educator Award from the IEEE Antennas & Propagation Society (2001).

George J. Fikioris is a lecturer at the National Technical University of Athens (NTUA) in Greece. After obtaining his Ph.D. at Harvard University in 1993 he worked for six years as an electronics engineer at the Air Force Research Laboratory, Hanscom AFB, USA before joining the NTUA in 1999.

Richard B. Mack is a consultant with his own company, Mack Consulting, specializing in electro-magnetic measurement techniques.

Cylindrical Antennas and Arrays

Revised and enlarged 2nd edition of
'Arrays of Cylindrical Dipoles'

Ronold W. P. King

Gordon McKay Professor of Applied Physics, Emeritus, Harvard University

George J. Fikioris

Lecturer, National Technical University of Athens, Greece

Richard B. Mack

Consultant in EM Measurement Techniques, Mack Consulting

CAMBRIDGE
UNIVERSITY PRESS

CAMBRIDGE UNIVERSITY PRESS
Cambridge, New York, Melbourne, Madrid, Cape Town, Singapore, São Paulo

Cambridge University Press
The Edinburgh Building, Cambridge CB2 2RU, UK

Published in the United States of America by Cambridge University Press, New York

www.cambridge.org
Information on this title: www.cambridge.org/9780521431071

First published 2002
This digitally printed first paperback version 2005

A catalogue record for this publication is available from the British Library

Library of Congress Cataloguing in Publication data

King, Ronold Wyeth Percival, 1905–
Cylindrical antennas and arrays / Ronold W. P. King, George J. Fikioris, Richard B. Mack.
    p.   cm.
Includes bibliographical references and index.
Previously published as: Arrays of cylindrical dipoles. London : Cambridge U.P., 1968.
ISBN 0 521 43107 7
1. Cylindrical antennas. 2. Antenna arrays. 3. Antennas, Dipole.
I. Fikioris, George J., 1962– II. Mack, Richard Bruce. III. King, Ronold Wyeth Percival, 1905–
Arrays of cylindrical dipoles. IV. Title.
TK7871.67.C95 K56 2001
621.384′135–dc21   2001025623

ISBN-13 978-0-521-43107-1 hardback
ISBN-10 0-521-43107-7 hardback

ISBN-13 978-0-521-01786-2 paperback
ISBN-10 0-521-01786-6 paperback

Good reasons must, of force, give place to better.
SHAKESPEARE

Contents

7   Planar and three-dimensional arrays   241

8   Vertical dipoles on and over the earth or sea   290

13   Direct numerical methods: a detailed discussion   452

14   Techniques and theory of measurements   475

Preface

Over three decades have passed since the publication, in 1968, of "Arrays of Cylindrical Dipoles" by R. W. P. King, R. B. Mack, and S. S. Sandler. The present volume is a revised and enlarged second edition of that work. The objectives of "Cylindrical Antennas and Arrays" are similar to those of "Arrays of Cylindrical Dipoles": to present approximate but efficient theoretical methods for determining current distributions, input admittances, and field patterns of arrays of cylindrical dipoles; to use such methods to analyze particular types of arrays; to describe experimental methods for determining current distributions, input admittances, and field patterns; and to correlate and compare theoretical and experimental results.

The most fundamental quantities, and the ones most difficult to determine theoretically, are the current distributions on the array elements. Rather than postulating the current distributions, perhaps the most common treatment in the literature of the 1960s, "Arrays of Cylindrical Dipoles" sought to determine the current distributions on the array elements by solving integral equations. Today's antenna and engineering literature is quite different from that of the 1960s: even elementary textbooks include discussions on determining current distributions from integral or similar equations.

As an example, consider the simplest configuration discussed in this book, the single, isolated cylindrical antenna of given length and radius. The integral equation treated in Chapter 2 is but one of the several integral or integro-differential equations that are encountered in the present-day literature. Although such equations were derived many years ago, the reasons for their increased popularity are the easy accessibility to high-speed computers and the availability of a large number of numerical methods. As a result of these developments, application of general-purpose numerical methods to the aforementioned equations is today much more common than in the 1960s.

Rather than discretizing the integral equations, the "two-term" and "three-term" theories developed in this book treat them by analytical means. These theories, which apply to elements that are not too long, are here presented as powerful alternatives to applying general-purpose numerical methods. Because the final two- and three-term formulas are quite simple in form, they require less running time when programmed in a computer. In addition, the analytical methods present a physical basis for understanding changes in the characteristics of the antenna as the parameters are changed. Furthermore, applying numerical methods to the integral equations in this book presents difficulties that are often overlooked in the literature.

Chapter 1, which has been completely rewritten, has the same purpose as the first chapter of the 1968 edition: to introduce the reader to the theory of antennas and arrays presented in subsequent chapters by discussing some commonly used methods of studying a single antenna in free space. The chapter concludes with an introductory discussion of integral equations and the application of numerical methods, a subject discussed in greater depth in Chapter 13.

Chapters 2–5 develop the two- and three-term theories for the single isolated antenna, the two-element array, the circular array, and the curtain array, respectively. In Chapters 3–5, the array elements are assumed to be identical, parallel, and non-staggered. In Chapters 6 and 7, the two- and three-term theories are extended and applied to certain types of arrays that do not satisfy the aforementioned conditions. Apart from editing changes, Chapters 2–7 are the same as those in the first edition.

Chapters 8–13 have no counterparts in "Arrays of Cylindrical Dipoles". Chapters 8 and 9 analyze vertical and horizontal dipoles and arrays over and on the surface of the conducting and dielectric earth or sea. Included are asphalt-coated earth and ice-coated water. A major new addition is long-distance propagation over the spherical surface of the sea.

In Chapter 10, arrays of identical, parallel, non-staggered elements are discussed once more, from the point of view of computer implementation of the two-term formulas. Some restrictions placed on the arrays of Chapters 3–5 are removed in this chapter, which also serves as an introduction to the study of the large circular array in Chapters 11 and 12.

In Chapters 11 and 12, a novel type of circular array is studied. The arrays under consideration differ from conventional circular arrays in that only one or two of the many array elements are excited and the entire array is tuned to spatial resonance. Both the integral equations and the two-term theory must be modified and extended to deal with the phenomena studied in this chapter. The modified theory is used to show that such arrays possess new, unusual and potentially useful resonant and directive properties. The analytical nature of the underlying theory is an important advantage for this study.

Although the two- and three-term theories are analytical in nature, they are not claimed to be mathematically rigorous. (Nonetheless, very good agreement between theory and experiment is obtained in Chapters 1–12.) Until shown otherwise [1], the lack of rigor is a necessity rather than a convenience: most of the integral equations dealt with in this book have no exact solutions, even in principle. From a theoretical point of view, then, one cannot do better than to find "reasonable solutions" that satisfy the integral equations approximately, and the aforementioned lack of rigor must be true for any method used to "solve" the integral equations. The main purpose of Chapter 13 is to discuss the consequences of this rather peculiar situation when general-purpose numerical methods are applied to the integral equations in question. The case of an isolated antenna is discussed in detail, and extensions to arrays are pointed out. The special numerical difficulties associated with the circular arrays of the previous two chapters are also discussed here.

The concluding Chapter 14 discusses experimental methods, with emphasis on the measurement of antenna impedance. This chapter corresponds to Chapter 8 of the first edition; it has been significantly revised to discuss modern measurement techniques.

R. B. Mack wrote Chapter 14, G. Fikioris wrote Chapters 1 and 10–13; R. W. P. King wrote Chapters 8 and 9, and organized the present edition as a whole. In addition to the contributions of the several individuals named in the Preface to the first edition, the authors gratefully acknowledge the contribution of Chapter 5 by Sheldon S. Sandler. Without the extensive and very thorough work of Margaret Owens, both in preparing and correcting the new manuscripts and in editing Chapters 2–7, the present arrangement would not be possible. Finally, we thank Tai Tsun Wu for providing the initial and most fundamental ideas for the work in Chapters 11–13, and for guiding and inspiring the ensuing researches.

R.W.P.K.
G.J.F.
R.B.M.

Preface to first edition

Studies of coupled antennas in arrays may be separated into two groups: those which postulate a single convenient distribution of current along all structurally identical elements regardless of their relative locations in the array and those which seek to determine the actual currents in the several elements. Virtually all of the early and most of the more recent analyses are in the first group in which both field patterns and impedances have been obtained for elements with assumed currents. Pioneer work in the determination of field patterns of arrays of elements with sinusoidally distributed currents was carried out for uniform arrays by Bontsch-Bruewitsch [1] in 1926, by Southworth [2] in 1930, by Sterba [3] and by Carter *et al.* [4] in 1931. Early studies of non-uniform arrays are by Schelkunoff [5] in 1943, by Dolph [6] in 1946, and by Taylor and Whinnery [7] in 1951. The self- and mutual impedances of arrays of elements with sinusoidally distributed currents were studied especially by Carter [8] in 1932, by Brown [9] in 1937, by Walkinshaw [10] in 1946, by Cox [11] in 1947, by Barzilai [12] in 1948, and by Starnecki and Fitch [13] in 1948. A thorough presentation of the basic theory of antennas with sinusoidal currents was given by Brückmann [14] in 1939. Actually, the current in any cylindrical antenna of length $2h$ and finite radius a is accurately sinusoidal only when it is driven by a continuous distribution of electromotive forces of proper amplitude and phase along its entire length. It is approximately sinusoidal in an isolated very thin antenna ($a \ll h$) driven by a single lumped generator primarily when the antenna is near resonance. When antennas are coupled in an array with each driven by a single generator or excited parasitically, it is generally assumed that (1) the phase of the current along each element is the same as at the driving point and (2) the amplitude is distributed sinusoidally. Both of these assumptions are reasonably well satisfied only for very thin antennas ($a \ll \lambda$) that are not too long ($h \leq \lambda/4$). Nevertheless, a very extensive theory of arrays has been developed based implicitly on one or both of these assumptions. Evidently it is correspondingly restricted in its generality.

The analysis of coupled antennas from the point of view of determining the actual distributions of current was studied for two antennas by Tai [15] in 1948 and extended to the N-element circular array by King [16] in 1950. A general analysis of arrays of coupled antennas has been given by King [17]. Unfortunately, the rigorous solution of the simultaneous integral equations for the distributions of current in the elements of an array of parallel elements is very complicated and no simple and practically useful set of formulas was obtained. As a consequence, the extensive study of the electromagnetic fields of antennas and arrays in this earlier work (Chapters 5 and 6 in King [17]) was limited to arrays with currents in the elements that satisfied the assumptions of constant phase angle and sinusoidal amplitude. Similar restrictions are implicit in the fields calculated, for example, by Aharoni [18], Stratton [19], Hansen [20] and many others.

A practical method for obtaining solutions of the simultaneous integral equations for the distributions of current in the elements of a parallel array in a form that combines simplicity with quantitative accuracy was proposed by King [21] in 1959. In this analysis an approximate procedure was developed which provided simple, two-term trigonometric formulas for the currents in all of the arbitrarily driven or parasitic elements in a circular array of N elements in a manner that took full account of the effects of mutual interactions on the distributions of current. These formulas applied to elements up to one and one-quarter wavelengths long. The application of this new procedure to actual arrays and the experimental verification of the results were carried out in an extensive series of investigations by Mack [22]. The generalization of the method to curtain arrays was developed by King and Sandler [23, 24] in 1963 and 1964. The extension of the method to parasitic elements in arrays of the Yagi type was verified experimentally by Mailloux [25] in 1966. A modification of the theory and its application to the optimization of Yagi arrays by the use of a high-speed computer were devised by Morris [26] in 1965. In 1967 Cheong [27] extended the theory to unequal and unequally spaced elements. (The several researches were supported in part by Joint Services Contract Nonr 1866(32), Air Force Contract AF19(604)-4118 and National Science Foundation Grants NSF-GP-851 and GK-273.)

A further improvement in the simplified trigonometric representation of the current in an isolated antenna was introduced by King and Wu [28] in 1965 and extended to arrays in the present work.

This book begins with an introductory chapter that reviews the foundations and limitations of conventional antenna theory. It then proceeds to derive the new two- and three-term formulas for the isolated antenna in Chapter 2 and for two coupled antennas in Chapter 3. Chapter 4 provides the complete formulation of the new theory for the N-element circular array; Chapter 5 for the N-element curtain array of identical elements. The more difficult problem of treating elements of different lengths—notably in the Yagi array and the log-periodic antenna—is treated in Chapter 6. Chapter 7 is devoted to planar and three-dimensional arrays that include staggered and collinear elements. Chapter 8[1] is concerned with the broad problems of measurement—currents, impedances, field patterns and the correlation of theory with experiment. In the appendices summaries of programs are given for the computational analysis of circular, curtain, and Yagi arrays.[2]

---

[1] The original Chapter 8 corresponds to Chapter 14 of the present (2nd) edition.
[2] Much of this material is omitted from the appendices in the 2nd edition.

In the preparation of the manuscript, S. S. Sandler was responsible for Chapters 1 and 5, R. B. Mack for Chapters 4 and 8, and R. W. P. King for Chapters 2, 3, 6, and 7 and for the co-ordination of the several parts.

The authors are happy to acknowledge the important contributions of Drs Robert J. Mailloux, I. Larry Morris, and W.-M. Cheong whose researches form the basis of Chapter 6; and of V. W. H. Chang whose work underlies Chapter 7. They are grateful to Professor Tai Tsun Wu for many valuable suggestions and to Mrs Dilla G. Tingley for continuing painstaking assistance with the preparation of the manuscript, the graphical representation, the computations, and the programs. Mrs Barbara Sandler, Mrs Evelyn Mack, and Mr Chang also assisted with the programs, Mrs S. R. Seshadri with the preparation of the manuscript. Miss Margaret Owens contributed greatly to the accuracy of the presentation with her meticulous reading of the proofs. She also had a major share in the preparation of the index.

R.W.P.K.
R.B.M.
S.S.S.

Cambridge, Mass.
January 1967

1 Introduction

1.1 Linear antennas

Wireless communication depends upon the interaction of oscillating electric currents in specially designed, often widely separated configurations of conductors known as antennas. Those considered in this book consist of thin metal wires, rods or tubes arranged in arrays. Electric charges in the conductors of a transmitting array are maintained in systematic accelerated motion by suitable generators that are connected to one or more of the elements by transmission lines. These oscillating charges exert forces on other charges located in the distant conductors of a receiving array of elements of which at least one is connected by a transmission line to a receiver. Fundamental quantities which describe such interactions are the electromagnetic field, the driving-point admittance, and the driving-point impedance. These can be easily determined if the distributions of current on the array elements are known. The determination of the currents on the array elements is the main concern of this book. In this first chapter, the basic electromagnetic equations are formulated and applied to a single antenna in free space. The simplest approach of assuming the current rather than actually determining it is reviewed first. Then, integral equations for the current distributions are derived, and determining the current by numerical methods is discussed. These discussions serve as an introduction to the analytical theory of antennas and arrays based on the solution of integral equations that is presented in subsequent chapters.

Figures 1.1a and 1.1b show two simple practical radiating systems. In Fig. 1.1a, a section at the open end of a two-wire transmission line has been bent outward to form a dipole antenna. In Fig. 1.1b, the inner conductor of a coaxial transmission line is extended above a ground plane. In both cases, the transmission lines are connected to generators which oscillate at a frequency $f = \omega/2\pi$. In a small region (comparable in extent with the distance between the two conductors of the transmission line), the antenna and line are coupled. Owing to the complications involved in this coupling, it is convenient to replace the actual generator/transmission line with an idealized so-called *delta-function* generator, which maintains an impressed electric field $\mathbf{E}^e(z) = \hat{\mathbf{z}} E_z^e(z) = V\delta(z)\hat{\mathbf{z}}$ at the surface of the antenna. This is the linear antenna of Fig. 1.1c. The impressed field is non-zero only at the center $z = 0$ of the cylindrical surface. The delta-function generator is an independent voltage source in the sense of ordinary

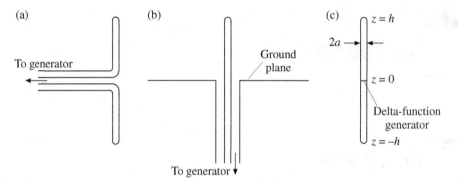

Figure 1.1 (a) Dipole antenna and two-wire transmission line. (b) Monopole antenna over a ground plane. (c) Simplified center-driven linear antenna.

circuit theory. The linear antenna of Fig. 1.1c can also serve as a model for other types of radiating systems. The simplifying assumption of studying the antenna in the absence of the connecting transmission line is particularly useful when the antenna is an array element.

The radius of the linear dipole antenna of Fig. 1.1c is a, and its half-length is h. It is assumed throughout this book that the radius is much smaller than both the wavelength λ and the length $2h$ of the antenna. Under such conditions, one can neglect the small currents on the capped ends of the antenna and assume that only a current $K_z(z) = I(z)/2\pi a$ is maintained on the cylindrical surface of the antenna. Other concepts of circuit theory can be introduced, and are particularly useful to the antenna engineer: the driving-point admittance Y_0 and driving-point impedance Z_0 are defined as

$$Y_0 = G_0 + jB_0 = \frac{I(0)}{V} = \frac{1}{Z_0}, \qquad Z_0 = R_0 + jX_0 = \frac{V}{I(0)} = \frac{1}{Y_0}. \tag{1.1}$$

G_0, B_0, R_0, and X_0 are respectively, the driving-point conductance, susceptance, resistance, and reactance. When h, a, and f are such that the antenna is at resonance, one has $X_0 = 0$ and $B_0 = 0$. As an example of the use of these quantities in a practical situation, consider the problem of designing the antenna so that, at a given frequency f, there is maximum power transfer from a transmission line of given characteristic impedance Z_c. With the assumption that the transmission line and the antenna can be studied separately, the problem is reduced to that of determining h and a so that Z_0 is equal to Z_c^*, the complex conjugate of Z_c.

The delta function $\delta(z)$ is zero except when $z = 0$. Additional, well-known properties of the delta function are

$$\delta(z) = \begin{cases} 0, & \text{if } z \neq 0 \\ \infty, & \text{if } z = 0 \end{cases}, \qquad \int_{-b}^{b} \delta(z)\,dz = 1 \tag{1.2a}$$

$$\delta(kz) = \frac{1}{|k|}\delta(z), \qquad f(z)\delta(z) = f(0)\delta(z) \tag{1.2b}$$

$$\int_{-b}^{b} f(z)\delta(z)\,dz = f(0) \tag{1.2c}$$

$$\frac{d}{dz}H(z) = \delta(z) \quad \text{where} \quad H(z) = \begin{cases} 1, & \text{if } z > 0 \\ 0, & \text{if } z < 0. \end{cases} \tag{1.2d}$$

In (1.2), b is any positive constant, k is any real constant, $f(z)$ is any smooth function of z, and $H(z)$ is the step function.

The next section introduces the fundamental equations of electromagnetic theory that are useful in the antenna problems considered in this book. More details can be found in [1], and in more concise form in [2, Chapter 1].

1.2   Maxwell's equations and the potential functions

The interaction of charges and currents is governed by Maxwell's equations which define the electromagnetic field. With an assumed time dependence $e^{j\omega t}$, they are

$$\nabla \times \mathbf{B} = \mu_0(\mathbf{J} + j\omega\epsilon_0\mathbf{E}), \qquad \nabla \cdot \mathbf{B} = 0 \tag{1.3a}$$

$$\nabla \times \mathbf{E} = -j\omega\mathbf{B}, \qquad \nabla \cdot \mathbf{E} = \rho/\epsilon_0, \tag{1.3b}$$

where the electric vector $\mathbf{E}$ is in volts per meter (V/m), the magnetic vector $\mathbf{B}$ in tesla (T). SI units are used throughout this book. The volume density of current $\mathbf{J}$ in amperes per square meter (A/m^2) is the charge crossing unit area per second. The volume density of charge ρ is in coulombs per cubic meter (C/m^3). $\mathbf{J}$ and ρ satisfy the equation of continuity,

$$\nabla \cdot \mathbf{J} + j\omega\rho = 0. \tag{1.3c}$$

In the interior of perfect conductors, $\mathbf{J} = 0$ and $\rho = 0$. In (1.3), ϵ_0 and μ_0 are the absolute permittivity and permeability of free space. They have the numerical values $\epsilon_0 = 8.854 \times 10^{-12}$ farads per meter (F/m) and $\mu_0 = 4\pi \times 10^{-7}$ henrys per meter (H/m), and are related to the velocity c of light and the characteristic impedance ζ_0 of free space by

$$c = \frac{1}{\sqrt{\mu_0\epsilon_0}}, \qquad \zeta_0 = \sqrt{\frac{\mu_0}{\epsilon_0}}. \tag{1.4}$$

Transmission lines and antennas are made from highly conducting materials such as brass or copper. In most cases, it is an excellent approximation to assume that the conductors are perfect. The relevant boundary conditions at an interface between a perfect conductor and air are

$$\hat{\mathbf{n}} \times \mathbf{E} = 0, \qquad \hat{\mathbf{n}} \cdot \mathbf{E} = \eta/\epsilon_0 \tag{1.5a}$$

$$\hat{\mathbf{n}} \times \mathbf{B} = \mu_0 \mathbf{K}, \qquad \hat{\mathbf{n}} \cdot \mathbf{B} = 0. \tag{1.5b}$$

In (1.5), $\hat{\mathbf{n}}$ is the unit normal to the conductor–air interface. Its direction is outward from the conductor to the air. $\mathbf{K}$ is the surface density of current in amperes per meter (A/m) and η is the surface density of charge in coulombs per square meter (C/m^2) on the perfect conductor. The left-hand equation in (1.5a) states that the component of the electric field in air tangent to the surface of the perfect conductor must be zero. The left-hand equation in (1.5b) states that the tangential magnetic field in air is proportional to the surface density of current on the conductor.

It is convenient to introduce the scalar and vector potentials ϕ, $\mathbf{A}$. The defining relationships between the potentials and the electromagnetic-field vectors are obtained with the aid of Maxwell's equations. With the vector identity $\nabla \cdot (\nabla \times \mathbf{C}) = 0$ (where $\mathbf{C}$ is any vector) and the equation $\nabla \cdot \mathbf{B} = 0$, the magnetic field may be expressed in the form

$$\mathbf{B} = \nabla \times \mathbf{A}. \tag{1.6}$$

If (1.6) is substituted in (1.3b), it follows that

$$\nabla \times (\mathbf{E} + j\omega\mathbf{A}) = 0. \tag{1.7}$$

The identity $\nabla \times (\nabla \psi) = 0$, where ψ is a scalar function, then permits the definition of ϕ in the form

$$-\nabla\phi = \mathbf{E} + j\omega\mathbf{A}. \tag{1.8}$$

The substitution of (1.6) and (1.8) into the remaining Maxwell equations leads to coupled partial differential equations for $\mathbf{A}$ and ϕ. They can be decoupled if the following condition relating $\mathbf{A}$ and ϕ is imposed:

$$\nabla \cdot \mathbf{A} = -j\omega\mu_0\epsilon_0\phi \quad \text{or} \quad \nabla \cdot \mathbf{A} = -j\frac{\beta_0^2}{\omega}\phi, \tag{1.9}$$

where the free-space wave number β_0 (also denoted by k in this book) is given by

$$\beta_0 = \omega\sqrt{\mu_0\epsilon_0} = \frac{\omega}{c} = \frac{2\pi}{\lambda} \tag{1.10}$$

and λ is the free-space wavelength. Equation (1.9) is known as the Lorentz condition. The resulting equations for $\mathbf{A}$ and ϕ are

$$(\nabla^2 + \beta_0^2)\mathbf{A} = -\mu_0\mathbf{J}, \qquad (\nabla^2 + \beta_0^2)\phi = -\rho/\epsilon_0. \tag{1.11}$$

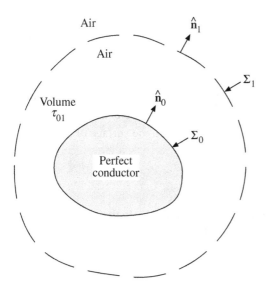

Figure 1.2 Perfect conductor in air.

The solutions to (1.11) can be derived with the use of the retarded Green's function. They are

$$\mathbf{A}(\mathbf{r}) = \frac{\mu_0}{4\pi} \int \mathbf{J}(\mathbf{r}') \frac{e^{-j\beta_0|\mathbf{r}-\mathbf{r}'|}}{|\mathbf{r}-\mathbf{r}'|} dV' \tag{1.12a}$$

and

$$\phi(\mathbf{r}) = \frac{1}{4\pi\epsilon_0} \int \rho(\mathbf{r}') \frac{e^{-j\beta_0|\mathbf{r}-\mathbf{r}'|}}{|\mathbf{r}-\mathbf{r}'|} dV', \tag{1.12b}$$

where the volume integrations extend over the entire region occupied by currents or charges. In most cases considered in this book, the conductors are perfect so that only surface current densities $\mathbf{K}$ and surface charge densities η are present. In such cases, the volume integrals in (1.12) reduce to surface integrals. In the limit of infinitely thin wire antennas, the surface integrals in turn reduce to line integrals.

1.3   Power and the Poynting vector

The complex Poynting vector is defined as

$$\mathbf{S} = \frac{1}{2\mu_0} \mathbf{E} \times \mathbf{B}^*, \tag{1.13}$$

where the asterisk denotes the complex conjugate. The integral of the normal component of Re{$\mathbf{S}$} over a closed surface Σ is the time-average, total power transferred from within Σ. The time average is over a period $T = 2\pi/\omega$. Several useful identities involving the Poynting vector are now derived. The geometry of interest is shown in Fig. 1.2. A perfect conductor surrounded by air is shown. The conductor–air interface is the closed surface Σ_0, and $\hat{\mathbf{n}}_0$ is the unit outward normal. Assume that there is an impressed electric field $\mathbf{E}^e$ tangent to the surface of the conductor. As a result, a surface current density $\mathbf{K}$ exists on the conductor's surface. This, in turn, maintains an electromagnetic field $\mathbf{E}$ and $\mathbf{B}$ in the air. The total electric field on the conductor's surface is $\mathbf{E} + \mathbf{E}^e$, and the boundary conditions on the surface of the perfect conductor are

$$\hat{\mathbf{n}}_0 \times (\mathbf{E} + \mathbf{E}^e) = 0, \qquad \hat{\mathbf{n}}_0 \times \mathbf{B} = \mu_0 \mathbf{K}. \tag{1.14}$$

Suppose that Σ_1 is a closed (mathematical) surface in the air surrounding the perfect conductor, and that $\hat{\mathbf{n}}_1$ is the corresponding unit normal vector. Let τ_{01} be the volume lying between Σ_0 and Σ_1, and consider the quantity

$$\int_{\tau_{01}} \nabla \cdot \mathbf{S} \, dV. \tag{1.15}$$

First, with (1.13), the vector identity

$$\nabla \cdot (\mathbf{E} \times \mathbf{B}^*) = \mathbf{B}^* \cdot (\nabla \times \mathbf{E}) - \mathbf{E} \cdot (\nabla \times \mathbf{B}^*) \tag{1.16}$$

and the Maxwell equations on the left in (1.3a, b), it is seen that

$$\int_{\tau_{01}} \nabla \cdot \mathbf{S} \, dV = -j\omega \int_{\tau_{01}} \left(\tfrac{1}{2} \mu_0^{-1} |\mathbf{B}|^2 - \tfrac{1}{2} \epsilon_0 |\mathbf{E}|^2 \right) dV. \tag{1.17}$$

The boundaries of the volume τ_{01} are the surfaces Σ_0 and Σ_1. Application of the divergence theorem to the quantity in (1.15) yields

$$\int_{\tau_{01}} \nabla \cdot \mathbf{S} \, dV = -\int_{\Sigma_0} (\hat{\mathbf{n}}_0 \cdot \mathbf{S}) \, d\Sigma + \int_{\Sigma_1} (\hat{\mathbf{n}}_1 \cdot \mathbf{S}) \, d\Sigma. \tag{1.18}$$

A comparison of (1.17) and (1.18) yields the identity

$$\int_{\Sigma_1} (\hat{\mathbf{n}}_1 \cdot \mathbf{S}) \, d\Sigma = \int_{\Sigma_0} (\hat{\mathbf{n}}_0 \cdot \mathbf{S}) \, d\Sigma - j\omega \int_{\tau_{01}} \left(\tfrac{1}{2} \mu_0^{-1} |\mathbf{B}|^2 - \tfrac{1}{2} \epsilon_0 |\mathbf{E}|^2 \right) dV. \tag{1.19a}$$

If one takes the real part of this equation, no volume integral appears:

$$P \equiv \int_{\Sigma_0} (\hat{\mathbf{n}}_0 \cdot \text{Re}\{\mathbf{S}\}) \, d\Sigma = \int_{\Sigma_1} (\hat{\mathbf{n}}_1 \cdot \text{Re}\{\mathbf{S}\}) \, d\Sigma. \tag{1.19b}$$

Equation (1.19b) states that P, the total time-average power entering Σ_0, is the same as the total time-average power leaving Σ_1.

The next identity of interest is obtained by expressing $\int_{\Sigma_0} (\hat{\mathbf{n}}_0 \cdot \mathbf{S}) \, d\Sigma$ in (1.19a) in terms of $\mathbf{E}^e$ and $\mathbf{K}$. With (1.13), the vector identity $\hat{\mathbf{n}}_0 \cdot (\mathbf{E} \times \mathbf{B}^*) = -\mathbf{E} \cdot (\hat{\mathbf{n}}_0 \times \mathbf{B}^*)$, and the boundary conditions (1.14), it is seen that $\int_{\Sigma_0} (\hat{\mathbf{n}}_0 \cdot \mathbf{S}) \, d\Sigma = \int_{\Sigma_0} \frac{1}{2} \mathbf{E}^e \cdot \mathbf{K}^* \, d\Sigma$ so that (1.19a) can be written as

$$\int_{\Sigma_0} \frac{1}{2} \mathbf{E}^e \cdot \mathbf{K}^* \, d\Sigma = j\omega \int_{\tau_{01}} (\frac{1}{2} \mu_0^{-1} |\mathbf{B}|^2 - \frac{1}{2} \epsilon_0 |\mathbf{E}|^2) \, dV + \int_{\Sigma_1} (\hat{\mathbf{n}}_1 \cdot \mathbf{S}) \, d\Sigma. \quad (1.20a)$$

The real part of this expression is

$$P \equiv \int_{\Sigma_0} \mathrm{Re}\{\frac{1}{2} \mathbf{E}^e \cdot \mathbf{K}^*\} \, d\Sigma = \int_{\Sigma_1} (\hat{\mathbf{n}}_1 \cdot \mathrm{Re}\{\mathbf{S}\}) \, d\Sigma. \quad (1.20b)$$

In (1.20), Σ_1 is any surface completely surrounding the air–conductor interface Σ_0. Equations (1.20a, b) can be extended to surfaces Σ_1 that pass through the surface of the perfect conductor, provided that $\mathbf{E}^e = 0$ on any part of Σ_0 excluded by Σ_1. This follows from the boundary condition $\hat{\mathbf{n}}_0 \times \mathbf{E} = 0$ on the part of Σ_0 excluded by Σ_1 and the fact that all fields are zero within the volume occupied by the perfect conductor.

Equation (1.20b) states that the time-average power transferred to the perfect conductor from the "generator" (i.e. the impressed electric field $\mathbf{E}^e$) is all radiated into free space. Equations (1.20a, b) possess analogues for the case of imperfect conductors; these involve a volume integral instead of a surface integral, and include a term due to the ohmic losses in the conductors. It is important to note that in both (1.19) and (1.20), only integrations of $\hat{\mathbf{n}} \cdot \mathbf{S}$ over closed surfaces appear; it is not mathematically justified to attach meaning to an integral of $\hat{\mathbf{n}} \cdot \mathbf{S}$ over only a part of a closed surface.

Consider the limiting case of an infinitely thin, perfectly conducting wire lying on the z-axis between $-h$ and h. The impressed electric field is $E_z^e(z)$, and the current on the wire is $I(z)$. In this limit, (1.20b) reduces to

$$P \equiv \int_{-h}^{h} \mathrm{Re}\{\frac{1}{2} E_z^e(z) I^*(z)\} \, dz = \int_{\Sigma_1} (\hat{\mathbf{n}}_1 \cdot \mathrm{Re}\{\mathbf{S}\}) \, d\Sigma. \quad (1.20c)$$

1.4   The field of thin linear antennas: general equations

Now consider the linear antenna of Fig. 1.1c and assume that $a \ll h$ and $\beta_0 a \ll 1$. Both cylindrical coordinates ρ, Φ, z and spherical coordinates r, Θ, Φ are to be used throughout this book. Rotational symmetry obtains, so that all cylindrical or spherical field components are independent of Φ. There is a surface current density $K_z(z)$ on the cylindrical surface $\rho = a$, and also a current on the small capped ends of the antenna. The latter currents can be neglected when calculating the field of the antenna. The total current $I(z)$ and the charge per unit length $q(z)$ are defined to be

$$I(z) = 2\pi a K_z(z), \qquad q(z) = 2\pi a \eta(z). \quad (1.21)$$

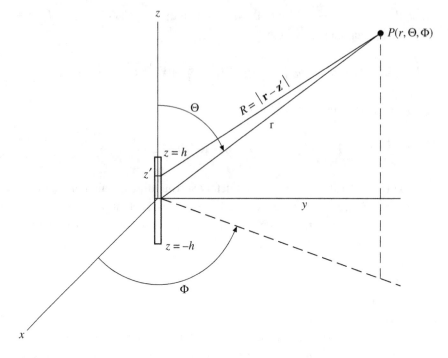

Figure 1.3 Coordinate system for calculations in the far zone.

They are related by the one-dimensional equation of continuity

$$\frac{dI(z)}{dz} = -j\omega q(z). \tag{1.22}$$

$I(z)$ is even with respect to z and $q(z)$ is odd.

When calculating the field of the antenna, one can assume that the current is located at the axis $z = 0$, which is the same as replacing the antenna of radius a by an infinitely thin antenna. With this assumption, but without reference to a particular current distribution $I(z)$, formulas for calculating the field are given in this section and some general characteristics of the field are discussed. The coordinate system is shown in Fig. 1.3.

It is seen from (1.12a) that $\mathbf{A} = \hat{\mathbf{z}}A_z(\rho, z)$. Equations (1.12a, b) reduce to

$$A_z = \frac{\mu_0}{4\pi} \int_{-h}^{h} I(z') \frac{e^{-j\beta_0 R}}{R} dz' \tag{1.23a}$$

and

$$\phi = \frac{1}{4\pi\epsilon_0} \int_{-h}^{h} q(z') \frac{e^{-j\beta_0 R}}{R} dz', \tag{1.23b}$$

where $R = |\mathbf{r} - \hat{\mathbf{z}}z'|$ is the distance from a point z' on the infinitely thin antenna to the observation point **r**. The one-dimensional Lorentz condition is

$$\frac{\partial A_z}{\partial z} = -j\frac{\beta_0^2}{\omega}\phi. \tag{1.23c}$$

The **E** and **B** fields are obtained from (1.6) and (1.8) with (1.23a) and (1.23c). In the cylindrical coordinates ρ, Φ, z, they are $\mathbf{B} = \hat{\boldsymbol{\Phi}} B_\Phi$ and $\mathbf{E} = \hat{\boldsymbol{\rho}} E_\rho + \hat{\mathbf{z}} E_z$, where

$$B_\Phi = \frac{-\partial A_z}{\partial \rho} \tag{1.24a}$$

$$E_\rho = \frac{-j\omega}{\beta_0^2}\frac{\partial^2 A_z}{\partial \rho \partial z} \tag{1.24b}$$

$$E_z = \frac{-j\omega}{\beta_0^2}\left(\frac{\partial^2 A_z}{\partial z^2} + \beta_0^2 A_z\right). \tag{1.24c}$$

In the spherical coordinates r, Θ, Φ with origin at the center of the antenna, the electric field is given by

$$E_r = E_z \cos\Theta + E_\rho \sin\Theta \tag{1.25a}$$

$$E_\Theta = -E_z \sin\Theta + E_\rho \cos\Theta. \tag{1.25b}$$

At sufficiently great distances from the antenna ($r^2 \gg h^2$ and $(\beta_0 r)^2 \gg 1$), the field reduces to a simple form known as the radiation or far field. It is given by

$$B_\Phi^r = E_\Theta^r/c, \tag{1.26a}$$

where

$$\mathbf{E}^r \doteq E_\Theta^r \hat{\boldsymbol{\Theta}}, \quad E_\Theta^r = \frac{j\omega\mu_0}{4\pi}\sin\Theta\int_{-h}^{h} I(z')\frac{e^{-j\beta_0 R}}{R}\,dz'. \tag{1.26b}$$

The distance R from an arbitrary point on the antenna to the field point is given in terms of r and z' by the cosine law, namely (Fig. 1.3),

$$R = \sqrt{r^2 + z'^2 - 2rz'\cos\Theta}. \tag{1.27a}$$

In the radiation zone, $r^2 \gg z'^2$. If the binomial expansion is applied to (1.27a) and only the linear term in z' is retained, the following approximate form is obtained for R:

$$R \doteq r - z'\cos\Theta, \quad (\beta_0 r)^2 \gg 1. \tag{1.27b}$$

The phase variation of $\exp(-j\beta_0 R)/R$ is replaced with the linear phase variation given by (1.27b), i.e. by $\exp(-j\beta_0 r + j\beta_0 z'\cos\Theta)$. The amplitude $1/R$ of $\exp(-j\beta_0 R)/R$ is a slowly varying function of z' and is replaced by $1/r$, where r is the distance to the center of the antenna. With these approximations, (1.26b) can be written as

$$E_\Theta^r = \frac{j\zeta_0 I(0)}{2\pi}\frac{e^{-j\beta_0 r}}{r} F_0(\Theta, \beta_0 h), \tag{1.28a}$$

where $\zeta_0 = \sqrt{\mu_0/\epsilon_0} \doteq 120\pi$ ohms and

$$F_0(\Theta, \beta_0 h) = \frac{\beta_0 \sin \Theta}{2I(0)} \int_{-h}^{h} I(z') e^{j\beta_0 z' \cos \Theta} \, dz'. \qquad (1.28b)$$

The term $F_0(\Theta, \beta_0 h)$ contains all the directional properties of a linear radiator of length $2h$. It is called the field characteristic, field factor, or element factor, and will be computed for some commonly used current distributions. The magnetic field $\mathbf{B}^r$ in the far zone is at right angles to $\mathbf{E}^r$ and also perpendicular to the direction of propagation $\mathbf{r}$. It is given by (1.26a). Thus

$$\mathbf{B}^r = \hat{\mathbf{\Phi}} B_\Phi^r, \quad B_\Phi^r = \frac{j\mu_0 I(0)}{2\pi} \frac{e^{-j\beta_0 r}}{r} F_0(\Theta, \beta_0 h). \qquad (1.28c)$$

Note that the field in the far zone depends on $F_0(\Theta, \beta_0 h)$ which is a function of the particular distribution of current in the antenna.

It is instructive to consider the instantaneous value of the field in (1.28a), which is obtained by multiplication with $e^{j\omega t}$ and selection of the real part. Except for a phase factor,

$$E_\Theta^r(\mathbf{r}, t) = \mathrm{Re}\, E_\Theta(\mathbf{r}) e^{j\omega t} \sim \frac{\sin(\omega t - \beta_0 r)}{r} = \frac{\sin \omega(t - r/c)}{r}. \qquad (1.29a)$$

Note that the field at the point r at the instant t is computed from the current at $r = 0$ at the earlier time $(t - r/c)$. This is a consequence of the finite velocity of propagation c.

The equiphase and equipotential surfaces of $\mathbf{E}$ and $\mathbf{B}$ are spherical shells on which r is equal to a constant. There are an infinite number of such shells that have the same phase (differ by an integral multiple of 2π) but only one that has both the same amplitude and the same phase. The velocity of propagation is the outward radial velocity of the surfaces of constant phase where the phase is represented by the argument of the sine term in (1.29a), that is

$$\text{phase} = \Psi = \omega t - \beta_0 r. \qquad (1.29b)$$

For a constant phase

$$\frac{d\Psi}{dt} = 0 = \omega - \frac{\beta_0 \, dr}{dt}. \qquad (1.29c)$$

It follows that

$$\frac{dr}{dt} = \frac{\omega}{\beta_0} = c = 3 \times 10^8 \text{ m/s}. \qquad (1.29d)$$

Since the phase repeats itself every 2π radians, a wavelength is the distance between two adjacent equiphase surfaces. For example, if one surface is defined by $r = r_1$ and the other by $r = r_2$, then

$$\omega t - \beta_0 r_1 = 2\pi \quad \text{and} \quad \omega t - \beta_0 r_2 = 4\pi \qquad (1.30a)$$

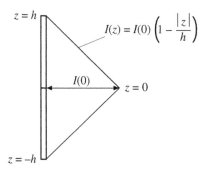

Figure 1.4 Linear antenna with triangular distribution of current.

or

$$r_2 - r_1 = \frac{2\pi}{\beta_0} = \lambda,$$ (1.30b)

where λ is the wavelength in air. The physical picture of the fields in the far zone is quite simple. The electric and magnetic vectors are mutually orthogonal and tangent to an outward traveling spherical shell. Thus, both components of the field are transverse to the radius vector $\mathbf{r}$; they have the same phase velocity $c = 3 \times 10^8$ m/s, the velocity of light.

1.5   The field of the electrically short antenna; directivity

If the current on a thin linear antenna is known, the far-field pattern can be easily determined from the equations in the previous section. When the antenna is electrically short, i.e. $\beta_0 a \ll \beta_0 h \ll 1$, the plausible assumption that the current distribution is triangular can be made. This assumed current distribution is adequate for calculating the field, even quite close to the antenna.

A diagram of the triangular distribution is shown in Fig. 1.4, where the magnitude of the current is plotted along an axis perpendicular to the antenna. In order to find a simple expression for the radiation field, the exponent in (1.28b) can be approximated by 1. Thus,

$$F_0(\Theta, \beta_0 h) \doteq \frac{\beta_0 \sin \Theta}{2} \int_{-h}^{h} \left(1 - \frac{|z'|}{h}\right) dz' = \frac{\beta_0 h \sin \Theta}{2}; \quad (\beta_0 h)^2 \ll 1.$$ (1.31)

Equation (1.31) shows that the radiation field of a short linear antenna is proportional to $\sin \Theta$. Polar and rectangular graphs of the field are shown in Figs. 1.5a and 1.5b, normalized with respect to the maximum at $\Theta = 90°$.

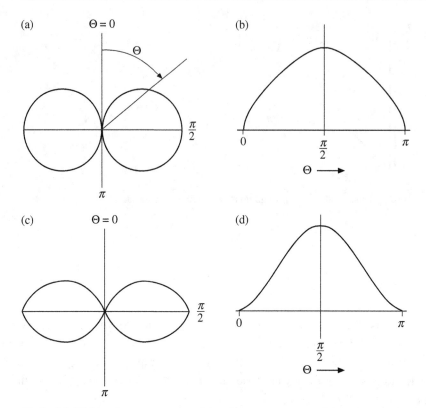

Figure 1.5 Field and power patterns of short linear antenna. (a) Field pattern, polar plot. (b) Field pattern, rectangular plot. (c) Power pattern, polar plot. (d) Power pattern, rectangular plot.

The field quite near an electrically short antenna is readily evaluated from (1.23a) with $I(z) = I(0)(1 - |z|/h)$ and $R \doteq r$. This gives

$$A_z \doteq \frac{\mu_0 h I(0)}{4\pi} \frac{e^{-j\beta_0 r}}{r}. \tag{1.32}$$

The components of the field can be evaluated in the spherical coordinates r, Θ, Φ from (1.6) and (1.3a). The results are

$$B_\Phi \doteq \frac{\mu_0 h I(0)}{4\pi} \left(\frac{j\beta_0}{r} + \frac{1}{r^2} \right) e^{-j\beta_0 r} \sin \Theta \tag{1.33a}$$

$$E_r \doteq \frac{\zeta_0 h I(0)}{4\pi} \left(\frac{2}{r^2} - \frac{j2}{\beta_0 r^3} \right) e^{-j\beta_0 r} \cos \Theta \tag{1.33b}$$

$$E_\Theta \doteq \frac{j\zeta_0 h I(0)}{4\pi} \left(\frac{\beta_0}{r} - \frac{j}{r^2} - \frac{1}{\beta_0 r^3} \right) e^{-j\beta_0 r} \sin \Theta. \tag{1.33c}$$

These may be expressed in terms of the dipole moment $p_z = I(0)h/j\omega$ if desired. The electromagnetic power transferred across a closed surface in the far zone is given by the integral of $\mathrm{Re}\{S_r\} \sim \sin^2 \Theta$. An angular graph of $\mathrm{Re}\{S_r\}$ is called a power pattern. Polar and rectangular graphs of the power pattern are shown in Figs. 1.5c and 1.5d. Note that because of symmetry, both the field pattern and power pattern are independent of the coordinate Φ.

The half-power beam width Θ_{hp} is defined as the angular distance between half-power points on the radiation pattern referred to the principal lobe. The value of Θ_{hp} for the short linear antenna is $90°$. Another parameter useful in defining the directive properties of an antenna is the absolute directivity D. This parameter is a measure of the total time-average power transferred across a closed surface in the direction of the principal lobe. The time-average power transferred across a closed surface Σ is the integral of the normal component of $\mathbf{S}$. Thus, in the far zone,

$$P = \int_{\Sigma} S_r \, d\Sigma. \tag{1.34}$$

The directivity D is the ratio of P with S_r set at its maximum value S_r^{max} to the actual value of P. For a short dipole with $|S_r| \sim \sin^2 \Theta$, the value of D is

$$D = \frac{4\pi}{\int_0^{2\pi} \int_0^{\pi} \sin^2 \Theta \sin \Theta \, d\Theta} = \frac{3}{2}. \tag{1.35}$$

A nearly omnidirectional pattern requires a large value of Θ_{hp} and a nearly unity value of D. A more directional pattern requires a smaller value of Θ_{hp} and a larger value of D.

1.6   The field of antennas with sinusoidally distributed currents; radiation resistance

It is customary to assume that the current distribution on a linear antenna is sinusoidal, i.e.

$$I(z) = \frac{I(0) \sin \beta_0 (h - |z|)}{\sin \beta_0 h} = I_m \sin \beta_0 (h - |z|). \tag{1.36}$$

For this current, the field characteristic $F_0(\Theta, \beta_0 h)$ is given by (1.28b) with (1.36),

$$F_0(\Theta, \beta_0 h) = \frac{\cos(\beta_0 h \cos \Theta) - \cos \beta_0 h}{\sin \beta_0 h \sin \Theta}. \tag{1.37a}$$

An alternative field characteristic $F_m(\Theta, \beta_0 h)$ is referred to the maximum value of the sinusoid, namely, $I_m = I(0)/\sin \beta_0 h$ which occurs at $h - \lambda/4$ when $\beta_0 h \geq \pi/2$.

$$F_m(\Theta, \beta_0 h) = \frac{\cos(\beta_0 h \cos \Theta) - \cos \beta_0 h}{\sin \Theta}. \tag{1.37b}$$

The function $F_m(\Theta, \beta_0 h)$ is shown graphically in Fig. 1.6 for several values of h. It is seen that the pattern corresponding to $\beta_0 h = \pi/2$ ($h = \lambda/4$) is only slightly narrower

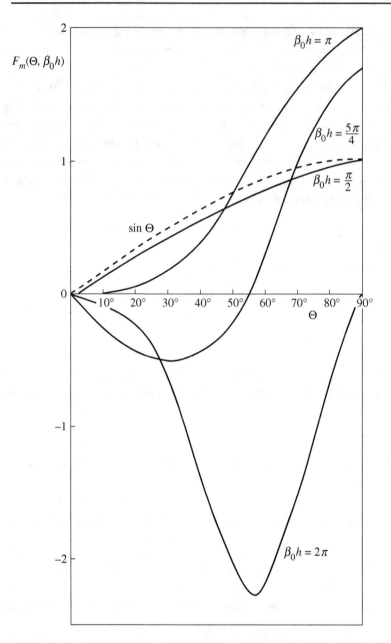

Figure 1.6 Field factor of linear antenna.

than the pattern for $(\beta_0 h)^2 \ll 1$ which is shown in Fig. 1.5b. Note that as $\beta_0 h$ is increased beyond π, minor lobes appear which successively become the major lobe and point in directions other than $\Theta = \pi/2$.

The theoretical model of an infinitely thin antenna with a sinusoidal distribution of current is a convenient one: the *complete* electromagnetic field can be evaluated exactly in terms of elementary functions, even for observation points arbitrarily close to the antenna. This is accomplished with the substitution of the current (1.36) in the general integral (1.23a) for the vector potential, and subsequent use of the resulting expression in (1.24). The indicated differentiations can be carried out directly without evaluating the integral. The result is

$$B_\Phi(\rho, z) = \frac{j I_m \mu_0}{4\pi\rho} [e^{-j\beta_0 R_{1h}} + e^{-j\beta_0 R_{2h}} - 2\cos\beta_0 h \, e^{-j\beta_0 r}] \tag{1.38a}$$

$$E_\rho(\rho, z) = \frac{j I_m \zeta_0}{4\pi\rho} \left[\frac{z-h}{R_{1h}} e^{-j\beta_0 R_{1h}} + \frac{z+h}{R_{2h}} e^{-j\beta_0 R_{2h}} - \frac{2z}{r}\cos\beta_0 h \, e^{-j\beta_0 r} \right] \tag{1.38b}$$

$$E_z(\rho, z) = \frac{-j I_m \zeta_0}{4\pi} \left[\frac{e^{-j\beta_0 R_{1h}}}{R_{1h}} + \frac{e^{-j\beta_0 R_{2h}}}{R_{2h}} - 2\cos\beta_0 h \, \frac{e^{-j\beta_0 r}}{r} \right] \tag{1.38c}$$

$$B_\rho(\rho, z) = B_z(\rho, z) = E_\phi(\rho, z) = 0, \tag{1.38d}$$

where

$$r = \sqrt{\rho^2 + z^2}, \quad R_{1h} = \sqrt{\rho^2 + (h-z)^2}, \quad R_{2h} = \sqrt{\rho^2 + (h+z)^2} \tag{1.38e}$$

are the distances from the observation point to the center and the two ends of the antenna, respectively.

When $\beta_0 h = \pi/2$, the interpretation of (1.38) in terms of spheroidal waves is available in [1, pp. 297–310] or [3, Chapter V]. It is easily checked that, when r is large, (1.38a–d) reduce to the radiation field given by (1.28a, c) with (1.37a). Furthermore, (1.38) are seen to reduce to the field (1.33) of the electrically short antenna when $\beta_0 h \ll 1$.

The total, time-average power P is equal to the integral of the normal component of $\text{Re}\{\mathbf{S}\} = (1/2\mu_0)\,\text{Re}\{\mathbf{E} \times \mathbf{B}^*\}$ over a closed surface surrounding the antenna, where $\mathbf{E}$ and $\mathbf{B}$ are given by (1.38). Although any closed surface completely surrounding the antenna will correctly give P, it is convenient to select a large sphere for the integration surface and use the expressions (1.28) and (1.37) for the radiation field in spherical coordinates. The complete formula for P determined in this manner can be found, for example, in [2, p. 140]. It is easy to see that the expression for P has the form

$$P = \tfrac{1}{2}|I_m|^2 R_m^e \quad \text{or} \quad P = \tfrac{1}{2}|I(0)|^2 R_0^e \tag{1.39}$$

where the quantities R_m^e and $R_0^e = R_m^e / \sin^2 \beta_0 h$ depend only on $\beta_0 h$. The units of R_m^e and R_0^e are ohms. By definition, R_m^e (R_0^e) is the radiation resistance referred to I_m ($I(0)$). R_m^e is equal to 73.1 ohms when $\beta_0 h = \pi/2$, and 199 ohms when $\beta_0 h = \pi$.

In general, R_0^e is *not* the driving-point resistance of a center-driven antenna. To see why this is true, let us examine in more detail the model of an infinitely thin antenna with a sinusoidal distribution of current. In particular, in what way can one maintain, at least in principle, the sinusoidal current distribution (1.36) on the infinitely thin, perfectly conducting wire?

From (1.38c), it is seen that the exact tangential electric field $E_z(0, z)$ on the wire's axis is non-zero along the entire length of the wire. This is the axial field maintained by the sinusoidal current distribution, and not the total axial field. Since the total axial field is zero, there must be an externally maintained field $E_z^e = -E_z(0, z)$ on the perfectly conducting wire. It is given by

$$E_z^e(z) = \frac{j I_m \zeta_0}{4\pi} \left[\frac{e^{-j\beta_0(h-z)}}{h-z} + \frac{e^{-j\beta_0(h+z)}}{h+z} - 2\cos\beta_0 h \frac{e^{-j\beta_0|z|}}{|z|} \right]; \quad -h < z < h$$

(1.40)

and is non-zero along the whole length of the wire. It follows that it is not possible to excite a sinusoidal current simply by a single delta-function generator with $E_z^e(z) = V\delta(z)$. Instead, a continuous distribution of electromotive forces is necessary.

Equations (1.40), (1.36), and the power identity (1.20c) provide another equivalent way to determine the time-average power P radiated by the infinitely thin antenna, by integrating $\text{Re}\{\frac{1}{2} E_z^e(z) I^*(z)\}$ along the length of the antenna. Note that the integrand is finite. As before, R_0^e is the coefficient of $\frac{1}{2} |I(0)|^2$ in the resulting expression.

The foregoing discussion clearly shows that R_0^e and the driving-point resistance R_0 of a center-driven antenna are two different quantities. In some cases, however, it is true that $R_0 \doteq R_0^e$. This will be seen in the next section.

1.7   Impedance of antenna: EMF method

In this section, the "induced EMF method" [4] is discussed. This is an approximate method used for calculating the impedance of a center-driven antenna with non-zero radius.

Let $I(z) = 2\pi a K_z(z)$ be the current on an antenna center-driven by a delta-function generator, and let $E_z(a, z)$ be the tangential electric field at the surface $\rho = a$. Consider the quantities

$$-\frac{1}{|I(0)|^2} \int_{-h}^{h} E_z(a, z) I^*(z) \, dz$$

(1.41a)

or

$$-\frac{1}{I^2(0)} \int_{-h}^{h} E_z(a, z) I(z) \, dz.$$

(1.41b)

These are both equal to Z_0, the driving-point impedance of the antenna. This is seen to be true by the substitution of the boundary condition $E_z(a, z) = -V\delta(z)$ in (1.41) and the subsequent use of the property (1.2c) of the delta function and the definition $Z_0 = V/I(0)$.

The "induced EMF method" consists of determining the driving-point impedance of the antenna from the formula

$$Z_0 \doteq -\frac{1}{|I(0)|^2} \int_{-h}^{h} E_z(a, z) I^*(z) \, dz \tag{1.41c}$$

where one uses the sinusoidal current distribution $I(z) = [I(0)/\sin \beta_0 h] \sin \beta_0 (h - |z|)$ on the right-hand side, and the associated value of $E_z(a, z)$ from (1.38c). It is easily seen from (1.38c) that the integral in (1.41c) is proportional to $|I(0)|^2$. Therefore, the final quantity obtained does not involve $I(0)$; it is an integral expression which depends only on $\beta_0 a$ and $\beta_0 h$. Since $I(z)$ is in phase with $I(0)$ for all z, the same result is obtained if (1.41b) is used instead of (1.41a).

The resulting integral expression for Z_0 can be evaluated by numerical integration, expressed [5] in terms of integrals tabulated in standard mathematical handbooks [6], or written in the form

$$Z_0 \doteq \frac{j\zeta_0}{2\pi} \frac{1}{\sin^2 \beta_0 h} \{\sin \beta_0 h \, [C_a(h, h) - \cos \beta_0 h \, C_a(h, 0)]$$
$$- \cos \beta_0 h \, [S_a(h, h) - \cos \beta_0 h \, S_a(h, 0)]\}, \tag{1.42a}$$

where the integrals $C_a(h, z)$ and $S_a(h, z)$, which occur frequently in antenna theory, are defined by

$$C_a(h, z) = \int_0^h \cos \beta_0 z' \left[\frac{e^{-j\beta_0 R_1}}{R_1} + \frac{e^{-j\beta_0 R_2}}{R_2} \right] dz' \tag{1.42b}$$

$$S_a(h, z) = \int_0^h \sin \beta_0 |z'| \left[\frac{e^{-j\beta_0 R_1}}{R_1} + \frac{e^{-j\beta_0 R_2}}{R_2} \right] dz' \tag{1.42c}$$

and where

$$R_1 = \sqrt{(z - z')^2 + a^2}, \qquad R_2 = \sqrt{(z + z')^2 + a^2}. \tag{1.42d}$$

A short table of these integrals for the case $a/\lambda = 0.007\,022$ is given in [2, Appendix 1].

Note that the value of Z_0 so obtained is infinite when $\beta_0 h = \pi, 2\pi, \ldots$. Therefore, the method cannot be used to determine the driving-point impedance of antennas with these lengths. Note also that, in the limit $\beta_0 a \to 0$, the value of $R_0 = \mathrm{Re}\{Z_0\}$ reduces to R_0^e, where R_0^e is the radiation resistance obtained in the previous section.

From a theoretical point of view, the valid objection can be raised that two different models are involved in (1.41). These are the antenna in which a sinusoidal current distribution is maintained (by a continuous distribution of electromotive forces), and the antenna center-driven by a delta-function generator. Only under special circumstances can the first model be regarded as being similar to the second, or, indeed, to the more practical antennas of Figs. 1.1a and 1.1b. For the first model, there

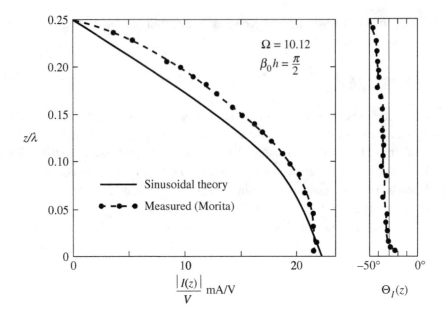

Figure 1.7 Distribution of amplitude and phase of current in half-wave dipole.

is no single pair of terminals, so that the quantity in (1.41) is not the driving-point impedance of an antenna. For the second model, the quantity in (1.41) is indeed the driving-point impedance but (1.41c) is an identity, and not a means of determining Z_0. The electric field maintained by the currents in the first case violates the boundary condition $E_z(a, z) = 0$ ($z \neq 0$) satisfied by the corresponding field in the case of the center-driven antenna.

In order to further understand the two models, it is instructive to consider the permissible choices of Σ_1 in (1.20b). In other words, for what types of surfaces Σ_1 does one correctly obtain the time-average power radiated? The answer is different for the two models: For the case of the center-driven antenna, Σ_1 can be any closed surface that encloses the delta-function generator at $z = 0$. It need not enclose the entire antenna. However, for the antenna with a sinusoidal distribution of current, it is necessary to enclose the entire antenna in order to correctly obtain P, the time-average power.

From an engineering point of view, the induced EMF method is best discussed by comparison with measurement. In order to obtain useful results, it is necessary that the assumed sinusoidal current distribution be close to the true current distribution, and that the antenna be electrically thin. Figures 1.7 and 1.8 show the measured amplitude and phase of the current for a base-driven monopole over a ground plane together with the sinusoidal current for $\beta_0 h = \pi/2$ and π, respectively. The parameter Ω is related to h/a by $\Omega = 2 \ln(2h/a)$. In Fig. 1.7, the experimental data are taken from [7]. The

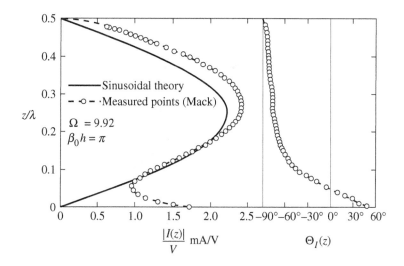

Figure 1.8 Distribution of amplitude and phase of current in full-wave dipole.

theoretical curve is in the form

$$\frac{I(z)}{V} = \left|\frac{I(z)}{V}\right| e^{j\Theta_I(z)} = \frac{I(0)}{V}\cos\beta_0 z = \frac{2}{Z_0}\cos\beta_0 z \qquad (1.43)$$

where Z_0 has been calculated from (1.41c) to be $Z_0 = 73 + j41$ ohms. The factor of 2 in the last equation in (1.43) is included so that $I(z)/V$ corresponds to that of a monopole over a ground plane. In Fig. 1.8, the measurements have been made by Mack. The value of $|I_m/V| = |I(\lambda/4)/V|$ in the theoretical curve is such that the total power radiated by the antenna with the sinusoidal current (as calculated from $P = R^e_m|I_m|^2$ with $R^e_m = 199$ ohms) is the same as the total power radiated by the base-driven monopole. The latter power can be found from the measured driving-point conductance $G_0 = 1.023$ millisiemens (mS) as $P = G_0|V|^2$.

In Fig. 1.7, the general agreement between the measured values and the sinusoidal approximation is fair, with more current near the top of the actual antenna than is indicated by the cosine curve. The driving-point admittance as calculated by the induced EMF method agrees quite well with the measured value. The phase differs somewhat from the constant required by the sinusoidal distribution of current. For the full-wave antenna of Fig. 1.8, the sinusoidal current fails completely near the driving point, where, instead of $|I(0)/V| = 0$, $|I(0)/V|$ is about three-quarters its maximum value along the antenna. The measured phase, instead of being constant, changes significantly along the antenna.

Some additional comments about the half-wave antenna are now made. More discussions along these lines can be found in [8]. For $\beta_0 h = \pi/2$, the measured current is fairly close to that predicted by the induced EMF method. It follows that the near-field B_Φ should also be fairly close. This is not true, however, for all near-field

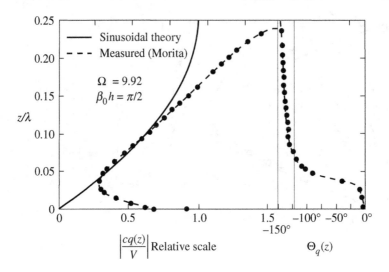

0.25

— Sinusoidal theory
- - - Measured (Morita)

$\Omega = 9.92$
$\beta_0 h = \pi/2$

z/λ

$\left|\dfrac{cq(z)}{V}\right|$ Relative scale

$\Theta_q(z)$

Figure 1.9 Normalized distribution of charge in amplitude and phase for a half-wave dipole.

quantities. It follows from preceding discussions that the E_z components are different. That the E_ρ components should also be different is illustrated in Fig. 1.9, where the measured [7] charge per unit length $q(z)/V = |q(z)/V|e^{j\Theta_q(z)}$ along the half-wave antenna is shown together with that predicted by the sinusoidal theory. The theoretical curve was calculated from (1.43) by the equation of continuity (1.22). Here, the agreement is quite poor.

The sinusoidal current distribution $\sin \beta_0(h - |z|)$ has been seen to be inadequate in many cases. Nevertheless, it is attractive because of its simplicity. In Chapter 2, linear antennas satisfying $\beta_0 a \ll \beta_0 h < 3\pi/2$ and $\beta_0 a \ll 1$ are considered. For such antennas, an improved representation of the current will be introduced. In this representation, $\sin \beta_0(h - |z|)$ is the first term, and the remaining terms are also simple trigonometric functions.

1.8   Integral equations for the current distribution

In the three preceding sections, the current distribution $I(z)$ along the length of the linear antenna has been assumed. A more scientific and more difficult method for investigating the properties of a center-driven linear antenna is to determine $I(z)$ from the boundary condition satisfied by E_z on the surface of the antenna. If this condition is imposed, an integral equation for $I(z)$ results. A history of the development of the integral equation, as well as many additional references, can be found in [3, 9, 10].

This section first introduces the model of the center-driven tubular dipole. Two integral equations will be derived, one of which is exact for this model and will be called the *exact* integral equation. The second integral equation is approximate and

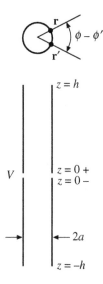

Figure 1.10 Center-driven tubular dipole: top view and cross section.

will be called the *approximate* integral equation. Both forms are referred to as *Hallén's* integral equation in recent literature.

Figure 1.10 shows a center-driven tubular dipole. It is a perfectly conducting, open-ended tube with walls of negligible thickness. A delta-function generator is located at an infinitesimal gap between $z = 0^-$ and $z = 0^+$, so that the scalar potential $\phi(\rho, z)$ satisfies $\phi(a, 0^+) - \phi(a, 0^-) = V$. From (1.8) and the boundary condition $E_z(a, z) = 0$ ($z \neq 0$), it is seen that, on the surface of the tube at $\rho = a$,

$$E_z(a, z) = -V\delta(z); \quad -h < z < h. \tag{1.44}$$

There is a rotationally symmetric surface current density $K_{z,\,\mathrm{out}}(z)$ on the outside of the tube, and a similar surface current density $K_{z,\,\mathrm{in}}(z)$ on the inside of the tube. The total current $I(z)$ is defined by $I(z) = 2\pi a[K_{z,\,\mathrm{out}}(z) + K_{z,\,\mathrm{in}}(z)]$. $I(z)$ is even in z; it vanishes at the ends $z = \pm h$ of the dipole by the continuity of $K_z(z)$.

Denote by $A_z(a, z)$ the rotationally symmetric vector potential on the surface of the tube. An integral equation for $I(z)$ can be derived by calculating $A_z(a, z)$ in two different ways and equating the results as follows.

The first step is to note that, at any observation point $\mathbf{r}$, $E_z(\mathbf{r})$ is related to $A_z(\mathbf{r})$ by equation (1.24c). (This equation holds as long as only z-directed currents are present.) When the observation point is on the surface of the dipole, (1.44) can be substituted in (1.24c). Thus, $A_z(a, z)$ satisfies the differential equation

$$\frac{\partial^2 A_z(a, z)}{\partial z^2} + \beta_0^2 A_z(a, z) = \frac{-j\beta_0^2}{\omega} V\delta(z); \quad -h < z < h. \tag{1.45}$$

The solution to this equation consists of a particular solution plus the general solution of the homogeneous equation. The latter is $C_1 \cos \beta_0 z + C_2 \sin \beta_0 z$ where C_1

and C_2 are constants to be determined. From the properties (1.2) of the delta function, a particular solution is readily verified to be $(-jV/2c) \sin \beta_0 |z|$. Since $A_z(a, z)$ is even in z, the coefficient C_2 is identically zero and the solution to (1.45) is

$$A_z(a, z) = \frac{-jV}{2c} \sin \beta_0 |z| + C_1 \cos \beta_0 z. \tag{1.46}$$

It is seen that the derivative of $A_z(a, z)$ is discontinuous. This is a consequence of the step-function behavior of the scalar potential at $z = 0$.

The second way to calculate $A_z(a, z)$ is from the integral (1.12a). At any observation point $\mathbf{r}$, the vector potential is given by integrating over the surface of the tube:

$$A_z(\mathbf{r}) = \frac{\mu_0}{4\pi} \int_{-h}^{h} \int_{-\pi}^{\pi} \frac{I(z')}{2\pi a} \frac{e^{-j\beta_0 |\mathbf{r} - \mathbf{r}'|}}{|\mathbf{r} - \mathbf{r}'|} \, a \, d\Phi' \, dz'. \tag{1.47}$$

When $\rho = a$, the distance between $\mathbf{r}$ and $\mathbf{r}'$ is given by (see Fig. 1.10)

$$|\mathbf{r} - \mathbf{r}'| = \sqrt{(z - z')^2 + 4a^2 \sin^2[(\Phi - \Phi')/2]}. \tag{1.48}$$

Although Φ appears in the integral for $A_z(a, z)$, it is apparent that $A_z(a, z)$ is independent of Φ, so that one can take $\Phi = 0$ with no loss of generality. Equating (1.46) with the integral expression for $A_z(a, z)$, the equation

$$\frac{4\pi}{\mu_0} A_z(a, z) \equiv \int_{-h}^{h} K(z - z') I(z') \, dz'$$

$$= \frac{-j2\pi V}{\zeta_0} \sin \beta_0 |z| + C \cos \beta_0 z; \quad -h < z < h \tag{1.49}$$

is obtained. In (1.49), $K(z)$ is given by

$$K(z) = K_{ex}(z) = \frac{1}{2\pi} \int_{-\pi}^{\pi} \frac{\exp\left(-j\beta_0 \sqrt{z^2 + 4a^2 \sin^2(\Phi/2)}\right)}{\sqrt{z^2 + 4a^2 \sin^2(\Phi/2)}} \, d\Phi; \quad |z| < 2h \tag{1.50}$$

and the constant $C = 4\pi C_1/\mu_0$ is to be determined from the condition that

$$I(h) = 0. \tag{1.51}$$

In equation (1.49), which is to hold for all values of z between $-h$ and h, the unknown current $I(z)$ appears inside the integral sign. This is the desired exact integral equation. The quantity $K(z - z') = K_{ex}(z - z')$ that multiplies the unknown is called the *kernel* of the integral equation. It depends only on the difference $z - z'$, and not on z and z' separately.

The same steps may be followed to derive the so-called *approximate* integral equation. On the basis that the antenna is electrically thin, $\beta_0 a \ll 1$, one makes the assumption that the current is located at the axis $z = 0$ of the antenna. Thus, in place of (1.47), one has the integral expression (1.23a) for the z-directed vector potential, where

$$R = |\mathbf{r} - \hat{\mathbf{z}}z'| = \sqrt{(z - z')^2 + a^2} \tag{1.52}$$

when the observation point $\mathbf{r}$ is on the antenna's surface. Equating this integral expression with (1.46) results in the approximate integral equation. It is the same as (1.49), but with the "approximate" or "reduced" kernel

$$K(z) = K_{\mathrm{ap}}(z) = \frac{\exp\left(-j\beta_0\sqrt{z^2 + a^2}\right)}{\sqrt{z^2 + a^2}}; \quad |z| < 2h \tag{1.53}$$

in place of the exact kernel of (1.50).

In this chapter, as well as in the related Chapter 13, the symbol $I_{\mathrm{ex}}(z)$ will denote the unknown current when the exact kernel $K_{\mathrm{ex}}(z)$ is used in (1.49). The corresponding quantity when $K_{\mathrm{ap}}(z)$ is used will be denoted by $I_{\mathrm{ap}}(z)$. A symbol $K(z)$ $[I(z)]$ with no subscripts can denote either $K_{\mathrm{ex}}(z)$ $[I_{\mathrm{ex}}(z)]$ or $K_{\mathrm{ap}}(z)$ $[I_{\mathrm{ap}}(z)]$.

The most pronounced advantage of the approximate integral equation is that $K_{\mathrm{ap}}(z)$ is simpler in form than $K_{\mathrm{ex}}(z)$, which involves an integration. The approximate integral equation will be used almost exclusively throughout this book. Despite the similarity of the two integral equations, their mathematical properties are very different. Such properties are discussed in detail in [10] and in Section 13.2 of this book.

Equations (1.49) and (1.51) can be written in various equivalent forms, one of which is

$$\int_{-h}^{h} K(z - z')I^{(1)}(z')\,dz' = \frac{-j2\pi V}{\zeta_0}\sin\beta_0|z|; \quad -h < z < h \tag{1.54a}$$

and

$$\int_{-h}^{h} K(z - z')I^{(2)}(z')\,dz' = \cos\beta_0 z; \quad -h < z < h, \tag{1.54b}$$

where the unknowns $I^{(1)}(z)$ and $I^{(2)}(z)$ are related to $I(z)$ and C by

$$I(z) = I^{(1)}(z) + CI^{(2)}(z), \qquad C = -\frac{I^{(1)}(h)}{I^{(2)}(h)}. \tag{1.55}$$

This form is slightly more convenient for the application of numerical methods. These are discussed in the next section and in Chapter 13.

1.9   Direct numerical methods

Methods for solving (1.49) approximately were proposed as early as the 1930s [11]. When high-speed computers appeared, solving (1.49) by numerical methods, especially moment methods [12], became popular. When applying moment methods to integral equations, the basic idea is to seek an approximate solution in the form of a linear combination of a finite number of basis functions. The coefficients of the basis functions are the unknowns. They are determined by approximating the integral equation by a system of algebraic equations which are then solved by computer.

In this section, numerical methods are introduced by describing the application of a particular method to (1.49). The method to be applied is Galerkin's [13] method with pulse functions, which is a form of the method of moments in [12]. Generalizations and additional information can be found in many standard antenna and engineering textbooks. A more critical discussion of the application of numerical methods to (1.49) is beyond the scope of an introductory chapter and is contained in Chapter 13.

As mentioned previously, it is convenient to deal first with equations (1.54a, b). One writes $I^{(1)}(z)$ and $I^{(2)}(z)$ as the sum of basis functions with unknown coefficients. Basis functions that are non-zero on only a part of the interval of interest [in our case $(-h, h)$] are called *subsectional* basis functions. Perhaps the simplest choice of subsectional basis functions are the pulse functions $u_n(z)$ which result by dividing $(-h, h)$ into $2N + 1$ segments of length z_p so that

$$(2N + 1)z_p = 2h. \tag{1.56}$$

The nth pulse function $u_n(z)$ is constant on the nth segment and zero elsewhere:

$$u_n(z) = \begin{cases} 1, & \text{if } (n - \tfrac{1}{2})z_p < z < (n + \tfrac{1}{2})z_p \\ 0, & \text{otherwise} \end{cases} \tag{1.57}$$

for $n = -N, -(N - 1), \dots, N$ (see Fig. 1.11). The choice of an odd number of segments is convenient, but not necessary.

"Staircase"-type approximate solutions to (1.54a, b) are sought by setting

$$I^{(1)}(z) \doteq \sum_{n=-N}^{N} I_n^{(1)} u_n(z), \qquad I^{(2)}(z) \doteq \sum_{n=-N}^{N} I_n^{(2)} u_n(z), \tag{1.58}$$

where $I_n^{(1)}$ and $I_n^{(2)}$ are coefficients to be determined. The substitution of (1.58) into (1.54) yields

$$\sum_{n=-N}^{N} I_n^{(1)} \int_{-h}^{h} K(z - z') u_n(z') \, dz' \doteq \frac{-j2\pi V}{\zeta_0} \sin \beta_0 |z|; \quad -h < z < h \tag{1.59a}$$

and

$$\sum_{n=-N}^{N} I_n^{(2)} \int_{-h}^{h} K(z - z') u_n(z') \, dz' \doteq \cos \beta_0 z; \quad -h < z < h. \tag{1.59b}$$

Note that the integrands in (1.59) are non-zero over the nth segment only.

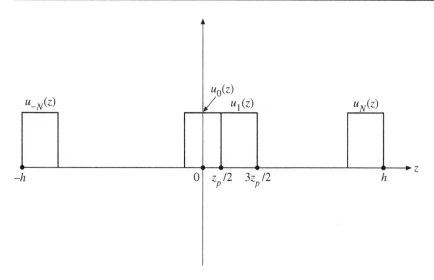

Figure 1.11 Pulse functions $u_n(z)$.

Equations (1.59) cannot be satisfied for all z, and one must satisfy them approximately. This can be done by selecting a second set of functions $v_l(z)$, $l = -N, \ldots, N$ (called *testing functions*), multiplying (1.59) by $v_l(z)$, and integrating from $z = -h$ to $z = h$. In the case of Galerkin's method, the set of testing functions is taken to be the same as the set of basis functions. Multiplication of (1.59) by $u_l(z)$ and integration from $z = -h$ to $z = h$ yield

$$\sum_{n=-N}^{N} A_{ln} I_n^{(1)} = B_l^{(1)}, \qquad \sum_{n=-N}^{N} A_{ln} I_n^{(2)} = B_l^{(2)}; \quad l = 0, \pm 1, \ldots, \pm N, \tag{1.60}$$

where

$$A_{ln} = \int_{(l-1/2)z_p}^{(l+1/2)z_p} \int_{(n-1/2)z_p}^{(n+1/2)z_p} K(z - z')\, dz'\, dz; \quad -N \le l, n \le N \tag{1.61}$$

$$B_l^{(1)} = \frac{-j2\pi V}{\zeta_0} \int_{(l-1/2)z_p}^{(l+1/2)z_p} \sin \beta_0 |z| \, dz$$

$$= \begin{cases} \dfrac{-j8\pi V}{\zeta_0 \beta_0} \sin^2(\beta_0 z_p/4), & \text{if } l = 0 \\[2mm] \dfrac{-j4\pi V}{\zeta_0 \beta_0} \sin(\beta_0 z_p/2) \sin(\beta_0 z_p |l|), & \text{if } l = \pm 1, \ldots, \pm N \end{cases} \tag{1.62a}$$

and

$$B_l^{(2)} = \int_{(l-1/2)z_p}^{(l+1/2)z_p} \cos \beta_0 z \, dz$$

$$= \frac{2}{\beta_0} \sin(\beta_0 z_p/2) \cos(\beta_0 z_p l); \quad l = 0, \pm 1, \ldots, \pm N. \tag{1.62b}$$

Equations (1.60) are two systems of algebraic equations with unknowns $I_{-N}^{(1)}$, $I_{-(N-1)}^{(1)}, \ldots, I_N^{(1)}$ and $I_{-N}^{(2)}, I_{-(N-1)}^{(2)}, \ldots, I_N^{(2)}$.

The procedure just described is Galerkin's method with pulse functions applied to the two integral equations (1.54). Having solved the systems in (1.60) by computer, one can determine C from [see (1.55)]

$$C \doteq -I_N^{(1)}/I_N^{(2)}. \tag{1.63a}$$

The final numerical solution is therefore

$$I(z) \doteq \sum_{n=-N}^{N} I_n u_n(z) = \sum_{n=-N}^{N} [I_n^{(1)} + C I_n^{(2)}] u_n(z). \tag{1.63b}$$

The symbols $I_{\text{ex},n}$ and $I_{\text{ap},n}$ will denote the values of I_n obtained with the exact kernel and the approximate kernel, respectively.

The double integral in (1.61) can be reduced to a single integral by setting $z - lz_p = x$, $z' - nz_p = x'$, and by using the identity

$$\int_{-z_p/2}^{z_p/2} \int_{-z_p/2}^{z_p/2} f(x - x') \, dx' \, dx = \int_0^{z_p} (z_p - z)[f(z) + f(-z)] \, dz. \tag{1.64}$$

From the resulting equation and from $K(z) = K(-z)$, it is seen that the A_{ln} depend on $|l - n|$ only, and not on l and n separately. Denoting $A_{ln} = A_{nl}$ by A_{l-n}, one has the simpler expression

$$A_l = A_{-l} = \int_0^{z_p} (z_p - z)[K(z + lz_p) + K(z - lz_p)] \, dz \tag{1.65}$$

for the matrix coefficients. In (1.65), the index l takes the values $0, \pm1, \pm2, \ldots, \pm2N$.

Thus, one must first solve the two $(2N + 1) \times (2N + 1)$ systems (1.60) for $I_n^{(1)}$ and $I_n^{(2)}$. These systems are symmetric since $A_{ln} = A_{nl}$ and Toeplitz [14] because A_{ln} depends only on the difference $l - n$. Whereas the vector elements on the right-hand side can be found from (1.62), it is necessary to compute the matrix elements A_l by numerical integration. This is a simple task for modern computers. Note that the expression (1.50) for the exact kernel is an integral, so that when the exact kernel is used, (1.65) is actually a double integral. Here, one can exploit the properties of the integrand and use standard techniques [15] to reduce the computer time required for the numerical integration. The solutions to the systems in (1.60) satisfy $I_n^{(1)} = I_{-n}^{(1)}$ and $I_n^{(2)} = I_{-n}^{(2)}$, so that each system is equivalent to a $(N + 1) \times (N + 1)$ system. It can also be shown that (1.60) and (1.63a) are equivalent (in the absence of roundoff errors in the computer) to one $(N + 1) \times (N + 1)$ system of equations with unknowns $I_0, I_1, \ldots, I_{N-1}$, and C.

When applying the method, one can scale h, a, and z_p by the wavelength $\lambda = 2\pi/\beta_0$. Figures 1.12a and 1.12b show the results obtained for $h/\lambda = 0.25$, $a/\lambda = 0.007\,022$,

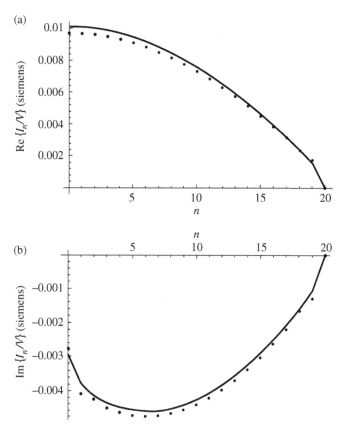

Figure 1.12 Numerical results obtained with method of Section 1.9; $h/\lambda = 0.25, a/\lambda = 0.007\,022$, $N = 20$. (a) $\mathrm{Re}\{I_{\mathrm{ex},n}/V\}$ (solid line) and $\mathrm{Re}\{I_{\mathrm{ap},n}/V\}$ (dots). (b) $\mathrm{Im}\{I_{\mathrm{ex},n}/V\}$ (solid line) and $\mathrm{Im}\{I_{\mathrm{ap},n}/V\}$ (dots).

and $N = 20$. Here, z_p/λ is about 0.012. In Fig. 1.12a, the component of current $\mathrm{Re}\{I_n/V\}$ in phase with the driving voltage is shown as a function of n, for both the approximate and the exact kernels. The values of $\mathrm{Re}\{I_{\mathrm{ap},n}/V\}$, $n = 0, 1, \ldots, 20$, are shown as dots. The corresponding values for the case of the exact kernel have been joined by straight lines. For this choice of N, the values agree quite well. Good agreement between the results obtained with the exact and the approximate kernels is also seen in Fig. 1.12b, where the components $\mathrm{Im}\{I_n/V\}$ are shown. The driving-point admittance obtained for this antenna with $N = 20$ is $Y_0 = I_{\mathrm{ex},0}/V = 10.1 - j2.91$ mS for the case of the exact kernel, and $Y_0 = I_{\mathrm{ap},0}/V = 9.72 - j2.76$ mS for the case of the approximate kernel.

When applying this numerical method to integral equations, the numerical solution ordinarily becomes closer to the true solution as N is increased, where $2N + 1$ is the number of pulse functions. In practice, the true solution is not known and one often resorts to the empirical criterion of making N larger until the solution has converged to a satisfactory final value.

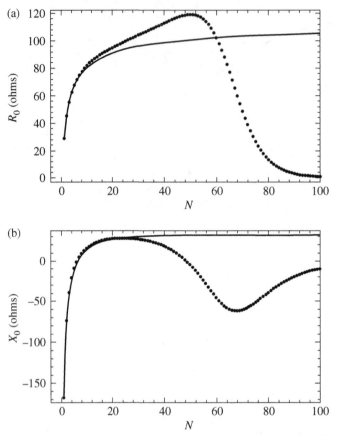

Figure 1.13 Numerical results obtained with method of Section 1.9; $h/\lambda = 0.25$, $a/\lambda = 0.007\,022$.
(a) $R_0 = \text{Re}\{V/I_{\text{ex},0}\}$ (solid line) and $R_0 = \text{Re}\{V/I_{\text{ap},0}\}$ (dots) as function of N.
(b) $X_0 = \text{Im}\{V/I_{\text{ex},0}\}$ (solid line) and $X_0 = \text{Im}\{V/I_{\text{ap},0}\}$ (dots) as function of N.

Figures 1.13a and 1.13b show the driving-point resistance R_0 and reactance X_0 obtained by the numerical method as a function of N for $N = 1, 2, \ldots, 100$. One may be surprised to see that the values of R_0 and X_0 agree for small values of N only. Whereas the values for the case of the exact kernel are relatively stable when N is large, a perhaps puzzling behavior is observed in the case of the approximate kernel: both R_0 and X_0 seem to converge to zero. Corresponding graphs for the driving-point conductance G_0 and susceptance B_0 (shown in Figs. 1.14a and 1.14b) reveal that the quantity B_0 is to blame: For large values of N in Fig. 1.14b, the values of B_0 are much larger than those obtained with the exact kernel. (In Fig. 1.14b, the values of B_0 for $N \geq 72$ in the case of the approximate kernel are not shown. These values continue to rise rapidly, with $B_0 = 0.1$ S when $N = 100$.) The values of G_0 in Fig. 1.14a, on the other hand, agree quite well with those obtained with the exact kernel.

It is seen from these figures that, in the case of the approximate kernel (at least for the specific parameters h/λ and a/λ under consideration), the choice of N is very

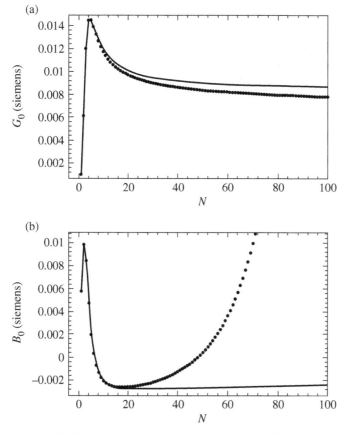

Figure 1.14 Numerical results obtained with method of Section 1.9; $h/\lambda = 0.25$, $a/\lambda = 0.007\,022$.
(a) $G_0 = \text{Re}\{I_{\text{ex},0}/V\}$ (solid line) and $G_0 = \text{Re}\{I_{\text{ap},0}/V\}$ (dots) as function of N.
(b) $B_0 = \text{Im}\{I_{\text{ex},0}/V\}$ (solid line) and $B_0 = \text{Im}\{I_{\text{ap},0}/V\}$ (dots) as function of N. Values of
$\text{Im}\{I_{\text{ap},0}/V\}$ for $N \geq 72$ are not shown.

important and the empirical criterion mentioned above cannot be used. These results
indicate that the application of numerical methods to (1.49) presents difficulties. Such
difficulties have been discussed in the literature from many points of view [16–21].
In Chapter 13 of this book, the interested reader will find detailed explanations
for the behavior observed in Figs. 1.13 and 1.14. The discussion there concerns
both integral equations. The situation is simpler in the case of the exact kernel; in
this case, the behavior of the numerical solutions can be readily inferred from the
mathematical properties of the integral equation. Knowledge of these properties also
leads to improvements of the numerical method. Even with the exact kernel, however,
difficulties exist.

   Despite the difficulties associated with solving (1.49) numerically, useful results can
be obtained by numerical methods. Many such results are available in the literature.
Chapter 2 introduces an alternative method for determining the current on a linear

antenna: $I(z)$ is represented as a linear combination of a *fixed* number (two or three) of trigonometric functions. The unknown coefficients are determined by exploiting the properties of the kernel and the right-hand side of the integral equation. Chapters 3–7 and 10–12 extend this method to arrays of cylindrical dipoles. In this case, one needs to solve a system of algebraic equations to obtain the unknown coefficients. Because the number of unknown coefficients per dipole is small, the systems of equations that result are generally much smaller than those resulting from the application of numerical methods.

2 An approximate analysis of the cylindrical antenna

2.1 The sinusoidal current

The distribution of current along a thin center-driven antenna of length $2h$ (or along a base-driven antenna of length h over an ideal ground plane) is assumed to have the sinusoidal form

$$I_z(z) = I_z(0) \frac{\sin \beta_0(h - |z|)}{\sin \beta_0 h} \tag{2.1}$$

in Sections 1.6 and 1.7. Actually, this is the correct distribution along a section of lossless coaxial line of length h that is short-circuited at $z = 0$ and terminated at $z = h$ in an infinite impedance. This is illustrated in Fig. 2.1a where the infinite impedance is obtained by means of an additional short-circuited quarter-wave section of coaxial line. In this case the current is entirely reactive, the electromagnetic field is completely confined within the coaxial shield in the form of axial standing waves and there is no radiation. When the ideal "open" end at $z = h$ is replaced by an actual one as shown in Fig. 2.1b, the distributions of current and charge are changed in a manner that resembles a crowding of the entire pattern toward the open end. In addition to a large reactive component, the current now also includes a very small resistive part. The associated electromagnetic field is still primarily a standing wave within the coaxial sleeve, but it does extend outside especially near the open end and there is some radiation. From the point of view of the transmission line the differences between currents and fields for Figs. 2.1a and 2.1b are interpreted as end-effects. If the outside shield is removed as in Fig. 2.1c these "end-effects" extend all the way to the generator and the distributions of current and charge are significantly changed over the entire length. The resistive component is now comparable in magnitude to the reactive part and the associated electromagnetic field includes a large radiation field that extends to infinity in the form of outward traveling waves. It is, of course, not at all surprising that the distributions of current along the conductors of radius a and length h are not the same in the three quite different situations represented in Figs. 2.1a–c. The boundary conditions are not alike except at $r = a$, $0 \leq z \leq h$, where the tangential electric field vanishes.

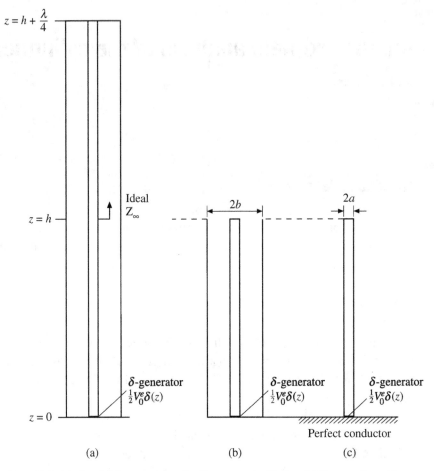

$z = h + \dfrac{\lambda}{4}$

$z = h$

Ideal
Z_∞

$2b$

$2a$

δ-generator
$\frac{1}{2}V_0^e\delta(z)$

δ-generator
$\frac{1}{2}V_0^e\delta(z)$

δ-generator
$\frac{1}{2}V_0^e\delta(z)$

$z = 0$

Perfect conductor

(a)   (b)   (c)

Figure 2.1   (a) Coaxial line terminated in Z_∞ at $z = h$. (b) Coaxial line terminated in open end. (c) Base-driven monopole over perfectly conducting ground screen.

2.2   The equation for the current

As discussed in Section 1.8, the determination of the actual current distribution along the antenna in Fig. 2.1c requires the derivation and solution of an integral equation. The derivation proceeds from the boundary condition $E_z(z) = -V_0^e\delta(z)$ on the surface $\rho = a$, $-h \leq z \leq h$, of the perfectly conducting, center-driven tubular antenna. The electric field $E_z(z)$ is expressed in terms of the vector potential defined in (1.23a) in the formula (1.24c). On the surface of the antenna where $E_z(z) = -V_0^e\delta(z)$, the following differential equation applies:

$$\left(\frac{d^2}{dz^2} + \beta_0^2\right) A_z(z) = \frac{-j\beta_0^2}{\omega} V_0^e \delta(z),$$ (2.2)

which has the solution

$$A_z(z) = \frac{-j}{c} (C_1 \cos \beta_0 z + \tfrac{1}{2} V_0^e \sin \beta_0 |z|)$$ (2.3)

if the symmetry conditions, $I_z(-z) = I_z(z)$, $A_z(-z) = A_z(z)$ are imposed. The second term in (2.3) is a particular solution to (2.2), and C_1 is a yet undetermined constant. With (1.23a) the integral equation for the current is

$$\frac{4\pi}{\mu_0} A_z(z) = \int_{-h}^{h} I_z(z') \frac{e^{-j\beta_0 R}}{R} dz' = \frac{-j4\pi}{\zeta_0} [C_1 \cos \beta_0 z + \tfrac{1}{2} V_0^e \sin \beta_0 |z|],$$ (2.4)

where V_0^e is the EMF of the delta-function generator, $\zeta_0 = \sqrt{\mu_0/\epsilon_0} \doteq 120\pi$ ohms, and $R = \sqrt{(z - z')^2 + a^2}$. By definition the driving voltage of the delta-function generator is $\lim_{z\to 0} [\phi(z) - \phi(-z)] = V_0^e$. The constant C_1 must be evaluated from the condition $I_z(\pm h) = 0$. Note that the "approximate kernel", discussed in Section 1.8, is used in (2.4).

Although it is not difficult to derive the integral equation (2.4), the problem of finding analytical solutions for the current is very complicated. What is needed is an approximate solution that is both sufficiently simple to be useful in the evaluation of the electromagnetic field and sufficiently accurate to provide quantitatively acceptable values not only of the details of the field but of the driving-point impedance. (In anticipation, it is well to note that a generalization of the method in order to make it useful in the solution of the simultaneous integral equations that occur in the analysis of arrays is also going to be required.)

The procedure to be followed in obtaining a useful approximate solution of (2.4) is straightforward and simple. It involves the replacement of the integral equation (2.4) by an approximately equivalent algebraic equation. In order to accomplish this a careful study must be made of the integral in (2.4).

2.3   Properties of integrals

The integrand in (2.4) consists of two parts: (1) the current $I_z(z)$ which is to be determined and about which nothing is known except that it vanishes at the ends $z = \pm h$, is continuous through the generator at $z = 0$, and satisfies the symmetry condition $I_z(-z) = I_z(z)$; (2) the kernel

$$K(z, z') = \frac{e^{-j\beta_0 R}}{R}, \qquad R = \sqrt{(z - z')^2 + a^2},$$ (2.5)

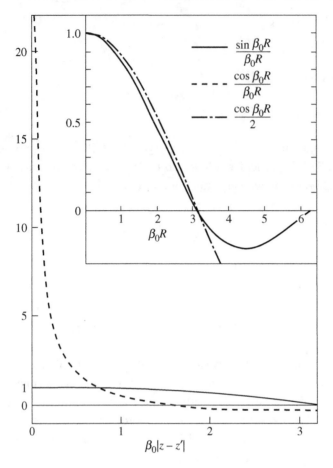

Figure 2.2 The functions $\dfrac{\sin \beta_0 R}{\beta_0 R}$, $\dfrac{\cos \beta_0 R}{\beta_0 R}$, and $\cos \dfrac{\beta_0 R}{2}$.

which may be separated into its real and imaginary parts,

$$K_R(z, z') = \frac{\cos \beta_0 R}{R}; \qquad K_I(z, z') = -\frac{\sin \beta_0 R}{R}. \tag{2.6}$$

The dimensionless quantities $K_R(z, z')/\beta_0$ and $K_I(z, z')/\beta_0$ are shown graphically in Fig. 2.2 as functions of $\beta_0|z - z'|$. A comparison in the lower figure shows that their behaviors are quite different. $K_R(z, z')/\beta_0$ has a sharp high peak precisely at $z' = z$; its magnitude $1/\beta_0 a$ is very large compared with 1 since it has been postulated that $\beta_0 a \ll 1$. On the other hand, $K_I(z, z')/\beta_0$ varies only slowly with $\beta_0|z - z'|$ and never exceeds the value 1. It is seen in the upper part of Fig. 2.2 that $\sin \beta_0 R/\beta_0 R$ is very well approximated by $\cos(\beta_0 R/2)$ in the range $0 \leq \beta_0|z - z'| \leq \pi$. Moreover, the value of $\cos(\beta_0 R/2)$ is hardly affected if the small quantity $\beta_0 a$ is neglected and $\beta_0 R$ is approximated by $\beta_0|z - z'|$.

These facts suggest the following approximations for the two parts of the integral in (2.4):

$$J_R(h, z) = \int_{-h}^{h} I_z(z') \frac{\cos \beta_0 R}{R} \, dz' = \Psi_1(z) I_z(z) \doteq \Psi_1 I_z(z) \tag{2.7}$$

$$J_I(h, z) = -\int_{-h}^{h} I_z(z') \frac{\sin \beta_0 R}{R} \, dz' = -\beta_0 \int_{-h}^{h} I_z(z') \cos \tfrac{1}{2}\beta_0(z - z') \, dz'. \tag{2.8}$$

The reasoning behind the approximation in (2.7) is simple. Since the kernel is quite small except at and very near $z' = z$, where it rises to a very large value, it is clear that the current near $z' = z$ is primarily significant in determining the value of the integral at z. In other words, the integral is approximately proportional to $I_z(z)$. The proportionality constant Ψ_1 is best determined where $I_z(z)$ is a maximum.

The integral in (2.8) may be transformed as follows:

$$J_I(h, z) = -\beta_0 \int_{-h}^{h} I_z(z') \cos \tfrac{1}{2}\beta_0(z - z') \, dz'$$

$$= -\beta_0 \int_{0}^{h} I_z(z')[\cos \tfrac{1}{2}\beta_0(z - z') + \cos \tfrac{1}{2}\beta_0(z + z')] \, dz'$$

$$= -2\beta_0 \cos \tfrac{1}{2}\beta_0 z \int_{0}^{h} I_z(z') \cos \tfrac{1}{2}\beta_0 z' \, dz'.$$

It follows that for antennas that do not greatly exceed $\beta_0 h = \pi$ in electrical half-length, specifically, $\beta_0 h \leq 5\pi/4$,

$$J_I(h, z) \doteq J_I(h, 0) \cos \tfrac{1}{2}\beta_0 z; \qquad J_I(h, 0) = -2\beta_0 \int_{0}^{h} I_z(z') \cos \tfrac{1}{2}\beta_0 z' \, dz'. \tag{2.9}$$

A further refinement in the approximation (2.7) is suggested by the fact that, while the integral on the left becomes quite small at the ends of the antenna where $z = \pm h$, the right-hand side vanishes identically at these points since $I_z(\pm h) = 0$. Evidently a better approximation than (2.7) is the following:

$$4\pi \mu_0^{-1}[A_z(z) - A_z(h)] = \int_{-h}^{h} I_z(z')[K_R(z, z') - K_R(h, z')] \, dz' \doteq \Psi_2 I_z(z), \tag{2.10}$$

where the left-hand side is simply the vector potential difference between the point (a, z) and the end (a, h) of the antenna; Ψ_2 is a new constant.

2.4   Rearranged equation for the current

In order to make use of (2.10), the integral equation (2.4) may be modified by subtracting $4\pi\mu_0^{-1}A_z(h)$ from both sides. The result is

$$4\pi\mu_0^{-1}[A_z(z) - A_z(h)] = \int_{-h}^{h} I_z(z')K_d(z, z')\,dz'$$

$$= \frac{-j4\pi}{\zeta_0}[C_1\cos\beta_0 z + \tfrac{1}{2}V_0^e\sin\beta_0|z| + U], \tag{2.11}$$

where

$$U = \frac{-j\zeta_0}{4\pi}\int_{-h}^{h} I_z(z')K(h, z')\,dz' \tag{2.12}$$

and the difference kernel is

$$K_d(z, z') = K(z, z') - K(h, z'). \tag{2.13}$$

The constant C_1 can now be expressed in terms of U and V_0^e by setting $z = h$. Since the left-hand side of (2.11) then vanishes, the right-hand side can be solved for C_1 to give

$$C_1 = -\frac{\tfrac{1}{2}V_0^e\sin\beta_0 h + U}{\cos\beta_0 h}. \tag{2.14}$$

If this value of C_1 is substituted into (2.11) the following equation is obtained:

$$\int_{-h}^{h} I_z(z')K_d(z, z')\,dz' = \frac{j4\pi}{\zeta_0\cos\beta_0 h}\left[\tfrac{1}{2}V_0^e\sin\beta_0(h - |z|) + U(\cos\beta_0 z - \cos\beta_0 h)\right]. \tag{2.15}$$

The integral equation (2.15) with (2.12) is a rearrangement of the original equation (2.4). No approximations are involved.

2.5   Reduction of integral equation to algebraic equation

The next and most important step is to make use of the information contained in (2.9) and (2.10) in order to reduce (2.15) to an approximately equivalent algebraic equation. The procedure is simple and straightforward. With (2.9) and (2.10) it is clear that the integral in (2.15) may be approximated as follows:

$$\int_{-h}^{h} I_z(z')K_d(z, z')\,dz' \doteq I_z(z)\Psi_2 + jJ_1(h, 0)\left(\cos\tfrac{1}{2}\beta_0 z - \cos\tfrac{1}{2}\beta_0 h\right). \tag{2.16}$$

If this is substituted in (2.15), the resulting equation can be solved explicitly for $I_z(z)$. It is seen to have the following zero-order form:

$$I_z(z) \doteq [I_z(z)]_0 = I_V\big[\sin\beta_0(h-|z|) + T_U(\cos\beta_0 z - \cos\beta_0 h)$$
$$+ T_D\big(\cos\tfrac{1}{2}\beta_0 z - \cos\tfrac{1}{2}\beta_0 h\big)\big], \tag{2.17a}$$

where I_V, T_U and T_D are complex coefficients. With the identity

$$\frac{\sin\beta_0(h-|z|)}{\cos\beta_0 h} = -(\sin\beta_0|z| - \sin\beta_0 h) + \tan\beta_0 h\,(\cos\beta_0 z - \cos\beta_0 h)$$

an alternative form of (2.17a) is

$$I_z(z) \doteq [I_z(z)]_0 = -I_V'[(\sin\beta_0|z| - \sin\beta_0 h) + T_U'(\cos\beta_0 z - \cos\beta_0 h)$$
$$- T_D'(\cos\tfrac{1}{2}\beta_0 z - \cos\tfrac{1}{2}\beta_0 h)], \tag{2.17b}$$

where I_V', T_U', and T_D' are complex coefficients.

This is a very significant result. It shows that an approximation of the current consists of three terms, each of which represents a different distribution. One of the terms is the simple sinusoid. As for the completely shielded transmission line, the sinusoidal component of the current is maintained directly by the generator; it does not include the components that are induced by coupling between different parts of the antenna. The currents induced by the interaction between charges moving in the more or less widely separated sections of the antenna appear in two parts. One of these, the shifted cosine, is maintained by that part of the interaction that is equivalent to a constant field acting in phase at all points along the antenna. The other part, the shifted cosine with half-angle arguments, is the correction that takes account of the phase lag introduced by the retarded instead of instantaneous interaction.

Thus, the new three-term approximation augments the conventionally assumed sinusoidal distribution with components represented by a shifted cosine and a shifted cosine with half-angle arguments, each with a complex coefficient.

It is quite possible to evaluate the coefficients Ψ_2, $J_I(h,0)$ and U that are involved in I_V, T_U, and T_D – obtained when (2.16) is substituted in (2.15). However, it is preferable to use the arguments and approximations introduced up to this point merely to determine the *form* of the distribution of current. The three new coefficients, I_V, T_U and T_D, may be evaluated directly if (2.17a) is substituted in the integral equation (2.15) and the principles involved in (2.9) and (2.10) are invoked.

The substitution of (2.17a) in the integral in (2.15) involves the following parts obtained from the real part $K_{dR}(z, z')$ of the difference kernel $K_d(z, z')$ defined in (2.13):

$$\int_{-h}^{h} \sin \beta_0(h - |z'|) K_{dR}(z, z') \, dz' \doteq \Psi_{dR} \sin \beta_0(h - |z|) \qquad (2.18a)$$

$$\int_{-h}^{h} [\cos \beta_0 z' - \cos \beta_0 h] K_{dR}(z, z') \, dz' \doteq \Psi_{dUR}(\cos \beta_0 z - \cos \beta_0 h) \qquad (2.18b)$$

$$\int_{-h}^{h} [\cos \tfrac{1}{2}\beta_0 z' - \cos \tfrac{1}{2}\beta_0 h] K_{dR}(z, z') \, dz' \doteq \Psi_{dDR}(\cos \tfrac{1}{2}\beta_0 z - \cos \tfrac{1}{2}\beta_0 h). \qquad (2.18c)$$

These expressions follow from (2.10). In order to enhance the accuracy, each part of the current is separately treated and supplied with its own coefficient. The evaluation of these coefficients is considered below.

The integrals obtained with the imaginary part $K_{dI}(z, z')$ of the difference kernel are easily approximated by the application of (2.9). Thus,

$$\int_{-h}^{h} \sin \beta_0(h - |z'|) K_{dI}(z, z') \, dz' \doteq \Psi_{dI}(\cos \tfrac{1}{2}\beta_0 z - \cos \tfrac{1}{2}\beta_0 h) \qquad (2.19a)$$

$$\int_{-h}^{h} (\cos \beta_0 z' - \cos \beta_0 h) K_{dI}(z, z') \, dz' \doteq \Psi_{dUI}(\cos \tfrac{1}{2}\beta_0 z - \cos \tfrac{1}{2}\beta_0 h) \qquad (2.19b)$$

$$\int_{-h}^{h} (\cos \tfrac{1}{2}\beta_0 z' - \cos \tfrac{1}{2}\beta_0 h) K_{dI}(z, z') \, dz' \doteq \Psi_{dDI}(\cos \tfrac{1}{2}\beta_0 z - \cos \tfrac{1}{2}\beta_0 h). \qquad (2.19c)$$

The three constants Ψ_{dI}, Ψ_{dUI} and Ψ_{dDI} are evaluated later. Finally, if the distribution (2.17a) is substituted in (2.12), the result is

$$U = \frac{-j\zeta_0 I_V}{4\pi} [\Psi_V(h) + T_U \Psi_U(h) + T_D \Psi_D(h)], \qquad (2.20)$$

where

$$\Psi_V(h) = \int_{-h}^{h} \sin \beta_0(h - |z'|) K(h, z') \, dz' \qquad (2.21a)$$

$$\Psi_U(h) = \int_{-h}^{h} (\cos \beta_0 z' - \cos \beta_0 h) K(h, z') \, dz' \qquad (2.21b)$$

$$\Psi_D(h) = \int_{-h}^{h} (\cos \tfrac{1}{2}\beta_0 z' - \cos \tfrac{1}{2}\beta_0 h) K(h, z') \, dz'. \qquad (2.21c)$$

With (2.18a–c) and (2.19a–c) the integral on the left-hand side in (2.15) is reduced to a mere sum of terms with suitable coefficients. And the integral equation as a whole has been replaced by an algebraic equation that involves the three distributions $\sin \beta_0 (h - |z|)$, $\cos \beta_0 z - \cos \beta_0 h$, and $\cos \frac{1}{2}\beta_0 z - \cos \frac{1}{2}\beta_0 h$. It is

$$\left(I_V \Psi_{dR} - \frac{j2\pi V_0^e}{\zeta_0 \cos \beta_0 h} \right) \sin \beta_0 (h - |z|)$$

$$+ \left(I_V T_U \Psi_{dUR} - \frac{j4\pi U}{\zeta_0 \cos \beta_0 h} \right)(\cos \beta_0 z - \cos \beta_0 h)$$

$$+ I_V (j\Psi_{dI} + j\Psi_{dUI} T_U + \Psi_{dD} T_D)(\cos \tfrac{1}{2}\beta_0 z - \cos \tfrac{1}{2}\beta_0 h) = 0, \quad (2.22)$$

where $\Psi_{dD} = \Psi_{dDR} + j\Psi_{dDI}$.

2.6   Evaluation of coefficients

The algebraic equation (2.22) is satisfied for all values of z when the coefficient of each of the three distributions vanishes. This step yields three equations for the determination of the coefficients I_V, T_U and T_D in (2.17a). They are:

$$I_V = \frac{j2\pi V_0^e}{\zeta_0 \Psi_{dR} \cos \beta_0 h} \quad (2.23a)$$

$$T_U[\Psi_{dUR} \cos \beta_0 h - \Psi_U(h)] - T_D \Psi_D(h) = \Psi_V(h) \quad (2.23b)$$

$$T_U \Psi_{dUI} - j T_D \Psi_{dD} = -\Psi_{dI}. \quad (2.23c)$$

The last two equations are easily solved for T_U and T_D. The results are:

$$T_U = Q^{-1}[\Psi_V(h)\Psi_{dD} - j\Psi_D(h)\Psi_{dI}] \quad (2.24a)$$

$$T_D = -jQ^{-1}\{\Psi_{dI}[\Psi_{dUR} \cos \beta_0 h - \Psi_U(h)] + \Psi_V(h)\Psi_{dUI}\} \quad (2.24b)$$

$$Q = \Psi_{dD}[\Psi_{dUR} \cos \beta_0 h - \Psi_U(h)] + j\Psi_D(h)\Psi_{dUI}. \quad (2.25)$$

The several Ψ functions in (2.24)–(2.25) are defined with (2.18a–c) and (2.19a–c) at the value of z that gives the maximum of the current distribution function. Since, in the range of interest, $\beta_0 h < 3\pi/2$, the maximum of $\sin \beta_0 (h - |z|)$ is at $z = 0$ when $\beta_0 h \leq \pi/2$ but at $z = h - \lambda/4$ when $\beta_0 h \geq \pi/2$, whereas the maxima of $(\cos \beta_0 z - \cos \beta_0 h)$ and $(\cos \frac{1}{2}\beta_0 z - \cos \frac{1}{2}\beta_0 h)$ are at $z = 0$, the following definitions are appropriate:

$$\Psi_{dR} = \Psi_{dR}(z_m), \quad \begin{cases} z_m = 0, & \beta_0 h \le \pi/2 \\ z_m = h - \lambda/4, & \beta_0 h > \pi/2 \end{cases} \tag{2.26}$$

$$\Psi_{dR}(z) = \csc \beta_0 (h - |z|) \int_{-h}^{h} \sin \beta_0 (h - |z'|)[K_R(z, z') - K_R(h, z')] \, dz' \tag{2.27}$$

$$\Psi_{dUR} = (1 - \cos \beta_0 h)^{-1} \int_{-h}^{h} (\cos \beta_0 z' - \cos \beta_0 h)[K_R(0, z') - K_R(h, z')] \, dz' \tag{2.28}$$

$$\Psi_{dD} = (1 - \cos \tfrac{1}{2}\beta_0 h)^{-1} \int_{-h}^{h} (\cos \tfrac{1}{2}\beta_0 z' - \cos \tfrac{1}{2}\beta_0 h)[K(0, z') - K(h, z')] \, dz' \tag{2.29}$$

$$\Psi_{dI} = (1 - \cos \tfrac{1}{2}\beta_0 h)^{-1} \int_{-h}^{h} \sin \beta_0 (h - |z'|)[K_I(0, z') - K_I(h, z')] \, dz' \tag{2.30}$$

$$\Psi_{dUI} = (1 - \cos \tfrac{1}{2}\beta_0 h)^{-1} \int_{-h}^{h} (\cos \beta_0 z' - \cos \beta_0 h)[K_I(0, z') - K_I(h, z')] \, dz'. \tag{2.31}$$

These integrals may be evaluated directly by high-speed computer or reduced to the tabulated generalized sine and cosine integral functions given by (1.42b–d) and the exponential integral,

$$E_a(h, z) = \int_{-h}^{h} \frac{e^{-j\beta_0 R_1}}{R_1} \, dz' = \int_{0}^{h} \left[\frac{e^{-j\beta_0 R_1}}{R_1} + \frac{e^{-j\beta_0 R_2}}{R_2} \right] dz'. \tag{2.32}$$

2.7   The approximate current and admittance

The final approximate expression for the current in an isolated cylindrical antenna for which $\beta_0 h < 3\pi/2$ and $\beta_0 a \ll 1$ is

$$I_z(z) = \frac{j2\pi V_0^e}{\zeta_0 \Psi_{dR} \cos \beta_0 h} [\sin \beta_0 (h - |z|) + T_U(\cos \beta_0 z - \cos \beta_0 h)$$

$$+ T_D(\cos \tfrac{1}{2}\beta_0 z - \cos \tfrac{1}{2}\beta_0 h)]. \tag{2.33}$$

The associated driving-point admittance is

$$Y_0 = \frac{j2\pi}{\zeta_0 \Psi_{dR} \cos \beta_0 h} [\sin \beta_0 h + T_U(1 - \cos \beta_0 h) + T_D(1 - \cos \tfrac{1}{2}\beta_0 h)]. \tag{2.34}$$

Since these formulas become indeterminate when $\beta_0 h = \pi/2$, it is convenient to use the alternative forms obtained from (2.17b) when $\beta_0 h$ is at or near $\pi/2$. They are

$$I_z(z) = \frac{-j2\pi V_0^e}{\zeta_0 \Psi_{dR}} [(\sin \beta_0 |z| - \sin \beta_0 h) + T_U'(\cos \beta_0 z - \cos \beta_0 h)$$

$$- T_D'(\cos \tfrac{1}{2}\beta_0 z - \cos \tfrac{1}{2}\beta_0 h)] \tag{2.35}$$

$$Y_0 = \frac{j2\pi}{\zeta_0 \Psi_{dR}} [\sin \beta_0 h - T_U'(1 - \cos \beta_0 h) + T_D'(1 - \cos \tfrac{1}{2}\beta_0 h)] \tag{2.36}$$

where

$$T_U' = -\frac{T_U + \sin \beta_0 h}{\cos \beta_0 h}, \qquad T_D' = \frac{T_D}{\cos \beta_0 h}. \tag{2.37}$$

T_U' and T_D' are both finite when $\beta_0 h = \pi/2$.

When the antenna is electrically short, so that $\beta_0 h < 1$, the trigonometric functions can be expanded in series and the leading terms retained. The current is then given by

$$I_z(z) = \frac{j2\pi V_0^e}{\zeta_0 \Psi_{dR}} \left[\beta_0 h \left(1 - \frac{|z|}{h}\right) + \tfrac{1}{2}\beta_0^2 h^2 T \left(1 - \frac{z^2}{h^2}\right) \right]. \tag{2.38}$$

This distribution includes triangular and parabolic components. The admittance is

$$Y_0 = \frac{j2\pi}{\zeta_0 \Psi_{dR}} [\beta_0 h + \tfrac{1}{2}\beta_0^2 h^2 T], \tag{2.39}$$

where $T = T_U + T_D/4$.

2.8   Numerical examples; comparison with experiment

Numerical computations have been made for typical antennas for which extensive measurements are available. For these antennas $a/\lambda = 7.022 \times 10^{-3}$. The parameters for the two critical lengths, $\beta_0 h = \pi/2$ with $\Omega = 2 \ln 2h/a = 8.54$ and $\beta_0 h = \pi$ with $\Omega = 9.92$ are listed below:

$$\beta_0 h = \frac{\pi}{2}: \qquad \Psi_{dR} = 6.218, \qquad T_U' = 3.085 + j3.581,$$

$$T_D' = 1.061 + j0.025 \tag{2.40a}$$

$$\beta_0 h = \pi: \qquad \Psi_{dR} = 5.737, \qquad T_U = -0.117 + j0.114,$$

$$T_D = -0.106 + j0.108. \tag{2.40b}$$

The corresponding normalized currents in amperes per volt, admittances in siemens and impedances in ohms are as follows.

For $\beta_0 h = \pi/2$,

$$\frac{I_z(z)}{V_0^e} = \{9.597 \cos \beta_0 z - 0.067 \cos \tfrac{1}{2}\beta_0 z + 0.047 - j[2.680(\sin \beta_0|z| - 1)$$

$$+ 8.269 \cos \beta_0 z - 2.843 \cos \tfrac{1}{2}\beta_0 z + 2.010]\} \times 10^{-3} \tag{2.41a}$$

$$Y_0 = (9.577 - j4.756) \times 10^{-3}, \qquad Z_0 = 83.76 + j41.60. \tag{2.41b}$$

For $\beta_0 h = \pi$,

$$\frac{I_z(z)}{V_0^e} = \{0.331(\cos \beta_0 z + 1) + 0.314 \cos \tfrac{1}{2}\beta_0 z$$

$$- j[2.905 \sin \beta_0|z| - 0.340(\cos \beta_0 z + 1) - 0.308 \cos \tfrac{1}{2}\beta_0 z]\} \times 10^{-3}$$

$$\tag{2.42a}$$

$$Y_0 = (0.976 + j0.988) \times 10^{-3}, \qquad Z_0 = 506.0 - j512.2. \tag{2.42b}$$

Note that when a sinusoidal distribution of current is assumed the corresponding impedances are for $\beta_0 h = \pi/2$, $Z_0 = 73.1 + j42.5$ (see Section 1.7); and for $\beta_0 h = \pi$, $Z_0 = \infty$.

Graphs of $I_z(z)/V_0^e = [I_z''(z) + jI_z'(z)]/V_0^e$ are presented in Figs. 2.3 and 2.4 for $\beta_0 h = \pi/2$ and π together with measured values. The approximate theoretical curves are seen to agree very well with measured values not only for $\beta_0 h = \pi/2$, but also for $\beta_0 h = \pi$.

As can be seen from Figs. 2.3 and 2.4, and especially from the latter, the theoretical currents at the driving point and, hence, the admittances differ somewhat from the measured values. In order to achieve a more accurate admittance, higher-order terms are required in the expressions for the current. Simple trigonometric functions cannot take adequate account of the rapid change in the current near the driving point when the antenna is not near resonance. Since higher-order terms are necessarily complicated, their introduction would defeat the primary purpose of this formulation, namely, to maintain a reasonably simple representation. Fortunately, there is a useful alternative. Since the only large error in the current occurs in the quadrature component of the current very near the driving point, it is possible to introduce a lumped susceptance B_c across the terminals which will correct the driving-point current and the susceptance while leaving the otherwise well-approximated current unchanged. Actually, since the use of a lumped corrective network is required in any case to take account of the local geometry of the junction between the feeding line and the antenna if quantitative accuracy is desired, the addition of B_c to the susceptance B_T of the terminal-zone network is no significant complication. In practice, it may be convenient to measure the apparent driving-point susceptance at $\beta_0 h = \pi$ and use the difference between this and the approximate theoretical value as the total lumped susceptance $B_T + B_c$ to be

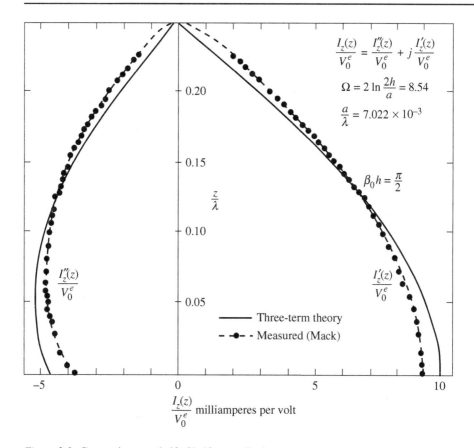

Figure 2.3 Current in upper half of half-wave dipole.

used with all theoretical values based on the approximate theory for any given ratio of a/λ.

2.9   The radiation field

The electric field in the radiation zone of an antenna with a distribution of current $I_z(z)$ is given by the integral

$$\mathbf{E}^r = \hat{\Theta} E_\Theta^r; \qquad E_\Theta^r = \frac{j\omega\mu_0}{4\pi} \sin\Theta \, \frac{e^{-j\beta_0 R_0}}{R_0} \int_{-h}^{h} I_z(z') e^{j\beta_0 z' \cos\Theta} \, dz'. \tag{2.43}$$

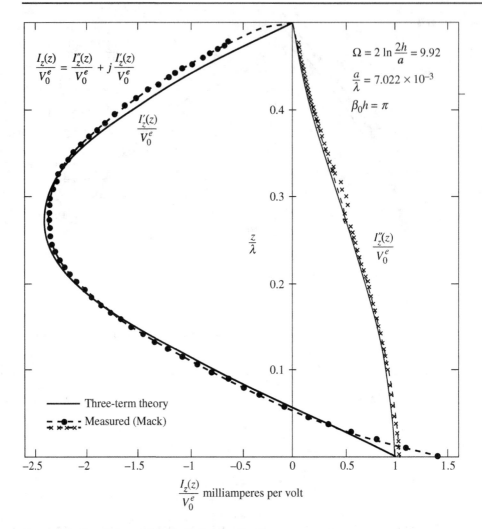

Figure 2.4 Current in upper half of full-wave dipole.

The far field maintained by the distribution (2.33) is obtained when

$$I_z(z) = \frac{j2\pi V_0^e}{\zeta_0 \Psi_{dR} \cos \beta_0 h} [\sin \beta_0 (h - |z|) + T_U (\cos \beta_0 z - \cos \beta_0 h)$$

$$+ T_D (\cos \tfrac{1}{2}\beta_0 z - \cos \tfrac{1}{2}\beta_0 h)] \tag{2.44}$$

is substituted in (2.43). The result may be expressed as follows:

$$E_\Theta^r = \frac{-V_0^e}{\Psi_{dR}} \frac{e^{-j\beta_0 R_0}}{R_0} f(\Theta, \beta_0 h), \tag{2.45a}$$

where

$$f(\Theta, \beta_0 h) = [F_m(\Theta, \beta_0 h) + T_U G_m(\Theta, \beta_0 h) + T_D D_m(\Theta, \beta_0 h)] \sec \beta_0 h. \tag{2.45b}$$

The several field functions are

$$
\begin{aligned}
F_m(\Theta, \beta_0 h) &= \frac{\beta_0}{2} \int_{-h}^{h} \sin \beta_0 (h - |z'|) e^{j\beta_0 z' \cos \Theta} \sin \Theta \, dz' \\
&= \frac{\cos(\beta_0 h \cos \Theta) - \cos \beta_0 h}{\sin \Theta}
\end{aligned}
\tag{2.46}
$$

$$
\begin{aligned}
G_m(\Theta, \beta_0 h) &= \frac{\beta_0}{2} \int_{-h}^{h} (\cos \beta_0 z' - \cos \beta_0 h) e^{j\beta_0 z' \cos \Theta} \sin \Theta \, dz' \\
&= \frac{\sin \beta_0 h \cos(\beta_0 h \cos \Theta) \cos \Theta - \cos \beta_0 h \sin(\beta_0 h \cos \Theta)}{\sin \Theta \cos \Theta}
\end{aligned}
\tag{2.47}
$$

$$
\begin{aligned}
D_m(\Theta, \beta_0 h) &= \frac{\beta_0}{2} \int_{-h}^{h} (\cos \tfrac{1}{2}\beta_0 z' - \cos \tfrac{1}{2}\beta_0 h) e^{j\beta_0 z' \cos \Theta} \sin \Theta \, dz' \\
&= \left[\frac{2\cos(\beta_0 h \cos \Theta) \sin \tfrac{1}{2}\beta_0 h - 4\sin(\beta_0 h \cos \Theta) \cos \tfrac{1}{2}\beta_0 h \cos \Theta}{1 - 4\cos^2 \Theta} \right. \\
&\quad \left. - \frac{\sin(\beta_0 h \cos \Theta) \cos \tfrac{1}{2}\beta_0 h}{\cos \Theta} \right] \sin \Theta .
\end{aligned}
\tag{2.48}
$$

For the alternative current

$$
\begin{aligned}
I_z(z) &= \frac{-j2\pi V_0^e}{\zeta_0 \Psi_{dR}} [(\sin \beta_0 |z| - \sin \beta_0 h) + T_U'(\cos \beta_0 z - \cos \beta_0 h) \\
&\quad - T_D'(\cos \tfrac{1}{2}\beta_0 z - \cos \tfrac{1}{2}\beta_0 h)],
\end{aligned}
\tag{2.49}
$$

which is useful when $\beta_0 h$ is at and near $\pi/2$, the far field is

$$
E_\Theta^r = \frac{V_0^e}{\Psi_{dR}} \frac{e^{-j\beta_0 R_0}}{R_0} f'(\Theta, \beta_0 h),
\tag{2.50a}
$$

where

$$
f'(\Theta, \beta_0 h) = H_m(\Theta, \beta_0 h) + T_U' G_m(\Theta, \beta_0 h) - T_D' D_m(\Theta, \beta_0 h).
\tag{2.50b}
$$

The new field function is

$$
\begin{aligned}
H_m(\Theta, \beta_0 h) &= \frac{\beta_0}{2} \int_{-h}^{h} (\sin \beta_0 |z'| - \sin \beta_0 h) e^{j\beta_0 z' \cos \Theta} \sin \Theta \, dz' \\
&= \frac{[1 - \cos \beta_0 h \cos(\beta_0 h \cos \Theta)] \cos \Theta - \sin \beta_0 h \sin(\beta_0 h \cos \Theta)}{\sin \Theta \cos \Theta} .
\end{aligned}
\tag{2.51}
$$

$G_m(\Theta, \beta_0 h)$ and $D_m(\Theta, \beta_0 h)$ are as in (2.47) and (2.48).

For the specific cases considered above, the coefficients are:

For $\beta_0 h = \pi/2$,

$$H_m\left(\Theta, \frac{\pi}{2}\right) = \frac{\cos\Theta - \sin\left(\frac{\pi}{2}\cos\Theta\right)}{\sin\Theta\cos\Theta} \tag{2.52a}$$

$$G_m\left(\Theta, \frac{\pi}{2}\right) = \frac{\cos\left(\frac{\pi}{2}\cos\Theta\right)}{\sin\Theta} = F_m\left(\Theta, \frac{\pi}{2}\right) \tag{2.52b}$$

$$D_m\left(\Theta, \frac{\pi}{2}\right) = \frac{\sqrt{2}}{2}\left\{ \frac{2\cos\left(\frac{\pi}{2}\cos\Theta\right) - 4\sin\left(\frac{\pi}{2}\cos\Theta\right)\cos\Theta}{1 - 4\cos^2\Theta} \right.$$

$$\left. - \frac{\sin\left(\frac{\pi}{2}\cos\Theta\right)}{\cos\Theta} \right\}\sin\Theta. \tag{2.52c}$$

For $\beta_0 h = \pi$,

$$F_m(\Theta, \pi) = \frac{\cos(\pi\cos\Theta) + 1}{\sin\Theta} \tag{2.53a}$$

$$G_m(\Theta, \pi) = \frac{\sin(\pi\cos\Theta)}{\sin\Theta\cos\Theta} \tag{2.53b}$$

$$D_m(\Theta, \pi) = \frac{2\cos(\pi\cos\Theta)\sin\Theta}{1 - 4\cos^2\Theta}. \tag{2.53c}$$

In the formulas (2.45a) and (2.50a), the field is referred to the driving voltage V_0^e. It can be referred to the current $I_z(0)$ at the driving point with the simple substitution of $I_z(0)/Y_0$ for V_0^e where Y_0 is the admittance given by (2.34) or (2.36). The field in (2.45a) is then expressed as follows:

$$E_\Theta^r = \frac{j\zeta_0 I_z(0)}{2\pi}\frac{e^{-j\beta_0 R_0}}{R_0} f_I(\Theta, \beta_0 h), \tag{2.54a}$$

where

$$f_I(\Theta, \beta_0 h) = \frac{F_m(\Theta, \beta_0 h) + T_U G_m(\Theta, \beta_0 h) + T_D D_m(\Theta, \beta_0 h)}{\sin\beta_0 h + T_U(1 - \cos\beta_0 h) + T_D(1 - \cos\frac{1}{2}\beta_0 h)}. \tag{2.54b}$$

The alternative form (2.50a) becomes

$$E_\Theta^r = \frac{j\zeta_0 I_z(0)}{2\pi}\frac{e^{-j\beta_0 R_0}}{R_0} f_I'(\Theta, \beta_0 h), \tag{2.55a}$$

where

$$f_I'(\Theta, \beta_0 h) = -\frac{H_m(\Theta, \beta_0 h) + T_U' G_m(\Theta, \beta_0 h) - T_D' D_m(\Theta, \beta_0 h)}{\sin\beta_0 h - T_U'(1 - \cos\beta_0 h) + T_D'(1 - \cos\frac{1}{2}\beta_0 h)}. \tag{2.55b}$$

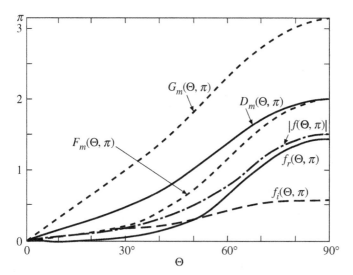

Figure 2.5 The functions $F_m(\Theta, \pi)$, $G_m(\Theta, \pi)$, $D_m(\Theta, \pi)$ and the field components $f(\Theta, \pi) = f_r(\Theta, \pi) + j f_i(\Theta, \pi)$ when $a/\lambda = 7.022 \times 10^{-3}$.

As a numerical illustration, the three functions $F_m(\Theta, \pi)$, $G_m(\Theta, \pi)$ and $D_m(\Theta, \pi)$ are shown graphically in Fig. 2.5 for a full-wave antenna. They all have nulls at $\Theta = 0$ and maxima at $\Theta = 90°$. However, $G_m(\Theta, \pi)$ and $D_m(\Theta, \pi)$ have relatively much greater values at small values of Θ than $F_m(\Theta, \pi)$.

If use is made of the numerical values of T_U and T_D given in (2.40b) for a cylindrical antenna with $a/\lambda = 7.022 \times 10^{-3}$ [for which the distribution of current is given in (2.42a) and the admittance and impedance in (2.42b)] the field factor

$$f(\Theta, \pi) = f_r(\Theta, \pi) + j f_i(\Theta, \pi) \tag{2.56}$$

may be evaluated. The real and imaginary parts $f_r(\Theta, \pi)$ and $f_i(\Theta, \pi)$ are shown in Fig. 2.5 together with the magnitude $|f(\Theta, \pi)|$. If $|f(\Theta, \pi)|$ and $F_m(\Theta, \pi)$ are divided by their respective maximum values at $\Theta = \pi/2$, two normalized functions are obtained. These resemble one another quite closely except for $\Theta < 30°$ where the first one is significantly greater. However, since the field is quite small when $\Theta < 30°$, no serious error is made in calculating the far field if the following approximations are used when $\beta_0 h \leq \pi$:

$$G_m(\Theta, \beta_0 h) \doteq \frac{G_m\left(\frac{\pi}{2}, \beta_0 h\right)}{F_m\left(\frac{\pi}{2}, \beta_0 h\right)} F_m(\Theta, \beta_0 h)$$

$$= \left(\frac{\sin \beta_0 h - \beta_0 h \cos \beta_0 h}{1 - \cos \beta_0 h}\right) F_m(\Theta, \beta_0 h) \tag{2.57a}$$

$$D_m(\Theta, \beta_0 h) \doteq \frac{D_m\left(\frac{\pi}{2}, \beta_0 h\right)}{F_m\left(\frac{\pi}{2}, \beta_0 h\right)} F_m(\Theta, \beta_0 h)$$

$$= \left(\frac{2 \sin \frac{1}{2}\beta_0 h - \beta_0 h \cos \frac{1}{2}\beta_0 h}{1 - \cos \beta_0 h}\right) F_m(\Theta, \beta_0 h) \tag{2.57b}$$

$$H_m(\Theta, \beta_0 h) \doteq \frac{H_m\left(\frac{\pi}{2}, \beta_0 h\right)}{F_m\left(\frac{\pi}{2}, \beta_0 h\right)} F_m(\Theta, \beta_0 h)$$

$$= \left(\frac{1 - \cos \beta_0 h - \beta_0 h \sin \beta_0 h}{1 - \cos \beta_0 h}\right) F_m(\Theta, \beta_0 h). \tag{2.57c}$$

These approximations are equivalent to the use of the far-field distribution associated with a sinusoidal current, but normalizing this to the value at $\Theta = \pi/2$ obtained from the three-term form of the current.

2.10 An approximate two-term theory

For all purposes when $\beta_0 h \leq \pi/2$ and for determining the far-field and driving-point impedance when $\beta_0 h \leq 5\pi/4$, the difference between the distribution functions $F_{0z} = \cos \beta_0 z - \cos \beta_0 h$ and $H_{0z} = \cos \frac{1}{2}\beta_0 z - \cos \frac{1}{2}\beta_0 h$ is small and the formulation may be simplified further by consolidating the two terms. If F_{0z} is substituted everywhere for H_{0z}, the current is well approximated as follows when $\beta_0 h \leq 5\pi/4$:

$$I_z(z) = \frac{j2\pi V_0^e}{\zeta_0 \Psi_{dR} \cos \beta_0 h} [\sin \beta_0(h - |z|) + T(\cos \beta_0 z - \cos \beta_0 h)] \tag{2.58}$$

or, in the form useful near $\beta_0 h = \pi/2$,

$$I_z(z) = \frac{-j2\pi V_0^e}{\zeta_0 \Psi_{dR}} [\sin \beta_0|z| - \sin \beta_0 h + T'(\cos \beta_0 z - \cos \beta_0 h)], \tag{2.59}$$

where T and T' are obtained by forming $T_U + T_D$ and $T_U' - T_D'$ but with the substitution $\Psi_{dD} = \Psi_{dU}$, $\Psi_D(h) = \Psi_U(h)$. The function T is simply

$$T = \frac{\Psi_V(h) - j\Psi_{dI} \cos \beta_0 h}{\Psi_{dU} \cos \beta_0 h - \Psi_U(h)}. \tag{2.60}$$

T' is given by

$$T' = -\frac{T + \sin \beta_0 h}{\cos \beta_0 h}$$

$$= \frac{[\Psi_V(h) - \Psi_U(h) \sin \beta_0 h] \sec \beta_0 h + \Psi_{dU} \sin \beta_0 h - j\Psi_{dI}}{\Psi_U(h) - \Psi_{dU} \cos \beta_0 h} \qquad (2.61a)$$

$$= \frac{[\Psi_{dU} + E_a(h,h)] \sin \beta_0 h - j\Psi_{dI} - S_a(h,h)}{C_a(h,h) - [\Psi_{dU} + E_a(h,h)] \cos \beta_0 h}. \qquad (2.61b)$$

Since $\Psi_U(h) = \Psi_V(h) = C_a(h,h)$ when $\beta_0 h = \pi/2$, this reduces simply to

$$T' = \frac{\Psi_{dU} - j\Psi_{dI} - S_a\left(\dfrac{\lambda}{4}, \dfrac{\lambda}{4}\right) + E_a\left(\dfrac{\lambda}{4}, \dfrac{\lambda}{4}\right)}{C_a\left(\dfrac{\lambda}{4}, \dfrac{\lambda}{4}\right)} \qquad (2.62)$$

when $\beta_0 h = \pi/2$.

For the numerical cases considered in Section 2.8 for $a/\lambda = 7.022 \times 10^{-3}$, the results for the two-term theory are:

$$\beta_0 h = \pi/2: \quad \left. \begin{array}{l} \Psi_{dR} = 6.218, \qquad T' = 2.65 + j3.79; \\[2mm] Y_0 = (10.17 - j4.43) \times 10^{-3} \text{ siemens} \end{array} \right\} \qquad (2.63)$$

$$\beta_0 h = \pi: \quad \left. \begin{array}{l} \Psi_{dR} = 5.737, \qquad T = -0.172 + j0.175; \\[2mm] Y_0 = (1.021 + j1.000) \times 10^{-3} \text{ siemens}. \end{array} \right\} \qquad (2.64)$$

These are seen to be in good agreement with the values obtained with the more accurate three-term theory. A more extensive list of numerical values of Ψ_{dR}, T, T' and $Y_0 = G_0 + jB_0$ is in Table 1 of Appendix I.

As with the three-term theory, the quadrature component of the current near the driving point is not adequately represented by simple trigonometric functions so that the same expedient previously described must be used in order to obtain quantitative agreement with measured values of the susceptance. The lumped value of B_c to be used with the two-term theory differs only slightly from that for the three-term theory. For $a/\lambda = 7.022 \times 10^{-3}$, it is $B_c = 0.72$ millisiemens. This value must be added to the two-term susceptance $B_0 + B_T$ (where B_T is the terminal-zone correction for a particular transmission-line-antenna junction) in order to obtain the measurable apparent susceptance $B_{sa} = B_0 + 0.72 + B_T$. It is seen in Fig. 2.6 that $B_0 + 0.72$ is in good agreement with the King–Middleton second-order values of B_0 and the apparent measured values corrected for the terminal-zone effects, $B_{sa} - B_T$.

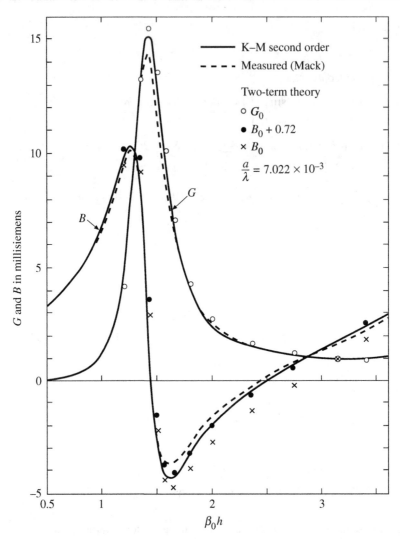

Figure 2.6 King–Middleton second-order admittance $Y = G + jB$. Two-term zero-order admittance $Y_0 = G_0 + jB_0$, and measured.

2.11   The receiving antenna

The general method of analysis introduced in this chapter as a means of analyzing the center-driven cylindrical antenna can be extended readily to the center-loaded receiving antenna in an incident plane-wave field. For the purposes of this book[1] – which includes the properties of receiving arrays – it is sufficient to treat only the simple case of normal incidence with the electric vector parallel to the z-axis which

---

[1] A more detailed analysis of the receiving antenna is in [1], Chapter 4.

is the axis of the antenna. The antenna is, therefore, in the plane wave front of the incident wave which may be assumed to travel in the positive x direction. That is $E_z^{\text{inc}}(x) = E_z^{\text{inc}} e^{-j\beta_0 x}$ where E_z^{inc} is the constant amplitude. The boundary condition that requires the total tangential electric field to vanish on the surface of the antenna gives

$$\left(\frac{d^2}{dz^2} + \beta_0^2\right) A_z(z) = -\frac{j\beta_0^2}{\omega} E_z^{\text{inc}} = -\beta_0^2 A_z^{\text{inc}} \tag{2.65}$$

instead of (2.2). In (2.65), $A_z(z)$ is the vector potential due to the currents in the receiving antenna itself. A_z^{inc} is the constant amplitude of the vector potential maintained on the surface of the antenna by the distant transmitter. Note that $E_z^{\text{inc}} = -j\omega A_z^{\text{inc}}$. Since the axis of the antenna lies in the wave front, even symmetry obtains with respect to z for both the current and the associated vector potential so that $A_z(-z) = A_z(z)$, $I_z(-z) = I_z(z)$. It follows that, on an unloaded receiving or scattering antenna, the vector potential on the surface of the antenna due to the currents in the antenna satisfies the equations

$$4\pi\mu_0^{-1} A_z(z) = \int_{-h}^{h} I_z(z') \frac{e^{-j\beta_0 R}}{R} \, dz' = \frac{-j4\pi}{\zeta_0} [C_1 \cos\beta_0 z + U^{\text{inc}}], \tag{2.66}$$

where C_1 is an arbitrary constant to be evaluated from the condition $I_z(h) = 0$ and

$$U^{\text{inc}} = \frac{E_z^{\text{inc}}}{\beta_0} = -\frac{j\omega A_z^{\text{inc}}}{\beta_0}. \tag{2.67}$$

This integral equation is like (2.4) with an added constant term on the right and with $V_0^e = 0$. If the same rearrangement is carried out as for (2.11), the result is

$$4\pi\mu_0^{-1}[A_z(z) - A_z(h)] = \int_{-h}^{h} I_z(z') K_d(z, z') \, dz'$$

$$= \frac{-j4\pi}{\zeta_0} [C_1 \cos\beta_0 z + U + U^{\text{inc}}], \tag{2.68}$$

where U, as defined in (2.12), is proportional to the vector potential at $z = h$, $\rho = a$ due to the currents in the antenna; U^{inc}, as defined in (2.67), is proportional to the vector potential maintained on the surface of the antenna by the distant transmitter. The sum $U + U^{\text{inc}}$ is proportional to the total vector potential on the surface of the antenna.

Since the integral equation (2.68) is just like (2.11) with $V_0^e = 0$, it follows that the rearranged equation corresponding to (2.15) is

$$\int_{-h}^{h} I_z(z') K_d(z, z') \, dz' = \frac{j4\pi(U + U^{\text{inc}})}{\zeta_0 \cos\beta_0 h} (\cos\beta_0 z - \cos\beta_0 h). \tag{2.69}$$

The approximate solution of this equation is like (2.33) with $V_0^e = 0$. It is obtained with $\Psi_{dD} = \Psi_{dU}$, $\Psi_D(h) = \Psi_U(h)$. It is

$$I_z(z) = \frac{j4\pi U^{\text{inc}}}{\zeta_0 Q} [\Psi_{dD}(\cos \beta_0 z - \cos \beta_0 h) - j\Psi_{dUI}(\cos \tfrac{1}{2}\beta_0 z - \cos \tfrac{1}{2}\beta_0 h)], \quad (2.70)$$

where Q is defined in (2.25), Ψ_{dD} in (2.29) and Ψ_{dUI} in (2.31). This solution for the unloaded receiving antenna corresponds to the three-term form (2.33) for the driven antenna. Corresponding to the simpler two-term approximation (2.58) for the driven antenna is the expression

$$I_z(z) = \frac{j4\pi E_z^{\text{inc}}}{\omega\mu_0} \left[\frac{\cos \beta_0 z - \cos \beta_0 h}{\Psi_{dU} \cos \beta_0 h - \Psi_U(h)} \right]. \quad (2.71)$$

In (2.71), U^{inc} has been set equal to E_z^{inc}/β_0, its value when the normally incident plane wave has its electric component parallel to the axis of the receiving antenna. When the axis of the antenna is oriented at an arbitrary angle with respect to the incident E-vector, the distribution of current is much more complicated. In particular, if the antenna does not lie in the plane wave front (surface of constant phase) of the incident field, the current and the vector potential have components with both even and odd symmetries with respect to z.

If the antenna is cut at $z = 0$ and a load Z_L is connected in series with the halves of the antenna, the current in the antenna is readily obtained. Note first that, if a generator with voltage V_0^e is connected across the terminals at $z = 0$ instead of the load, the resulting current in the antenna is

$$I_z(z) = V_0^e v(z) + U^{\text{inc}} u(z), \quad (2.72)$$

where $v(z)$ is $I_z(z)/V_0^e$ as obtained from (2.44) and $u(z)$ is $I_z(z)/U^{\text{inc}}$ as obtained from (2.71). If now V_0^e is replaced by the negative of the voltage drop across a load Z_L that is connected across the terminals of the antenna, that is,

$$V_0^e = -I_z(0)Z_L, \quad (2.73)$$

V_0^e is readily eliminated between (2.73) and (2.72). With Z_0, the driving-point impedance of the same antenna when driven, the result can be expressed as follows:

$$I_z(z) = U^{\text{inc}} \left[u(z) - v(z)u(0) \frac{Z_L Z_0}{Z_L + Z_0} \right]. \quad (2.74)$$

This is the current at any point z in the center-loaded receiving antenna. The current in the load at $z = 0$ is simply

$$I_z(0) = U^{\text{inc}} u(0) \frac{Z_0}{Z_0 + Z_L} \quad (2.75)$$

since $v(0) = 1/Z_0$. When $Z_L = 0$, this gives $I_z(0) = U^{\text{inc}}u(0)$. The voltage drop across the load is

$$I_z(0)Z_L = U^{\text{inc}}u(0)\frac{Z_0 Z_L}{Z_0 + Z_L}. \tag{2.76}$$

When $Z_L \to \infty$, this is the open-circuit voltage across the terminals at $z = 0$. That is

$$V(Z_L \to \infty) = \lim_{Z_L \to \infty} I_z(0)Z_L = U^{\text{inc}}u(0)Z_0 = [I_z(0)Z_0]_{Z_L=0}. \tag{2.77}$$

It is now clear that with (2.67) and (2.75) the current in the load is given by

$$I_z(0) = \frac{V(Z_L \to \infty)}{Z_0 + Z_L} = \frac{E_z^{\text{inc}}u(0)Z_0}{\beta_0(Z_0 + Z_L)}. \tag{2.78}$$

This is the current in a simple series circuit that consists of a generator with EMF equal to the open-circuit voltage across the terminals of the receiving antenna in series with the impedance of the antenna and the impedance of the load. The same conclusion is readily obtained by the application of Thévenin's theorem.

The quantity

$$u(0)Z_0/\beta_0 = 2h_e\left(\frac{\pi}{2}\right), \tag{2.79}$$

which occurs in (2.78) and is dimensionally a length, is known as the complex effective length of the receiving antenna with actual length $2h$. With (2.79), the current in the load is

$$I_z(0) = \frac{2h_e\left(\frac{\pi}{2}\right)E_z^{\text{inc}}}{Z_0 + Z_L}. \tag{2.80}$$

Note that (2.78), (2.79) and (2.80) apply only when the axis of the receiving antenna is parallel to the incident electric vector and, therefore, also perpendicular to the direction of propagation of the incident wave. Similar but more general expressions that involve the orientation of the antenna relative to the incident wave and the direction of the electric vector in the plane wave front are available in the literature.[2]

---

[2] See, for example, [1], Chapter 4, Section 4.

3 The two-element array

3.1 The method of symmetrical components

An array is a configuration of two or more antennas so arranged that the superposition of the electromagnetic fields maintained at distant points by the currents in the individual elements yields a resultant field that fulfils certain desirable directional properties. Since the individual elements in an array are quite close together – the distance between adjacent elements is often a half-wavelength or less – the currents in them necessarily interact. It follows that the distributions of both the amplitude and the phase of the current along each element depend not only on the length, radius, and driving voltage of that element, but also on the distributions in amplitude and phase of the currents along all elements in the array. Since these currents are the primary unknowns from which the radiation field and the driving-point admittance are computed, it is essential that they be determined accurately and not arbitrarily assumed to have identical distributions, as in uniform array theory.

In order to introduce the properties of arrays in a simple and direct manner, it is advantageous to study first the two-element array in some detail. The integral equation (2.15) for the current in a single isolated antenna is readily generalized to apply to the two identical parallel and non-staggered elements shown in Fig. 3.1. It is merely necessary to add to the vector potential on the surface of each element the contributions by the current in the other element. Thus, for element 1, the vector potential difference is

$$4\pi \mu_0^{-1}[A_{1z}(z) - A_{1z}(h)]$$

$$= \int_{-h}^{h} [I_{1z}(z')K_{11d}(z, z') + I_{2z}(z')K_{12d}(z, z')]\,dz'$$

$$= \frac{j4\pi}{\zeta_0 \cos \beta_0 h} \left[\tfrac{1}{2}V_{10} \sin \beta_0(h - |z|) + U_1(\cos \beta_0 z - \cos \beta_0 h)\right]. \tag{3.1}$$

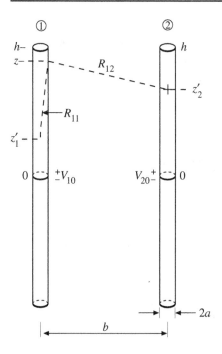

Figure 3.1 Two identical parallel antennas.

Similarly, for element 2:

$$4\pi\mu_0^{-1}[A_{2z}(z) - A_{2z}(h)]$$

$$= \int_{-h}^{h} [I_{1z}(z')K_{21d}(z, z') + I_{2z}(z')K_{22d}(z, z')]\,dz'$$

$$= \frac{j4\pi}{\zeta_0 \cos\beta_0 h} [\tfrac{1}{2}V_{20}\sin\beta_0(h - |z|) + U_2(\cos\beta_0 z - \cos\beta_0 h)]. \tag{3.2}$$

In these expressions

$$K_{11d}(z, z') = \frac{e^{-j\beta_0 R_{11}}}{R_{11}} - \frac{e^{-j\beta_0 R_{11h}}}{R_{11h}} = K_{11}(z, z') - K_{11}(h, z') \tag{3.3a}$$

$$K_{12d}(z, z') = \frac{e^{-j\beta_0 R_{12}}}{R_{12}} - \frac{e^{-j\beta_0 R_{12h}}}{R_{12h}} = K_{12}(z, z') - K_{12}(h, z') \tag{3.3b}$$

with

$$R_{11} = \sqrt{(z - z')^2 + a^2}, \qquad R_{11h} = \sqrt{(h - z')^2 + a^2} \tag{3.4a}$$

$$R_{12} = \sqrt{(z - z')^2 + b^2}, \qquad R_{12h} = \sqrt{(h - z')^2 + b^2}. \tag{3.4b}$$

$K_{22d}(z, z')$ and $K_{21d}(z, z')$ are obtained from the above formulas when 1 is substituted for 2 and 2 for 1 in the subscripts.

The two simultaneous integral equations (3.1) and (3.2) can be reduced to a *single* equation in two special cases. These are (a) the so-called *zero-phase sequence* when the two driving voltages are identical so that the two currents are the same, and (b) the *first-phase sequence* when the two driving voltages and the resulting two currents are equal in magnitude but 180° out of phase. Specifically, for the zero-phase sequence,

$$V_{10} = V_{20} = V^{(0)}, \qquad I_{1z}(z) = I_{2z}(z) = I_z^{(0)}(z), \tag{3.5}$$

so that the equations (3.1) and (3.2) both become

$$\int_{-h}^{h} I_z^{(0)}(z') K_d^{(0)}(z, z') \, dz'$$

$$= \frac{j4\pi}{\zeta_0 \cos \beta_0 h} \left[\tfrac{1}{2} V^{(0)} \sin \beta_0 (h - |z|) + U^{(0)} (\cos \beta_0 z - \cos \beta_0 h) \right], \tag{3.6}$$

where

$$U^{(0)} = \frac{-j\zeta_0}{4\pi} \int_{-h}^{h} I_z^{(0)}(z') K^{(0)}(h, z') \, dz' \tag{3.7}$$

and

$$K^{(0)}(z, z') = \frac{e^{-j\beta_0 R_{11}}}{R_{11}} + \frac{e^{-j\beta_0 R_{12}}}{R_{12}}, \left.\begin{array}{c} \\ \\ \end{array}\right\}$$
$$K_d^{(0)}(z, z') = K^{(0)}(z, z') - K^{(0)}(h, z'). \tag{3.8}$$

Similarly, for the first-phase sequence,

$$V_{10} = -V_{20} = V^{(1)}, \qquad I_{1z}(z) = -I_{2z}(z) = I_z^{(1)}(z), \tag{3.9}$$

so that the two equations again become alike and equal to

$$\int_{-h}^{h} I_z^{(1)}(z') K_d^{(1)}(z, z') \, dz'$$

$$= \frac{j4\pi}{\zeta_0 \cos \beta_0 h} \left[\tfrac{1}{2} V^{(1)} \sin \beta_0 (h - |z|) + U^{(1)} (\cos \beta_0 z - \cos \beta_0 h) \right], \tag{3.10}$$

where

$$U^{(1)} = \frac{-j\zeta_0}{4\pi} \int_{-h}^{h} I_z^{(1)}(z') K^{(1)}(h, z') \, dz' \tag{3.11}$$

and

$$K^{(1)}(z, z') = \frac{e^{-j\beta_0 R_{11}}}{R_{11}} - \frac{e^{-j\beta_0 R_{12}}}{R_{12}}, \left.\begin{array}{c} \\ \\ \end{array}\right\}$$
$$K_d^{(1)}(z, z') = K^{(1)}(z, z') - K^{(1)}(h, z'). \tag{3.12}$$

Note that the two phase sequences differ only in the sign in $K^{(0)}(z, z')$ and $K^{(1)}(z, z')$.

If (3.6) can be solved for the zero-phase-sequence current $I_z^{(0)}(z)$ and (3.10) for the first-phase-sequence current $I_z^{(1)}(z)$, the currents $I_{1z}(z)$ and $I_{2z}(z)$ maintained by the arbitrary voltages V_{10} and V_{20} can be obtained simply by superposition. This follows directly if $V^{(0)}$ and $V^{(1)}$ are so chosen that

$$V^{(0)} = \tfrac{1}{2}[V_{10} + V_{20}], \qquad V^{(1)} = \tfrac{1}{2}[V_{10} - V_{20}]. \tag{3.13}$$

In this case,

$$V_{10} = V^{(0)} + V^{(1)}, \qquad V_{20} = V^{(0)} - V^{(1)} \tag{3.14}$$

so that

$$I_{1z}(z) = I_z^{(0)}(z) + I_z^{(1)}(z), \qquad I_{2z}(z) = I_z^{(0)}(z) - I_z^{(1)}(z). \tag{3.15}$$

3.2   Properties of integrals

The two integral equations (3.6) and (3.10) for the phase-sequence currents are formally exactly like the equation (2.15) for the isolated antenna. They differ only in the kernels of the integrals on the left and in the definitions (3.7) and (3.11) of the functions U. Each of these is now the algebraic sum of two terms that are identical except that the radius a appears in the first term, the distance b between the elements in the second term.

The two elements may be considered close together when $\beta_0 b < 1$ and $b < h$. In this case, since b satisfies substantially the same conditions as a, the behavior of the integrals that contain b corresponds closely to that of the integrals that contain a. These are discussed in the preceding chapter. When the separation of the two elements is such that $\beta_0 b > 1$ but not so great that $\beta_0\sqrt{b^2 + h^2}$ differs negligibly from $\beta_0 b$, the vector potentials maintained by the currents on the one antenna at points along the surface of the other differ significantly from one another in phase due to retarded action. A detailed investigation[1] has been made of the four current distributions involved in (2.33) and (2.35) which lead to the integrals

$$S_1(z) = \sin \beta_0 h \; C_\rho(h, z) - \cos \beta_0 h \; S_\rho(h, z) \tag{3.16a}$$

$$S_2(z) = -S_\rho(h, z) + \sin \beta_0 h \; E_\rho(h, z) \tag{3.16b}$$

$$C(z) = C_\rho(h, z) - \cos \beta_0 h \; E_\rho(h, z) \tag{3.16c}$$

$$D(z) = D_\rho(h, z) - \cos \tfrac{1}{2}\beta_0 h \; E_\rho(h, z), \tag{3.16d}$$

---

[1] See [1], pp. 1456–1458.

where $C_\rho(h, z)$ and $S_\rho(h, z)$ are defined in (1.42b–d) with a replaced by ρ, $E_\rho(h, z)$ is defined in (2.32) with a replaced by ρ, and

$$D_\rho(h, z) = \int_0^h \cos \tfrac{1}{2}\beta_0 z' \left(\frac{e^{-j\beta_0 R_1}}{R_1} + \frac{e^{-j\beta_0 R_2}}{R_2} \right) dz', \tag{3.17a}$$

where

$$R_1 = \sqrt{(z - z')^2 + b^2}, \qquad R_2 = \sqrt{(z + z')^2 + b^2}. \tag{3.17b}$$

Accurately calculated three-dimensional graphs[2] of $S_1(z)$, $S_2(z)$, $C(z)$, and $D(z)$ for $\beta_0 h = \pi$, $\pi/2$, and 1.2 show that the real parts of all four of these functions quite accurately correspond to cylindrical waves when $b \geq h$ and $-h \leq z \leq h$ and that they are reasonable approximations when $\beta_0 b \geq 1$, $-h \leq z \leq h$. The respective amplitudes are $S_1(0)$, $S_2(0)$, $C(0)$, and $D(0)$. It follows that for most purposes the distribution of current induced by the real part of any of the four functions (3.16a–d) in a parallel dipole when this is at distances b that satisfy the inequality $\beta_0 b \geq 1$ can be assumed to be like that induced by a normally incident plane wave, namely $\cos \beta_0 z - \cos \beta_0 h$. In special cases, such as that of the resonant circular array analyzed in Chapter 11, the more severe condition $b \geq h$ may have to be enforced. The distribution of current induced by the imaginary part of any of the four functions is closely approximated by $\cos \tfrac{1}{2}\beta_0 z - \cos \tfrac{1}{2}\beta_0 h$ for all values of $\beta_0 b$ when $\beta_0 h \leq 5\pi/4$.

In order to verify the correctness of these conclusions the difference integral

$$S_b(h, z) - S_b(h, h) = \int_{-h}^h \sin \beta_0 |z'| \, K_d(z, z') \, dz' \tag{3.18}$$

has been evaluated for $\beta_0 h = \pi$ over a range of values of $\beta_0 b$ extending from 0.04 to 4.5. The real and imaginary parts are shown in Fig. 3.2 together with the three trigonometric functions, $\sin \beta_0 z$, $(\cos \beta_0 z + 1)$ and $\cos \tfrac{1}{2}\beta_0 z$, to which the sine, shifted cosine and shifted cosine with half-angle arguments reduce when $\beta_0 h = \pi$. For convenience in the graphical comparison, $-(\cos \beta_0 z + 1)$ and $-\cos \tfrac{1}{2}\beta_0 z$ are shown. It is evident from Fig. 3.2 that the real part of the difference integral approximates $\sin \beta_0 z$ when $\beta_0 b < 1$, $1 + \cos \beta_0 z$ when $\beta_0 b \geq 1$. On the other hand, the imaginary part resembles the shifted cosine with half-angle arguments, in this case $\cos \tfrac{1}{2}\beta_0 z$, for all values of $\beta_0 b$.

As a consequence of these observations, the following approximate representation of the integrals in (3.6) and (3.10) is indicated. For $\beta_0 b < 1$,

$$\int_{-h}^h I_z(z') \left(\frac{\cos \beta_0 R_{12}}{R_{12}} - \frac{\cos \beta_0 R_{12h}}{R_{12h}} \right) dz' = \Psi_{12}(z) I_z(z) \doteq \Psi_{12} I_z(z), \tag{3.19a}$$

---

[2] For examples, see [1], Figs. 1, 2, and 3.

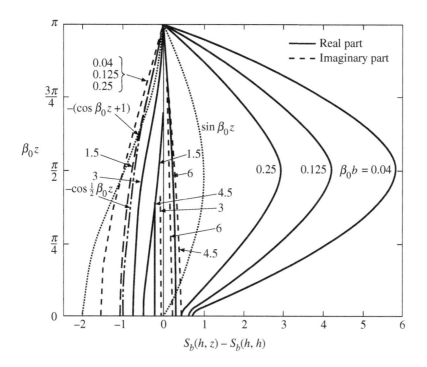

Figure 3.2 The functions $S_b(h, z) - S_b(h, h)$ compared with three trigonometric functions.

where Ψ_{12} is a constant. For $\beta_0 b \geq 1$,

$$\int_{-h}^{h} I_z(z') \left(\frac{\cos \beta_0 R_{12}}{R_{12}} - \frac{\cos \beta_0 R_{12h}}{R_{12h}} \right) dz' \sim \cos \beta_0 z - \cos \beta_0 h. \tag{3.19b}$$

For all values of $\beta_0 b$

$$\int_{-h}^{h} I_z(z') \left(\frac{\sin \beta_0 R_{12}}{R_{12}} - \frac{\sin \beta_0 R_{12h}}{R_{12h}} \right) dz' \sim \cos \tfrac{1}{2}\beta_0 z - \cos \tfrac{1}{2}\beta_0 h. \tag{3.19c}$$

3.3   Reduction of integral equations for phase sequences to algebraic equations

The relations (3.19a, b, c), combined with the results of Chapter 2, indicate that the current in each of the two coupled elements in both phase sequences must have leading terms that are well approximated by the following zero-order, three-term formula:

$$I_z^{(m)}(z) \doteq [I_z^{(m)}(z)]_0 = I_V^{(m)}[\sin \beta_0(h - |z|) + T_U^{(m)}(\cos \beta_0 z - \cos \beta_0 h)$$

$$+ T_D^{(m)}(\cos \tfrac{1}{2}\beta_0 z - \cos \tfrac{1}{2}\beta_0 h)], \tag{3.20}$$

where $m = 0$ or 1 and $I_V^{(m)}$, $T_U^{(m)}$ and $T_D^{(m)}$ are complex coefficients that must be determined.

The substitution of (3.20) into the integral in (3.6) and (3.10) involves the following parts obtained from the real part $K_{dR}^{(m)}(z, z')$ of the difference kernel $K_d^{(m)}(z, z')$ defined in (3.8) and (3.12) for $m = 0$ and 1:

$$\beta_0 b < 1, \qquad \int_{-h}^{h} \sin \beta_0 (h - |z'|) K_{dR}^{(m)}(z, z') \, dz' \doteq \Psi_{dR}^{(m)} \sin \beta_0 (h - |z|) \qquad (3.21a)$$

$$\beta_0 b \geq 1, \qquad \int_{-h}^{h} \sin \beta_0 (h - |z'|) K_{dR}^{(m)}(z, z') \, dz' \doteq \Psi_{dR} \sin \beta_0 (h - |z|)$$

$$+ \Psi_{d\Sigma R}^{(m)}(\cos \beta_0 z - \cos \beta_0 h). \qquad (3.21b)$$

For all values of $\beta_0 b$,[3]

$$\int_{-h}^{h} (\cos \beta_0 z' - \cos \beta_0 h) K_{dR}^{(m)}(z, z') \, dz' \doteq \Psi_{dUR}^{(m)}(\cos \beta_0 z - \cos \beta_0 h) \qquad (3.22a)$$

$$\int_{-h}^{h} (\cos \tfrac{1}{2}\beta_0 z' - \cos \tfrac{1}{2}\beta_0 h) K_{dR}^{(m)}(z, z') \, dz' \doteq \Psi_{dDR}^{(m)}(\cos \tfrac{1}{2}\beta_0 z - \cos \tfrac{1}{2}\beta_0 h). \qquad (3.22b)$$

The corresponding integrals obtained with the imaginary part $K_{dI}^{(m)}(z, z')$ of the difference kernel $K_d^{(m)}(z, z')$ are valid for all values of $\beta_0 b$. They are:

$$\int_{-h}^{h} \sin \beta_0 (h - |z'|) K_{dI}^{(m)}(z, z') \, dz' \doteq \Psi_{dI}^{(m)}(\cos \tfrac{1}{2}\beta_0 z - \cos \tfrac{1}{2}\beta_0 h) \qquad (3.23a)$$

$$\int_{-h}^{h} (\cos \beta_0 z' - \cos \beta_0 h) K_{dI}^{(m)}(z, z') \, dz' \doteq \Psi_{dUI}^{(m)}(\cos \tfrac{1}{2}\beta_0 z - \cos \tfrac{1}{2}\beta_0 h) \qquad (3.23b)$$

$$\int_{-h}^{h} (\cos \tfrac{1}{2}\beta_0 z' - \cos \tfrac{1}{2}\beta_0 h) K_{dI}^{(m)}(z, z') \, dz' \doteq \Psi_{dDI}^{(m)}(\cos \tfrac{1}{2}\beta_0 z - \cos \tfrac{1}{2}\beta_0 h). \qquad (3.23c)$$

The several Ψ functions introduced in the above expressions are defined as follows:

$$\beta_0 b < 1: \begin{cases} \Psi_{dR}^{(m)} = \Psi_{dR}^{(m)}(z_m); & \begin{cases} z_m = 0, & \beta_0 h \leq \pi/2 \\ z_m = h - \lambda/4, & \beta_0 h > \pi/2 \end{cases} \qquad (3.24a) \\[2ex] \Psi_{dR}^{(m)}(z) = \csc \beta_0 (h - |z|) \int_{-h}^{h} \sin \beta_0 (h - |z'|) K_{dR}^{(m)}(z, z') \, dz' \qquad (3.24b) \end{cases}$$

---

[3] Strictly according to (3.19b) the integral in (3.22b) should be treated separately with different behaviors when $\beta_0 b < 1$ and $\beta_0 b \geq 1$. However, since the distributions $\cos \beta_0 z - \cos \beta_0 h$ and $\cos(\beta_0 z/2) - \cos(\beta_0 h/2)$ are quite similar when $\beta_0 h \leq 5\pi/4$, and since considerable complication is avoided by not making this separation, the relation (3.22b) is used for both real and imaginary parts of the kernel and for all spacings.

$$\beta_0 b \geq 1: \begin{cases} \Psi_{dR} \text{ is defined in (2.26) and (2.27)} & (3.25a) \\[1em] \Psi_{d\Sigma R}^{(m)} = (-1)^m (1 - \cos \beta_0 h)^{-1} \int_{-h}^{h} \sin \beta_0 (h - |z'|) \\[1em] \qquad \times \left[\dfrac{\cos \beta_0 R_{12}}{R_{12}} - \dfrac{\cos \beta_0 R_{12h}}{R_{12h}} \right] dz'. & (3.25b) \end{cases}$$

The following apply for all values of $\beta_0 b$:

$$\Psi_{dI}^{(m)} = (1 - \cos \tfrac{1}{2}\beta_0 h)^{-1} \int_{-h}^{h} \sin \beta_0 (h - |z'|) K_{dI}^{(m)}(0, z') \, dz' \tag{3.26}$$

$$\Psi_{dUR}^{(m)} = (1 - \cos \beta_0 h)^{-1} \int_{-h}^{h} (\cos \beta_0 z' - \cos \beta_0 h) K_{dR}^{(m)}(0, z') \, dz' \tag{3.27a}$$

$$\Psi_{dUI}^{(m)} = (1 - \cos \tfrac{1}{2}\beta_0 h)^{-1} \int_{-h}^{h} (\cos \beta_0 z' - \cos \beta_0 h) K_{dI}^{(m)}(0, z') \, dz' \tag{3.27b}$$

$$\Psi_{dD}^{(m)} = (1 - \cos \tfrac{1}{2}\beta_0 h)^{-1} \int_{-h}^{h} (\cos \tfrac{1}{2}\beta_0 z' - \cos \tfrac{1}{2}\beta_0 h) K_d^{(m)}(0, z') \, dz'. \tag{3.28}$$

For each pair of real and imaginary parts, the notation $\Psi_d = \Psi_{dR} + j\Psi_{dI}$ will be used.

When (3.20) is substituted in $U^{(m)}$ as defined in (3.7) and (3.11), the notation of (2.20)–(2.21c) applies in the form

$$U^{(m)} = \frac{-j\zeta_0}{4\pi} I_V^{(m)} [\Psi_V^{(m)}(h) + T_U^{(m)} \Psi_U^{(m)}(h) + T_D^{(m)} \Psi_D^{(m)}(h)], \tag{3.29}$$

where

$$\Psi_V^{(m)}(h) = \int_{-h}^{h} \sin \beta_0 (h - |z'|) K^{(m)}(h, z') \, dz' \tag{3.30}$$

$$\Psi_U^{(m)}(h) = \int_{-h}^{h} (\cos \beta_0 z' - \cos \beta_0 h) K^{(m)}(h, z') \, dz' \tag{3.31}$$

$$\Psi_D^{(m)}(h) = \int_{-h}^{h} (\cos \tfrac{1}{2}\beta_0 z' - \cos \tfrac{1}{2}\beta_0 h) K^{(m)}(h, z') \, dz' \tag{3.32}$$

with $m = 0, 1$.

If the approximate formulas for the several parts of the integrals when $\beta_0 b < 1$ are substituted in (3.6) and (3.10), an algebraic equation is obtained that is just like (2.22) for the single antenna but with superscripts (m) on I, T, Ψ, V and U. It follows that (2.23a), (2.24), (2.25) and (2.26) give the solutions for $I_V^{(m)}, T_U^{(m)}$ and $T_D^{(m)}$ if superscripts (m) are affixed to all Ψ's and $V^{(m)}$ replaces V_0^e.

When $\beta_0 b \geq 1$, the equation corresponding to (2.22) has the following slightly different form:

$$\left(I_V^{(m)} \Psi_{dR} - \frac{j2\pi V^{(m)}}{\zeta_0 \cos \beta_0 h} \right) \sin \beta_0 (h - |z|)$$

$$+ \left(I_V^{(m)} \Psi_{d\Sigma R}^{(m)} + I_V^{(m)} T_U^{(m)} \Psi_{dUR}^{(m)} - \frac{j4\pi U^{(m)}}{\zeta_0 \cos \beta_0 h} \right) (\cos \beta_0 z - \cos \beta_0 h)$$

$$+ I_V^{(m)} (j\Psi_{dI}^{(m)} + j\Psi_{dUI}^{(m)} T_U^{(m)} + \Psi_{dD}^{(m)} T_D^{(m)}) (\cos \tfrac{1}{2}\beta_0 z - \cos \tfrac{1}{2}\beta_0 h) = 0. \quad (3.33)$$

If the coefficients of the trigonometric functions are individually equated to zero and (3.29) is substituted for $U^{(m)}$, three relations corresponding to (2.23a–c) are obtained. They are readily solved to give

$$I_V^{(m)} = \frac{j2\pi V^{(m)}}{\zeta_0 \Psi_{dR} \cos \beta_0 h} \quad (3.34)$$

$$T_U^{(m)} = \{ \Psi_{dD}^{(m)} [\Psi_V^{(m)}(h) - \Psi_{d\Sigma R}^{(m)} \cos \beta_0 h] - j\Psi_D^{(m)}(h) \Psi_{dI}^{(m)} \} / Q^{(m)} \quad (3.35)$$

$$T_D^{(m)} = -j\{ \Psi_{dI}^{(m)} [\Psi_{dUR}^{(m)} \cos \beta_0 h - \Psi_U^{(m)}(h)]$$

$$+ \Psi_{dUI}^{(m)} [\Psi_V^{(m)}(h) - \Psi_{d\Sigma R}^{(m)} \cos \beta_0 h] \} / Q^{(m)} \quad (3.36)$$

$$Q^{(m)} = \Psi_{dD}^{(m)} [\Psi_{dUR}^{(m)} \cos \beta_0 h - \Psi_U^{(m)}(h)] + j\Psi_D^{(m)}(h) \Psi_{dUI}^{(m)}. \quad (3.37)$$

As throughout this chapter, $m = 0, 1$.

3.4   The phase-sequence currents and admittances

With the three coefficients $I_V^{(m)}$, $T_U^{(m)}$ and $T_D^{(m)}$ determined, the phase-sequence currents and the admittances may be written down directly. When $\beta_0 b < 1$, they are

$$I_z^{(m)}(z) = \frac{j2\pi V^{(m)}}{\zeta_0 \Psi_{dR}^{(m)} \cos \beta_0 h} [\sin \beta_0 (h - |z|) + T_U^{(m)} (\cos \beta_0 z - \cos \beta_0 h)$$

$$+ T_D^{(m)} (\cos \tfrac{1}{2}\beta_0 z - \cos \tfrac{1}{2}\beta_0 h)] \quad (3.38)$$

$$Y^{(m)} = \frac{j2\pi}{\zeta_0 \Psi_{dR}^{(m)} \cos \beta_0 h} [\sin \beta_0 h + T_U^{(m)} (1 - \cos \beta_0 h) + T_D^{(m)} (1 - \cos \tfrac{1}{2}\beta_0 h)]$$

$$\quad (3.39)$$

with

$$T_U^{(m)} = [\Psi_V^{(m)}(h)\Psi_{dD}^{(m)} - j\Psi_D^{(m)}(h)\Psi_{dI}^{(m)}]/Q^{(m)} \tag{3.40}$$

$$T_D^{(m)} = -j\{\Psi_{dI}^{(m)}[\Psi_{dUR}^{(m)}\cos\beta_0 h - \Psi_U^{(m)}(h)] + \Psi_V^{(m)}(h)\Psi_{dUI}^{(m)}\}/Q^{(m)} \tag{3.41}$$

$$Q^{(m)} = \Psi_{dD}^{(m)}[\Psi_{dUR}^{(m)}\cos\beta_0 h - \Psi_U^{(m)}(h)] + j\Psi_D^{(m)}(h)\Psi_{dUI}^{(m)}. \tag{3.42}$$

The functions $\Psi^{(m)}$ are defined in (3.24), (3.26)–(3.28) and (3.30)–(3.32). For the zero-phase sequence, $m = 0$; for the first-phase sequence, $m = 1$.

As for the single antenna, these formulas for $I_z^{(m)}(z)$ and $Y_0^{(m)}$ become indeterminate when $\beta_0 h = \pi/2$. Convenient alternative forms when $\beta_0 h$ is at or near $\pi/2$ are

$$I_z^{(m)}(z) = \frac{-j2\pi V^{(m)}}{\zeta_0 \Psi_{dR}^{(m)}} [(\sin\beta_0|z| - \sin\beta_0 h) + T_U'^{(m)}(\cos\beta_0 z - \cos\beta_0 h)$$

$$- T_D'^{(m)}(\cos\tfrac{1}{2}\beta_0 z - \cos\tfrac{1}{2}\beta_0 h)] \tag{3.43}$$

$$Y^{(m)} = \frac{j2\pi}{\zeta_0 \Psi_{dR}^{(m)}} [\sin\beta_0 h - T_U'^{(m)}(1 - \cos\beta_0 h) + T_D'^{(m)}(1 - \cos\tfrac{1}{2}\beta_0 h)] \tag{3.44}$$

where, as in (2.37),

$$T_U'^{(m)} = -\frac{T_U^{(m)} + \sin\beta_0 h}{\cos\beta_0 h}, \qquad T_D'^{(m)} = \frac{T_D^{(m)}}{\cos\beta_0 h}. \tag{3.45}$$

$T_U^{(m)}$ and $T_D^{(m)}$ are given in (3.40) and (3.41).

When $\beta_0 b \geq 1$, the general form of the expressions for the phase-sequence current and admittance are similar to those for $\beta_0 b < 1$. They are

$$I_z^{(m)}(z) = \frac{j2\pi V^{(m)}}{\zeta_0 \Psi_{dR}\cos\beta_0 h} [\sin\beta_0(h - |z|) + T_U^{(m)}(\cos\beta_0 z - \cos\beta_0 h)$$

$$+ T_D^{(m)}(\cos\tfrac{1}{2}\beta_0 z - \cos\tfrac{1}{2}\beta_0 h)] \tag{3.46}$$

$$Y^{(m)} = \frac{j2\pi}{\zeta_0 \Psi_{dR}\cos\beta_0 h} [\sin\beta_0 h + T_U^{(m)}(1 - \cos\beta_0 h) + T_D^{(m)}(1 - \cos\tfrac{1}{2}\beta_0 h)] \tag{3.47}$$

with $T_U^{(m)}$ and $T_D^{(m)}$ given by (3.35) and (3.36) with (3.37). Similarly, when $\beta_0 h$ is near $\pi/2$ and $\beta_0 b \geq 1$,

$$I_z^{(m)}(z) = \frac{-j2\pi V^{(m)}}{\zeta_0 \Psi_{dR}} [(\sin\beta_0|z| - \sin\beta_0 h) + T_U'^{(m)}(\cos\beta_0 z - \cos\beta_0 h)$$

$$- T_D'^{(m)}(\cos\tfrac{1}{2}\beta_0 z - \cos\tfrac{1}{2}\beta_0 h)] \tag{3.48}$$

$$Y^{(m)} = \frac{j2\pi}{\zeta_0 \Psi_{dR}} [\sin\beta_0 h - T_U'^{(m)}(1 - \cos\beta_0 h) + T_D'^{(m)}(1 - \cos\tfrac{1}{2}\beta_0 h)]. \tag{3.49}$$

The parameters $T_U'^{(m)}$ and $T_D'^{(m)}$ are defined as in (3.45); $T_U^{(m)}$ and $T_D^{(m)}$ are given by (3.35) and (3.36).

Note that the currents and admittances when $\beta_0 b \geq 1$ differ from those when $\beta_0 b < 1$ not only in the T (or T') parameters but also in the appearance of Ψ_{dR} for the isolated antenna instead of $\Psi_{dR}^{(m)}$ for the coupled pair.

3.5 Currents for arbitrarily driven antennas; self- and mutual admittances and impedances

With the phase-sequence currents $I_z^{(0)}(z)$ and $I_z^{(1)}(z)$ determined, it is straightforward to obtain the expressions for the currents $I_{1z}(z)$ and $I_{2z}(z)$ in the two antennas when they are driven by the arbitrary voltages V_{10} and V_{20}. If

$$V^{(0)} = \tfrac{1}{2}(V_{10} + V_{20}), \qquad V^{(1)} = \tfrac{1}{2}(V_{10} - V_{20}) \tag{3.50}$$

it follows that, when $\beta_0 b \geq 1$,

$$I_{1z}(z) = I_z^{(0)}(z) + I_z^{(1)}(z) = V_{10}v(z) + V_{20}w(z) \tag{3.51a}$$

$$I_{2z}(z) = I_z^{(0)}(z) - I_z^{(1)}(z) = V_{10}w(z) + V_{20}v(z), \tag{3.51b}$$

where

$$v(z) = \frac{j2\pi}{\zeta_0 \Psi_{dR} \cos \beta_0 h} [\sin \beta_0(h - |z|) + \tfrac{1}{2}(T_U^{(0)} + T_U^{(1)})(\cos \beta_0 z - \cos \beta_0 h)$$

$$+ \tfrac{1}{2}(T_D^{(0)} + T_D^{(1)})(\cos \tfrac{1}{2}\beta_0 z - \cos \tfrac{1}{2}\beta_0 h)] \tag{3.52a}$$

$$w(z) = \frac{j2\pi}{\zeta_0 \Psi_{dR} \cos \beta_0 h} [\tfrac{1}{2}(T_U^{(0)} - T_U^{(1)})(\cos \beta_0 z - \cos \beta_0 h)$$

$$+ \tfrac{1}{2}(T_D^{(0)} - T_D^{(1)})(\cos \tfrac{1}{2}\beta_0 z - \cos \tfrac{1}{2}\beta_0 h)]. \tag{3.52b}$$

Alternatively, when $\beta_0 h$ is near $\pi/2$,

$$v(z) = \frac{-j2\pi}{\zeta_0 \Psi_{dR}} [(\sin \beta_0 |z| - \sin \beta_0 h) + \tfrac{1}{2}(T_U'^{(0)} + T_U'^{(1)})(\cos \beta_0 z - \cos \beta_0 h)$$

$$- \tfrac{1}{2}(T_D'^{(0)} + T_D'^{(1)})(\cos \tfrac{1}{2}\beta_0 z - \cos \tfrac{1}{2}\beta_0 h)] \tag{3.52c}$$

$$w(z) = \frac{-j2\pi}{\zeta_0 \Psi_{dR}} [\tfrac{1}{2}(T_U'^{(0)} - T_U'^{(1)})(\cos \beta_0 z - \cos \beta_0 h)$$

$$- \tfrac{1}{2}(T_D'^{(0)} - T_D'^{(1)})(\cos \tfrac{1}{2}\beta_0 z - \cos \tfrac{1}{2}\beta_0 h)]. \tag{3.52d}$$

The corresponding expressions when $\beta_0 b < 1$ are easily obtained from (3.38). The driving-point currents may be expressed in the form

$$I_{1z}(0) = V_{10}Y_{s1} + V_{20}Y_{12} \tag{3.53a}$$

$$I_{2z}(0) = V_{10}Y_{21} + V_{20}Y_{s2}, \tag{3.53b}$$

where Y_{s1} and Y_{s2} are the self-admittances, Y_{12} and Y_{21} are the mutual admittances. They are given by

$$Y_{s1} = Y_{s2} = v(0) = \tfrac{1}{2}(Y^{(0)} + Y^{(1)}) \tag{3.54}$$

$$Y_{21} = Y_{12} = w(0) = \tfrac{1}{2}(Y^{(0)} - Y^{(1)}). \tag{3.55}$$

Specifically,

$$Y_{s1} = Y_{s2} = v(0) = \frac{j\pi}{\zeta_0 \Psi_{dR} \cos \beta_0 h} [2 \sin \beta_0 h + (T_U^{(0)} + T_U^{(1)})(1 - \cos \beta_0 h)$$

$$+ (T_D^{(0)} + T_D^{(1)})(1 - \cos \tfrac{1}{2}\beta_0 h)] \tag{3.56}$$

$$Y_{12} = Y_{21} = w(0) = \frac{j\pi}{\zeta_0 \Psi_{dR} \cos \beta_0 h} [(T_U^{(0)} - T_U^{(1)})(1 - \cos \beta_0 h)$$

$$+ (T_D^{(0)} - T_D^{(1)})(1 - \cos \tfrac{1}{2}\beta_0 h)]. \tag{3.57}$$

When $\beta_0 h = \pi/2$, the self- and mutual admittances are

$$Y_{s2} = Y_{s1} = \frac{j\pi}{\zeta_0 \Psi_{dR}} [2 \sin \beta_0 h - (T_U'^{(0)} + T_U'^{(1)})(1 - \cos \beta_0 h)$$

$$+ (T_D'^{(0)} + T_D'^{(1)})(1 - \cos \tfrac{1}{2}\beta_0 h)] \tag{3.58}$$

$$Y_{21} = Y_{12} = \frac{-j\pi}{\zeta_0 \Psi_{dR}} [(T_U'^{(0)} - T_U'^{(1)})(1 - \cos \beta_0 h) - (T_D'^{(0)} - T_D'^{(1)})(1 - \cos \tfrac{1}{2}\beta_0 h)]. \tag{3.59}$$

The associated self- and mutual impedances are the coefficients of the currents in the equations

$$V_{10} = I_{1z}(0)Z_{s1} + I_{2z}(0)Z_{12} \tag{3.60a}$$

$$V_{20} = I_{1z}(0)Z_{21} + I_{2z}(0)Z_{s2}. \tag{3.60b}$$

It is readily shown that

$$Z_{s1} = Y_{s2}/D = \tfrac{1}{2}(Z^{(0)} + Z^{(1)}); \qquad Z_{s2} = Y_{s1}/D = \tfrac{1}{2}(Z^{(0)} + Z^{(1)}) \tag{3.61a}$$

$$Z_{12} = -Y_{21}/D = \tfrac{1}{2}(Z^{(0)} - Z^{(1)}); \qquad Z_{21} = -Y_{12}/D = \tfrac{1}{2}(Z^{(0)} - Z^{(1)}), \tag{3.61b}$$

where $D = Y_{s1}Y_{s2} - Y_{12}Y_{21} = (Z_{s1}Z_{s2} - Z_{12}Z_{21})^{-1}$. If lumped impedances Z_1 and Z_2 are connected in series with V_{10} and V_{20}, Z_{s1} and Z_{s2} in (3.60a, b) are replaced by $Z_{11} = Z_{s1} + Z_1$ and $Z_{22} = Z_{s2} + Z_2$.

This completes the general formulation for the currents and admittances of two parallel antennas driven by the arbitrary voltages V_{10} and V_{20}.

3.6   Currents for one driven, one parasitic antenna

If antenna 2 is parasitic instead of driven and is center-loaded by an arbitrary impedance Z_2, the driving voltage V_{20} may be replaced by the negative of the voltage drop across the load. Thus,

$$V_{20} = -I_{2z}(0)Z_2 = -I_{2z}(0)/Y_2. \tag{3.62}$$

If (3.62) is substituted in (3.53b), the result is

$$I_{2z}(0) = V_{10}\frac{Y_{21}}{1 + Y_{s2}Z_2} = -V_{10}\frac{Z_{21}}{Z_{s2}(Z_{s1} + Z_2) + Z_{12}Z_{21}}, \tag{3.63}$$

so that

$$V_{20} = -V_{10}\left(\frac{Y_{21}}{Y_2 + Y_{s2}}\right) = V_{10}\frac{Z_{21}Z_2}{Z_{s2}(Z_{s1} + Z_2) + Z_{12}Z_{21}}. \tag{3.64}$$

It follows from (3.51a, b) that

$$I_{1z}(z) = V_{10}\left[v(z) - \left(\frac{Y_{21}}{Y_2 + Y_{s2}}\right)w(z)\right] \tag{3.65a}$$

$$I_{2z}(z) = V_{10}\left[w(z) - \left(\frac{Y_{21}}{Y_2 + Y_{s2}}\right)v(z)\right]. \tag{3.65b}$$

The driving-point admittance and impedance are

$$Y_{1in} = \frac{I_{1z}(0)}{V_{10}} = Y_{s1} - \frac{Y_{21}Y_{12}}{Y_2 + Y_{s2}} \tag{3.66a}$$

$$Z_{1in} = \frac{1}{Y_{1in}} = \frac{Z_{s1}(Z_{s2} + Z_2) + Z_{12}Z_{21}}{Z_{s2} + Z_2}. \tag{3.66b}$$

Note that when $Z_2 = 0$ or $Y_2 = \infty$,

$$Y_{1in} = Y_{s1}; \qquad I_{1z}(z) = V_{10}v(z); \qquad I_{2z}(z) = V_{10}w(z). \tag{3.67a}$$

Alternatively, when $Z_2 = \infty$ or $Y_2 = 0$,

$$Z_{1in} = Z_{s1}; \qquad I_{1z}(z) = V_{10}\left[v(z) - \frac{Y_{21}}{Y_{s2}}w(z)\right];$$

$$I_{2z}(z) = V_{10}\left[w(z) - \frac{Y_{21}}{Y_{s2}}v(z)\right]. \qquad \qquad (3.67b)$$

The parasitic element is tuned to resonance when $Y_2 = jB_2$ and $B_2 = -B_{s2}$ in $Y_{s2} = G_{s2} + jB_{s2}$. With this choice, $Y_{21}/(Y_2 + Y_{s2})$ is maximized so that

$$I_{1z}(z) = V_{10}\left[v(z) - \frac{Y_{21}}{G_{s2}}w(z)\right] \qquad \qquad (3.68a)$$

$$I_{2z}(z) = V_{10}\left[w(z) - \frac{Y_{21}}{G_{s2}}v(z)\right]. \qquad \qquad (3.68b)$$

Since the coefficient Y_{21}/G_{s2} is of the order of magnitude of one, the coefficients of $v(z)$ and $w(z)$ are comparable. It follows that the distributions of $I_{1z}(z)$ and $I_{2z}(z)$ are roughly similar, whereas when $Z_2 = 0$ as in (3.67a), they are quite different unless $\beta_0 h$ is near $\pi/2$.

3.7   The couplet

Perhaps the most interesting two-element array is the couplet in which the distance between the elements is $\lambda/4$ and the currents at the driving points are equal in amplitude but differ in phase by a quarter period. That is

$$I_{2z}(0) = jI_{1z}(0). \qquad \qquad (3.69)$$

It follows from (3.60a, b) that with $Z_{12} = Z_{21}$ and $Z_{s2} = Z_{s1}$,

$$V_{10} = I_{1z}(0)[Z_{s1} + jZ_{12}] \qquad \qquad (3.70a)$$

$$V_{20} = I_{2z}(0)[Z_{s1} - jZ_{12}] = I_{1z}(0)[Z_{12} + jZ_{s1}]. \qquad \qquad (3.70b)$$

Hence,

$$Z_{1in} = Z_{s1} + jZ_{12}, \qquad Z_{2in} = Z_{s1} - jZ_{12}. \qquad \qquad (3.70c)$$

The distributions of current are obtained from (3.51a, b). Thus

$$I_{1z}(z) = V_{10}\left[v(z) + \frac{Z_{12} + jZ_{s1}}{Z_{s1} + jZ_{12}}w(z)\right] \qquad \qquad (3.71a)$$

$$I_{2z}(z) = V_{10}\left[w(z) + \frac{Z_{12} + jZ_{s1}}{Z_{s1} + jZ_{12}}v(z)\right]. \qquad \qquad (3.71b)$$

Instead of specifying the driving-point currents $I_{1z}(0)$ and $I_{2z}(0)$ as in (3.69), the driving voltages may be assigned so that

$$V_{20} = jV_{10}. \tag{3.72}$$

It then follows from (3.53a, b) that

$$I_{1z}(0) = V_{10}(Y_{s1} + jY_{12}) \tag{3.73a}$$

$$I_{2z}(0) = V_{20}(Y_{s1} - jY_{12}) = V_{10}(Y_{12} + jY_{s1}). \tag{3.73b}$$

The driving-point admittances are

$$Y_{1in} = Y_{s1} + jY_{12}, \qquad Y_{2in} = Y_{s1} - jY_{12}. \tag{3.74}$$

The currents are obtained from (3.51a, b) with (3.61a, b). Thus,

$$I_{1z}(z) = V_{10}\left[v(z) + \frac{Y_{12} + jY_{s1}}{Y_{s1} - jY_{12}}w(z)\right]$$

$$= V_{10}\left[v(z) - \frac{Z_{12} - jZ_{s1}}{Z_{s1} + jZ_{12}}w(z)\right] \tag{3.75a}$$

$$I_{2z}(z) = V_{10}\left[w(z) + \frac{Y_{12} + jY_{s1}}{Y_{s1} - jY_{12}}v(z)\right]$$

$$= V_{10}\left[w(z) - \frac{Z_{12} - jZ_{s1}}{Z_{s1} + jZ_{12}}v(z)\right]. \tag{3.75b}$$

The currents are not the same when $I_{1z}(0)$ and $I_{2z}(0)$ are specified as when V_{10} and V_{20} are assigned. Note that

$$[I_{1z}(z)]_I - [I_{1z}(z)]_V = 2V_{10}Z_{12}w(z) \tag{3.76a}$$

$$[I_{2z}(z)]_I - [I_{2z}(z)]_V = 2V_{10}Z_{12}v(z). \tag{3.76b}$$

If the currents differ significantly, the field patterns cannot be the same.

3.8   Field patterns

The radiation field of an array of two parallel elements is the vector sum of the fields maintained by the currents in the individual elements. In terms of the spherical coordinates R, Θ, Φ, that have their origin midway between the centers of the two elements, the individual electric fields are readily expressed in the form (2.45a, b) for the currents (3.51a, b). Thus

$$E^r_{\Theta 1} = -\frac{1}{\Psi_{dR}} \frac{e^{-j\beta_0 R_1}}{R} [V_{10} f(\Theta, \beta_0 h) + V_{20} g(\Theta, \beta_0 h)] \tag{3.77a}$$

$$E^r_{\Theta 2} = -\frac{1}{\Psi_{dR}} \frac{e^{-j\beta_0 R_2}}{R} [V_{10} g(\Theta, \beta_0 h) + V_{20} f(\Theta, \beta_0 h)], \tag{3.77b}$$

where

$$R_1 = R + \frac{b}{2} \cos \Phi \sin \Theta \tag{3.78a}$$

$$R_2 = R - \frac{b}{2} \cos \Phi \sin \Theta \tag{3.78b}$$

$$f(\Theta, \beta_0 h) = [F_m(\Theta, \beta_0 h) + \tfrac{1}{2}(T_U^{(0)} + T_U^{(1)}) G_m(\Theta, \beta_0 h)$$

$$+ \tfrac{1}{2}(T_D^{(0)} + T_D^{(1)}) D_m(\Theta, \beta_0 h)] \sec \beta_0 h \tag{3.79}$$

$$g(\Theta, \beta_0 h) = [\tfrac{1}{2}(T_U^{(0)} - T_U^{(1)}) G_m(\Theta, \beta_0 h) + \tfrac{1}{2}(T_D^{(0)} - T_D^{(1)}) D_m(\Theta, \beta_0 h)] \sec \beta_0 h. \tag{3.80}$$

The field functions $F_m(\Theta, \beta_0 h)$, $G_m(\Theta, \beta_0 h)$ and $D_m(\Theta, \beta_0 h)$ are defined in (2.46), (2.47) and (2.48). Alternatively, when $\beta_0 h$ is near $\pi/2$, the fields for the currents (3.52c, d) are:

$$E^r_{\Theta 1} = \frac{1}{\Psi_{dR}} \frac{e^{-j\beta_0 R_1}}{R} [V_{10} f'(\Theta, \beta_0 h) + V_{20} g'(\Theta, \beta_0 h)] \tag{3.81a}$$

$$E^r_{\Theta 2} = \frac{1}{\Psi_{dR}} \frac{e^{-j\beta_0 R_2}}{R} [V_{10} g'(\Theta, \beta_0 h) + V_{20} f'(\Theta, \beta_0 h)] \tag{3.81b}$$

$$f'(\Theta, \beta_0 h) = H_m(\Theta, \beta_0 h) + \tfrac{1}{2}(T_U^{\prime(0)} + T_U^{\prime(1)}) G_m(\Theta, \beta_0 h)$$

$$- \tfrac{1}{2}(T_D^{\prime(0)} + T_D^{\prime(1)}) D_m(\Theta, \beta_0 h) \tag{3.82}$$

$$g'(\Theta, \beta_0 h) = \tfrac{1}{2}(T_U^{\prime(0)} - T_U^{\prime(1)}) G_m(\Theta, \beta_0 h) - \tfrac{1}{2}(T_D^{\prime(0)} - T_D^{\prime(1)}) D_m(\Theta, \beta_0 h). \tag{3.83}$$

The function $H_m(\Theta, \beta_0 h)$ is defined in (2.51).

The resultant radiation field of the arbitrarily driven two-element array is

$$E^r_{\Theta} = E^r_{\Theta 1} + E^r_{\Theta 2} = \frac{-1}{\Psi_{dR}} \frac{e^{-j\beta_0 R}}{R} \{[V_{10} f(\Theta, \beta_0 h)$$

$$+ V_{20} g(\Theta, \beta_0 h)] e^{-j(\beta_0 b/2) \cos \Phi \sin \Theta}$$

$$+ [V_{10} g(\Theta, \beta_0 h) + V_{20} f(\Theta, \beta_0 h)] e^{j(\beta_0 b/2) \cos \Phi \sin \Theta}\}. \tag{3.84}$$

When $\beta_0 h$ is near $\pi/2$, $-f'(\Theta, \beta_0 h)$ and $-g'(\Theta, \beta_0 h)$ may be substituted, respectively, for $f(\Theta, \beta_0 h)$ and $g(\Theta, \beta_0 h)$.

3.9   The two-term approximation

As pointed out in Section 2.10, the difference between the distribution functions $F_{0z} = \cos \beta_0 z - \cos \beta_0 h$ and $H_{0z} = \cos \frac{1}{2}\beta_0 z - \cos \frac{1}{2}\beta_0 h$ is relatively unimportant in the determination of the far-field and the driving-point admittance of an isolated antenna when $\beta_0 h \leq 5\pi/4$. This is also true of the far-field and driving-point admittances of two coupled antennas provided the interaction between them is not sensitive to small changes in the current distributions. When both elements are driven by comparable voltages and when the distance between them is sufficiently great so that $\beta_0 b \geq 1$, it may be assumed that the substitution of $\cos \beta_0 z - \cos \beta_0 h$ for $\cos \frac{1}{2}\beta_0 z - \cos \frac{1}{2}\beta_0 h$ can produce no important change in the admittances or the far-field. When one element is parasitic and unloaded, the three-term approximation is automatically reduced to two terms since the distribution $\sin \beta_0 (h - |z|)$ is excited only by a generator or an equivalent voltage drop across a load. Correspondingly, the two-term approximation is reduced to a single term. However, this is quite adequate for many purposes. In anticipation, it may be added at this point that when an array consists of one driven antenna and many parasitic elements, at least two terms are desirable in the representation of the current distributions. This is considered in a later chapter.

As for the single antenna, the two-term approximations are quickly obtained from the three-term formulas by the simple substitution of $\cos \beta_0 z - \cos \beta_0 h$ for $\cos \frac{1}{2}\beta_0 z - \cos \frac{1}{2}\beta_0 h$ and the representation of the resulting coefficient $(T_U + T_D)$ by T. It is implicit that $\Psi_{dD} \to \Psi_{dU}$, $\Psi_D(h) \to \Psi_U(h)$. Thus, the phase-sequence currents and admittances (3.46) and (3.47) become, for $\beta_0 b \geq 1$,

$$I_z^{(m)}(z) = \frac{j2\pi V^{(m)}}{\zeta_0 \Psi_{dR} \cos \beta_0 h} [\sin \beta_0 (h - |z|) + T^{(m)}(\cos \beta_0 z - \cos \beta_0 h)] \tag{3.85}$$

$$Y^{(m)} = \frac{j2\pi}{\zeta_0 \Psi_{dR} \cos \beta_0 h} [\sin \beta_0 h + T^{(m)}(1 - \cos \beta_0 h)], \tag{3.86}$$

where

$$T^{(m)} = T_U^{(m)} + T_D^{(m)} = -\frac{\Psi_V^{(m)}(h) - (\Psi_{d\Sigma R}^{(m)} + j\Psi_{dI}^{(m)})\cos \beta_0 h}{\Psi_U^{(m)}(h) - \Psi_{dU}^{(m)} \cos \beta_0 h}. \tag{3.87}$$

Similarly, when $\beta_0 h$ is near $\pi/2$, (3.48) and (3.49) reduce to

$$I_z^{(m)}(z) = \frac{-j2\pi V^{(m)}}{\zeta_0 \Psi_{dR}} [(\sin \beta_0 |z| - \sin \beta_0 h) + T'^{(m)}(\cos \beta_0 z - \cos \beta_0 h)] \tag{3.88}$$

$$Y^{(m)} = \frac{j2\pi}{\zeta_0 \Psi_{dR}} [\sin \beta_0 h - T'^{(m)}(1 - \cos \beta_0 h)], \tag{3.89}$$

where

$$T'^{(m)} = -\frac{T^{(m)} + \sin \beta_0 h}{\cos \beta_0 h}$$

$$= \frac{[\Psi_V^{(m)}(h) - \Psi_U^{(m)}(h) \sin \beta_0 h] \sec \beta_0 h + \Psi_{dU}^{(m)} \sin \beta_0 h - j\Psi_{dI}^{(m)} - \Psi_{d\Sigma R}^{(m)}}{\Psi_U^{(m)}(h) - \Psi_{dU}^{(m)} \cos \beta_0 h}$$

$$= \frac{[\Psi_{dU}^{(m)} + E_a(h, h)] \sin \beta_0 h - j\Psi_{dI}^{(m)} - S_a(h, h) - \Psi_{d\Sigma R}^{(m)}}{C_a(h, h) - [\Psi_{dU} + E_a(h, h)] \cos \beta_0 h}. \tag{3.90}$$

Note that when $\beta_0 h = \pi/2$, $\Psi_U^{(m)}(\lambda/4) = \Psi_V^{(m)}(\lambda/4)$.

As an example, the phase-sequence currents have been evaluated specifically for two antennas for which $\Omega = 2 \ln(2h/a) = 10$, $\beta_0 h = \pi$ and $\beta_0 b = 1.5$. For these

$$\Psi_{dR} = 5.834, \qquad \Psi_{d\Sigma R}^{(0)} = -0.245, \qquad \Psi_{d\Sigma R}^{(1)} = 0.245 \tag{3.91a}$$

$$\left.\begin{array}{l} \Psi_{dI}^{(0)} = -0.633 - 0.524 = -1.157; \\ \Psi_{dI}^{(1)} = -0.633 + 0.524 = -0.109 \end{array}\right\} \tag{3.91b}$$

$$\Psi_{dU}^{(0)} = 7.848 - j3.939, \qquad \Psi_{dU}^{(1)} = 7.352 - j0.661. \tag{3.91c}$$

The amplitude functions are

$$T^{(0)}\left(\frac{\lambda}{2}\right) = -0.216 + j0.274, \qquad T^{(1)}\left(\frac{\lambda}{2}\right) = -0.177 + j0.066.$$

With these values the two-term zero-phase-sequence and first-phase-sequence currents (in amperes when V_0 is in volts) in the two antennas are

$$I_{2z}^{(0)}(z) = I_{1z}^{(0)}(z) = V^{(0)}\{0.783(\cos \beta_0 z + 1)$$

$$- j[2.805 \sin \beta_0 |z| - 0.617(\cos \beta_0 z + 1)]\} \times 10^{-3} \tag{3.92a}$$

$$-I_{2z}^{(1)}(z) = I_{1z}^{(1)}(z) = V^{(1)}\{0.189(\cos \beta_0 z + 1)$$

$$- j[2.805 \sin \beta_0 |z| - 0.506(\cos \beta_0 z + 1)]\} \times 10^{-3}. \tag{3.92b}$$

These currents are shown graphically in Fig. 3.3 in the form $I_z = I_z'' + jI_z'$, where I_z'' is in phase, I_z' in phase quadrature with V_0. The corresponding driving-point admittances and impedances are

$$\left.\begin{array}{l} Y^{(0)} = (1.566 + j1.234) \text{ millisiemens}, \\ Y^{(1)} = (0.378 + j1.012) \text{ millisiemens}, \end{array}\right\} \tag{3.93a}$$

$$Z^{(0)} = 394 - j310 \text{ ohms}, \qquad Z^{(1)} = 324 - j867 \text{ ohms}. \tag{3.93b}$$

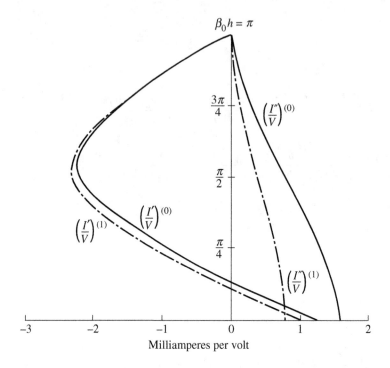

$$\beta_0 h = \pi$$

Figure 3.3 Zero- and first-phase-sequence currents on two-element array. $\Omega = 10$, $\beta_0 b = 1.5$.

The two-term approximations of the general formulas (3.51a, b) are

$$I_{1z}(z) = V_{10}v(z) + V_{20}w(z) \tag{3.94a}$$

$$I_{2z}(z) = V_{10}w(z) + V_{20}v(z), \tag{3.94b}$$

where now

$$v(z) = \frac{j2\pi}{\zeta_0 \Psi_{dR} \cos \beta_0 h} \left[\sin \beta_0(h - |z|) + \tfrac{1}{2}(T^{(0)} + T^{(1)})(\cos \beta_0 z - \cos \beta_0 h) \right] \tag{3.95a}$$

$$w(z) = \frac{j\pi}{\zeta_0 \Psi_{dR} \cos \beta_0 h} (T^{(0)} - T^{(1)})(\cos \beta_0 z - \cos \beta_0 h). \tag{3.95b}$$

When $\beta_0 h$ is near $\pi/2$,

$$v(z) = \frac{-j2\pi}{\zeta_0 \Psi_{dR}} [(\sin \beta_0 |z| - \sin \beta_0 h) + \tfrac{1}{2}(T'^{(0)} + T'^{(1)})(\cos \beta_0 z - \cos \beta_0 h)] \tag{3.95c}$$

$$w(z) = \frac{-j\pi}{\zeta_0 \Psi_{dR}} (T'^{(0)} - T'^{(1)})(\cos \beta_0 z - \cos \beta_0 h). \tag{3.95d}$$

The self- and mutual admittances (3.56) and (3.57) become

$$Y_{s1} = Y_{s2} = \frac{j\pi}{\zeta_0 \Psi_{dR} \cos \beta_0 h} [2 \sin \beta_0 h + (T^{(0)} + T^{(1)})(1 - \cos \beta_0 h)] \qquad (3.96)$$

$$Y_{21} = Y_{12} = \frac{j\pi}{\zeta_0 \Psi_{dR} \cos \beta_0 h} (T^{(0)} - T^{(1)})(1 - \cos \beta_0 h). \qquad (3.97)$$

Similarly, when $\beta_0 h$ is near $\pi/2$, (3.58) and (3.59) reduce to

$$Y_{s1} = Y_{s2} = \frac{j\pi}{\zeta_0 \Psi_{dR}} [2 \sin \beta_0 h - (T'^{(0)} + T'^{(1)})(1 - \cos \beta_0 h)] \qquad (3.98)$$

$$Y_{21} = Y_{12} = \frac{-j\pi}{\zeta_0 \Psi_{dR}} (T'^{(0)} - T'^{(1)})(1 - \cos \beta_0 h). \qquad (3.99)$$

The two-term self- and mutual admittances for the special case $a/\lambda = 7.022 \times 10^{-3}$, $\beta_0 h = \pi$ are shown in Fig. 3.4 as a function of b/λ. The self-susceptance is expressed in the corrected form $B_{11} + 0.72$. Agreement with measured values is seen to be very good. Numerical values of Ψ_{dR}, $T^{(m)}$, $T'^{(m)}$, $Y^{(m)}$, $Y_{si} = Y_{11}$ and Y_{12} are in Tables 2–4 of Appendix I for three values of $\beta_0 h$ and a range of $b/\lambda = d/\lambda$.

For the special case $\Omega = 10$, $\beta_0 h = \pi$, $\beta_0 b = 1.5$, the two-term self- and mutual impedances defined in (3.61) with the two-term expressions (3.96) and (3.97) are

$$Z_{s2} = Z_{s1} = \tfrac{1}{2}(Z^{(0)} + Z^{(1)}) = 359 - j588 \text{ ohms} \qquad (3.100a)$$

$$Z_{21} = Z_{12} = \tfrac{1}{2}(Z^{(0)} - Z^{(1)}) = 35 + j278 \text{ ohms}. \qquad (3.100b)$$

If antenna 1 is driven and antenna 2 is an unloaded parasitic element, (3.67a) applies. The two-term formulas for the currents may be obtained directly from (3.94a, b) with $V_{20} = 0$. Then, in the special case $\Omega = 10$, $\beta_0 h = \pi$, $\beta_0 b = 1.5$,

$$I_{1z}(z) = V_{10}\{0.486(\cos \beta_0 z + 1) - j[2.805 \sin \beta_0|z| - 0.566(\cos \beta_0 z + 1)]\} \times 10^{-3}$$
$$(3.101a)$$

$$I_{2z}(z) = V_{10}(0.287 + j0.055)(\cos \beta_0 z + 1) \times 10^{-3}. \qquad (3.101b)$$

The corresponding driving-point admittance and impedance are

$$Y_{1in} = (0.972 + j1.33) \text{ millisiemens}, \qquad Z_{1in} = 436 - j508 \text{ ohms}. \qquad (3.102)$$

The currents in the driven and parasitic antennas are shown in Fig. 3.5a. They differ from each other greatly in both distribution and amplitude. Indeed, contributions to the far-field by the currents in the parasitic element are insignificant and the horizontal field pattern is almost circular. Note that this behavior is entirely different from what it would be if the two elements were half-wave instead of full-wave dipoles. In the former, the current in the parasitic element is comparable and essentially similar in

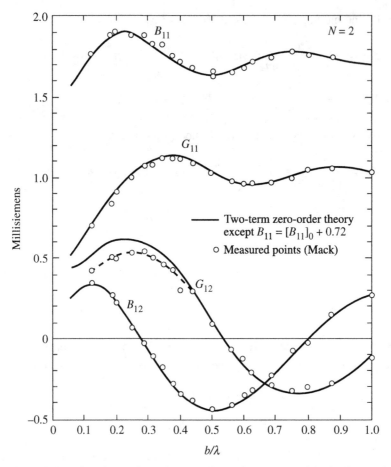

Figure 3.4 Self- and mutual admittances of two-element array; b is the distance between elements; $\beta_0 h = \pi$.

distribution to that in the driven element. The reason for this difference is that the half-wave elements are near resonance, the full-wave elements near anti-resonance. This condition can be changed by inserting a lumped susceptance B_2 (or an equivalent transmission line) in series with the full-wave parasitic element at its center and tuning this susceptance to make the entire circuit resonant. When this is done the distribution functions $v(z)$ and $w(z)$ given by (3.95a, b) give

$$I_{1z}(z) = V_{10}\{0.369(\cos \beta_0 z + 1) - j[2.805 \sin \beta_0 |z| - 0.494(\cos \beta_0 z + 1)]\} \times 10^{-3}$$

$$(3.103a)$$

$$I_{2z}(z) = V_{10}\{[0.064(\cos \beta_0 z + 1) - 0.320 \sin \beta_0 |z|]$$

$$+ j[1.712 \sin \beta_0 |z| - 0.343(\cos \beta_0 z + 1)]\} \times 10^{-3}. \qquad (3.103b)$$

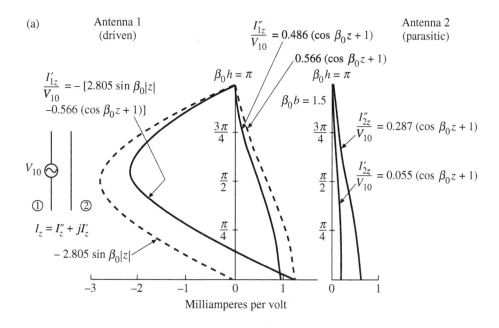

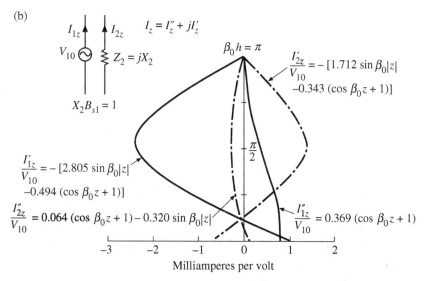

Figure 3.5 Currents on full-wave antenna with (a) $a/\lambda = 7.022 \times 10^{-3}$ untuned parasite, $\Omega = 10$; (b) tuned parasite, $\Omega = 10$.

These currents are shown in Fig. 3.5b. They are very nearly alike in both distribution and amplitude, so that the horizontal field pattern of the tuned full-wave parasitic array must correspond closely to that of the half-wave array with an unloaded parasitic element.

The two-term formulas for the currents in the couplet are given by (3.71a, b) with $v(z)$ and $w(z)$ as in (3.95a, b). For the special case $\Omega = 10$, $\beta_0 h = \pi$, $\beta_0 b = 1.5$, (3.70a, b) give

$$V_{20} = V_{10}\left(\frac{Z_{12} + jZ_{s1}}{Z_{s1} + jZ_{12}}\right) = (-0.966 + j1.267)V_{10} = 1.59V_{10}e^{-j146.3°}. \qquad (3.104)$$

With this value, the explicit formulas for the current in an array are

$$I_{1z} = V_{10}\{0.129(\cos\beta_0 z + 1) - j[2.805\sin\beta_0|z| - 0.884(\cos\beta_0 z + 1)]\} \times 10^{-3} \qquad (3.105a)$$

$$I_{2z} = V_{20}\{0.400(\cos\beta_0 z + 1) - j[2.805\sin\beta_0|z| - 0.397(\cos\beta_0 z + 1)]\} \times 10^{-3}. \qquad (3.105b)$$

In order to obtain expressions for the current that are comparable from the point of view of maintaining an electromagnetic field, it is necessary to use the same reference for amplitude and phase. If I_{2z} is referred to V_{10} instead of V_{20}, the following formula is obtained in place of (3.105b):

$$I_{2z} = V_{10}\{[3.554\sin\beta_0|z| - 0.884(\cos\beta_0 z + 1)]$$
$$+ j[2.710\sin\beta_0|z| + 0.129(\cos\beta_0 z + 1)]\} \times 10^{-3}. \qquad (3.105c)$$

The corresponding driving-point admittances and impedances are

$$Y_{10} = (0.258 + j1.768)\text{ millisiemens}, \qquad Y_{20} = (0.801 + j0.784)\text{ millisiemens} \qquad (3.106a)$$

$$Z_{10} = 80.8 - j554\text{ ohms}, \qquad Z_{20} = 638 - j624\text{ ohms}. \qquad (3.106b)$$

The ratio of the power supplied to antenna 1 to that supplied to antenna 2 is $|V_{20}|^2 G_{20}/|V_{10}|^2 G_{10} = 7.9$. The currents represented by (3.105a) and (3.105b) are shown in the upper diagram in Fig. 3.6 in the form I_{1z}/V_{10} and I_{2z}/V_{20}. The distribution of I_{2z}/V_{10} is shown in the bottom diagram in Fig. 3.6. It differs greatly from I_{1z}/V_{10} (shown in the upper graph) even though the input currents at $z = 0$ satisfy the assigned relation, $I_{20} = jI_{10}$.

The radiation field of the full-wave couplet may be expressed as follows:

$$E_\Theta^r = E_{\Theta 1}^r + E_{\Theta 2}^r = K[A_1 e^{j(\beta_0 b/2)\cos\Phi} + A_2 e^{-j(\beta_0 b/2)\cos\Phi}], \qquad (3.107)$$

where

$$A_1 = V_{10}[(0.129 + j0.884)G_m(\Theta, \pi) - j2.805F_m(\Theta, \pi)] \qquad (3.108a)$$

$$A_2 = V_{10}[(-0.884 + j0.129)G_m(\Theta, \pi) + (3.554 + j2.710)F_m(\Theta, \pi)] \qquad (3.108b)$$

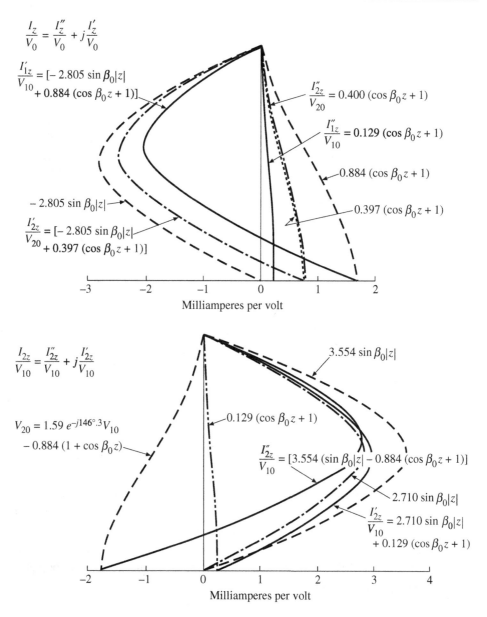

$$\frac{I_z}{V_0} = \frac{I''_z}{V_0} + j\frac{I'_z}{V_0}$$

$$\frac{I'_{1z}}{V_{10}} = [-2.805 \sin \beta_0|z| + 0.884 (\cos \beta_0 z + 1)]$$

$$\frac{I''_{2z}}{V_{20}} = 0.400 (\cos \beta_0 z + 1)$$

$$\frac{I''_{1z}}{V_{10}} = 0.129 (\cos \beta_0 z + 1)$$

$$0.884 (\cos \beta_0 z + 1)$$

$$-2.805 \sin \beta_0|z|$$

$$0.397 (\cos \beta_0 z + 1)$$

$$\frac{I'_{2z}}{V_{20}} = [-2.805 \sin \beta_0|z| + 0.397 (\cos \beta_0 z + 1)]$$

Milliamperes per volt

$$\frac{I_{2z}}{V_{10}} = \frac{I''_{2z}}{V_{10}} + j\frac{I'_{2z}}{V_{10}}$$

$$3.554 \sin \beta_0|z|$$

$$V_{20} = 1.59\, e^{-j146°.3} V_{10} - 0.884 (1 + \cos \beta_0 z)$$

$$0.129 (\cos \beta_0 z + 1)$$

$$\frac{I''_{2z}}{V_{10}} = [3.554 (\sin \beta_0|z| - 0.884 (\cos \beta_0 z + 1)]$$

$$2.710 \sin \beta_0|z|$$

$$\frac{I'_{2z}}{V_{10}} = 2.710 \sin \beta_0|z| + 0.129 (\cos \beta_0 z + 1)$$

Milliamperes per volt

Figure 3.6 Currents in full-wave couplet; $I_{20} = j I_{10}$; $\beta_0 b = 2\pi b/\lambda = 1.5$; $\Omega = 10$.

and where K is a constant. Note that in the equatorial plane, $\Theta = \pi/2$, and $G_m(\pi/2, \pi) = \pi$, $F_m(\pi/2, \pi) = 2$. The field pattern calculated from the magnitude of (3.107) with (3.108a, b) for the couplet of full-wave elements is shown in Fig. 3.7 together with the corresponding pattern for the ideal couplet with identical distributions of current in the two elements. (This latter is quite closely approximated by the pattern of a couplet of half-wave elements.) Both patterns are normalized to

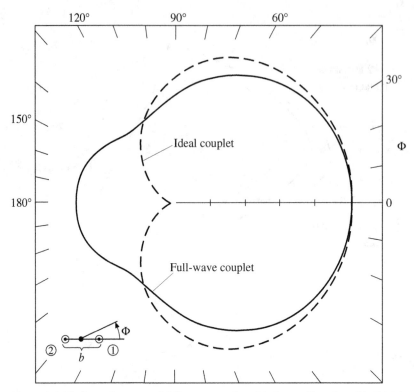

Figure 3.7 Horizontal pattern of full-wave couplet with $I_{20} = j I_{10}$; $\Omega = 2 \ln(2h/a) = 10$, $\beta_0 h = \pi$, $\beta_0 b = 1.5$.

unity at $\Phi = 0$. It is seen that the deep minimum at $\Phi = 180°$ in the ideal pattern (this would be a null if $\beta_0 b = \pi/2$ had been used instead of $\beta_0 b = 1.5$) is replaced by a minor maximum with an amplitude that is about one-half that of the principal maximum at $\Phi = 0$. Thus, the characteristic property of the ideal couplet of providing a null in one direction does not exist in actual couplets when $\beta_0 h = \pi$ or, in fact, for any other value of $\beta_0 h$ that is not near $\pi/2$ or that is not an odd multiple thereof. Significantly, this makes the cardioid pattern of the half-wave couplet a relatively narrow-band property!

4 The circular array

The two-element array, which was investigated in the preceding chapter, may be regarded as the special case $N = 2$ of an array of N elements arranged either at the vertices of a regular polygon inscribed in a circle, or along a straight line to form a curtain. Owing to its greater geometrical symmetry, the circular array is advantageously treated next. Indeed, the basic assumptions which underlie the subsequent study of the curtain array (Chapter 5) depend for their justification on the prior analysis of the circular array.

The real difficulty in analyzing an array of N arbitrarily located elements is that the solution of N simultaneous integral equations for N unknown distributions of current is involved. Although the same set of equations applies to the circular array, they may be replaced by an equivalent set of N independent integral equations in the manner illustrated in Chapter 3 for the two-element array. Since the N elements are geometrically indistinguishable, it is only necessary to make them electrically identical as well. One way is to drive them all with generators that maintain voltages that are equal in amplitude and in phase. When this is done all N currents must also be equal in amplitude and in phase at corresponding points. But this is only one of N possibilities. If the N voltages are all equal in magnitude but made to increase equally and progressively in phase from element 1 to element N, a condition may be achieved such that each element is in exactly the same environment as every other element. There are N such possibilities since the phase sequence closes around the circle when the phase shift from element to element is an integral multiple of $2\pi/N$. Any increment in phase given by $2\pi m/N$ with $m = 0, 1, 2, \ldots, N - 1$ may be used. Specifically, when $N = 2$, the two possibilities are 0 and π. This means that the two driving voltages and the two currents may be equal in magnitude and in phase $(0, 0)$ or equal in magnitude and 180° out of phase $(0, 180°)$. Similarly, when $N = 3$, there are three possibilities, 0, $2\pi/3$ and $4\pi/3$. The voltages and currents around the circle may now be equal in magnitude with phases $(0, 0, 0)$, $(0, 120°, 240°)$ or $(0, 240°, 480°)$.

The analysis of the circular array involves the solution of N simultaneous equations similar in form to (2.15). The case $N = 2$ is solved in Chapter 3 by rearranging the two simultaneous equations for the currents $I_{1z}(z)$ and $I_{2z}(z)$ into two independent equations. These were derived by adding and subtracting the two original equations. When the elements were driven by voltages which were equal in magnitude and either in phase or 180° out of phase, the resulting currents were independent and named, respectively, the zero- and first-phase-sequence currents. The solution for the phase-sequence currents was then carried out and, after a simple algebraic transformation, the actual currents in the elements were derived. The solution for arbitrary driving conditions could also be obtained from the two phase-sequence solutions. A generalization of this procedure is followed in the analysis of the circular array.

The arrays considered here consist of N identical, parallel, non-staggered, center-driven elements that are equally spaced about the circumference of a circle. This means that the elements are at the vertices of an N-sided regular polygon. Arrays of this type are frequently called single-ring arrays in the literature. Their analysis formally parallels step-by-step the analysis of the two-element array in Chapter 3. However, contributions to the vector potential on the surface of each antenna by the currents in all of the elements must be included and this leads to a set of N coupled integral equations for the N currents in the elements. The complete geometrical symmetry of the array permits the use of the method of symmetrical components to reduce the coupled set of integral equations to a single integral equation for each of N possible phase-sequence currents. All other quantities that are required to design and describe the array can be calculated from the solution of essentially one equation with N somewhat different kernels.

The coordinate system and parameters that are used to specify an array are shown in Fig. 4.1 for five elements. The diameter of each element is $2a$, its length is $2h$, the distance between the kth and the ith elements is b_{ki}, the distance between adjacent elements – the length of the side of a regular polygon with the elements at its vertices – is d, and the radius of the circle is ρ.

As indicated in Sections 3.9 and 2.10, the two-term approximation is generally adequate when $h \leq 5\lambda/8$ and $b \geq \lambda/2\pi$. Since it is much simpler and has been used to compute the theoretical results discussed in this chapter, the currents, admittances, and fields in the following sections are determined in the two-term form. Later when matrix notation is introduced, both the two- and the three-term forms of the theory are presented in compact form. This serves both as a summary of the theory of circular arrays and as an introduction to the analysis of more general arrays in Chapters 5 and 6.

4.1   Integral equations for the sequence currents

The vector potential difference at the surface of each element in a circular array of N elements is easily obtained as a generalization of (3.1). Since all elements are thin and parallel to the z-axis, only z-components of the current and the associated vector

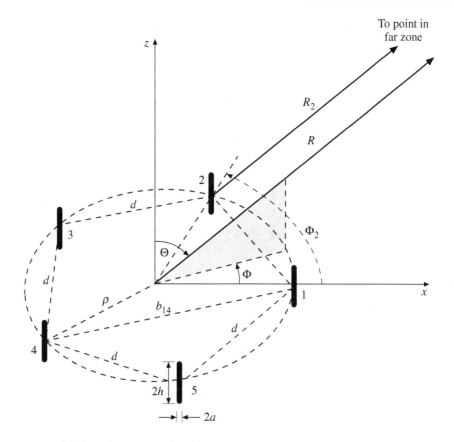

Figure 4.1 Coordinate system for circular arrays.

potential at the surface of each element are significant. Thus, the vector potential difference on the surface of element 1 is

$$4\pi \mu_0^{-1}[A_{1z}(z) - A_{1z}(h)]$$

$$= \int_{-h}^{h}[I_{1z}(z')K_{11d}(z, z') + I_{2z}(z')K_{12d}(z, z') + \cdots + I_{Nz}(z')K_{1Nd}(z, z')]\,dz'$$

$$= \frac{j4\pi}{\zeta_0 \cos \beta_0 h}[\tfrac{1}{2}V_{10} \sin \beta_0(h - |z|) + U_1(\cos \beta_0 z - \cos \beta_0 h)], \qquad (4.1a)$$

where

$$U_1 = \frac{-j\zeta_0}{4\pi} \int_{-h}^{h}[I_{1z}(z')K_{11}(h, z') + I_{2z}(z')K_{12}(h, z') + \cdots$$

$$+ I_{Nz}(z')K_{1N}(h, z')]\,dz'. \qquad (4.1b)$$

Similarly, the vector potential difference on the surface of element 2 is

$$4\pi\mu_0^{-1}[A_{2z}(z) - A_{2z}(h)] = \int_{-h}^{h} [I_{1z}(z')K_{21d}(z, z') + I_{2z}(z')K_{22d}(z, z') + \cdots$$

$$+ I_{Nz}(z')K_{2Nd}(z, z')] dz'$$

$$= \frac{j4\pi}{\zeta_0 \cos\beta_0 h} [\tfrac{1}{2}V_{20} \sin\beta_0(h - |z|)$$

$$+ U_2(\cos\beta_0 z - \cos\beta_0 h)], \tag{4.1c}$$

where

$$U_2 = \frac{-j\zeta_0}{4\pi} \int_{-h}^{h} [I_{1z}(z')K_{21}(h, z') + I_{2z}(z')K_{22}(h, z') + \cdots$$

$$+ I_{Nz}(z')K_{2N}(h, z')] dz'. \tag{4.1d}$$

The vector potential difference on the kth element is

$$4\pi\mu_0^{-1}[A_{kz}(z) - A_{kz}(h)] = \int_{-h}^{h} \sum_{i=1}^{N} I_{iz}(z')K_{kid}(z, z') dz'$$

$$= \frac{j4\pi}{\zeta_0 \cos\beta_0 h} [\tfrac{1}{2}V_{k0} \sin\beta_0(h - |z|)$$

$$+ U_k(\cos\beta_0 z - \cos\beta_0 h)], \tag{4.1e}$$

$$k = 1, 2, \ldots, N$$

where

$$U_k = \frac{-j\zeta_0}{4\pi} \int_{-h}^{h} \sum_{i=1}^{N} I_{iz}(z')K_{ki}(h, z') dz'. \tag{4.1f}$$

In these expressions the kernels are

$$K_{kid}(z, z') = K_{ki}(z, z') - K_{ki}(h, z') = \frac{e^{-j\beta_0 R_{ki}}}{R_{ki}} - \frac{e^{-j\beta_0 R_{kih}}}{R_{kih}} \tag{4.1g}$$

with

$$R_{ki} = \sqrt{(z_k - z_i')^2 + b_{ki}^2} \tag{4.1h}$$

$$R_{kih} = \sqrt{(h - z_i')^2 + b_{ki}^2}, \qquad b_{kk} = a. \tag{4.1i}$$

V_{k0} is the applied driving voltage at the center of element k (or the voltage of an equivalent generator if the element is parasitic with an impedance connected across its terminals), and U_k is the effective driving function characteristic of that part of all the currents that maintains a vector potential of constant amplitude equal to that at $z = h$ along the entire length of the antenna. To reduce the set of N simultaneous equations in (4.1e) to N independent equations, the symmetry conditions characteristic of a circular array must be imposed and the phase-sequence voltages and currents introduced.

Assume that all of the driving voltages are equal in magnitude and have a uniformly progressive phase such that the total phase change around the circle is an integral multiple of 2π. Each multiple of 2π is one of the N phase sequences designated by a superscript (m); these range from zero to $N-1$. In the zero phase sequence, all driving voltages are the same; in the first phase sequence, the driving voltages of adjacent elements differ by $\exp(j2\pi/N)$; in the mth phase sequence, the driving voltages of adjacent elements differ by $\exp(j2\pi m/N)$, and the voltages of the kth and the ith elements are related by

$$V_i = V_k e^{j2\pi(i-k)m/N}. \tag{4.2a}$$

Because of the symmetry of a circular array, the currents in the elements must be related in the same manner as are the driving voltages. That is,

$$I_i(z') = I_k(z')e^{j2\pi(i-k)m/N}. \tag{4.2b}$$

Note that with these driving voltages both the geometric and the electrical environments of each element in the array are identical. Therefore, when (4.2a) and (4.2b) are substituted into the set of coupled integral equations, $I_i(z')$ can be removed from the summation, the remaining kernel is the same regardless of the element to which it is referred, and each equation in the set reduces to

$$\int_{-h}^{h} I^{(m)}(z')K_d^{(m)}(z, z')\, dz'$$

$$= \frac{j4\pi}{\zeta_0 \cos\beta_0 h} \left[\tfrac{1}{2}V^{(m)}\sin\beta_0(h - |z|) + U^{(m)}(\cos\beta_0 z - \cos\beta_0 h)\right], \tag{4.3a}$$

where $m = 0, 1, \ldots N - 1$ and

$$U^{(m)} = \frac{-j\zeta_0}{4\pi}\int_{-h}^{h} I^{(m)}(z')K^{(m)}(h, z')\, dz' \tag{4.3b}$$

$$K^{(m)}(h, z') = \sum_{i=1}^{N} e^{j2\pi(i-1)m/N}\left[\frac{e^{-j\beta_0 R_{1ih}}}{R_{1ih}}\right] \tag{4.3c}$$

$$K_d^{(m)}(z, z') = \sum_{i=1}^{N} e^{j2\pi(i-1)m/N}\left[\frac{e^{-j\beta_0 R_{1i}}}{R_{1i}} - \frac{e^{-j\beta_0 R_{1ih}}}{R_{1ih}}\right]. \tag{4.3d}$$

For later use, it is convenient to separate this difference kernel into two parts that depend, respectively, on the real and imaginary parts of the exponential functions.

That is

$$K_d^{(m)}(z, z') = K_{dR}^{(m)}(z, z') + j K_{dI}^{(m)}(z, z'), \qquad (4.3e)$$

where

$$K_{dR}^{(m)}(z, z') = \sum_{i=1}^{N} e^{j2\pi(i-1)m/N} \operatorname{Re} \left[\frac{e^{-j\beta_0 R_{1i}}}{R_{1i}} - \frac{e^{-j\beta_0 R_{1ih}}}{R_{1ih}} \right] \qquad (4.3f)$$

$$K_{dI}^{(m)}(z, z') = \sum_{i=1}^{N} e^{j2\pi(i-1)m/N} \operatorname{Im} \left[\frac{e^{-j\beta_0 R_{1i}}}{R_{1i}} - \frac{e^{-j\beta_0 R_{1ih}}}{R_{1ih}} \right]. \qquad (4.3g)$$

The method of solution for (4.3a) parallels that of (3.6) and (3.10); the discussion of Section 3.2 and the steps of Section 3.3 are applicable if note is taken of Section 3.9, which relates the two-term to the three-term theory. In fact, the solution is formally given by (3.85) and (3.86) with $m = 0, 1, 2, \ldots, N-1$. This is discussed in somewhat greater detail in a later section (Section 4.6) rather than at this point in order to avoid complications in these initial stages of the analysis. Thus, the mth phase-sequence current in the two-term form is given by

$$I^{(m)}(z) = \frac{j2\pi V^{(m)}}{\zeta_0 \Psi_{dR} \cos \beta_0 h} [\sin \beta_0 (h - |z|) + T^{(m)} (\cos \beta_0 z - \cos \beta_0 h)], \quad \beta_0 h \neq \frac{\pi}{2} \qquad (4.4a)$$

$$I^{(m)}(z) = \frac{j2\pi V^{(m)}}{\zeta_0 \Psi_{dR}} [1 - \sin \beta_0 |z| - T'^{(m)} \cos \beta_0 z], \qquad \beta_0 h = \frac{\pi}{2}. \qquad (4.4b)$$

The Ψ and T functions which occur in (4.4a, b) are defined as follows when $\beta_0 d \geq 1$:

$$T^{(m)} = \frac{\Psi_V^{(m)}(h) - [\Psi_{d\Sigma}^{(m)} + j\Psi_{dI}^{(m)}] \cos \beta_0 h}{\Psi_{dU}^{(m)} \cos \beta_0 h - \Psi_U^{(m)}(h)} \qquad (4.5a)$$

$$T'^{(m)} = \frac{\Psi_{dR} + E_\Sigma \left(\frac{\lambda}{4}, \frac{\lambda}{4} \right) - S_\Sigma \left(\frac{\lambda}{4}, \frac{\lambda}{4} \right)}{C_\Sigma \left(\frac{\lambda}{4}, \frac{\lambda}{4} \right)} \qquad (4.5b)$$

$$\Psi_{dR} = \operatorname{Re} \left[\sin \beta_0 h \, C_{d\Sigma 1} \left(h, h - \frac{\lambda}{4} \right) - \cos \beta_0 h \, S_{d\Sigma 1} \left(h, h - \frac{\lambda}{4} \right) \right], \qquad h \geq \frac{\lambda}{4} \qquad (4.6a)$$

$$= \operatorname{Re} [C_{d\Sigma 1}(h, 0) - \cot \beta_0 h \, S_{d\Sigma 1}(h, 0)], \qquad h < \frac{\lambda}{4} \qquad (4.6b)$$

$$\Psi_V^{(m)}(h) = \sin \beta_0 h \, C_{\Sigma}^{(m)}(h, h) - \cos \beta_0 h \, S_{\Sigma}^{(m)}(h, h) \tag{4.7}$$

$$\Psi_U^{(m)}(h) = C_{\Sigma}^{(m)}(h, h) - \cos \beta_0 h \, E_{\Sigma}^{(m)}(h, h) \tag{4.8}$$

$$\Psi_{d\Sigma}^{(m)} = (1 - \cos \beta_0 h)^{-1} [\sin \beta_0 h \, C_{d\Sigma2}^{(m)}(h, 0) - \cos \beta_0 h \, S_{d\Sigma2}^{(m)}(h, 0)] \tag{4.9}$$

$$\Psi_{dI}^{(m)} = \mathrm{Im}\{(1 - \cos \beta_0 h)^{-1} [\sin \beta_0 h \, C_{d\Sigma1}^{(m)}(h, 0) - \cos \beta_0 h \, S_{d\Sigma1}^{(m)}(h, 0)]\} \tag{4.10}$$

$$\Psi_{dU}^{(m)} = (1 - \cos \beta_0 h)^{-1} [C_{d\Sigma}^{(m)}(h, 0) - \cos \beta_0 h \, E_{d\Sigma}^{(m)}(h, 0)] \tag{4.11}$$

$$C_{d\Sigma}^{(m)}(h, z) = C_{\Sigma}^{(m)}(h, z) - C_{\Sigma}^{(m)}(h, h), \qquad S_{d\Sigma}^{(m)}(h, z) = S_{\Sigma}^{(m)}(h, z) - S_{\Sigma}^{(m)}(h, h)$$

$$E_{d\Sigma}^{(m)}(h, z) = E_{\Sigma}^{(m)}(h, z) - E_{\Sigma}^{(m)}(h, h) \tag{4.12}$$

$$C_{\Sigma}^{(m)}(h, z) = \sum_{i=1}^{N} e^{j2\pi(i-1)m/N} C_{bi}, \qquad C_{bi} = \int_{-h}^{h} \cos \beta_0 z' \, \frac{e^{-j\beta_0 R_{bi}}}{R_{bi}} \, dz' \tag{4.13a}$$

$$S_{\Sigma}^{(m)}(h, z) = \sum_{i=1}^{N} e^{j2\pi(i-1)m/N} S_{bi}, \qquad S_{bi} = \int_{-h}^{h} \sin \beta_0 |z'| \, \frac{e^{-j\beta_0 R_{bi}}}{R_{bi}} \, dz' \tag{4.13b}$$

$$E_{\Sigma}^{(m)}(h, z) = \sum_{i=1}^{N} e^{j2\pi(i-1)m/N} E_{bi}, \qquad E_{bi} = \int_{-h}^{h} \frac{e^{-j\beta_0 R_{bi}}}{R_{bi}} \, dz' \tag{4.13c}$$

$$R_{bi} = \sqrt{(z - z')^2 + b_i^2}, \qquad b_i = a \text{ for } i = 1. \tag{4.14}$$

The subscript $d\Sigma 1$ indicates that only element 1 ($i = 1$) is to be included and effects of all other elements are ignored; the subscript $d\Sigma 2$ indicates that only the effects of elements other than element number 1 are to be included ($i = 2, \ldots, N$).

In order to evaluate (4.4a, b) it is convenient to lump the various coefficients into new parameters defined as follows:

$$s^{(m)} = \frac{j2\pi}{\zeta_0 \Psi_{dR} \cos \beta_0 h}, \qquad c^{(m)} = s^{(m)} T^{(m)} \tag{4.15a}$$

$$s'^{(m)} = \frac{j2\pi}{\zeta_0 \Psi_{dR}}, \qquad c'^{(m)} = s'^{(m)} T'^{(m)} \tag{4.15b}$$

so that, when normalized to $V^{(m)}$, (4.4a, b) become

$$\frac{I^{(m)}(z)}{V^{(m)}} = s^{(m)} \sin \beta_0 (h - |z|) + c^{(m)} (\cos \beta_0 z - \cos \beta_0 h), \qquad \beta_0 h \neq \frac{\pi}{2} \tag{4.16a}$$

$$= s'^{(m)} (1 - \sin \beta_0 |z|) - c'^{(m)} \cos \beta_0 z, \qquad \beta_0 h = \frac{\pi}{2}. \tag{4.16b}$$

The sequence admittances are given by the normalized sequence currents in amperes per volt evaluated at $z = 0$. Thus,

$$Y^{(m)} = \frac{I^{(m)}(0)}{V^{(m)}} = s^{(m)} \sin \beta_0 h + c^{(m)}(1 - \cos \beta_0 h), \qquad \beta_0 h \neq \frac{\pi}{2} \qquad (4.17a)$$

$$= s'^{(m)} - c'^{(m)}, \qquad \beta_0 h = \frac{\pi}{2}. \qquad (4.17b)$$

For a circular array of N elements there are N sequences but only $(N + 1)/2$ are different if N is odd or $(N/2) + 1$ if N is even. This is the same as the number of different self- and mutual admittances.

The sequence currents form a set of functions that are characteristic of the geometrical and electrical properties of the array. Thus, Ψ_{dR} and the $T^{(m)}$ or $T'^{(m)}$ function depend upon the number of elements in the array, their spacing, and the length and thickness of the elements. Once these parameters have been specified, the set of sequence currents can be calculated. Distributions of current in the elements, their driving-point admittances, and the far-zone fields of the arrays with arbitrary driving conditions can be determined from the set of sequence currents with the relations given in Section 4.2. Short tables of Ψ_{dR} and $T^{(m)}$ or $T'^{(m)}$ are given in Appendix I; additional values are available [1]. It may be noted parenthetically that in the notation of [1], the terms 'quasi-zeroth-order' and 'zeroth-order admittances' refer identically to what is called the 'two-term approximation' in this book.

4.2   Sequence functions and array properties

Imagine the array to be excited simultaneously with currents in all of the N possible phase sequences. Then the driving voltage and current for the kth element are

$$V_k = \sum_{m=0}^{N-1} V^{(m)} e^{j2\pi(k-1)m/N} \qquad (4.18a)$$

$$I_k(z) = \sum_{m=0}^{N-1} I^{(m)}(z) e^{j2\pi(k-1)m/N}, \qquad (4.18b)$$

where $V^{(m)}$ is the mth phase-sequence voltage and $I^{(m)}(z)$ is the corresponding phase-sequence current. Similarly, from (4.18b) and (4.17), the self- and mutual admittances are

$$Y_{1k} = \frac{1}{N} \sum_{m=0}^{N-1} Y^{(m)} e^{j2\pi(k-1)m/N}. \qquad (4.19)$$

If the elements of the array are driven by arbitrary voltages V_i which produce corresponding currents $I_i(z)$ along the elements, the sequence voltages and currents are readily obtained from the relations

$$V^{(m)} = \frac{1}{N} \sum_{i=1}^{N} V_i e^{-j2\pi(i-1)m/N} \tag{4.20a}$$

$$I^{(m)}(z) = \frac{1}{N} \sum_{i=1}^{N} I_i(z) e^{-j2\pi(i-1)m/N}. \tag{4.20b}$$

With (4.18b) and (4.16) the normalized current distribution along the kth element can conveniently be expressed as follows:

$$\frac{I_k(z)}{V_1} = s_k \sin \beta_0(h - |z|) + c_k(\cos \beta_0 z - \cos \beta_0 h), \qquad \beta_0 h \neq \frac{\pi}{2} \tag{4.21a}$$

$$= s_k'[1 - \sin \beta_0|z|] - c_k' \cos \beta_0 z, \qquad \beta_0 h = \frac{\pi}{2} \tag{4.21b}$$

where the complex amplitude functions s_k and c_k are

$$s_k = \sum_{m=0}^{N-1} \frac{V^{(m)}}{V_1} s^{(m)} e^{j2\pi(k-1)m/N} \tag{4.22a}$$

$$c_k = \sum_{m=0}^{N-1} \frac{V^{(m)}}{V_1} c^{(m)} e^{j2\pi(k-1)m/N}. \tag{4.22b}$$

The corresponding expressions for s_k' and c_k' are similar. The radiation-zone electric field for each element is given by (2.43); the total field is a superposition of the fields maintained by each element. When the currents in the form (4.21a, b) are substituted in (2.43), the resulting expressions for the field are

$$\frac{E_\Theta^r}{KK_1V_1} = F(\Theta, \beta_0 h) \sum_{i=1}^{N} s_i e^{j\beta_0\rho \sin\Theta \cos(\phi_i - \Phi)}$$

$$+ G(\Theta, \beta_0 h) \sum_{i=1}^{N} c_i e^{j\beta_0\rho \sin\Theta \cos(\phi_i - \Phi)}, \qquad \beta_0 h \neq \frac{\pi}{2} \tag{4.23a}$$

$$= -H\left(\Theta, \frac{\pi}{2}\right) \sum_{i=1}^{N} s_i' e^{j\beta_0\rho \sin\Theta \cos(\phi_i - \Phi)}$$

$$- G\left(\Theta, \frac{\pi}{2}\right) \sum_{i=1}^{N} c_i' e^{j\beta_0\rho \sin\Theta \cos(\phi_i - \Phi)}, \qquad \beta_0 h = \frac{\pi}{2}, \tag{4.23b}$$

with

$$K_1 = \frac{e^{-j\beta_0 R}}{R}, \quad K = \frac{j\zeta_0}{2\pi}, \quad \beta_0\rho = \frac{\pi d/\lambda}{\sin(\pi/N)}, \quad \phi_i = (i-1)2\pi/N.$$

$F(\Theta, \beta_0 h)$, $G(\Theta, \beta_0 h)$ and $H(\Theta, \pi/2)$ are given by (2.46), (2.47), and (2.52a), respectively. These are the so-called element factors and there is one for each type of current distribution. The sums in (4.23a, b) are the array factors. The complex amplitude coefficients are not simply related to one another and the array factors generally cannot be summed in a closed form to yield something equivalent to the familiar $\sin Nx / \sin x$ patterns. In (4.23) the driving voltage V_1 appears since the other driving voltages have been referred to the voltage of element 1. Any other element could have been used for this normalization.

The steps required to make use of this theory in the analysis of a particular array can now be summarized. If the driving voltages are specified, sequence voltages are computed from (4.20a), s_k and c_k from (4.22a, b), the current distributions from (4.21a, b), and far-zone fields from (4.23a, b). Driving-point admittances are found either from the current evaluated at $z = 0$, namely

$$Y_{kin} = \frac{I_k(0)}{V_k} = \frac{I_k(0)}{V_1} \frac{V_1}{V_k} \tag{4.24a}$$

or from the coupled circuit equations and the self- and mutual admittances

$$Y_{kin} = \sum_{i=1}^{N} \frac{V_i}{V_k} Y_{ki}. \tag{4.24b}$$

If the driving-point currents are specified, sequence currents can be found from (4.20b), (4.16a) or (4.16b) solved for $V^{(m)}$, and the remaining steps carried out as when the driving voltages are specified. Numerical results for a particular array can be obtained from the tables of Appendix I or [1].

4.3 Self- and mutual admittances

For a circular array with uniformly-spaced elements, self- and mutual admittances are defined in terms of the sequence admittances by (4.19). The more general definition (discussed in Chapter 14) of self- and mutual admittances as the coefficients of the driving-point voltages in the coupled circuit equations also applies. For the pth element,

$$I_p(0) = \sum_{i=1}^{N} V_i Y_{pi} \tag{4.25}$$

from which it follows that the self-admittance Y_{pp} of the pth element is the driving-point admittance of that element when all other elements are present and short-circuited at their driving points. The mutual admittance Y_{pk} ($p \neq k$) between element p and element k is the driving-point current of element p per unit driving voltage of element k with all other elements present and short-circuited at their driving points. Thus, the mutual admittances characterize the degree in which power that is fed to one element of the array is transferred to the remaining elements.

Among the properties of circular arrays that are revealed by a study of their self- and mutual admittances are resonant spacings at which all of the elements interact vigorously and, in larger arrays, spacings at which at least some of the mutual admittances are very small compared with the self-admittance. In arrays containing only a few elements, the resonant spacings are most important for elements with lengths near $h = \lambda/4$; in larger arrays they are most important for elements with somewhat greater lengths. When the elements in an array are at the resonant spacings, their currents are essentially all in phase and their properties are very sensitive to small changes in frequency. Although calculations of the driving-point admittances generally must include all of the mutual admittances when the array consists of only a few elements, there are ranges of spacings in larger arrays over which at least some of the mutual admittances are much smaller than the self-admittance. In larger arrays there is also a range of spacings over which many of the mutual admittances are nearly the same in magnitude and phase. These properties are illustrated in Figs. 4.2–4.7, which show graphically examples of self- and mutual admittances in millisiemens for a range of values of d/λ, the distance between adjacent elements.

With the exception of the self-susceptance shown in Fig. 4.3b, the theoretical results are all evaluated from the two-term theory and were computed from (4.19), (4.17b), and the functions in (4.5)–(4.14). The theoretical self-susceptance in Fig. 4.3b is shown in the corrected form $B_{11} + 1.16$ with B_{11} calculated from the two-term theory. The correcting susceptance 1.16 includes the term 0.72 needed to correct the two-term susceptance and an additional susceptance that takes account of the particular end-effects of the coaxial measuring line. The measured results in Figs. 4.2, 4.3 and 4.4 were obtained from load admittances apparently terminating the coaxial line. They were measured by the distribution-curve method discussed in Chapter 14. The experimental apparatus consisted of combined slotted measuring lines and monopoles driven over a ground-screen. The actual measured results have been divided by two and an approximate terminal-zone correction of $Y_T = j0.286$ millisiemens as obtained from Fig. 14.3b has been combined with B_{11} so that the final results apply to an ideal center-driven dipole with all contributions to the admittance by an associated driving mechanism eliminated.

An array of four elements of length $h = \lambda/4$ (Fig. 4.2) has a resonant spacing near $d/\lambda = 0.54$. At this spacing all conductances have sharp positive maxima while the suceptances are all essentially zero. If the length of the elements is increased to $h = 3\lambda/8$, a similar resonance occurs in the range between $d/\lambda = 0.37$ and 0.40, but the maxima are not as sharp. With eight elements (Fig. 4.5) there are several

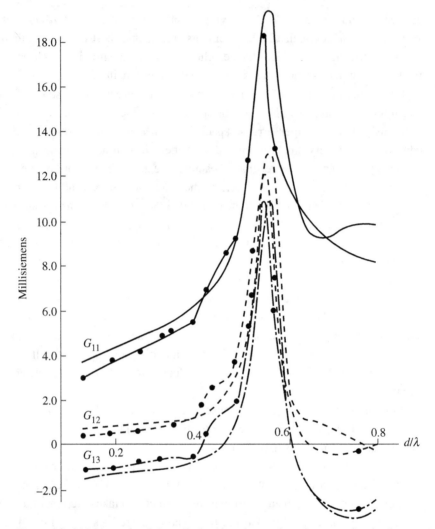

Figure 4.2 (a) Measured and theoretical self- and mutual conductances for circular array; $N = 4$, $h/\lambda = 1/4$, $a/\lambda = 0.007\,022$.

resonances but only the first two, which occur near $d/\lambda = 0.35$ and 0.50, are sharply defined. Also, from Fig. 4.5 it is seen that the conductances all have the same sign at the first resonance but not at the second. For twenty elements with length $h = \lambda/4$ it is seen from Fig. 4.6 that a number of resonances occur, but that they no longer have large amplitudes. On the other hand, when the length of the elements is near $h = 3\lambda/8$, the resonances are sharply defined and a small change in spacing (or frequency) produces large changes in the admittances as shown in Fig. 4.7.

Note also that, whereas the four- and eight-element arrays have only one spacing each at which some of the mutual conductances or susceptances are small compared

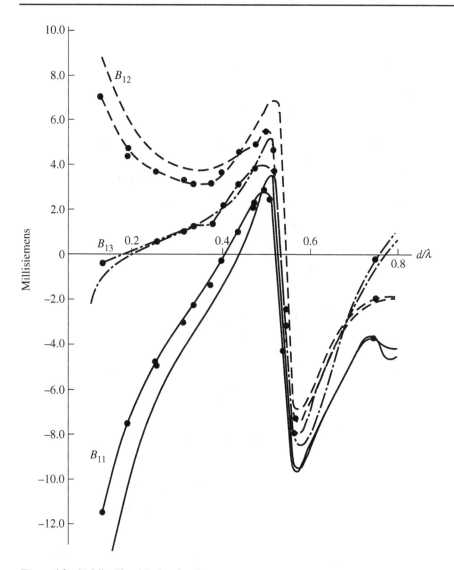

Figure 4.2 (b) Like Fig. 4.2a but for the susceptances.

to the self-conductance or susceptance, there is a considerable range of spacings for a twenty-element array over which only Y_{12} is important and all other mutual admittances are small compared to Y_{11}. For close spacings, many of the mutual admittances have essentially the same value in Figs. 4.5 and 4.6. Also, at small spacings the self-susceptance and the mutual susceptance between adjacent elements become very large compared to either the remaining susceptances or the conductances. This indicates that it is these quantities which cause difficulties in matching arrays of closely-spaced elements. These susceptances can be controlled at least partially by an adjustment of the lengths of the elements. Additional, more extensive graphs

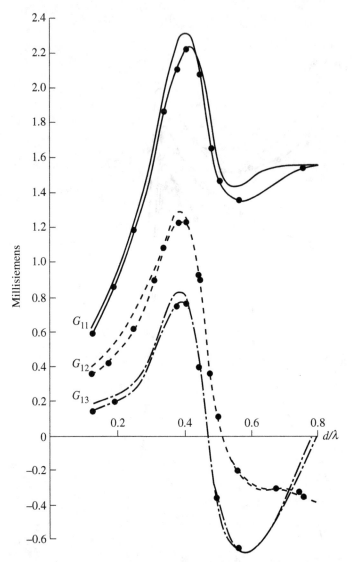

Figure 4.3 (a) Measured and theoretical self- and mutual conductances for circular array; $N = 4$, $h/\lambda = 3/8, a/\lambda = 0.007\,022$.

and tables of self- and mutual admittances are in the literature [2]. All of the results discussed here are for elements with the radius $a/\lambda = 0.007$, but since the parameters of an array change quite slowly with the thickness of the elements, the qualitative behavior should be the same for thicknesses that do not violate the requirement of 'thin', i.e. $\beta_0 a \leq 0.10$. Note, however, that the self-impedances of the individual elements change significantly with their radius – especially when h is not near $\lambda/4$.

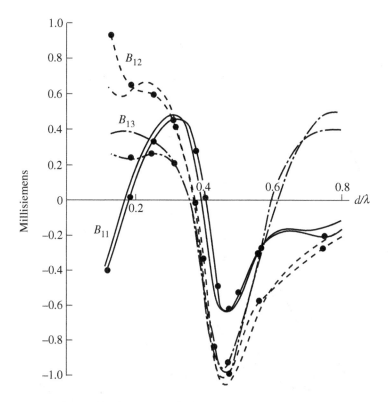

Figure 4.3 (b) Like Fig. 4.3a but for the susceptances.

Figures 4.2 and 4.3 indicate that, except near the sharp resonances, the results from the two-term theory are in good agreement with the measured values for all conductances and mutual susceptances. Part of the differences between measured and computed conductances at the sharp resonant maxima for $h = \lambda/4$ may be due to the difficulty encountered in obtaining accurate measurements over this region. The self-susceptance and its correction were discussed in Chapter 2. In Fig. 4.2, no correction has been applied to B_{11}; the use of the correction 0.72 millisiemens that was indicated in Chapter 2 would yield better agreement for $d/\lambda < 0.40$. In Fig. 4.3, the correction applied to B_{11} is 1.16 millisiemens. As previously discussed, this includes both the term 0.72 and an additional empirically determined susceptance that takes account of the end-correction for the coaxial measuring line actually used. It was determined from a comparison of theoretical and measured results for a single element (Fig. 2.6). Since the correction to B_{11} is a constant, it is evident that the correct variation of B_{11} with d/λ is given by the theory. In a practical application, the characteristics of a given array are determined from the theory, a single model of the elements of the array is constructed, and its driving-point admittance measured. The difference between theoretical and measured driving-point susceptances for the single

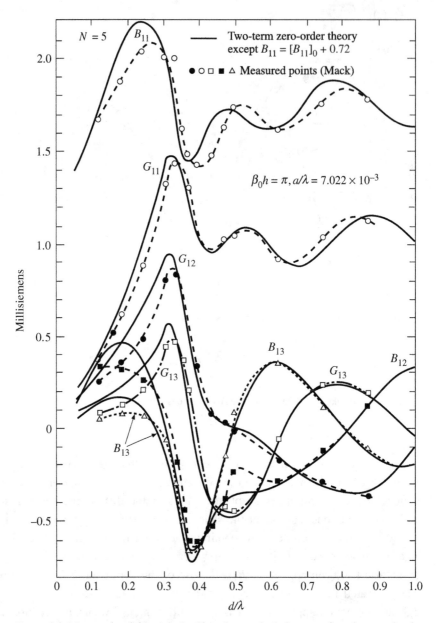

Figure 4.4 Measured and theoretical self- and mutual admittances, five-element circular array; $h = \lambda/2$.

element may be used as a correction for the computed driving-point admittances in the array.

It is sometimes convenient to characterize element intercoupling by self- and mutual impedances instead of admittances. For a general array, the conversion from an

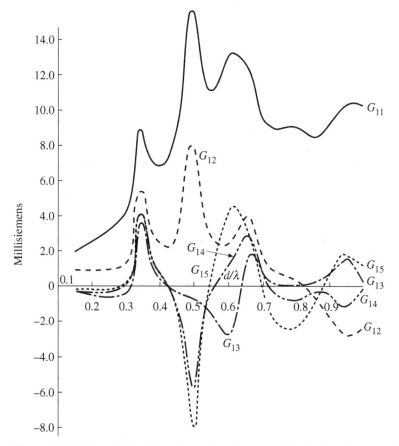

Figure 4.5 (a) Theoretical self- and mutual conductances for circular array; $N = 8$, $h = \lambda/4$, $a/\lambda = 0.007\,022$.

admittance basis to an impedance basis requires an inversion of the admittance matrix. For a circular array, the sequence admittances and impedances are reciprocals, that is,

$$Z^{(m)} = 1/Y^{(m)} \tag{4.26}$$

$$Z_{1i} = \frac{1}{N} \sum_{m=0}^{N-1} e^{-j2\pi(i-1)m/N} Z^{(m)} \tag{4.27}$$

so that the reciprocal of only one complex number is required for each sequence.

4.4   Currents and fields; arrays with one driven element

One of the simplest examples of the application of the two-term theory is provided by ring arrays with one element driven and the remaining elements short-circuited at their driving points. In the following examples, the radius of the elements is taken to be

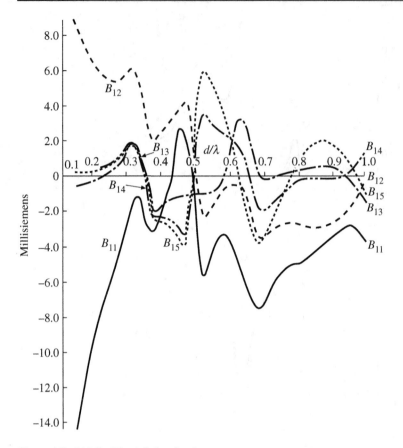

Figure 4.5 (b) Like Fig. 4.5a but for the susceptances.

$a/\lambda = 0.007$ and the radiation patterns are all measured or computed in the equatorial plane, $\Theta = \pi/2$. The relative radiation patterns are computed from

$$P_{dB} = 10 \log_{10} \left| \frac{E_\Theta^r(\Theta, \Phi) \cdot E_\Theta^{r*}(\Theta, \Phi)}{E_{\Theta m}^r \cdot E_{\Theta m}^{r*}} \right|, \tag{4.28}$$

where $E_{\Theta m}^r$ is the maximum value of the field in the plane $\Theta = \pi/2$. An asterisk indicates the complex conjugate, $E_\Theta^r(\Theta, \Phi)$ is given by (4.23), and P_{dB} is the relative magnitude of the Poynting vector in decibels (dB).

Figure 4.8a contains two examples of the radiation patterns of five-element arrays. One pattern is for $d = \lambda/4$ and $h = \lambda/4$ and has a back-to-front ratio of about -14 dB with half-power beam widths of about $100°$. The second pattern is also for $d = \lambda/4$ but $h = 3\lambda/8$; it has a very smooth angular variation with a back-to-front ratio of about -20 dB and beam widths of about $140°$. Agreement between the theoretical and measured results is well within 1 dB except near the deeper minimum in the backward direction near $\Phi = 180°$. Similar patterns for $h/\lambda = 0.5$ and two values of d/λ are in Fig. 4.8b.

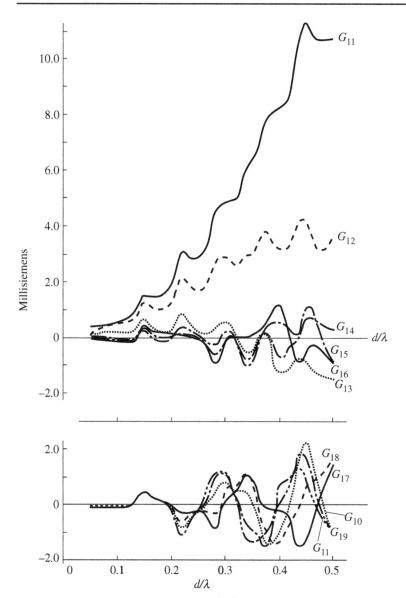

Figure 4.6 (a) Theoretical self- and mutual conductances for circular array; $N = 20$, $h = \lambda/4$, $a/\lambda = 0.007\,022$.

Corresponding currents in the elements of the two arrays with the patterns given in Fig. 4.8a are shown in Figs. 4.9 and 4.10. As a consequence of the symmetry, only three of the currents are different for each five-element array. The radiation patterns depend only on the relative distributions of current. If the currents in Figs. 4.9 and 4.10 were simply normalized to their maximum values, it is evident that agreement between theoretical and measured results would be very good and, therefore, measured patterns well represented by the theory. In order to permit detailed comparison of the

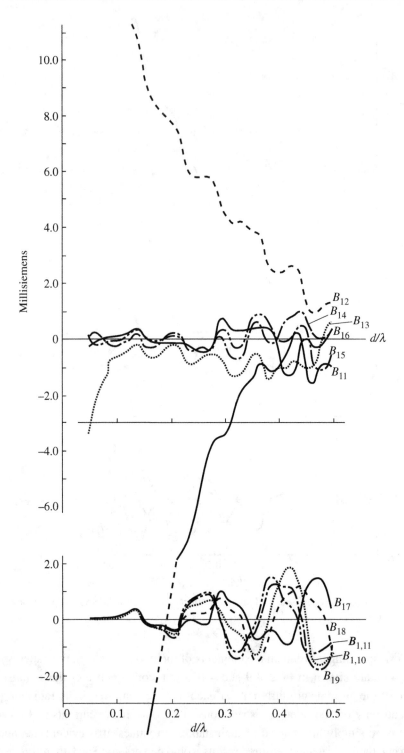

Figure 4.6 (b) Like Fig. 4.6a but for the susceptances.

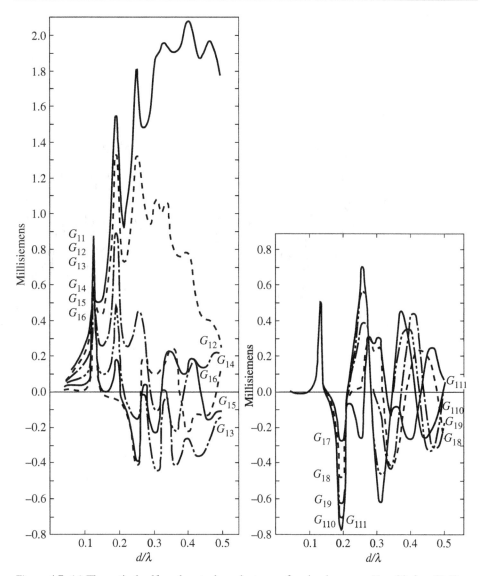

Figure 4.7 (a) Theoretical self- and mutual conductances for circular array; $N = 20$, $h = 3\lambda/8$, $a/\lambda = 0.007\,022$.

experimental and theoretical models, the relative amplitude and phase of the current along each element were measured and normalized to the measured self- and mutual admittances. Thus,

$$\frac{I_k(z)}{V_1} = \frac{|I_k(z)|}{V_1} e^{j\Psi_k(z)}$$

$$= \frac{|I_k(z)|}{V_1} [\cos \Psi_k(z) + j \sin \Psi_k(z)] = \frac{\operatorname{Re} I_k(z) + j \operatorname{Im} I_k(z)}{V_1}, \tag{4.29}$$

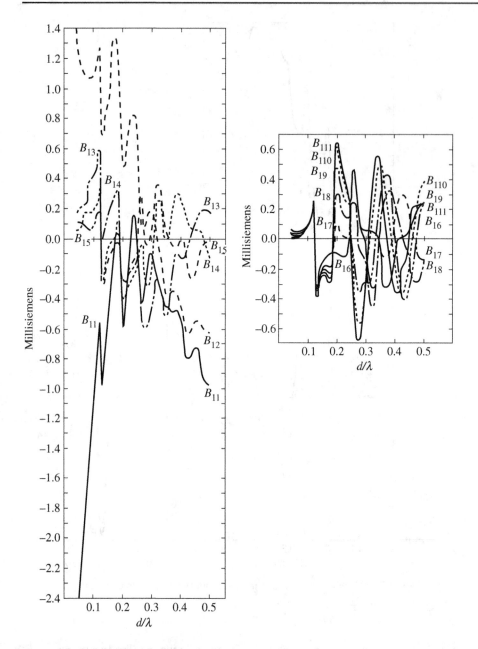

Figure 4.7 (b) Like Fig. 4.7a but for the susceptances.

where the real and imaginary parts are, respectively, in phase and in phase quadrature with the driving voltage. The relatively small amplitude of the current $|I_3(z)|/V_1$ in Fig. 4.9 prevented an accurate measurement of phase in this case.

The experimental model that was used for the measurement of both field patterns and currents consisted of five monopoles over a ground plane combined with a

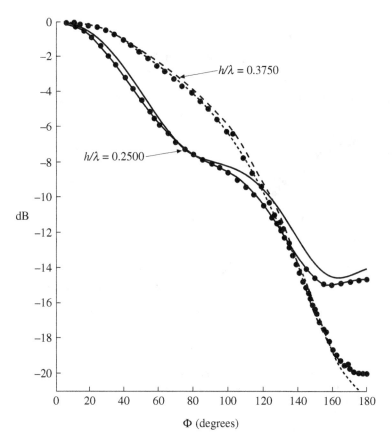

Figure 4.8 (a) Radiation patterns for five-element arrays with one driven element; $h/\lambda = 0.25$ and 0.375.

measuring line for each. The equipment and procedures for measuring amplitude and phase are discussed in Chapter 14. The s_k and c_k coefficients for use in (4.21) and (4.23) can be computed from the values of Ψ_{dR} and T in the tables of Appendix I with the use of (4.15a, b) and (4.22a, b). Numerical data for the two five-element arrays under discussion are

$$N = 5, \quad d = \lambda/4, \quad h = \lambda/4:$$

$$\left. \begin{array}{l} s_1' = j2.6824; \quad s_2' = s_3' = s_4' = s_5' = 0; \\ c_1' = -4.2084 + j9.2159; \quad c_2' = c_5' = -0.8072 - j4.6906; \\ c_3' = c_4' = 0.6835 - j0.3656; \end{array} \right\} \quad (4.30)$$

$$N = 5, \quad d = \lambda/4, \quad h = 3\lambda/8:$$

$$\left. \begin{array}{l} s_1 = -j3.6571; \quad s_2 = s_3 = s_4 = s_5 = 0; \\ c_1 = 0.7504 + j1.1391; \quad c_2 = c_5 = 0.4492 + j0.4419; \\ c_3 = c_4 = 0.1890 + j0.1678. \end{array} \right\} \quad (4.31)$$

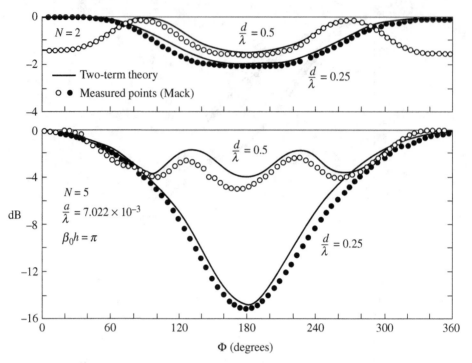

Figure 4.8 (b) Horizontal pattern of parasitic arrays of two and five elements in a circle; $h/\lambda = 0.50$; d is the distance between adjacent elements.

Note that the currents in the parasitic elements are represented by shifted cosine components only.

The radiation patterns in Fig. 4.8a suggest that spacings can be found at which the pattern is a smooth function of Φ and has a deep minimum near $\Phi = 180°$. Examples of such patterns are shown in Fig. 4.11 for $N = 4, 5, 10$ and 20 and $h = 3\lambda/8$. As N increases, such patterns occur when the circumference of the circle containing the array approaches 2λ. For them, the phase of the electric field is also a smooth slowly changing function of the azimuth angle Φ as shown in Fig. 4.11. The phase was computed from (4.23a) in the form

$$\frac{E_\Theta^r(\pi/2, \Phi)}{K K_1 V_1} = \text{Re}\left[\frac{E_\Theta^r(\pi/2, \Phi)}{K K_1 V_1}\right] + j\,\text{Im}\left[\frac{E_\Theta^r(\pi/2, \Phi)}{K K_1 V_1}\right]$$

$$= \left|\frac{E_\Theta^r(\pi/2, \Phi)}{K K_1 V_1}\right| e^{j\Psi(\pi/2, \Phi)} \qquad (4.32a)$$

$$\Psi(\pi/2, \Phi) = \tan^{-1}\frac{\text{Im}\,E_\Theta^r(\pi/2, \Phi)}{\text{Re}\,E_\Theta^r(\pi/2, \Phi)}. \qquad (4.32b)$$

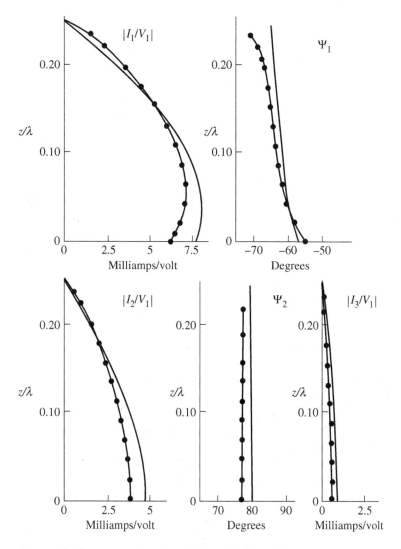

Figure 4.9 Element currents; $N = 5$, $h = \lambda/4$, $d = \lambda/4$.

4.5   Matrix notation and the method of symmetrical components

In the preceding sections the N simultaneous integral equations for the N currents in a circular array were replaced by N independent integral equations by a procedure known as the method of symmetrical components. This procedure was introduced as a generalization of the corresponding treatment of the two coupled equations analyzed in Chapter 3. It is now appropriate to systematize the general formulation with the compact notation of matrices.

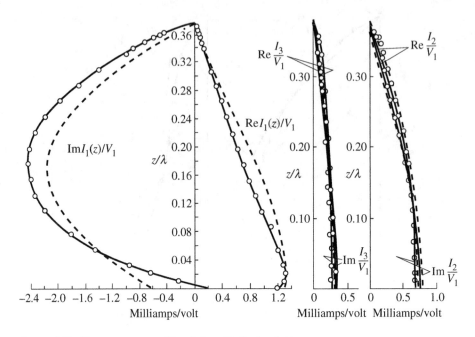

Figure 4.10 Element currents; $N = 5$, $h = 3\lambda/8$, $d = \lambda/4$.

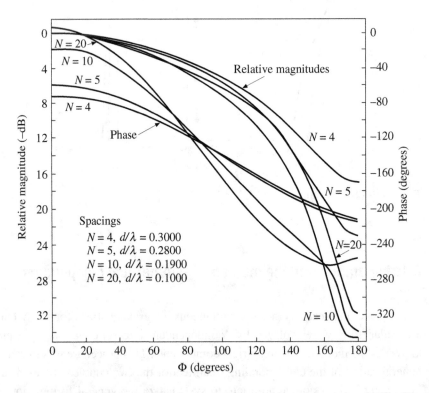

Figure 4.11 Power pattern of circular arrays with one driven element; $h = 3\lambda/8$, $a/\lambda = 0.007\,022$.

The general method of symmetrical components became well known in its application to problems in multi-phase electric circuits. Loads on three-phase power systems, for example, must generally be balanced to give equal currents in all three branches. Under some conditions unequal loads are placed across the supply lines. The calculation of the resulting branch currents is usually made in terms of three phase-sequence components. The zero phase-sequence currents are all in phase. The first sequence contains three equal phasors which have $120°$ progressive phase shifts. These phasors rotate in the counter-clockwise direction in the complex plane as time increases. The angular velocity is ω. The second phase sequence has three phasors with equal magnitude and a progressive $-120°$ phase shift. Since the currents generated by the three sets of phase-sequence voltages do not interact with one another, they may be calculated separately and later combined to give the actual currents. A similar procedure applies to an N-phase system.

The equations which relate the currents and voltages in N coupled circuits have the following matrix form:

$$[Z]\{I\} = \{V\}, \tag{4.33}$$

where

$$\{I\} = \begin{Bmatrix} I_1 \\ I_2 \\ \vdots \\ I_N \end{Bmatrix}, \qquad \{V\} = \begin{Bmatrix} V_1 \\ V_2 \\ \vdots \\ V_N \end{Bmatrix} \tag{4.34}$$

$$[Z] = \begin{bmatrix} Z_{11} & Z_{12} & Z_{13} & \cdots & Z_{1N} \\ Z_{21} & Z_{22} & & \cdots & Z_{2N} \\ \vdots & & & & \\ Z_{N1} & & & \cdots & Z_{NN} \end{bmatrix}. \tag{4.35}$$

The usual reciprocity of off-diagonal impedances is assumed, i.e., $Z_{ij} = Z_{ji}$. In addition, $[Z]$ is a circulant matrix, so that all rows are cyclic permutations of the first row.

In order to illustrate the application of the method of symmetrical components, this set of equations will not be solved in the usual manner by setting $\{I\} = [Z]^{-1}\{V\}$. Instead, the phase-sequence voltages and impedances will be calculated first by means of the following transformation matrices:

$$\begin{Bmatrix} V^{(0)} \\ V^{(1)} \\ \vdots \\ V^{(N-1)} \end{Bmatrix} = \frac{1}{N} \begin{bmatrix} 1 & 1 & 1 & \cdots & 1 \\ 1 & p^{-1} & p^{-2} & \cdots & p^{-(N-1)} \\ 1 & p^{-2} & p^{-4} & \cdots & p^{-2(N-1)} \\ \vdots & \vdots & \vdots & & \vdots \\ 1 & p^{-(N-1)} & p^{-2(N-1)} & \cdots & p^{-(N-1)(N-1)} \end{bmatrix} \begin{Bmatrix} V_1 \\ V_2 \\ V_3 \\ \vdots \\ V_N \end{Bmatrix}, \tag{4.36}$$

where $p = e^{j2\pi/N}$, or

$$\{V^{(m)}\}[P]^{-1}\{V\}. \tag{4.37}$$

Similarly, for the impedances

$$\{Z^{(m)}\} = [P]\{Z\}, \tag{4.38}$$

where

$$\{Z^{(m)}\} = \begin{Bmatrix} Z^{(0)} \\ Z^{(1)} \\ \vdots \\ Z^{(N-1)} \end{Bmatrix} \tag{4.39}$$

$$\{Z\} = \begin{Bmatrix} Z_{11} \\ Z_{12} \\ Z_{13} \\ \vdots \\ Z_{1N} \end{Bmatrix} \tag{4.40}$$

$$[P] = \begin{bmatrix} 1 & 1 & 1 & \cdots & 1 \\ 1 & p & p^2 & \cdots & p^{(N-1)} \\ 1 & p^2 & p^4 & \cdots & p^{2(N-1)} \\ \vdots & \vdots & & & \\ 1 & p^{(N-1)} & p^{2(N-1)} & \cdots & p^{(N-1)(N-1)} \end{bmatrix}. \tag{4.41}$$

The phase-sequence currents are given by the algebraic equations

$$I^{(m)} = V^{(m)}/Z^{(m)}, \qquad m = 0, 1, \ldots (N-1). \tag{4.42}$$

The original currents I_i, $i = 1, 2, 3, \ldots N$ are given by

$$\{I\} = [P]\{I^{(m)}\}, \tag{4.43}$$

where

$$\{I\} = \begin{Bmatrix} I_1 \\ I_2 \\ \vdots \\ I_N \end{Bmatrix}. \tag{4.44}$$

As a trivial example of the method, consider two coupled circuits with the same self-impedances. The matrix equation is

$$\begin{bmatrix} Z_{11} & Z_{12} \\ Z_{12} & Z_{11} \end{bmatrix} \begin{Bmatrix} I_1 \\ I_2 \end{Bmatrix} = \begin{Bmatrix} V_1 \\ V_2 \end{Bmatrix}. \tag{4.45}$$

With $p = e^{j2\pi/N} = e^{j\pi} = -1$, the matrix P^{-1} defined by (4.36) is

$$[P]^{-1} = \tfrac{1}{2} \begin{bmatrix} 1 & 1 \\ 1 & -1 \end{bmatrix}. \tag{4.46}$$

The phase-sequence voltages and impedances as obtained from (4.37) and (4.38) are

$$\begin{Bmatrix} V^{(0)} \\ V^{(1)} \end{Bmatrix} = \tfrac{1}{2} \begin{bmatrix} 1 & 1 \\ 1 & -1 \end{bmatrix} \begin{Bmatrix} V_1 \\ V_2 \end{Bmatrix} = \tfrac{1}{2} \begin{Bmatrix} V_1 + V_2 \\ V_1 - V_2 \end{Bmatrix} \tag{4.47}$$

and

$$\begin{Bmatrix} Z^{(0)} \\ Z^{(1)} \end{Bmatrix} = \begin{bmatrix} 1 & 1 \\ 1 & -1 \end{bmatrix} \begin{Bmatrix} Z_{11} \\ Z_{12} \end{Bmatrix} = \begin{Bmatrix} Z_{11} + Z_{12} \\ Z_{11} - Z_{12} \end{Bmatrix}. \tag{4.48}$$

The resulting phase-sequence currents $I^{(m)}$, $m = 1, 2$, are given by

$$I^{(0)} = \tfrac{1}{2}(V_1 + V_2)/(Z_{11} + Z_{12}) \tag{4.49}$$

and

$$I^{(1)} = \tfrac{1}{2}(V_1 - V_2)/(Z_{11} - Z_{12}). \tag{4.50}$$

The desired currents I_i, $i = 1, 2$ (which are generated by the actual driving voltages V_1 and V_2) are given by (4.43) with (4.49) and (4.50). They are

$$\begin{Bmatrix} I_1 \\ I_2 \end{Bmatrix} = \begin{bmatrix} 1 & 1 \\ 1 & -1 \end{bmatrix} \begin{Bmatrix} I^{(0)} \\ I^{(1)} \end{Bmatrix} = \begin{Bmatrix} (V_1 Z_{11} - V_2 Z_{12})/(Z_{11}^2 - Z_{12}^2) \\ (V_2 Z_{11} - V_1 Z_{12})/(Z_{11}^2 - Z_{12}^2) \end{Bmatrix}. \tag{4.51}$$

These equations are, of course, the same as those obtained directly from (4.45). Note that in the method of symmetrical components the matrix inversion is performed in a number of straightforward steps. In the analysis of circular arrays it allows a large matrix to be inverted for each phase sequence by obtaining the reciprocal of one complex number.

4.6   General formulation and solution

In Section 4.1, the solutions for the N independent integral equations for the phase-sequence currents in a circular array of identical elements were obtained by a logical generalization of the parallel analysis for the two-element array in Chapter 3. A more complete formulation and solution with special reference to the complications of the N-element array is now in order.

With the matrix notation introduced in Section 4.5, the integral equations (4.3a) for the N phase-sequence currents may be expressed as follows

$$\int_{-h}^{h} I_z^{(m)}(z') K_d^{(m)}(z, z') \, dz'$$

$$= \frac{j4\pi}{\zeta_0 \cos \beta_0 h} \left[\tfrac{1}{2} V^{(m)} \sin \beta_0 (h - |z|) + U^{(m)}(\cos \beta_0 z - \cos \beta_0 h) \right], \tag{4.52}$$

where

$$\{I_z^{(m)}\} = [P]^{-1}\{I_z\} \tag{4.53a}$$

$$\{V^{(m)}\} = [P]^{-1}\{V\} \tag{4.53b}$$

$$\{K_d^{(m)}(z, z')\} = [P]\{K_d(z, z')\} \tag{4.54}$$

and

$$U^{(m)} = \frac{-j\zeta_0}{4\pi} \int_{-h}^{h} I_z^{(m)}(z') K^{(m)}(h, z') \, dz'. \tag{4.55}$$

In order to reduce the integral equation (4.52) to an approximately equivalent algebraic equation in the manner described in Chapter 3, it is necessary to introduce approximate expressions for the several parts of the integral. The procedure and the reasoning behind it are in principle the same as described in Sections 3.2 and 3.3 for two elements. However, for N elements in a circle the kernel consists of a sum of N instead of two terms. In the interest of simplicity, the introductory discussion in Section 4.1 assumed that all elements are separated by distances sufficiently great so that $\beta_0 b_{ki} \geq 1$ for all values of k and i. Although this condition is satisfied in most circular arrays, there are exceptions. One is the cage antenna in which the N parallel elements are distributed around an electrically small circle so that the condition $\beta_0 b_{ki} < 1$ is satisfied for all k and i. An intermediate case arises when the circle is electrically large, but the elements are quite closely spaced so that one or more on each side of every element satisfies the inequality $\beta_0 b_{ki} < 1$, but all of the others are far enough away so that $\beta_0 b_{ki} \geq 1$. Since the behavior of the parts of the integrals that relate closely spaced elements is different from the parts that represent widely spaced ones, it is necessary to treat them separately. Since for each phase sequence all elements are in identical environments, element no. 1 is conveniently selected for reference. Let it be assumed that n elements on each side of element 1 are sufficiently near so that for them $\beta_0 b_{1i} < 1$, $1 \leq i \leq n$, $N - n + 1 \leq i \leq N$ and that for all other elements, $\beta_0 b_{1i} \geq 1$, $n < i < N - n + 1$. Let the sum over all the $2n + 1$ elements for which $\beta_0 b_{1i} < 1$ be denoted by $\Sigma 1$, the sum over all other elements in the circle by $\Sigma 2$. Similarly, let $K_{d\Sigma1}^{(m)}(z, z')$ be the part of the sum in (4.3d) which includes the $2n + 1$ elements for which $\beta_0 b_{1i} < 1$, $K_{d\Sigma2}^{(m)}(z, z')$ the rest of the sum. It now follows by analogy with (3.21a, b) that

$$\int_{-h}^{h} \sin \beta_0(h - |z'|) K_{d\Sigma1R}^{(m)}(z, z') \, dz' \doteq \Psi_{d\Sigma1R}^{(m)} \sin \beta_0(h - |z|) \tag{4.56a}$$

$$\int_{-h}^{h} \sin \beta_0(h - |z'|) K_{d\Sigma2R}^{(m)}(z, z') \, dz' \doteq \Psi_{d\Sigma2R}^{(m)}(\cos \beta_0 z - \cos \beta_0 h), \tag{4.56b}$$

where $K_{d\Sigma 1R}^{(m)}(z, z')$ and $K_{d\Sigma 2R}^{(m)}(z, z')$ are the appropriate parts of $K_{dR}^{(m)}(z, z')$ as defined in (4.3f). On the other hand, all remaining parts of the integral are independent of $\beta_0 b_{1i}$, so that they are the same as in (3.22a)–(3.23c) but with $K_{dR}^{(m)}(z, z')$ and $K_{dI}^{(m)}(z, z')$ as given in (4.3f) and (4.3g).

The Ψ-functions introduced in (4.56a, b) are defined as follows:

$$\Psi_{dR}^{(m)} \equiv \Psi_{d\Sigma 1R}^{(m)} = \Psi_{d\Sigma 1R}^{(m)}(z_m); \quad \begin{cases} z_m = 0, & \beta_0 h \leq \pi/2 \\ z_m = h - \lambda/4, & \beta_0 h > \pi/2 \end{cases} \tag{4.57a}$$

$$\Psi_{d\Sigma 1R}^{(m)}(z) = \csc \beta_0 (h - |z|) \int_{-h}^{h} \sin \beta_0 (h - |z'|) K_{d\Sigma 1R}^{(m)}(z, z') \, dz' \tag{4.57b}$$

$$\Psi_{d\Sigma R}^{(m)} \equiv \Psi_{d\Sigma 2R}^{(m)} = (1 - \cos \beta_0 h)^{-1} \int_{-h}^{h} \sin \beta_0 (h - |z'|) K_{d\Sigma 2R}^{(m)}(0, z') \, dz'. \tag{4.58}$$

These are generalizations of (3.24a, b) and (3.25a, b). The other Ψ-functions, specifically $\Psi_{dU}^{(m)} = \Psi_{dUR}^{(m)} + j\Psi_{dUI}^{(m)}$, $\Psi_{dD}^{(m)} = \Psi_{dDR}^{(m)} + j\Psi_{dDI}^{(m)}$ and $\Psi_{dI}^{(m)}$ are the same as defined in (3.26)–(3.28) but with the N-term kernel given in (4.3e). Note that when all elements are sufficiently far apart to satisfy the inequality $\beta_0 b_{1i} > 1$, $1 < i \leq N$, only $i = 1$ with $b_{11} = a$ contributes to $\Psi_{dR}^{(m)}$ which is then equal to Ψ_{dR} for the isolated element.

With the notation introduced in (4.57a) and (4.58), the equation (3.33) applies directly to the N-element array. The same equation with $\Psi_{dR}^{(m)}$ substituted for Ψ_{dR} is correct when some elements are sufficiently close together so that $\beta_0 b_{1i} < 1$, $i > 1$. It follows that the entire formal solution in Sections 3.3 and 3.4 is valid for the phase sequences of the N-element array. The N independent phase-sequence currents $I^{(m)}(z)$, $m = 0, 1, \ldots N$, may be expressed as the solution of a column matrix equation.

A summary of the relevant equations is given below.

Phase-sequence currents

$$\{I_z^{(m)}(z)\} = \frac{j2\pi}{\zeta_0 \Psi_{dR} \cos \beta_0 h} [\{V^{(m)} M_{0z}\} + \{V^{(m)} T_U^{(m)} F_{0z}\} + \{V^{(m)} T_D^{(m)} H_{0z}\}], \tag{4.59}$$

where $M_{0z} = \sin \beta_0 (h - |z|)$, $F_{0z} = \cos \beta_0 z - \cos \beta_0 h$, and $H_{0z} = \cos \frac{1}{2}\beta_0 z - \cos \frac{1}{2}\beta_0 h$.

$$\begin{Bmatrix} T_U^{(m)} \\ T_D^{(m)} \end{Bmatrix} = [\Phi_T^{(m)}]^{-1} \begin{Bmatrix} \Psi_V^{(m)}(h) - \Psi_{d\Sigma R}^{(m)} \cos \beta_0 h \\ -j\Psi_{dI}^{(m)} \end{Bmatrix} \tag{4.60a}$$

$$[\Phi_T^{(m)}] = \begin{bmatrix} \Phi_{T11}^{(m)} & \Phi_{T12}^{(m)} \\ \Phi_{T21}^{(m)} & \Phi_{T22}^{(m)} \end{bmatrix}. \tag{4.60b}$$

$[\Phi_T^{(m)}]^{-1}$ is the reciprocal of $[\Phi_T^{(m)}]$.

$$
\left.
\begin{aligned}
\Phi_{T11}^{(m)} &= \Psi_{dUR}^{(m)} \cos \beta_0 h - \Psi_U^{(m)}(h) \\
\Phi_{T12}^{(m)} &= -\Psi_D^{(m)}(h) \\
\Phi_{T21}^{(m)} &= j\Psi_{dUI}^{(m)} \\
\Phi_{T22}^{(m)} &= \Psi_{dD}^{(m)}
\end{aligned}
\right\}.
\tag{4.61}
$$

The phase-sequence admittance is given by setting $z = 0$ in (4.59), thus:

$$
Y^{(m)} = \frac{I_z^{(m)}(0)}{V^{(m)}}.
\tag{4.62}
$$

The phase-sequence impedance is the reciprocal of the phase-sequence admittance,

$$
Z^{(m)} = \frac{1}{Y^{(m)}}.
\tag{4.63}
$$

The mutual impedances Z_{1i}, $1 < i \leq N$, may be calculated from the phase-sequence impedances by multiplying by the inverse (4.36) of the phase-sequence matrix P. Thus,

$$
\{Z\} = [P]^{-1}\{Z^{(m)}\}.
\tag{4.64}
$$

When the identical elements of a circular array are equally spaced around a circle, symmetry reduces the number of different admittances or impedances to $(N+1)/2$ if N is odd and $(N/2)+1$ if N is even. For example,

$$
Z_{12} = Z_{1N}; \qquad Z_{13} = Z_{1(N-1)}; \qquad Z_{14} = Z_{1(N-2)} \text{ etc.}
\tag{4.65}
$$

When the expression for the phase-sequence currents becomes indeterminate for $\beta_0 h = \pi/2$ and for a range near this value, the alternative form given in (3.43)–(3.45) is useful. It is

$$
\{I_z^{(m)}(z)\} = \frac{-j2\pi}{\zeta_0 \Psi_{dR}} \{V^{(m)} S_{0z} + V^{(m)} T_U^{\prime(m)} F_{0z} - V^{(m)} T_D^{\prime(m)} H_{0z}\}, \qquad \beta_0 h \sim \frac{\pi}{2},
\tag{4.66}
$$

where

$$
\begin{aligned}
S_{0z} &= \sin \beta_0 |z| - \sin \beta_0 h \\
&= \sin \beta_0 |z| - 1 \qquad \text{when } \beta_0 h = \frac{\pi}{2}
\end{aligned}
\tag{4.67}
$$

and

$$
\left.
\begin{aligned}
\{T_U^{\prime(m)}\} &= -\{(T_U^{(m)} + \sin \beta_0 h)/\cos \beta_0 h\} \\
\{T_D^{\prime(m)}\} &= \{T_D^{(m)}/\cos \beta_0 h\}
\end{aligned}
\right\}.
\tag{4.68}
$$

The two-term approximation used earlier in this chapter is quickly obtained from the three-term formulas. As stated in Section 3.9, the procedure involves the substitution of F_{0z} for H_{0z} and $T^{(m)}$ for $T_U^{(m)} + T_D^{(m)}$. This implies that $\Psi_{dD} \to \Psi_{dU}$, $\Psi_D(h) \to \Psi_U(h)$. The two-term forms for the phase-sequence currents and admittances (cf. (3.85)–(3.87)) are

$$\{I_z^{(m)}(z)\} = \frac{j2\pi}{\zeta_0 \Psi_{dR} \cos \beta_0 h} \{V^{(m)} M_{0z} + V^{(m)} T^{(m)} F_{0z}\} \tag{4.69}$$

and

$$\{Y^{(m)}\} = \left\{ \frac{I_z^{(m)}(0)}{V^{(m)}} \right\}, \tag{4.70}$$

where

$$\{T^{(m)}\} = -\left\{ \frac{\Psi_V^{(m)}(h) - (\Psi_{d\Sigma R}^{(m)} + j\Psi_{dI}^{(m)}) \cos \beta_0 h}{\Psi_U^{(m)}(h) - \Psi_{dU}^{(m)} \cos \beta_0 h} \right\}. \tag{4.71}$$

When $\beta_0 h$ is at or near $\pi/2$ the alternative formulas (3.88)–(3.90) are applicable. They are

$$\{I_z^{(m)}(z)\} = \frac{-j2\pi}{\zeta_0 \Psi_{dR}} \{V^{(m)} S_{0z} + V^{(m)} T'^{(m)} F_{0z}\} \tag{4.72}$$

$$\{Y^{(m)}\} = \left\{ \frac{I_z^{(m)}(0)}{V^{(m)}} \right\}, \tag{4.73}$$

where

$$\{T'^{(m)}\} = -\left\{ \frac{T^{(m)} + \sin \beta_0 h}{\cos \beta_0 h} \right\}, \qquad \beta_0 h \sim \frac{\pi}{2}. \tag{4.74}$$

Note that (4.69) and (4.72) are the same as (4.4a) and (4.4b).

5 The circuit and radiating properties of curtain arrays

In Chapters 2 and 3 an accurate theory is presented for a single antenna and for a two-element array. The present chapter is concerned with the analysis of the general N-element curtain array. This is a linear array with the centers of all elements along a straight line and with their axes all perpendicular to and in a plane containing the line.

5.1 Comparison of conventional and two-term theories

The analysis of arrays is conventionally formulated under the implicit assumption that distributions of current along all elements are identical. It follows that self- and mutual impedances depend only on the geometry of the elements. Circuit equations can then be written to relate the driving-point voltages and currents through an impedance matrix. Thus,

$$\{V\} = [Z]\{I\} \tag{5.1}$$

where

$$\{V\} = \begin{Bmatrix} V_{01} \\ V_{02} \\ \vdots \\ V_{0N} \end{Bmatrix}, \qquad \{I\} = \begin{Bmatrix} I_{01} \\ I_{02} \\ \vdots \\ I_{0N} \end{Bmatrix} \tag{5.2}$$

and

$$[Z] = \begin{bmatrix} Z_{11} & Z_{12} & Z_{13} & \cdots & Z_{1N} \\ Z_{21} & Z_{22} & Z_{23} & \cdots & Z_{2N} \\ \vdots & & & & \\ Z_{N1} & Z_{N2} & Z_{N3} & \cdots & Z_{NN} \end{bmatrix}. \tag{5.3}$$

The bracketed terms are $N \times N$ matrices; the terms in braces are column matrices. The usual reciprocity of off-diagonal impedances in (5.3) holds (i.e. $Z_{12} = Z_{21}$ etc.).

Equation (5.1) relates the quantities that can be assigned at the driving point of the antenna, namely the voltages and currents. The simple matrix relation between V's and I's shows that it is *immaterial* whether the voltages or the currents are specified, since the ratio between each voltage and current is unchanged. The phase and magnitude of the currents in the individual elements are normally specified so as to produce a particular radiation pattern. The assumption of identical distributions of current on all elements involves the tacit assumption that the phase and amplitude of the current at all points in each element are completely determined by their values at the driving point.

The preceding remarks may, at first glance, seem like a repetition of well-known facts. However, the assumptions implied in the conventional formulation are not satisfactory approximations for actual arrays except when the elements are very thin and have lengths near $\lambda/2$. Even for this special case difficulties arise when the elements are very closely spaced. Fortunately, a more realistic theory can be developed that is generally applicable to arrays with elements that are less than $3\lambda/4$ in half-length. The new theory is somewhat more complicated than the conventional approach. However, for engineering purposes it is more important that a theory agree with experiment than that it be mathematically simple. As with most new approaches, much of the complexity disappears with continued use and understanding. At the outset the fundamental processes will be explained without reference to the details of the theory.

An example of the notation of a three-element array is shown in Fig. 5.1. The conventional assumption is that regardless of the driving conditions each element has the same distribution of current. For example,

$$I_i(z) = I_{0i} \frac{\sin \beta_0(h - |z|)}{\sin \beta_0 h}, \qquad i = 1, 2, 3. \tag{5.4}$$

Equation (5.4) shows that once the currents are assigned at any point, e.g. at $z = 0$, the entire current is completely specified. The more accurate theory requires the individual currents to have distributions determined by their electrical environments. Specifically, they are represented by the following formula:

$$I_i(z) = j A_i \sin \beta_0(h - |z|) + B_i(\cos \beta_0 z - \cos \beta_0 h), \tag{5.5}$$

where $i = 1, 2, 3$, and A_i is real and B_i is complex. In (5.5) the A coefficients are directly proportional to the respective driving voltages. That is,

$$A_i = C V_{0i}, \qquad i = 1, 2, 3, \tag{5.6}$$

where C is a constant. On the other hand, the complex B coefficients depend on contributions not only from the individual element but also from all of the remaining elements. For example, there are contributions to B_1 from V_{01} and also from V_{02} and V_{03}.

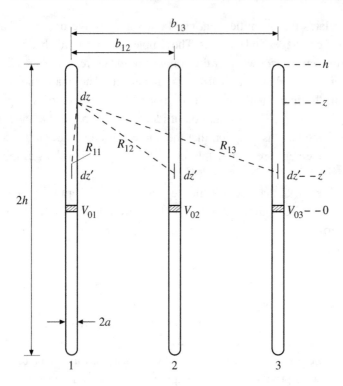

Figure 5.1 Three-element array.

5.2   Two-term theory of curtain arrays

The theoretical solution of the general problem of the curtain array will now be examined in detail. The essential basis for this theory was given in Chapter 2. Since the two-term theory described in Section 2.10 yields results of sufficient accuracy, it will be used for the curtain array to reduce the complexity of the formulation. However, the more accurate three-term representation involves only added algebraic complications.

The integral equation (2.4) may be written as follows for the kth antenna of an N-element array:

$$4\pi \mu_0^{-1} A_{zk}(z) = \int_{-h}^{h} \sum_{i=1}^{N} I_{zi}(z') K_{ki}(z, z') \, dz'$$

$$= -j \frac{4\pi}{\zeta_0} (C_k \cos \beta_0 z + \tfrac{1}{2} V_{0k} \sin \beta_0 |z|), \tag{5.7}$$

where

$$K_{ki}(z, z') = \frac{e^{-j\beta_0 R_{ki}}}{R_{ki}},$$

(5.8)

$$R_{ki} = \sqrt{(z_k - z_i')^2 + b_{ki}^2}, \qquad b_{kk} = a, \qquad \zeta_0 \doteq 120\pi \text{ ohms.}$$

(5.9)

The notation is illustrated in Fig. 5.1 for a three-element array. If the array has N elements, it is necessary to solve N simultaneous integral equations of the form (5.7), where $k = 1, 2, 3, \ldots N$. Following the procedure used in Chapter 2, an approximate zero-order solution will be obtained for the general linear array. That is, given the N driving voltages, a solution will be obtained for the currents in the N elements. Alternatively, given the N driving-point currents, the N driving voltages will be determined.

As a first step in the solution, the constant part of the vector potential is removed from the right-hand side of (5.7) by the introduction of the vector potential difference

$$W_{zk}(z) = A_{zk}(z) - A_{zk}(h).$$

The result is

$$4\pi \mu_0^{-1} W_{zk}(z) = \int_{-h}^{h} \sum_{i=1}^{N} I_{zi}(z') K_{kid}(z, z') \, dz'$$

(5.10)

$$= -j \frac{4\pi}{\zeta_0} [C_k \cos \beta_0 z + \tfrac{1}{2} V_{0k} \sin \beta_0 |z|]$$

$$- \int_{-h}^{h} \sum_{i=1}^{N} I_{zi}(z') K_{ki}(h, z') \, dz',$$

(5.11)

where

$$K_{kid}(z, z') = \frac{e^{-j\beta_0 R_{ki}}}{R_{ki}} - \frac{e^{-j\beta_0 R_{kih}}}{R_{kih}}.$$

(5.12)

The constants of integration C_k are expressed in terms of quantities U_k that are proportional to the $A_{zk}(h)$ by means of the relation $W_{zk}(h) = 0$. The final rearrangement of the integral equation (5.7) is [cf. (2.15)]

$$\int_{-h}^{h} \sum_{i=1}^{N} I_{zi}(z') K_{kid}(z, z') \, dz' = j \frac{4\pi}{\zeta_0 F_0(h)} (U_k F_{0z} + \tfrac{1}{2} V_{0k} M_{0z}),$$

(5.13)

where

$$F_{0z} = F_0(z) - F_0(h) = \cos \beta_0 z - \cos \beta_0 h$$

(5.14)

$$M_{0z} = F_0(z) \sin \beta_0 h - \sin \beta_0 |z| F_0(h) = \sin \beta_0 (h - |z|)$$

(5.15)

$$U_k = \sum_{i=1}^{N} U_{ki} = -j \frac{\zeta_0}{4\pi} \int_{-h}^{h} \sum_{i=1}^{N} I_{zi}(z') K_{ki}(h, z') \, dz'.$$

(5.16)

The difference kernel (5.12) may be separated into its real and imaginary parts as follows [cf. (2.5) et seq.]:

$$K_{kidR}(z, z') + jK_{kidI}(z, z') = K_{kid}(z, z'), \tag{5.17}$$

where

$$\left. \begin{array}{l} K_{kidR} = K_{kiR}(z, z') - K_{kiR}(h, z') \\ K_{kidI} = K_{kiI}(z, z') - K_{kiI}(h, z') \end{array} \right\}. \tag{5.18}$$

For the single element, the integrals corresponding to those in (5.13) were separated into two groups depending on the manner in which their leading terms varied as functions of z. The same principle of separation may be applied to the integrals for the curtain array. As before, one group varies approximately as M_{0z}, the other as F_{0z}. The following functional forms for the integrals in (5.13) are important general criteria for the separation:

$$\int_{-h}^{h} I_{zi}(z') K_{kiR}(z, z') \, dz' \sim I_{zi}(z) \qquad \text{when } \beta_0 b_{ki} < 1 \tag{5.19}$$

$$\int_{-h}^{h} I_{zi}(z') K_{kiR}(z, z') \, dz' \sim F_{0z} \qquad \text{when } \beta_0 b_{ki} \geq 1 \tag{5.20}$$

$$\int_{-h}^{h} I_{zi}(z') K_{kiI}(z, z') \, dz' \sim F_{0z} \qquad \text{for any } I(z') \text{ and all } \beta_0 b_{ki}. \tag{5.21}$$

The current in each element can now be expressed in two parts in the form

$$I_{zi}(z) = I_{ui}(z) + I_{vi}(z) \tag{5.22}$$

where, by definition, the leading terms behave approximately as follows:

$$I_{vi}(z) \sim M_{0z}, \qquad I_{ui}(z) \sim F_{0z}. \tag{5.23}$$

Some appreciation of the importance of the general functional forms in (5.23) may be obtained from an investigation of the integral equation (5.13). If attention is directed to the right-hand side of (5.13), it is seen that the equation contains two apparent sources, the coefficients of F_{0z} and M_{0z}. The function U_k has a constant amplitude over the entire length of the kth element and is generated primarily by the distributed currents on each element in the array. The other source function is the potential difference V_{0k}, as in a transmission line or in an isolated antenna; it is localized at $z = 0$. The form of the integral equation (5.13) suggests that the current on each element can be separated into two parts, the one apparently generated by the U_k, the other by the V_{0k}. The part of the current due to U_k is closely related to the current in an unloaded receiving antenna that is located in the wave front of an incident plane-wave field that has the same amplitude and phase over the entire length of the element. For this the leading term varies as F_{0z}. Except when the elements are very closely spaced (as in an open-wire line), the sinusoidal parts of the currents (i.e. M_{0z}) are maintained primarily by the individual driving potentials V_{0k}. Thus, the current due to each of the V_{0k} is essentially the same as in an isolated antenna.

When (5.22) is substituted in (5.13), groups of integrals occur that may be expressed as follows for all k and i in the ranges 1 to N:

$$\int_{-h}^{h} I_{ui}(z') K_{kid}(z, z') \, dz' = \left(\frac{B_i}{B_k}\right) \Psi_{kidu} I_{uk}(z) - D_{kidu}(z) \tag{5.24}$$

$$\int_{-h}^{h} I_{vi}(z') K_{kid}(z, z') \, dz' = \left(\frac{jA_i}{B_k}\right) \Psi_{kidv} I_{uk}(z) - D_{kidv}(z); \qquad \beta_0 b_{ki} \geq 1 \tag{5.25}$$

$$\int_{-h}^{h} I_{vi}(z') K_{kidR}(z, z') \, dz' = \left(\frac{A_i}{A_k}\right) \Psi_{kidR} I_{vk}(z) - D_{kidR}(z); \qquad \beta_0 b_{ki} < 1 \tag{5.26}$$

$$\int_{-h}^{h} I_{vi}(z') K_{kidI}(z, z') \, dz' = \left(\frac{jA_i}{B_k}\right) \Psi_{kidI} I_{uk}(z) - D_{kidI}(z); \qquad \beta_0 b_{ki} < 1. \tag{5.27}$$

It is assumed that the functions Ψ_{ki} are defined so that the difference terms $D_{ki}(z)$ are small enough to be negligible in a solution of zero order. The coefficient (jA_i/B_k) in (5.27) is the ratio of the amplitude of $I_{vi}(z)$ to that of $I_{uk}(z)$. When (5.24)–(5.27) are substituted in (5.13) and only the leading terms are retained, the following separation into two groups of equations may be carried out:

$$\sum_{i=k-m}^{k+m} \left(\frac{A_i}{A_k}\right) \Psi_{kidR} I_{vk}(z) = j \frac{2\pi}{\zeta_0 F_0(h)} V_{0k} M_{0z} \tag{5.28}$$

$$\sum_{i=1}^{N} \left(\frac{B_i}{B_k}\right) \Psi_{kidu} I_{uk}(z) + \left[\sum_{i=1}^{k-m-1} + \sum_{k+m+1}^{N}\right] \left(\frac{jA_i}{B_k}\right) \Psi_{kidv} I_{uk}(z)$$

$$+ j \sum_{i=k-m}^{k+m} \left(\frac{jA_i}{B_k}\right) \Psi_{kidI} I_{uk}(z) = j \frac{4\pi}{\zeta_0 F_0(h)} U_k F_{0z}. \tag{5.29}$$

The index m in the sums is defined by

$$\beta_0 b_{km} < 1, \qquad \beta_0 b_{k,m+1} \geq 1 \tag{5.30}$$

where b_{km} is the distance between the centers of the elements m and k. In most curtain arrays the spacing of the elements is sufficiently great so that all elements with $m \neq k$

satisfy the right-hand inequality in (5.30) and only $\beta_0 b_{kk} = \beta_0 a < 1$. When this is true, (5.28) and (5.29) reduce to

$$I_{vk}(z) = j \frac{2\pi V_{0k}}{\zeta_0 \Psi_{dR} F_0(h)} M_{0z} \qquad (5.31)$$

and

$$\sum_{i=1}^{N} \left\{ \left(\frac{B_i}{B_k} \right) \Psi_{kidu} + \left(\frac{jA_i}{B_k} \right) [\Psi_{kidv}(1 - \delta_{ik}) + j\Psi_{kidI}\delta_{ik}] \right\} I_{uk}(z) = \frac{j4\pi U_k}{\zeta_0 F_0(h)} F_{0z}, \qquad (5.32)$$

where

$$\delta_{ik} = \begin{cases} 0, & i \neq k \\ 1, & i = k. \end{cases}$$

The notation $\Psi_{dR} = \Psi_{kkdR}$ is used, since with identical elements all the Ψ_{kkdR} are identical and equal to Ψ_{dR} for the isolated antenna.

It follows directly from (5.31) that the leading term in $I_{vk}(z)$ is always M_{0z} for each value of k. Similarly from (5.32) the leading term in $I_{uk}(z)$ is of the form F_{0z}. Hence, it is possible to set

$$I_{vi}(z) = jA_i M_{0z}, \qquad I_{ui}(z) = B_i F_{0z} \qquad (5.33)$$

or

$$I_{zi}(z) = jA_i M_{0z} + B_i F_{0z}. \qquad (5.34)$$

Since Ψ_{dR} is real, it is clear from (5.31) that A_i is real when V_{0k} is real and from (5.32) that B_i is in general complex, or

$$B_i = B_{iR} + jB_{iI}. \qquad (5.35)$$

Note that the constant (jA_i/B_k), introduced in (5.25) and (5.27), is the ratio of the coefficients of the two terms in (5.34).

With the zero-order current formally determined, the constant U_k may be obtained from the substitution of (5.34) in (5.16). It is given by

$$U_k = -j \frac{\zeta_0}{4\pi} \sum_{i=1}^{N} [jA_i \Psi_{kiv}(h) + B_i \Psi_{kiu}(h)], \qquad (5.36)$$

where

$$\Psi_{kiv}(h) = \int_{-h}^{h} M_{0z'} K_{ki}(h, z') \, dz' \qquad (5.37)$$

$$\Psi_{kiu}(h) = \int_{-h}^{h} F_{0z'} K_{ki}(h, z') \, dz'. \qquad (5.38)$$

If (5.36), (5.33), and (5.34) are substituted in (5.31) and (5.32) the result is

$$A_k = \frac{2\pi}{\zeta_0 \Psi_{dR} F_0(h)} V_{0k} \tag{5.39}$$

$$\sum_{i=1}^{N} B_i [\Psi_{kidu} F_0(h) - \Psi_{kiu}(h)]$$

$$= j \sum_{i=1}^{N} A_i \{\Psi_{kiv}(h) - [\Psi_{kidv}(1 - \delta_{ik}) + j\Psi_{kidI}\delta_{ik}]F_0(h)\} \tag{5.40}$$

where $k = 1, 2, 3, \ldots N$. The physical significance of the zero-order solution is evident from (5.39) and (5.40). The coefficients of the 'transmitting part' of the current are given by (5.39). The N driving voltages generate the expected sinusoidal distribution of current on each element. The coefficients of the 'receiving part' of the current are given by (5.40). The N currents act as distributed sources to generate distributions of the receiving type which are present in all the elements of the array. Equation (5.40) permits the prediction in each driven element of the shifted-cosine component of the current that is due to coupling between currents distributed along the element itself and along all other elements in the array. Conventional array theory is concerned only with (5.39), since all currents are assumed to have the same sinusoidal distribution. In the special case of an array with thin half-wave elements, the real and imaginary parts of the current in each element do have approximately the same distribution. It follows that conventional array theory should work quite well for an array of very thin half-wave elements. On the other hand, in the more general case, the real and imaginary parts of the current in each element have different distributions so that (5.40) is needed along with (5.39) to determine the actual currents.

An important case to which conventional theory has no application is the array of full-wave elements in which the currents are near anti-resonance, and their real and imaginary parts have quite different distributions. Before some particular parallel arrays are analyzed, (5.40) is best expressed in matrix form. A general expression will be given for the $\Psi_{ki}(z)$ functions, and rigorous expressions will be derived for the radiation field.

Equation (5.40) is a set of linear algebraic equations with N unknowns that may be solved for the B_i in terms of the A_i. The N values of the A_i are expressed in terms of the N driving voltages V_{0i} by (5.39). In order to express (5.40) in matrix form, let the following quantities be defined:

$$\Phi_{kiu} = \Psi_{kidu} F_0(h) - \Psi_{kiu}(h) \tag{5.41}$$

$$\Phi_{kiv} = \Psi_{kiv}(h) - \Psi_{kidv} F_0(h)(1 - \delta_{ik}) - j\Psi_{kidI} F_0(h)\delta_{ik}. \tag{5.42}$$

Also let

$$[\Phi_u] = \begin{bmatrix} \Phi_{11u} & \Phi_{12u} & \cdots & \Phi_{1Nu} \\ \Phi_{21u} & \Phi_{22u} & \cdots & \Phi_{2Nu} \\ \vdots & & & \\ \Phi_{N1u} & \Phi_{N2u} & \cdots & \Phi_{NNu} \end{bmatrix} \tag{5.43}$$

$$[\Phi_v] = \begin{bmatrix} \Phi_{11v} & \Phi_{12v} & \cdots & \Phi_{1Nv} \\ \Phi_{21v} & \Phi_{22v} & \cdots & \Phi_{2Nv} \\ \vdots & & & \\ \Phi_{N1v} & \Phi_{N2v} & \cdots & \Phi_{NNv} \end{bmatrix} \tag{5.44}$$

$$\{A\} = \begin{Bmatrix} A_1 \\ A_2 \\ \vdots \\ A_N \end{Bmatrix}, \qquad \{B\} = \begin{Bmatrix} B_1 \\ B_2 \\ \vdots \\ B_N \end{Bmatrix}. \tag{5.45}$$

The bracketed terms are $N \times N$ matrices; the terms in braces are column matrices. From the substitution of (5.41)–(5.45) in (5.40), it follows that

$$[\Phi_u]\{B\} = [\Phi_v]\{jA\} \tag{5.46}$$

and from (5.39)

$$\{A\} = \frac{2\pi}{\zeta_0 \Psi_{dR} F_0(h)} \{V_0\} \tag{5.47}$$

with $\{V_0\}$ defined as in (5.2).

The solutions of two important problems in linear array theory are readily obtained from (5.46) and (5.47). Case I is concerned with specifying the driving-point[1] currents and determining the N potentials V_{0k} required to maintain these currents. In Case II the N potentials V_{0k} are specified and the corresponding driving-point currents are determined.

In the zero-order current distribution (5.34), the coefficients B_i are the amplitudes of the shifted cosine currents due to the distributed interaction of all elements of current in the array. The A_i coefficients are determined completely by the voltages of the

---

[1] The term 'base current' is also used for driving-point current.

individual generators. The distribution of the current in the ith element (5.34) may be separated into its real and imaginary parts as follows:

$$I_{zi}(z) = j\{A_i \sin \beta_0(h - |z|) + B_{iI}(\cos \beta_0 z - \cos \beta_0 h)\}$$

$$+ B_{iR}(\cos \beta_0 z - \cos \beta_0 h) \tag{5.48}$$

$$= I_{zi}''(z) + j I_{zi}'(z). \tag{5.49}$$

At $z = 0$, the real and imaginary parts of the driving-point current are

$$I_{zi}''(0) = B_{iR}(1 - \cos \beta_0 h) \tag{5.50a}$$

$$I_{zi}'(0) = A_i \sin \beta_0 h + B_{iI}(1 - \cos \beta_0 h). \tag{5.50b}$$

The driving-point impedance and admittance under the two driving conditions can be computed from the following general formulas obtained by combining (5.46)–(5.50). (Note: A special form is convenient when $\beta_0 h$ is at or near $\pi/2$.)

Case I   Input currents specified

$$\{V_0\} = \frac{1}{c_1(1 - \cos \beta_0 h)} [\Phi_w]^{-1}[\Phi_u]\{I_z(0)\}, \tag{5.51}$$

where

$$c_1 = j2\pi/(\zeta_0 \Psi_{dR} \cos \beta_0 h);$$

$$[\Phi_w] = [\Phi_v + \Phi_u \sin \beta_0 h/(1 - \cos \beta_0 h)]. \tag{5.52}$$

Case II   Driving voltages specified

$$\{I_z(0)\} = c_1(1 - \cos \beta_0 h)[\Phi_u]^{-1}[\Phi_w]\{V_0\}. \tag{5.53}$$

The matrix components in (5.51) and (5.53) as well as numerical values of the driving-point impedances and admittances under different driving conditions are given in tables in Appendices II and III. These tables were extracted from a more complete table [1]. The forms of the current for specified driving-point voltages and currents are not generally the same since the A_i and B_i coefficients differ for the two cases.

The symmetry properties of the impedance matrix in (5.1) and its counterpart in (5.51) are not identical. The assumption of identical current distributions implies that the mutual impedance between any two elements in an array is only a function of the size and spacing of the elements. Thus, with identical elements in an array, elements with the same center-to-center spacing have the same value of mutual impedance. For example, in an array with elements equally spaced, $Z_{12} = Z_{23} = Z_{34}$ and $Z_{13} =$

$Z_{24} = Z_{46}$. The mutual impedance for elements near the center of the array is then the same as for corresponding elements near the ends of the array. The more accurate theory correctly shows that the coupling properties of an element in the array depend on the distribution of current and the location of every element in the array. Elements near the edges of an array are coupled differently from elements near the center.

The general $\Psi(z)$ functions, obtained from the defining integrals (5.24)–(5.27), are

$$\Psi_{kidu}(z) = \frac{1}{\cos \beta_0 z - \cos \beta_0 h} \{[C_b(h, z) - C_b(h, h)]$$

$$- \cos \beta_0 h \, [E_b(h, z) - E_b(h, h)]\} \tag{5.54}$$

$$\Psi_{kidR}(z) = \frac{1}{\sin \beta_0 (h - |z|)} \, \mathrm{Re}\{[C_b(h, z) - C_b(h, h)] \sin \beta_0 h$$

$$- [S_b(h, z) - S_b(h, h)] \cos \beta_0 h\} \tag{5.55}$$

$$\Psi_{kidI}(z) = \frac{1}{\cos \beta_0 z - \cos \beta_0 h} \, \mathrm{Im}\{[C_b(h, z) - C_b(h, h)] \sin \beta_0 h$$

$$- [S_b(h, z) - S_b(h, h)] \cos \beta_0 h\} \tag{5.56}$$

$$\Psi_{kidv}(z) = \frac{1}{\cos \beta_0 z - \cos \beta_0 h} \{[C_b(h, z) - C_b(h, h)] \sin \beta_0 h$$

$$- [S_b(h, z) - S_b(h, h)] \cos \beta_0 h\} \tag{5.57}$$

$$\Psi_{kiv}(h) = C_b(h, h) \sin \beta_0 h - S_b(h, h) \cos \beta_0 h \tag{5.58}$$

$$\Psi_{kiu}(h) = C_b(h, h) - E_b(h, h) \cos \beta_0 h, \tag{5.59}$$

where in subscripts $b = b_{ki}$, and

$$S_b(h, z) = \int_0^h \sin \beta_0 z' \left[\frac{e^{-j\beta_0 R_1}}{R_1} + \frac{e^{-j\beta_0 R_2}}{R_2} \right] dz' \tag{5.60}$$

$$C_b(h, z) = \int_0^h \cos \beta_0 z' \left[\frac{e^{-j\beta_0 R_1}}{R_1} + \frac{e^{-j\beta_0 R_2}}{R_2} \right] dz' \tag{5.61}$$

$$E_b(h, z) = \int_0^h \left[\frac{e^{-j\beta_0 R_1}}{R_1} + \frac{e^{-j\beta_0 R_2}}{R_2} \right] dz' \tag{5.62}$$

$$R_1 = \sqrt{(z - z')^2 + b_{ki}^2}, \qquad R_2 = \sqrt{(z + z')^2 + b_{ki}^2}. \tag{5.63}$$

The functions S_b, C_b and E_b are found in King[2] and are tabulated for a wide range of values of h, z and b by Mack [3]. In order to obtain satisfactory overall agreement, the Ψ functions are evaluated at the point of maximum current. This ensures a good approximation for the determination of both the far field and the input power. However,

---

[2] [2], p. 94.

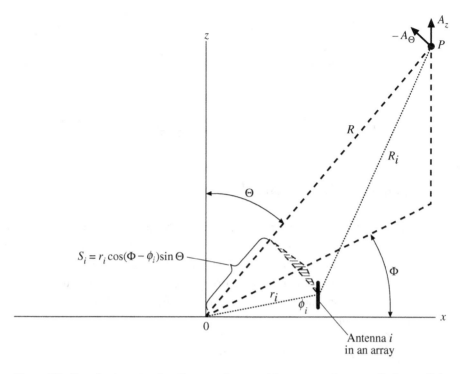

Figure 5.2 Coordinate system locating one element with respect to the center 0 of a parallel array.

the input susceptance may be somewhat in error. This does not present any practical difficulty since appropriate corrections may be applied at the driving point (cf. Section 2.8).

The far-zone electric field depends upon the location of each element and its current distribution. Thus, for the geometry of Fig. 5.2, which defines R_i,

$$E_\Theta(\Theta, \Phi) = j \, \frac{\omega\mu_0 \sin \Theta}{4\pi} \sum_{i=1}^{N} \frac{e^{-j\beta_0 R_i}}{R_i} \int_{-h}^{h} I_i(z') e^{j\beta_0 z' \cos \Theta} \, dz'. \tag{5.64}$$

For the conventional sinusoidal distribution of current, the electric field is given by

$$E_\Theta(\Theta, \Phi) = j \, \frac{\zeta_0}{2\pi} \sum_{i=1}^{N} \frac{e^{-j\beta_0 R_i}}{R_i} I_i(0) F_m(\Theta, \beta_0 h), \tag{5.65}$$

where

$$F_m(\Theta, \beta_0 h) = \frac{\cos(\beta_0 h \cos \Theta) - \cos \beta_0 h}{\sin \Theta}. \tag{5.66}$$

The more accurate far-zone electric field is

$$E_\Theta(\Theta, \Phi) = j \frac{\zeta_0}{2\pi} \sum_{i=1}^{N} \frac{e^{-j\beta_0 R_i}}{R_i} [j A_i F_m(\Theta, \beta_0 h) + B_i G_m(\Theta, \beta_0 h)], \qquad (5.67)$$

where

$$G_m(\Theta, \beta_0 h) = \frac{\sin \beta_0 h \cos(\beta_0 h \cos \Theta) \cos \Theta - \cos \beta_0 h \sin(\beta_0 h \cos \Theta)}{\sin \Theta \cos \Theta}. \qquad (5.68)$$

5.3   Example: the three-element array

Consider a three-element array with elements that are a full wavelength long ($2h = \lambda$) and separated by a quarter wavelength ($b_{i,i+1} = \lambda/4$). The conventional approach to this problem is doomed to failure when $\beta_0 h = \pi$, since an assumed sinusoidal current (i.e. $I_z(z) \sim \sin \beta_0|z|$) is zero at the driving point. This gives rise to a zero admittance or an infinite impedance for each element in the array. This difficulty does not exist with the current obtained from (5.5) with $\beta_0 h = \pi$. This is

$$I_i = j A_i \sin \beta_0 |z| + B_i (\cos \beta_0 z + 1). \qquad (5.69)$$

At $z = 0$, the current is finite and is given by the coefficient B_i for each element in the array. In order to demonstrate the difference between the two antenna theories, the conventional approach will be used for $\beta_0 h = 3.157$ and compared to the results of the two-term theory for $\beta_0 h = \pi$.

Consider now the three-element array shown in Fig. 5.1. Either the driving-point voltages or the driving-point currents may be specified. Conventionally the phases of the equal driving-point currents are specified to produce a radiation pattern. The electric field E_Θ in the far zone can be expressed in the simple form

$$E_\Theta = \frac{j\zeta_0 I_{z0}(0)}{2\pi} \frac{e^{-j\beta_0 R}}{R} F_0(\Theta, \beta_0 h) A(\Theta, \Phi), \qquad (5.70)$$

where $F_0(\Theta, \beta_0 h)$ is the vertical field function of an isolated element and $A(\Theta, \Phi)$ is the array factor. A uniform array with equally spaced elements and with $|I_{zi}(0)| = |I_{z0}(0)|$ for all values of i has the array factor

$$A(\Theta, \Phi) = \frac{\sin Nx}{\sin x} \equiv A(\Theta, \Phi; N, n, t), \qquad (5.71)$$

where $x = \pi(n \sin \Theta \cos \Phi - t)$, n is the distance between elements in fractions of a wavelength, and t is the time delay from element to element in fractions of a cycle. The expanded form of (5.71) in which all five variables and parameters are explicit is useful when several array factors are superimposed. For the array of Fig. 5.1, the number of elements is $N = 3$, and n is chosen to be $\frac{1}{4}$. Now let attention be directed to the

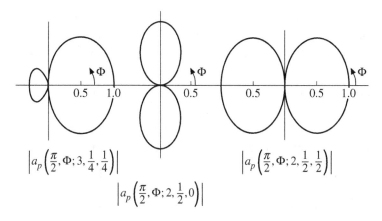

$$\left|a_p\left(\frac{\pi}{2},\Phi;3,\frac{1}{4},\frac{1}{4}\right)\right|$$

$$\left|a_p\left(\frac{\pi}{2},\Phi;2,\frac{1}{2},\frac{1}{2}\right)\right|$$

$$\left|a_p\left(\frac{\pi}{2},\Phi;2,\frac{1}{2},0\right)\right|$$

Figure 5.3 Arrays factors which comprise actual three-element array factor.

horizontal pattern in the equatorial or H-plane defined by $\Theta = \pi/2$. The three driving-point currents are equal in magnitude but the phase delay t between elements may be varied to produce a particular pattern. For example, a value of $t = \frac{1}{4}$ will produce an endfire radiation pattern with the maximum value of the directivity D toward the right in Fig. 5.1.

The actual radiation pattern of the three-element endfire array with specified base currents differs from the ideal pattern shown on the left of Fig. 5.3. The several components of current on the elements which are discussed later in this section, are equivalent to separate sources producing different radiation patterns. The additional patterns in the middle and right of Fig. 5.3 fill in the deep nulls and reduce the back-to-front ratio of the ideal pattern. The electric field for the three-element array with the expanded form of the array factor in (5.71) is given by the sum

$$E_\Theta \sim A(\pi/2, \Phi; 3, \tfrac{1}{4}, \tfrac{1}{4}) + (-0.53 + j0.57)A(\pi/2, \Phi; 2, \tfrac{1}{2}, 0)$$

$$+ (-0.07 + j0.50)A(\pi/2, \Phi; 2, \tfrac{1}{2}, \tfrac{1}{2}).$$

The individual normalized field patterns

$$a_p(\Theta, \Phi; N, n, t) = \frac{A(\Theta, \Phi; N, n, t)}{N} = \frac{\sin Nx}{N \sin x}$$

are shown in Fig. 5.3.

The ideal radiation pattern as determined from (5.70) depends on the vertical field factor $F_0(\Theta, \beta_0 h)$ of an isolated element. Consider now an array with full-wave elements ($\beta_0 h = \pi$). The particular value of $F_0(\Theta, \beta_0 h)$ is given by (5.66). Thus,

$$F_0(\Theta, \pi) = \frac{\cos(\pi \cos \Theta) + 1}{\sin \Theta} \tag{5.72}$$

$$F_0\left(\frac{\pi}{2}, \pi\right) = 2. \tag{5.73}$$

As an illustrative comparison of the results of the present theory with other methods, an examination of the driving-point impedances for the three-element array with $\beta_0 h = \pi/2$ and $\beta_0 h = \pi$ is given. An extension of the "induced EMF method", discussed in Section 1.7 for the case of a single isolated antenna, is frequently used for the calculation of mutual impedance.[3] As in the case of a single antenna, the method assumes sinusoidal currents. The resulting formula for the mutual impedance is the same as in (1.42) but with a replaced by b. For $\beta_0 h = \pi/2$, and with $\zeta_0 = 120\pi$ ohms, the self- and mutual impedances obtained with the EMF method are

$$
\left.
\begin{aligned}
Z_{11} &= 73.12 + j41.28 \text{ ohms}, & \Omega &= 2\ln(2h/a) = 10 \\
Z_{12} &= 40.79 - j28.35 \text{ ohms}, & \beta_0 b &= \pi/2 \\
Z_{13} &= -12.53 - j29.93 \text{ ohms}, & \beta_0 b &= \pi
\end{aligned}
\right\}
\tag{5.74}
$$

$$
Z_{12} = Z_{21} = Z_{23} = Z_{32}, \quad Z_{13} = Z_{31}, \quad Z_{11} = Z_{22} = Z_{33}.
\tag{5.75}
$$

The driving-point currents are specified in the following way to produce an endfire pattern:

$$
\{I\} =
\left\{
\begin{matrix}
I_{01} \\
I_{02} \\
I_{03}
\end{matrix}
\right\}
= I_{01}
\left\{
\begin{matrix}
1 \\
-j \\
-1
\end{matrix}
\right\}.
\tag{5.76}
$$

The substitution of (5.76) and (5.74) into (5.1) yields the following driving-point impedances for the three elements:

$$
\left.
\begin{aligned}
Z_{01} &= Z_{11} - jZ_{12} - Z_{13} = 57.3 + j30.42 \text{ ohms} = V_{01}/I_{01} \\
Z_{02} &= Z_{11} = 73.12 + j41.28 \text{ ohms} = V_{02}/I_{02} \\
Z_{03} &= Z_{11} + jZ_{12} - Z_{13} = 114.0 + j112.0 \text{ ohms} = V_{03}/I_{03}
\end{aligned}
\right\}.
\tag{5.77}
$$

The same results are, of course, obtained when the driving voltages are assigned instead of the currents by the substitution of V for I in (5.76), since no changes are possible in the assumed distributions of current and, hence, in the mutual coupling.

It is now in order to examine the results obtained by the two-term theory which takes full account of the changes in the distributions of current due to the presence of any number of coupled elements. The driving-point impedances for the three elements are readily computed.[4] They are

$$
\left.
\begin{aligned}
Z_{01} &= 67.51 + j24.14 \text{ ohms}, & \Omega &= 2\ln\frac{2h}{a} = 10 \\
Z_{02} &= 78.47 + j31.23 \text{ ohms} \\
Z_{03} &= 145.61 + j96.91 \text{ ohms}
\end{aligned}
\right\}.
\tag{5.78}
$$

[3] See, for example, [4], pp. 535–556.   [4] [1], p. 84.

These values are comparable with those in (5.77) (with differences not exceeding about 30%) simply because the current in half-wave dipoles is predominantly sinusoidal with only relatively small changes due to finite radius and mutual coupling.

The situation is quite different when the elements are not near resonance. This is well illustrated with the same three-element array but now with $\beta_0 h$ near π instead of $\pi/2$. The conventional application of the EMF method with assumed sinusoidal currents on all elements yields meaningless results. Since the currents at all three driving points are identically zero all driving-point impedances are infinite – which is, of course, absurd.

Once again it is in order to introduce the results from the two-term theory which actually determines the distributions of the currents on all three elements and the associated driving-point impedances. The following values are readily calculated[5] for the driving-point currents specified in (5.76):

$$\left.\begin{array}{ll} Z_{01} = 612 - j591 \text{ ohms} & Y_{01} = (0.845 + j0.817) \times 10^{-3} \text{ siemens} \\ Z_{02} = 160 - j590 \text{ ohms} & Y_{02} = (0.429 + j1.578) \times 10^{-3} \text{ siemens} \\ Z_{03} = 61.5 - j435 \text{ ohms} & Y_{03} = (0.318 + j2.252) \times 10^{-3} \text{ siemens} \end{array}\right\}. \qquad (5.79)$$

When normalized to I_{01}, the voltages required to maintain the currents specified in (5.76) when $\beta_0 h = \pi$ and $\beta_0 b = \pi/2$, i.e. $I_{02} = -jI_{01}$, $I_{03} = -I_{01}$, are $V_{01}/I_{01} = 612 - j591$ volt/ampere, $V_{02}/I_{01} = -590 - j160$ volt/ampere, and $V_{03}/I_{01} = -61.5 + j435$ volt/ampere. The power supplied to element k by its generator is

$$P_k = |I_{0k}|^2 R_{0k} = |V_{0k}|^2 G_{0k}. \qquad (5.80)$$

The ratios of the powers supplied to the three-element array are

$$P_1/P_3 = 9.82, \qquad P_2/P_3 = 2.51. \qquad (5.81)$$

Evidently element 1 receives almost ten times the power that is supplied to the terminals of element 3.

The two-term theory gives the following values[6] of Z_{0i} and Y_{0i}, $i = 1, 2, 3$ when the driving-point voltages are specified ($V_{02} = -jV_{01}$, $V_{03} = -V_{01}$) instead of the currents:

$$\left.\begin{array}{ll} Z_{01} = 675 - j484 \text{ ohms} & Y_{01} = (0.979 + j0.701) \times 10^{-3} \text{ siemens} \\ Z_{02} = 359 - j479 \text{ ohms} & Y_{02} = (1.003 + j1.336) \times 10^{-3} \text{ siemens} \\ Z_{03} = 170 - j426 \text{ ohms} & Y_{03} = (0.808 + j2.024) \times 10^{-3} \text{ siemens} \end{array}\right\}. \qquad (5.82)$$

---

[5] [1], p. 203.   [6] [1], p. 221.

Clearly, the results for Cases I and II are not the same as seen from a comparison of (5.79) and (5.82). This difference is due to the unequal distributions of the currents in the elements which cause non-uniform coupling. This effect will become clearer when the currents in the individual elements are examined.

The conventional currents in the three-element endfire array with $\beta_0 h = 3.157$ are

$$I_i(z) = I_{0i} \sin(3.157 - \beta_0|z|), \qquad i = 1, 2, 3$$

$$\text{driving-point currents specified} \tag{5.83}$$

$$I_i(z) = V_i Y_i \sin(3.157 - \beta_0|z|), \qquad i = 1, 2, 3$$

$$\text{driving voltages specified.} \tag{5.84}$$

The form of the currents in (5.83) and (5.84) is identical for each element. Both the real and imaginary parts have the same distribution. The currents in the two-term theory are given by (5.69) with (5.46) and (5.47). They are shown in Figs. 5.4 and 5.5 for the two different driving conditions. When the currents at $z = 0$ are specified, the distributions differ widely in form from element to element. Note that the currents are shown both with respect to the individual driving voltages and with respect to V_{02}. In the computation of radiation patterns the currents must all be normalized with respect to a single driving voltage. The large differences in the real and imaginary parts of the currents in Fig. 5.4 practically disappear when the driving voltages are specified in Fig. 5.5.

5.4   Electronically scanned arrays

Previous sections of this book have demonstrated the general invalidity of the assumption of equal current distributions in the elements of an array. A most significant result of the two-term theory is that the expected conventional radiation pattern is not achieved since the contributions by the individual elements to the radiation pattern are different. The results of the two-term theory for the broadside and endfire arrays show an appreciable difference not only between the driving-point impedances for the broadside and endfire arrays, but between the conventional and two-term theories. The experimental determination of the individual driving-point impedances is a complicated problem and a theoretical prediction of the individual circuit properties would certainly be an aid in the efficient operation of an array.

A comparison of the corresponding expressions for the far fields based on the conventional method and the more accurate two-term approach helps to illustrate some of the problems in the theory of scanned arrays. Consider an array in which the currents at the driving points of the elements are specified in both amplitude and phase. For the present, let the amplitudes be equal and the phases required to change linearly from

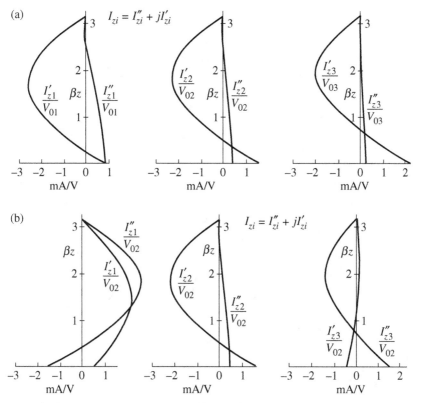

Figure 5.4 Three-element endfire array; driving-point currents specified. Drawn with respect to (a) individual driving voltages, (b) V_{02} ($\lambda/4$ spacing, $\beta_0 h = \pi$, $\Omega = 10$).

element to element across the array. For example, the base current might increase in phase by $30°$ toward one end of the array. Expressed in general terms

$$I_i = I_0 e^{-j\delta_i} = I_0 e^{-j2\pi it} \tag{5.85}$$

where t is the time delay between elements in fractions of a period.

With the currents at $z = 0$ specified in (5.85) and under the assumption of identical distributions of current, the far-zone electric field has the form

$$E_\Theta^r = \left\{ \frac{j\zeta_0 I_0}{2\pi} \frac{e^{-j\beta_0 R_0}}{R_0} F_0(\Theta, \beta_0 h) \right\} \left\{ 1 + \sum_{i=1}^{\frac{1}{2}(N-1)} [e^{-j(\delta_i - \beta_0 S_i)} + e^{j(\delta_i - \beta_0 S_i)}] \right\}.$$
$$\tag{5.86}$$

The second term in braces in (5.86) is the familiar array factor given by (5.71). If the distance between the elements is small enough, the radiation pattern has only one principal maximum in the visible ranges of Θ and Φ. The first maximum of (5.70) occurs when $x = 0$. Thus, to direct the main beam in a specific direction (Θ_m, Φ_m)

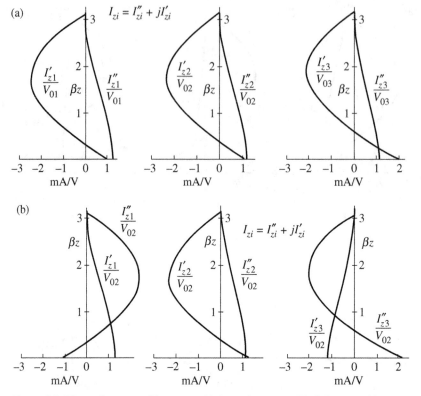

Figure 5.5 Three-element endfire array; driving voltages specified. Drawn with respect to (a) individual driving voltages, (b) V_{02} ($\lambda/4$ spacing, $\beta_0 h = \pi$, $\Omega = 10$).

in space, the time delay between the currents in the elements must be set equal to a particular value t_m such that

$$n \sin \Theta_m \cos \Phi_m - t_m = 0,$$

or

$$t_m = n \sin \Theta_m \cos \Phi_m. \tag{5.87}$$

For example, in an array with half-wave spacing ($n = \frac{1}{2}$) for which the main beam is to point in the direction $\Theta_m = \pi/2$ (H-plane) and $\Phi_m = 60°$, the required phase shift given by (5.87) is $t_m = \frac{1}{4}$ of a period. In a single curtain array it is not possible with ordinary elements to have any control over the beam pointing in the Θ direction. The control of the main beam in the Θ direction could be achieved by a planar array formed by an array of collinear elements.

Now let the conventional requirement, that the distributions of current be equal, be removed. Let the currents at $z = 0$ again be specified so that, on the basis of the conventional theory, the main beam will point in a desired direction. However, and this obvious fact is often overlooked, the specification of the currents at each driving point usually does not determine the entire current along each element. A variety of distributions of current may be associated with any given value at $z = 0$. In general, the radiation pattern can be considered the superposition of two parts. One part is the pattern of an array of elements with equal distributions of current; the other the pattern of the same array with dissimilar distributions of current. Conventional theories assume that the first part is the entire pattern.

The beam-pointing properties of a scanning array are affected by the interaction between the currents in the elements. The simple array factor in (5.71) characterizes an ideal array in which the exact phase and amplitude of the current are specified for each element. This specification applies not only at $z = 0$ but all along each element. The phase of the current is of primary importance in the determination of the direction of the main beam. In an actual array the variation in phase along the length of each individual element differs from element to element. In practice, this variation is responsible for a beam-pointing error of non-negligible value. Furthermore, with this phase variation perfect phase cancellation and addition are impossible. Perfect nulls in the radiation pattern will disappear and side-lobe levels will be modified significantly. The side-lobe level and the angular width of the main beam are also affected by changes in the magnitudes of the currents from element to element across the array.

As a specific example, consider the three-element array with full-wave elements ($\beta_0 h = \pi$) and half-wave spacing ($\beta_0 b = \pi/2$). The driving-point currents are specified to produce a maximum field in the direction indicated by the conventional theory. The driving-point impedance is to be calculated for each element as a function of the scanning angle. The actual position of the maximum, as given by the two-term theory, is to be compared with the corresponding angular position predicted by the conventional theory. The difference is the beam-pointing error Δ.

The general matrix relation (5.51) between the driving-point voltages and currents may be reduced to the following symbolic form:

$$\{V_0\} = [\Phi_T]\{I_0\}, \tag{5.88}$$

where

$$\{V_0\} = \begin{Bmatrix} V_{01} \\ V_{02} \\ \vdots \\ V_{0N} \end{Bmatrix}, \qquad \{I_0\} = \begin{Bmatrix} I_{01} \\ I_{02} \\ \vdots \\ I_{0N} \end{Bmatrix} \tag{5.89}$$

and

$$[\Phi_T] = \begin{bmatrix} \Phi_{T11} & \Phi_{T12} & \cdots & \Phi_{T1N} \\ \Phi_{T21} & \Phi_{T22} & \cdots & \Phi_{T2N} \\ \vdots & & & \\ \Phi_{TN1} & \Phi_{TN2} & \cdots & \Phi_{TNN} \end{bmatrix} = \frac{1}{c_1(1 - \cos \beta_0 h)}[\Phi_w]^{-1}[\Phi_u], \tag{5.90}$$

where [cf. (5.52)]

$$[\Phi_w] = [\Phi_v + \Phi_u \sin \beta_0 h/(1 - \cos \beta_0 h)]$$

$$c_1 = j2\pi/\zeta_0 \Psi_{dR} \cos \beta_0 h.$$

The elements of Φ_T may be derived from the machine-tabulated values of Φ_u and Φ_v in the tables [Appendix II]. From the tabulated values for the two different sets of driving conditions [Appendix III] and a knowledge of the symmetry properties of the Φ_u and Φ_v values, the Φ_T values may also be calculated. In the present example calculations similar to those given in Appendix III yield the following information for the case $\beta_0 h = \pi$, $\beta_0 b = \pi/2$:

When

$$\{I_0\} = I_{01} \left\{ \begin{array}{c} 1 \\ -j \\ -1 \end{array} \right\} \tag{5.91}$$

then

$$\{V_0\} = I_{01} \left\{ \begin{array}{c} 612 - j591 \\ -590 - j160 \\ -61.5 + j435 \end{array} \right\}. \tag{5.92}$$

Also when

$$\{I_0\} = I_{01} \left\{ \begin{array}{c} 1 \\ 1 \\ 1 \end{array} \right\} \tag{5.93}$$

then

$$\{V_0\} = I_{01} \left\{ \begin{array}{c} 435 - j346 \\ 309 - j37.9 \\ 435 - j346 \end{array} \right\}. \tag{5.94}$$

The specifications in (5.91) and (5.93) are the conventional ones for the endfire and broadside arrays. For $\beta_0 h = \pi$, $\beta_0 b = \pi/2$, the time delay between elements as given by (5.87) is

$$t_m = n \cos \Phi_m = \tfrac{1}{4} \cos \Phi_m. \tag{5.95}$$

Table 5.1. *Relative values of driving-point currents I_i, $i = 1, 2, 3$ for different values of Φ_m*

Φ_m	0	$30°$	$60°$	$75°$	$90°$
I_1	1	$1 + j0$	$1 + j0$	$1 + j0$	1
I_2/I_1	$-j$	$0.209 - j0.978$	$0.707 - j0.707$	$0.919 - j0.395$	1
I_3/I_1	-1	$-0.913 - j0.409$	$-0.707 - j0.707$	$0.687 - j0.726$	1

The driving-point currents in the N elements can now be expressed in terms of the angle Φ_m with the aid of (5.85). The result is

$$I_i = I_1 e^{-j2\pi(i-1)n \cos \Phi_m} = I_1 e^{-j(\pi/2)(i-1)\cos \Phi_m}, \qquad n = \tfrac{1}{4}, \qquad i = 1, 2, 3. \tag{5.96}$$

Table 5.1 is useful for the computation of the driving-point impedances for different values of the angle Φ_m.

Before the driving-point impedances can be computed the elements of the matrix Φ_T must be found. They can be computed directly from the basic matrix equations in terms of Φ_u and Φ_v, or they may be computed from the tables of driving-point impedances for different driving conditions. For example, from the two sets of information contained in (5.92) and (5.94), the symbolic matrix multiplication (5.88) yields

$$\left.\begin{aligned}
\Phi_{T11} + \Phi_{T12} + \Phi_{T13} &= 435 - j346 \\
2\Phi_{T21} + \Phi_{T22} &= 309 - j37.9 \\
\Phi_{T11} - j\Phi_{T12} - \Phi_{T13} &= 612 - j591 \\
-j\Phi_{T22} &= -590 - j160 \\
-\Phi_{T11} - j\Phi_{T12} + \Phi_{T13} &= -61.5 + j435
\end{aligned}\right\}, \tag{5.97}$$

where

$$[\Phi_T] = \begin{bmatrix} \Phi_{T11} & \Phi_{T12} & \Phi_{T13} \\ \Phi_{T21} & \Phi_{T22} & \Phi_{T21} \\ \Phi_{T13} & \Phi_{T12} & \Phi_{T11} \end{bmatrix}. \tag{5.98}$$

The symmetry properties of the elements of (5.98) were deduced from those of the component matrices involved in (5.88). The elements of (5.98) may be compared to the impedance matrix whose elements are the self- and mutual impedances computed under the conventional assumptions. For example, Φ_{T11} could be compared to Z_{11}, the self-impedance of the first antenna. The result shown symbolically in (5.98) indicates that the off-diagonal terms are not necessarily equal (e.g. $\Phi_{T12} \neq \Phi_{T21}$) and that the

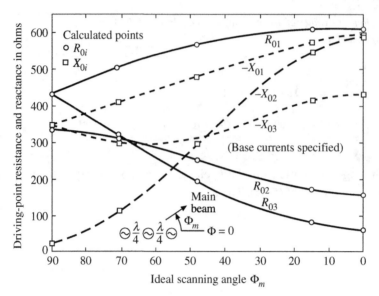

Figure 5.6 Variation of driving-point resistance and reactance with beam-pointing angle Φ_m for three-element array ($\lambda/4$ spacing, $\beta_0 h = \pi$, $\Omega = 10$).

diagonal terms may differ (e.g. $\Phi_{T11} \neq \Phi_{T22}$). The numerical values of the matrix elements of Φ_T are

$$
\left.
\begin{aligned}
\Phi_{T11} &= 347 - j567 \\
\Phi_{T22} &= 160 - j590 \\
\Phi_{T12} &= 77.9 + j275 \\
\Phi_{T21} &= 74.3 + j276 \\
\Phi_{T13} &= 10.4 - j53.7
\end{aligned}
\right\} .
\tag{5.99}
$$

Consider the specific case $\Phi_m = 75°$, where the driving-point currents are given in Table 5.1 and the elements of the Φ_T matrix are given by (5.99). Thus,

$$
\left\{
\begin{aligned}
V_{01} \\
V_{02} \\
V_{03}
\end{aligned}
\right\}
=
\begin{bmatrix}
\Phi_{T11} & \Phi_{T12} & \Phi_{T13} \\
\Phi_{T21} & \Phi_{T22} & \Phi_{T21} \\
\Phi_{T13} & \Phi_{T12} & \Phi_{T11}
\end{bmatrix}
\left\{
\begin{aligned}
1 + j0 \\
0.919 - j0.395 \\
0.687 - j0.726
\end{aligned}
\right\} I_1.
\tag{5.100}
$$

To compute, for example, $Z_{02} = (V_{02}/I_2)$, the quantity (V_{02}/I_1) is computed from (5.100) or $(V_{02}/I_1) = \Phi_{T21}(1 + j0) + \Phi_{T22}(0.919 - j0.395) + \Phi_{T21}(0.687 - j0.726) = 240 - j193$ ohms. The driving-point impedance Z_{02} is found from the substitution of the relation $I_2 = (0.919 - j0.395)I_1$ in this expression with the result, $Z_{02} = 297 - j82.8$ ohms. The variation of the driving-point resistance and reactance with the beam-pointing angle Φ_m is shown in Fig. 5.6. It is seen that even if the beam-pointing angle is restricted to moderate departures from a normal position, significant changes in the impedance function occur. These will be apparent in the mismatch between the generator and the antenna. Note also that from symmetry, the continuation of R_{01} and X_{01} in the range $90° \leq \Phi_m \leq 180°$ is the mirror image of R_{03} and X_{03} about $\Phi_m = 90°$.

The driving-point currents have been specified according to the criteria of the conventional theory. This specification does not control the distribution of either the phase or the amplitude of the current away from the driving point. As a result, the location of the maximum of the main beam may differ from that predicted by the ideal angle in (5.87). This difference is the beam-pointing error Δ and represents the difference between the ideal scanning angle Φ_m and the actual angle Φ_a.

The far-zone electric field is given by (5.67). The computation of this field requires all currents to be normalized with respect to a single driving voltage. Thus, with the kth element as a reference, (5.67) may be rearranged to give

$$E_\Theta^r(\Theta) = \frac{j\zeta_0}{2\pi} \frac{e^{-j\beta_0 R_0}}{R_0} \sum_{i=1}^{N} C_i e^{-j\beta_0 b[(N-2i+1)/2]\cos\Phi\sin\Theta} \tag{5.101}$$

where

$$C_1 = \xi_i \left[\frac{-j}{60\Psi_{dR}} F_m(\Theta, \beta_0 h) + \frac{Y_i}{2} G_m(\Theta, \beta_0 h) \right] \tag{5.102}$$

and

$$\xi_i = V_{0i}/V_{0k}. \tag{5.103}$$

The conventional theory equates the C coefficient in (5.101) to the driving-point currents [cf. (5.65)]. These, in turn, are chosen to produce a given radiation pattern. The two-term theory has shown that the currents in the elements as well as the radiation pattern cannot be specified merely by adjusting the currents at the driving point. Moreover, the direction of the main beam may differ considerably from the value predicted by the conventional theory.

The true location of the principal lobe is found from the location of the major maximum of $|E_\Theta^r(\Theta)|$ or of $|E_\Theta^r(\Theta)|^2$. For the special case $\Theta = \pi/2$, $\beta_0 b = \pi/2$, $N = 3$, and $\beta_0 h = \pi/2$, the electric field in the far zone is

$$E_\Theta^r(\Theta) = K(C_1 e^{-ju} + C_2 + C_3 e^{ju}), \tag{5.104}$$

where

$$K = \frac{j\zeta_0}{2\pi} \frac{e^{-j\beta_0 R_0}}{R_0}$$

$$u = \beta_0 b \cos\Phi = \frac{\pi}{2}\cos\Phi$$

$$\Theta = \frac{\pi}{2} \ (H\text{-plane}).$$

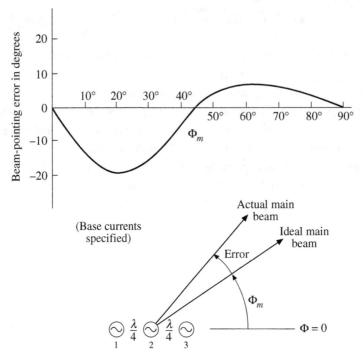

Figure 5.7 Variation of beam-pointing error with beam-pointing angle for three-element array ($\lambda/4$ spacing, $\beta_0 h = \pi$, $\Omega = 10$).

The square of the absolute value of $E_\Theta^r(\Theta)$ is formed from (5.104) with the result

$$E_\Theta^r E_\Theta^{r*} = KK^* C_2 C_2^* [C_{12}^* C_{32} e^{j2u} + (C_{12}^* + C_{32}) e^{ju} + (1 + C_{12} C_{12}^* + C_{32} C_{32}^*)$$

$$+ (C_{12} + C_{32}^*) e^{-ju} + C_{12} C_{32}^* e^{-j2u}], \tag{5.105}$$

where

$$C_{12} = \frac{C_1}{C_2} \quad \text{and} \quad C_{32} = \frac{C_3}{C_2}. \tag{5.106}$$

With the substitution $x = e^{ju}$, (5.105) is seen to be an algebraic equation of fourth degree. Thus,

$$|E_\Theta^r(x)|^2 = C_a x^4 + C_b x^3 + C_c x^2 + C_d x + C_e. \tag{5.107}$$

The true location of the principal lobe is determined from the equation obtained when (5.107) is differentiated with respect to x and equated to zero. The computed beam-pointing error for the three-element array as determined from the conventional theory is shown in Fig. 5.7. This graph shows an appreciable plus and minus variation over most of the visible range of Φ.

The expression for the square of the absolute value of the far-field for the N-element array is

$$|E_\Theta^r(x)|^2 = KK^* \sum_i \sum_n C_i C_n^* x^{(i-n)}. \tag{5.108}$$

The location of the extremes of (5.108) is given by

$$\frac{\partial}{\partial x} |E_\Theta^r(x)|^2 = KK^* j \sum_i \sum_n (i-n) C_i C_n^* x^{(i-n)} = 0, \qquad x = x_0, x_1, x_2, \ldots. \tag{5.109}$$

5.5   Examples of the general theory for large arrays

Thus far the simple array with $N = 3$ and $\beta_0 b = \pi/2$ has been examined for a variety of driving-point conditions. Calculations have also been made for arrays with a larger number of elements. For these the lengths $2h$ of the elements were varied from a quarter to a full wavelength. The driving-point voltages or currents were specified according to conventional array theory to produce a broadside or endfire radiation pattern.

The driving-point currents required for an ideal broadside array are

$$I_{z1}(0) = I_{z2}(0) = I_{z3}(0), \quad \text{etc.} \tag{5.110}$$

or, in matrix form,

$$\{I_z(0)\} = I_{z1}(0) \begin{Bmatrix} 1 \\ 1 \\ \vdots \end{Bmatrix}. \tag{5.111}$$

Alternatively, the driving voltages may be assigned as follows:

$$\{V\} = V_1 \begin{Bmatrix} 1 \\ 1 \\ 1 \\ \vdots \end{Bmatrix}. \tag{5.112}$$

The relatively large sinusoidal parts of the currents on the antennas are determined directly from (5.47) by the specification of the voltages. However, the relations (5.50a) and (5.50b) between the coefficients A, B_R, B_I and the currents at $z = 0$ do not in general suffice to determine the distributions of current along the elements.

The driving-point impedances for broadside arrays are shown in Figs. 5.8–5.10. Driving-point currents and voltages are specified for arrays of up to 25 elements ($N \leq 25$) for quarter- and half-wavelength spacings ($\beta_0 b = \pi/2$ and $\beta_0 b = \pi$).

In Figs. 5.8a–d are shown graphs of the resistances and reactances of the individual elements of a broadside array when the driving-point currents are specified. In Figs. 5.8a and 5.8b the distance between the adjacent antennas is one-quarter wavelength; the lengths of the elements are, respectively, a quarter and a half wavelength. In Figs. 5.8c and 5.8d the spacing of the elements has been increased to a half wavelength.

Since the main beam of a broadside array is at right angles to the curtain of antennas, it is to be expected that the effect of mutual coupling will be much less than for an endfire array. However, when the elements are separated by only a quarter wavelength, differences in the interactions between the currents in differently situated elements are sufficient to produce small but significant changes in the resistances even when the elements are as short as a quarter wavelength (Fig. 5.8a). In this case there is only a very small variation in the reactance. When the length of the elements is increased to a half wavelength with the same quarter wavelength spacing, both resistance and reactance vary greatly from element to element (Fig. 5.8b). Note that the change in the reactance from the central element in the array to one at the extremities may be as large as from near 100 ohms to near zero.

As is to be expected, an increase in the spacing of the elements to a half wavelength substantially reduces the changes in resistance and reactance due to differences in mutual interaction. When $2h = \lambda/4$ both resistance and reactance are substantially constant across the array (Fig. 5.8c). When $2h = \lambda/2$ significant differences in both resistance and reactance exist, but they are much smaller than for the more closely spaced array (Fig. 5.8b). In all cases, the obviously different environment of elements at the extremities of the array is responsible for the largest differences in the impedances. For the two lengths, $2h = \lambda/4$ and $2h = \lambda/2$, there is little difference between the results obtained with specified voltages and with specified driving-point currents.

Graphs of the resistances of the individual antennas in a broadside array of three-quarter and full wavelength elements are shown in Figs. 5.9a–d when the driving-point currents are specified. Similar curves for the same array with the voltages specified are in Figs. 5.10a–d. Especially noteworthy when $2h = 3\lambda/4$ are the large differences between the resistances and reactances of the elements when the driving-point voltages are specified instead of the driving-point currents (Figs. 5.9a, c and 5.10a, c). When $2h = \lambda$ the reactance and to a lesser extent the resistance of the elements at the extremities of the array differ greatly from the others (Figs. 5.9b, d and 5.10b, d). As an example of typical digital results prepared for this study, a table of impedances is given in Appendix III.

The radiation patterns in the equatorial or H-plane are shown in Fig. 5.11 for a broadside array of 15 elements. The ideal patterns are fairly well approximated when the amplitude and phase of the current along each antenna are specified near the point of maximum amplitude. For the array of half-wave dipoles this occurs essentially

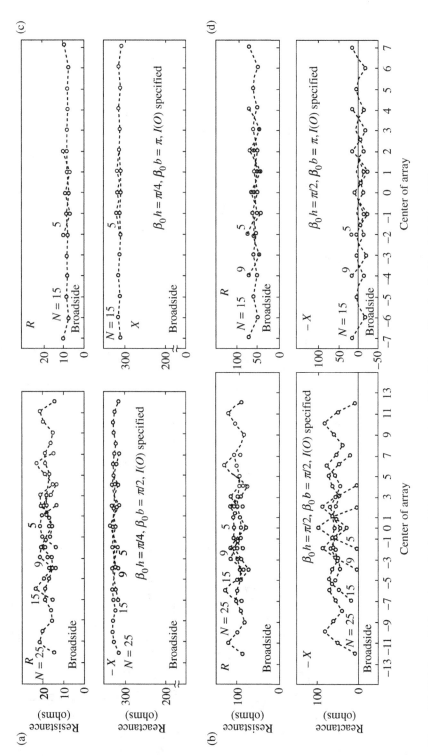

Figure 5.8 Driving-point impedances for broadside arrays, current specified; $\beta_0 h = \pi/4$ and $\pi/2$, $N = 5$, 9, 15 and 25.

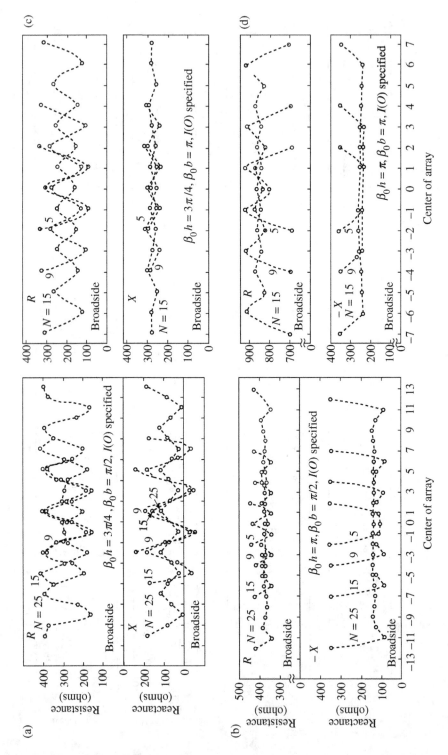

Figure 5.9 Driving-point impedances for broadside arrays, current specified; $\beta_0 h = 3\pi/4$ and π, $N = 5$, 9, 15 and 25.

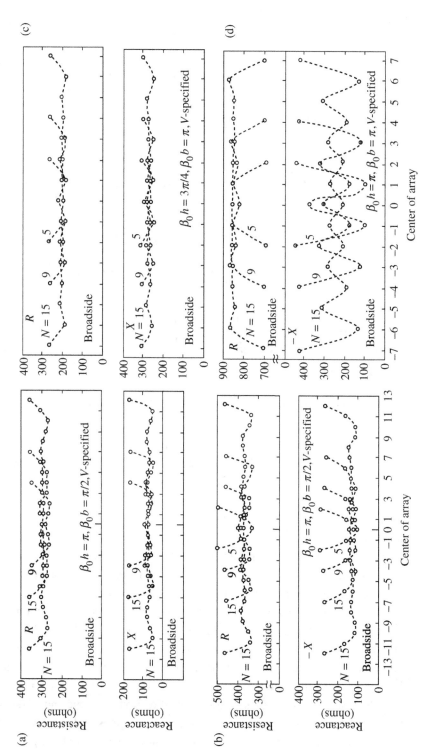

Figure 5.10 Driving-point impedances for broadside arrays, voltage specified; $\beta_0 h = 3\pi/4$ and π, $N = 5, 9, 15$ and 25.

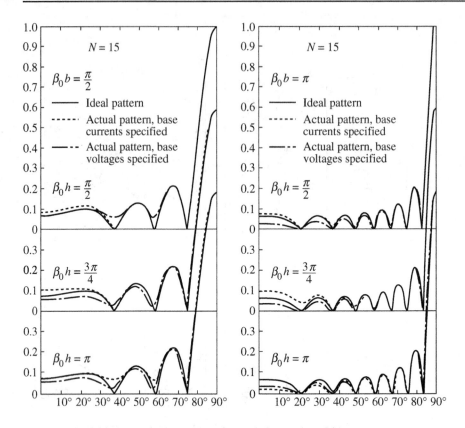

Figure 5.11 Field patterns in the *H*-plane for a 15-element broadside array.

when the driving-point currents are specified, for the full-wave dipoles when the voltages are specified. On the other hand, when the current is not specified at the maximum, the actual pattern differs considerably from the ideal especially in the region of the minima (nulls). This is true when the driving-point currents are specified for the full-wave elements and when the voltages are specified for the half-wave elements.

In endfire arrays the currents are adjusted to produce the main beam of the radiation pattern along the line of the elements. For the unilateral endfire array there is a single major lobe in the direction $\Phi = 0$; for the bilateral endfire array there are two major lobes, one in the direction $\Phi = 0$, the other in the opposite direction, $\Phi = 180°$. Whereas in the broadside array the interaction between all but the next adjacent elements is quite small owing to extensive cancellation of the fields of the several elements in both directions along the line of the array, exactly the opposite is true for the endfire array. In the unilateral endfire array there is a cumulative reinforcement of the fields due to the several elements in one direction from one end of the array to the other, a more or less complete cancellation in the opposite direction. In the bilateral array the cumulative reinforcement is in both directions. It is to be expected, therefore, that mutual coupling between neighboring and even quite widely separated elements must play a major role in determining the amplitude, phase, and distribution of each current.

In an ideal endfire array the currents must all be equal in amplitude and vary progressively in phase by an amount equal to the electrical distance between the elements. The specifications for a unilateral endfire array are

$$\{I_z(0)\} = I_{z1}(0) \begin{Bmatrix} 1 \\ -j \\ -1 \\ \vdots \end{Bmatrix}, \qquad \beta_0 b = \frac{\pi}{2}. \tag{5.113}$$

For the bilateral array,

$$\{I_z(0)\} = I_{z1}(0) \begin{Bmatrix} 1 \\ -1 \\ 1 \\ \vdots \end{Bmatrix}, \qquad \beta_0 b = \pi. \tag{5.114}$$

Alternatively, the voltages may be specified in the same manner. Thus, for the unilateral array

$$\{V\} = V_1 \begin{Bmatrix} 1 \\ -j \\ -1 \\ \vdots \end{Bmatrix}, \qquad \beta_0 b = \frac{\pi}{2}. \tag{5.115}$$

For the bilateral array,

$$\{V\} = V_1 \begin{Bmatrix} 1 \\ -1 \\ 1 \\ \vdots \end{Bmatrix}, \qquad \beta_0 b = \pi. \tag{5.116}$$

The resistances and reactances of the individual elements in a unilateral endfire array are shown in Figs. 5.12a and 5.12b, respectively with $2h = \lambda/4$ and $2h = \lambda/2$. The driving-point currents were specified according to (5.113). Corresponding values for the bilateral array are in Figs. 5.12c and 5.12d. Note that these are symmetrical with respect to the center of the array. For the shorter elements ($2h = \lambda/4$), the reactances of all elements are reasonably alike; the resistances also vary little except for the two elements at the ends of the unilateral array. When the elements are a half-wavelength long, the resistances and reactances both vary greatly along the unilateral array (Fig. 5.12b), moderately along the bilateral array (Fig. 5.12d). It is interesting to note that in the unilateral array the impedance of the forward element (in the direction of the beam) is greatest, that of the rear element smallest. Since the amplitudes of the driving-point currents are all the same, the power supplied to each element is proportional to its resistance. It follows from Fig. 5.12b that the power supplied to and radiated from the forward element is approximately five times that supplied to and radiated from the rear element. Note that the resistance and the reactance of all but the last two elements in each array are significantly greater than for an isolated antenna. In effect, each element after the forward one acts partly as a driven element, partly as a parasitic reflector for the element in front of it.

The resistances of the antennas in a unilateral endfire array with elements of length $2h = 3\lambda/4$ (Fig. 5.13a) decrease continually from the forward element to that in the rear in a manner resembling that for the half-wave elements (Fig. 5.12b). However, the range of magnitudes is much greater. The corresponding values for the same array but constructed of full-wave elements ($2h = \lambda$) are in Fig. 5.13b. They are startlingly different. The resistances of all elements are now reasonably alike except for that of the rear element, which is much greater. Evidently, the rear element is supplied and radiates the most power – approximately four to six times as much as any other element. This suggests that all but the rear element act in part as driven radiators and in part as parasitic directors for the elements behind them, especially the rear one. Note that for the bilateral array of full-wave elements (Fig. 5.13d) the resistances of the elements increase from the center outward, whereas for the corresponding array of half-wave elements (Fig. 5.12d) the resistances decrease from the center outward. If the voltages are specified according to (5.115) and (5.116) instead of the driving-point currents, the graphs of Figs. 5.13a–d are replaced by those of Figs. 5.14a–d. The two sets are seen to differ considerably.

The radiation patterns in the equatorial or H-plane are shown in Figs. 5.15 and 5.16 respectively for the unilateral and bilateral endfire arrays. The ideal pattern is fairly well approximated when the current along each antenna is specified near its point of maximum according to the criteria for an ideal array. For the half-wave dipoles this is true essentially when the driving-point currents $I(0)$ are specified, for the full-wave dipoles when the voltages are assigned. On the other hand, when the current is not specified at its maximum value, the actual pattern differs considerably from the ideal, especially in its minor lobe structure and the region of the minima (nulls). This is true when the driving-point currents are specified for the full-wave dipoles, when the voltages are specified for the half-wave dipoles. In general, the departure from the ideal

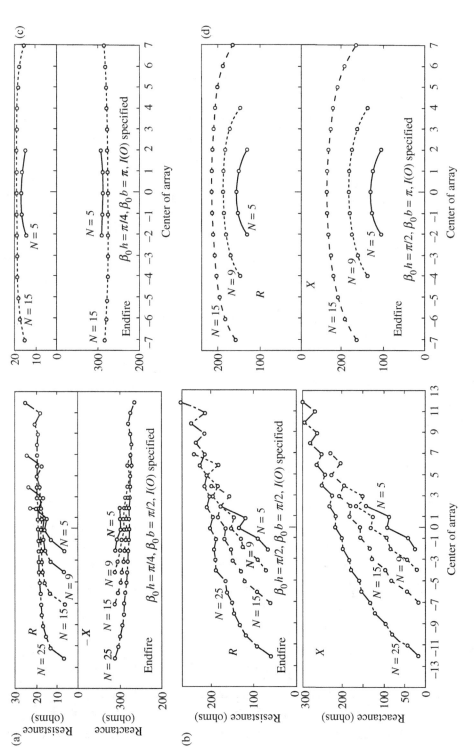

Figure 5.12 Driving-point impedances for endfire arrays, current specified; $\beta_0 h = \pi/4$ and $\pi/2$, $N = 5$, 9, 15 and 25.

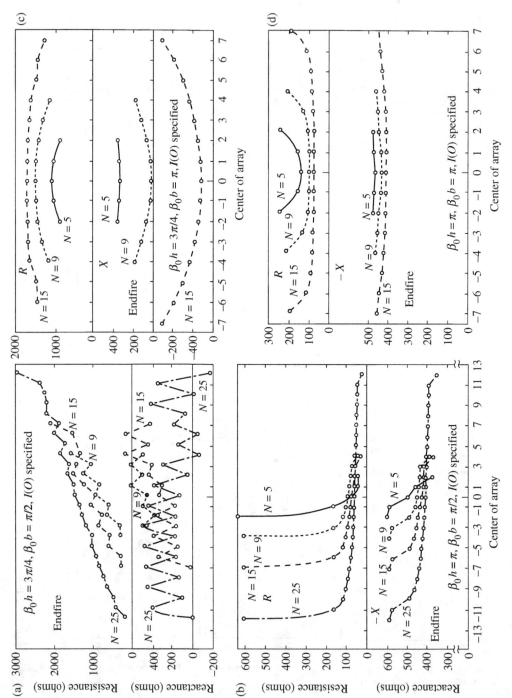

Figure 5.13 Driving-point impedances for endfire arrays, current specified; $\beta_0 h = 3\pi/4$ and π, $N = 5$, 9, 15 and 25.

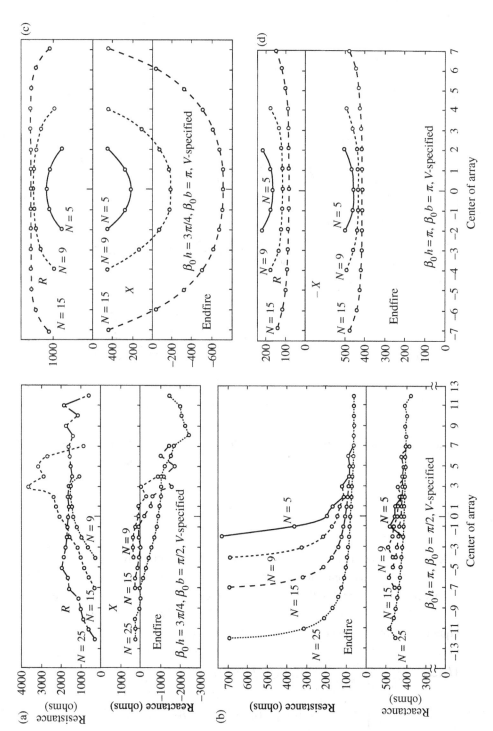

Figure 5.14 Driving-point impedances for endfire arrays, voltage specified; $\beta_0 h = 3\pi/4$ and π, $N = 5, 9, 15$ and 25.

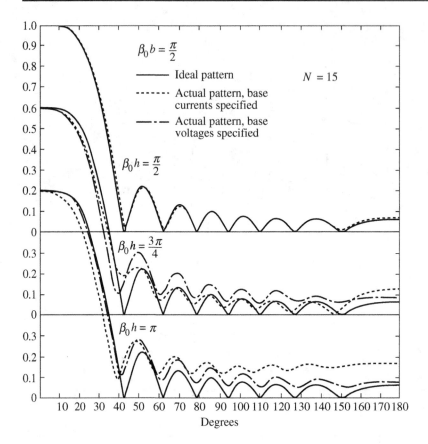

Figure 5.15 Field patterns in the *H*-plane for a 15-element unilateral endfire array.

patterns is greater for the unilateral endfire array (Fig. 5.15) than for the broadside array (Fig. 5.11) since the effect of mutual interaction is greater.

5.6   The special case when $\beta_0 h = \pi/2$

The general functional form for the currents in the elements given by (5.34) with (5.46) and (5.47) presents some difficulties when $\beta_0 h = \pi/2$. For both circular and curtain arrays the expression for the currents becomes indeterminate in the form 0/0 when $\beta_0 h = \pi/2$. This behavior is illustrated for the curtain array in the following matrix equation for the currents:

$$\{I_z(z)\} = \frac{j2\pi}{\zeta_0 \Psi_{dR} \cos \beta_0 h} \{V_0\} \sin \beta_0 (h - |z|)$$

$$+ \frac{j2\pi}{\zeta_0 \Psi_{dR} \cos \beta_0 h} [\Phi_u]^{-1} [\Phi_v] \{V_0\} (\cos \beta_0 z - \cos \beta_0 h). \tag{5.117}$$

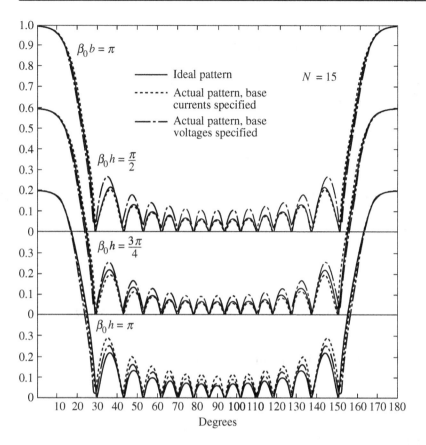

Figure 5.16 Radiation patterns in the H-plane for a 15-element bilateral endfire array.

From the form of the Ψ functions at $\beta_0 h = \pi/2$ it follows that

$$\lim_{\beta_0 h \to \pi/2} [\Phi_u]^{-1} [\Phi_v] = -[\mathcal{I}], \tag{5.118}$$

where $[\mathcal{I}]$ is the identity matrix. The indeterminate form for the currents in the elements follows directly when (5.118) is used in (5.117). It is

$$\{I_z(z)\} = \frac{j 2\pi \cos \beta_0 z}{\zeta_0 \Psi_{dR} \cdot 0} \{V_0\} - j \frac{2\pi \cos \beta_0 z}{\zeta_0 \Psi_{dR} \cdot 0} \{V_0\} = \frac{0}{0}, \qquad \beta_0 h = \frac{\pi}{2}. \tag{5.119}$$

Two alternatives are available for avoiding this difficulty: (a) the formula for the currents may be rearranged as in Section 2.7; or (b) a special formulation for $\beta_0 h = \pi/2$ may be used. The former method has the advantage that it is applicable over a range near $\beta_0 h = \pi/2$, whereas the latter method is valid only at $\beta_0 h = \pi/2$. Both methods are presented here, although the numerical results were calculated based on the special form for $\beta_0 h = \pi/2$. Numerical calculations have shown the results of the two approximate forms that are useful when $\beta_0 h = \pi/2$ to be approximately the same.

The expression for the currents when $\beta_0 h$ is near $\pi/2$ follows directly from the results of Section 2.7. In matrix form

$$\{I_z(z)\} = \frac{-j2\pi}{\zeta_0 \Psi_{dR}} (\{V_0\}(\sin \beta_0 |z| - \sin \beta_0 h) + [\Phi_u']^{-1}[\Phi_v']\{V_0\}(\cos \beta_0 z - \cos \beta_0 h)),$$

(5.120a)

where the elements of the matrices are

$$\Phi_{kiu}' = -\Phi_{kiu} \cos \beta_0 h \qquad (5.120b)$$

$$\Phi_{kiv}' = \Phi_{kiv} + \Phi_{kiu} \sin \beta_0 h. \qquad (5.120c)$$

When $\beta_0 h = \pi/2$,

$$\{I_z(z)\} = \frac{-j2\pi}{\zeta_0 \Psi_{dR}} (\{V_0\}(\sin \beta_0 |z| - 1) + [\Phi_u']^{-1}[\Phi_v']\{V_0\} \cos \beta_0 z). \qquad (5.121)$$

For $\beta_0 h = \pi/2$, the elements of the Φ_u' and Φ_v' matrices are

$$\Phi_{kiu}' = \Psi_{kiu}(h) \qquad (5.122a)$$

and

$$\Phi_{kiv}' = \Psi_{kidu} - \Psi_{kidv}(1 - \delta_{ik}) - j\Psi_{kidI}\delta_{ik}. \qquad (5.122b)$$

The alternative approach begins with the special form for the integral equation valid at $\beta_0 h = \pi/2$. It was this latter method which was used for the original curtain-array calculations [5]. The final form is similar to (5.121) with slightly different values for the constant Ψ_{dR} and the Φ_u' and Φ_v' matrices. In this method the Ψ functions are computed with the following cosine and shifted-sine currents:

$$I_{zi}(z) = -jA_i S_{0z} + B_i F_{0z}, \qquad (5.123)$$

where $S_{0z} = \sin \beta_0 |z| - \sin \beta_0 h$ and $F_{0z} = \cos \beta_0 z - \cos \beta_0 h$. The final expression for the current with $\beta_0 h = \pi/2$ is

$$\{I_z(z)\} = \frac{-j2\pi}{\zeta_0 \Psi_{dR}^h} \{V_0\}(\sin \beta_0 |z| - 1) - j \frac{2\pi}{\zeta_0 \Psi_{dR}^h} [\Phi_u]^{-1}[\Phi_v^h]\{V_0\} \cos \beta_0 z, \qquad (5.124)$$

where

$$\Phi^h_{kiv} = \Psi^h_{kidv}(1 - \delta_{ik}) + j\Psi^h_{kidI}\delta_{ik} - \Psi^h_{kiv}(0) \qquad (5.125a)$$

$$\Phi_{kiu} = -\Psi_{kidu} + \Psi_{kiu}(0) \qquad (5.125b)$$

and

$$\Psi^h_{dR} = -\,\text{Re}\{[S_b(h,0) - E_b(h,0)] - [S_b(h,h) - E_b(h,h)]\} \qquad (5.126a)$$

$$\Psi^h_{kidI} = \text{Im}\{[S_b(h,0) - E_b(h,0)] - [S_b(h,h) - E_b(h,h)]\} \qquad (5.126b)$$

$$\Psi^h_{kidv} = [S_b(h,0) - E_b(h,0)] - [S_b(h,h) - E_b(h,h)] \qquad (5.126c)$$

$$\Psi_{kidu} = C_b(h,0) - C_b(h,h) \qquad (5.126d)$$

$$\Psi^h_{kiv}(0) = S_b(h,0) - E_b(h,0) \qquad (5.126e)$$

$$\Psi_{kiu}(0) = C_b(h,0) \qquad (5.126f)$$

$$\beta_0 h = \pi/2, \qquad b \equiv b_{ki}, \qquad b_{kk} = a.$$

Numerical calculations show that the results obtained with (5.124) are comparable with those obtained with (5.121).

5.7   Summary

In this chapter a complete theory of curtain arrays of practical antennas has been presented. Mutual coupling among all elements is included in a manner that takes account of changes in the amplitudes and the phases of the currents along all elements as determined by their locations in an array. The theory is quantitatively useful for cylindrical elements with electrical half-lengths in the range $\beta_0 h \leq 5\pi/4$ and electric radii with values $\beta_0 a \leq 0.02$. This includes lengths over the full range in which the principal lobe in the vertical field pattern is in the equatorial plane; it provides a 5 to 1 frequency band for electrical half-lengths included in the range $\pi/4 \leq \beta_0 h \leq 5\pi/4$.

In this chapter no measurements have been cited to verify the quantitative correctness of the two-term theory in determining distributions of current, driving-point impedances or admittances, and field patterns of typical curtain arrays. This is due in part to the relative difficulty in carrying out accurate measurements of the self- and mutual impedances for curtain arrays owing to the lack of the symmetry which underlies the corresponding measurements with the circular array. The primary reason, however, is the adequacy of the experimental verification of all phases of the theory as applied to the two-element array – the simplest curtain array (Chapter 3), general circular arrays (Chapter 4) and to curtain arrays of parasitic elements (Chapter 6). As in the case of the circular array, the most sensitive and, at the same time, the most convenient experimental verification of the theory is in its application to an array in which only one element is driven while all others are parasitic. The first section in the next chapter is concerned specifically with the application of the theory developed in this chapter to a curtain array of twenty elements of which only one is driven and a comparison of theoretically and experimentally determined currents, admittances, and field patterns.

6 Arrays with unequal elements: parasitic and log-periodic antennas

The general theory of curtain arrays which is developed in the preceding chapter requires all N elements to be identical geometrically, but allows them to be driven by arbitrary voltages or loaded by arbitrary reactors at their centers. If some of these voltages are zero, the corresponding elements are parasitic and their currents are maintained entirely by mutual interaction. In arrays of the well-known Yagi–Uda type, only one element is driven, so that the importance of an accurate analytical treatment of the inter-element coupling is increased. In a long array the possible cumulative effect of a small error in the interaction between the currents in adjacent elements must not be overlooked. As an added complication, the tuning of the individual parasitic elements is accomplished by adjustments in their lengths and spacings. This introduces the important problem of arrays with elements that are different in length and that are separated by different distances. In the Yagi–Uda array the range of these differences is relatively small. On the other hand, in frequency-independent arrays of the log-periodic type the range of lengths and distances between adjacent elements is very great.

In this chapter the analytical treatment of arrays with elements that are different in length and unequally spaced is carried out successively for parasitic arrays of the conventional Yagi–Uda type and for driven log-periodic arrays. However, the formulation is sufficiently general to permit its extension to arrays of other types, both parasitic and driven, that involve geometrically different elements.

6.1 Application of the two-term theory to a simple parasitic array

The simplest parasitic array consists of N geometrically identical antennas each of length $2h$ and radius a arranged in a curtain of parallel non-staggered elements with spacing b. Element 1 is driven, all others are parasitic. Such an array is illustrated in Fig. 6.1. The directional properties of the electromagnetic field maintained by the array depend on the relative amplitudes and phases of the currents in all of the elements. The currents in the parasitic elements are all induced by their mutual interaction. The current in the driven antenna is determined in part by the driving generator, in part by the mutual interaction with the currents in the other elements. The coupling between

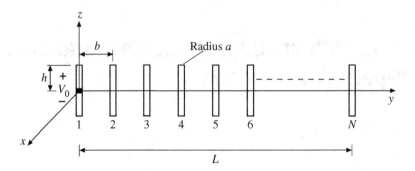

Figure 6.1 Parasitic array of identical elements.

the currents in any pair of elements of given length depends primarily on the distance between them.

The general theory of curtain arrays formulated in the preceding chapter may be applied directly by setting $V_{0i} = 0$, $1 < i \leq N$. The currents in the N elements are given by (5.34). They are

$$I_{z1}(z) = jA_1 \sin \beta_0(h - |z|) + B_1(\cos \beta_0 z - \cos \beta_0 h) \tag{6.1}$$

$$I_{zi}(z) = B_i(\cos \beta_0 z - \cos \beta_0 h), \qquad i = 2, 3, \ldots N, \tag{6.2}$$

where from (5.47)

$$A_1 = \frac{2\pi}{\zeta_0 \Psi_{dR} \cos \beta_0 h} V_{01} \tag{6.3}$$

and the B_i are obtained from (5.46). With V_{01} specified, the currents at the centers of the elements are obtained from (5.53). The driving-point admittance of element 1 is

$$Y_{01} = I_{z1}(0)/V_{01}. \tag{6.4}$$

The field pattern of the array is obtained from (5.67) with the appropriate values of A_i and B_i. As only A_1 differs from zero the applicable formula is

$$E_{\Theta}(\Theta, \Phi) = \frac{j\zeta_0}{2\pi} \left\{ jA_1 \frac{e^{-j\beta_0 R_1}}{R_1} F_m(\Theta, \beta_0 h) + \sum_{i=1}^{N} B_i \frac{e^{-j\beta_0 R_i}}{R_i} G_m(\Theta, \beta_0 h) \right\}, \tag{6.5}$$

where $F_m(\Theta, \beta_0 h)$ and $G_m(\Theta, \beta_0 h)$ are defined in (5.66) and (5.68). In (6.5) the field is evaluated in the far zone of each element so that the distances R_i are measured to the centers of the elements. The far field of the array implies in addition that $R_i \doteq R_1$ in amplitudes and $R_i = R_1 - (i - 1)b \sin \Theta \cos \Phi$ in phase angles.

Numerical computations have been made by Mailloux [1] for an array of 20 elements with $a/\lambda = 0.006\,35$ and $b/\lambda = 0.20$. Several values of h/λ were chosen in the range for endfire operation between 0.16 and 0.204.

(a)

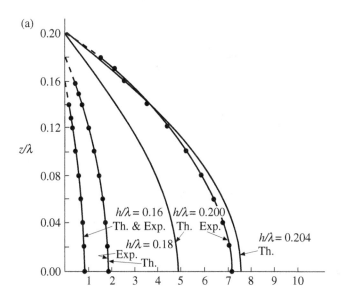

(b)

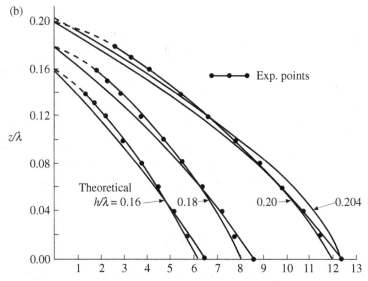

Figure 6.2 Components of current (mA) on driven dipole in 20-element parasitic array (Mailloux), (a) in phase with driving voltage, (b) in phase quadrature with driving voltage. $b/\lambda = 0.20$, $a/\lambda = 0.006\,35$.

The calculated distributions of current along the driven element are shown in Fig. 6.2 together with measured values. The agreement is excellent for $h/\lambda = 0.16$ and 0.18. The agreement when $h/\lambda = 0.20$ is not so close. However, the theoretical curves for antennas with h/λ increased by only 0.004 – a distance of less than $a/\lambda = 0.006\,35$ – are in excellent agreement with the experimental data for $h/\lambda = 0.20$. Evidently, as resonance is approached the current amplitude becomes increasingly sensitive to

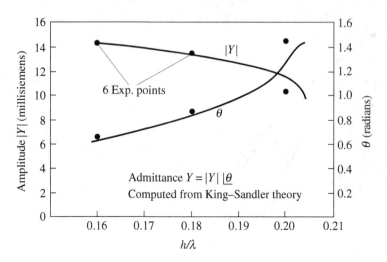

Figure 6.3 Driving-point admittance of 20-element parasitic array (Mailloux). $b/\lambda = 0.20$, $a/\lambda = 0.006\,35$.

small changes in length. The theoretical and experimental driving-point admittances are shown in Fig. 6.3. As for the current distribution in general, the agreement is very good for $h/\lambda = 0.16$ and 0.18, but the theoretical value at $h/\lambda = 0.204$ is in better agreement with the measured value for $h/\lambda = 0.20$ than is the theoretical value for $h/\lambda = 0.20$.

The normalized theoretical distributions of current along all parasitic dipoles are the same. The experimental values were also found to be remarkably alike. Theoretical and experimental distributions of the magnitude of the current along a typical parasitic element are shown in Fig. 6.4. It is seen that the theoretical currents differ somewhat from the measured values. Measured changes in the phase of the current along the parasitic elements were very small.

The amplitudes of the currents at $z = 0$ along each of the twenty elements are shown in Fig. 6.5. The agreement with measured values is again excellent for $h/\lambda = 0.16$ and 0.18. As before, the theoretical curve for $h/\lambda = 0.204$ is in much better agreement with the measured curve for $h/\lambda = 0.20$ than is the theoretical curve for $h/\lambda = 0.20$. The corresponding phases are shown in Fig. 6.6.

It is interesting to note that when $h/\lambda = 0.16$ and 0.18 the amplitudes of the currents in all of the parasitic elements except those nearest the driven antenna are quite small and substantially equal and the phase shift from element to element is linear. On the other hand, as h/λ approaches resonance the amplitudes of the currents increase greatly and they oscillate in magnitude from element to element. The small constant amplitude and linear phase shift that are characteristic of the shorter elements suggest a traveling wave along the array; the large oscillating amplitudes near resonance are characteristic of a standing wave.

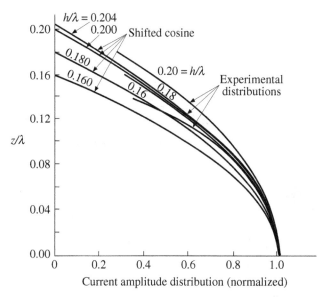

Figure 6.4 Normalized current amplitudes on a typical parasitic element in a 20-element array (Mailloux). $b/\lambda = 0.20$, $a/\lambda = 0.006\,35$.

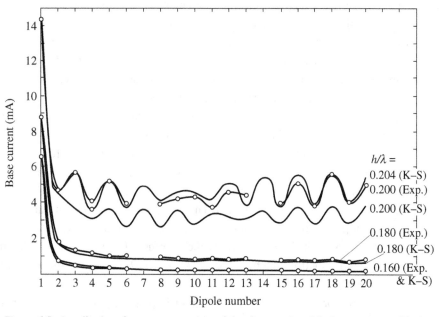

Figure 6.5 Amplitudes of currents at centers of the elements in a 20-element array with element No. 1 driven; comparison of King–Sandler theory with experiment (Mailloux). $b/\lambda = 0.20$, $a/\lambda = 0.006\,35$, frequency 600 MHz.

The theoretical and experimental field patterns are shown in Fig. 6.7 for the three values of h/λ. Although the measurements were made in the far zone of each element

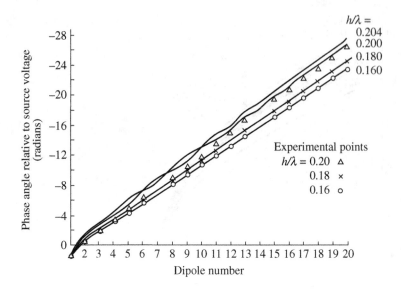

Figure 6.6 Same as Fig. 6.5 but for phases of currents (Mailloux).

(*E* far zone), the length *L* of the 20-element array was such that the true far-zone approximations $R_i \doteq R_1$ in amplitudes and $R_i \doteq R_1 - (i-1)b \sin \Theta \cos \Phi$ in phases were not sufficiently well satisfied. Accordingly, the field was evaluated from (6.5) with the actual distances to the elements for comparison with the measured values. The true far field was also computed for comparison. The former is designated '*E* far zone' in the figures, the latter is labelled 'far zone'. The agreement between theory and experiment is seen to be quite good even in the details of the minor lobe structure.

It may be concluded that the two-term theory of curtain arrays developed in Chapter 5 provides remarkably accurate results even for parasitic arrays for which one of the terms vanishes for each of the $N-1$ parasitic elements. This is somewhat surprising since the single term provides no flexibility in the representation of the distribution of the currents in the parasitic elements. They are all assumed to be the same and given by $I(z) \sim \cos \beta_0 z - \cos \beta_0 h$. Moreover, the phase of the current $I(z)$ along each element is assumed to be the same as that of the current $I(0)$ at the center. This means that the current distribution function $f(z)$ in $I(z) = I(0)f(z)$ is assumed to be real for all parasitic elements.

It is unreasonable to suppose that these implied assumptions are generally valid when longer elements are involved. After all, the investigation in this section has been limited to relatively short elements with $h/\lambda \leq 0.2$. It would appear that a more accurate representation of the currents in the parasitic elements is required – this is suggested in Fig. 6.4 where the actual distributions of current even on the relatively short elements were not very accurately represented by the single shifted-cosine term.

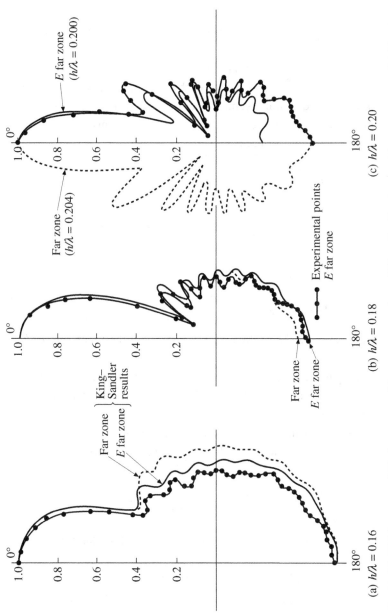

Figure 6.7 Field patterns of 20-element parasitic array; comparison of King–Sandler theory with measurements (Mailloux). (Far zone is far zone referred to length of array, *E* far zone is far zone referred to length of the elements.)

(a) $h/\lambda = 0.16$ (b) $h/\lambda = 0.18$ (c) $h/\lambda = 0.20$

Far zone
($h/\lambda = 0.204$)

E far zone
($h/\lambda = 0.200$)

Far zone $\left.\begin{array}{c} \\ \\ \end{array}\right\}$ King–Sandler results
E far zone

Far zone
E far zone

Experimental points
E far zone

6.2   The problem of arrays with parasitic elements of unequal lengths

In order to provide a more accurate representation of the current in the parasitic elements of an array, use may be made of the three-term approximation given in (3.20). This is known to be an improvement over the two-term theory used in Chapter 5 and, when applied to parasitic elements, it provides two terms with complex coefficients instead of only a single term. Specifically let

$$I_{zk}(z_k) = A_k M_{0zk} + B_k F_{0zk} + D_k H_{0zk}, \tag{6.6}$$

where

$$M_{0zk} = \sin \beta_0 (h_k - |z_k|) \tag{6.7a}$$

$$F_{0zk} = \cos \beta_0 z_k - \cos \beta_0 h_k \tag{6.7b}$$

$$H_{0zk} = \cos \tfrac{1}{2} \beta_0 z_k - \cos \tfrac{1}{2} \beta_0 h_k. \tag{6.7c}$$

In parasitic elements the coefficient A_k is zero, but the two terms $B_k F_{0zk} + D_k H_{0zk}$ remain.

It is anticipated that the distribution (6.6) provides sufficient flexibility to represent the currents in elements of different lengths when each element is allowed to have its own length $2h_k$.

When the several antennas in an array are not all equal in length so that the h_i differ, the problem of solving the N simultaneous integral equations

$$\sum_{i=1}^{N} \int_{-h_i}^{h_i} I_{zi}(z_i') K_{kid}(z_k, z_i') \, dz_i' = \frac{j 4\pi}{\zeta_0 \cos \beta_0 h_k} [\tfrac{1}{2} V_{0k} M_{0zk} + U_k F_{0zk}] \tag{6.8}$$

with $k = 1, 2, \ldots N$, is more complicated. The kernel has the form

$$K_{kid}(z_k, z_i') = K_{ki}(z_k, z_i') - K_{ki}(h_k, z_i') = \frac{e^{-j\beta_0 R_{ki}}}{R_{ki}} - \frac{e^{-j\beta_0 R_{kih}}}{R_{kih}}, \tag{6.9}$$

where

$$R_{ki} = \sqrt{(z_k - z_i')^2 + b_{ki}^2}, \qquad R_{kih} = \sqrt{(h_k - z_i')^2 + b_{ki}^2}. \tag{6.10}$$

Note that $b_{kk} = a$. The function U_k is

$$U_k = \frac{-j\zeta_0}{4\pi} \sum_{i=1}^{N} \int_{-h_k}^{h_k} I_{zi}(z_i') K_{ki}(h_k, z_i') \, dz_i'. \tag{6.11}$$

In a parasitic antenna l the driving voltage $V_{0l} = 0$, so that

$$\sum_{i=1}^{N} \int_{-h_i}^{h_i} I_{zi}(z_i') K_{kid}(z_l, z_i') \, dz_i' = \frac{j4\pi}{\zeta_0 \cos \beta_0 h_l} U_l F_{0zl}. \tag{6.12}$$

In order to obtain approximate solutions of the N simultaneous integral equations (6.8) by the procedure developed in the earlier chapters, use may be made of the properties of the real and imaginary parts of the kernel. As shown in Chapter 2,

$$\int_{-h_k}^{h_k} G_{0z'k} K_{kkdR}(z_k, z_k') \, dz_k' \sim G_{0zk}, \tag{6.13}$$

where $G_{0z'k}$ stands for $M_{0z'k}$, $F_{0z'k}$ or $H_{0z'k}$ and $K_{kkdR}(z_k, z_k')$ is the real part of the kernel. On the other hand,

$$\int_{-h_k}^{h_k} G_{0z'k} K_{kkdI}(z_k, z_k') \, dz_k' \sim H_{0zk} \tag{6.14}$$

for any distribution $G_{0z'k}$. It follows that

$$W_{kkV}(z_k) \equiv \int_{-h_k}^{h_k} M_{0z'k} K_{kkd}(z_k, z_k') \, dz_k' \doteq \Psi_{kkdV}^m M_{0zk} + \Psi_{kkdV}^h H_{0zk} \tag{6.15}$$

$$W_{kkU}(z_k) \equiv \int_{-h_k}^{h_k} F_{0z'k} K_{kkd}(z_k, z_k') \, dz_k' \doteq \Psi_{kkdU}^f F_{0zk} + \Psi_{kkdU}^h H_{0zk} \tag{6.16}$$

$$W_{kkD}(z_k) \equiv \int_{-h_k}^{h_k} H_{0z'k} K_{kkd}(z_k, z_k') \, dz_k' \doteq \Psi_{kkdD}^f F_{0zk} + \Psi_{kkdD}^h H_{0zk}, \tag{6.17}$$

where the Ψ's are complex coefficients yet to be determined. Actually, (6.13) with $G = H$ and (6.14) suggest that the term $\Psi_{kkdD}^h H_{0zk}$ should be an adequate approximation. The term $\Psi_{kkdD}^f F_{0zk}$ is added in order to provide greater flexibility and symmetry.

When $i \neq k$ and $\beta_0 b \geq 1$, it has been shown by direct comparison in Chapter 3 that

$$\int_{-h_i}^{h_i} G_{0z'i} K_{kidR}(z_k, z_i') \, dz_i' \sim F_{0zk} \tag{6.18}$$

$$\int_{-h_i}^{h_i} G_{0z'i} K_{kidI}(z_k, z_i') \, dz_i' \sim H_{0zk}, \tag{6.19}$$

where $G_{0z'i}$ stands for $M_{0z'i}$, $F_{0z'i}$ or $H_{0z'i}$. It follows that with $i \neq k$,

$$W_{kiV}(z_k) \equiv \int_{-h_i}^{h_i} M_{0z'i} K_{kid}(z_k, z'_i)\, dz'_i \doteq \Psi_{kidV}^f F_{0zk} + \Psi_{kidV}^h H_{0zk} \tag{6.20}$$

$$W_{kiU}(z_k) \equiv \int_{-h_i}^{h_i} F_{0z'i} K_{kid}(z_k, z'_i)\, dz'_i \doteq \Psi_{kidU}^f F_{0zk} + \Psi_{kidU}^h H_{0zk} \tag{6.21}$$

$$W_{kiD}(z_k) \equiv \int_{-h_i}^{h_i} H_{0z'i} K_{kid}(z_k, z'_i)\, dz'_i \doteq \Psi_{kidD}^f F_{0zk} + \Psi_{kidD}^h H_{0zk}, \tag{6.22}$$

where the Ψ's are complex coefficients yet to be determined.

In the formulation developed in the earlier chapters for driven elements of equal lengths, the coefficients Ψ were defined individually in terms of the two integrals obtained from the real and imaginary parts of the kernel. In order to take account of the more varied distributions that may be obtained when the elements are neither all driven nor all equal in length, the separation into two parts is not made. Instead the entire integral is represented by a linear combination of the two distributions that best represent the parts of the integral. The complex coefficients of these distributions are to be determined by matching the integral and its approximation at two points along the antenna, instead of at only one such point. It is anticipated that by fitting the trigonometric approximations to the integrals at $z = 0$, $h_k/2$, and h_k (where both must vanish) a good representation may be achieved in reasonably simple form of all of the different distributions that may occur along antennas of unequal lengths. It is, of course, assumed that $\beta_0 h_i \leq 5\pi/4$ for all h_i.

6.3   Application to the Yagi–Uda array

In order to clarify the description of the procedure used to solve the N simultaneous integral equations for a parasitic array, it will be carried out in detail for the specific and practically useful Yagi–Uda array. In general, this consists of a curtain of N antennas of which No. 1 is parasitic and adjusted in length to function as a reflector, No. 2 is driven by a voltage V_{02} and Nos. 3 to N are also parasitic and adjusted to act as directors. Such an array is shown in Fig. 6.8 for the special case (treated later) with $2h_1 = 0.51\lambda$; $2h_2 = 0.50\lambda$; $2h_i = 2h$, $i > 2$; $b_{21} = 0.25\lambda$; $b_{i,i\pm1} = b$, $i > 2$. The details of these adjustments are examined later.

On the basis of the three-term approximation, the current in the single driven element has the form

$$I_{z2}(z_2) = A_2 M_{0z2} + B_2 F_{0z2} + D_2 H_{0z2}. \tag{6.23}$$

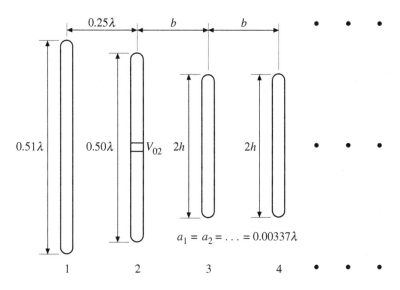

Figure 6.8 A Yagi array with directors of constant length, radius and spacing.

The currents in the parasitic elements are

$$I_{zi}(z_i) = B_i F_{0zi} + D_i H_{0zi}, \qquad i = 1, 3, 4, \ldots N, \tag{6.24}$$

where the constants A_2, B_i and D_i must be evaluated ultimately in terms of V_{02}. The integral equation for the driven element is

$$A_2 \int_{-h_2}^{h_2} M_{0z'2} K_{22d}(z_2, z_2') \, dz_2' + \sum_{i=1}^{N} B_i F_{0z'i} K_{2id}(z_2, z_i') \, dz_i'$$

$$+ \sum_{i=1}^{N} D_i H_{0z'i} K_{2id}(z_2, z_i') \, dz_i'$$

$$= \frac{j4\pi}{\zeta_0 \cos \beta_0 h_2} [\tfrac{1}{2} V_{02} M_{0z2} + U_2 F_{0z2}]. \tag{6.25}$$

The remaining $N - 1$ integral equations are

$$A_2 \int_{-h_2}^{h_2} M_{0z'2} K_{k2d}(z_k, z_2') \, dz_2' + \sum_{i=1}^{N} B_i F_{0z'i} K_{kid}(z_k, z_i') \, dz_i'$$

$$+ \sum_{i=1}^{N} D_i H_{0z'i} K_{kid}(z_k, z_i') \, dz_i'$$

$$= \frac{j4\pi}{\zeta_0 \cos \beta_0 h_k} U_k F_{0zk}, \qquad k = 1, 3, 4, \ldots N. \tag{6.26}$$

With (6.15)–(6.17) and (6.20)–(6.22) these may be expressed in terms of the parameters Ψ. Thus, for (6.25)

$$A_2[\Psi^m_{22dV} M_{0z2} + \Psi^h_{22dV} H_{0z2}] + \sum_{i=1}^{N} B_i[\Psi^f_{2idU} F_{0z2} + \Psi^h_{2idU} H_{0z2}]$$

$$+ \sum_{i=1}^{N} D_i[\Psi^f_{2idD} F_{0z2} + \Psi^h_{2idD} H_{0z2}]$$

$$= \frac{j4\pi}{\zeta_0 \cos \beta_0 h_2} [\tfrac{1}{2} V_{02} M_{0z2} + U_2 F_{0z2}]. \tag{6.27}$$

For (6.26), the $N - 1$ equations are

$$A_2[\Psi^f_{k2dV} F_{0zk} + \Psi^h_{k2dV} H_{0zk}] + \sum_{i=1}^{N} B_i[\Psi^f_{kidU} F_{0zk} + \Psi^h_{kidU} H_{0zk}]$$

$$+ \sum_{i=1}^{N} D_i[\Psi^f_{kidD} F_{0zk} + \Psi^h_{kidD} H_{0zk}]$$

$$= \frac{j4\pi}{\zeta_0 \cos \beta_0 h_k} U_k F_{0zk}, \qquad k = 1, 3, 4, \ldots N. \tag{6.28}$$

These equations will be satisfied if the coefficient of each of the three distribution functions is individually required to vanish. That is, in (6.27):

$$A_2 = \frac{j2\pi V_{02}}{\zeta_0 \Psi^m_{22dV} \cos \beta_0 h_2} \tag{6.29}$$

$$\sum_{i=1}^{N} [B_i \Psi^f_{2idU} + D_i \Psi^f_{2idD}] \cos \beta_0 h_2 - \frac{j4\pi}{\zeta_0} U_2 = 0 \tag{6.30a}$$

$$A_2 \Psi^h_{22dV} + \sum_{i=1}^{N} [B_i \Psi^h_{2idU} + D_i \Psi^h_{2idD}] = 0. \tag{6.30b}$$

Similarly in (6.28) with $k = 1, 3, \ldots N$

$$\left\{ A_2 \Psi^f_{k2dV} + \sum_{i=1}^{N} [B_i \Psi^f_{kidU} + D_i \Psi^f_{kidD}] \right\} \cos \beta_0 h_k - \frac{j4\pi}{\zeta_0} U_k = 0 \tag{6.30c}$$

$$A_2 \Psi^h_{k2dV} + \sum_{i=1}^{N} [B_i \Psi^h_{kidU} + D_i \Psi^h_{kidD}] = 0. \tag{6.30d}$$

Actually, the single equations in (6.30a) and (6.30b) may be combined with the $N - 1$ equations in (6.30c) and (6.30d) with the aid of the Kronecker δ defined by

$$\delta_{ik} = \begin{cases} 0 & i \neq k \\ 1 & i = k \end{cases}.$$

The $2N$ equations are

$$\left\{ A_2(1 - \delta_{k2}) \Psi^f_{k2dV} + \sum_{i=1}^{N} [B_i \Psi^f_{kidU} + D_i \Psi^f_{kidD}] \right\} \cos \beta_0 h_k - \frac{j4\pi}{\zeta_0} U_k = 0 \quad (6.31a)$$

$$A_2 \Psi^h_{k2dV} + \sum_{i=1}^{N} [B_i \Psi^h_{kidU} + D_i \Psi^h_{kidD}] = 0 \quad (6.31b)$$

with $k = 1, 2, \ldots N$. These equations, together with (6.29), determine the $2N + 1$ constants A_2, B_i and D_i, $i = 1, 2, \ldots N$.

Before these two sets of equations can be solved, it is necessary to evaluate the functions U_k. This is readily done in terms of the following integrals:

$$\Psi_{kiV}(h_k) = \int_{-h_i}^{h_i} M_{0z'i} K_{ki}(h_k, z'_i) \, dz'_i \quad (6.32)$$

$$\Psi_{kiU}(h_k) = \int_{-h_i}^{h_i} F_{0z'i} K_{ki}(h_k, z'_i) \, dz'_i \quad (6.33)$$

$$\Psi_{kiD}(h_k) = \int_{-h_i}^{h_i} H_{0z'i} K_{ki}(h_k, z'_i) \, dz'_i, \quad (6.34)$$

where

$$K_{ki}(h_k, z'_i) = \frac{e^{-j\beta_0 R_{kih}}}{R_{kih}}, \qquad R_{kih} = \sqrt{(h_k - z'_i)^2 + b^2_{ki}}. \quad (6.35)$$

It follows from the definition in (6.11) that

$$U_k = \frac{-j\zeta_0}{4\pi} \sum_{i=1}^{N} [A_i \Psi_{kiV}(h_k) + B_i \Psi_{kiU}(h_k) + D_i \Psi_{kiD}(h_k)]. \quad (6.36a)$$

Since only antenna 2 is driven, $A_i = 0$, $i \neq 2$ so that

$$U_k = \frac{-j\zeta_0}{4\pi} \left\{ A_2 \Psi_{k2V}(h_k) + \sum_{i=1}^{N} [B_i \Psi_{kiU}(h_k) + D_i \Psi_{kiD}(h_k)] \right\}. \quad (6.36b)$$

The substitution of (6.36b) in (6.31a) gives for these equations

$$A_2[\Psi_{k2V}(h_k) - (1 - \delta_{k2}) \Psi^f_{k2dV} \cos \beta_0 h_k] + \sum_{i=1}^{N} B_i [\Psi_{kiU}(h_k)$$

$$- \Psi^f_{kidU} \cos \beta_0 h_k] + \sum_{i=1}^{N} D_i [\Psi_{kiD}(h_k) - \Psi^f_{kidD} \cos \beta_0 h_k] = 0, \quad (6.37)$$

with $k = 1, 2, \ldots N$. These equations can be simplified formally by the introduction of the notation

$$\Phi_{k2V} = \Psi_{k2V}(h_k) - (1 - \delta_{k2}) \Psi^f_{k2dV} \cos \beta_0 h_k \quad (6.38)$$

$$\Phi_{kiU} = \Psi_{kiU}(h_k) - \Psi_{kidU}^{f} \cos \beta_0 h_k \tag{6.39a}$$

$$\Phi_{kiD} = \Psi_{kiD}(h_k) - \Psi_{kidD}^{f} \cos \beta_0 h_k. \tag{6.39b}$$

With this notation, (6.37) together with (6.31b) gives the following set of $2N$ equations for determining the $2N$ coefficients B_i and D_i in terms of A_2:

$$\sum_{i=1}^{N} [\Phi_{kiU} B_i + \Phi_{kiD} D_i] = -\Phi_{k2V} A_2; \qquad k = 1, 2, \ldots N \tag{6.40}$$

$$\sum_{i=1}^{N} [\Psi_{kidU}^{h} B_i + \Psi_{kidD}^{h} D_i] = -\Psi_{k2dV}^{h} A_2; \qquad k = 1, 2, \ldots N. \tag{6.41}$$

These equations may be expressed in matrix form after the introduction of the following notation:

$$[\Phi_U] = \begin{bmatrix} \Phi_{11U} & \Phi_{12U} & \cdots & \Phi_{1NU} \\ \vdots & & & \\ \Phi_{N1U} & \cdots & & \Phi_{NNU} \end{bmatrix} \tag{6.42a}$$

$$[\Phi_D] = \begin{bmatrix} \Phi_{11D} & \Phi_{12D} & \cdots & \Phi_{1ND} \\ \vdots & & & \\ \Phi_{N1D} & \cdots & & \Phi_{NND} \end{bmatrix} \tag{6.42b}$$

$$[\Psi_{dU}^{h}] = \begin{bmatrix} \Psi_{11dU}^{h} & \Psi_{12dU}^{h} & \cdots & \Psi_{1NdU}^{h} \\ \vdots & & & \\ \Psi_{N1dU}^{h} & \cdots & & \Psi_{NNdU}^{h} \end{bmatrix} \tag{6.43a}$$

$$[\Psi_{dD}^{h}] = \begin{bmatrix} \Psi_{11dD}^{h} & \Psi_{12dD}^{h} & \cdots & \Psi_{1NdD}^{h} \\ \vdots & & & \\ \Psi_{N1dD}^{h} & \cdots & & \Psi_{NNdD}^{h} \end{bmatrix} \tag{6.43b}$$

$$\{\Phi_{2V}\} = \begin{Bmatrix} \Phi_{12V} \\ \Phi_{22V} \\ \vdots \\ \Phi_{N2V} \end{Bmatrix} \qquad \{\Psi_{2dV}^{h}\} = \begin{Bmatrix} \Psi_{12dV}^{h} \\ \Psi_{22dV}^{h} \\ \vdots \\ \Psi_{N2dV}^{h} \end{Bmatrix} \tag{6.44}$$

$$\{B\} = \begin{Bmatrix} B_1 \\ B_2 \\ \vdots \\ B_N \end{Bmatrix} \qquad \{D\} = \begin{Bmatrix} D_1 \\ D_2 \\ \vdots \\ D_N \end{Bmatrix}. \tag{6.45}$$

The matrix forms of (6.40) and (6.41) are

$$[\Phi_U]\{B\} + [\Phi_D]\{D\} = -\{\Phi_{2V}\}A_2 \tag{6.46}$$

$$[\Psi^h_{dU}]\{B\} + [\Psi^h_{dD}]\{D\} = -\{\Psi^h_{2dV}\}A_2. \tag{6.47}$$

The N coefficients B_i and the N coefficients D_i must be determined from these equations for substitution in the equations (6.23) and (6.24) for the currents in the N elements. The coefficient A_2, which is a common factor, is obtained from (6.29) in terms of the single driving voltage V_{02}.

It remains to evaluate the parameters Ψ that occur in the Φ's in (6.46) and explicitly in (6.47).

6.4   Evaluation of coefficients for the Yagi–Uda array

The equations (6.46) and (6.47) involve the elements of the $N \times N$ matrices $[\Phi_U]$, $[\Phi_D]$, $[\Psi^h_{dU}]$ and $[\Psi^h_{dD}]$. These, in turn, depend on the parameters Ψ introduced in (6.15)–(6.17) and (6.20)–(6.22) and the parameters $\Psi(h)$ defined in (6.32)–(6.34). Since each integral is approximated by a linear combination of two terms with arbitrary coefficients, these can be evaluated by equating both sides in (6.15)–(6.17) and (6.20)–(6.22) at two values of z. The values chosen are $z = 0$, and $z = h_k/2$ in addition to $z = h_k$ where both sides must vanish.

Specific formulas for the two values of each of the integrals W defined in (6.15)–(6.17) and (6.20)–(6.22) are as follows:

$$W_{kiV}(0) \equiv A_i^{-1} \int_{-h_i}^{h_i} I_{Vi}(z_i') K_{kid}(0, z_i') \, dz_i' \doteq \int_{-h_i}^{h_i} M_{0z'i} K_{kid}(0, z_i') \, dz_i' \tag{6.48a}$$

$$W_{kiV}\left(\frac{h_k}{2}\right) \equiv A_i^{-1} \int_{-h_i}^{h_i} I_{Vi}(z_i') K_{kid}\left(\frac{h_k}{2}, z_i'\right) dz_i'$$

$$\doteq \int_{-h_i}^{h_i} M_{0z'i} K_{kid}\left(\frac{h_k}{2}, z_i'\right) dz_i' \tag{6.48b}$$

$$W_{kiU}(0) \equiv B_i^{-1} \int_{-h_i}^{h_i} I_{Ui}(z_i') K_{kid}(0, z_i') \, dz_i' \doteq \int_{-h_i}^{h_i} F_{0z'i} K_{kid}(0, z_i') \, dz_i' \tag{6.49a}$$

$$W_{kiU}\left(\frac{h_k}{2}\right) \equiv B_i^{-1} \int_{-h_i}^{h_i} I_{Ui}(z_i') K_{kid}\left(\frac{h_k}{2}, z_i'\right) dz_i'$$

$$\doteq \int_{-h_i}^{h_i} F_{0z'i} K_{kid}\left(\frac{h_k}{2}, z_i'\right) dz_i' \tag{6.49b}$$

$$W_{kiD}(0) \equiv D_i^{-1} \int_{-h_i}^{h_i} I_{Di}(z_i') K_{kid}(0, z_i') \, dz_i' \doteq \int_{-h_i}^{h_i} H_{0z'i} K_{kid}(0, z_i') \, dz_i' \tag{6.50a}$$

$$W_{kiD}\left(\frac{h_k}{2}\right) \equiv D_i^{-1} \int_{-h_i}^{h_i} I_{Di}(z_i') K_{kid}\left(\frac{h_k}{2}, z_i'\right) \, dz_i'$$

$$\doteq \int_{-h_i}^{h_i} H_{0z'i} K_{kid}\left(\frac{h_k}{2}, z_i'\right) \, dz_i'. \tag{6.50b}$$

In all of the above, $k = 1, 2, 3, \ldots N$. These are a set of complex numbers which give the values of the integrals (6.20)–(6.22) at the two points $z = 0$ and $z = h_k/2$. They are readily evaluated numerically by high-speed computer, or they may be expressed in terms of the tabulated generalized sine and cosine integral functions. Once the W's in (6.48a)–(6.50b) have been obtained for all values of i and k, the coefficients Ψ may be determined from the equations (6.15)–(6.17) and (6.20)–(6.22). At $z = 0$ these become:

$$\Psi_{kkdV}^m \sin \beta_0 h_k + \Psi_{kkdV}^h [1 - \cos(\beta_0 h_k/2)] = W_{kkV}(0) \tag{6.51a}$$

$$\Psi_{kidV}^f (1 - \cos \beta_0 h_k) + \Psi_{kidV}^h [1 - \cos(\beta_0 h_k/2)] = W_{kiV}(0) \qquad i \neq k \tag{6.51b}$$

$$\Psi_{kidU}^f (1 - \cos \beta_0 h_k) + \Psi_{kidU}^h [1 - \cos(\beta_0 h_k/2)] = W_{kiU}(0) \tag{6.51c}$$

$$\Psi_{kidD}^f (1 - \cos \beta_0 h_k) + \Psi_{kidD}^h [1 - \cos(\beta_0 h_k/2)] = W_{kiD}(0). \tag{6.51d}$$

At $z = h_k/2$, they are

$$\Psi_{kkdV}^m \sin(\beta_0 h_k/2) + \Psi_{kkdV}^h [\cos(\beta_0 h_k/4) - \cos(\beta_0 h_k/2)] = W_{kkV}\left(\frac{h_k}{2}\right) \tag{6.52a}$$

$$\Psi_{kidV}^f [\cos(\beta_0 h_k/2) - \cos \beta_0 h_k] + \Psi_{kidV}^h [\cos(\beta_0 h_k/4) - \cos(\beta_0 h_k/2)]$$

$$= W_{kiV}\left(\frac{h_k}{2}\right) \qquad i \neq k \tag{6.52b}$$

$$\Psi_{kidU}^f [\cos(\beta_0 h_k/2) - \cos \beta_0 h_k] + \Psi_{kidU}^h [\cos(\beta_0 h_k/4) - \cos(\beta_0 h_k/2)]$$

$$= W_{kiU}\left(\frac{h_k}{2}\right) \tag{6.52c}$$

$$\Psi_{kidD}^f [\cos(\beta_0 h_k/2) - \cos \beta_0 h_k] + \Psi_{kidD}^h [\cos(\beta_0 h_k/4) - \cos(\beta_0 h_k/2)]$$

$$= W_{kiD}\left(\frac{h_k}{2}\right). \tag{6.52d}$$

The solutions of these equations for the Ψ's are obtained directly. They are

$$\Psi_{kkdV}^m = \Delta_1^{-1}\left\{W_{kkV}(0)\left[\cos\left(\frac{\beta_0 h_k}{4}\right) - \cos\left(\frac{\beta_0 h_k}{2}\right)\right]\right.$$

$$\left. - W_{kkV}(h_k/2)[1 - \cos(\beta_0 h_k/2)]\right\} \tag{6.53}$$

$$\Psi_{kkdV}^h = \Delta_1^{-1}\{W_{kkV}(h_k/2)\sin\beta_0 h_k - W_{kkV}(0)\sin(\beta_0 h_k/2)\} \tag{6.54}$$

$$\Psi_{kidV}^f = \Delta_2^{-1}\{W_{kiV}(0)[\cos(\beta_0 h_k/4) - \cos(\beta_0 h_k/2)]$$

$$- W_{kiV}(h_k/2)[1 - \cos(\beta_0 h_k/2)]\} \qquad i \neq k \tag{6.55}$$

$$\Psi_{kidV}^h = \Delta_2^{-1}\{W_{kiV}(h_k/2)[1 - \cos\beta_0 h_k]$$

$$- W_{kiV}(0)[\cos(\beta_0 h_k/2) - \cos\beta_0 h_k]\} \qquad i \neq k \tag{6.56}$$

$$\Psi_{kidU}^f = \Delta_2^{-1}\{W_{kiU}(0)[\cos(\beta_0 h_k/4) - \cos(\beta_0 h_k/2)]$$

$$- W_{kiU}(h_k/2)[1 - \cos(\beta_0 h_k/2)]\} \tag{6.57}$$

$$\Psi_{kidU}^h = \Delta_2^{-1}\{W_{kiU}(h_k/2)[1 - \cos\beta_0 h_k] - W_{kiU}(0)[\cos(\beta_0 h_k/2) - \cos(\beta_0 h_k)]\} \tag{6.58}$$

$$\Psi_{kidD}^f = \Delta_2^{-1}\{W_{kiD}(0)[\cos(\beta_0 h_k/4) - \cos(\beta_0 h_k/2)]$$

$$- W_{kiD}(h_k/2)[1 - \cos(\beta_0 h_k/2)]\} \tag{6.59}$$

$$\Psi_{kidD}^h = \Delta_2^{-1}\{W_{kiD}(h_k/2)[1 - \cos\beta_0 h_k]$$

$$- W_{kiD}(0)[\cos(\beta_0 h_k/2) - \cos\beta_0 h_k]\}, \tag{6.60}$$

where

$$\Delta_1 = \sin\beta_0 h_k [\cos(\beta_0 h_k/4) - \cos(\beta_0 h_k/2)]$$

$$- \sin(\beta_0 h_k/2)[1 - \cos(\beta_0 h_k/2)] \tag{6.61}$$

and

$$\Delta_2 = [1 - \cos\beta_0 h_k][\cos(\beta_0 h_k/4) - \cos(\beta_0 h_k/2)]$$

$$- [\cos(\beta_0 h_k/2) - \cos\beta_0 h_k][1 - \cos(\beta_0 h_k/2)]. \tag{6.62}$$

All of the Ψ's have been determined. The $\Psi(h)$ coefficients are given in (6.32)–(6.34). The elements of the Φ matrices are obtained from (6.38)–(6.39b). This completes the solution for all of the currents in the elements of the Yagi–Uda array.

6.5   Arrays with half-wave elements

When an array includes half-wave parasitic elements the formulation in Sections 6.3 and 6.4 is directly applicable. Specifically, when $\beta_0 h_i = \pi/2$ and element i is parasitic, the current (6.24) has the form

$$I_{zi}(z_i) = B_i \cos \beta_0 z_i + D_i[\cos(\beta_0 z_i/2) - \sqrt{2}/2]. \tag{6.63}$$

If the length of the driven element 2 is such that $\beta_0 h_2$ is near or exactly $\pi/2$ (as in Fig. 6.8), the alternative form for the current given in (2.35) for the isolated antenna is more convenient since it does not yield an indeterminate form at $\beta_0 h_2 = \pi/2$. That is, in the notation of (6.23),

$$I_{z2}(z_2) = A_2' S_{0z2} + B_2' F_{0z2} + D_2 H_{0z2}, \tag{6.64}$$

where

$$S_{0z2} = \sin \beta_0 |z_2| - \sin \beta_0 h_2 \tag{6.65}$$

and

$$A_2' = -A_2 \cos \beta_0 h_2 = -j(2\pi V_{02}/\zeta_0 \Psi_{22dV}^m) \tag{6.66a}$$

$$B_2' = B_2 + A_2 \sin \beta_0 h_2 = B_2 - A_2' \tan \beta_0 h_2. \tag{6.66b}$$

Note that A_2' and B_2' are finite when $\beta_0 h_2 = \pi/2$. In this case

$$S_{0z2} = \sin \beta_0 |z_2| - 1, \qquad B_2' = B_2 + A_2. \tag{6.67}$$

Since (6.64) is not actually a different distribution from the original in (6.23) but merely a rearrangement that is more convenient when $\beta_0 h_2$ is at or near $\pi/2$, it is not necessary to repeat the formulation in the preceding sections with S_{0z2} substituted for M_{0z2}. A simple rearrangement of the $2N$ equations in (6.40) and (6.41) is all that is required. This is accomplished by the substitutions (6.66a) and (6.66b) for A_2 and B_2. Specifically, let

$$A_2 = -A_2' \sec \beta_0 h_2, \qquad B_2 = B_2' + A_2' \tan \beta_0 h_2 \tag{6.68}$$

$$\Phi_{k2V}' = [\Phi_{k2V} - \Phi_{k2U} \sin \beta_0 h_2] \sec \beta_0 h_2 \tag{6.69}$$

$$\Psi_{k2dV}'^h = [\Psi_{k2dV}^h - \Psi_{k2dU}^h \sin \beta_0 h_2] \sec \beta_0 h_2. \tag{6.70}$$

Also let B_i' stand for $B_1, B_2', B_3, \ldots B_N$. With this notation, the equations (6.40) and (6.41) become:

$$\sum_{i=1}^{N}[\Phi_{kiU} B_i' + \Phi_{kiD} D_i] = \Phi_{k2V}' A_2'; \qquad k = 1, 2, \ldots N \qquad (6.71)$$

$$\sum_{i=1}^{N}[\Psi_{kidU}^{h} B_i' + \Psi_{kidD}^{h} D_i] = \Psi_{k2dV}'^{h} A_2'; \qquad k = 1, 2, \ldots N. \qquad (6.72)$$

In matrix form these are

$$[\Phi_U]\{B'\} + [\Phi_D]\{D\} = \{\Phi_{2V}'\} A_2' \qquad (6.73)$$

$$[\Psi_{dU}^{h}]\{B'\} + [\Psi_{dD}^{h}]\{D\} = \{\Psi_{2dV}'^{h}\} A_2', \qquad (6.74)$$

where the four square matrices and the column matrix $\{D\}$ are defined in (6.42a, b), (6.43a, b) and (6.45). The other column matrices are

$$\{B'\} = \begin{Bmatrix} B_1 \\ B_2' \\ B_3 \\ \vdots \\ B_N \end{Bmatrix}, \qquad \{\Phi_{2V}'\} = \begin{Bmatrix} \Phi_{12V}' \\ \Phi_{22V}' \\ \Phi_{32V}' \\ \vdots \\ \Phi_{N2V}' \end{Bmatrix}, \qquad \{\Psi_{2dV}'^{h}\} = \begin{Bmatrix} \Psi_{12dV}'^{h} \\ \Psi_{22dV}'^{h} \\ \Psi_{32dV}'^{h} \\ \vdots \\ \Psi_{N2dV}'^{h} \end{Bmatrix}. \qquad (6.75)$$

These equations are to be solved for the $2N$ coefficients B_i' and D_i in terms of $A_2' = -j(2\pi V_{02}/\zeta_0 \Psi_{22dV}^{m})$. The Ψ functions that occur in these equations are defined in the same manner as in Sections 6.3 and 6.4. This is illustrated below for $\beta_0 h_k = \pi/2$.

When $\beta_0 h_k = \pi/2$, $F_{0zk} = M_{0zk} = \cos \beta_0 z_k$. It follows from (6.48) and (6.49) that $W_{k2V}(0) = W_{k2U}(0)$ and $W_{k2V}(h_k/2) = W_{k2U}(h_k/2)$. From (6.32) and (6.33), (6.61) and (6.62), it follows that $\Psi_{k2V}(h_k) = \Psi_{k2U}(h_k)$ and $\Delta_1 = \Delta_2$. Hence, from (6.53) and (6.57), (6.54) and (6.58), it follows that $\Psi_{22dV}^{m} = \Psi_{22dU}^{f}$, $\Psi_{22dV}^{h} = \Psi_{22dU}^{h}$ when $k = 2$. Similarly, (6.55) and (6.57), (6.56) and (6.58) give $\Psi_{k2dV}^{f} = \Psi_{k2dU}^{f}$, $\Psi_{k2dV}^{h} = \Psi_{k2dU}^{h}$ when $k \neq 2$. As a consequence, Φ_{k2V}' and $\Psi_{k2dV}'^{h}$ become indeterminate in the form $0/0$ when $k \neq 2$. However, the limiting value for each as $\beta_0 h_2 \to \pi/2$ is finite. Thus, (6.69) and (6.70) may be expanded as follows. When $k = 2$,

$$\Phi_{22V}' = -S_a(h_2, h_2) + E_a(h_2, h_2) + \Psi_{22dU}^{f}; \qquad (6.76a)$$

when $k \neq 2$,

$$
\begin{aligned}
\Phi'_{k2V} = & -S_{bk2}(h_2, h_k) + E_{bk2}(h_2, h_k) \\
& + \frac{\cos \beta_0 h_k}{\Delta_2} \Big\{ [\cos \tfrac{1}{4}\beta_0 h_k - \cos \tfrac{1}{2}\beta_0 h_k][-E_{bk2}(h_2, 0) \\
& + E_{bk2}(h_2, h_k) + S_{bk2}(h_2, 0) - S_{bk2}(h_2, h_k)] \\
& + (1 - \cos \tfrac{1}{2}\beta_0 h_k) \Big[E_{bk2}\Big(h_2, \frac{h_k}{2}\Big) \\
& - E_{bk2}(h_2, h_k) - S_{bk2}\Big(h_2, \frac{h_k}{2}\Big) + S_{bk2}(h_2, h_k) \Big] \Big\},
\end{aligned}
\tag{6.76b}
$$

where Δ_2 is defined in (6.62). Similarly, when $k = 2$,

$$
\begin{aligned}
\Psi'^h_{22dV} = & \frac{1 - \sqrt{2}}{\Delta_2} \Big\{ \Big[C_a\Big(h_2, \frac{h_2}{2}\Big) - C_a(h_2, h_2) \Big] \Big[1 - \frac{1}{\sqrt{2}} \Big] \\
& - [C_a(h_2, 0) - C_a(h_2, h_2)] \Big[\cos \frac{\pi}{8} - \frac{1}{\sqrt{2}} \Big] \Big\} \\
& + \frac{1}{\Delta_2} \Big\{ \Big[-S_a\Big(h_2, \frac{h_2}{2}\Big) + S_a(h_2, h_2) + E_a\Big(h_2, \frac{h_2}{2}\Big) - E_a(h_2, h_2) \Big] \\
& + \frac{1}{\sqrt{2}} [S_a(h_2, 0) - S_a(h_2, h_2) - E_a(h_2, 0) + E_a(h_2, h_2)] \Big\};
\end{aligned}
\tag{6.76c}
$$

when $k \neq 2$,

$$
\begin{aligned}
\Psi'^h_{k2dV} = & \frac{1}{\Delta_2} \Big\{ [1 - \cos \beta_0 h_k] \Big[-S_{bk2}\Big(h_2, \frac{h_k}{2}\Big) + S_{bk2}(h_2, h_k) \\
& + E_{bk2}\Big(h_2, \frac{h_k}{2}\Big) - E_{bk2}(h_2, h_k) \Big] \\
& + [\cos \tfrac{1}{2}\beta_0 h_k - \cos \beta_0 h_k][S_{bk2}(h_2, 0) - S_{bk2}(h_2, h_k) \\
& - E_{bk2}(h_2, 0) + E_{bk2}(h_2, h_k)] \Big\}.
\end{aligned}
\tag{6.76d}
$$

The coefficients B'_i and D_i obtained for $\beta_0 h_2 = \pi/2$ from (6.73) and (6.74) with (6.76) are to be used in the current distributions

$$
I_{z2}(z_2) = A'_2 S_{0z2} + B'_2 F_{0z2} + D_2 H_{0z2}
\tag{6.77}
$$

$$
I_{zi}(z_i) = B'_i F_{0zi} + D_i H_{0zi}, \qquad i = 1, 3, \ldots N.
\tag{6.78}
$$

In the original analysis of arrays with half-wave elements [2] and in its application to arrays of the Yagi type [3], a somewhat different procedure was used. In effect, this treated the alternative form (6.64) of the distribution of current along a driven half-wave element as an independent representation. The entire procedure carried through in Sections 6.3 and 6.4 was repeated with the distribution function M_{0z2} replaced by S_{0z2}. This also involved a simple rearrangement of the integral equations (6.8) so that when $k = 2$, the right-hand member is $(j4\pi/\zeta_0)[\frac{1}{2}V_{02}S_{0z2} + C_2 F_{0z2}]$.

The alternative procedure is basically equivalent to that outlined in Section 6.5 but the two are not identical and involve small quantitative differences when applied to a particular array. In particular, the values of $W_{22V}(0)$ and $W_{22V}(h_2/2)$ from (6.48a, b) are necessarily somewhat different when, with $\beta_0 h_2 = \pi/2$, $S_{0z2} = \sin \beta_0|z_2| - 1$ is substituted for $M_{0z2} = \cos \beta_0 z_2$ in the integrals. It follows that the two values of Ψ_{22dV}^m as defined in (6.53) are also not quite the same when S_{0z2} is used instead of M_{0z2}. These differences are small and either procedure should give satisfactory results, although in the interest of simplicity and consistency the generalization in Section 6.3 is to be preferred.

Reference is here made to the alternative procedure primarily because it was used by Morris in an extensive quantitative study of the Yagi–Uda array. The results of his work, described later in this chapter, differ negligibly from those actually given.

6.6   The far field of the Yagi–Uda array; gain

The electric field maintained at distant points by the currents in the N elements of the Yagi–Uda array is readily determined. For the currents

$$I_{z2}(z_2) = A_2 \sin \beta_0(h_2 - |z_2|) + B_2(\cos \beta_0 z_2 - \cos \beta_0 h_2)$$

$$+ D_2(\cos \tfrac{1}{2}\beta_0 z_2 - \cos \tfrac{1}{2}\beta_0 h_2) \tag{6.79a}$$

$$I_{zi}(z_i) = B_i(\cos \beta_0 z_i - \cos \beta_0 h_i) + D_i(\cos \tfrac{1}{2}\beta_0 z_i - \cos \tfrac{1}{2}\beta_0 h_i), \qquad i \neq 2 \quad (6.79b)$$

the electromagnetic field is

$$E_\Theta(R_2, \Theta, \Phi) = \frac{j\zeta_0}{2\pi}\left\{A_2 \frac{e^{-j\beta_0 R_2}}{R_2} F_m(\Theta, \beta_0 h_2)\right.$$

$$\left. + \sum_{i=1}^{N} \frac{e^{-j\beta_0 R_i}}{R_i}[B_i G_m(\Theta, \beta_0 h_i) + D_i D_m(\Theta, \beta_0 h_i)]\right\}, \tag{6.80}$$

where $F_m(\Theta, \beta_0 h)$, $G_m(\Theta, \beta_0 h)$ and $D_m(\Theta, \beta_0 h)$ are defined in (2.46)–(2.48) and R_i is the distance from the point of calculation to the center of element i. This may be rearranged as follows:

$$E_{\Theta N}(R_2, \Theta, \Phi) = -\frac{V_{02}}{\Psi}\frac{e^{-j\beta_0 R_2}}{R_2} f_{VN}(\Theta, \Phi). \tag{6.81a}$$

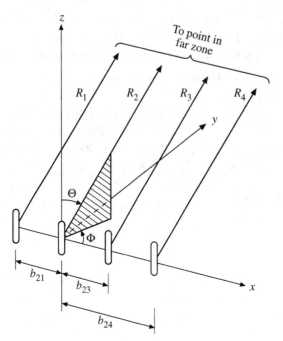

Figure 6.9 Coordinates for four-element array referred to origin at center of element 2. $b_{21} = b_{23} = b_{34} = b$.

Since no ambiguity can arise the symbol Ψ without subscripts and superscripts is used for Ψ^m_{22dV} as defined in (6.53). The field factor in (6.81a) for the N-element array is given by

$$f_{VN}(\Theta, \Phi) = \left\{ F_m(\Theta, \beta_0 h_2) + \sum_{i=1}^{N} e^{-j\beta_0(R_i - R_2)} [T_{Ui} G_m(\Theta, \beta_0 h_i) \right.$$

$$\left. + T_{Di} D_m(\Theta, \beta_0 h_i)] \right\} \sec \beta_0 h_2. \qquad (6.81b)$$

In obtaining (6.81a, b) the far-field approximation, $R_i \doteq R_2$, in amplitudes has been made. In the spherical coordinates R_2, Θ, Φ (Fig. 6.9), and with $b_{i,i\pm1} = b$,

$$R_i - R_2 = -(i - 2)b \sin \Theta \cos \Phi. \qquad (6.81c)$$

The following set of parameters has been introduced:

$$T_{Ui} = B_i/A_2, \qquad T_{Di} = D_i/A_2, \qquad (6.82)$$

where $A_2 = j2\pi V_{02}/\zeta_0 \Psi \cos \beta_0 h_2$. The quantity $E_\Theta(\Theta, \Phi)/V_{02}$ is the far field per unit voltage driving element 2.

An alternative expression for the field per unit input current to the driven antenna 2, i.e. $E_\Theta(\Theta, \Phi)/I_{z2}(0)$, is readily obtained with the substitution of $V_{02} = I_{z2}(0)/Y_{2N}$ where, from (6.23), the input admittance of antenna 2 when driving the N-element array is

$$Y_{2N} = \frac{I_{z2}(0)}{V_{02}} = \frac{j2\pi}{\zeta_0 \Psi \cos \beta_0 h_2} [\sin \beta_0 h_2 + T_{U2}(1 - \cos \beta_0 h_2)$$

$$+ T_{D2}(1 - \cos \tfrac{1}{2}\beta_0 h_2)]. \tag{6.83}$$

The result is

$$E_{\Theta N}(R_2, \Theta, \Phi) = \frac{j\zeta_0 I_{z2}(0)}{2\pi} \frac{e^{-j\beta_0 R_2}}{R_2} f_{IN}(\Theta, \Phi), \tag{6.84a}$$

where

$$f_{IN}(\Theta, \Phi)$$

$$= \left\{ \frac{F_m(\Theta, \beta_0 h_2) + \sum\limits_{i=1}^{N} e^{-j\beta_0(R_i - R_2)}[T_{Ui} G_m(\Theta, \beta_0 h_i) + T_{Di} D_m(\Theta, \beta_0 h_i)]}{\sin \beta_0 h_2 + T_{U2}(1 - \cos \beta_0 h_2) + T_{D2}(1 - \cos \tfrac{1}{2}\beta_0 h_2)} \right\}.$$

$$\tag{6.84b}$$

If the driven element is near a half wavelength long, the more convenient alternative form of the current is

$$I_{z2}(z_2) = A_2'(\sin \beta_0 |z_2| - \sin \beta_0 h_2) + B_2'(\cos \beta_0 z_2 - \cos \beta_0 h_2)$$

$$+ D_2(\cos \tfrac{1}{2}\beta_0 z_2 - \cos \tfrac{1}{2}\beta_0 h_2), \tag{6.85}$$

where $A_2' = -j2\pi V_{02}/\zeta_0 \Psi$. The currents in the parasitic elements are given by (6.79b). With the notation

$$T_{Ui}' = B_i'/A_2, \qquad B_i' = B_1, B_2', B_3, \ldots B_N \tag{6.86}$$

the formula for the distant field is

$$E_{\Theta N}(R_2, \Theta, \Phi) = \frac{V_{02}}{\Psi} \frac{e^{-j\beta_0 R_2}}{R_2} f_{VN}'(\Theta, \Phi), \tag{6.87a}$$

where

$$f_{VN}'(\Theta, \Phi) = H_m(\Theta, \beta_0 h_2) + \sum_{i=1}^{N} e^{-j\beta_0(R_i - R_2)}$$

$$\times [T_{Ui}' G_m(\Theta, \beta_0 h_i) + T_{Di} D_m(\Theta, \beta_0 h_i)]. \tag{6.87b}$$

$H_m(\Theta, \beta_0 h)$ is defined in (2.51) and, specifically for $\beta_0 h = \pi/2$, in (2.52a). As before, $G_m(\Theta, \beta_0 h)$ and $D_m(\Theta, \beta_0 h)$ are given in (2.47) and (2.48). Special values for $\beta_0 h =$

$\pi/2$ are in (2.52b, c). If desired $E_{\Theta N}(R_2, \Theta, \Phi)$ as given in (6.87a, b) may be referred to the current $I_{z2}(0)$ instead of the voltage V_{02}. In this case

$$Y_{2N} = \frac{-j2\pi}{\zeta_0 \Psi} [-\sin \beta_0 h_2 + T'_{U2}(1 - \cos \beta_0 h_2) + T_{D2}(1 - \cos \tfrac{1}{2}\beta_0 h_2)] \qquad (6.88)$$

so that

$$E_{\Theta N}(R_2, \Theta, \Phi) = \frac{j\zeta_0 I_{z2}(0)}{2\pi} \frac{e^{-j\beta_0 R_2}}{R_2} f'_{IN}(\Theta, \Phi), \qquad (6.89a)$$

where

$$f'_{IN}(\Theta, \Phi)$$

$$= \frac{H_m(\Theta, \beta_0 h_2) + \sum_{i=1}^{N} e^{-j\beta_0(R_i - R_2)}[T'_{Ui} G_m(\Theta, \beta_0 h_i) + T_{Di} D_m(\Theta, \beta_0 h_i)]}{-\sin \beta_0 h_2 + T'_{U2}(1 - \cos \beta_0 h_2) + T_{D2}(1 - \cos \tfrac{1}{2}\beta_0 h_2)}.$$

$$(6.89b)$$

The graphical representations of the normalized field factors $|f_N(\Theta, \Phi)|/|f_N(\pi/2, 0)|$ or $|f'_N(\Theta, \Phi)|/|f'_N(\pi/2, 0)|$ in appropriate planes are the field patterns. The field pattern in the equatorial (horizontal) plane is given by $|f_N(\pi/2, \Phi)|/|f_N(\pi/2, 0)|$ as a function of Φ. Important field patterns in planes perpendicular to the equatorial plane are with $\Phi = 0$ and π. In this case $|f_N(\Theta, \{ {0 \atop \pi} \})|/|f_N(\pi/2, 0)|$ is shown graphically as a function of Θ. The ratio of the field in the forward direction (i.e. toward the directors, $\Phi = 0$) to the field in the backward direction (i.e. toward the reflector, $\Phi = \pi$) in the equatorial plane $\Theta = \pi/2$ is known as the front-to-back ratio. It is given by

$$R_{FB} = \frac{\left| f_N \left(\frac{\pi}{2}, 0 \right) \right|}{\left| f_N \left(\frac{\pi}{2}, \pi \right) \right|}. \qquad (6.90a)$$

The front-to-back ratio in decibels is

$$r_{FB} = 20 \log_{10} \left| f_N \left(\frac{\pi}{2}, 0 \right) \Big/ f_N \left(\frac{\pi}{2}, \pi \right) \right|. \qquad (6.90b)$$

Note that in all of the ratios involving $f_N(\Theta, \Phi)$ either $f_{VN}(\Theta, \Phi)$ or $f_{IN}(\Theta, \Phi)$ may be used.

Since the total power radiated by an array is given by the integral over a great sphere of the normal component of the Poynting vector

$$|S_R(R, \Theta, \Phi)| = |E_{\Theta}(R, \Theta, \Phi)|^2/2\zeta_0 \qquad (6.91)$$

the distribution as a function of Θ and Φ of $|S_R(R, \Theta, \Phi)|$ is of interest. The total power supplied to the N-element array is

$$P_{2N} = \tfrac{1}{2}|I_{z2}(0)|^2 R_{2N} = \tfrac{1}{2}|V_{02}|^2 G_{2N}, \tag{6.92}$$

where R_{2N} and G_{2N} are, respectively, the driving-point resistance and conductance of the element 2 when driving the N-element parasitic array. With (6.81a), (6.84a) and (6.92), (6.91) becomes

$$|S_R(R_2, \Theta, \Phi)| = \frac{P_{2N}}{\zeta_0 \Psi^2 R_2^2} \frac{1}{G_{2N}} |f_{VN}(\Theta, \Phi)|^2 \tag{6.93a}$$

$$= \frac{P_{2N}}{4\pi^2 R_2^2} \frac{\zeta_0}{R_{2N}} |f_{IN}(\Theta, \Phi)|^2. \tag{6.93b}$$

A graphical representation of $|f_N(\Theta, \Phi)/f_N(\pi/2, 0)|^2$ is known as a power pattern. (Note that R_2 is a distance, R_{2N} a resistance.)

If ohmic losses in the conductors of the antennas and in the surrounding dielectric medium (air) are neglected, the total power radiated by an array outside a great sphere of radius R_2 is the same as the total power supplied at the terminals of the driven element 2. That is

$$P_{2N} = \tfrac{1}{2}|V_{02}|^2 G_{2N} = \tfrac{1}{2}|I_{z2}(0)|^2 R_{2N}$$

$$= \int_0^{2\pi} \int_0^{\pi} |S_R(R_2, \Theta, \Phi)| R_2^2 \sin\Theta \, d\Theta \, d\Phi. \tag{6.94}$$

With (6.93) and (6.94), formulas are obtained for R_{2N} and G_{2N} in terms of the far field. They are

$$R_{2N} = \frac{\zeta_0}{4\pi^2} \int_0^{2\pi} \int_0^{\pi} |f_{IN}(\Theta, \Phi)|^2 \sin\Theta \, d\Theta \, d\Phi \tag{6.95a}$$

$$G_{2N} = \frac{1}{\zeta_0 \Psi^2} \int_0^{2\pi} \int_0^{\pi} |f_{VN}(\Theta, \Phi)|^2 \sin\Theta \, d\Theta \, d\Phi. \tag{6.95b}$$

Actually, both R_{2N} and G_{2N} are already known from

$$I_{z2}(0)/V_{02} = G_{2N} + jB_{2N}$$

when the medium in which the array is immersed is lossless.

The absolute directivity of the N-element Yagi array is defined in terms of the power radiated by a fictitious omnidirectional antenna that maintains the same field in *all* directions as the Yagi array does in the one direction of its maximum, namely, $\Theta = \pi/2$, $\Phi = 0$. This power is

$$P_{N \, \text{omni}} = 4\pi R_2^2 \left| S_R\left(R_2, \frac{\pi}{2}, 0\right)\right|$$

$$= P_{2N} \frac{\zeta_0}{\pi R_{2N}} \left|f_{IN}\left(\frac{\pi}{2}, 0\right)\right|^2 = P_{2N} \frac{4\pi}{\zeta_0 \Psi^2 G_{2N}} \left|f_{VN}\left(\frac{\pi}{2}, 0\right)\right|^2. \tag{6.96}$$

The ratio $P_{N\,\text{omni}}/P_{2N}$ is the absolute directivity. Thus

$$D_N\left(\frac{\pi}{2},0\right) = \frac{P_{N\,\text{omni}}}{P_{2N}} = \frac{\zeta_0}{\pi R_{2N}}\left|f_{IN}\left(\frac{\pi}{2},0\right)\right|^2 = \frac{4\pi}{\zeta_0\Psi^2 G_{2N}}\left|f_{VN}\left(\frac{\pi}{2},0\right)\right|^2.$$

(6.97)

This formula is often written with R_{2N} expressed explicitly as given in (6.95a). The quantity

$$G_N\left(\frac{\pi}{2},0\right) = 10\log_{10} D_N\left(\frac{\pi}{2},0\right)$$

(6.98)

is the absolute gain in decibels.

The absolute directivity of the driven element 2 when isolated is

$$D_1\left(\frac{\pi}{2},0\right) = \frac{P_{1\,\text{omni}}}{P_{21}} = \frac{\zeta_0}{\pi R_{21}}\left|f_{I1}\left(\frac{\pi}{2},0\right)\right|^2 = \frac{4\pi}{\zeta_0\Psi^2 G_{21}}\left|f_{V1}\left(\frac{\pi}{2},0\right)\right|^2.$$

(6.99)

The relative directivity at constant power of the array referred to the isolated driven element is

$$D_r(0) = \frac{D_N\left(\frac{\pi}{2},0\right)}{D_1\left(\frac{\pi}{2},0\right)} = \frac{R_{21}\left|f_{IN}\left(\frac{\pi}{2},0\right)\right|^2}{R_{2N}\left|f_{I1}\left(\frac{\pi}{2},0\right)\right|^2} = \frac{G_{21}\left|f_{VN}\left(\frac{\pi}{2},0\right)\right|^2}{G_{2N}\left|f_{V1}\left(\frac{\pi}{2},0\right)\right|^2}.$$

(6.100)

The corresponding relative gain in decibels is

$$G_r(0) = G_N\left(\frac{\pi}{2},0\right) - G_1\left(\frac{\pi}{2},0\right) = 10\left[\log_{10} D_N\left(\frac{\pi}{2},0\right) - \log_{10} D_1\left(\frac{\pi}{2},0\right)\right].$$

(6.101)

The relative directivity (6.100) is readily expressed in terms of the electric field in (6.84a). Thus,

$$D_r(0) = \frac{\left|E_{\Theta N}\left(R_2,\frac{\pi}{2},0\right)\right|^2}{\left|E_{\Theta 1}\left(R_2,\frac{\pi}{2},0\right)\right|^2}\frac{P_{21}}{P_{2N}}.$$

(6.102)

The relative directivity at constant power, $P_{21} = P_{2N}$, is

$$D_r(0) = \frac{\left|E_{\Theta N}\left(R_2,\frac{\pi}{2},0\right)\right|^2}{\left|E_{\Theta 1}\left(R_2,\frac{\pi}{2},0\right)\right|^2}.$$

(6.103)

This is equivalent to (6.100).

The relative directivity (6.100) or (6.103) is also the relative forward directivity in the direction $\Theta = \pi/2$, $\Phi = 0$. The relative directivity at constant power $P_{21} = P_{2N}$ in the backward direction $\Theta = \pi/2$, $\Phi = \pi$ is defined by

$$D_r(\pi) = \frac{\left|E_{\Theta N}\left(R_2, \frac{\pi}{2}, \pi\right)\right|^2}{\left|E_{\Theta 1}\left(R_2, \frac{\pi}{2}, \pi\right)\right|^2} = \frac{R_{21}}{R_{2N}} \frac{\left|f_{IN}\left(\frac{\pi}{2}, \pi\right)\right|^2}{\left|f_{I1}\left(\frac{\pi}{2}, \pi\right)\right|^2} = \frac{G_{21}}{G_{2N}} \frac{\left|f_{VN}\left(\frac{\pi}{2}, \pi\right)\right|^2}{\left|f_{V1}\left(\frac{\pi}{2}, \pi\right)\right|^2}.$$

(6.104)

The relative backward gain in decibels is

$$G_r(\pi) = 10 \log_{10} D_r(\pi).$$

(6.105)

Since for a single element rotational symmetry with respect to Φ gives $f_1(\pi/2, 0) = f_1(\pi/2, \pi)$, it follows that

$$\frac{D_r(0)}{D_r(\pi)} = \frac{\left|f_N\left(\frac{\pi}{2}, 0\right)\right|^2}{\left|f_N\left(\frac{\pi}{2}, \pi\right)\right|^2}$$

(6.106)

and

$$r_{FB} = G_r(0) - G_r(\pi)$$

(6.107)

in decibels. Note that R_2 is a distance, R_{21} and R_{2N} resistances.

6.7   Simple applications of the modified theory; comparison with experiment

The theory of arrays developed in the preceding sections is like that formulated in the earlier chapters in that the complicated simultaneous integral equations for the currents in the elements are replaced by a set of algebraic equations. This is accomplished by approximating the integrals with an appropriate combination of trigonometric functions. In dealing with arrays of driven elements of equal length it was convenient to use different trigonometric functions for different parts of the integrals and to match these to the integrals at the point of maximum current, $z = z_m$, and at the ends, $z = \pm h$. For use with parasitic elements of unequal length this procedure is modified. Each integral is approximated by a sum of trigonometric terms with coefficients matched to the integral at $z = 0$, $\pm h/2$ and $\pm h$. In order to illustrate the application of the modified theory and at the same time verify its accuracy it is convenient to consider the simplest cases, the isolated antenna and the two-element parasitic array. Since conventional (sinusoidal) theory fails completely

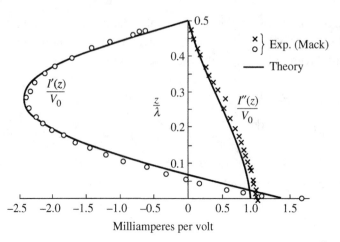

Figure 6.10 Distribution of current on full-wave antenna; $I(z) = I''(z) + jI'(z)$; $a/\lambda = 0.007\,022$, $h/\lambda = 0.5$.

when full-wave elements are involved, the examples are selected deliberately to include such elements.

In Fig. 6.10 are the distributions of current along a full-wave isolated antenna as computed from the modified theory, and as measured. They may be compared with the three-term approximation in Fig. 2.4 where the same experimental data are also shown. The two theoretical representations, while not identical, are nevertheless both very good approximations of the current. The modified theory does not provide quite as good an overall fit, but is somewhat better in specifying the susceptance – as would be expected since all integrals are matched at $z = 0$ and not only at the maximum of current. The admittance in the modified theory is $Y_0 = (0.926 + j1.350) \times 10^{-3}$ siemens; the value obtained previously is $Y_0 = (0.976 + j0.988) \times 10^{-3}$ siemens. The measured value after correction for end effects is $(1.025 + j1.676) \times 10^{-3}$ siemens. As indicated in conjunction with Fig. 2.6 a lumped susceptance $B_0 = 0.72 \times 10^{-3}$ siemens must be added to the three-term admittance to give $Y_0 = (0.976 + j1.708) \times 10^{-3}$ siemens. A similar lumped correction is also required with the modified theory, but it is smaller, namely $B_0 = 0.35 \times 10^{-3}$ siemens. It is clear that when suitably corrected to give the right susceptance, either theory provides a very acceptable approximation of the current in a dipole.

The distributions of current in an array of two full-wave elements in which element 1 is center driven and element 2 is parasitic are shown in Fig. 6.11 for four values of b, the distance between the parallel antennas. The corresponding field patterns in the equatorial plane are in Fig. 6.12. The distributions of current in Fig. 6.11 may be compared with measured values in Fig. 6.13. The agreement is seen to be very good. Equally good agreement has been observed for the field patterns.

As an illustration of the computations for the currents in a two-element array with elements differing greatly in length, the graphs in Fig. 6.14 are provided. The

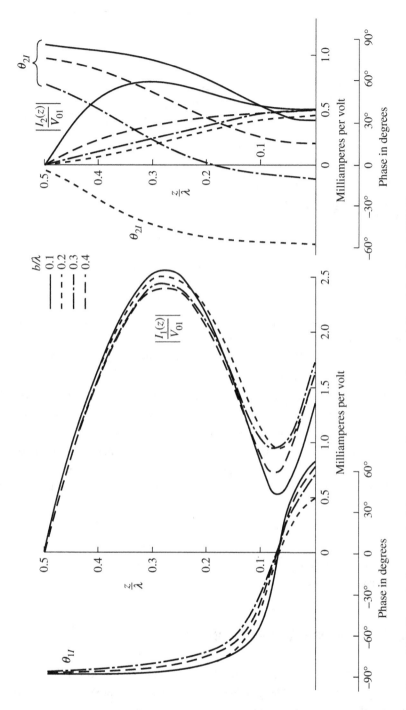

Figure 6.11 Theoretical currents in full-wave two-element parasitic array. $h/\lambda = 0.5$, $a/\lambda = 0.007\,022$.

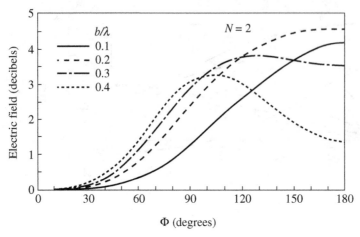

Figure 6.12 Horizontal field patterns of full-wave two-element parasitic array. $h/\lambda = 0.5$, $a/\lambda = 0.007\,022$.

associated horizontal field patterns are in Fig. 6.15. In the case considered, the driven element is a wavelength long, the parasitic element has successively the three lengths $h_2 = 0.2\lambda$, 0.4λ, and 0.65λ. Large changes in the distributions of current are seen to occur in the parasitic element as its length is changed while fixed at the specified distance $b = 0.2\lambda$ from the driven element. Note that except for the shortest length, the currents in the parasitic element differ significantly from the sinusoidal. The current in the driven antenna is only slightly affected by the changes in length of the coupled parasitic antenna, the largest changes occurring near the driving point so that the admittance is noticeably modified. Specifically, for the values $h_2/\lambda = 0.2, 0.4,$ 0.65 the admittances are $(0.916 + j1.041) \times 10^{-3}$, $(0.790 + j1.480) \times 10^{-3}$, and $(0.805 + j1.510) \times 10^{-3}$ siemens.

A typical computer printout for a two-element parasitic array is in Table 6.1. The coefficients of the trigonometric components of the current, the admittance, the impedance, the current distributions, the horizontal and vertical field patterns, the forward gain, the backward gain and the front-to-back ratio are all given.

6.8   The three-element Yagi–Uda array[1]

The computed distributions of current and the field pattern for a three-element array consisting of a reflector of length $2h_1 = 0.51\lambda$, a driven element of length $2h_2 = 0.50\lambda$ and a single director of length $2h_3 = 0.45\lambda$ are shown in Figs. 6.16 and 6.17. For this array the radius of all elements was taken as $a = 0.003\,369\lambda$. The driving-point impedance of element 2 is $Z_2 = 27.4 + j1.27$ ohms. The computed values of the

---

[1] This section is based on the work of Dr I. L. Morris [3].

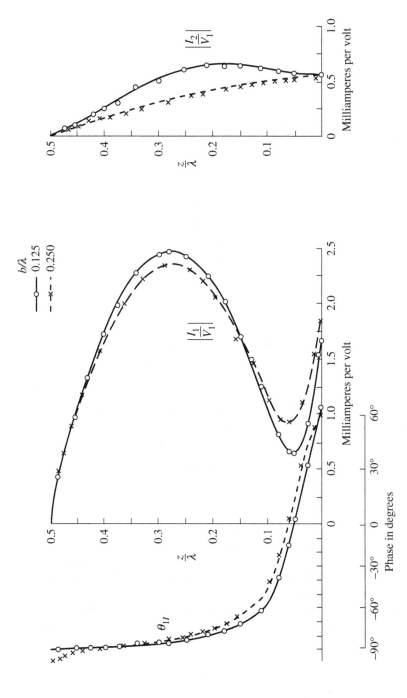

Figure 6.13 Measured currents in full-wave array of two elements (Mack). $h/\lambda = 0.5$, $a/\lambda = 0.007\,022$.

Table 6.1. *Computer printout for two-element parasitic array*

No. of elements = 2
Half-length of driving antenna = 0.500 0000E   00
Half-length of parasitic antenna = 0.650 0000E   00
Radius = 0.702 2000E−02
Element spacing = 0.200 0000E   00

Coefficients for current distributions

Element No. 1

AR	AI	BR
−0.249 034E−04	−0.318 019E−02	0.182 919E−03
BI	DR	DI
0.441 346E−03	0.439 517E−03	0.627 672E−03

Element No. 2

		BR
		0.707 925E−04
BI	DR	DI
−0.493 231E−03	0.221 251E−03	0.456 011E−03

Current distributions and input admittances

Element No. 1

	Real	Imaginary	Magnitude	Argument
Input admittance =				
	0.805 356E−03	0.151 036E−02	0.171 167E−02	61.8473
Input impedance =				
	0.274 884E 03	−0.515 518E 03	0.584 226E 03	−61.8473

Z/H	Real	Imaginary	Magnitude	Argument
0.	0.805 356E−03	0.151 036E−02	0.171 167E−02	61.8473
0.1	0.783 297E−03	0.498 302E−03	0.928 363E−03	32.4182
0.2	0.734 272E−03	−0.473 916E−03	0.873 929E−03	−32.7939
0.3	0.661 902E−03	−0.131 281E−02	0.147 023E−02	−63.1562
0.4	0.571 337E−03	−0.193 902E−02	0.202 144E−02	−73.4809
0.5	0.468 802E−03	−0.229 502E−02	0.234 241E−02	−78.3470
0.6	0.361 052E−03	−0.235 064E−02	0.237 821E−02	−81.1558
0.7	0.254 792E−03	−0.210 594E−02	0.212 130E−02	−82.9870
0.8	0.156 115E−03	−0.159 102E−02	0.159 866E−02	−84.2797
0.9	0.700 128E−04	−0.862 943E−03	0.865 779E−03	−85.2440

Table 6.1. – *continued*

Element No. 2

Z/H	Real	Imaginary	Magnitude	Argument
0.	0.434 100E−03	−0.120 109E−03	0.450 410E−03	−15.4447
0.1	0.423 681E−03	−0.890 176E−04	0.432 931E−03	−11.8492
0.2	0.393 571E−03	−0.202 264E−05	0.393 577E−03	−0.2940
0.3	0.347 053E−03	0.123 120E−03	0.368 245E−03	19.5057
0.4	0.289 068E−03	0.260 242E−03	0.388 955E−03	41.9382
0.5	0.225 521E−03	0.379 298E−03	0.441 278E−03	59.1837
0.6	0.162 456E−03	0.451 620E−03	0.479 950E−03	70.1187
0.7	0.105 249E−03	0.455 010E−03	0.467 024E−03	76.8698
0.8	0.579 298E−04	0.377 819E−03	0.382 234E−03	81.1709
0.9	0.227 404E−04	0.221 326E−03	0.222 491E−03	84.0178

Horizontal field pattern

Phi	E	E dB
0.	1.000 000	−0.
10.00	0.999 009	−0.0086
20.00	0.997 528	−0.0215
30.00	0.999 831	−0.0015
40.00	1.012 188	0.1052
50.00	1.041 196	0.3507
60.00	1.091 405	0.7597
70.00	1.163 272	1.3136
80.00	1.252 706	1.9570
90.00	1.352 388	2.6220
100.00	1.453 907	3.2507
110.00	1.549 639	3.8046
120.00	1.633 938	4.2647
130.00	1.703 568	4.6272
140.00	1.757 561	4.8982
150.00	1.796 660	5.0893
160.00	1.822 565	5.2137
170.00	1.837 157	5.2829
180.00	1.841 848	5.3051

F gain = 0.4079 dB   B gain = 5.7130 dB   FTBR = −5.3051 dB

Vertical field pattern

Theta	E	E dB
10.00	0.068 390	−23.3002
20.00	0.151 670	−16.3820
30.00	0.245 067	−12.2143
40.00	0.345 650	−9.2273
50.00	0.460 228	−6.7405
60.00	0.606 571	−4.3424
70.00	0.782 575	−2.1295
80.00	0.937 397	−0.5615

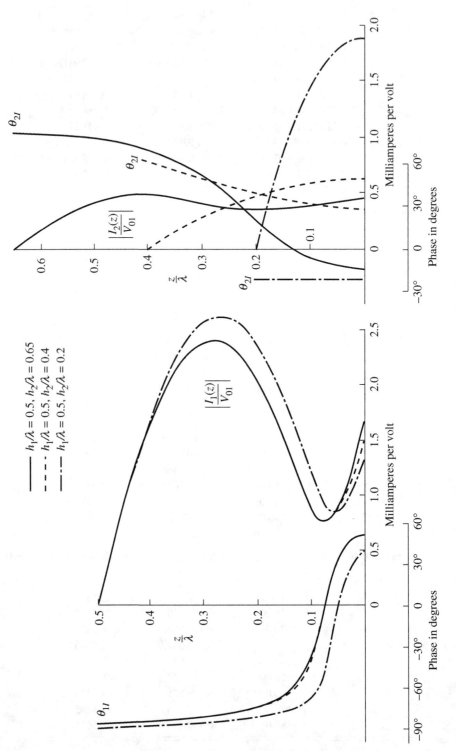

Figure 6.14 Currents in arrays of two elements of different lengths. $b/\lambda = 0.2$, $a/\lambda = 0.007\,022$, $N = 2$.

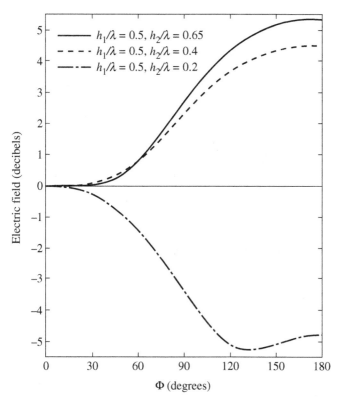

Figure 6.15 Horizontal field patterns of arrays of two elements of different lengths. $b/\lambda = 0.2$, $a/\lambda = 0.007\,022$, $N = 2$.

phase angle of the current along the reflector are nearly constant; it decreases from $74°.5$ at $z/h_1 = 0$ to $72°.7$ at $z/h_1 = 0.9$. The phase angle of the current along the driven element decreases from $-2°.66$ at $z/h_2 = 0$ to $-8°.47$ at $z/h_2 = 0.9$. The phase angle of the current along the director is almost exactly constant, changing only from $-154°.3$ at $z/h_3 = 0$ to $-154°.0$ at $z/h_3 = 0.9$. It is clear from Fig. 6.16 that the current in the reflector is so small that it actually contributes negligibly to the field.

In order to determine whether the particular length h_3 and spacing b_{23} are the best values to maintain the largest forward gain $G(0)$ or the maximum front-to-back ratio, the quantities h_3/λ and b_{23}/λ can be varied over a suitable range and the associated forward gain or front-to-back ratio computed. A computer printout for the front-to-back ratio is shown in Fig. 6.18. The ordinates are $2h_3/\lambda = 2H/L$, in a range from 0.50 to 0.36 in steps of 0.01; the abscissae are $b_{23}/\lambda = B/L$ in the range from 0.02 to 0.30 in steps of 0.02. The contours are drawn along estimated lines of constant front-to-back ratio ranging from 1 to 19. It is seen that the maximum value of front-to-back ratio is close to $b_{23}/\lambda = 0.12$ with $2h_3/\lambda = 0.44$ Thus, the distributions of current and the field pattern in Figs. 6.16 and 6.17 do not quite correspond to those

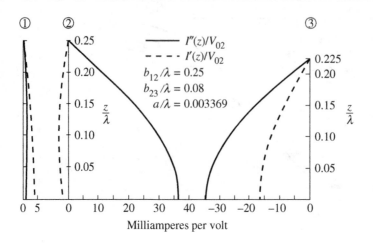

Figure 6.16 Currents in three-element Yagi–Uda array.

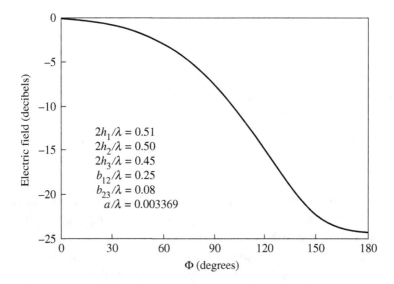

Figure 6.17 Field pattern of three-element Yagi–Uda array with element No. 2 driven.

for maximum front-to-back ratio. A small readjustment in the length of the director from $2h_3 = 0.45\lambda$ to $2h_3 = 0.44\lambda$ and an increase in its spacing b_{23} from 0.08λ to 0.12λ produce an increase in front-to-back ratio from 24.14 to 30.70. If the parameters $2h_3/\lambda$ and b_{23}/λ were varied in steps smaller than 0.01 and 0.02, respectively, an even higher ratio might be obtained within the narrow ranges $2h_3/\lambda = 0.44 \pm 0.01$, $b_{23}/\lambda = 0.12 \pm 0.02$. A more extended set of contours of the front-to-back ratios is shown in Fig. 6.19b in which the computed numbers have been deleted and only the contours of constant r_{FB} are shown. It is clear that a number of successive maxima in front-to-back ratio are obtained as the distance b_{23} between the director and the driven

1-Director Yagi antenna with constant director length and spacing
front-to-back ratio in dB

Reflector-feeder spacing = 0.250000
Radius of each element = 0.003369

Reflector length = 0.510000
Feeder length = 0.500000

$2h/\lambda$ →	0.52	0.33	0.22	0.11	0.02	−0.07	−0.16	−0.19	−0.21	−0.21	−0.18	−0.11	0.01	0.17	0.39
0.500000	10.94	4.95	2.72	1.71	1.15	0.81	0.58	0.41	0.30	0.24	0.22	0.24	0.30	0.41	0.58
0.490000	15.19	13.23	8.72	5.98	4.31	3.24	2.52	2.02	1.66	1.40	1.23	1.12	1.08	1.10	1.18
0.480000	13.42	17.39	16.01	11.96	9.02	7.02	5.61	4.60	3.85	3.29	2.88	2.58	2.37	2.24	2.20
0.470000	12.24	15.72	20.17	19.66	15.18	11.85	9.55	7.89	6.65	5.71	5.00	4.45	4.03	3.73	3.52
0.460000	11.53	14.03	17.81	24.14	25.08	18.38	14.34	11.70	9.82	8.42	7.34	6.51	5.86	5.35	4.98
0.450000	11.08	12.93	15.54	19.66	28.40	30.70	20.39	15.93	13.14	11.16	9.68	8.54	7.65	6.95	6.40
0.440000	10.76	12.19	14.08	16.70	20.58	26.07	25.03	19.71	16.10	13.60	11.76	10.35	9.24	8.36	7.67
0.430000	10.53	11.66	13.09	14.93	17.28	20.01	21.75	20.41	17.68	15.23	13.27	11.73	10.49	9.50	8.71
0.420000	10.36	11.27	12.40	13.77	15.39	17.13	18.49	18.66	17.48	15.74	14.05	12.58	11.34	10.32	9.48
0.410000	10.22	10.98	11.88	12.94	14.14	15.38	16.40	16.82	16.43	15.42	14.17	12.93	11.81	10.84	10.01
0.400000	10.11	10.74	11.48	12.33	13.26	14.19	14.97	15.40	15.32	14.76	13.90	12.93	11.98	11.11	10.34
0.390000	10.02	10.55	11.17	11.86	12.60	13.33	13.95	14.33	14.37	14.06	13.47	13.00	11.96	11.20	10.51
0.380000	9.94	10.40	10.91	11.49	12.09	12.68	13.17	13.50	13.58	13.41	13.00	12.74	12.45	11.82	11.19
0.370000	9.88	10.27	10.71	11.19	11.69	12.16	12.57	12.85	12.95	12.85	12.56	12.14	11.64	11.10	10.58

b/λ: 0.020000 0.040000 0.060000 0.080000 0.100000 0.120000 0.140000 0.160000 0.180000 0.200000 0.220000 0.240000 0.260000 0.280000 0.300000

Figure 6.18 Typical printout for three-element Yagi–Uda array.

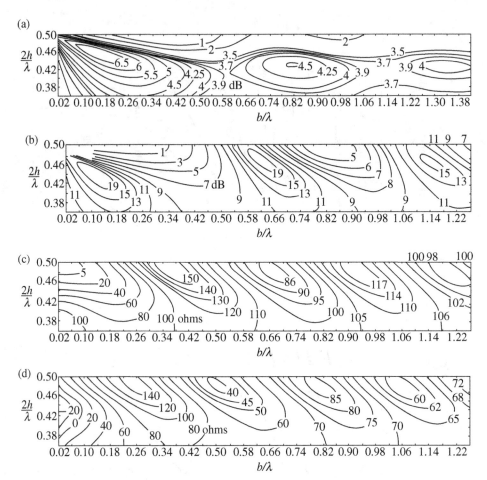

Figure 6.19 Contour diagrams constructed with computer printouts for a one-director Yagi array, (a) forward gain, (b) front-to-back ratio, (c) input resistance, (d) input reactance.

element is increased. These occur substantially at intervals of $\lambda/2$ with $2h_3$ between 0.44λ and 0.46λ. Similar computer printouts for forward gain, driving-point resistance and reactance are also shown in Fig. 6.19.

6.9   The four and eight director Yagi–Uda arrays[2]

The theory developed and illustrated with simple examples in the preceding sections can be applied to analyze the properties of longer Yagi–Uda arrays. For such arrays the quantities of principal interest include the distributions of current along all elements (since these determine the field), the admittance or impedance of the single

---

[2] This section is based on the work of Dr I. L. Morris [3].

driven element, the far-field pattern, the forward gain and the front-to-back ratio. For many purposes the determination of conditions that yield a maximum in the forward gain or in the front-to-back ratio is important. The parameters that may be varied are the length $2h_i$ and radius a_i of each element i, the distances b_{ij} between the elements i and j and their number N. Thus, there are a total of $3N$ parameters.

Because of the large number of possible combinations, an exhaustive study of the Yagi array would be very costly in both time and money even when a high-speed digital computer is available. An investigation of reasonable proportions must be restricted to a choice and range of parameters that is appropriate to a particular purpose.

In general, the purpose of the Yagi–Uda array is to obtain a highly directive field pattern with large values of the forward gain and front-to-back ratio. It has been shown implicitly that these desired properties can be achieved with the array pictured in Fig. 6.8. It consists of the following components:

1. A single driven element No. 2 that is a typical half-wave dipole of length $2h_2 = 0.5\lambda$ but with a finite radius a_2 and a distribution of current that is not assumed in advance to be sinusoidal, but remains to be determined.
2. A single reflecting element No. 1 that is slightly longer ($2h_1 = 0.51\lambda$) than the driven antenna is placed at a distance $b_{12} = 0.25\lambda$ from it. The field maintained by the currents induced in a parasitic element of this length and relative location tends to reinforce the field maintained by the currents in the driven element in the forward direction (from 1 to 2) and to reduce or cancel it in the opposite or backward direction (from 2 to 1).
3. The balance of the array consists of $N-2$ directors that all have the same half-length $h_i = h$ and that are separated by the same distance $b_{i-1,i} = b$ with $3 \leq i \leq N$. In order to function as directors, the length h of the $N-2$ parasitic elements must satisfy the inequality $h < h_2 = 0.5\lambda$ if the field maintained by the currents in them is to reinforce in the forward direction the field maintained by the currents in the driven element and in the reflector. If it is required that all antennas have the same radius, $a_i = a$, the $3N$ parameters have been reduced to three; h, b and N.

Contour diagrams constructed from computer printouts of the forward gain, the front-to-back ratio, the input resistance and the input reactance are shown in Figs. 6.20 and 6.21 for an array with four identical directors. The parameters are $2h/\lambda$ and b/λ where $h = h_3 = h_4 = h_5 = h_6$ and $b = b_{23} = b_{34} = b_{45} = b_{56}$. From these, combinations of h and b may be selected for which the forward gain or the front-to-back ratio is a maximum. For example, the following pairs of values are obtained from Fig. 6.20 to give a maximum front-to-back ratio: $2h/\lambda = 0.413$, 0.420, 0.426, 0.424; $b/\lambda = 0.033, 0.139, 0.248, 0.360$. These four sets all give a maximum front-to-back ratio, but the field patterns are quite different. These are shown in Fig. 6.22 together with corresponding patterns for similarly optimized one- and

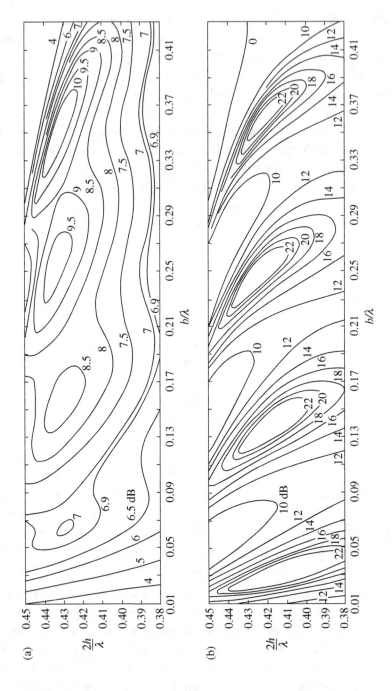

Figure 6.20 Forward gain (a) and front-to-back ratio (b) for a four-director Yagi array.

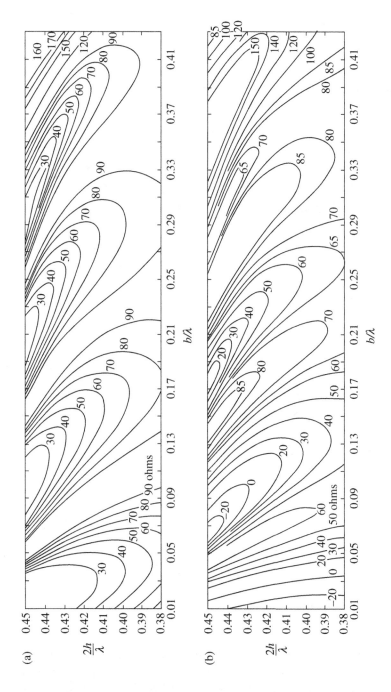

Figure 6.21 Input resistance (a) and reactance (b) for a four-director Yagi array.

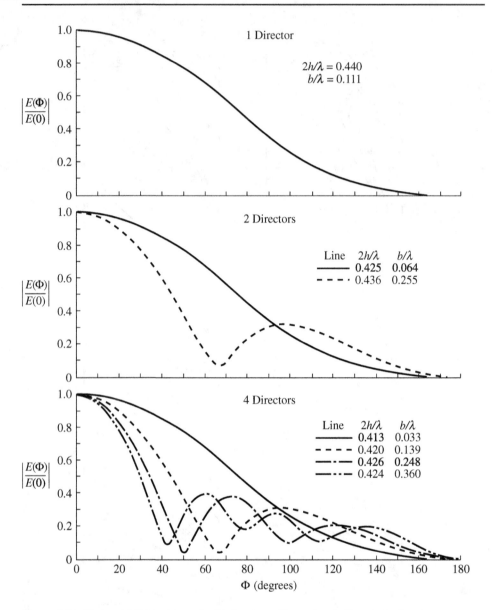

Figure 6.22 Horizontal field patterns for Yagi arrays with maxima in front-to-back ratio.

two-director arrays. From these it is seen that the field patterns for the most closely spaced condition for maximum front-to-back ratio are practically identical regardless of the number of directors. This is due to the fact that the directors are all so close to the driven element that no minor lobes are possible. As the distance between directors is increased, but limited to values that yield maxima in the front-to-back ratio, minor lobes appear and the beam width is reduced. The currents at the centers of the elements for the arrays that maintain the field patterns in Fig. 6.22 are represented in the form

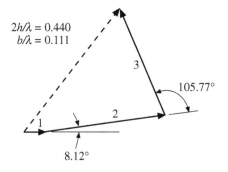

Element	$|I(0)|$
1	4.28
2	24.47
3	23.08

$2h/\lambda = 0.440$
$b/\lambda = 0.111$

$105.77°$

$8.12°$

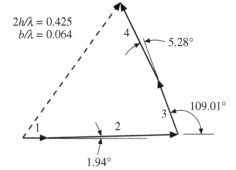

Element	$|I(0)|$
1	4.35
2	27.34
3	12.55
4	16.68

$2h/\lambda = 0.425$
$b/\lambda = 0.064$

$5.28°$

$109.01°$

$1.94°$

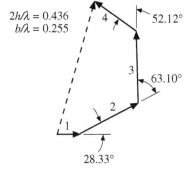

Element	$|I(0)|$
1	4.55
2	13.32
3	14.17
4	11.05

$2h/\lambda = 0.436$
$b/\lambda = 0.255$

$52.12°$

$63.10°$

$28.33°$

Figure 6.23 (a) Phasor diagrams for Yagi arrays with maxima in front-to-back ratio; one and two directors.

of phasor diagrams in Figs. 6.23a, b. The magnitude and angle of $I(0)$ in each element are shown. Note that for the very closely spaced four-director array with $b/\lambda = 0.033$, the currents in the directors are almost equal and in phase and much smaller than the current in the driven element. On the other hand, for the largest spacing shown $b/\lambda = 0.36$, the currents in the directors are comparable in magnitude with the current in the driven element and their phase differences are close to the progressive phase difference $360°b/\lambda = 130°$ of a wave traveling with the velocity of light from element to element.

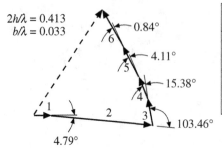

Element	$|I(0)|$
1	4.49
2	30.66
3	7.91
4	7.34
5	8.79
6	12.21

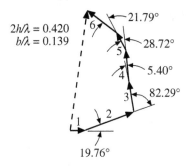

Element	$|I(0)|$
1	4.48
2	15.54
3	9.64
4	9.14
5	5.70
6	9.73

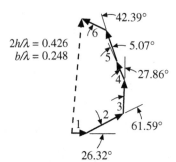

Element	$|I(0)|$
1	4.60
2	12.04
3	9.96
4	6.37
5	8.05
6	7.89

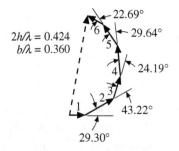

Element	$|I(0)|$
1	4.74
2	9.59
3	7.46
4	8.14
5	7.58
6	5.29

Figure 6.23 (b) Like Fig. 6.23a but for four directors.

Composite diagrams showing the forward gain, the front-to-back ratio, the input resistance and the input reactance as functions of b/λ for 1-, 2-, 4- and 8-director

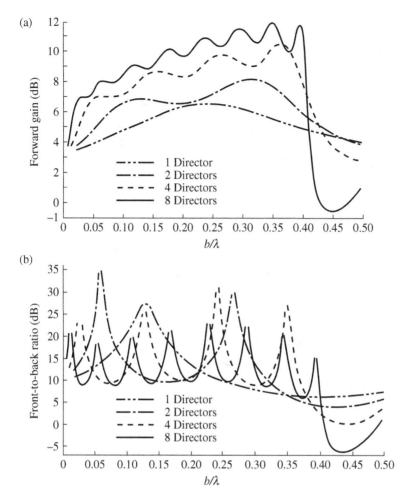

Figure 6.24 Forward gain (a), and front-to-back ratio (b), for a Yagi array with directors of constant length, radius and spacing (0.43λ, $0.003\,37\lambda$ and b, respectively).

arrays with $2h/\lambda = 0.43$, $a/\lambda = 0.003\,37$ are shown in Figs. 6.24a, b and 6.25a, b. From these the major quantities of interest are readily obtained.

A computer printout of an 8-director Yagi–Uda array[3] with $2h/\lambda = 0.4$ and $b/\lambda = 0.3$ is given in the accompanying Table 6.2. The impedance of the driven element when isolated is $Z_0 = 88.94 + j39.11$ ohms. Graphs of the currents in all of the elements are shown in Fig. 6.26. The phase angle along each parasitic element is essentially constant. It is represented in Fig. 6.27 as a function of the distance of the element from the driven antenna No. 2. The curve drawn through the points has no physical significance; it serves merely to interrelate the discrete points and thus reveal how nearly constant the phase change from director to director actually is. The electrical separation of adjacent directors is $108°$; the average phase difference of the currents

---

[3] The numerical evaluation for the 8-director Yagi array was done by V. W. H. Chang.

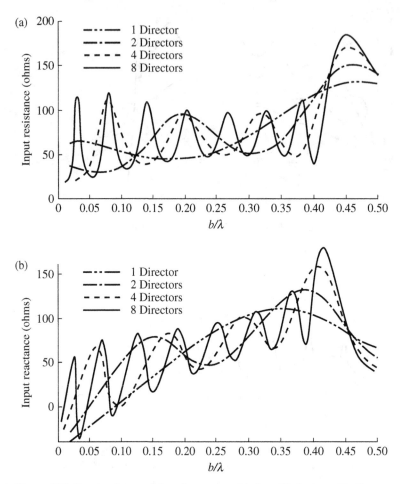

Figure 6.25 Input resistance (a), and reactance (b), for a Yagi array with directors of constant length, radius and spacing (0.43λ, $0.003\,37\lambda$ and b, respectively).

is $115°.6$. The horizontal field pattern maintained by the currents in the ten-element array is shown in Fig. 6.28.

6.10   Receiving arrays

The study of arrays of cylindrical antennas in all of the earlier sections of the book has been directed specifically to the problem of transmission, which involves the determination of distributions of current, driving-point admittances and field patterns. Arrays of antennas are also used to secure desired directional properties for receivers.

   In a transmitting array a single element may be driven, as in parasitic arrays of the Yagi–Uda type, or all the elements may be active as in the broadside or endfire arrays. In these latter the driving voltage is usually supplied from a single power oscillator

Table 6.2. *Computer printout for eight-director Yagi–Uda array*

No. of elements = 10
Half-length of driving antenna = 0.250 0000E−00
Half-length of parasitic antennas = 0.200 0000E−00
Half-length of reflector antenna = 0.255 0000E−00
Radius = 0.336 9000E−02
Spacing between reflector and driving antennas = 0.250 0000E−00
Spacing between parasitic antennas = 0.300 0000E−00

Coefficients for current distributions

Element No.	AR	AI	BR
1	0	0	−0.261 108E−03
2	0.260 791E−04	−0.128 598E−02	0.603 188E−03
3	0	0	0.373 823E−02
4	0	0	−0.247 170E−02
5	0	0	−0.117 657E−02
6	0	0	0.302 159E−02
7	0	0	−0.130 879E−02
8	0	0	−0.176 549E−02
9	0	0	0.264 461E−02
10	0	0	−0.391 583E−03

Element No.	BI	DR	DI
1	0.443 744E−03	0.626 141E−02	0.121 838E−01
2	0.204 197E−01	0.200 499E−01	−0.917 549E−01
3	−0.956 042E−03	−0.339 069E−01	0.757 068E−02
4	−0.290 843E−02	0.207 297E−01	0.252 446E−01
5	0.282 252E−02	0.102 848E−01	−0.238 580E−01
6	0.397 791E−03	−0.256 008E−01	−0.357 119E−02
7	−0.284 188E−02	0.109 612E−01	0.241 588E−01
8	0.176 994E−02	0.149 901E−01	−0.149 562E−01
9	0.139 167E−02	−0.223 427E−01	−0.119 466E−01
10	−0.278 379E−02	0.390 606E−02	0.234 452E−01

Table 6.2. – *continued*

Current distributions and input admittances

Element No. 1

Z/H	Real	Imaginary	Magnitude	Argument
0.	0.163 470E−02	0.416 262E−02	0.447 210E−02	68.4651
0.1	0.161 797E−02	0.411 786E−02	0.442 432E−02	68.4550
0.2	0.156 780E−02	0.398 398E−02	0.428 136E−02	68.4247
0.3	0.148 433E−02	0.376 216E−02	0.404 439E−02	68.3743
0.4	0.136 779E−02	0.345 436E−02	0.371 530E−02	68.3042
0.5	0.121 849E−02	0.306 328E−02	0.329 672E−02	68.2146
0.6	0.103 685E−02	0.259 232E−02	0.279 199E−02	68.1060
0.7	0.823 442E−03	0.204 556E−02	0.220 508E−02	67.9789
0.8	0.578 922E−03	0.142 768E−02	0.154 059E−02	67.8340
0.9	0.304 115E−03	0.743 922E−03	0.803 683E−03	67.6719

Element No. 2

Z/H	Real	Imaginary	Magnitude	Argument
0.	0.644 960E−02	−0.516 874E−02	0.826 517E−02	−38.6555
0.1	0.638 444E−02	−0.533 846E−02	0.832 227E−02	−39.8463
0.2	0.618 129E−02	−0.543 588E−02	0.823 147E−02	−41.2718
0.3	0.584 171E−02	−0.544 297E−02	0.798 446E−02	−42.9171
0.4	0.536 841E−02	−0.533 362E−02	0.756 752E−02	−44.7520
0.5	0.476 516E−02	−0.507 442E−02	0.696 107E−02	−46.7358
0.6	0.403 674E−02	−0.462 572E−02	0.613 943E−02	−48.8223
0.7	0.318 893E−02	−0.394 290E−02	0.507 107E−02	−50.9645
0.8	0.222 841E−02	−0.297 779E−02	0.371 928E−02	−53.1176
0.9	0.116 268E−02	−0.168 029E−02	0.204 333E−02	−55.2423

Element No. 3

Z/H	Real	Imaginary	Magnitude	Argument
0.	−0.389 258E−02	0.785 263E−03	0.397 100E−02	168.6103
0.1	−0.385 515E−02	0.777 863E−03	0.393 285E−02	168.6082
0.2	−0.374 266E−02	0.755 602E−03	0.381 817E−02	168.6018
0.3	−0.355 451E−02	0.718 302E−03	0.362 636E−02	168.5912
0.4	−0.328 973E−02	0.665 672E−03	0.335 640E−02	168.5765
0.5	−0.294 700E−02	0.597 315E−03	0.300 693E−02	168.5580
0.6	−0.252 470E−02	0.512 741E−03	0.257 624E−02	168.5358
0.7	−0.202 096E−02	0.411 377E−03	0.206 240E−02	168.5102
0.8	−0.143 372E−02	0.292 589E−03	0.146 327E−02	168.4815
0.9	−0.760 799E−03	0.155 698E−03	0.776 567E−03	168.4500

Table 6.2. *– continued*

Element No. 4

Z/H	Real	Imaginary	Magnitude	Argument
0.	0.225 112E−02	0.281 162E−02	0.360 177E−02	51.2469
0.1	0.222 970E−02	0.278 474E−02	0.356 740E−02	51.2456
0.2	0.216 531E−02	0.270 393E−02	0.346 408E−02	51.2416
0.3	0.205 751E−02	0.256 871E−02	0.329 114E−02	51.2350
0.4	0.190 559E−02	0.237 827E−02	0.304 753E−02	51.2259
0.5	0.170 859E−02	0.213 152E−02	0.273 178E−02	51.2144
0.6	0.146 531E−02	0.182 713E−02	0.234 212E−02	51.2006
0.7	0.117 439E−02	0.146 354E−02	0.187 647E−02	51.1848
0.8	0.834 289E−03	0.103 904E−02	0.133 254E−02	51.1671
0.9	0.443 383E−03	0.551 817E−03	0.707 878E−03	51.1476

Element No. 5

Z/H	Real	Imaginary	Magnitude	Argument
0.	0.115 124E−02	−0.260 615E−02	0.284 910E−02	−66.0760
0.1	0.114 022E−02	−0.258 133E−02	0.282 195E−02	−66.0770
0.2	0.110 710E−02	−0.250 670E−02	0.274 030E−02	−66.0799
0.3	0.105 169E−02	−0.238 177E−02	0.260 363E−02	−66.0847
0.4	0.973 653E−03	−0.220 574E−02	0.241 108E−02	−66.0913
0.5	0.872 566E−03	−0.197 751E−02	0.216 147E−02	−66.0997
0.6	0.747 887E−03	−0.169 575E−02	0.185 335E−02	−66.1096
0.7	0.598 995E−03	−0.135 890E−02	0.148 506E−02	−66.1212
0.8	0.425 205E−03	−0.965 221E−03	0.105 473E−02	−66.1340
0.9	0.225 787E−03	−0.512 883E−03	0.560 383E−03	−66.1481

Element No. 6

Z/H	Real	Imaginary	Magnitude	Argument
0.	−0.280 144E−02	−0.407 170E−03	0.283 087E−02	−171.7418
0.1	−0.277 475E−02	−0.403 259E−03	0.280 390E−02	−171.7424
0.2	−0.269 450E−02	−0.391 507E−03	0.272 279E−02	−171.7442
0.3	−0.256 017E−02	−0.371 848E−03	0.258 703E−02	−171.7473
0.4	−0.237 090E−02	−0.344 178E−03	0.239 575E−02	−171.7516
0.5	−0.212 552E−02	−0.308 354E−03	0.214 777E−02	−171.7569
0.6	−0.182 261E−02	−0.264 202E−03	0.184 166E−02	−171.7633
0.7	−0.146 050E−02	−0.211 518E−03	0.147 573E−02	−171.7707
0.8	−0.103 734E−02	−0.150 081E−03	0.104 814E−02	−171.7790
0.9	−0.551 178E−03	−0.796 550E−04	0.556 904E−03	−171.7880

Table 6.2. *– continued*

Element No. 7

Z/H	Real	Imaginary	Magnitude	Argument
0.	0.118 905E−02	0.265 022E−02	0.290 474E−02	65.7455
0.1	0.117 774E−02	0.262 496E−02	0.287 706E−02	65.7450
0.2	0.114 373E−02	0.254 901E−02	0.279 384E−02	65.7437
0.3	0.108 680E−02	0.242 187E−02	0.265 454E−02	65.7414
0.4	0.100 657E−02	0.224 275E−02	0.245 827E−02	65.7383
0.5	0.902 524E−03	0.201 056E−02	0.220 384E−02	65.7344
0.6	0.774 037E−03	0.172 395E−02	0.188 974E−02	65.7297
0.7	0.620 375E−03	0.138 136E−02	0.151 427E−02	65.7243
0.8	0.440 727E−03	0.981 069E−03	0.107 552E−02	65.7182
0.9	0.234 232E−03	0.521 244E−03	0.571 454E−03	65.7116

Element No. 8

Z/H	Real	Imaginary	Magnitude	Argument
0.	0.164 294E−02	−0.163 338E−02	0.231 672E−02	−44.7710
0.1	0.162 728E−02	−0.161 782E−02	0.229 464E−02	−44.7712
0.2	0.158 020E−02	−0.157 105E−02	0.222 828E−02	−44.7718
0.3	0.150 140E−02	−0.149 275E−02	0.211 720E−02	−44.7727
0.4	0.139 038E−02	−0.138 243E−02	0.196 068E−02	−44.7741
0.5	0.124 645E−02	−0.123 940E−02	0.175 777E−02	−44.7757
0.6	0.106 878E−02	−0.106 281E−02	0.150 727E−02	−44.7777
0.7	0.856 408E−03	−0.851 691E−03	0.120 781E−02	−44.7800
0.8	0.608 252E−03	−0.604 957E−03	0.857 871E−03	−44.7826
0.9	0.323 173E−03	−0.321 454E−03	0.455 822E−03	−44.7854

Element No. 9

Z/H	Real	Imaginary	Magnitude	Argument
0.	−0.243 969E−02	−0.131 999E−02	0.277 389E−02	−151.6237
0.1	−0.241 646E−02	−0.130 739E−02	0.274 746E−02	−151.6242
0.2	−0.234 660E−02	−0.126 951E−02	0.266 799E−02	−151.6258
0.3	−0.222 965E−02	−0.120 611E−02	0.253 497E−02	−151.6285
0.4	−0.206 487E−02	−0.111 680E−02	0.234 754E−02	−151.6321
0.5	−0.185 124E−02	−0.100 106E−02	0.210 457E−02	−151.6367
0.6	−0.158 748E−02	−0.858 237E−03	0.180 462E−02	−151.6422
0.7	−0.127 214E−02	−0.687 575E−03	0.144 607E−02	−151.6486
0.8	−0.903 608E−03	−0.488 242E−03	0.102 708E−02	−151.6557
0.9	−0.480 150E−03	−0.259 353E−03	0.545 718E−03	−151.6634

Element No. 10

Z/H	Real	Imaginary	Magnitude	Argument
0.	0.475 414E−03	0.255 409E−02	0.259 796E−02	79.3463
0.1	0.470 794E−03	0.252 977E−02	0.257 321E−02	79.3483
0.2	0.456 916E−03	0.245 667E−02	0.249 880E−02	79.3545
0.3	0.433 725E−03	0.233 429E−02	0.237 425E−02	79.3647
0.4	0.401 134E−03	0.216 185E−02	0.219 875E−02	79.3787
0.5	0.359 024E−03	0.193 825E−02	0.197 122E−02	79.3965
0.6	0.307 248E−03	0.166 218E−02	0.169 033E−02	79.4177
0.7	0.245 641E−03	0.133 207E−02	0.135 453E−02	79.4421
0.8	0.174 023E−03	0.946 231E−03	0.962 100E−03	79.4695
0.9	0.922 045E−04	0.502 831E−03	0.511 215E−03	79.4994

Table 6.2. – *continued*

	Real	Imaginary	Magnitude	Argument
Input admittance =				
	0.644 960E−02	−0.516 874E−02	0.826 517E−02	−38.6555
Input impedance =				
	0.944 123E 02	0.756 624E 02	0.120 990E 03	38.6555

Horizontal field pattern

Phi	E	E dB
0.	1.000 000	−0.
5.00	0.986 511	−0.1180
10.00	0.944 099	−0.4996
15.00	0.867 684	−1.2328
20.00	0.751 469	−2.4818
25.00	0.593 565	−4.5306
30.00	0.404 522	−7.8611
35.00	0.232 860	−12.6581
40.00	0.225 985	−12.9184
45.00	0.344 538	−9.2553
50.00	0.415 399	−7.6307
55.00	0.390 921	−8.1582
60.00	0.298 334	−10.5060
65.00	0.247 244	−12.1375
70.00	0.299 673	−10.4671
75.00	0.334 813	−9.5039
80.00	0.292 586	−10.6749
85.00	0.225 493	−12.9373
90.00	0.229 328	−12.7909
95.00	0.260 556	−11.6820
100.00	0.241 345	−12.3473
105.00	0.183 770	−14.7145
110.00	0.156 452	−16.1124
115.00	0.176 931	−15.0439
120.00	0.187 247	−14.5517
125.00	0.166 602	−15.5664
130.00	0.130 073	−17.7163
135.00	0.107 503	−19.3716
140.00	0.118 454	−18.5290
145.00	0.146 355	−16.6919
150.00	0.171 587	−15.3103
155.00	0.187 308	−14.5489
160.00	0.193 479	−14.2673
165.00	0.192 924	−14.2923
170.00	0.189 266	−14.4586
175.00	0.185 684	−14.6245
180.00	0.184 254	−14.6917

F gain = 11.5646 dB   B gain = −3.1270 dB   FTBR = 14.6917 dB

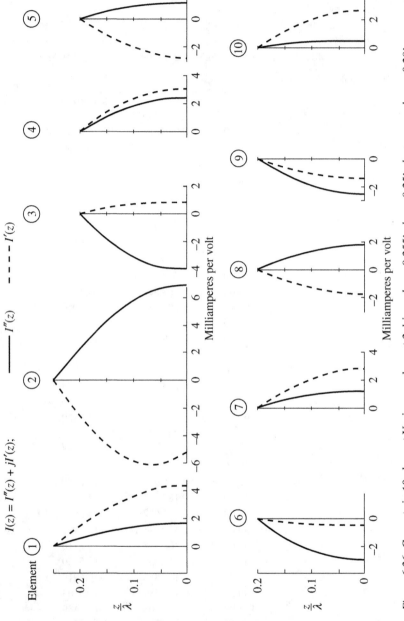

Figure 6.26 Currents in 10-element Yagi array, element 2 driven; $h_1 = 0.255\lambda$, $h_2 = 0.25\lambda$, $h_3 = \cdots = h_{10} = 0.20\lambda$; $b_{12} = 0.25\lambda$, $b_{23} = \cdots = b_{9,10} = 0.3\lambda$; $a_1 = a_2 = \cdots = a_{10} = 0.003\,37\lambda$; $\Omega = 10$ for $h = 0.25\lambda$.

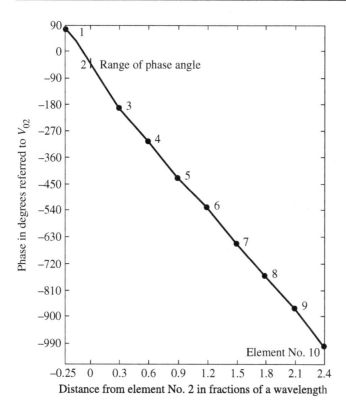

Figure 6.27 Phases of currents in elements referred to V_{02}.

by way of a suitable network of transmission lines, transformers and phase shifters. The design of such a feeding system of transmission lines is beyond the scope of this book. However, most transmitting arrays with their associated networks have a single pair of terminals across which the driving voltage is maintained. Since this pair of terminals is directly obvious in the parasitic arrays which have only a single driven element, attention in the following discussion is focused specifically on arrays of this type. Note that all references to the terminals of the driven element in a parasitic array apply equally to the single pair of input terminals of the transmission-line network that drives any other array.

Consider a receiving array of antennas in the incident plane-wave field of a distant transmitter. For convenience let the array be that shown in Fig. 6.8 with a load impedance Z_L instead of the generator connected across the terminals of antenna 2. In order to determine all of the properties of this system including, for example, the distributions of current in the elements and the reradiated or scattered field, it is necessary to formulate the coupled integral equations from the boundary condition that requires the tangential component of the total electric field to vanish on the perfectly conducting surface of each element. Fortunately, if interest is restricted to

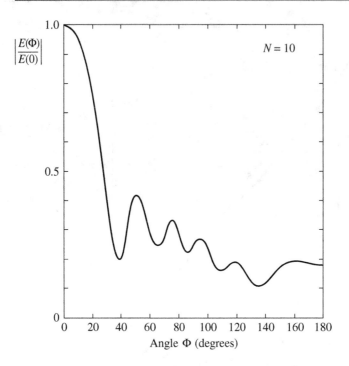

Figure 6.28 Field of 10-element Yagi array.

the transmission of information from a distant transmitter to the load Z_L, this elaborate analysis is unnecessary since the current in the load between the given terminals can be determined by the application of the reciprocal theorem[4] to the identical array when driven by the voltage V_0^e across the same terminals.

The reciprocal theorem applies to two arbitrarily located pairs of terminals, the one, for example, in an array A, the other in a simple dipole D. First, let the array be used for transmission, the dipole for reception. A generator with EMF V_0^e and internal impedance Z_g is connected across the terminals of the array; a load Z_L is connected across the terminals of the dipole. The center of the driven element 2 in the array is located at the origin of the spherical coordinates r, Θ, Φ; the receiving dipole is used to measure the field pattern of the array. For this purpose it is moved along the surface of a great sphere so that its axis is always tangent to the electric field maintained by the transmitter. The current $I_D(\Theta, \Phi)$ in Z_L at the center of the dipole varies as the dipole is moved. From (2.78) with (2.79), it is given by

$$I_D(\Theta, \Phi) = \frac{2h_e\left(\frac{\pi}{2}\right) E_z^{\text{inc}}}{Z_0 + Z_L} = \frac{-2h_e\left(\frac{\pi}{2}\right) E_\Theta(R_2, \Theta, \Phi)}{Z_0 + Z_L}, \tag{6.108}$$

4   See, for example, [4], p. 690 and [5], p. 216.

where $2h_e(\pi/2)$ is the effective length of the dipole when its axis is parallel to the incident electric field and perpendicular to the direction of propagation. Note that when the axis of the receiving dipole is tangent to the surface of the great sphere parallel to E^{inc}, the positive direction of the spherical coordinate Θ is opposite to the positive direction z along the antenna.

The far-zone electric field maintained by the N-element Yagi array driven by a generator at the center of element No. 2 is given by (6.84a). It is

$$E_\Theta(R_2, \Theta, \Phi) = \frac{j\zeta_0 I_{z2}(0)}{2\pi} \frac{e^{-j\beta_0 R_2}}{R_2} f_{IN}(\Theta, \Phi), \tag{6.109}$$

where R_2 is measured from the center of element No. 2 and the field factor of the array, $f_{IN}(\Theta, \Phi)$, is given by (6.84b). If the driving-point impedance of the array at the terminals of element No. 2 is Z_{02} and the internal impedance of the generator is Z_g, it follows that

$$I_{z2}(0) = \frac{V_0^e}{Z_{02} + Z_g}. \tag{6.110}$$

With (6.109) and (6.110), (6.108) becomes

$$I_D(\Theta, \Phi) = -\frac{2h_e\left(\frac{\pi}{2}\right)}{Z_0 + Z_L} \cdot \frac{j\zeta_0 V_0^e}{Z_{02} + Z_g} \frac{e^{-j\beta_0 R_2}}{2\pi R_2} f_{IN}(\Theta, \Phi). \tag{6.111}$$

Now let the generator with its EMF V_0^e and internal impedance Z_g be interchanged with the load Z_L so that the dipole is the transmitter, the array the receiver. The dipole is again moved over the surface of the same great sphere; the array remains fixed at the origin of coordinates. The current $I_A(\Theta, \Phi)$ in the load Z_L in the array varies as the location of the transmitter is changed.

The reciprocal theorem states that if the same voltage V_0^e is applied successively to both antennas and provided $Z_g = Z_L$, then

$$I_D(\Theta, \Phi) = I_A(\Theta, \Phi) \tag{6.112}$$

for all values of Θ and Φ. It follows by a rearrangement of (6.111) and with (6.112) that the current in the load Z_L of the Yagi array when used for reception is given by

$$I_A(\Theta, \Phi) = -\frac{2 f_{IN}(\Theta, \Phi)}{\beta_0(Z_{02} + Z_L)} \cdot \frac{j\zeta_0 V_0^e}{Z_0 + Z_g} \frac{e^{-j\beta_0 R_2}}{2\pi R_2} \beta_0 h_e\left(\frac{\pi}{2}\right) \tag{6.113}$$

provided $Z_L = Z_g$. Since it has been proved[5] in general that

$$\beta_0 h_e \left(\frac{\pi}{2} \right) = f_I \left(\frac{\pi}{2}, \beta_0 h \right), \tag{6.114}$$

where $f_I(\pi/2, \beta_0 h)$ is the field factor of the dipole given in (2.54b) and evaluated at $\Theta = \pi/2$, it follows that (6.113) can be expressed as follows:

$$I_A(\Theta, \Phi) = -\frac{2h_{eN}(\Theta, \Phi) E_\Theta^r}{Z_{02} + Z_L}, \tag{6.115}$$

where

$$E_\Theta^r = \frac{j \zeta_0 I_z(0)}{2\pi} \frac{e^{-j\beta_0 R_2}}{R_2} f_I \left(\frac{\pi}{2}, \beta_0 h \right) \tag{6.116}$$

is the field maintained by the dipole at the center of element No. 2 of the array and where

$$2h_{eN}(\Theta, \Phi) = 2 f_{IN}(\Theta, \Phi)/\beta_0 \tag{6.117}$$

is by definition the effective length of the Yagi array. It follows that the directional properties of the Yagi (or any other array) are the same for reception as for transmission.

The preceding discussion has been concerned with reciprocity with constant applied voltage. If reciprocity is to be preserved with constant power somewhat different conditions must be fulfilled. This problem is considered elsewhere.[6]

6.11   Driven arrays of elements that differ greatly in length

The procedure outlined in Section 6.2 for approximating the integrals in the simultaneous integral equations (6.8) for the currents in a parasitic array of unequal elements is quite adequate when the elements do not differ greatly in length. In the Yagi–Uda array the lengths $2h_i$ of the individual elements $i = 1, \ldots N$ always lie in a range that extends from slightly greater than $\lambda/2$ to approximately $\lambda/3$. Unfortunately, when elements have lengths that encompass the full range permitted by the present theory, namely, $0 \le \beta_0 h_i \le 5\pi/4$, the representations (6.20)–(6.22) for the several integrals are not adequate under certain conditions. In particular the two-term approximations on the right in (6.20)–(6.22) do not adequately represent the integrals $W_{ki}(z_k)$ on the left whenever element k is quite long ($\beta_0 h_k \sim \pi$) but element i is short ($\beta_0 h_i \lesssim \pi/4$). Extensive computations and measurements by W.-M. Cheong [6] have shown that the two-term approximations in (6.20)–(6.22) with the two-point fitting used in

---

[5] [4], pp. 568–570.   [6] [4], p. 694.

(6.53)–(6.60) are especially unsatisfactory for points on the longer element in the range $|z| > h/2$.

A better representation of all of the integrals (6.15)–(6.17) and (6.20)–(6.22) is obtained when full advantage is taken of the three-term distribution of current given in (6.6) to approximate the integrals. Specifically, let

$$W_{kiV}(z_k) \equiv \int_{-h_i}^{h_i} M_{0z'i} K_{kid}(z_k, z_i') \, dz_i'$$

$$\dot{=} \Psi_{kidV}^m M_{0zk} + \Psi_{kidV}^f F_{0zk} + \Psi_{kidV}^h H_{0zk} \tag{6.118}$$

$$W_{kiU}(z_k) \equiv \int_{-h_i}^{h_i} F_{0z'i} K_{kid}(z_k, z_i') \, dz_i'$$

$$\dot{=} \Psi_{kidU}^m M_{0zk} + \Psi_{kidU}^f F_{0zk} + \Psi_{kidU}^h H_{0zk} \tag{6.119}$$

$$W_{kiD}(z_k) \equiv \int_{-h_i}^{h_i} H_{0z'i} K_{kid}(z_k, z_i') \, dz_i'$$

$$\dot{=} \Psi_{kidD}^m M_{0zk} + \Psi_{kidD}^f F_{0zk} + \Psi_{kidD}^h H_{0zk}. \tag{6.120}$$

The inclusion of the distribution M_{0z} in the approximate representation of the integrals $W_{kiU}(z_k)$ and $W_{kiD}(z_k)$ is a new departure. In all previous discussions it has been pointed out that the part of the integral that depends on the real part of the kernel is approximately proportional to the distribution in the integrand when the distance $\beta_0 b_{ki} < 1$ (which usually occurs only when $i = k$ and $b_{kk} = a$) and that otherwise the entire integral is proportional to combinations of $F_{0z} = \cos \beta_0 z - \cos \beta_0 h$ and $H_{0z} = \cos(\beta_0 z/2) - \cos(\beta_0 h/2)$. This means that the distribution $M_{0z} = \sin \beta_0(h - |z|)$ can appear on the right only when $M_{0z'}$ appears in the integrand. These statements are still correct. However, the investigations of Cheong [6] have shown that the current induced in the relatively long antenna ($h \sim \lambda/2$) by a very short one ($h < \lambda/4$) is not well represented by combinations of F_{0z} and H_{0z} alone. These distributions are excellent when the amplitude and phase of the inducing field are approximately constant along the entire length of an antenna. Clearly, this is not at all true of the field maintained, for example, along a full-wave antenna by the current in an adjacent quite short element. By including the term in M_{0z}, Cheong has obtained an improved overall representation of the amplitudes of the currents, especially at points at some distance from the centers of the longer elements. On the other hand, since M_{0z} has a discontinuous slope at $z = 0$ (except when $\beta_0 h = (2n + 1)\pi/2$), which the actual induced current cannot have, the slope of an approximate representation that makes use of M_{0z} is necessarily somewhat in error near $z = 0$ even though the amplitude is quite well described. The slope of the current is, of course, proportional to the charge per unit length. Fortunately, an incorrect slope with a discontinuity at $z = 0$ does not significantly affect the admittance or the far field. These are determined by the magnitude and phase of the current alone.

Since combinations of F_{0z} and H_{0z} are excellent approximations of the two integrals in (6.119) and (6.120) except in the special situations just described, it is to be anticipated that the coefficients Ψ_{kidU}^m and Ψ_{kidD}^m will be small except under those conditions. In any event, the three-term representation of the current for all elements including the M_{0z} terms in (6.119) and (6.120), can only serve to improve the representation of the amplitudes of the currents at the expense of a small error in their slopes near $z = 0$.

In order to determine the complex parameters Ψ in (6.118)–(6.120), the approximate expressions on the right are made exactly equal to the integrals at the three points $z_k = 0$, $z_k = h_k/3$ and $z_k = 2h_k/3$ instead of only at the two points $z_k = 0$ and $z_k = h_k/2$ used in Section 6.4. That is, three equations are obtained from each of the relations (6.118)–(6.120) in the form:

$$W_{kiV}(0) = \int_{-h_i}^{h_i} M_{0z'i} K_{kid}(0, z_i') \, dz_i'$$

$$= \Psi_{kidV}^m \sin \beta_0 h_k + \Psi_{kidV}^f (1 - \cos \beta_0 h_k) + \Psi_{kidV}^h [1 - \cos(\beta_0 h_k/2)] \tag{6.121}$$

$$W_{kiV}(h_k/3) = \int_{-h_i}^{h_i} M_{0z'i} K_{kid}(h_k/3, z_i') \, dz_i'$$

$$= \Psi_{kidV}^m \sin(2\beta_0 h_k/3) + \Psi_{kidV}^f [\cos(\beta_0 h_k/3) - \cos \beta_0 h_k]$$

$$+ \Psi_{kidV}^h [\cos(\beta_0 h_k/6) - \cos(\beta_0 h_k/2)] \tag{6.122}$$

$$W_{kiV}(2h_k/3) = \int_{-h_i}^{h_i} M_{0z'i} K_{kid}(2h_k/3, z_i') \, dz_i'$$

$$= \Psi_{kidV}^m \sin(\beta_0 h_k/3) + \Psi_{kidV}^f [\cos(2\beta_0 h_k/3) - \cos \beta_0 h_k]$$

$$+ \Psi_{kidV}^h [\cos(\beta_0 h_k/3) - \cos(\beta_0 h_k/2)]. \tag{6.123}$$

Each integral when evaluated is a complex number. There are, then, three simultaneous complex algebraic equations to evaluate the three complex parameters Ψ_{kidV}^m, Ψ_{kidV}^f, and Ψ_{kidV}^h for each pair of values i and k. A similar second set of three equations is obtained with the different complex numbers $W_{kiU}(0)$, $W_{kiU}(h_k/3)$ and $W_{kiU}(2h_k/3)$ on the left. These are obtained from the same integrals when $M_{0z'i}$ is replaced by $F_{0z'i}$. The simultaneous solution of these three equations for each pair of values i and k yields the complex parameters Ψ_{kidU}^m, Ψ_{kidU}^f and Ψ_{kidU}^h. A third set of three equations is obtained with the quantities $W_{kiD}(0)$, $W_{kiD}(h_k/3)$ and $W_{kiD}(2h_k/3)$ appearing on the left in (6.121)–(6.123). These quantities are defined by the integrals in (6.121)–(6.123) with $M_{0z'i}$ replaced by $H_{0z'i}$. For each pair of values of i and k this third set of three equations yields Ψ^m_{kidD}, Ψ^f_{kidD} and Ψ^h_{kidD}. In this manner all values of the parameters Ψ_{kid} are determined. They have the following forms for each of the subscripts V, U and D on the Ψ's and W's:

$$\Psi^m_{ki} = \Delta^{-1} \begin{vmatrix} W_{ki}(0) & 1 - \cos\beta_0 h_k & 1 - \cos(\beta_0 h_k/2) \\ W_{ki}(h_k/3) & \cos(\beta_0 h_k/3) - \cos\beta_0 h_k & \cos(\beta_0 h_k/6) - \cos(\beta_0 h_k/2) \\ W_{ki}(2h_k/3) & \cos(2\beta_0 h_k/3) - \cos\beta_0 h_k & \cos(\beta_0 h_k/3) - \cos(\beta_0 h_k/2) \end{vmatrix}$$

(6.124)

$$\Psi^f_{ki} = \Delta^{-1} \begin{vmatrix} \sin\beta_0 h_k & W_{ki}(0) & 1 - \cos(\beta_0 h_k/2) \\ \sin(2\beta_0 h_k/3) & W_{ki}(h_k/3) & \cos(\beta_0 h_k/6) - \cos(\beta_0 h_k/2) \\ \sin(\beta_0 h_k/3) & W_{ki}(2h_k/3) & \cos(\beta_0 h_k/3) - \cos(\beta_0 h_k/2) \end{vmatrix}$$

(6.125)

$$\Psi^h_{ki} = \Delta^{-1} \begin{vmatrix} \sin\beta_0 h_k & 1 - \cos\beta_0 h_k & W_{ki}(0) \\ \sin(2\beta_0 h_k/3) & \cos(\beta_0 h_k/3) - \cos\beta_0 h_k & W_{ki}(h_k/3) \\ \sin(\beta_0 h_k/3) & \cos(2\beta_0 h_k/3) - \cos\beta_0 h_k & W_{ki}(2h_k/3) \end{vmatrix},$$

(6.126)

where

$$\Delta = \begin{vmatrix} \sin\beta_0 h_k & 1 - \cos\beta_0 h_k & 1 - \cos(\beta_0 h_k/2) \\ \sin(2\beta_0 h_k/3) & \cos(\beta_0 h_k/3) - \cos\beta_0 h_k & \cos(\beta_0 h_k/6) - \cos(\beta_0 h_k/2) \\ \sin(\beta_0 h_k/3) & \cos(2\beta_0 h_k/3) - \cos\beta_0 h_k & \cos(\beta_0 h_k/3) - \cos(\beta_0 h_k/2) \end{vmatrix}.$$

(6.127)

The N simultaneous integral equations for the currents in the elements are

$$\sum_{i=1}^{N} \left\{ A_i \int_{-h_i}^{h_i} M_{0z'i} K_{kid}(z_k, z'_i) \, dz'_i + B_i \int_{-h_i}^{h_i} F_{0z'i} K_{kid}(z_k, z'_i) \, dz'_i \right.$$

$$\left. + D_i \int_{-h_i}^{h_i} H_{0z'i} K_{kid}(z_k, z'_i) \, dz'_i \right\}$$

$$= \frac{j4\pi}{\zeta_0 \cos\beta_0 h_k} [\tfrac{1}{2} V_{0k} M_{0zk} + U_k F_{0zk}]; \qquad k = 1, 2, \dots N,$$

(6.128)

where

$$U_k = \frac{-j\zeta_0}{4\pi} \sum_{i=1}^{N} \int_{-h_k}^{h_k} I_{zi}(z'_i) K_{ki}(h_k, z'_i) \, dz'_i$$

$$= \frac{-j\zeta_0}{4\pi} \sum_{i=1}^{N} [A_i \Psi_{kiV}(h_k) + B_i \Psi_{kiU}(h_k) + D_i \Psi_{kiD}(h_k)]$$

(6.129)

with $\Psi_{kiV}(h_k)$, $\Psi_{kiU}(h_k)$ and $\Psi_{kiD}(h_k)$ defined in (6.32)–(6.34). If the integrals in (6.128) are replaced by their approximate algebraic equivalents, the following set of algebraic equations for the coefficients A_i, B_i and D_i is obtained:

$$
\sum_{i=1}^{N}\{A_i[\Psi_{kidV}^m M_{0zk} + \Psi_{kidV}^f F_{0zk} + \Psi_{kidV}^h H_{0zk}]
$$

$$
+ B_i[\Psi_{kidU}^m M_{0zk} + \Psi_{kidU}^f F_{0zk} + \Psi_{kidU}^h H_{0zk}]
$$

$$
+ D_i[\Psi_{kidD}^m M_{0zk} + \Psi_{kidD}^f F_{0zk} + \Psi_{kidD}^h H_{0zk}]\}
$$

$$
= \frac{j4\pi}{\zeta_0 \cos\beta_0 h_k} [\tfrac{1}{2} V_{0k} M_{0zk} + U_k F_{0zk}]. \tag{6.130}
$$

Finally, if (6.129) is substituted for U_k, the set of equations may be arranged as follows:

$$
M_{0zk} \sum_{i=1}^{N}\left[(A_i \Psi_{kidV}^m + B_i \Psi_{kidU}^m + D_i \Psi_{kidD}^m)\cos\beta_0 h_k - \frac{j2\pi}{\zeta_0} V_{0k}\right]
$$

$$
+ F_{0zk} \sum_{i=1}^{N}[(A_i \Psi_{kidV}^f + B_i \Psi_{kidU}^f + D_i \Psi_{kidD}^f)\cos\beta_0 h_k
$$

$$
- A_i \Psi_{kiV}(h_k) - B_i \Psi_{kiU}(h_k) - D_i \Psi_{kiD}(h_k)]
$$

$$
+ H_{0zk} \sum_{i=1}^{N}[A_i \Psi_{kidV}^h + B_i \Psi_{kidU}^h + D_i \Psi_{kidD}^h]\cos\beta_0 h_k = 0 \tag{6.131}
$$

with $k = 1, 2, \ldots N$. These equations are satisfied if the coefficient of each of the three distribution functions is allowed to vanish. The result is a set of $3N$ simultaneous equations for the $3N$ unknown coefficients A, B and D. They are:

$$
\sum_{i=1}^{N}[A_i \Psi_{kidV}^m + B_i \Psi_{kidU}^m + D_i \Psi_{kidD}^m] = \frac{j2\pi}{\zeta_0} \frac{V_{0k}}{\cos\beta_0 h_k} \tag{6.132}
$$

$$
\sum_{i=1}^{N}[A_i \Phi_{kiV} + B_i \Phi_{kiU} + D_i \Phi_{kiD}] = 0 \tag{6.133}
$$

$$
\sum_{i=1}^{N}[A_i \Psi_{kidV}^h + B_i \Psi_{kidU}^h + D_i \Psi_{kidD}^h] = 0 \tag{6.134}
$$

with $k = 1, 2, \ldots N$. In (6.133) the following notation has been introduced:

$$
\Phi_{kiV} \equiv \Psi_{kiV}(h_k) - \Psi_{kidV}^f \cos\beta_0 h_k \tag{6.135}
$$

$$\Phi_{kiU} \equiv \Psi_{kiU}(h_k) - \Psi_{kidU}^f \cos\beta_0 h_k \tag{6.136}$$

$$\Phi_{kiD} \equiv \Psi_{kiD}(h_k) - \Psi_{kidD}^f \cos\beta_0 h_k. \tag{6.137}$$

These equations can be expressed in matrix notation. Let

$$[\Phi] = \begin{bmatrix} \Phi_{11} & \Phi_{12} & \cdots & \Phi_{1N} \\ \vdots & & & \\ \Phi_{N1} & \cdots & & \Phi_{NN} \end{bmatrix}, \tag{6.138}$$

where the matrix elements Φ_{ki} are defined in (6.135)–(6.137) for each subscript V, U and D. Also let

$$[\Psi^h] = \begin{bmatrix} \Psi_{11}^h & \Psi_{12}^h & \cdots & \Psi_{1N}^h \\ \vdots & & & \\ \Psi_{N1}^h & \cdots & & \Psi_{NN}^h \end{bmatrix}, \qquad [\Psi^m] = \begin{bmatrix} \Psi_{11}^m & \Psi_{12}^m & \cdots & \Psi_{1N}^m \\ \vdots & & & \\ \Psi_{N1}^m & \cdots & & \Psi_{NN}^m \end{bmatrix},$$
$$\tag{6.139}$$

where the Ψ_{ki}^h are obtained from (6.126). The following column matrices are needed:

$$\{A\} = \begin{Bmatrix} A_1 \\ A_2 \\ \vdots \\ A_N \end{Bmatrix}, \qquad \{B\} = \begin{Bmatrix} B_1 \\ B_2 \\ \vdots \\ B_N \end{Bmatrix}, \qquad \{D\} = \begin{Bmatrix} D_1 \\ D_2 \\ \vdots \\ D_N \end{Bmatrix} \tag{6.140}$$

$$\left\{ \frac{j2\pi}{\zeta_0} \frac{V_0}{\cos\beta_0 h} \right\} = \frac{j2\pi}{\zeta_0} \begin{Bmatrix} V_{01}/\cos\beta_0 h_1 \\ V_{02}/\cos\beta_0 h_2 \\ \vdots \\ V_{0N}/\cos\beta_0 h_N \end{Bmatrix}. \tag{6.141}$$

With this notation, the equivalent matrix equations for determining the coefficients A_i, B_i and D_i are

$$[\Psi_{dV}^m]\{A\} + [\Psi_{dU}^m]\{B\} + [\Psi_{dD}^m]\{D\} = \left\{ \frac{j2\pi}{\zeta_0} \frac{V_0}{\cos\beta_0 h} \right\} \tag{6.142a}$$

$$[\Phi_V]\{A\} + [\Phi_U]\{B\} + [\Phi_D]\{D\} = 0 \tag{6.142b}$$

$$[\Psi_{dV}^h]\{A\} + [\Psi_{dU}^h]\{B\} + [\Psi_{dD}^h]\{D\} = 0. \tag{6.142c}$$

These equations correspond to (6.29) with (6.46) and (6.47) in the simpler case of the Yagi array with two-term fitting of the integrals.

The solutions of (6.132)–(6.134) or (6.142a, b, c) express each of the coefficients A_i, B_i and D_i as a sum of terms in the N voltages V_{0k}, $k = 1, 2, \ldots N$. That is

$$A_i = j \frac{2\pi}{\zeta_0} \sum_{k=1}^{N} \frac{V_{0k}}{\cos \beta_0 h_k} \alpha_{ik} \tag{6.143}$$

$$B_i = j \frac{2\pi}{\zeta_0} \sum_{k=1}^{N} \frac{V_{0k}}{\cos \beta_0 h_k} \beta_{ik} \tag{6.144}$$

$$D_i = j \frac{2\pi}{\zeta_0} \sum_{k=1}^{N} \frac{V_{0k}}{\cos \beta_0 h_k} \gamma_{ik}, \tag{6.145}$$

where the α_{ik}, β_{ik} and γ_{ik} are the appropriate cofactors divided by the determinant of the system.

It follows that with the coefficients A_i, B_i and D_i evaluated, the currents in all elements are available in the form:

$$I_{zi}(z) = j \frac{2\pi}{\zeta_0} \sum_{k=1}^{N} \frac{V_{0k}}{\cos \beta_0 h_k} \{\alpha_{ik} \sin \beta_0 (h_k - |z|) + \beta_{ik} (\cos \beta_0 z - \cos \beta_0 h_k)$$

$$+ \gamma_{ik} [\cos(\beta_0 z/2) - \cos(\beta_0 h_k/2)]\} \tag{6.146}$$

$$I_{zi}(0) = j \frac{2\pi}{\zeta_0} \sum_{k=1}^{N} \frac{V_{0k}}{\cos \beta_0 h_k} \{\alpha_{ik} \sin \beta_0 h_k + \beta_{ik}(1 - \cos \beta_0 h_k)$$

$$+ \gamma_{ik}[1 - \cos(\beta_0 h_k/2)]\}$$

$$= \sum_{k=1}^{N} V_{0k} Y_{ik}. \tag{6.147}$$

In these relations $i = 1, 2, \ldots N$, and

$$Y_{ik} = j \frac{2\pi}{\zeta_0 \cos \beta_0 h_k} \{\alpha_{ik} \sin \beta_0 h_k + \beta_{ik}(1 - \cos \beta_0 h_k) + \gamma_{ik}[1 - \cos(\beta_0 h_k/2)]\}. \tag{6.148}$$

The quantities Y_{ik}, with $k = i$, are the self-admittances of the N elements in the array; the Y_{ik}, with $k \neq i$, are the mutual admittances. They are readily determined from (6.148). Note that in general the self-admittance of an element when coupled to other antennas is not the same as the self-admittance of the same element when isolated.

In matrix form, the equations for the N driving-point currents are

$$\{I_z(0)\} = [Y_A]\{V_0\}, \tag{6.149}$$

where

$$\{I_z(0)\} = \begin{Bmatrix} I_{z1}(0) \\ I_{z2}(0) \\ \vdots \\ I_{zN}(0) \end{Bmatrix}, \qquad \{V_0\} = \begin{Bmatrix} V_{01} \\ V_{02} \\ \vdots \\ V_{0N} \end{Bmatrix} \qquad (6.150)$$

and

$$[Y_A] = \begin{bmatrix} Y_{11} & Y_{12} & \cdots & Y_{1N} \\ \vdots & & & \\ Y_{N1} & \cdots & & Y_{NN} \end{bmatrix}. \qquad (6.151)$$

The solution for the currents in the N elements of the array is thus completed in terms of arbitrary voltages. When these are specified, the complete distributions of current are given in the form (6.146). The driving-point admittances Y_{0i} and impedances Z_{0i} are given by

$$Y_{0i} = \frac{I_{zi}(0)}{V_{0i}} = \frac{1}{Z_{0i}}. \qquad (6.152)$$

6.12   The log-periodic dipole array

An interesting and important example of a curtain of driven elements that all have different lengths and radii and that are unequally spaced is the so-called log-periodic dipole array illustrated in Fig. 6.29. In spite of the fact that in this array all elements are connected directly to an active transmission line, its operation when suitably designed is closely related to that of the Yagi–Uda antenna in which only one element is driven and all others are parasitic. However, unlike the Yagi antenna, the log-periodic array has important broad-band properties. These are best introduced in terms of an array of an infinite number of center-driven dipoles arranged as shown in Fig. 6.29. Let the half-length of a typical element i be h_i, let its radius be a_i. The distance between element i and the next adjacent element to the right is $b_{i,i+1}$ where $i = 1, 2, 3, \ldots$. The array is constructed so that the following parameters

$$\frac{h_i}{h_{i+1}} = \tau, \qquad \frac{h_{i+1}}{b_{i,i+1}} = \sigma, \qquad 2 \ln \frac{2h_i}{a_i} = \Omega \qquad (6.153)$$

are treated as constants independent of i. As throughout this book, it is assumed that $h_i \gg a_i$.

If the dipoles individually approximate perfect conductors, the electrical properties of the array (such as the driving-point admittances of the elements and the field pattern of the array) at an angular frequency ω_0 depend only on the electrical dimensions

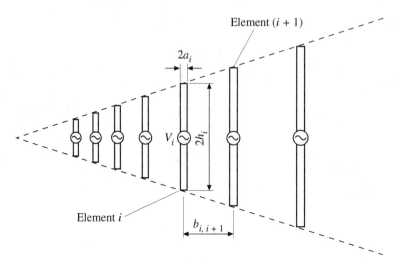

Figure 6.29 Seven elements of an infinite log-periodic array.

$\beta_0 h_i$, $\beta_0 b_{i,i+1}$, and $\beta_0 a_i$ where $\beta_0 = \omega_0/c = 2\pi/\lambda_0$ and c is the velocity of light. If the angular frequency is changed to $\omega_n = \tau^n \omega_0$, where n is a positive or negative integer, the original electrical properties are determined by $\tau^{-n}\beta_n h_i$, $\tau^{-n}\beta_n b_{i,i+1}$ and $\tau^{-n}\beta_n a_i$ where $\beta_n = \omega_n/c$. However, there are along the array antennas with half-lengths $h_{i+n} = \tau^{-n} h_i$ for which $(h_{i+n+1}/b_{i+n,i+n+1}) = \sigma$ and $2\ln(2h_{i+n}/a_{i+n}) = \Omega$. Since $\beta_0 h_i = \tau^{-n}\beta_n h_i = \beta_n h_{i+n}$, it follows that all properties of the array at the angular frequency ω_0 referred to element i are repeated at the angular frequency ω_n but referred to the element $i+n$. This periodicity of the properties with respect to frequency is linear with respect to the logarithm of the frequency. That is, since $\log \omega_n = \log \omega_0 + n \log \tau$, it is clear that any property shown graphically on a logarithmic frequency scale is periodic with period $\log \tau$. Accordingly, arrays with this construction are known as log-periodic dipole arrays [7–10]. Such arrays are generally driven from a two-wire line in the manner illustrated in Figs. 6.30a, b. The arrangement with reversed connections in Fig. 6.30b is the one required for endfire operation.

Actual arrays are, of course, never infinite so that the ideal frequency-independent properties of the infinite array are modified by asymmetries near the ends. These may be modified by the use of a terminating impedance Z_T as shown in Fig. 6.30 which provides an additional parameter. The value $Z_T = Z_c$, where Z_c is the characteristic impedance of the line, is an obvious choice.

6.13   Analysis of the log-periodic dipole array

The theory developed in Section 6.11 for arrays of antennas with unequal lengths, spacings and radii can be applied directly to the log-periodic dipole array. It is

(a)

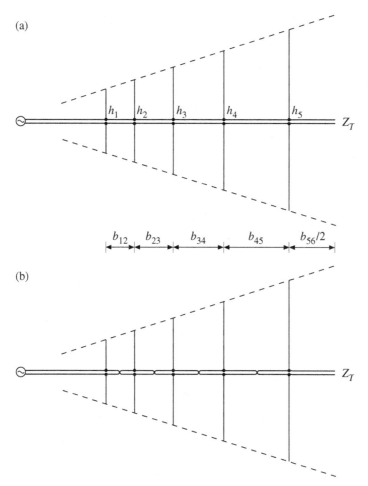

(b)

Figure 6.30 Log-periodic array driven from a two-wire line with (a) direct connections, (b) reversed connections.

only necessary to specify the driving-point voltages to the elements in order to obtain a complete solution for the distributions of current along the elements and their individual input admittances. The driving-point admittance of the array and the complete field pattern are then readily obtained over any frequency range for which the condition $\beta_0 h_i \leq 5\pi/4$ is satisfied for all elements. Such quantities as the beam width, the directivity, front-to-back ratio and side-lobe level can, of course, be obtained from the field pattern.

Consider specifically the array shown in Fig. 6.30b. The driving voltage is applied to a transmission-line that is connected successively to all of the elements beginning with the shortest. Between each adjacent pair of elements the connections are reversed by crossing the conductors of the transmission line in order to achieve the desired phase relations. The analysis of this circuit is conveniently carried out following the

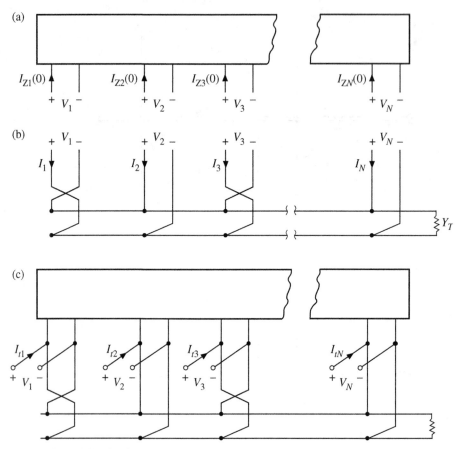

Figure 6.31 Schematic diagram of (a) the antenna circuit, (b) the transmission-line circuit, and (c) the antenna and transmission-line circuits connected in parallel.

method introduced by Carrel [9]. The procedure is simply to determine first the matrix equation for the antenna circuit shown in Fig. 6.31a, then the matrix equation for the transmission-line circuit shown in Fig. 6.31b, and finally the matrix equation for the two circuits in parallel. Note that in Fig. 6.31, a generator is connected across each of the N terminals.

The matrix equation for the antenna circuit in Fig. 6.31a has already been given in (6.149). The elements of the admittance matrix $[Y_A]$ are the self- and mutual admittances of the antenna array.

The matrix equation for the transmission-line circuit in Fig. 6.31b is readily derived. Consider a typical section of the line between the terminal pairs i and $i + 1$ which are separated by a length of line $b_{i,i+1}$ as shown in Fig. 6.32. The relations between the current and voltage at terminals i and those at terminals $i + 1$ are readily obtained.[7] For temporary convenience let $d_i = b_{i,i+1}$; also let ϕ be any constant

---

[7] [11], p. 83, equations (6) and (7).

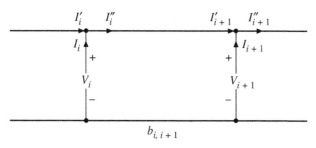

Figure 6.32 Section of transmission line between the terminal pairs i and $i+1$ when the voltages V_i and V_{i+1} are maintained.

phase-shift introduced between adjacent elements in addition to the value $\beta_0 d_i$ which is determined by the length of line between elements i and $i+1$.

$$V_i = V_{i+1}\cos(\beta_0 d_i + \phi) + jI'_{i+1}R_c \sin(\beta_0 d_i + \phi) \tag{6.154a}$$

$$I''_i R_c = jV_{i+1}\sin(\beta_0 d_i + \phi) + I'_{i+1}R_c \cos(\beta_0 d_i + \phi), \tag{6.154b}$$

where R_c is the characteristic resistance of the lossless line.

These equations can be rearranged in the form

$$I''_i = -jG_c[V_i \cot(\beta_0 d_i + \phi) - V_{i+1}\csc(\beta_0 d_i + \phi)] \tag{6.155a}$$

$$I'_{i+1} = -jG_c[V_i \csc(\beta_0 d_i + \phi) - V_{i+1}\cot(\beta_0 d_i + \phi)], \tag{6.155b}$$

where $G_c = R_c^{-1}$ is the characteristic conductance of the lossless line. It follows that

$$I''_{i+1} = -jG_c[V_{i+1}\cot(\beta_0 d_{i+1} + \phi) - V_{i+2}\csc(\beta_0 d_{i+1} + \phi)] \tag{6.156a}$$

$$I'_{i+2} = -jG_c[V_{i+1}\csc(\beta_0 d_{i+1} + \phi) - V_{i+2}\cot(\beta_0 d_{i+1} + \phi)]. \tag{6.156b}$$

The total current in the generator at the terminals $i+1$ is

$$I_{i+1} = I''_{i+1} - I'_{i+1} = jG_c\{V_i \csc(\beta_0 d_i + \phi) - V_{i+1}[\cot(\beta_0 d_i + \phi)$$

$$+ \cot(\beta_0 d_{i+1} + \phi)] + V_{i+2}\csc(\beta_0 d_{i+1} + \phi)\}. \tag{6.157}$$

In particular, when $\phi = \pi$ as in Fig. 6.31b,

$$I_{i+1} = -jG_c\{V_i \csc\beta_0 b_{i,i+1}$$

$$+ V_{i+1}(\cot\beta_0 b_{i,i+1} + \cot\beta_0 b_{i+1,i+2}) + V_{i+2}\csc\beta_0 b_{i+1,i+2}\}. \tag{6.158}$$

Also

$$I_1 = I''_1 = -jG_c[V_1 \cot\beta_0 b_{12} + V_2 \csc\beta_0 b_{12}] \tag{6.159}$$

and

$$I_N = -jG_c[V_{N-1} \csc \beta_0 b_{N-1,N} + V_N(\cot \beta_0 b_{N-1,N} + jy_N)] \qquad (6.160a)$$

since

$$I_N'' = V_N Y_N = V_N y_N G_c, \qquad (6.160b)$$

where

$$y_N = Y_N/G_c = \left[\frac{Y_T + jG_c \tan \beta_0 b_T}{G_c + jY_T \tan \beta_0 b_T} \right] \qquad (6.161)$$

is the normalized admittance in parallel with element N. $Y_T = 1/Z_T$ is the admittance terminating the final section of line of length $b_T = b_{N,N+1}/2$.

With (6.158), (6.159) and (6.160), the matrix equation for the transmission line has the form

$${I} = [Y_L]{V}, \qquad (6.162)$$

where

$${I} = \left\{ \begin{array}{c} I_1 \\ I_2 \\ \vdots \\ I_N \end{array} \right\}, \qquad {V} = \left\{ \begin{array}{c} V_1 \\ V_2 \\ \vdots \\ V_N \end{array} \right\} \qquad (6.163)$$

and

$$[Y_L] = -jG_c$$

$$\times \left[\begin{array}{ccccc} \cot \beta_0 b_{12} & \csc \beta_0 b_{12} & 0 & 0 & \cdots \\ \csc \beta_0 b_{12} & (\cot \beta_0 b_{12} + \cot \beta_0 b_{23}) & \csc \beta_0 b_{23} & 0 & \cdots \\ 0 & \csc \beta_0 b_{23} & (\cot \beta_0 b_{23} + \cot \beta_0 b_{34}) & \csc \beta_0 b_{34} & \cdots \\ \vdots & \vdots & \vdots & \vdots & \\ 0 & 0 & 0 & 0 & \cdots \\ 0 & 0 & 0 & 0 & \cdots \end{array} \right.$$

$$\left. \begin{array}{c} 0 \\ 0 \\ 0 \\ \vdots \\ \csc \beta_0 b_{N-2,N-1} \quad (\cot \beta_0 b_{N-2,N-1} + \cot \beta_0 b_{N-1,N}) \quad \csc \beta_0 b_{N-1,N} \\ 0 \qquad\qquad\qquad \csc \beta_0 b_{N-1,N} \qquad\qquad (\cot \beta_0 b_{N-1,N} + jy_N) \end{array} \right]$$

$$(6.164)$$

The final step in the analysis of the array in Fig. 6.30b is to connect the transmission-line circuit in Fig. 6.31b in parallel with the antenna circuit in Fig. 6.31a as shown schematically in Fig. 6.31c. The same driving voltages are maintained across the N input terminals. Let the total currents in the generators be represented by $I_{ti} = I_{zi}(0) + I_i$ where $I_{zi}(0)$ is the current entering antenna i and I_i is the current into the transmission line at terminals i. The matrix equation for the total current is

$$\{I_t\} = ([Y_A] + [Y_L])\{V_0\} = [Y]\{V_0\}. \tag{6.165}$$

This gives the N currents supplied by N generators connected across the N sets of terminals in Fig. 6.31c. In the actual circuit in Fig. 6.30b, there is only one generator, V_{01}, and all of the total currents I_{ti} are zero except I_{t1}. Hence, in (6.165)

$$\{I_t\} = \begin{Bmatrix} I_{t1} \\ 0 \\ 0 \\ \vdots \\ 0 \end{Bmatrix}, \qquad \{V_0\} = \begin{Bmatrix} V_{01} \\ \vdots \\ V_{0N} \end{Bmatrix} \tag{6.166}$$

$$[Y] = [Y_A] + [Y_L]. \tag{6.167}$$

The voltages V_{0i} driving the N elements are, therefore, given by

$$\{V_0\} = [Y]^{-1}\{I_t\} \tag{6.168}$$

in terms of the total current I_{t1}. The driving-point admittances of the N elements can be determined as follows. The substitution of

$$\{V_0\} = [Y_A]^{-1}\{I_z(0)\} \tag{6.169}$$

in (6.165) yields

$$\{I_t\} = [U + [Y_L][Y_A]^{-1}]\{I_z(0)\}, \tag{6.170}$$

where U is the unit matrix. Note that $[Z_A] = [Y_A]^{-1}$ is the impedance matrix of the array. The equation (6.170) can be solved for the driving-point currents of the several elements in terms of the driving-point current in element 1. Thus,

$$\{I_z(0)\} = [U + [Y_L][Y_A]^{-1}]^{-1}\{I_t\}. \tag{6.171}$$

These currents with a common phase and amplitude reference value are convenient for calculating the field pattern and for comparing relative amplitudes. The admittances of the N elements are

$$Y_{0i} = G_{0i} + jB_{0i} = I_{zi}(0)/V_{0i}, \qquad i = 1, 2, \ldots N, \tag{6.172}$$

where V_{0i} and $I_{zi}(0)$ are given, respectively, by (6.168) and (6.171). The driving-point admittance of the array at the terminals $i = 1$ of the first element is

$$Y_1 = G_1 + jB_1 \doteq I_{t1}/V_{01}. \tag{6.173}$$

6.14   Characteristics of a typical log-periodic dipole array[8]

A complete determination of the properties of the log-periodic dipole array involves a systematic study in which the several parameters that characterize its operation are varied progressively over adequately wide ranges. These include the degree of taper of the array ($\tau = h_i / h_{i+1}$), the relative spacing of the elements ($\sigma = h_{i+1}/b_{i,i+1}$), the relative thickness of the elements ($\Omega = 2\ln(2h_i/a_i)$), the total number of elements N, the normalized admittance ($y_T = Y_T R_c$) terminating the transmission line beyond the Nth element, and the phase shift ϕ introduced between successive elements in addition to that specified by the electrical distance $\beta_0 b_{i,i+1}$ between adjacent elements. Such an investigation could also make use of optimization procedures for the forward gain, front-to-back ratio, band width, and other properties of practical interest in a manner similar to that used earlier in this chapter for the Yagi–Uda array. Use of the formulation of Sections 6.11–6.13, which takes full account of the coupling among all elements in determining the different distributions of current and the individual driving-point admittances, should lead to results of considerable quantitative accuracy to supplement those of earlier, more approximate investigations [7–10]. A complete analysis of a typical log-periodic dipole array has been made by Cheong [6] with a high-speed computer. The parameters for this array are $\tau = 0.93$, $\sigma = 0.70$, $\Omega = 11.4$, $N = 12$, $Y_T R_c = 1$, and $\phi = \pi$. The results obtained serve admirably to illustrate both the detailed operation of the log-periodic dipole array and the power of the theory.

Consider first the operation of the array at a frequency[9] such that an element k near its center is a half wavelength long. At this frequency the admittances of the 12 elements when individually isolated lie on a curve in the complex admittance plane that is very nearly an arc of a circle that extends on both sides of the axis $B_0 = 0$ as shown in Fig. 6.33. Note that element 7 is nearest to resonance with only a small negative susceptance. The actual admittances $Y_{0i} = G_{0i} + jB_{0i}$ of the same elements when driven as parts of the log-periodic array lie on a curve that departs significantly from the circle for the isolated admittances.[10] It is roughly circular for the group of elements from No. 3 to No. 9, but the circle has a much greater radius than that for the isolated elements. Indeed, it is so great that the conductances of a number of elements (Nos. 2 and 3) are negative. This large difference in the driving-point admittances is due to coupling; it indicates a strong interaction between the currents in this group of elements. Note that element 7 is still very nearly resonant. Since the admittance curve near its ends bends inward and comes quite close to the circle for the isolated elements, it must be concluded that the elements near the ends of the array behave much as if

[8] This section is based on Chapter 9 of [6]. Parts of Sections 6.14–6.16 were first published in Radio Science [12].

[9] Designated as f_{14} in a notation described in Section 6.15.

[10] Note that only the plotted points are physically meaningful; the continuous curve serves only to guide the eye.

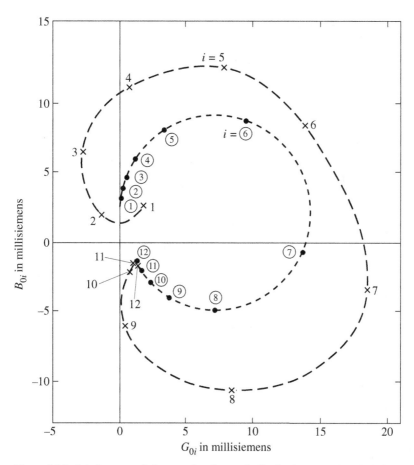

Figure 6.33 Admittances of elements in a log-periodic dipole array when individually isolated and when in an array with $\tau = 0.93$, $\sigma = 0.7$, $\Omega = 11.4$, $Y_T R_c = 1$, $\phi = \pi$ (operating frequency f_{14}). •, Isolated admittances; ×, admittances in array.

they were individually isolated. This is possible only if their currents are relatively small and contribute little to the properties of the array.

In Fig. 6.34 are shown the magnitudes and relative phase angles[11] of the complex voltages V_{0i} that obtain across the input terminals of the elements in the array. The amplitudes are fairly constant for the shorter capacitive elements but they decrease rapidly as soon as the elements are long enough to pass through resonance and become inductive. The phase of the voltages is seen to shift continuously from element to element along the line. Corresponding curves for the driving-point currents $I_{zi}(0)$ are also in Fig. 6.34. Note particularly that elements 4, 5 and 6 all carry larger currents than element 7 which is nearest resonance. Note also that the phase curve for the current crosses that for the voltage at resonance. The shorter elements have leading (capacitive) currents, the longer elements lagging (inductive) currents. The relative

---

[11] Note that only the plotted points are physically meaningful; the continuous curve serves only to guide the eye.

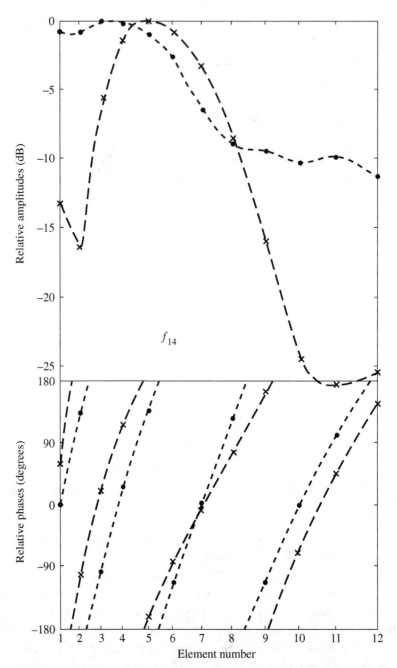

Figure 6.34 Relative amplitudes and phases of driving-point currents and voltages for a log-periodic dipole array; $\tau = 0.93$, $\sigma = 0.7$, $\Omega = 11.4$, $Y_T R_c = 1$, $\phi = \pi$. •, Driving-point voltages, V_{0i}; ×, driving-point currents, $I_{zi}(0)$.

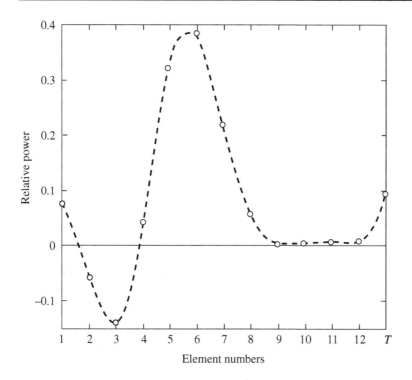

Figure 6.35 Relative power in the 12 elements and the termination. (Operating frequency f_{14}.)

powers[12] P_{0i} in each element and in the termination are given in Fig. 6.35. Note that in the elements 2 and 3, which have a negative input conductance, the power is negative. This means that power is transferred from the other elements to Nos. 2 and 3 by radiation coupling and then from these back to the feeder. The small rise in voltage shown in Fig. 6.34 at elements 3 and 4 may be ascribed to elements 2 and 3 acting as generators and not as loads. It is significant that the maximum power per element is not in the resonant element 7 but in the shorter elements 5 and 6 which also have larger currents. This is a consequence of the very much smaller voltage maintained across the terminals of element 7 as compared with the voltages across the terminals of elements 5 and 6.

The roles played by the several elements in the array may be seen most clearly from their currents. The distributions of current $I_{zi}(z)$ along all 12 elements are shown in Fig. 6.36a referred to the driving voltage V_{01} at the input terminals of the array. Note these distributions differ greatly from element to element – they are not simple sinusoids. The quantity $I_{zi}(z)/V_{01}$ is represented in its real and imaginary parts; it provides the relative currents that together maintain the electromagnetic field. It is seen that (as predicted from the admittance curves in Fig. 6.33) the currents in the

---

[12] Note that only the plotted points are physically significant. The continuous curve serves merely to guide the eye.

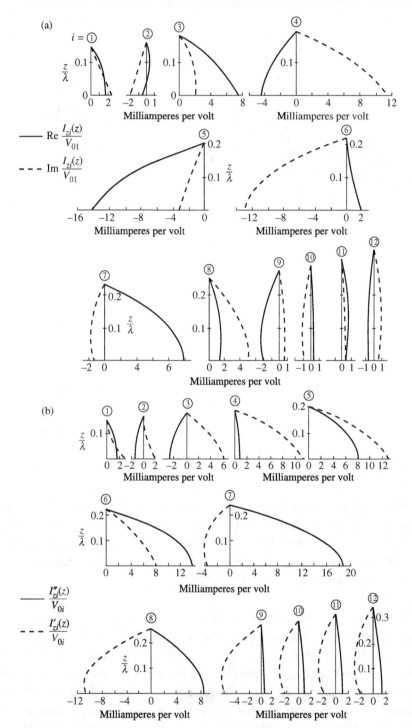

Figure 6.36 Normalized currents in the elements of a log-periodic dipole array; $\tau = 0.93$, $\sigma = 0.7$, $\Omega = 11.4$, $Y_T R_c = 1$, $\phi = \pi$. (a) $I_{zi}(z)/V_{01}$. (b) $I_{zi}(z)/V_{0i} = [I''_{zi}(z) + j I'_{zi}(z)]/V_{0i}$; $I_{zi}(0)/V_{0i} = Y_{0i} = G_{0i} + j B_{0i}$.

outer elements 1, 2, 9, 10, 11, 12 are extremely small so that their contributions are negligible. Clearly, the distant electromagnetic field is determined essentially by the currents in elements 3 to 8 and of these elements 4, 5, 6 and 7 predominate. Note in particular that the currents in the shorter-than-resonant elements 4, 5 and 6 actually exceed the current in the practically resonant element 7.

The current distributions are also shown in Fig. 6.36b but each current is now referred to its own driving voltage. Thus, the quantities represented are $I_{zi}(z)/V_{0i} = [I''_{zi}(z) + jI'_{zi}(z)]/V_{0i}$ where $I''_{zi}(z)$ is the component in phase with V_{0i}, $I'_{zi}(z)$ the component in phase quadrature. Note that $I_{zi}(0)/V_{0i} = Y_{0i}$ so that $I''_{zi}(0)/V_{0i} = G_{0i}$ and $I'_{zi}(0)/V_{0i} = B_{0i}$. The power in antenna i is $P_{0i} = |V_{0i}|^2 G_{0i}$, but since the value of V_{0i} differs greatly from element to element as seen in Fig. 6.34, the relative powers in the several elements are not proportional simply to the real parts of the currents $I''_{zi}(0)$ in the terminals. However, the distributions in Fig. 6.36b are instructive since they show the negative real parts for elements 2 and 3 that transfer power to the feeding line. They also show that the imaginary parts of the currents in elements 1 to 6 are capacitive, those in elements 7 to 12 inductive. This means that each of the elements 1 to 6 acts as a director for the elements to its right, whereas each of the elements 7 to 12 acts as a reflector for all elements to its left. Actually, the capacitive components of current in elements 3, 4 and 5 exceed the conductive components so that relatively little power is supplied to them from the line, and they behave substantially like parasitic directors. The inductive component of current predominates in elements 8 to 12 and these act in major part like parasitic reflectors. However, since the amplitudes of the currents in elements 9 to 12 are quite small, it is clear that the principal reflector action comes from element 8. In summary, Figs. 6.36a, b indicate that of the 12 elements numbers 1, 2, 9, 10, 11 and 12 may be ignored since their currents are small; elements 5, 6, 7 are supplied most of the power from the feeder and behave primarily like driven antennas in an endfire array; elements 3 and 4 act predominantly like parasitic directors; and element 8 is essentially a parasitic reflector. Thus, the log-periodic antenna is very much like a somewhat generalized Yagi–Uda array when driven at a frequency for which the antenna closest to resonance is not too near the ends and the array is long enough to include relatively inactive elements at each end. A lengthening of the array by the addition of one or two or even a great many more elements at either end or at both ends cannot significantly modify the circuit or field properties of the array at the particular frequency since these are determined by the active group.

The normalized far-field pattern in the equatorial or H-plane (variable Φ with $\Theta = \pi/2$) is shown in Fig. 6.37. Note the smoothness of the pattern and the very small minor lobes. As is to be expected this low minor-lobe level is achieved at the expense of the beam width. A comparison with the field pattern in Fig. 6.28 for a 10-element Yagi–Uda array shows that the latter has larger minor lobes but a much narrower beam. However, the Yagi–Uda array does not have the important frequency-independent properties of the log-periodic dipole array.

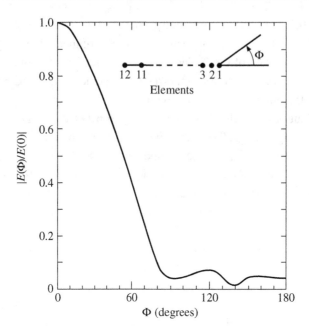

Figure 6.37 Normalized far field of log-periodic array with currents shown in Fig. 6.36a.

6.15   Frequency-independent properties of the log-periodic dipole array

The principle underlying the properties of the log-periodic dipole array when driven at the terminals of the shortest element as shown in Fig. 6.30b and operated as illustrated in the preceding section depends upon the following:

1. A small group of about seven dipoles constitutes the active or radiating part of the array. These may be described approximately as including: (a) three strongly driven and radiating elements near resonance; (b) three shorter elements each of which combines the functions of a rather weakly driven antenna and a highly active parasitic director; and (c) one longer antenna that acts both as a weakly driven element and a strong parasitic reflector.
2. All other elements in the array and the terminating admittance Y_T have such small currents and so little power that they may be ignored both as loads on the feeding line and as contributing radiators of the far-zone field.
3. The driving-point admittance of the array at the terminals of the shortest element is approximately equal to the characteristic conductance G_c of the transmission line.
4. The currents in the active elements maintain a unilateral endfire field pattern with very small minor lobes.

The effect of a change in frequency is to shift the active group toward the terminated end with longer elements when the frequency is lowered. As long as the frequency

Table 6.3. *Relation between the relative heights of the elements, h/λ, and the frequencies f_i.*

i in f_i	h_1/λ	h_{12}/λ	f_i when $h_i = 1$ m
1	0.0962	0.2138	28.86 MHz
2	0.0998	0.2217	29.94
3	0.1035	0.2299	31.05
4	0.1073	0.2384	32.19
5	0.1113	0.2473	33.39
6	0.1154	0.2564	34.62
7	0.1197	0.2659	35.91
8	0.1241	0.2757	37.23
9	0.1287	0.2859	38.61
10	0.1335	0.2966	40.05
11	0.1384	0.3075	41.52
12	0.1435	0.3188	43.05
13	0.1488	0.3306	44.64
14	0.1543	0.3428	46.29
15	0.1600	0.3555	48.00
16	0.1659	0.3686	49.77
17	0.1721	0.3824	51.63
18	0.1785	0.3966	53.55
19	0.1850	0.4110	55.50
20	0.1918	0.4261	57.54
21	0.1989	0.4419	59.67
22	0.2063	0.4583	61.89
23	0.2139	0.4752	64.17
24	0.2218	0.4928	66.54
25	0.2300	0.5110	69.00
26	0.2385	0.5299	71.55
27	0.2473	0.5494	74.19

range is bounded so that neither the shortest nor the longest element in the array is a part of the active group, there can be no significant change in either the circuit or the field properties. The array must behave substantially as if infinitely long. On the other hand, as the frequency is increased or decreased sufficiently to make the element at either end of the array a member of the active group, all of the properties of the array must begin to change. This change becomes drastic when the frequency is varied so much that none of the N elements is near resonance.

The general behavior of the 12-element log-periodic dipole array as a function of frequency has been investigated by Cheong [6] using a discrete set of frequencies f_i, $i = 1, \ldots, 27$. These are chosen so that the lowest frequency f_1 is below the resonant frequency of the longest element No. 12 and the highest frequency f_{27} is above the resonant value for the shortest element No. 1 as shown in Table 6.3. In

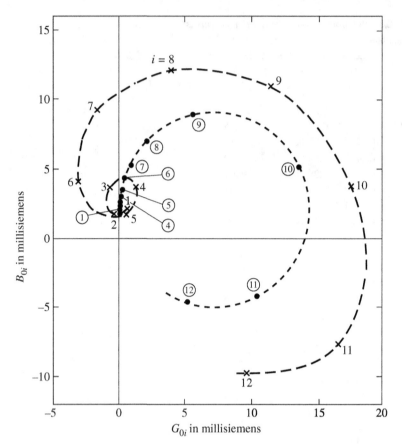

Figure 6.38 Like Fig. 6.33 but for lower frequency with resonance near element 10. (Operating frequency f_7.) •, Isolated admittances; ×, admittances in array.

order to distribute the frequencies according to the log-periodic scheme of lengths and spacings, the ratio factor $\sqrt{0.93}$ was chosen so that $f_{j+2}/f_j = 0.93$ where j is an integer. This provides an intermediate frequency step $f_{j+1}/f_j = \sqrt{0.93}$ to achieve a closer approximation of a continuous spectrum. The properties of the array described in the preceding section and represented in Figs. 6.33–6.37 are obtained specifically at the center frequency f_{14} in this set for which an element (No. 7) near the middle of the array is most nearly resonant.

Consider first a decrease in frequency from f_{14} to f_7 so that resonance is moved from approximately element 7 to approximately element 10. The corresponding driving-point admittances are shown in the complex admittance plane in Fig. 6.38 together with the admittances of the elements when these are individually isolated. The admittance circle for the isolated antennas and the admittance curve[13] for the array resemble those in Fig. 6.33 but appear to have been moved in a counter-clockwise

---

[13] Note that only the plotted points are physically significant. The continuous curve serves merely to guide the eye.

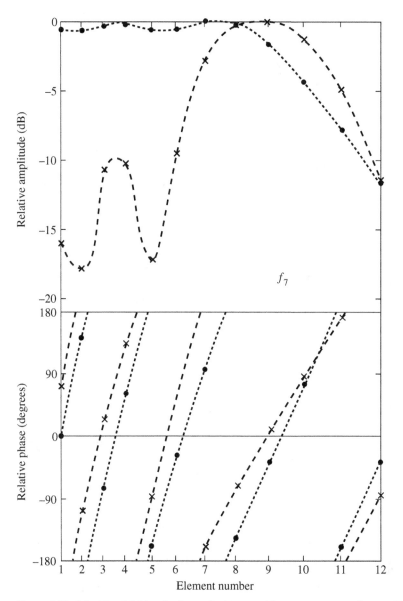

Figure 6.39 Like Fig. 6.34 but for lower frequency with resonance near element 10. •, Driving-point voltages, V_{0i}; ×, driving-point currents, $I_{zi}(0)$.

direction. The admittances of the short elements from 1 to 6 now form a small spiral around the values for the same elements when isolated. The previous tight little spiral of admittances for the longer elements in Fig. 6.33 is completely unwound and the admittance curve for the array no longer comes near to the circle for the admittances of the isolated elements. It is clear that in Fig. 6.38 elements 6 to 12 instead of 3 to 9 as in Fig. 6.33 form the active group. This is further confirmed in Fig. 6.39

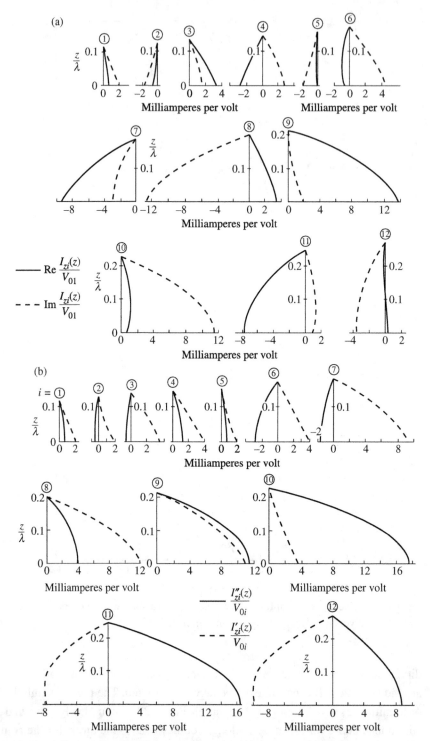

Figure 6.40 (a) Like Fig. 6.36a but for lower frequency with resonance near element 10. (b) Like Fig. 6.36b but for lower frequency with resonance near element 10.

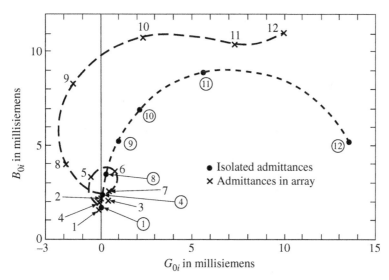

Figure 6.41 Like Fig. 6.38 but for lower frequency with resonance beyond element 12. (Operating frequency f_3.)

which shows the voltages and currents at the driving points of the elements. The voltage amplitudes are quite constant from elements 1 to 8, then decrease rapidly. The associated current amplitudes are small for elements 1 to 6, large for elements 7 to 11 and again small for element 12. Evidently, with reference to Fig. 6.39 (and Fig. 6.34), the group consisting of director-radiators 6, 7 and 8 (instead of 3, 4, 5), radiators 9, 10, 11 (instead of 6, 7, 8) and reflector-radiator 12 (instead of 9) is primarily responsible for the properties of the array. These conclusions may also be reached from a study of the current-distribution curves for $I_{zi}(z)/V_{01}$ in Fig. 6.40a and for $I_{zi}(z)/V_{0i}$ in Fig. 6.40b. The former show clearly that the amplitudes of the currents in elements 1 through 5 are negligibly small. The latter indicate the following: the capacitive currents dominate in elements 6, 7 and 8, in element 9 the capacitive and conductive currents are practically equal, element 10 is nearly resonant with a very small capacitive current, element 11 has large inductive and conductive components, and in element 12 the inductive current exceeds the conductive component. It may be concluded, therefore, that the decrease in frequency which moved resonance from near element 7 to near element 10 has not significantly changed the properties of the active group and, hence, of the array.

If the frequency is decreased still further to f_3 at which even element No. 12 is too short to be resonant, the admittance curve is that shown in Fig. 6.41. The counter-clockwise rotation of the curves has been increased beyond that in Fig. 6.38 so that now none of the elements is either inductive or resonant. The small spiral formed by the admittances of the short elements around the circle of their isolated values has two complete turns. It is to be expected, therefore, that elements 1 through

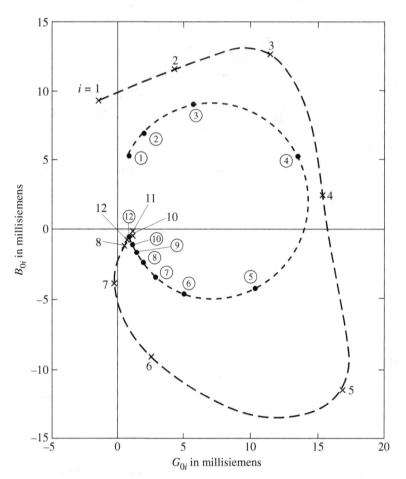

Figure 6.42 Like Fig. 6.33 but for higher frequency with resonance near element 4. (Operating frequency f_{19}.) •, Isolated admittances; ×, admittances in array.

7 must have negligible currents. The active group in Fig. 6.41 includes dipoles 8 to 12. However, none of these is resonant and there are no inductive reflectors. Moreover, since there must be a significant voltage across the terminals of element No. 12, considerable power must be dissipated in the terminating admittance Y_T. Under these conditions the properties of the array must differ significantly from those existing for the frequencies determining Figs. 6.33 and 6.38. The frequency-independent behavior requires at least two radiating and reflecting elements longer than the one nearest resonance.

If the frequency is increased to f_{19} so that element No. 4 is most nearly resonant, the admittance curve takes the form shown in Fig. 6.42. As compared with Fig. 6.33, the curves have been rotated clockwise with respect to the axis $B_0 = 0$. The admittances of the longer elements Nos. 8 through 12 in the array are all clustered close to one end of the circular arc formed by the admittances of the isolated elements. On the

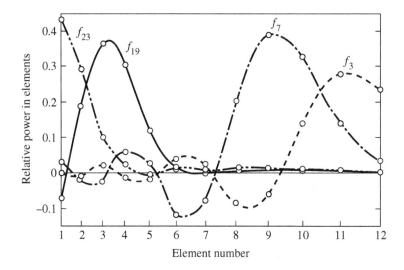

Figure 6.43 Like Fig. 6.35 but for frequencies f_3, f_7, f_{19}, and f_{23}.

other hand, not even the shortest element No. 1 is near the other end of the circular arc. Since a detailed study (in conjunction with Figs. 6.33 and 6.36a) of the currents and power in the elements longer than resonance has shown that at most two elements longer than the one nearest resonance carry significant currents, it follows that all elements from No. 12 down through No. 7 play no significant role in the array. On the other hand, it is clear from Fig. 6.42 that the admittance of element No. 1 does not produce a curve that bends inward toward the circular arc of isolated admittances, but rather outward away from the arc. This is a consequence of the fact that the region of active elements has been moved too close to the end of the array. It is clear from Fig. 6.36a that the active region includes at least four elements shorter than the one nearest resonance. For the frequency f_{19} leading to Fig. 6.42 there are only three such elements available. This means that the frequency responsible for Fig. 6.42 is already somewhat higher than acceptable for the frequency-independent properties of the array and that the currents in element No. 1 must differ from the expected since one of the required director-radiators is missing.

The useful range for a frequency-independent behavior lies between the frequencies at which elements 5 and 10 (or, in general, $N - 2$) are resonant. In the scale of discrete frequencies used for the 12-element array this range is approximately $f_7 \leq f \leq f_{17}$. The power in the several elements at the frequencies f_3, f_7, f_{14}, f_{19} and f_{23} is shown in Figs. 6.35 and 6.43. Note that in these figures only the plotted points are significant. The connecting curves serve only to guide the eye.

A detailed study of the operation of the 12-element array over the full range of frequencies from f_1 to f_{27} has been made by Cheong [6]. Important results in addition to those already discussed are contained in Figs. 6.44–6.46. They may be summarized as follows:

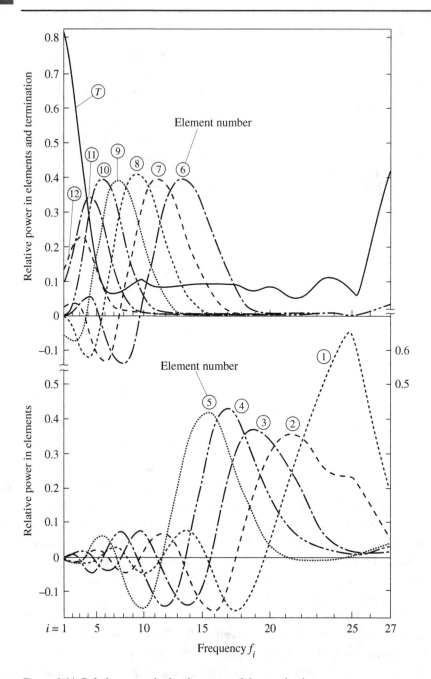

Figure 6.44 Relative power in the elements and the termination.

1. As shown in Fig. 6.44, curve T, a large fraction of the total power is dissipated in the terminating admittance $Y_T = G_c$, in the ranges $f < f_5$ and $f > f_{26}$. As a consequence only a small fraction of power appears in the dipoles so that little is radiated. It is also clear from Fig. 6.44 that in the range $f_5 \leq f \leq f_{26}$ only a

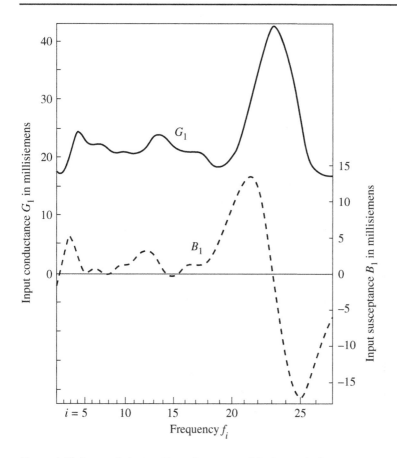

Figure 6.45 Input admittance $Y_1 = G_1 + jB_1$ of the log-periodic array.

small part of the power is dissipated in the terminating admittance, most of it appears in and is radiated from a relatively small group of active dipoles near resonance.

2. In the range $f_5 \leq f \leq f_{17}$ elements which have half-lengths h_i in the range $0.18 \leq h_i/\lambda \leq 0.255$ form the active group. Resonance occurs with $h_i/\lambda \doteq 0.216$. Elements which have half-lengths h_i less than 0.18λ or greater than 0.255λ play an insignificant part in the operation of the array. On the other hand, outside this range of frequencies the shorter and longer elements cannot be ignored.

3. As shown in Fig. 6.45 the driving-point admittance of the array, Y_1, is reasonably constant at a value very near the characteristic conductance G_c of the transmission line over the range $f_5 \leq f \leq f_{17}$. Specifically $Y_1 \doteq (23.0 + j0.0) \times 10^{-3}$ siemens with $G_c = 20 \times 10^{-3}$ siemens. Outside this range of frequencies Y_1 varies widely in both real and imaginary parts.

4. The band of frequencies $f_5 \leq f \leq f_{17}$ is characterized by a very stable main lobe in the forward direction, i.e. toward the shorter elements and the driving point,

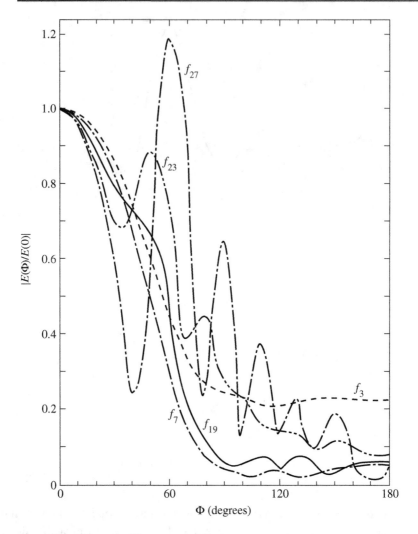

Figure 6.46 Like Fig. 6.37 but for a number of different frequencies.

and very small side and back lobes. This is clear from Fig. 6.37 and Fig. 6.46. Figure 6.47 shows that the ratio of the forward field to the largest side- or back-lobe level is roughly constant near 15 and that the 3 dB forward beam width remains quite stable at about 38° in the range $f_5 \leq f \leq f_{17}$. Outside this band of frequencies large side and back lobes appear.

It is important to note that all of the computed data apply to a particular array with a single set of values of the basic parameters τ, σ, Ω, Y_T, R_c and ϕ. A numerical study of the effects of changes in these parameters and of optimum designs based on the three-term theory can readily be made if required. Additional information is given in Cheong [6], Cheong and King [12], and Carrel [9].

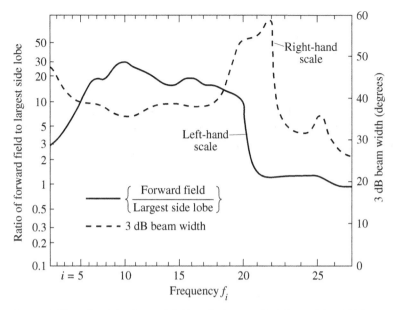

Figure 6.47 Ratio of the forward field to the largest side lobe and the 3 dB beam width over the frequency range f_1 to f_{27}.

6.16 Experimental verification of the theory for arrays of unequal dipoles

In order to verify experimentally the predictions of the general theory developed in Section 6.11 for arrays of dipoles with a wide range of lengths and spacings, a series of measurements on the 12-element log-periodic dipole array would be appropriate. However, arrays of this type are driven from two-wire lines in a manner that makes accurate measurements of current distributions, admittances, voltages and field patterns very difficult – especially over a two-to-one or greater range of frequencies. For this reason a less elaborate array arranged to permit precision measurements was preferred by Cheong [6].

As a first step, an extensive experimental study was made of two coupled dipoles over wide ranges of lengths and spacings in order to verify the adequacy of the three-term representation of the currents. When this had been established, a complete array of five elements was constructed after the log-periodic design with the longest element approximately twice as long as the shortest element. This array consisted of monopoles over a very large ground screen. Each element was the extension of the inner conductor of a coaxial line of which the outer conductor pierced the metal ground screen. In order to provide an equivalent for the reversal of the connections between adjacent pairs of elements, provision was made to permit the insertion of an arbitrary length of coaxial line in addition to a length equal to the spacing of the elements. Since the added phase shift had to be exactly π for each different frequency, it was necessary to readjust the length of the sections of coaxial line between the elements for each frequency. Careful measurements were made of the driving-point admittance, the currents and voltages in amplitude and phase at the base of each element, and the field pattern over a range of some 17 different frequencies that included resonance for the longest and the shortest elements. The agreement between theory and measurement was remarkable in all details, thus confirming the adequacy of the theory for use not only on the five-element array but on an array of any type that satisfies the requirements of the theory. Details and extensive graphs are in the work of Cheong [6] and Cheong and King [13].

7   Planar and three-dimensional arrays

The study of dipole arrays in Chapters 3 through 6 has proceeded from simpler to more complicated configurations. In Chapters 3 and 4 all elements are physically alike and arranged to be parallel with their centers uniformly spaced around a circle so that when driven in suitable phase sequences all elements are geometrically and electrically identical. Chapter 5 is also concerned with parallel elements that are structurally alike, but they lie in a curtain with their centers along a straight line of finite length; consequently the electromagnetic environments of the several elements are not all the same. In Chapter 6 the requirement that the elements in a curtain array be equal in length is omitted and consideration is given first to arrays of elements that differ only moderately in length, then to arrays in which not only the lengths but also the radii of the elements and the distances between them vary widely. The lifting of each restriction introduces additional complications in the approximate representation of the currents on the elements by simple trigonometric functions and in the reduction of the integrals in the simultaneous integral equations to sums of such functions with suitably defined complex coefficients.

The final generalization, which is carried out in this chapter, is the omission of the requirement maintained throughout the book until this point, that all elements be non-staggered. The removal of this condition leads to the discussion of arrays of parallel elements that are arranged in a plane as in Fig. 7.1 and in three dimensions as shown in Fig. 7.2. Note that such arrays include arbitrarily staggered elements and collinear elements which do not occur in the circular and curtain arrays considered in Chapters 3 through 6. When the centers of the elements are displaced from a common plane, the halves of many antennas are in different electrical environments so that an even symmetry with respect to their individual centers no longer obtains for the distributions of current. An important new complication is thus introduced: components of current with odd symmetries in addition to those with even symmetries.

7.1   Vector potentials and integral equations for the currents

Four typical elements in an array of N parallel dipoles are shown in Fig. 7.3. All antennas have their axes parallel to the Z-axis of a system of rectangular coordinates X, Y, Z. The center of the kth element is at X_k, Y_k, Z_k; its radius is a_k, its half-length

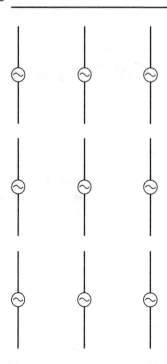

Figure 7.1   Planar array of nine identical elements.

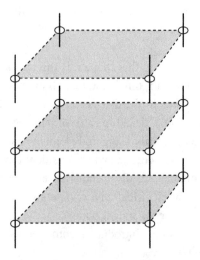

Figure 7.2   Three-dimensional array of 12 identical elements.

h_k, and it is center driven by a delta-function generator with EMF V_{0k}. As before, the antennas are assumed to be perfectly conducting and electrically thin so that $\beta_0 a_k \ll 1$ for $k = 1, 2, \ldots, N$. A local axial coordinate z_k has its origin at the center of element k.

The vector potential on the surface of antenna k no longer has the simple form given in (2.3), since the even symmetry conditions $I_{zk}(-z_k) = I_{zk}(z_k)$ for the current and

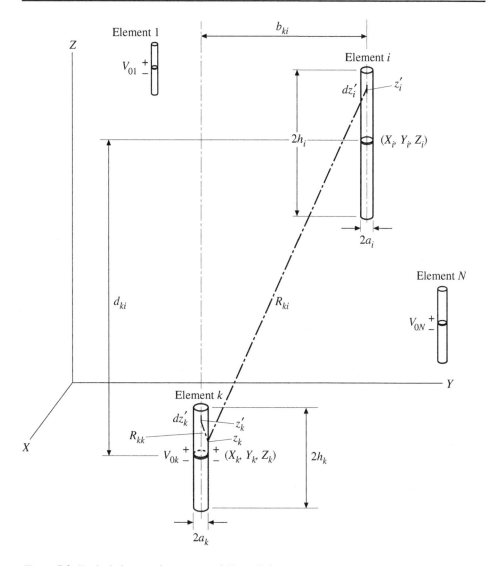

Figure 7.3 Typical elements in an array of N parallel antennas.

$A_{zk}(-z_k) = A_{zk}(z_k)$ for the vector potential no longer apply. However, the vector potential can be resolved into two parts, one with even symmetry, the other with odd symmetry. Thus

$$A_{zk}(z_k) = A_{zk}^{\text{even}}(z_k) + A_{zk}^{\text{odd}}(z_k), \tag{7.1}$$

where, in the range $-h_k \leq z_k \leq h_k$,

$$A_{zk}^{\text{even}}(z_k) = \frac{-j}{c}\left[C_{k1}\cos\beta_0 z_k + \tfrac{1}{2}V_{0k}\sin\beta_0|z_k|\right] \tag{7.2}$$

as in (2.3), and

$$A_{zk}^{\text{odd}}(z_k) = \frac{-j}{c} C_{k2} \sin \beta_0 z_k. \tag{7.3}$$

The vector potential on the surface of antenna k is also given by the sum of integrals,

$$A_{zk}(z_k) = \sum_{i=1}^{N} \frac{\mu_0}{4\pi} \int_{-h_i}^{h_i} I_{zi}(z_i') G_{ki}(d_{ki}, z_k, z_i') \, dz_i', \tag{7.4}$$

where

$$G_{ki}(d_{ki}, z_k, z_i') = \frac{e^{-j\beta_0 R_{ki}}}{R_{ki}} \tag{7.5a}$$

with

$$R_{ki} = \sqrt{(d_{ki} + z_i' - z_k)^2 + b_{ki}^2}. \tag{7.5b}$$

As shown in Fig. 7.3, $d_{ki} = |Z_k - Z_i|$ is the axial distance between the transverse planes containing the centers of elements k and i, $d_{kk} = 0$; $b_{ki} = \sqrt{(X_k - X_i)^2 + (Y_k - Y_i)^2}$, $i \neq k$, is the distance between the center of element k and the projection of the center of element i onto the plane $z_k = 0$; $b_{kk} = a_k$. The currents $I_{zi}(z_i)$ in the N elements that generate the vector potential on the surface of antenna k as given in (7.4) include even and odd parts with respect to the centers of the respective elements. That is,

$$I_{zi}(z_i) = I_{zi}^{\text{even}}(z_i) + I_{zi}^{\text{odd}}(z_i), \tag{7.6}$$

where

$$I_{zi}^{\text{even}}(z_i) = \tfrac{1}{2}[I_{zi}(z_i) + I_{zi}(-z_i)], \quad I_{zi}^{\text{odd}}(z_i) = \tfrac{1}{2}[I_{zi}(z_i) - I_{zi}(-z_i)].$$

In order to separate the even and the odd parts of the vector potential in (7.4), the kernel $G_{ki}(d_{ki}, z_k, z_i')$ in (7.4) must be separated into its even and odd parts. Thus,

$$G_{ki}(d_{ki}, z_k, z_i') = G_{ki}^{\text{even}}(d_{ki}, z_k, z_i') + G_{ki}^{\text{odd}}(d_{ki}, z_k, z_i'), \tag{7.7}$$

where, as is readily shown,

$$G_{ki}^{\text{even}}(d_{ki}, z_k, z_i') = \tfrac{1}{2}[K_{ki}(z_k - d_{ki}, z_i') + K_{ki}(z_k + d_{ki}, z_i')] \tag{7.8}$$

$$G_{ki}^{\text{odd}}(d_{ki}, z_k, z_i') = \tfrac{1}{2}[K_{ki}(z_k - d_{ki}, z_i') - K_{ki}(z_k + d_{ki}, z_i')]. \tag{7.9}$$

The function K occurring in (7.8) and (7.9) is the kernel previously used for non-staggered arrays, namely,

$$K_{ki}(z_k, z_i') = K_{kiR}(z_k, z_i') + j K_{kiI}(z_k, z_i') = \frac{e^{-j\beta_0 \sqrt{(z_k - z_i')^2 + b_{ki}^2}}}{\sqrt{(z_k - z_i')^2 + b_{ki}^2}}. \tag{7.10}$$

Note that when $d_{ki} = 0$, $G_{ki}^{\text{even}}(0, z_k, z_i') = K_{ki}(z_k, z_i')$ and $G_{ki}^{\text{odd}}(0, z_k, z_i') = 0$, as required for the previously analyzed non-staggered array. By means of the obvious relation,

$$K_{ki}(z_k, z_i') = K_{ki}(-z_k, -z_i'),$$

it is readily shown that, when (7.6) and (7.7) are substituted in (7.4), the parts of the integral that involve the products $I^{\text{even}}G^{\text{even}}$, $I^{\text{odd}}G^{\text{odd}}$ are themselves even in z, the parts that contain $I^{\text{even}}G^{\text{odd}}$, $I^{\text{odd}}G^{\text{even}}$ are themselves odd in z. It follows that the even part of the vector potential is given by

$$
\begin{aligned}
4\pi \mu_0^{-1} A_{zk}^{\text{even}}(z_k) &= \int_{-h_k}^{h_k} I_{zk}^{\text{even}}(z_k') G_{kk}^{\text{even}}(0, z_k, z_k') \, dz_k' \\
&\quad + \sum_{i=1}^{N} {}' \int_{-h_i}^{h_i} I_{zi}^{\text{even}}(z_i') G_{ki}^{\text{even}}(d_{ki}, z_k, z_i') \, dz_i' \\
&\quad + \sum_{i=1}^{N} {}' \int_{-h_i}^{h_i} I_{zi}^{\text{odd}}(z_i') G_{ki}^{\text{odd}}(d_{ki}, z_k, z_i') \, dz_i' \\
&= \frac{-j4\pi}{\zeta_0} [C_{k1} \cos \beta_0 z_k + \tfrac{1}{2} V_{0k} \sin \beta_0 |z_k|],
\end{aligned}
\tag{7.11}
$$

where $k = 1, 2, \ldots N$; $\zeta_0 \doteq 120\pi$ ohms; and $\sum'$ is the sum with $i = k$ omitted. The odd part of the vector potential is contained in

$$
\begin{aligned}
4\pi \mu_0^{-1} A_{zk}^{\text{odd}}(z_k) &= \int_{-h_k}^{h_k} I_{zk}^{\text{odd}}(z_k') G_{kk}^{\text{even}}(0, z_k, z_k') \, dz_k' \\
&\quad + \sum_{i=1}^{N} {}' \int_{-h_i}^{h_i} I_{zi}^{\text{even}}(z_i') G_{ki}^{\text{odd}}(d_{ki}, z_k, z_i') \, dz_i' \\
&\quad + \sum_{i=1}^{N} {}' \int_{-h_i}^{h_i} I_{zi}^{\text{odd}}(z_i') G_{ki}^{\text{even}}(d_{ki}, z_k, z_i') \, dz_i' \\
&= -(j4\pi/\zeta_0) C_{k2} \sin \beta_0 z_k
\end{aligned}
\tag{7.12}
$$

where $k = 1, 2, \ldots N$.

The relations on the right in (7.11) and (7.12) are $2N$ simultaneous integral equations for the even and odd parts of the currents in the N elements.

7.2   Vector potential differences and integral equations

In order to determine approximate distributions of current from the two sets of N simultaneous integral equations in the general manner described in earlier chapters, it is convenient to introduce the vector potential differences. This is quite straightforward for the even part of $A_{zk}(z_k)$. Thus, if $4\pi\mu_0^{-1}A_{zk}^{\text{even}}(h_k)$ is subtracted from both sides of (7.11), the result is

$$4\pi\mu_0^{-1}[A_{zk}^{\text{even}}(z_k) - A_{zk}^{\text{even}}(h_k)] = \int_{-h_k}^{h_k} I_{zk}^{\text{even}}(z_k')G_{kkd}^{\text{even}}(0, z_k, z_k')\,dz_k'$$

$$+ \sum_{i=1}^{N}{}' \int_{-h_i}^{h_i} I_{zi}^{\text{even}}(z_i')G_{kid}^{\text{even}}(d_{ki}, z_k, z_i')\,dz_i'$$

$$+ \sum_{i=1}^{N}{}' \int_{-h_i}^{h_i} I_{zi}^{\text{odd}}(z_i')G_{kid}^{\text{odd}}(d_{ki}, z_k, z_i')\,dz_i'$$

$$= -(j4\pi/\zeta_0)[\tfrac{1}{2}V_{0k}(\sin\beta_0|z_k| - \sin\beta_0 h_k)$$

$$+ C_{k1}(\cos\beta_0 z_k - \cos\beta_0 h_k)]$$

$$= (j4\pi/\zeta_0\cos\beta_0 h_k)[\tfrac{1}{2}V_{0k}\sin\beta_0(h_k - |z_k|)$$

$$+ U_k(\cos\beta_0 z_k - \cos\beta_0 h_k)], \tag{7.13}$$

where $k = 1, 2, \ldots N$ and

$$U_k = \frac{-j\zeta_0}{\mu_0}A_{zk}^{\text{even}}(h_k), \qquad C_{k1} = \frac{-U_k + \tfrac{1}{2}V_{0k}\sin\beta_0 h_k}{\cos\beta_0 h_k} \tag{7.14}$$

as in the corresponding equation with $d_{ki} = 0$ for the curtain array. The difference kernel (with extra subscript d) is defined by

$$G_{kid}^{\text{even}}(d_{ki}, z_k, z_i') = G_{ki}^{\text{even}}(d_{ki}, z_k, z_i') - G_{ki}^{\text{even}}(d_{ki}, h_k, z_i'), \tag{7.15}$$

and a similar equation for $G_{kid}^{\text{odd}}(d_{ki}, z_k, z_i')$.

It is not possible to form an equation like (7.13) with $A_{zk}^{\text{odd}}(z_k)$ since this is an odd function of z_k so that if $A_{zk}^{\text{odd}}(z_k) - A_{zk}^{\text{odd}}(h_k)$ is zero at $z_k = h_k$, it is $-2A_{zk}^{\text{odd}}(h_k)$ at $z_k = -h_k$. A convenient alternative[1] is to subtract the odd function $(z_k/h_k)A_{zk}^{\text{odd}}(h_k)$ which is equal to the vector potential at both $z_k = h_k$ and $z_k = -h_k$. Thus, with (7.12),

$$4\pi\mu_0^{-1}[A_{zk}^{\text{odd}}(z_k) - (z_k/h_k)A_{zk}^{\text{odd}}(h_k)]$$

$$= \int_{-h_k}^{h_k} I_{zk}^{\text{odd}}(z_k')\mathcal{G}_{kkd}^{\text{even}}(0, z_k, z_k')\,dz_k'$$

$$+ \sum_{i=1}^{N}{}' \int_{-h_i}^{h_i} I_{zi}^{\text{even}}(z_i')\mathcal{G}_{kid}^{\text{odd}}(d_{ki}, z_k, z_i')\,dz_i'$$

[1] See [1].

$$+ \sum_{i=1}^{N}{}' \int_{-h_i}^{h_i} I_{zi}^{\text{odd}}(z_i') \mathcal{G}_{kid}^{\text{even}}(d_{ki}, z_k, z_i') \, dz_i'$$

$$= -(j4\pi C_{k2}/\zeta_0)[\sin \beta_0 z_k - (z_k/h_k) \sin \beta_0 h_k] \tag{7.16}$$

where $k = 1, 2, \ldots N$ and the difference kernels are given by

$$\mathcal{G}_{kid}^{\text{even}}(d_{ki}, z_k, z_i') = G_{ki}^{\text{even}}(d_{ki}, z_k, z_i') - (z_k/h_k)G_{ki}^{\text{even}}(d_{ki}, h_k, z_i') \tag{7.17a}$$

$$\mathcal{G}_{kid}^{\text{odd}}(d_{ki}, z_k, z_i') = G_{ki}^{\text{odd}}(d_{ki}, z_k, z_i') - (z_k/h_k)G_{ki}^{\text{odd}}(d_{ki}, h_k, z_i'). \tag{7.17b}$$

For each superscript, the kernel may be expanded into its real and imaginary parts as follows:

$$\mathcal{G}_{kid}(d_{ki}, z_k, z_i') = \mathcal{G}_{kidR}(d_{ki}, z_k, z_i') + j\mathcal{G}_{kidI}(d_{ki}, z_k, z_i'). \tag{7.18}$$

The desired alternative set of $2N$ simultaneous integral equations for the even and odd parts of the currents in the N elements is contained in (7.13) and (7.16).

7.3   Approximate distribution of current

It has been shown in earlier chapters that the first integral in (7.13) is well approximated by

$$\int_{-h_k}^{h_k} I_{zk}^{\text{even}}(z_k') G_{kkdR}^{\text{even}}(0, z_k, z_k') \, dz_k' = \int_{-h_k}^{h_k} I_{zk}^{\text{even}}(z_k') K_{kkdR}(z_k, z_k') \, dz_k'$$

$$\sim I_{zk}^{\text{even}}(z_k) \tag{7.19}$$

and

$$\int_{-h_k}^{h_k} I_{zk}^{\text{even}}(z_k') G_{kkdI}^{\text{even}}(0, z_k, z_k') \, dz_k' = \int_{-h_k}^{h_k} I_{zk}^{\text{even}}(z_k') K_{kkdI}(z_k, z_k') \, dz_k'$$

$$\sim H_{0zk}, \tag{7.20}$$

where

$$H_{0zk} = \cos(\beta_0 z_k/2) - \cos(\beta_0 h_k/2) \tag{7.21}$$

provided $\beta_0 h_k \leq 5\pi/4$. By the same procedure it is readily shown that the first integral in (7.16) can be separated into analogous parts for which the following relations are good approximations:

$$\int_{-h_k}^{h_k} I_{zk}^{\text{odd}}(z_k') \mathcal{G}_{kkdR}^{\text{even}}(0, z_k, z_k') \, dz_k' \sim I_{zk}^{\text{odd}}(z_k) \tag{7.22}$$

$$\int_{-h_k}^{h_k} I_{zk}^{\text{odd}}(z_k') \mathcal{G}_{kkdI}^{\text{even}}(0, z_k, z_k') \, dz_k' \sim E_{0zk}, \tag{7.23}$$

where

$$E_{0zk} = \sin(\beta_0 z_k/2) - (z_k/h_k)\sin(\beta_0 h_k/2). \tag{7.24}$$

As a consequence of (7.19) and (7.22) it follows that the trigonometric functions that occur on the right-hand side of (7.13) and (7.16) must also occur as leading terms in the approximate expressions for the currents, together with (7.21) and (7.24). That is, appropriate approximate formulas for the even and odd currents in antenna k are given below. For the even currents

$$I_{zk}^{\text{even}}(z_k) = A_k M_{0zk} + B_k F_{0zk} + D_k H_{0zk} \tag{7.25a}$$

or the alternative equivalent form:

$$I_{zk}^{\text{even}}(z_k) = A_k' S_{0zk} + B_k' F_{0zk} + D_k H_{0zk}, \tag{7.25b}$$

where A_k, A_k', B_k, B_k' and D_k are complex coefficients and

$$M_{0zk} = \sin\beta_0(h_k - |z_k|) \tag{7.26}$$

$$S_{0zk} = \sin\beta_0|z_k| - \sin\beta_0 h_k \tag{7.27}$$

$$F_{0zk} = \cos\beta_0 z_k - \cos\beta_0 h_k \tag{7.28}$$

$$H_{0zk} = \cos(\beta_0 z_k/2) - \cos(\beta_0 h_k/2). \tag{7.29}$$

For the odd currents

$$I_{zk}^{\text{odd}}(z_k) = Q_k P_{0zk} + R_k E_{0zk}, \tag{7.30}$$

where Q_k and R_k are complex coefficients and

$$P_{0zk} = \sin\beta_0 z_k - (z_k/h_k)\sin\beta_0 h_k. \tag{7.31}$$

E_{0zk} is defined in (7.24). The above formulas are for $k = 1, 2, \ldots N$. The approximate formulas (7.25a, b) and (7.30) are obtained specifically from the first integrals in (7.13) and (7.16). When there are no staggered elements ($d_{ki} = 0$), it is known that the induced currents are well represented by a linear combination of F_{0zk} and H_{0zk}. It may be argued that a similar linear combination must also be an acceptable representation of the even parts of the currents induced in staggered elements. This follows from the theoretical and experimental studies, referred to in Chapter 6, of currents in non-staggered elements that differ greatly in length. If the current induced on a relatively long element (but with $\beta_0 h_k \leq 5\pi/4$) by an adjacent very short antenna is well represented by (7.25a, b), it may be concluded that the same must be true of the current induced in antenna k by other coupled elements which maintain a vector potential with an even part that varies less in amplitude and phase along antenna k

than the vector potential generated by the currents in a very short element. Since no measurements were available of the currents induced by coupled staggered elements, a numerical check was made of the degree in which the assumed current distributions satisfy the integral equation. The results were quite satisfactory.

It may be concluded that the current along any element k in an array of N parallel antennas is approximately

$$I_{zk}(z_k) = A_k \sin \beta_0(h_k - |z_k|) + B_k(\cos \beta_0 z_k - \cos \beta_0 h_k) + D_k[\cos(\beta_0 z_k/2)$$
$$- \cos(\beta_0 h_k/2)] + Q_k[\sin \beta_0 z_k - (z_k/h_k) \sin \beta_0 h_k]$$
$$+ R_k[\sin(\beta_0 z_k/2) - (z_k/h_k) \sin(\beta_0 h_k/2)]. \tag{7.32}$$

If more convenient, the first two terms may be replaced by those in (7.25b). The first three terms for the even part of the current are the same in form as for arrays of parallel, non-staggered elements. They include the term $\sin \beta_0(h_k - |z_k|)$ which represents that part of the current excited directly by the generator voltage V_{0k}. No such term is possible for the odd part of the current in a center-driven dipole. The remaining problem is to determine the coefficients in (7.32).

7.4   Evaluation of coefficients

The coefficients in the approximate formula (7.32) for the current in a typical element k in an array of N arbitrarily located parallel elements may be evaluated in various ways. The method outlined here is the one selected by V. W. H. Chang in his study of planar and three-dimensional arrays. He preferred to use the following alternative form for the current:

$$I_{zk}(z_k) = A'_k(\sin \beta_0 |z_k| - \sin \beta_0 h_k) + B'_k(\cos \beta_0 z_k - \cos \beta_0 h_k)$$
$$+ D_k[\cos(\beta_0 z_k/2) - \cos(\beta_0 h_k/2)]$$
$$+ Q_k[\sin \beta_0 z_k - (z_k/h_k) \sin \beta_0 h_k]$$
$$+ R_k[\sin(\beta_0 z_k/2) - (z_k/h_k) \sin(\beta_0 h_k/2)], \tag{7.33}$$

where $k = 1, 2, \ldots N$. Instead of substituting the even and odd parts into the integral equations (7.13) and (7.16) he used the simpler integral equation for the total current obtained when (7.4) is equated to (7.1) with (7.2) and (7.3). That is,

$$\sum_{i=1}^{N} \int_{-h_i}^{h_i} I_{zi}(z'_i) \frac{e^{-j\beta_0 R_{ki}}}{R_{ki}} dz'_i = -(j4\pi/\zeta_0)[C_{k1} \cos \beta_0 z_k + C_{k2} \sin \beta_0 z_k$$
$$+ \tfrac{1}{2} V_{0k} \sin \beta_0 |z_k|]. \tag{7.34}$$

The substitution of (7.33) in the integral in (7.34) yields N equations with $7N$ unknowns, namely the $5N$ coefficients in (7.33) and the $2N$ constants C_{k1} and C_{k2} with $k = 1, 2, \ldots N$. The required $7N$ equations can be obtained by satisfying (7.34) exactly at seven points along each antenna. The points chosen for z_k are h_k, $2h_k/3$, $h_k/3$, 0, $-h_k/3$, $-2h_k/3$, and $-h_k$. These correspond to the values used in the evaluation of the coefficients for the array of unequal elements in the last sections of Chapter 6, but since the currents are now not even functions of z_k, the negative values $-h_k$, $-2h_k/3$ and $-h_k/3$ must also be used.

The number of unknowns can be reduced by the elimination of the constants C_{k1} and C_{k2}. The former is conveniently evaluated at $z_k = h_k$ where the current vanishes; the latter can be obtained from the equation at $z_k = 2h_k/3$. Thus, with the notation

$$U_{k1} = \sum_{i=1}^{N} \int_{-h_i}^{h_i} I_{zi}(z_i')G_{ki}(d_{ki}, h_k, z_i')\, dz_i' \tag{7.35}$$

$$U_{k2} = \sum_{i=1}^{N} \int_{-h_i}^{h_i} I_{zi}(z_i')G_{ki}(d_{ki}, 2h_k/3, z_i')\, dz_i' \tag{7.36}$$

(7.34) evaluated at $z_k = h_k$ and $2h_k/3$ yields

$$(j4\pi/\zeta_0)C_{k1} = [U_{k1}\sin(2\beta_0 h_k/3) - U_{k2}\sin\beta_0 h_k]\csc(\beta_0 h_k/3) \tag{7.37}$$

$$(j4\pi/\zeta_0)(C_{k2} + V_{0k}/2) = [U_{k2}\cos\beta_0 h_k - U_{k1}\cos(2\beta_0 h_k/3)]\csc(\beta_0 h_k/3). \tag{7.38}$$

Note that in the range $\beta_0 h_k \le 5\pi/4$, these expressions remain finite.

With (7.37) and (7.38), C_{k1} and C_{k2} can be eliminated from (7.34) to obtain

$$\sum_{i=1}^{N}\sin(\beta_0 h_k/3)\int_{-h_i}^{h_i} I_{zi}(z_i')G_{ki}(d_{ki}, z_k, z_i')\, dz_i'$$

$$+ \sum_{i=1}^{N}\sin\beta_0(\tfrac{2}{3}h_k - z_k)\int_{-h_i}^{h_i} I_{zi}(z_i')G_{ki}(d_{ki}, h_k, z_i')\, dz_i'$$

$$- \sum_{i=1}^{N}\sin\beta_0(h_k - z_k)\int_{-h_i}^{h_i} I_{zi}(z_i')G_{ki}(d_{ki}, \tfrac{2}{3}h_k, z_i')\, dz_i'$$

$$= \frac{j4\pi V_{0k}}{\zeta_0}\sin(\beta_0 h_k/3)\sin\beta_0 z_k\, H(-z_k), \tag{7.39}$$

where $H(-z_k)$ is the Heaviside function defined by $H(-z_k) = 0$, $z_k > 0$; $H(-z_k) = 1$, $z_k \le 0$.

The next step is to substitute the current (7.33) in the integrals in (7.39). This leads to quantities of the following form:

$$\xi_{kij}(z_k) = \sin(\beta_0 h_k/3) \int_{-h_i}^{h_i} J_{zi}^j(z_i') G_{ki}(d_{ki}, z_k, z_i') \, dz_i'$$

$$+ \sin \beta_0 (\tfrac{2}{3} h_k - z_k) \int_{-h_i}^{h_i} J_{zi}^j(z_i') G_{ki}(d_{ki}, h_k, z_i') \, dz_i'$$

$$+ \sin \beta_0 (h_k - z_k) \int_{-h_i}^{h_i} J_{zi}^j(z_i') G_{ki}(d_{ki}, 2h_k/3, z_i') \, dz_i', \qquad (7.40)$$

where $k = 1, 2, \ldots N$, $i = 1, 2, \ldots N$ and $j = 1, 2, \ldots 5$. The notation

$$J_{zi}^1(z_i') = S_{0zi} = \sin \beta_0 |z_i| - \sin \beta_0 h_i \qquad (7.41a)$$

$$J_{zi}^2(z_i') = F_{0zi} = \cos \beta_0 z_i - \cos \beta_0 h_i \qquad (7.41b)$$

$$J_{zi}^3(z_i') = H_{0zi} = \cos(\beta_0 z_i/2) - \cos(\beta_0 h_i/2) \qquad (7.41c)$$

$$J_{zi}^4(z_i') = P_{0zi} = \sin \beta_0 z_i - (z_i/h_i) \sin \beta_0 h_i \qquad (7.41d)$$

$$J_{zi}^5(z_i') = E_{0zi} = \sin(\beta_0 z_i/2) - (z_i/h_i) \sin(\beta_0 h_i/2) \qquad (7.41e)$$

is used. Note that for any specified value of z_k in a fixed array, (7.40) defines a set of N complex numbers that can be evaluated by high-speed computer. With (7.40) and (7.41a–e), (7.39) becomes:

$$\sum_{i=1}^{N} [A_i' \xi_{ki1}(z_k) + B_i' \xi_{ki2}(z_k) + D_i \xi_{ki3}(z_k) + Q_i \xi_{ki4}(z_k) + R_i \xi_{ki5}(z_k)]$$

$$= j(4\pi V_{0k}/\zeta_0) \sin(\beta_0 h_k/3) \sin \beta_0 z_k \, H(-z_k), \qquad (7.42)$$

with $k = 1, 2, \ldots N$. Five sets of N equations can be obtained from (7.42) if z_k is successively made equal to the five values $h_k/3$, 0, $-h_k/3$, $-2h_k/3$ and $-h_k$. These contain the $M = 5N$ unknown coefficients given by the column matrix

$$\{A\} = \text{tr}(A_1, \ldots A_N, B_1, \ldots B_N, D_1, \ldots D_N, Q_1, \ldots Q_N, R_1, \ldots R_N) \qquad (7.43)$$

where tr indicates the transpose. Let

$$[\Phi] = \begin{bmatrix} \Phi_{11} & \cdots & \Phi_{1M} \\ \vdots & & \vdots \\ \Phi_{M1} & \cdots & \Phi_{MM} \end{bmatrix}, \qquad (7.44)$$

where $M = 5N$ and

$$\Phi_{k+(m-1)N, i+(j-1)N} = \xi_{kij}(z_k^m) \qquad (7.45)$$

with $j = 1, 2, \ldots 5$; $m = 1, 2, \ldots 5$; $k = 1, 2, \ldots N$; $i = 1, 2, \ldots N$. The notation $z_k^1 = h_k/3$, $z_k^2 = 0$, $z_k^3 = -h_k/3$, $z_k^4 = -2h_k/3$, $z_k^5 = -h_k$ is used. Also let the following column matrix of $5N$ terms be defined:

$$\{W\} = \text{tr}(0 \ldots 0, 0 \ldots 0, W_1 \ldots W_N, T_1 \ldots T_N, S_1 \ldots S_N), \tag{7.46}$$

where

$$W_k = -(j4\pi V_{0k}/\zeta_0)\sin^2(\beta_0 h_k/3) \tag{7.47a}$$

$$T_k = -(j4\pi V_{0k}/\zeta_0)\sin(\beta_0 h_k/3)\sin(2\beta_0 h_k/3) \tag{7.47b}$$

$$S_k = -(j4\pi V_{0k}/\zeta_0)\sin(\beta_0 h_k/3)\sin\beta_0 h_k \tag{7.47c}$$

with $k = 1, 2, \ldots N$.

With this matrix notation the $5N$ equations for the N coefficients of the currents in terms of the N driving voltages V_{0k} with $k = 1, 2, \ldots N$ are given by the single matrix equation

$$[\Phi]\{A\} = \{W\}. \tag{7.48}$$

If (7.48) is solved for the $5N$ coefficients given by (7.43), the N currents $I_{zk}(z_k)$, $k = 1, 2, \ldots N$ given in (7.33) are known in terms of the N voltages V_{0k}. The currents at the driving points are then given by the matrix equation

$$\{I_z(0)\} = [Y]\{V_0\}, \tag{7.49a}$$

where

$$\{I_z(0)\} = \left\{ \begin{array}{c} I_{z1}(0) \\ \vdots \\ I_{zN}(0) \end{array} \right\}, \qquad \{V_0\} = \left\{ \begin{array}{c} V_{01} \\ \vdots \\ V_{0N} \end{array} \right\}. \tag{7.49b}$$

The square matrix

$$[Y] = \left[\begin{array}{cccc} Y_{11} & Y_{12} & \cdots & Y_{1N} \\ \vdots & & & \vdots \\ Y_{N1} & \cdots & & Y_{NN} \end{array} \right] \tag{7.49c}$$

is the admittance matrix. The terms Y_{ii} are the self-admittances, the terms Y_{ij}, $i \neq j$ are the mutual admittances.

The N driving voltages can be expressed in terms of the currents at the driving points in the form

$$\{V_0\} = [Z]\{I_z(0)\}, \tag{7.49d}$$

where $[Z] = [Y]^{-1}$ is the impedance matrix.

The driving-point admittance of element k is defined by

$$Y_{0k} = I_{zk}(0)/V_{0k}. \tag{7.49e}$$

7.5 The field patterns

The radiation field of an array of arbitrarily located parallel elements is the superposition of the fields generated by the individual elements. The far field of element i in such an array is given by

$$E^r_{\Theta i} = \frac{j\omega\mu_0}{4\pi} \frac{e^{-j\beta_0 R_i}}{R_i} \int_{-h_i}^{h_i} I_{zi}(z_i')e^{j\beta_0 z_i' \cos\Theta} \sin\Theta \, dz_i', \tag{7.50a}$$

where R_i is the distance from the center of the antenna i at X_i, Y_i, Z_i to the point of calculation P, and Θ is the angle between the Z-axis and the line $0P$ from the origin of coordinates near the center of the array.

If the distribution of current (7.33) is substituted in (7.50a), the field of element i can be expressed in the following integrated form:

$$E^r_{\Theta i} = \frac{j\zeta_0}{4\pi} \frac{e^{-j\beta_0 R_i}}{R_i} [A_i' H_m(\Theta, \beta_0 h_i) + B_i' G_m(\Theta, \beta_0 h_i)$$
$$+ D_i D_m(\Theta, \beta_0 h_i) + Q_i Q_m(\Theta, \beta_0 h_i) + R_i R_m(\Theta, \beta_0 h_i)], \tag{7.50b}$$

where the individual field factors are as follows:

$$H_m(\Theta, \beta_0 h_i) = \frac{\beta_0 \sin\Theta}{2} \int_{-h_i}^{h_i} (\sin\beta_0|z_i'| - \sin\beta_0 h_i)e^{j\beta_0 z_i' \cos\Theta} \, dz_i'$$
$$= \{\cos\Theta - [1 - \cos\beta_0 h_i \cos(\beta_0 h_i \cos\Theta)]\} \sec\Theta \csc\Theta \tag{7.51a}$$

$$H_m(0, \beta_0 h_i) = H_m(\pi, \beta_0 h_i) = 0 \tag{7.51b}$$

$$H_m\left(\frac{\pi}{2}, \beta_0 h_i\right) = 1 - \cos\beta_0 h_i - \beta_0 h_i \sin\beta_0 h_i \tag{7.51c}$$

$$G_m(\Theta, \beta_0 h_i) = \frac{\beta_0 \sin\Theta}{2} \int_{-h_i}^{h_i} (\cos\beta_0 z_i' - \cos\beta_0 h_i)e^{j\beta_0 z_i' \cos\Theta} \, dz_i'$$
$$= [\cos\Theta \sin\beta_0 h_i \cos(\beta_0 h_i \cos\Theta)$$
$$- \cos\beta_0 h_i \sin(\beta_0 h_i \cos\Theta)] \sec\Theta \csc\Theta \tag{7.52a}$$

$$G_m(0, \beta_0 h_i) = G_m(\pi, \beta_0 h_i) = 0 \tag{7.52b}$$

$$G_m\left(\frac{\pi}{2}, \beta_0 h_i\right) = \sin\beta_0 h_i - \beta_0 h_i \cos\beta_0 h_i \tag{7.52c}$$

$$D_m(\Theta, \beta_0 h_i) = \frac{\beta_0 \sin \Theta}{2} \int_{-h_i}^{h_i} [\cos(\beta_0 z_i'/2) - \cos(\beta_0 h_i/2)] e^{j\beta_0 z_i' \cos \Theta} \, dz_i'$$

$$= \left[\frac{\sin \Theta}{\cos \Theta \, (1 - 4 \cos^2 \Theta)} \right] [2 \cos \Theta \sin(\beta_0 h_i/2)$$

$$\times \cos(\beta_0 h_i \cos \Theta) - \cos(\beta_0 h_i/2) \sin(\beta_0 h_i \cos \Theta)] \qquad (7.53a)$$

$$D_m\left(\frac{\pi}{2}, \beta_0 h_i\right) = 2 \sin(\beta_0 h_i/2) - \beta_0 h_i \cos(\beta_0 h_i/2) \qquad (7.53b)$$

$$D_m\left(\frac{\pi}{3}, \beta_0 h_i\right) = D_m\left(\frac{2\pi}{3}, \beta_0 h_i\right) = \frac{\sqrt{3}}{4} (\beta_0 h_i - \sin \beta_0 h_i) \qquad (7.53c)$$

$$Q_m(\Theta, \beta_0 h_i) = \frac{\beta_0 \sin \Theta}{2} \int_{-h_i}^{h_i} [\sin \beta_0 z_i' - (z_i'/h_i) \sin \beta_0 h_i] e^{j\beta_0 z_i' \cos \Theta} \, dz_i'$$

$$= (j/\beta_0 h_i)[-\beta_0 h_i \cos^2 \Theta \cos \beta_0 h_i \sin(\beta_0 h_i \cos \Theta)$$

$$- \sin^2 \Theta \sin \beta_0 h_i \sin(\beta_0 h_i \cos \Theta)$$

$$+ \beta_0 h_i \cos \Theta \sin \beta_0 h_i \cos(\beta_0 h_i \cos \Theta)] \csc \Theta \sec^2 \Theta \qquad (7.54a)$$

$$Q_m(0, \beta_0 h_i) = Q_m\left(\frac{\pi}{2}, \beta_0 h_i\right) = Q_m(\pi, \beta_0 h_i) = 0 \qquad (7.54b)$$

$$R_m(\Theta, \beta_0 h_i) = \frac{\beta_0 \sin \Theta}{2} \int_{-h_i}^{h_i} [\sin(\beta_0 z_i'/2) - (z_i'/h_i) \sin(\beta_0 h_i/2)] e^{j\beta_0 z_i' \cos \Theta} \, dz_i'$$

$$= \left[\frac{j \sin \Theta}{\beta_0 h_i \cos^2 \Theta \, (1 - 4 \cos^2 \Theta)} \right] \{\sin(\beta_0 h_i \cos \Theta)$$

$$\times [-2\beta_0 h_i \cos^2 \Theta \cos(\beta_0 h_i/2) - \sin(\beta_0 h_i/2)$$

$$+ 4 \sin(\beta_0 h_i/2) \cos^2 \Theta] + \beta_0 h_i \sin(\beta_0 h_i/2)$$

$$\times \cos \Theta \cos(\beta_0 h_i \cos \Theta)\} \qquad (7.55a)$$

$$R_m\left(\frac{\pi}{2}, \beta_0 h_i\right) = 0 \qquad (7.55b)$$

$$R_m\left(\frac{\pi}{3}, \beta_0 h_i\right) = -R_m\left(\frac{2\pi}{3}, \beta_0 h_i\right) = j\frac{\sqrt{3}}{4} [\beta_0 h_i + \sin \beta_0 h_i$$

$$- (8/\beta_0 h_i) \sin^2(\beta_0 h_i/2)]. \qquad (7.55c)$$

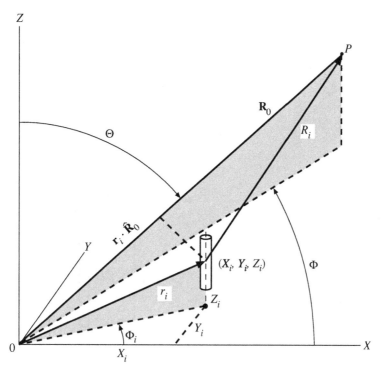

Figure 7.4 Point P in the far field of an array of parallel elements of which element i at X_i, Y_i, Z_i is typical.

The radiation field of N parallel antennas is the sum of the contributions from each element. In the far-field approximation

$$\frac{e^{-j\beta_0 R_i}}{R_i} = \frac{e^{-j\beta_0 R_0}}{R_0} e^{j\beta_0(\mathbf{r}_i \cdot \hat{\mathbf{R}}_0)}, \tag{7.56}$$

where $\mathbf{r}_i$ is the vector drawn from the origin near the center of the array to the center of antenna i and $\hat{\mathbf{R}}_0$ is the unit vector along the line $0P$ where P is the point of calculation as shown in Fig. 7.4. With (7.56) the far field of the array is

$$E_\Theta^r = \frac{j\zeta_0}{4\pi} \frac{e^{-j\beta_0 R_0}}{R_0} \sum_{i=1}^{N} e^{j\beta_0(\mathbf{r}_i \cdot \hat{\mathbf{R}}_0)} [A_i' H_m(\Theta, \beta_0 h_i) + B_i' G_m(\Theta, \beta_0 h_i)$$

$$+ D_i D_m(\Theta, \beta_0 h_i) + Q_i Q_m(\Theta, \beta_0 h_i) + R_i R_m(\Theta, \beta_0 h_i)]. \tag{7.57}$$

If the point P where the field is calculated is located by the spherical coordinates R_0, Θ, Φ and the center of element i is at X_i, Y_i, Z_i, then

$$\mathbf{r}_i \cdot \hat{\mathbf{R}}_0 = X_i \sin \Theta \cos \Phi + Y_i \sin \Theta \sin \Phi + Z_i \cos \Theta. \tag{7.58}$$

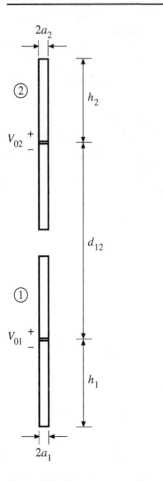

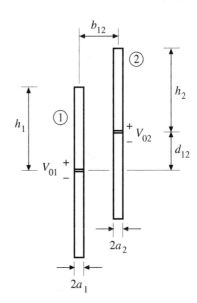

Figure 7.5 Two-element collinear array.   *Figure 7.6* Two-element staggered array.

7.6   The general two-element array[2]

In the introductory analysis of the two-element array in Chapter 2 only parallel non-staggered antennas are considered. As a consequence of the resulting even symmetry for the currents in the elements and the vector potentials, a three- or even two-term representation is adequate. The more general five-term approximation of the currents introduced in this chapter includes the previous three terms to describe the even currents and two additional terms to represent the odd currents generated by asymmetrical coupling when the elements are collinear as shown in Fig. 7.5 or staggered as in Fig. 7.6.

The general formulas for the currents, driving-point admittances and field patterns derived in the preceding sections are readily specialized for the two-element array.

---

[2] The computations in this section were planned and programmed by V. W. H. Chang.

Table 7.1a. *Symmetrical and anti-symmetrical admittances in millisiemens of parallel, non-staggered array of two elements; $a/\lambda = 0.007\,022$*

b_{12}/λ	$h/\lambda = 0.50$		$h/\lambda = 0.25$	
	Y^s	Y^a	Y^s	Y^a
0.05	$0.813 + j1.397$	$0.071 + j0.941$	$4.939 - j0.820$	$4.831 - j53.897$
0.10	$1.028 + j1.668$	$0.146 + j1.122$	$5.572 - j0.290$	$4.344 - j24.150$
0.15	$1.197 + j1.749$	$0.230 + j1.286$	$6.258 + j0.112$	$4.385 - j15.386$
0.25	$1.448 + j1.627$	$0.408 + j1.531$	$7.853 + j0.809$	$4.758 - j8.655$
0.50	$1.079 + j0.932$	$0.865 + j1.774$	$14.321 - j1.496$	$6.129 - j3.340$
0.75	$0.635 + j1.374$	$1.244 + j1.570$	$8.960 - j7.101$	$8.318 - j1.318$
1.00	$0.872 + j1.657$	$1.066 + j1.141$	$7.211 - j3.839$	$11.577 - j2.517$
1.25	$1.157 + j1.537$	$0.728 + j1.359$	$8.543 - j2.095$	$9.503 - j5.707$
1.50	$1.053 + j1.226$	$0.877 + j1.602$	$10.725 - j2.875$	$7.777 - j3.918$

Table 7.1b. *Symmetrical and anti-symmetrical admittances in millisiemens of collinear array of two elements; $a/\lambda = 0.007\,022$*

$\dfrac{d_{12} - 2h}{\lambda}$	$h/\lambda = 0.50$		$h/\lambda = 0.25$	
	Y^s	Y^a	Y^s	Y^a
0	$1.050 + j1.581$	$0.816 + j1.139$	$5.549 - j1.953$	$15.281 - j6.078$
0.05	$1.042 + j1.505$	$0.808 + j1.338$	$7.508 - j1.868$	$11.013 - j6.483$
0.10	$1.035 + j1.465$	$0.844 + j1.409$	$8.444 - j1.913$	$9.529 - j5.821$
0.15	$1.026 + j1.437$	$0.875 + j1.440$	$9.100 - j2.104$	$8.858 - j5.208$
0.25	$0.999 + j1.403$	$0.919 + j1.461$	$9.850 - j2.764$	$8.412 - j4.288$
0.50	$0.941 + j1.409$	$0.966 + j1.446$	$9.521 - j4.026$	$8.832 - j3.243$
0.75	$0.945 + j1.435$	$0.965 + j1.422$	$8.923 - j3.800$	$9.415 - j3.404$
1.00	$0.958 + j1.435$	$0.950 + j1.422$	$9.042 - j3.445$	$9.300 - j3.785$
1.25	$0.959 + j1.426$	$0.950 + j1.431$	$9.295 - j3.518$	$9.045 - j3.699$
1.50	$0.953 + j1.425$	$0.956 + j1.432$	$9.238 - j3.708$	$9.103 - j3.517$

Comparative examples of the admittances of coupled full-wave and half-wave elements when driven symmetrically with $V_{01} = V_{02} = 1$ volt and anti-symmetrically with $V_{01} = -V_{02} = 1$ volt as a function of the distance between centers are given in Tables 7.1a, b, c for antennas with $a/\lambda = 0.007\,022$. These tables give the symmetrical admittance $Y^s = G^s + jB^s$ and the anti-symmetrical admittance $Y^a = G^a + jB^a$. The associated self- and mutual admittances are $Y_{1s} = (Y^s + Y^a)/2$ and $Y_{12} = -(Y^s - Y^a)/2$.

Table 7.1a applies to the non-staggered antennas considered in Chapter 2; the variable parameter is b_{12}/λ, the normalized distance between centers. Table 7.1b is

Table 7.1c. *Symmetrical and anti-symmetrical admittances in millisiemens of parallel, staggered array of two elements; $a/\lambda = 0.007\,022$*

$\dfrac{b_{12}}{\lambda} = \dfrac{d_{12}}{\lambda}$	$h/\lambda = 0.50$		$h/\lambda = 0.25$	
	Y^s	Y^a	Y^s	Y^a
0.05	$0.825 + j1.318$	$0.086 + j0.608$	$4.692 - j0.775$	$11.923 - j76.274$
0.10	$1.040 + j1.720$	$0.786 + j0.801$	$5.456 - j0.470$	$8.631 - j29.120$
0.15	$1.189 + j1.832$	$0.236 + j0.984$	$6.336 - j0.297$	$7.399 - j16.489$
0.25	$1.363 + j1.681$	$0.420 + j1.379$	$8.372 - j0.097$	$6.669 - j\ 8.084$
0.50	$1.035 + j1.110$	$0.965 + j1.673$	$11.077 - j4.641$	$7.804 - j\ 2.856$
0.75	$0.785 + j1.474$	$1.089 + j1.414$	$8.235 - j3.925$	$10.159 - j\ 3.106$
1.00	$1.008 + j1.489$	$0.906 + j1.350$	$9.340 - j2.908$	$8.903 - j\ 4.302$
1.25	$0.976 + j1.380$	$0.939 + j1.484$	$9.483 - j4.115$	$8.862 - j\ 3.166$
1.50	$0.918 + j1.400$	$0.989 + j1.419$	$8.729 - j3.561$	$9.650 - j\ 3.650$

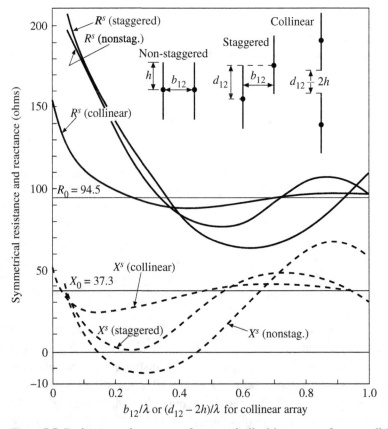

Figure 7.7 Resistance and reactance of symmetrically driven array of two parallel half-wave dipoles when non-staggered, staggered with $b_{12} = d_{12}$, and collinear; $a/\lambda = 0.007\,022$, $h/\lambda = 0.25$, $V_{02} = V_{01}$.

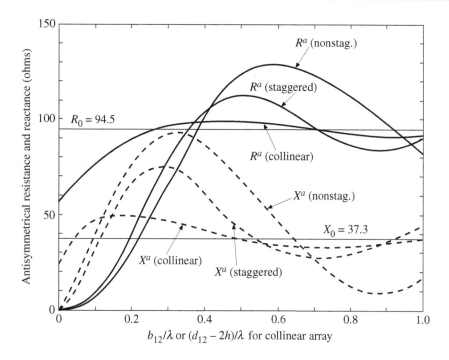

Figure 7.8 Like Fig. 7.7 but for anti-symmetrically driven elements, $V_{02} = -V_{01}$.

for the collinear pair with the distance $(d_{12} - 2h)$ between the adjacent ends as the parameter (d_{12} is the distance between centers). The admittances in Table 7.1c are for the staggered pair as the center of element 2 is moved along a $45°$ line so that $b_{12} = d_{12}$. The impedances $Z^s = 1/Y^s$ and $Z^a = 1/Y^a$ corresponding to the admittances in Tables 7.1a, b, c are shown graphically in Figs. 7.7 and 7.8, respectively, for the symmetrically and anti-symmetrically driven pairs. In these figures R_0 and X_0 are the resistance and reactance for infinite separation. The interaction between the elements is seen to be greatest for the non-staggered pair, smallest for the collinear arrangement. The self- and mutual impedances are given by $Z_{s1} = (Z^s + Z^a)/2$ and $Z_{12} = (Z^s - Z^a)/2$; they are listed in Tables 7.2a, b.

The current distribution along the lower element of a collinear pair when the adjacent ends are separated by a distance $d_{12} - 2h = 0.1\lambda$ is shown in Fig. 7.9 for both symmetrically and anti-symmetrically driven full-wave elements. Note that in both cases the currents are asymmetrical with respect to the center of the element. When the excitation is symmetrical ($V_{01} = V_{02}$), the current in the outer half is the greater; when the excitation is anti-symmetrical ($V_{01} = -V_{02}$), the current in the inner half is the greater.

The current distribution for a pair of coupled full-wave antennas in the staggered position with $b_{12}/\lambda = d_{12}/\lambda = 0.1$ is shown in Fig. 7.10 for symmetric excitation ($V_{01} = V_{02}$) in broken line. Since the two elements are very close together, the

Table 7.2a. *Self- and mutual impedances of coupled half-wave dipoles; $h/\lambda = 0.25$, $a/\lambda = 0.007\,022$*

| | Parallel, non-staggered | | Staggered | | Collinear | |
| | $x = b_{12}/\lambda$ | | $x = b_{12}/\lambda = d_{12}/\lambda$ | | $x = (d_{12} - 2h)/\lambda$ | |
x	$Z_{s1} = R_{s1} + jX_{s1}$	$Z_{12} = R_{12} + jX_{12}$	$Z_{s1} = R_{s1} + jX_{s1}$	$Z_{12} = R_{12} + jX_{12}$	$Z_{s1} = R_{s1} + jX_{s1}$	$Z_{12} = R_{12} + jX_{12}$
0					$108.4 + j39.5$	$51.9 - j17.0$
0.05	$99.4 + j25.6$	$97.7 + j7.15$	$104.7 + j23.5$	$102.7 + j10.7$	$96.4 + j35.4$	$29.0 - j4.2$
0.10	$93.1 + j24.7$	$85.9 - j15.4$	$95.6 + j23.6$	$86.3 - j7.9$	$94.5 + j36.1$	$18.1 - j10.6$
0.15	$88.4 + j28.6$	$71.3 - j31.5$	$90.1 + j28.9$	$67.4 - j21.5$	$94.1 + j36.7$	$10.2 - j12.6$
0.25	$87.4 + j37.9$	$38.6 - j50.8$	$90.1 + j37.5$	$29.4 - j36.1$	$94.2 + j37.3$	$-0.1 - j10.8$
0.50	$97.4 + j37.9$	$-28.4 - j30.7$	$94.1 + j39.1$	$-1.7 - j30.9$	$94.4 + j37.2$	$-5.3 + j0.5$
0.75	$92.9 + j36.4$	$-24.4 + j17.9$	$94.9 + j36.8$	$-18.1 - j4.6$	$94.4 + j37.2$	$0.5 + j3.2$
1.00	$95.3 + j37.7$	$12.8 + j19.8$	$94.5 + j37.3$	$4.5 + j9.8$	$94.4 + j37.2$	$2.2 - j0.4$
1.25	$93.9 + j37.8$	$16.5 - j9.7$	$94.3 + j37.2$	$3.3 - j6.8$	$94.4 + j37.2$	$-0.3 - j1.6$
1.50	$94.8 + j37.5$	$-7.8 - j14.2$	$94.4 + j37.1$	$-5.7 + j1.4$	$94.4 + j37.2$	$-1.2 + j0.2$

Table 7.2b. *Self- and mutual impedances of coupled full-wave dipoles; $h/\lambda = 0.5$, $a/\lambda = 0.007\,022$*

| | Parallel, non-staggered | | Staggered | | Collinear | |
| | $x = b_{12}/\lambda$ | | $x = b_{12}/\lambda = d_{12}/\lambda$ | | $x = (d_{12} - 2h)/\lambda$ | |
x	$Z_{s1} = R_{s1} + jX_{s1}$	$Z_{12} = R_{12} + jX_{12}$	$Z_{s1} = R_{s1} + jX_{s1}$	$Z_{12} = R_{12} + jX_{12}$	$Z_{s1} = R_{s1} + jX_{s1}$	$Z_{12} = R_{12} + jX_{12}$
0					$353.7 - j509.5$	$-62.1 + j70.6$
0.05	$195.5 - j795.5$	$115.6 + j260.8$	$285.1 - j1078.8$	$56.2 + j533.7$	$320.8 - j498.4$	$-9.8 + j49.2$
0.10	$190.9 - j655.5$	$76.8 + j221.0$	$250.5 - j825.2$	$7.0 + j399.4$	$317.3 - j488.9$	$4.4 + j33.7$
0.15	$200.5 - j571.3$	$66.0 + j182.0$	$239.9 - j672.3$	$9.4 + j288.3$	$318.6 - j484.1$	$10.5 + j23.1$
0.25	$233.8 - j476.4$	$71.4 + j133.5$	$246.6 - j511.3$	$44.4 + j152.4$	$322.6 - j481.8$	$14.1 + j8.7$
0.50	$376.4 - j457.1$	$154.2 - j1.6$	$354.2 - j465.1$	$95.4 - j16.6$	$323.5 - j484.5$	$4.1 - j6.4$
0.75	$293.7 - j495.5$	$-16.4 - j104.2$	$311.8 - j486.2$	$-30.2 - j42.2$	$323.3 - j483.7$	$-3.4 - j2.4$
1.00	$342.9 - j470.3$	$-94.4 - j2.3$	$327.3 - j485.7$	$-15.6 - j25.1$	$323.4 - j484.0$	$-1.6 + j2.1$
1.25	$309.3 - j493.5$	$3.2 + j78.2$	$323.1 - j482.0$	$18.4 - j0.8$	$323.3 - j483.9$	$1.3 + j1.1$
1.50	$333.0 - j474.8$	$70.0 + j5.4$	$322.5 - j484.5$	$-8.1 - j10.2$	$323.4 - j484.0$	$0.8 - j0.9$

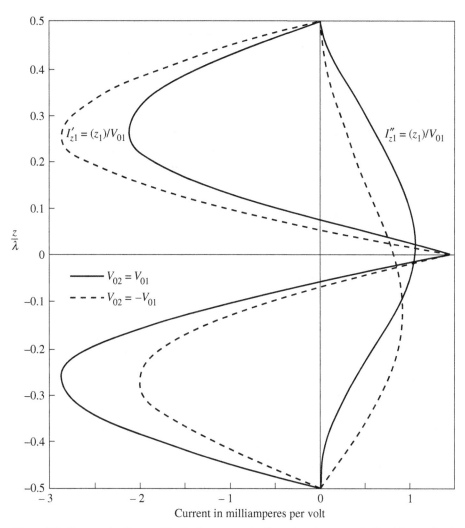

$\frac{z}{\lambda}$

Figure 7.9 Currents in element No. 1 of a symmetrically and anti-symmetrically driven pair of collinear antennas. $I_{z1}(z) = I''_{z1}(z) + j I'_{z1}(z)$, $V_{02} = \pm V_{01} = \pm 1$ volt, $a/\lambda = 0.007\,022$, $h/\lambda = 0.5$, $d_{12}/\lambda = 1.1$. (Element 1 is below element 2.)

interaction is great. When center driven with equal and opposite voltages, the two conductors form a slightly displaced two-wire line with a large and only slightly asymmetrical reactive current $I'_{z1}(z)$ that is almost sinusoidal, and a very small in-phase component $I''_{z1}(z)$. Since the current induced in each element by that in the other is essentially 180° out of phase, the coupling reinforces the currents excited by the generator voltages. When center driven by equal and co-directional generators the distribution of current is extremely asymmetrical. The half of each element that is removed from the other has a very large approximately sinusoidal reactive current $I'_{zi}(z)$, whereas the adjacent halves have only a small and oppositely

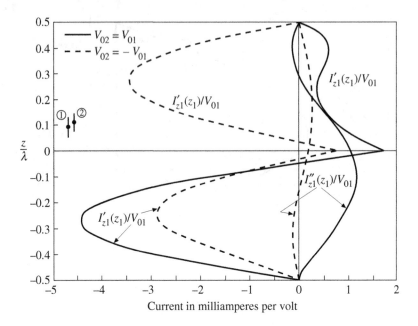

Figure 7.10 Like Fig. 7.9 but for two staggered elements with $b_{12}/\lambda = d_{12}/\lambda = 0.1$.

directed reactive component. The in-phase component $I_{zi}''(z)$ is much greater than when the antennas are anti-symmetrically driven and the asymmetry is in the opposite directions.

The distributions of current for the symmetrically and anti-symmetrically driven staggered pair are sketched approximately to scale in Figs. 7.11a, b. Note that the more closely coupled adjacent halves of the elements have the greater current when the excitation is asymmetrical, very much the smaller when the excitation is symmetrical. In the former case the coupling between the elements reinforces the generators, in the latter it opposes them.

It is interesting to note that the distribution along the symmetrically driven pair in Fig. 7.11a resembles that along a sleeve dipole.[3] This is to be expected since the two elements are very closely coupled.

7.7   A simple planar array[4]

The application of the general theory developed earlier in this chapter to planar arrays is conveniently illustrated with the three by three nine-element array shown

---

[3] See, for example, R. W. P. King, [2], p. 413, Fig. 30.7e.
[4] The computations in this section are those of V. W. H. Chang. Parts of Sections 7.7–7.10 were first published in *Radio Science* [3].

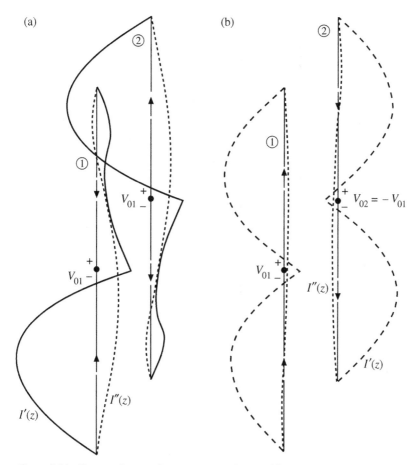

Figure 7.11 Currents in two-element staggered array driven (a) symmetrically with $V_{02} = V_{01}$ and (b) anti-symmetrically with $V_{02} = -V_{01}$. The distributions of current are taken from Fig. 7.10.

in Fig. 7.12. This involves non-staggered, staggered and collinear elements, so that the effects of the different types of coupling on otherwise identical elements can be studied.

Consider first the broadside array in which all elements are driven with equal voltages, that is, $V_{0i} = 1$ volt, $i = 1, 2, \ldots 9$. Since conventional theory is unable to treat full-wave elements, let an array of nine elements with $h/\lambda = 0.5$ be analyzed. Let the lateral distances between elements be $b/\lambda = 0.25$ and the axial distance between adjacent ends be $(d - 2h)/\lambda = 0.1$ where d is the distance between centers of adjacent collinear elements. With this symmetric excitation, elements 1, 2 and 3 are like elements 7, 8 and 9 in the even parts of these currents, but the algebraic sign of the odd parts is reversed. The coefficients of the five trigonometric functions in the current distribution given in (7.33) are listed in Table 7.3 for all of the elements. The associated driving-point admittances and impedances are also listed. Note that these differ significantly. The four different distributions of current are shown in Fig.

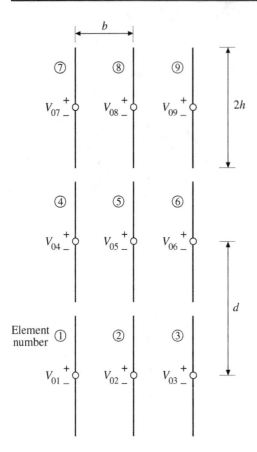

Figure 7.12 Planar array of nine identical, equally-spaced elements.

7.13 in the form $I_{zi}(z_i) = I''_{zi}(z_i) + jI'_{zi}(z_i)$ where $I''_{zi}(z_i)$ is in phase with V_{0i}, $I'_{zi}(z_i)$ in phase quadrature. Note that the currents for elements 7, 8 and 9 are like those for 1, 2 and 3 but with $-z_i$ substituted for z_i. Elements 4, 5 and 6 have even currents. The contribution by the odd currents in elements 1, 2 and 3 is seen to be large.

When the same nine-element array is driven to obtain a unilateral endfire pattern with $V_{01} = V_{04} = V_{07} = 1$, $V_{02} = V_{05} = V_{08} = -j$, $V_{03} = V_{06} = V_{09} = -1$ volt, the coefficients for the trigonometric functions in the current distribution (7.33) are listed in Table 7.4. Note that there are now six different sets of coefficients since elements 1 and 3 and their counterparts are no longer electrically identical. The driving-point admittances and impedances are also given in Table 7.4. They are seen to have a wider range of values than in the broadside array. Note that the resistances of the elements in the collinear trio in the backward direction (1, 4, 7) are much greater than the corresponding resistances in the forward trio (3, 6, 9). This is characteristic of endfire arrays of full-wave elements. The six different currents are shown in Figs. 7.14 and

Table 7.3. *Nine-element planar array – broadside; $a/\lambda = 0.007\,022$, $h/\lambda = 0.5$, $b/\lambda = 0.25$, $d/\lambda = 1.1$, $V_{0i} = 1$ volt, $i = 1, 2, \ldots 9$*

Coefficients of trigonometric functions in milliamperes per volt

i	A_i'	B_i'	D_i	Q_i [a]	R_i [a]
1, 3, 7, 9	$0.006 - j3.626$	$0.763 - j0.739$	$0.287 + j2.849$	$0.197 + j0.444$	$-0.008 + j0.134$
2, 8	$0.010 - j3.580$	$1.065 - j0.643$	$0.197 + j3.164$	$0.317 + j0.604$	$0.279 + j0.199$
4, 6	$0.007 - j3.615$	$0.832 - j0.590$	$0.176 + j2.596$	$0 + j0$	$0 + j0$
5	$0.010 - j3.567$	$1.176 - j0.443$	$0.087 + j2.821$	$0 + j0$	$0 + j0$

[a] Reverse signs for $i = 7, 8, 9$.

Admittances in millisiemens and impedances in ohms

i	$Y_{0i} = G_{0i} + jB_{0i}$	$Z_{0i} = R_{0i} + jX_{0i}$
1, 3, 7, 9	$1.759 + j1.371$	$353.7 - j275.7$
2, 8	$2.328 + j1.878$	$260.2 - j209.9$
4, 6	$1.840 + j1.416$	$341.3 - j262.8$
5	$2.440 + j1.955$	$249.6 - j200.0$

Table 7.4. *Nine-element planar array – endfire; $a/\lambda = 0.007\,022$, $h/\lambda = 0.5$, $b/\lambda = 0.25$, $d/\lambda = 1.1$, $V_{01} = V_{04} = V_{07} = 1$, $V_{02} = V_{05} = V_{08} = -j$, $V_{03} = V_{06} = V_{09} = -1$ volt*

Coefficients of trigonometric functions in milliamperes per volt

i	A_i'	B_i'	D_i	Q_i [a]	R_i [a]
1, 7	$0.053 - j3.674$	$0.321 - j0.426$	$0.391 - j2.060$	$0.206 + j0.463$	$0.288 + j0.213$
2, 8	$-3.675 - j0.006$	$-0.282 - j0.501$	$2.375 - j0.027$	$0.500 - j0.151$	$0.196 - j0.168$
3, 9	$0.042 + j3.668$	$-0.724 - j0.152$	$0.480 - j2.201$	$-0.119 - j0.562$	$-0.223 - j0.124$
4	$0.053 - j3.664$	$0.381 - j0.299$	$0.321 + j1.845$	$0 + j0$	$0 + j0$
5	$-3.665 - j0.005$	$-0.129 - j0.546$	$2.137 + j0.010$	$0 + j0$	$0 + j0$
6	$0.044 + j3.657$	$-0.735 - j0.344$	$0.463 - j1.918$	$0 + j0$	$0 + j0$

[a] Reverse signs for $i = 7, 8, 9$.

Admittances in millisiemens and impedances in ohms

i	$Y_{0i} = G_{0i} + jB_{0i}$	$Z_{0i} = R_{0i} + jX_{0i}$
1, 7	$1.034 + j1.208$	$409.0 - j477.8$
2, 8	$1.030 + j1.811$	$237.2 - j417.2$
3, 9	$0.966 + j2.506$	$134.0 - j347.4$
4	$1.084 + j1.250$	$396.0 - j456.7$
5	$1.083 + j1.879$	$230.2 - j399.5$
6	$1.008 + j2.607$	$129.0 - j333.7$

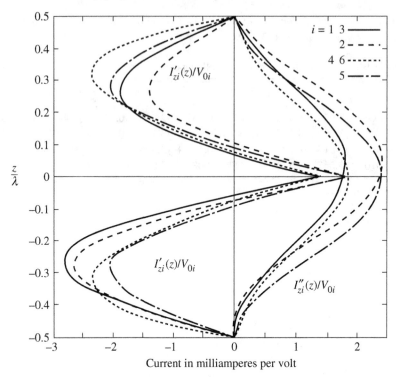

Figure 7.13 Currents in a planar array of nine elements in broadside. $I_{zi}(z_i) = I''_{zi}(z_i) + jI'_{zi}(z_i)$, $V_{0i} = 1$ volt; $a/\lambda = 0.007\,022$, $h/\lambda = 0.5$, $b/\lambda = 0.25$, $d/\lambda = 1.1$.

7.15. The currents in elements 7, 8 and 9 are like those in 1, 2 and 3 but with $-z_i$ substituted for z_i. Note that the currents in the rear collinear trio (1, 4, 7) are greater and contribute more to the far field than the currents in the forward trio of the elements (3, 6, 9). The far-field patterns in the horizontal or H-plane and the vertical or E-plane are shown in Fig. 7.16 for both the broadside and the endfire arrays. The horizontal pattern of the broadside array is bidirectional with maxima at $\Phi = 90°$ and $270°$, the endfire pattern is unidirectional with a broad maximum in the direction $\Phi = 0°$. The vertical patterns in the direction $\Phi = 0$ are seen to be very sharp as would be expected when three full-wave elements (which correspond to six half-wave elements) are stacked. (Note that the vertical pattern shown for the broadside array is not in the direction of the maximum at $\Phi = 90°$.)

When the length of the elements is a half wavelength instead of a full wavelength, it is usually desirable to assign the driving-point currents $I_{zi}(0)$ instead of the voltages V_{0i}. If the array shown in Fig. 7.12 is constructed of half-wave elements with $h/\lambda = 0.25$, $b/\lambda = 0.25$, $(d - 2h)/\lambda = 0.1$, and the currents are assigned for a broadside pattern with $I_{zi}(0) = 2.5$ milliamperes for $i = 1, 2, \ldots 9$, the coefficients for the trigonometric functions in the expression (7.33) for the currents in the elements are those given in Table 7.5. The required driving voltages V_{0i} are also

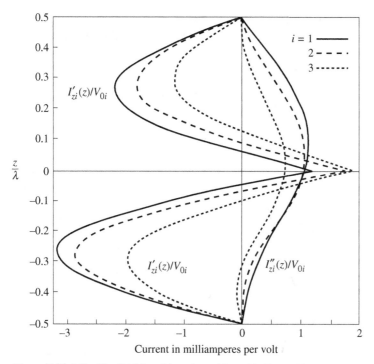

Figure 7.14 Like Fig. 7.13 but driven in endfire with $V_{01} = V_{04} = V_{07} = 1$, $V_{02} = V_{05} = V_{08} = -j$, $V_{03} = V_{06} = V_{09} = -1$ volt. Currents in elements 1, 2, 3.

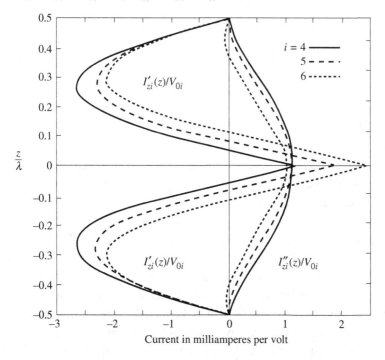

Figure 7.15 Like Fig. 7.14 but for elements 4, 5 and 6.

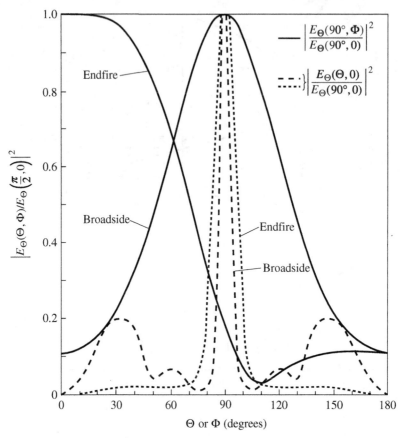

Figure 7.16 Horizontal ($\Theta = 90°$) and vertical ($\Phi = 0°$) patterns of a nine-element planar array of full-wave antennas; broadside and endfire excitation.

listed together with the associated driving-point admittances and impedances. Note that the voltages differ considerably, as do the impedances. This is due entirely to mutual coupling.

The fact that the driving-point currents are all maintained equal and in phase by a suitable choice of voltages does not mean that the several distributions of current are therefore equal and in phase. The very different interactions among the several elements necessarily lead to distributions of current that are quite dissimilar in both amplitude and phase. This is shown graphically in Fig. 7.17 for the real and imaginary parts of the currents. The real parts are seen to be more nearly triangular than cosinusoidal; the imaginary parts are quite large and distributed so differently from the real part that the phase angle is very far from constant. This means that even for half-wave elements the conventional assumption that all currents are cosinusoidally distributed and constant in phase along each element is of questionable validity for determining impedances and minor lobe structures.

Table 7.5. *Nine-element planar array – broadside; $a/\lambda = 0.007\,022$, $h/\lambda = 0.25$, $b/\lambda = 0.25$, $d/\lambda = 0.6$, $I_{zi}(0) = 2.5 \times 10^{-3}$ amperes; $i = 1, 2, \ldots 9$*

Coefficients of trigonometric functions in milliamperes

i	A'_i	B'_i	D_i	$Q_i{}^{\text{a}}$	$R_i{}^{\text{a}}$
1, 3, 7, 9	$-0.810 - j1.304$	$-4.422 - j2.913$	$20.869 + j5.492$	$-0.392 + j0.650$	$2.679 - j6.099$
2, 8	$-1.233 - j2.300$	$-5.529 - j5.054$	$23.204 + j9.404$	$-0.202 + j1.214$	$0.897 - j11.409$
4, 6	$-1.194 - j1.213$	$-5.759 - j2.193$	$24.120 + j3.342$	$0 + j0$	$0 + j0$
5	$-1.188 - j2.419$	$-7.570 - j4.620$	$27.976 + j7.513$	$0 + j0$	$0 + j0$

^{a} Reverse signs for $i = 7, 8, 9$.

Admittances in millisiemens, impedances in ohms, EMF's in volts

i	$Y_{0i} = G_{0i} + j B_{0i}$	$Z_{0i} = R_{0i} + j X_{0i}$	V_{0i}
1, 3, 7, 9	$8.203 + j4.310$	$95.5 - j50.2$	$0.239 - j0.125$
2, 8	$4.817 + j2.320$	$168.6 - j81.1$	$0.421 - j0.203$
4, 6	$6.369 + j5.538$	$89.4 - j77.7$	$0.223 - j0.194$
5	$3.713 + j2.663$	$177.8 - j127.5$	$0.445 - j0.319$

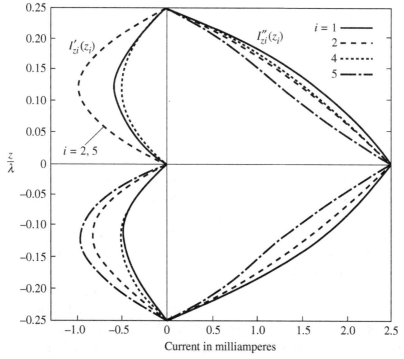

Figure 7.17 Normalized currents in the nine elements of a planar array.
$I_{zi}(z_i) = I''_{zi}(z_i) + j I'_{zi}(z_i)$, $I_{zi}(0) = 2.5$ milliamperes with $i = 1, 2, \ldots 9$. $a/\lambda = 0.007\,022$, $h/\lambda = 0.25$, $b/\lambda = 0.25$, $d/\lambda = 0.6$.

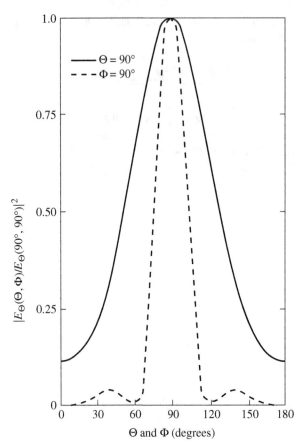

Figure 7.18 Horizontal ($\Theta = 90°$) and vertical ($\Phi = 90°$) patterns of a planar array of nine elements with currents shown in Fig. 7.17; $a/\lambda = 0.007\,022$, $h/\lambda = 0.25$, $b/\lambda = 0.25$, $d/\lambda = 0.6$; $I_{zi}(0) = 2.5$ milliamperes, $i = 1, 2, \ldots 9$.

The far-field pattern of the nine-element broadside planar array of half-wave elements is shown in Fig. 7.18 in the horizontal plane ($\Theta = 90°$) and the vertical plane in the direction of the maximum horizontal field ($\Phi = 90°$).

The general five-term theory is also valid for arrays that include parasitic elements. For example, in the nine-element planar array, the upper and lower rows may be parasitic with only elements 4, 5 and 6 driven and all constants the same as for the array described in Table 7.5 except that $V_{01} = V_{02} = V_{03} = V_{07} = V_{08} = V_{09} = 0$, $V_{04} = V_{05} = V_{06} = 1$ volt. The coefficients of the trigonometric functions in the distribution of current (7.33) are as given in Table 7.6. The admittances and impedances for the three driven elements are also tabulated. The distributions of the real and imaginary parts of the current referred to the driving voltage are shown in Fig. 7.19. The currents in the collinear parasitic elements are, of course, much smaller than in the driven elements and their distributions are quite different. Note, however, that the current

Table 7.6. *Nine-element planar array, with three elements driven; $a/\lambda = 0.007\,022$, $h/\lambda = 0.25$,*
$b/\lambda = 0.25$, $d/\lambda = 0.6$; $V_{01} = V_{02} = V_{03} = V_{07} = V_{08} = V_{09} = 0$, $V_{04} = V_{05} = V_{06} = 1$ volt

Coefficients of trigonometric functions in milliamperes per volt

i	A_i'	B_i'	D_i	$Q_i{}^a$	$R_i{}^a$
1, 3, 7, 9	$0.020 - j0.038$	$-0.554 - j0.197$	$-1.305 - j3.260$	$-2.225 - j3.406$	$16.467 - j33.106$
2, 8	$0.026 - j0.045$	$-0.519 - j0.340$	$-1.735 - j4.588$	$-2.145 - j2.471$	$16.336 - j27.315$
4, 6	$-0.386 - j5.485$	$-8.106 - j14.193$	$54.503 + j29.666$	$0 + j0$	$0 + j0$
5	$-0.309 - j5.731$	$-7.211 - j18.733$	$45.860 + j62.562$	$0 + j0$	$0 + j0$

^{a} Reverse signs for $i = 7, 8, 9$.

Admittances in millisiemens and impedances in ohms

i	$Y_{0i} = G_{0i} + jB_{0i}$	$Z_{0i} = R_{0i} + jX_{0i}$
4, 6	$8.244 - j0.019$	$121.3 + j0.3$
5	$6.530 + j5.322$	$92.0 - j75.0$

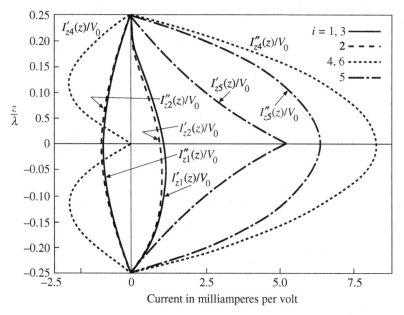

Figure 7.19 Currents in the nine elements of a planar array. $I_{zi}(z_i) = I_{zi}''(z_i) + j I_{zi}'(z_i)$,
$V_{01} = V_{02} = V_{03} = V_{07} = V_{08} = V_{09} = 0$, $V_{04} = V_{05} = V_{06} = 1$ volt. $a/\lambda = 0.007\,022$,
$h/\lambda = 0.25$, $b/\lambda = 0.25$, $d/\lambda = 0.6$.

in the middle element No. 5 has quite a different distribution from that of the other two
driven elements Nos. 4 and 6.

7.8   A three-dimensional array of twenty-seven elements[5]

As a final example of the application of the five-term theory consider a three-dimensional array consisting of three stacked, three-element, broadside curtains arranged in endfire as shown in Fig. 7.20. Let the lateral distances between the adjacent identical elements be $b_x/\lambda = b_y/\lambda = 0.25$, the axial distance between adjacent ends $(d - 2h)/\lambda = 0.1$; also let $a/\lambda = 0.007\,022$. If the antennas are individually a full wavelength long ($h/\lambda = 0.5$), the desired unidirectional endfire pattern is well realized when the driving voltages (which directly excite the large sinusoidal components of the currents) are assigned the following values: $V_{1+3n} = 1$, $V_{2+3n} = -j$, $V_{3+3n} = -1$ volt with $n = 0, 1, 2, \ldots 8$. The unidirectional beam is to be in the positive x direction. With this choice of parameters the five coefficients for the trigonometric components of the current in (7.33) have been computed and listed in Table 7.7. The associated driving-point admittances and impedances are also given. These are seen to vary widely as a necessary consequence of differences in the induced currents. Since the power in each element is given by $P_{0i} = \frac{1}{2} V_{0i}^2 G_{0i}$, and $V_{0i}^2 = V_{0i} V_{0i}^* = 1$, the relative powers are proportional to the driving-point conductances. It is seen from Table 7.7 that the nine elements in the plane $x = -b_x$ (which are the rear elements if the forward direction along the positive x-axis is that of the maximum beam), receive the largest amount of power from the generators $(9.78 V_0^2)$; the nine elements in the plane $x = 0$ the next largest amount $(5.78 V_0^2)$; and the nine forward elements in the plane $x = b_x$ the smallest amount $(4.51 V_0^2)$. However, the power is reasonably well divided among the elements. It is greatest in the middle elements 10, 13, 16 of the rear plane $(3.89 V_0^2)$ where induced currents are relatively small; it is least in the middle elements 12, 15, 18 of the forward plane $(1.26 V_0^2)$ where induced currents are relatively large.

The computed currents in 18 of the elements are shown graphically in Figs. 7.21a, b, c. The currents in elements 7, 8, 9, 16, 17, 18, 25, 26, 27 are obtained, respectively, from those in elements 1, 2, 3, 10, 11, 12, 19, 20, 21 with the substitution of $-z_i$ for z_i. Both the real and imaginary parts of the currents on differently situated but otherwise identical elements are seen to vary widely. Those in the outer tiers of elements with centers in the planes $z = \pm d$ exhibit a large asymmetry owing to the one-sidedness of the coupling.

If the full-wave elements in the array are replaced by half-wave elements ($h/\lambda = 0.25$) with the same axial distance $(d - 2h)/\lambda = 0.1$ between adjacent ends and all other conditions, including the driving voltages unchanged, the coefficients for the trigonometric components of the currents are computed to have the values listed in Table 7.8. The associated driving-point admittances and impedances are also given in Table 7.8. Note their very wide range.

---

[5] The computations in this section are those of V. W. H. Chang.

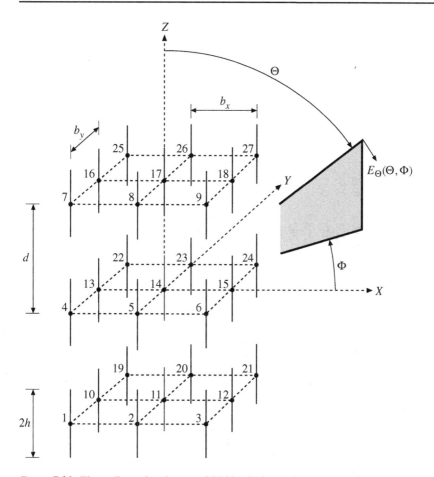

Figure 7.20 Three-dimensional array of 27 identical, equally-spaced elements.

As with the full-wave elements, the power in the nine elements in the rear plane $(x = -b_x)$ is greatest $(35.56 V_0^2)$, in the nine elements in the middle plane next greatest $(4.90 V_0^2)$, and in the nine elements in the forward plane least $(0.94 V_0^2)$. The distribution of power is seen to be very uneven. Indeed, the currents induced in the central forward elements 12, 15, 18 are now so great that these act as negative resistances or generators. The assigned voltage at the terminals of these elements can be maintained only if loads are connected across their terminals instead of generators. This is also true of the central element 14 in the middle plane. In evaluating the powers in the elements in the three planes, the negative values were subtracted since they represent power dissipated in a load, not radiated power. Note that the powers in elements 11 and 17 are not negative but very small. The entire admittance is very low, the input impedance correspondingly high. It might be supposed that these elements contribute negligibly to the radiation field. But this is not necessarily true. The fact that

Table 7.7. *Twenty-seven-element three-dimensional endfire array, with main beam in direction* $\Phi = 0$; $a/\lambda = 0.007022$, $h/\lambda = 0.5$, $b_x/\lambda = 0.5$, $b_y/\lambda = b_Y/\lambda = 0.25$, $d/\lambda = 1.1$; $V_{1+3n} = 1$, $V_{2+3n} = -j$, $V_{3+3n} = -1$ *volt*, $n = 0, 1, 2, \ldots 8$

Coefficients of trigonometric functions in milliamperes per volt

i	A'_i	B'_i	D_i	$Q_i{}^a$	$R_i{}^a$
1, 7, 19, 25	$0.062 - j3.633$	$0.442 - j0.687$	$1.033 - j2.590$	$0.157 + j0.529$	$0.102 + j0.430$
2, 8, 20, 26	$-3.639 - j0.017$	$-0.394 - j0.815$	$3.419 - j0.333$	$0.593 - j0.120$	$0.225 - j0.015$
3, 9, 21, 27	$0.033 + j3.613$	$-1.355 - j0.336$	$1.635 - j2.430$	$-0.110 - j0.651$	$-0.309 + j0.062$
4, 22	$0.062 - j3.623$	$0.522 - j0.544$	$0.963 - j2.358$	$0 + j0$	$0 + j0$
5, 23	$-3.628 - j0.017$	$-0.189 - j0.869$	$2.314 + j0.382$	$0 + j0$	$0 + j0$
6, 24	$0.037 + j3.599$	$-1.348 - j0.608$	$1.601 - j2.041$	$0 + j0$	$0 + j0$
10, 16	$0.075 - j3.590$	$0.744 - j0.361$	$1.063 + j2.997$	$0.242 + j0.686$	$0.306 + j0.474$
11, 17	$-0.359 - j0.025$	$-0.157 - j0.934$	$3.637 + j0.656$	$0.762 + j0.006$	$0.494 - j0.002$
12, 18	$0.037 + j3.559$	$-1.381 - j0.642$	$1.915 - j2.338$	$0.060 - j0.746$	$-0.126 - j0.165$
13	$0.074 - j3.578$	$0.846 - j0.422$	$0.985 + j2.675$	$0 + j0$	$0 + j0$
14	$-2.358 - j0.023$	$0.097 - j0.945$	$3.310 - j0.646$	$0 + j0$	$0 + j0$
15	$-0.041 + j3.545$	$-1.330 - j0.931$	$1.828 - j1.936$	$0 + j0$	$0 + j0$

[a] Reverse signs for $i = 7, 8, 9, 16, 17, 18, 25, 26, 27$.

Admittances in millisiemens and impedances in ohms

i	$Y_{0i} = G_{0i} + jB_{0i}$	$Z_{0i} = R_{0i} + jX_{0i}$	i	$Y_{0i} = G_{0i} + jB_{0i}$	$Z_{0i} = R_{0i} + jX_{0i}$
1, 7, 19, 25	$1.917 + j1.217$	$371.9 - j236.1$	10, 16	$2.552 + j1.735$	$268.0 - j182.2$
2, 8, 20, 26	$1.297 + j2.630$	$150.9 - j305.9$	11, 17	$1.212 + j3.322$	$96.9 - j265.6$
3, 9, 21, 27	$1.076 + j3.102$	$99.8 - j287.7$	12, 18	$0.847 + j3.623$	$61.2 - j261.7$
4, 22	$2.008 + j1.269$	$355.9 - j225.0$	13	$2.678 + j1.830$	$254.6 - j174.0$
5, 23	$1.357 + j2.761$	$143.4 - j291.8$	14	$1.245 + j3.504$	$90.0 - j253.4$
6, 24	$1.096 + j3.257$	$92.8 - j275.8$	15	$0.831 + j3.798$	$55.0 - j251.3$

Table 7.8. *Twenty-seven-element three-dimensional endfire array, with main beam in direction* $\Phi = 0$; $a/\lambda = 0.007\,022$, $h/\lambda = 0.25$, $b_x/\lambda = b_y/\lambda = 0.25$, $d/\lambda = 0.6$; $V_{1+3n} = 1$, $V_{2+3n} = -j$, $V_{3+3n} = -1$ volt, $n = 0, 1, 2, \ldots 8$

Coefficients of trigonometric functions in milliamperes per volt

i	A_i'	B_i'	D_i	$Q_i{}^{a}$	$R_i{}^{a}$
1, 7, 19, 25	$-0.432 - j5.436$	$-9.874 - j12.025$	$62.110 + j20.189$	$-2.912 + j2.092$	$24.552 - j20.379$
2, 8, 20, 26	$-5.365 + j0.084$	$-11.235 + j1.644$	$12.666 - j11.274$	$2.203 + j0.779$	$-19.840 - j6.877$
3, 9, 21, 27	$0.033 + j5.467$	$0.526 + j12.608$	$-3.857 - j24.434$	$0.807 - j1.716$	$-7.485 + j16.453$
4, 22	$-0.411 - j5.509$	$-10.672 - j12.054$	$61.129 + j27.424$	$0 + j0$	$0 + j0$
5, 23	$-5.415 + j0.079$	$-11.015 + j1.880$	$17.355 - j11.122$	$0 + j0$	$0 + j0$
6, 24	$0.039 + j5.511$	$1.022 + j12.365$	$-5.463 - j28.184$	$0 + j0$	$0 + j0$
10, 16	$-0.373 - j5.669$	$-9.514 - j16.649$	$56.729 + j51.632$	$-2.990 + j0.788$	$26.280 - j10.367$
11, 17	$-5.461 - j0.004$	$-13.163 + j0.023$	$25.922 - j0.547$	$1.665 + j0.330$	$-15.317 - j3.610$
12, 18	$-0.024 + j5.556$	$-0.643 + j14.403$	$3.868 - j37.307$	$0.504 - j1.226$	$-5.271 + j12.429$
13	$-0.332 - j5.754$	$-10.184 - j17.271$	$53.761 + j60.828$	$0 + j0$	$0 + j0$
14	$-5.505 - j0.014$	$-13.073 - j0.071$	$30.191 + j1.477$	$0 + j0$	$0 + j0$
15	$-0.026 + j5.595$	$-0.359 + j14.166$	$3.450 - j39.822$	$0 + j0$	$0 + j0$

[a] Reverse signs for $i = 7, 8, 9, 16, 17, 18, 25, 26, 27$.

Admittances in millisiemens and impedances in ohms

i	$Y_{0i} = G_{0i} + jB_{0i}$	$Z_{0i} = R_{0i} + jX_{0i}$	i	$Y_{0i} = G_{0i} + jB_{0i}$	$Z_{0i} = R_{0i} + jX_{0i}$
1, 7, 19, 25	$8.749 - j0.675$	$113.6 + j8.8$	10, 16	$7.475 + j4.143$	$102.3 - j56.7$
2, 8, 20, 26	$1.742 - j2.160$	$226.2 + j280.5$	11, 17	$0.141 - j0.109$	$4432.9 + j3430.8$
3, 9, 21, 27	$0.637 + j0.015$	$1570.1 - j38.0$	12, 18	$-0.513 + j2.087$	$-111.1 - j451.9$
4, 22	$7.643 + j1.488$	$126.1 - j24.5$	13	$5.895 + j6.299$	$79.2 - j84.6$
5, 23	$1.456 - j0.517$	$609.8 + j216.6$	14	$-0.367 + j1.275$	$-208.7 - j724.4$
6, 24	$0.617 + j1.400$	$263.5 - j598.0$	15	$-0.678 + j3.093$	$-67.6 - j308.5$

(a)

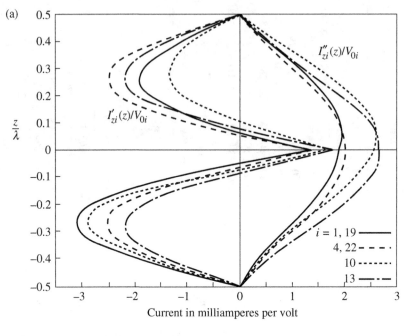

(b)

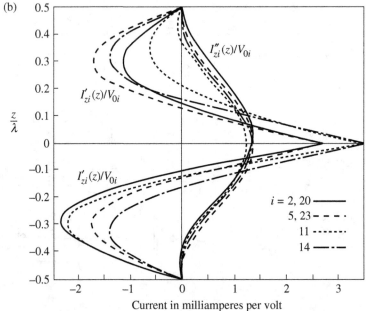

Figure 7.21 (a) Currents in elements Nos. 1, 4, 10 and 13 of the 27-element array shown in Fig. 7.20. $V_{1+3n} = 1$, $V_{2+3n} = -j$, $V_{3+3n} = -1$ volt, $n = 0, 1, 2, \ldots 8$; $a/\lambda = 0.007\,022$, $h/\lambda = 0.5$, $b_x/\lambda = b_y/\lambda = 0.25$, $d/\lambda = 1.1$. (b) Like (a) but for elements 2, 5, 11 and 14.

$I_{z11}(0)$ is near zero does not mean that $I_{z11}(z_{11})$ is everywhere equally small. It may be quite large.

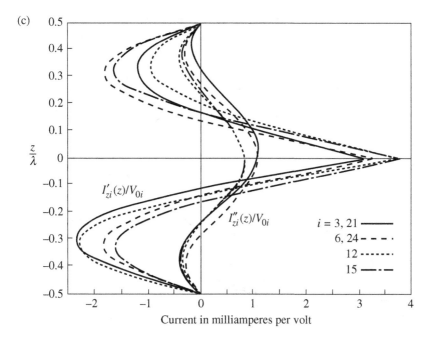

Figure 7.21 (c) Like (a) but for elements 3, 6, 12 and 15.

    The currents in the elements of the 27-element array of half-wave dipoles when the driving voltages are assigned to be $V_{1+3n} = 1$, $V_{2+3n} = -j$, $V_{3+3n} = -1$ volt with $n = 0, 1, \ldots 8$ are shown in Fig. 7.22. Note that the currents on the elements in the rear plane (top figure) are greater than those in the middle plane (lower left) and still greater than those in the forward plane (lower right). Specifically, the current in element 11 is very small at $z = 0$, but quite comparable with the other currents out along the antenna. It is seen from Fig. 7.22 that even with half-wave elements the conventional assumption that the distributions of current along all elements are identical and cosinusoidal is not well satisfied. Since this assumption also implies that the phase of each current is the same along the antenna as at the driving point, it is of interest to examine the relative phases referred to a common reference, namely V_{01}. This is done in Fig. 7.23 where the phase angles of the currents along all elements are shown. For the elements in the rear plane where induced currents are not of major significance, the phase angle varies relatively little from $z = 0$ to $z = \pm h$, much as in an isolated antenna. On the other hand, when induced currents constitute the major parts of the currents in an element, the phase angle varies very widely – as much as 153° in the middle element 14. It is clear that when large currents are induced in some elements of an array, as in endfire arrangements which maintain a maximum field along the antennas, an assumed current with constant phase cannot be expected to represent even approximately the actual currents in an array.

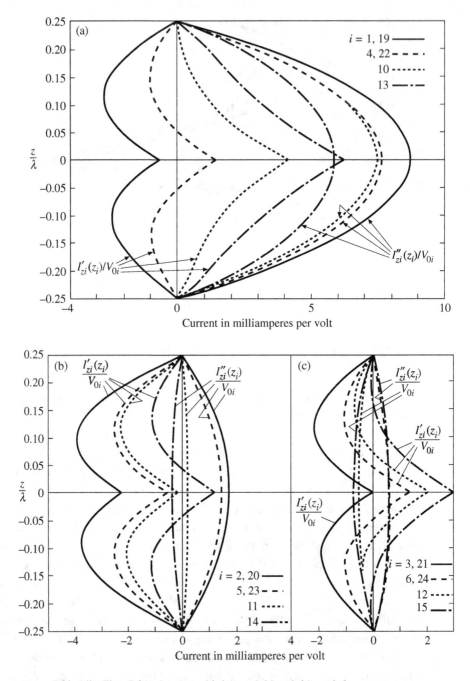

Figure 7.22 Like Figs. 7.21a, b, c but with $h/\lambda = 0.25$ and $d/\lambda = 0.6$.

Since with half-wave antennas the principal component of the current has its maximum value at $z = 0$, the progressive phases in the currents required for a specified field pattern can be approximated more closely when the maxima of the

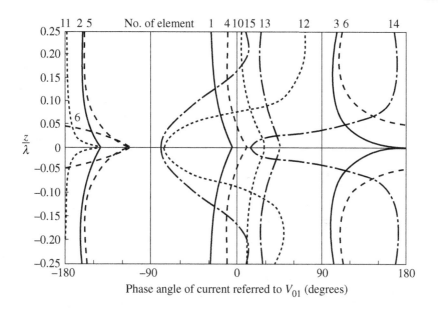

Figure 7.23 Phases of the currents in Fig. 7.22 all referred to V_{01}.

currents, i.e. the values $I_{zi}(0)$, are assigned instead of the voltages. Let the values $I_{1+3n}(0) = 2.5 \times 10^{-3}$, $I_{2+3n}(0) = -j2.5 \times 10^{-3}$, $I_{3+3n}(0) = -2.5 \times 10^{-3}$ amperes be specified for the same 27-element array. The corresponding coefficients for the currents as evaluated by computer are in Table 7.9 together with the required driving voltages and the associated admittances and impedances for the elements. Note that the voltages range from $V_{02} = -0.006 - j0.236$ to $V_{15} = -1.173 + j0.403$ volts. The complete distributions of current are in Figs. 7.24a, b, c in the normalized form: $I_{zi}(z_i)/I_{zi}(0) = I''_{zi}(z_i)/I_{zi}(0) + jI'_{zi}(z_i)/I_{zi}(0)$.

With the driving-point currents specified, the power in each element is conveniently determined from $P_{0i} = \frac{1}{2}|I_{zi}(0)|^2 R_{0i}$. It is seen to be proportional to R_{0i} as given in Table 7.9. The distribution of power to the elements with the driving-point currents assigned is quite different from when the voltages are specified. Note that the nine elements in the rear plane $(x = -b_x)$, which with voltages specified received the greatest power, now receive the least $(411.8I_0^2)$, the middle plane $(x = 0)$ is again intermediate $(523.6I_0^2)$, and the elements in the forward plane $(x = b_x)$, which with voltages assigned received the least power, now receive the greatest $(1484.8I_0^2)$. However, the division of power is not as extreme as before and there are no elements that have negative resistances and, therefore, feed power into a load instead of receiving power from a generator. A comparison of the relative powers in all of the elements is shown schematically in Fig. 7.25 in which boxes are located in the three-dimensional pattern of the array. The upper number in each box is the conductance G_{0i} when the conditions of Table 7.8 with voltages specified obtain; it is proportional to the power P_{0i} in each element. The lower number in each box is

Table 7.9. *Twenty-seven-element three-dimensional endfire array, with main beam in direction $\Phi = 0$;*
$a/\lambda = 0.007022$, $h/\lambda = 0.25$, $b_x/\lambda = b_y/\lambda = 0.25$, $d/\lambda = 0.6$; $I_{1+3n}(0) = 2.5$, $I_{2+3n}(0) = -j2.5$,
$I_{3+3n}(0) = -2.5$ milliamperes, $n = 0, 1, 2, \ldots 8$

Coefficients of trigonometric functions in milliamperes

i	A'_i	B'_i	D_i	$Q_i{}^a$	$R_i{}^a$
1, 7, 19, 25	$-0.981 - j1.034$	$-4.633 - j1.870$	$21.004 + j2.854$	$-0.096 + j0.670$	$-0.252 - j6.390$
2, 8, 20, 26	$-1.289 + j0.157$	$-2.825 - j2.993$	$5.244 - j18.220$	$0.702 + j0.802$	$-6.607 - j6.543$
3, 9, 21, 27	$1.088 + j3.730$	$4.704 + j8.473$	$-20.880 - j16.196$	$-0.066 - j1.859$	$1.803 + j17.612$
4, 22	$-1.615 - j1.016$	$-6.550 - j1.345$	$25.384 + j1.124$	$0 + j0$	$0 + j0$
5, 23	$-1.362 + j0.257$	$-2.582 + j3.699$	$4.165 - j20.288$	$0 + j0$	$0 + j0$
6, 24	$2.298 + j4.128$	$8.353 + j8.561$	$-29.207 - j15.136$	$0 + j0$	$0 + j0$
10, 16	$-1.524 - j1.647$	$-5.991 - j3.109$	$23.784 + j4.991$	$0.194 + j1.044$	$-2.972 - j9.951$
11, 17	$-2.062 + j0.445$	$-4.486 + j3.812$	$8.275 - j20.031$	$1.141 + j0.698$	$-10.697 - j5.579$
12, 18	$0.907 + j5.463$	$4.470 + j12.196$	$-20.701 - j22.986$	$0.010 - j2.896$	$1.103 + j26.825$
13	$-2.408 - j1.738$	$-8.521 - j2.671$	$29.404 + j3.185$	$0 + j0$	$0 + j0$
14	$-2.273 + j0.712$	$-4.393 + j4.988$	$7.235 - j23.137$	$0 + j0$	$0 + j0$
15	$2.341 + j0.639$	$8.894 + j13.302$	$-30.910 - j23.585$	$0 + j0$	$0 + j0$

[a] Reverse signs for $i = 7, 8, 9, 16, 17, 18, 25, 26, 27$.

Admittances in millisiemens, impedances in ohms, driving EMF's in volts

i	$Y_{0i} = G_{0i} + jB_{0i}$	$Z_{0i} = R_{0i} + jX_{0i}$	V_{0i}	i	$Y_{0i} = G_{0i} + jB_{0i}$	$Z_{0i} = R_{0i} + jX_{0i}$	V_{0i}
1, 7, 19, 25	$7.810 + j6.470$	$75.9 - j62.9$	$0.190 - j0.157$	10, 16	$4.804 + j4.082$	$120.9 - j102.7$	$0.302 - j0.257$
2, 8, 20, 26	$10.580 + j0.277$	$94.4 - j2.5$	$-0.006 - j0.236$	11, 17	$6.459 + j1.007$	$151.1 - j23.6$	$-0.059 - j0.378$
3, 9, 21, 27	$3.430 + j0.888$	$273.2 - j70.8$	$-0.683 + j0.177$	12, 18	$2.447 + j0.351$	$400.4 - j57.5$	$-1.001 + j0.144$
4, 22	$4.299 + j6.231$	$75.0 - j108.7$	$0.186 - j0.272$	13	$2.896 + j3.771$	$128.1 - j166.8$	$0.320 - j0.417$
5, 23	$9.898 + j0.910$	$100.2 - j9.2$	$-0.023 - j0.250$	14	$5.622 + j1.427$	$167.1 - j42.4$	$-0.106 - j0.418$
6, 24	$2.591 + j1.353$	$303.2 - j158.3$	$-0.758 + j0.396$	15	$1.905 + j0.654$	$469.5 - j161.2$	$-1.173 + j0.403$

the resistance R_{0i} for the same array when the conditions of Table 7.9 are maintained with input currents assigned; it is proportional to the power P_{0i} in each element. The relative distribution of power is seen to be reversed.

The distributions of current in Fig. 7.24a, b, c all have the same value at $z_i = 0$ and the components in phase with the input current are similarly distributed along the antenna in a rough sense. They range from a flattened cosine to a triangle. However, the quadrature currents are by no means negligible (they are presumed not to exist in conventional array theory). Indeed, they are of major significance in those elements which radiate most of the power. Note in particular the very large quadrature currents in all of the elements in the forward plane $x = b_x$ which are shown in Fig. 7.24c. (The currents in elements 9, 18 and 24 are like those in 3, 12 and 21 with $-z_i$ substituted for z_i.) These have distributions quite different from the conventionally assumed cosine curve. Evidently the phases are also as far from constant as those shown in Fig. 7.23 for the same array with assigned voltages.

The purpose of an array is to maintain a useful far field. The computed far-field patterns of the 27-element endfire array shown in Fig. 7.20 are in Fig. 7.26 for all the cases considered in this section, that is, for $h/\lambda = 0.5$ with voltages assigned, $h/\lambda = 0.25$ with voltages and currents assigned. The horizontal patterns in the equatorial plane $\Theta = 90°$ all have the principal maximum in the desired direction, $\Phi = 0, \Theta = 90°$. They also have a minor maximum in the backward direction, $\Phi = 180°, \Theta = 90°$. This is smallest with the array of half-wave elements with specified input currents, it is largest with the half-wave elements with voltages specified. The array of full-wave elements with voltages specified has a backward lobe of intermediate height. The vertical patterns for the array of half-wave elements are essentially the same when currents or voltages are specified. The former has a very slightly broader main beam and a correspondingly slightly lower minor lobe level. The array of full-wave elements has the narrowest main beam in the vertical pattern – the array is, of course, axially twice as long. On the other hand, its minor lobe level is correspondingly somewhat higher. Note that since very good approximations of actual currents on all of the elements are used, there are no nulls as would have been obtained with assumed sinusoidal currents with constant phase along each antenna. The details of the minor lobe structure derived from the five-term approximations of the several currents should have an accuracy comparable to that of the major lobe.

If all of the 27 elements are driven in phase, an approximately circular pattern with some undulations is obtained as would be expected; of interest is the fact that in this case, too, a number of the elements have negative driving-point conductances and resistances. This indicates that the induced currents in these elements predominate so that they act as generators and not as loads when connected to a transmission line. Elements with negative resistances are likely to occur in most arrays with large numbers of rather closely coupled elements.

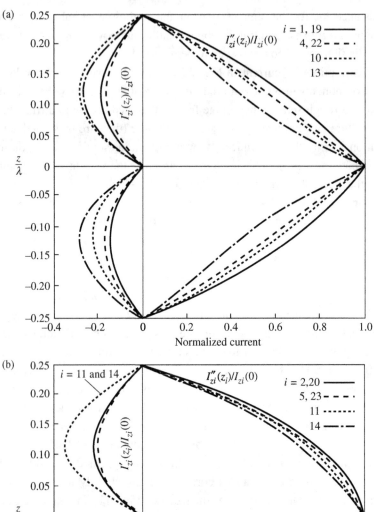

Figure 7.24 (a) Currents in elements Nos. 1, 4, 10 and 13 of the 27-element array shown in Fig. 7.20. $I_{1+3n}(0) = 2.5\,\text{mA}$, $I_{2+3n}(0) = -j2.5\,\text{mA}$, $I_{3+3n}(0) = -2.5\,\text{mA}$, with $n = 0, 1, 2, \ldots 8$. $a/\lambda = 0.007\,022$, $h/\lambda = 0.25$, $b_x/\lambda = b_y/\lambda = 0.25$, $d/\lambda = 0.6$. (b) Like (a) but for elements 2, 5, 11 and 14.

(c)

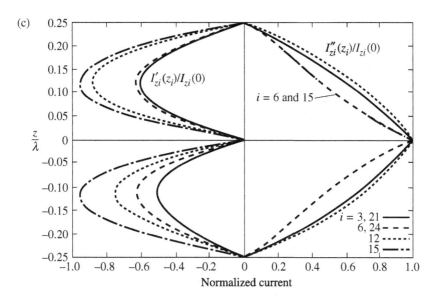

Figure 7.24 (c) Like (a) but for elements 3, 6, 12 and 15.

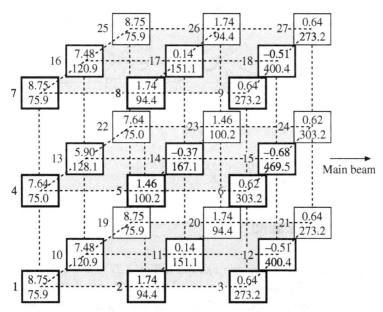

Figure 7.25 Schematic diagram showing the relative powers supplied to the half-wave ($h/\lambda = 0.25$) elements in a 27-element endfire array. The upper number in each box is G_{0i} (in millisiemens) which is proportional to power supplied when the V_{0i} are specified. The lower number is R_{0i} (in ohms) which is proportional to power when the driving-point currents $I_{zi}(0)$ are specified.

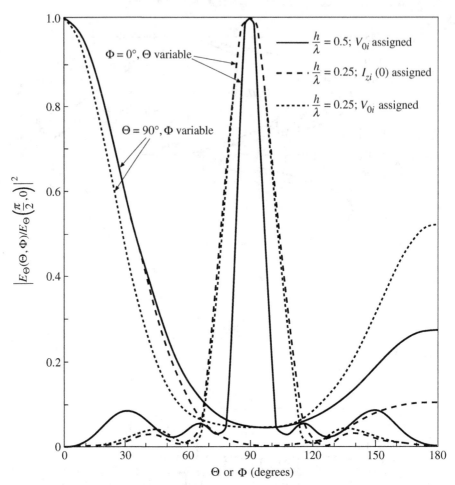

Figure 7.26 Horizontal ($\Theta = 90°$) and vertical ($\Phi = 0°$) patterns of a three-dimensional endfire array of 27 elements; $a/\lambda = 0.007\,022$, $b_x/\lambda = b_y/\lambda = 0.25$, $(d - 2h)/\lambda = 0.1$; $V_{1+3n} = 1$, $V_{2+3n} = -j$, $V_{3+3n} = -1$ volt or $I_{1+3n}(0) = 2.5$ mA, $I_{2+3n}(0) = -j2.5$ mA, $I_{3+3n}(0) = -2.5$ mA, with $n = 0, 1, 2, \ldots 8$.

7.9 Electrical beam scanning

The major lobe in the endfire patterns shown in Fig. 7.27 is in the direction $\Theta = 90°$, $\Phi = 0°$. This is readily switched electrically to the direction $\Theta = 90°$, $\Phi = 90°$ simply by interchanging the phases of the voltages or currents in the broadside rows (parallel to the y-axis in Fig. 7.20) and the endfire rows (parallel to the x-axis). For example, the assigned voltages would be $V_{0i} = 1$ volt, $1 \leq i \leq 9$; $V_{0i} = -j$ volt, $10 \leq i \leq 18$; $V_{0i} = -1$ volt, $19 \leq i \leq 27$ or, if the driving-point currents are assigned, $I_{zi}(0) = 2.5$ mA, $1 \leq i \leq 9$; $I_{zi}(0) = -j2.5$ mA, $10 \leq i \leq 18$; $I_{zi}(0) = -2.5$ mA, $19 \leq i \leq 27$.

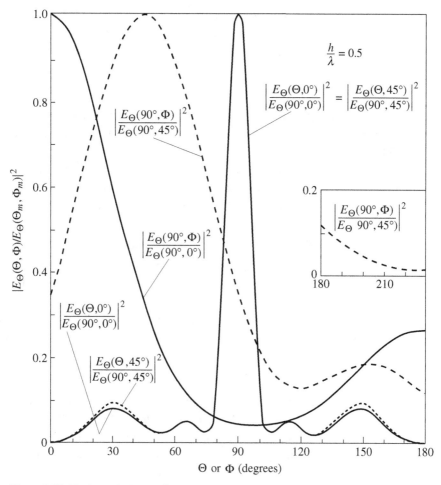

Figure 7.27 Horizontal ($\Theta = 90°$) and vertical patterns of 27-element three-dimensional endfire array with beam in the directions $\Phi = 0°$ and $\Phi = 45°$. Driving voltages specified as in Table 7.10; $a/\lambda = 0.007\,022$, $h/\lambda = 0.5$, $b_x/\lambda = b_y/\lambda = 0.25$, $d/\lambda = 1.1$.

More generally, the direction of the beam is specified by the far-field formula (7.57) in which the field factor of each individual antenna i is given by the square bracket, and the combination of these into a pattern for the array is determined by the phase factors $\exp(j\beta_0 \mathbf{r}_i \cdot \hat{\mathbf{R}}_0)$. The contribution to the pattern by each element is greatest when the amplitudes A'_i, B'_i, D_i, Q_i and R_i all include the common factor $\exp(-j\beta_0 \mathbf{r}_i \cdot \hat{\mathbf{R}}_0)$. When this is true the contributions from all the elements arrive in phase in the direction specified by particular values of Θ, Φ in $(\mathbf{r}_i \cdot \hat{\mathbf{R}}_0)$ as given in (7.58). This is, evidently, a necessary condition for a maximum in the field pattern. However, it is not a sufficient condition since the directional properties of the individual elements, as given by the square bracket in (7.57) for element i, are also involved. These may differ considerably from element to element owing to differences in the distributions of current so that no simple formula for the direction Θ_m, Φ_m of the main lobe on the field pattern can be written down. In the special case of maxima in the equatorial plane ($\Theta_m = 90°$) the presence of the common phase factor

$$\exp(-j\beta_0 \mathbf{r}_i \cdot \hat{\mathbf{R}}_0) = \exp[-j\beta_0(X_i \cos \Phi_m + Y_i \sin \Phi_m)],$$

is sufficient to fix the main beam in the direction Φ_m. For example, when $\Phi_m = 0$, the factor is $\exp(-j\beta_0 X_i)$. This means that when voltages are assigned, these must have the relative phases $\exp(j\beta_0 b_x)$, 1, $\exp(-j\beta_0 b_x)$, respectively, for the elements in the planes $X_i = -b_x$, $X_i = 0$, and $X_i = b_x$. When $b_x = \lambda/4$ as in the arrays considered in this chapter, the phases are $\exp[j(\pi/2)] = j$, 1, and $\exp[-j(\pi/2)] = -j$; or, if j is removed as a common factor, the relative phases are given by 1, $-j$, -1, which are the values used in Table 7.7. When the beam is switched to $\Phi_m = 90°$, the phase factor is $\exp(-j\beta_0 Y_i)$ so that the voltages in the planes $Y_i = -b_y = -\lambda/4$, $Y_i = 0$, $Y_i = b_y = \lambda/4$ must have the relative phases $\exp[j(\pi/2)]$, 1, and $\exp[-j(\pi/2)]$. When driving-point currents instead of voltages are assigned, the coefficients apply to them unchanged.

If the direction of the maximum beam is to be $\Theta_m = 90°$, $\Phi_m = 45°$, the coefficients are given by

$$\exp(-j\beta_0 \mathbf{r}_i \cdot \hat{\mathbf{R}}_0) = \exp[-j\beta_0(X_i + Y_i)\sqrt{2}/2].$$

For the three-dimensional array of 27 elements shown in Fig. 7.20, the elements are located at $X_i = -b_x = -\lambda/4$, $X_i = 0$, $X_i = b_x = \lambda/4$ and $Y_i = -b_y = -\lambda/4$, $Y_i = 0$, $Y_i = b_y = \lambda/4$. Thus, the following relative phases must be assigned to the driving voltages (or currents if these are specified instead of the voltages):

$$X_i = Y_i = -\frac{\lambda}{4}: \qquad \exp(j\pi\sqrt{2}/2)$$

$$X_i = -\frac{\lambda}{4}, \; Y_i = 0; \; X_i = 0, \; Y_i = -\frac{\lambda}{4}: \qquad \exp(j\pi\sqrt{2}/4)$$

$$X_i = \frac{\lambda}{4}, \; Y_i = -\frac{\lambda}{4}; \; X_i = 0, \; Y_i = 0; \; X_i = -\frac{\lambda}{4}, \; Y_i = \frac{\lambda}{4}: \qquad 1$$

$$X_i = 0, \; Y_i = \frac{\lambda}{4}; \; X_i = \frac{\lambda}{4}, \; Y_i = 0: \qquad \exp(-j\pi\sqrt{2}/4)$$

$$X_i = Y_i = \frac{\lambda}{4}: \qquad \exp(-j\pi\sqrt{2}/2).$$

Alternatively, if $\exp(j\pi\sqrt{2}/2)$ is removed as a common factor, the five phase coefficients are, respectively, 1, $\exp(-j\pi\sqrt{2}/4)$, $\exp(-j\pi\sqrt{2}/2)$, $\exp(-j3\pi\sqrt{2}/4)$ and $\exp(-j\pi\sqrt{2})$. With reference to Fig. 7.20, the required assigned voltages are listed in exponential form near the top of Table 7.10 and in complex numerical form later in the table. If these assigned voltages are used in the computer program, the coefficients for

Table 7.10. *Twenty-seven-element three-dimensional endfire array – beam direction* $\Phi = 45°$; $a/\lambda = 0.007\,022$,
$h/\lambda = 0.5$, $b_x/\lambda = b_y/\lambda = 0.25$, $d/\lambda = 1.1$; $V_{1+3n} = 1$, $V_{2+3n} = V_{10+3n} = \exp(-j\pi\sqrt{2}/4)$,
$V_{3+3n} = V_{11+3n} = V_{19+3n} = \exp(-j\pi\sqrt{2}/2)$, $V_{12+3n} = V_{20+3n} = \exp(-j3\pi\sqrt{2}/4)$,
$V_{21+3n} = \exp(-j\pi\sqrt{2})$ *volt*, $n = 0, 1, 2$

Coefficients of trigonometric functions in milliamperes per volt

i	A_i'	B_i'	D_i	Q_i [a]	R_i [a]
1, 7	$0.096 - j3.641$	$0.147 - j0.558$	$1.204 + j1.871$	$0.184 + j0.572$	$0.334 + j0.528$
2, 8, 10, 16	$-3.219 - j1.657$	$-0.308 - j0.769$	$3.264 + j0.528$	$0.613 + j0.097$	$0.424 + j0.058$
3, 9, 19, 25	$-2.898 - j2.189$	$-0.974 - j0.365$	$2.792 - j2.055$	$0.300 - j0.442$	$0.104 + j0.025$
4	$0.096 - j3.631$	$0.216 - j0.427$	$1.142 + j1.641$	$0 + j0$	$0 + j0$
5, 13	$-3.210 - j1.652$	$-0.108 - j0.767$	$2.994 + j0.473$	$0 + j0$	$0 + j0$
6, 22	$-2.889 + j2.181$	$-0.878 - j0.552$	$2.632 - j0.181$	$0 + j0$	$0 + j0$
11, 17	$-2.878 + j2.160$	$-0.793 - j0.646$	$3.415 - j1.816$	$0.613 - j0.473$	$0.398 - j0.321$
12, 18, 20, 26	$0.701 + j3.532$	$-1.307 + j0.030$	$1.112 - j3.167$	$-0.141 - j0.697$	$-0.906 - j0.147$
14	$-2.867 + j2.154$	$-0.598 - j0.817$	$3.147 - j1.620$	$0 + j0$	$0 + j0$
15, 23	$0.702 + j3.519$	$-1.346 - j0.236$	$1.152 - j2.800$	$0 + j0$	$0 + j0$
21, 27	$3.500 + j0.888$	$-1.364 + j0.883$	$-0.930 - j2.181$	$-0.695 - j0.160$	$-0.207 + j0.471$
24	$3.491 + j0.879$	$-2.161 + j0.730$	$-0.596 - j1.941$	$0 + j0$	$0 + j0$

[a] Reverse signs for $i = 7, 8, 9, 16, 17, 18, 25, 26, 27$.

Admittances in millisiemens, impedances in ohms, voltages in volts

i	$Y_{0i} = G_{0i} + jB_{0i}$	$Z_{0i} = R_{0i} + jX_{0i}$	V_{0i}	i	$Y_{0i} = G_{0i} + jB_{0i}$	$Z_{0i} = R_{0i} + jX_{0i}$	V_{0i}
1, 7	$1.498 + j0.755$	$532.3 - j268.2$	$1.0000 + j0.0000$	11, 17	$1.366 + j3.339$	$105.0 - j256.6$	$-0.6057 - j0.7957$
2, 8, 10, 16	$2.081 + j1.925$	$258.9 - j239.6$	$0.4440 + j0.8960$	12, 18, 20, 26	$0.886 + j3.335$	$74.4 - j280.1$	$-0.9819 + j0.1894$
3, 9, 19, 25	$1.704 + j2.358$	$201.3 - j278.6$	$-0.6057 - j0.7957$	14	$1.408 + j3.524$	$97.7 - j244.7$	$-0.6057 - j0.7957$
4	$1.574 + j0.788$	$508.1 - j254.3$	$1.0000 + j0.0000$	15, 23	$0.893 + j3.505$	$68.3 - j267.9$	$-0.9819 + j0.1894$
5, 13	$2.186 + j2.019$	$246.9 - j228.0$	$0.4440 + j0.8960$	21, 27	$0.573 + j3.637$	$42.3 - j268.3$	$-0.2662 + j0.9639$
6, 22	$1.789 + j2.464$	$192.9 - j265.7$	$-0.6057 - j0.7957$	24	$0.550 + j3.779$	$37.4 - j257.8$	$-0.2662 + j0.9639$

the currents in the 27 elements of the same array analyzed in Table 7.7, but with the beam rotated 45°, are as listed in Table 7.10. Note that the number of different currents is greater than when the main beam is in the direction $\Phi = 0$ or $\Phi = 90°$. The currents in the elements with centers in the plane $Z = d$ are, of course, the same as those in the plane $Z = -d$ with z_i replaced by $-z_i$. The associated admittances and impedances are also given together with the numerical values of the assigned voltages.

Since the voltage magnitudes are $|V_{0i}| = 1$ volt, the relative powers to the elements are proportional to the driving-point conductances. In general, these are quite comparable in magnitude and range to those in Table 7.7 but the larger values are shifted to the new elements in the backward direction ($\Phi = 225°$), the smaller values to the new elements in the forward direction ($\Phi = 45°$).

The horizontal field pattern in the plane $\Theta = 90°$ and the vertical pattern in the plane $\Phi = 45°$ are shown in Fig. 7.27 together with the corresponding patterns from Fig. 7.26. It is seen that in the horizontal plane the main lobe has been rotated substantially unchanged through 45°, but that the minor lobe structure is somewhat different. The change in the vertical pattern is so small that it can be distinguished only near the peak of a minor lobe. Evidently, the rather narrow beam of a three-dimensional array of full-wave elements in collinear, broadside, and endfire combinations is readily rotated by appropriate changes in the phases of the driving voltages. A similar rotation of the corresponding array of half-wave elements is readily achieved with precisely the same changes in the phases of the assigned driving-point currents.

7.10   Problems with practical arrays

The theory developed in this and the preceding chapters provides a complete, practical tool for the quantitative determination of the properties of very general arrays when the active elements are driven by a concentrated EMF at their centers. In practice, antennas are driven from transmission lines that maintain the desired voltage across the terminals of the antennas, but also introduce the complications that accompany transmission-line end-effects and the coupling between the antenna and the line. There is also the possibility of unbalanced currents on open-wire lines or on the outside surfaces of coaxial lines. These latter can be excited by asymmetrical conditions at the junctions of antennas and feeding lines, or by the intense near fields in an array whenever transmission lines are not in a neutral plane of these fields. Since such currents induced along transmission lines usually contribute significantly to the radiation field and can, therefore, constitute a non-negligible part of the load, both the circuit and field properties of an array can be modified greatly whenever they are excited. Important aspects of the problems relating to end-effects and coupling effects between antennas and transmission lines as well as techniques of measurement are considered in Chapter 14. However, questions relating to the maintenance of the required voltages for antennas with positive conductances and loads for those with negative conductances in large arrays are not analyzed since they involve the specific geometry of each array. A problem of this sort in which elements with both positive and negative resistances play important roles is treated in detail at the end of Chapter 6 where the log-periodic array is analyzed. This antenna includes not only radiating elements with specified geometrical properties but also a feeding line with definite electrical characteristics. Since it is in the neutral plane, the problems of unbalanced currents are avoided, but those relating to the transfer of power from the radiating elements to the line and vice versa constitute a major aspect of the analysis.

8 Vertical dipoles on and over the earth or sea

8.1 Introduction

In their practical application in radio broadcasting, communicating with ships and submerged submarines, and in cellular telephone transmission, dipole antennas are in proximity with the surface of the earth or sea. More generally, the earth may be coated with a layer of asphalt or concrete and the water may have a layer of ice. In a common engineering approximation, the earth or sea is treated as a perfectly conducting reflector or, in the far field, the earth-reflected field is assumed to be correctly given by the plane-wave reflection coefficient. As shown in Fig. 8.1, the far field of a short dipole at the height d over the earth or sea [1] is represented as the superposition of a direct field and an earth-reflected field. The former travels the distance $r_1 = [\rho^2 + (z - d)^2]^{1/2}$ from the source to the point of observation at the point ρ, z. The latter is reflected as a plane wave from the surface so that it travels the distance $r_2 = [\rho^2 + (z + d)^2]^{1/2}$. The plane-wave reflection coefficient is

$$f_{er}(\Theta) = \frac{N^2 \cos \Theta - (N^2 - \sin^2 \Theta)^{1/2}}{N^2 \cos \Theta + (N^2 - \sin^2 \Theta)^{1/2}}, \tag{8.1}$$

where, with the time dependence $e^{-i\omega t}$,

$$N^2 = \frac{k_2^2}{k_0^2} = \epsilon_{2r} + \frac{i\sigma_2}{\omega\epsilon_0}$$

$$k_2^2 = k_0^2 \left(\epsilon_{2r} + \frac{i\sigma_2}{\omega\epsilon_0} \right), \qquad k_0^2 = \omega^2 \mu_0 \epsilon_0 = \frac{\omega^2}{c^2}. \tag{8.2}$$

Here, k_0 is the wave number of air and $k_2 = \beta_2 + i\alpha_2$ is the complex wave number of the earth or sea. Note that when $k_2 \to \infty$ as for a perfect conductor, $f_{er}(\Theta) = 1$. In the spherical coordinates r_0, Θ, Φ, the far field in the air of a dipole with the current I and the effective length $2h_e$ is given by

$$E_{0\Theta}^r(r_0, \Theta) = cB_{0\Phi}^r(r_0, \Theta)$$

$$= -\frac{i\omega\mu_0(2h_e I)}{2\pi} \frac{e^{ik_0 r_0}}{r_0} A(\Theta) \sin \Theta, \tag{8.3}$$

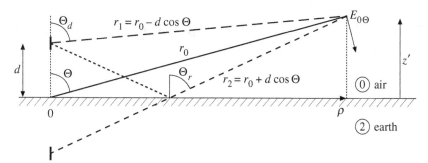

Figure 8.1 Vertical electric dipole at height d in air (region 0) over the earth (region 2). Taken from King and Sandler [1, Fig. 1]. © 1994 I.E.E.E.

where

$$A(\Theta) = \tfrac{1}{2}[e^{-ik_0 d \cos \Theta} + f_{er}(\Theta)e^{ik_0 d \cos \Theta}]. \tag{8.4}$$

When $f_{er}(\Theta) = 1$ for a perfect conductor,

$$A(\Theta) = \cos(k_0 d \cos \Theta). \tag{8.5}$$

The field on the surface of the earth or sea is obtained with $\Theta = \pi/2$, where

$$
\begin{aligned}
A(\pi/2) &= \tfrac{1}{2}[1 + f_{er}(\pi/2)] \\
&= \lim_{\Theta \to \pi/2} \left[\frac{N^2 \cos \Theta}{N^2 \cos \Theta + (N^2 - \sin^2 \Theta)^{1/2}} \right] \\
&= \begin{cases} 0, & N < \infty, \\ 1, & N = \infty. \end{cases}
\end{aligned}
\tag{8.6}
$$

Since $N = \infty$ only when $\sigma_2 = \infty$, i.e. the earth is a perfect conductor, it follows that the field on the surface of the earth is zero for all physically available types of earth or water. But this is contrary to fact. Vertical receiving antennas on the surface of the earth or sea receive strong signals at large distances from the source!

The reason that (8.3) with (8.4) gives incorrect results when Θ is at and near $\pi/2$ is that the assumption that the far field of a vertical dipole is a plane wave and is reflected from the earth or sea as such is not true. The electromagnetic field near and along the air–earth or air–sea boundary actually includes an inhomogeneous wave known as the Norton surface wave or, more generally, as a lateral wave. It is not included in (8.3) with (8.4) so that the correct answer is not obtained when Θ is at or near $\pi/2$.

8.2   The complete electromagnetic field of a vertical dipole over the earth or sea with or without a coating

The electromagnetic field of a center-driven electric dipole with the length $2h$ and the approximate current distribution

$$I_z(z) = I_z(0) \frac{\sin k_0(h - |z|)}{\sin k_0 h}, \qquad k_0 h < \pi \tag{8.7}$$

is conveniently expressed in terms of a dipole with the constant current I and a length $2h_e$ – called the effective length – that has the same electric moment. This is defined as follows:

$$2h_e I = \int_{-h}^{h} I_z(z)\, dz = \frac{2I_z(0)}{\sin k_0 h} \int_0^h \sin k_0(h - z)\, dz$$

$$= 2I_z(0)k_0^{-1} \frac{1 - \cos k_0 h}{\sin k_0 h}. \tag{8.8}$$

With $I_z(0) = I$, the effective length is

$$2h_e = 2k_0^{-1} \frac{1 - \cos k_0 h}{\sin k_0 h}. \tag{8.9}$$

When $k_0^2 h^2 \ll 9$, $\sin k_0 h \sim k_0 h$ and $1 - \cos k_0 h \sim k_0^2 h^2 / 2$, so that

$$2h_e = h. \tag{8.10}$$

With the dipole represented by $(2h_e I)$, the complete field of any dipole with an electrical length $k_0 h < \pi$ can be expressed in terms of the field of an infinitesimal unit dipole with $(2h_e I) = 1$ ampere meter (A m).

The analytical determination of the electromagnetic field in the air (region 0, wave number k_0) over the earth or sea (region 2, wave number k_2) when this is coated with a thin layer of dielectric (region 1, wave number k_1, thickness l) as shown in Fig. 8.2 is complicated and is not carried out here. It is available in [2]. Subject only to the conditions

$$k_0^2 \ll |k_1^2| \ll |k_2^2|, \qquad |k_1 l| \le 0.6 \tag{8.11}$$

and with the definition of the small quantity ϵ (not a permittivity when written without a subscript)

$$\epsilon = \frac{k_0}{k_2} - i k_0 l \tag{8.12}$$

the following formulas give the electromagnetic field of a short vertical dipole located at any height d over the air–dielectric surface. The cylindrical components of the field

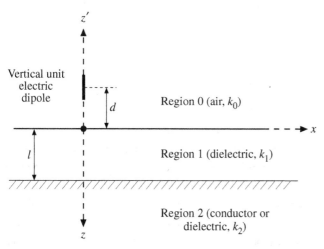

Figure 8.2 Unit vertical electric dipole at height d over plane boundary $z = -z' = 0$ between air and a sheet of dielectric with thickness l over a conducting (or dielectric) half-space. Taken from King and Sandler [2, Fig. 1]. © 1994 American Geophysical Union.

are defined at any radial distance ρ and height z' from the source. The electric moment of the dipole with the length $2h$ is $2h_e I$ where I is the current at its center. The field is

$$B_{0\phi'}(\rho, z') = -\frac{\mu_0(2h_e I)}{4\pi}\left[e^{ik_0 r_1}\left(\frac{\rho}{r_1}\right)\left(\frac{ik_0}{r_1} - \frac{1}{r_1^2}\right) + e^{ik_0 r_2}\left(\frac{\rho}{r_2}\right)\left(\frac{ik_0}{r_2} - \frac{1}{r_2^2}\right)\right.$$

$$\left. - 2k_0^2 \epsilon\, e^{ik_0 r_2}\left(\frac{\pi}{k_0 r_2}\right)^{1/2} e^{-iP}\mathcal{F}(P)\right] \qquad (8.13)$$

$$E_{0\rho}(\rho, z') = -\frac{\omega\mu_0(2h_e I)}{4\pi k_0}\left\{e^{ik_0 r_1}\left(\frac{\rho}{r_1}\right)\left(\frac{z'-d}{r_1}\right)\left(\frac{ik_0}{r_1} - \frac{3}{r_1^2} - \frac{3i}{k_0 r_1^3}\right)\right.$$

$$+ e^{ik_0 r_2}\left(\frac{\rho}{r_2}\right)\left(\frac{z'+d}{r_2}\right)\left(\frac{ik_0}{r_2} - \frac{3}{r_2^2} - \frac{3i}{k_0 r_2^3}\right)$$

$$\left. - 2\epsilon\, e^{ik_0 r_2}\left[\left(\frac{\rho}{r_2}\right)\left(\frac{ik_0}{r_2} - \frac{1}{r_2^2}\right) - k_0^2\epsilon\left(\frac{\pi}{k_0 r_2}\right)^{1/2} e^{-iP}\mathcal{F}(P)\right]\right\} \qquad (8.14)$$

$$E_{0z'}(\rho, z') = \frac{\omega\mu_0(2h_e I)}{4\pi k_0}$$

$$\times \left\{e^{ik_0 r_1}\left[\frac{ik_0}{r_1} - \frac{1}{r_1^2} - \frac{i}{k_0 r_1^3} - \left(\frac{z'-d}{r_1}\right)^2\left(\frac{ik_0}{r_1} - \frac{3}{r_1^2} - \frac{3i}{k_0 r_1^3}\right)\right]\right.$$

$$+ e^{ik_0 r_2}\left[\frac{ik_0}{r_2} - \frac{1}{r_2^2} - \frac{i}{k_0 r_2^3} - \left(\frac{z'+d}{r_2}\right)^2\left(\frac{ik_0}{r_2} - \frac{3}{r_2^2} - \frac{3i}{k_0 r_2^3}\right)\right]$$

$$\left. - 2k_0^2 \epsilon\, e^{ik_0 r_2}\left(\frac{\pi}{k_0 r_2}\right)^{1/2}\left(\frac{\rho}{r_2}\right)e^{-iP}\mathcal{F}(P)\right\}. \qquad (8.15)$$

In these formulas, $r_1 = [\rho^2 + (z' - d)^2]^{1/2}$ and $r_2 = [\rho^2 + (z' + d)^2]^{1/2}$. Also,

$$P = \frac{k_0 r_2}{2} \left(\frac{\epsilon r_2 + z' + d}{\rho} \right)^2, \qquad \epsilon = \frac{k_0}{k_2} - i k_0 l \qquad (8.16)$$

$$\mathcal{F}(P) = \tfrac{1}{2}(1 + i) - \int_0^P \frac{e^{it}}{(2\pi t)^{1/2}} \, dt. \qquad (8.17)$$

The integral in (8.17) is the Fresnel integral. Note that all three components of the electromagnetic field in the air, $z' \geq 0$, are independent of the wave number k_1 of the dielectric layer. Only its thickness is involved in the small parameter ϵ. This is a consequence of the fact that all but one of the so-called "trapped" waves in the dielectric layer are cut off by the condition $|k_1 l| \leq 0.6$ and the one possible mode, TM_{01}, has the same components as the lateral wave in the air above the dielectric layer. Furthermore, the trapped waves are plane waves that are not strongly excited by a vertical dipole as are lateral waves. Note that when $l = 0$ and $\epsilon = k_0/k_2$, the formulas (8.13)–(8.15) for the three-layered region reduce to formulas for the two-layered region consisting of regions 0 and 2.

The complete field in the air, given by (8.13)–(8.15), is conveniently studied in ranges, each with simple characteristic properties. These can be visualized with the help of Fig. 8.3, which shows the radial dependence of the electric field of a vertical electric dipole over sea water [3] in the important special case when both the dipole and the point of observation are in the air very close to the boundary surface, i.e. with $z' \sim 0$ and $d \sim 0$ in (8.14) and (8.15). With these values, the formulas reduce to

$$E_{0\rho}(\rho, 0) = -\frac{\omega \epsilon}{k_0} B_{0\phi'}(\rho, 0)$$

$$= \frac{\omega \mu_0 \epsilon (2h_e I)}{2\pi k_0} e^{ik_0\rho} \left[\frac{ik_0}{\rho} - \frac{1}{\rho^2} - k_0^2 \epsilon \left(\frac{\pi}{k_0 \rho} \right)^{1/2} e^{-iP_0} \mathcal{F}(P_0) \right] \qquad (8.18)$$

$$E_{0z'}(\rho, 0) = \frac{\omega \mu_0 (2h_e I)}{2\pi k_0} e^{ik_0\rho} \left[\frac{ik_0}{\rho} - \frac{1}{\rho^2} - \frac{i}{k_0 \rho^3} - k_0^2 \epsilon \left(\frac{\pi}{k_0 \rho} \right)^{1/2} e^{-iP_0} \mathcal{F}(P_0) \right], \qquad (8.19)$$

where now

$$P_0 = \frac{k_0 \rho \epsilon^2}{2}. \qquad (8.20)$$

In their radial dependence, these formulas can be separated into four ranges. They include the near field defined by

$$0 < k_0 \rho < 1: \qquad E_{0\rho}^n(\rho, 0) = -\frac{\omega \epsilon}{k_0} B_{0\phi'}^n(\rho, 0) = -\frac{\omega \mu_0 \epsilon (2h_e I)}{2\pi k_0} \frac{e^{ik_0\rho}}{\rho^2} \qquad (8.21)$$

$$E_{0z'}^n(\rho, 0) = -\frac{\omega \mu_0 (2h_e I)}{2\pi k_0} e^{ik_0\rho} \left(\frac{1}{\rho^2} + \frac{i}{k_0 \rho^3} \right). \qquad (8.22)$$

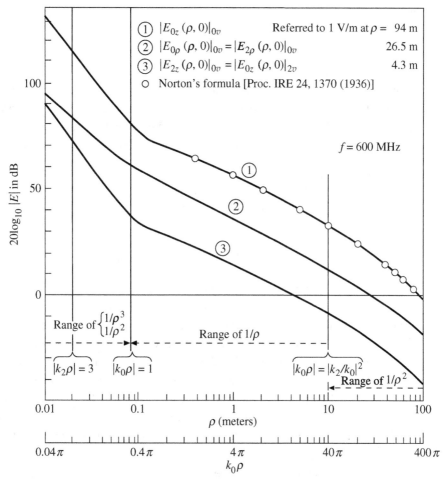

Figure 8.3 Components of electric field $E(\rho, 0)$ in V/m of unit vertical electric dipole ($I\,\Delta l = 1$ A m) at boundary between region 2 (sea water, $\epsilon_{2r} = 80$, $\sigma_2 = 3.5$ S/m) and region 0 (air); $z' = -z$. Taken from King [3, Fig. 2]. © 1990 American Geophysical Union.

The intermediate range is defined by $k_0\rho \geq 1$ and $|P_0| \leq 1$ or

$$1 \leq k_0\rho \leq \left|\frac{2}{\epsilon^2}\right|: \quad E^i_{0\rho}(\rho, 0) = -\frac{\omega\epsilon}{k_0} B^i_{0\phi'}(\rho, 0) = \frac{i\omega\mu_0\epsilon(2h_eI)}{2\pi} \frac{e^{ik_0\rho}}{\rho} \quad (8.23)$$

$$E^i_{0z'}(\rho, 0) = \frac{i\omega\mu_0(2h_eI)}{2\pi} \frac{e^{ik_0\rho}}{\rho}. \quad (8.24)$$

The transition range is bounded by the conditions

$$1 < |P_0| \leq 4 \quad \text{or} \quad \left|\frac{2}{\epsilon^2}\right| < k_0\rho \leq \left|\frac{8}{\epsilon^2}\right|. \quad (8.25)$$

The far field is defined by

$$4 < |P_0| < \infty \quad \text{or} \quad \left| \frac{8}{\epsilon^2} \right| < k_0\rho < \infty. \tag{8.26}$$

When (8.26) is satisfied, the Fresnel-integral terms reduce to a remarkably simple asymptotic form. This cancels the factor e^{-iP_0} and then cancels the $1/\rho$ term in (8.18) and (8.19). As shown in detail in [4, Eq. (3.4.35)], the entire bracketed terms in (8.18) and (8.19) reduce simply to $-1/\epsilon^2\rho^2$. It follows that the far field for the surface wave is defined by

$$\left| \frac{8}{\epsilon^2} \right| \le k_0\rho < \infty: \quad E^r_{0\rho}(\rho, 0) = -\frac{\omega\epsilon}{k_0} B^r_{0\phi'}(\rho, 0) = -\frac{\omega\mu_0(2h_eI)}{2\pi k_0} \frac{e^{ik_0\rho}}{\epsilon\rho^2} \tag{8.27}$$

$$E^r_{0z'}(\rho, 0) = -\frac{\omega\mu_0(2h_eI)}{2\pi k_0} \frac{e^{ik_0\rho}}{\epsilon^2\rho^2}. \tag{8.28}$$

In the transition range between the intermediate and far fields, i.e. when $1 \le |P_0| \le 4$ or $|2/\epsilon^2| \le k_0\rho \le |8/\epsilon^2|$, the amplitude of the field curves smoothly from the $1/\rho$ dependence to the $1/\rho^2$ dependence as the lateral wave increases from negligible to dominant. These ranges are shown in later sections for specific applications.

It is now appropriate to examine the properties of the field at all points in the air, i.e. for all values of ρ and z'. The condition $d^2 \ll \rho^2$ will be retained. A schematic diagram of the four ranges of the field is shown in Fig. 8.4. The contour $P = 1$ bounds the $1/r_0$ intermediate field and the transition range; the contour $P = 4$ bounds the transition range and the $1/r_0^2$ far field. For most purposes, it is convenient to eliminate specific reference to the transition range and divide it equally between the intermediate and far fields. When this is done, the $1/r_0$ decrease characteristic of the intermediate range is extended to include half of the transition range. Similarly, the remainder of the transition region is made part of the far field where the dependence on distance is $1/r_0^2$. With this approximation, the extended intermediate and far fields are

$$1 \le k_0\rho \le \left| \frac{5}{\epsilon^2} \right|, \quad \left| \frac{5}{\epsilon^2} \right| \le k_0\rho < \infty. \tag{8.29}$$

8.3   The field in the air in the intermediate range

In the intermediate range defined by $k_0r_0 \ge 1$ and $|P| \le 1$, the near-field terms are negligible and the Fresnel-integral terms are small compared to the $1/r_0$ terms. In this range, the components of the field are more convenient in the spherical coordinates r_0, Θ, Φ than in the cylindrical coordinates ρ, ϕ', z'. The relations between them are $\rho = r_0 \sin \Theta$ and $z' = r_0 \cos \Theta$. When the dipole is at $d = 0$ on the surface of the

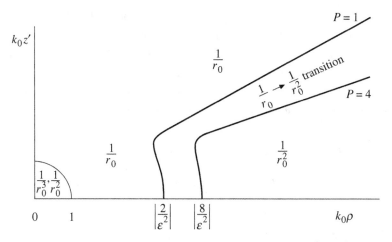

Figure 8.4 Schematic diagram of the four ranges: near field, $1/r_0^3$, $1/r_0^2$; intermediate field, $1/r_0$; transition range, $1/r_0 \rightarrow 1/r_0^2$; and far field, $1/r_0^2$. $\epsilon = (k_0/k_2) - ik_0l$.

layered region or directly on the earth or sea, or if it is at heights d that are small compared to the radial distance to the point of observation so that $d^2 \ll r_0^2$, and when, in addition, $dz' \ll r_0^2$ where $r_0 = (\rho^2 + z'^2)^{1/2}$, it follows that

$$r_1 = [\rho^2 + (z' - d)^2]^{1/2} \sim (r_0^2 - 2z'd)^{1/2} = r_0 - d \cos \Theta \tag{8.30a}$$

$$r_2 = [\rho^2 + (z' + d)^2]^{1/2} \sim (r_0^2 + 2z'd)^{1/2} = r_0 + d \cos \Theta. \tag{8.30b}$$

These values are used in phases. In amplitudes,

$$r_1 \sim r_2 \sim r_0$$

is an adequate approximation. With these formulas and with the near-field and the Fresnel-integral terms omitted because they are negligibly small in the intermediate range, (8.13)–(8.15) become

$$\overset{\cdot}{B}{}^i_{0\Phi}(r_0, \Theta) = -\frac{\mu_0(2h_e I)}{2\pi} e^{ik_0r_0} \frac{ik_0}{r_0} \sin \Theta \cos(k_0d \cos \Theta) \tag{8.31}$$

$$E^i_{0\Theta}(r_0, \Theta) = E^i_{0\rho}(\rho, z') \cos \Theta - E^i_{0z'}(\rho, z') \sin \Theta$$

$$= -\frac{i\omega\mu_0(2h_e I)}{2\pi} \frac{e^{ik_0r_0}}{r_0} \sin \Theta \left[\cos(k_0d \cos \Theta) \right.$$

$$\left. + \frac{id}{r_0} \cos \Theta \sin(k_0d \cos \Theta) - \epsilon \cos \Theta \, e^{ik_0d \cos \Theta} \right]. \tag{8.32}$$

Here the terms multiplied by the small quantities ϵ and d/r_0 can usually be neglected. This leaves the leading term

$$E^i_{0\Theta}(r_0, \Theta) = cB^i_{0\Phi}(r_0, \Theta) = -\frac{i\omega\mu_0(2h_e I)}{2\pi} \frac{e^{ik_0r_0}}{r_0} \sin \Theta \cos(k_0d \cos \Theta). \tag{8.33}$$

This is the same as the far field of a vertical dipole at the height d over a perfectly conducting plane as given by (8.3) with $k_1 = \infty$ and $f_{er}(\Theta) = 1$. In the intermediate range, the properties of the layered region are evidently irrelevant so long as the conditions (8.11) are satisfied. However, whereas the radial component of the electric field $E_{0r}^i(r_0, \Theta) = 0$ when the dipole is over a perfect conductor, it has a finite value when the dipole is over a region that satisfies (8.11). Specifically,

$$
E_{0r}^i(r_0, \Theta) = E_{0\rho}^i(\rho, z') \sin \Theta + E_{0z'}^i(\rho, z') \cos \Theta
$$

$$
= \frac{\omega \mu_0 (2h_e I)}{2\pi} \frac{e^{ik_0 r_0}}{r_0} \left[\frac{d}{r_0} \sin(k_0 d \cos \Theta) + i\epsilon\, e^{ik_0 d \cos \Theta} \right] \sin^2 \Theta. \quad (8.34)
$$

When both d/r_0 and $k_0 d$ are small, the significant radial field is

$$
E_{0r}^i(r_0, \Theta) = \frac{i\omega \mu_0 (2h_e I)}{2\pi} \frac{e^{ik_0(r_0 + d \cos \Theta)}}{r_0} \epsilon \sin^2 \Theta, \quad (8.35)
$$

where ϵ is defined in (8.16). Note that the radial component also decreases as $1/r_0$, and has the factor $\sin^2 \Theta$, not $\sin \Theta$ as does the transverse component.

With $d \sim 0$,

$$
E_{0\Theta}^i(r_0, \Theta) = -\frac{i\omega \mu_0 (2h_e I)}{2\pi} \frac{e^{ik_0 r_0}}{r_0} \sin \Theta \quad (8.36)
$$

$$
E_{0r}^i(r_0, \Theta) = \frac{i\omega \mu_0 (2h_e I)}{2\pi} \frac{e^{ik_0 r_0}}{r_0} \epsilon \sin^2 \Theta. \quad (8.37)
$$

8.4   The far field in the air

When the condition $|P| \geq 4$ is satisfied, the far-field terms of the direct and image fields can be combined with the far field of the Fresnel-integral term. In its asymptotic form with

$$
P = \frac{k_0 r_2}{2} \left(\frac{\epsilon r_2 + z' + d}{\rho} \right)^2 \quad (8.38)
$$

$$
T = k_0^2 \epsilon \left(\frac{\pi}{k_0 r_2} \right)^{1/2} e^{-iP} \mathcal{F}(P)
$$

$$
\rightarrow \frac{ik_0 \rho}{r_2^2} \frac{\epsilon}{\epsilon + (z' + d)/r_2} + \frac{\epsilon}{r_2^2} \left(\frac{\rho/r_2}{\epsilon + (z' + d)/r_2} \right)^3. \quad (8.39)
$$

With (8.39), (8.13)–(8.15) have the following far-field forms:

$$B_{0\phi'}^r(\rho, z') = -\frac{\mu_0(2h_e I)}{4\pi}\left\{e^{ik_0 r_1}\frac{ik_0\rho}{r_1^2} + e^{ik_0 r_2}\left[\frac{ik_0\rho}{r_2^2} - 2T\right]\right\}$$

$$= -\frac{\mu_0(2h_e I)}{4\pi}\left\{e^{ik_0 r_1}\frac{ik_0\rho}{r_1^2} + e^{ik_0 r_2}\left[\frac{ik_0\rho}{r_2^2}\left(\frac{(z'+d)/r_2 - \epsilon}{(z'+d)/r_2 + \epsilon}\right)\right.\right.$$

$$\left.\left. - \frac{2\epsilon}{r_2^2}\left(\frac{\rho/r_2}{\epsilon + (z'+d)/r_2}\right)^3\right]\right\}. \tag{8.40}$$

Since the angle of incidence on the surface $z' = 0$ in the far field is Θ^i with

$$\cos \Theta^i = \frac{z'+d}{r_2}, \qquad \sin \Theta^i = \frac{\rho}{r_2} \tag{8.41}$$

the plane-wave reflection coefficient for the layered surface at $z' = 0$ is

$$f_{er}(\Theta^i) = \frac{\cos\Theta^i - \epsilon(1 - \epsilon^2 \sin^2\Theta^i)^{1/2}}{\cos\Theta^i + \epsilon(1 - \epsilon^2 \sin^2\Theta^i)^{1/2}} \sim \frac{\cos\Theta^i - \epsilon}{\cos\Theta^i + \epsilon}, \tag{8.42}$$

where $\epsilon = (k_0/k_2) - ik_0 l$ and $|\epsilon|^2 \ll 1$. When $l = 0$, $\epsilon = N^{-1} = k_0/k_2$.
With (8.41) and (8.42), (8.40) becomes

$$B_{0\phi'}^r(\rho, z') = -\frac{\mu_0(2h_e I)}{4\pi}\left\{e^{ik_0 r_1}\frac{ik_0\rho}{r_1^2}\right.$$

$$\left. + e^{ik_0 r_2}\left[\frac{ik_0\rho}{r_2^2}f_{er}(\Theta^i) - \frac{2\epsilon}{r_2^2}\left(\frac{\rho/r_2}{\epsilon + (z'+d)/r_2}\right)^3\right]\right\}. \tag{8.43}$$

Similarly,

$$E_{0\rho}^r(\rho, z') = -\frac{\omega\mu_0(2h_e I)}{4\pi k_0}\left\{e^{ik_0 r_1}\frac{ik_0\rho}{r_1^2}\left(\frac{z'-d}{r_1}\right)\right.$$

$$\left. + e^{ik_0 r_2}\left[\frac{ik_0\rho}{r_2^2}\left(\frac{z'+d}{r_2} - 2\epsilon\right) + 2\epsilon T\right]\right\}$$

$$= -\frac{\omega\mu_0(2h_e I)}{4\pi k_0}\left\{e^{ik_0 r_1}\frac{ik_0\rho}{r_1^2}\left(\frac{z'-d}{r_1}\right)\right.$$

$$\left. + e^{ik_0 r_2}\left[\frac{ik_0\rho}{r_2^2}\left(\frac{z'+d}{r_2}\right)f_{er}(\Theta^i) + \frac{2\epsilon^2}{r_2^2}\left(\frac{\rho/r_2}{\epsilon + (z'+d)/r_2}\right)^3\right]\right\} \tag{8.44}$$

$$E_{0z'}^r(\rho, z') = \frac{\omega\mu_0(2h_e I)}{4\pi k_0}\left\{e^{ik_0 r_1}\frac{ik_0}{r_1}\left(\frac{\rho}{r_1}\right)^2 + e^{ik_0 r_2}\left[\frac{ik_0}{r_2}\left(\frac{\rho}{r_2}\right)^2 - 2T\left(\frac{\rho}{r_2}\right)\right]\right\}$$

$$= \frac{\omega\mu_0(2h_e I)}{4\pi k_0}\left\{e^{ik_0 r_1}\frac{ik_0}{r_1}\left(\frac{\rho}{r_1}\right)^2 + e^{ik_0 r_2}\left[\frac{ik_0}{r_2}\left(\frac{\rho}{r_2}\right)^2 f_{er}(\Theta^i)\right.\right.$$

$$\left.\left. - \frac{2\epsilon}{r_2^2}\left(\frac{\rho/r_2}{\epsilon + (z'+d)/r_2}\right)^3\left(\frac{\rho}{r_2}\right)\right]\right\}. \tag{8.45}$$

The formulas (8.43), (8.44), and (8.45) give the complete far field at any height z' of a vertical electric dipole located at any height d subject only to the conditions

$$k_0^2 \ll |k_1|^2 \ll |k_2|^2, \qquad |P| \geq 4, \qquad |k_1 l| \leq 0.6. \tag{8.46}$$

The field and the dipole are over a layered region consisting of a dielectric with the thickness l (region 1) on a half-space (region 2). It is significant to note that each of these far-field formulas consists of three terms. They are the direct field, the plane-wave reflected field, and the lateral-wave field.

It is possible to express the field in the spherical coordinates r_0, Θ, Φ with the substitutions

$$r_0 = (\rho^2 + z'^2)^{1/2}, \qquad \sin \Theta = \frac{\rho}{r_0}, \qquad \cos \Theta = \frac{z'}{r_0} \tag{8.47}$$

if the additional restriction

$$d^2 \ll r_0^2 \tag{8.48}$$

is made. With it,

$$r_1 \sim r_0 - d \cos \Theta, \qquad r_2 \sim r_0 + d \cos \Theta \tag{8.49}$$

in phases, and

$$r_1 \sim r_2 \sim r_0 \tag{8.50}$$

in amplitudes. It follows that

$$\frac{\rho}{r_1} \sim \frac{\rho}{r_2} \sim \frac{\rho}{r_0} = \sin \Theta \tag{8.51}$$

$$\frac{z'-d}{r_1} \sim \frac{z'-d}{r_0} = \cos \Theta - \frac{d}{r_0}$$

$$\cos \Theta^i = \frac{z'+d}{r_2} \sim \frac{z'+d}{r_0} = \cos \Theta + \frac{d}{r_0}. \tag{8.52}$$

With these substitutions and

$$\tfrac{1}{2}[1 + f_{er}(\Theta^i)] = \frac{\cos \Theta + d/r_0}{\epsilon + \cos \Theta + d/r_0}, \qquad \tfrac{1}{2}[1 - f_{er}(\Theta^i)] = \frac{\epsilon}{\epsilon + \cos \Theta + d/r_0} \tag{8.53}$$

(8.43)–(8.45) become

$$B_{0\phi'}^r(r_0, \Theta) = -\frac{\mu_0(2h_e I)}{2\pi} e^{ik_0 r_0} \left\{ \frac{ik_0}{r_0} \sin \Theta \left[\cos(k_0 d \cos \Theta) \right. \right.$$

$$\times \left(\frac{\cos \Theta + d/r_0}{\epsilon + \cos \Theta + d/r_0} \right) - i \sin(k_0 d \cos \Theta) \left(\frac{\epsilon}{\epsilon + \cos \Theta + d/r_0} \right) \Bigg]$$

$$\left. - \frac{\epsilon}{r_0^2} \left(\frac{\sin \Theta}{\epsilon + \cos \Theta + d/r_0} \right)^3 e^{ik_0 d \cos \Theta} \right\} \tag{8.54}$$

$$
E_{0\rho}^r(r_0, \Theta) = -\frac{\omega\mu_0(2h_e I)}{2\pi k_0} e^{ik_0 r_0} \left\{ \frac{ik_0}{r_0} \sin\Theta \left[\cos\Theta \cos(k_0 d \cos\Theta) \right. \right.
$$

$$
\left. + \frac{id}{r_0} \sin(k_0 d \cos\Theta) \right] \left(\frac{\cos\Theta + d/r_0}{\epsilon + \cos\Theta + d/r_0} \right) - \frac{ik_0}{r_0} \sin\Theta
$$

$$
\times \left[i \cos\Theta \sin(k_0 d \cos\Theta) + \frac{d}{r_0} \cos(k_0 d \cos\Theta) \right]
$$

$$
\times \left(\frac{\epsilon}{\epsilon + \cos\Theta + d/r_0} \right) + \frac{\epsilon^2}{r_0^2} \left(\frac{\sin\Theta}{\epsilon + \cos\Theta + d/r_0} \right)^3 e^{ik_0 d \cos\Theta} \right\}
$$

$$
(8.55)
$$

$$
E_{0z'}^r(r_0, \Theta) = \frac{\omega\mu_0(2h_e I)}{2\pi k_0} e^{ik_0 r_0} \left\{ \frac{ik_0}{r_0} \sin^2\Theta \right.
$$

$$
\times \left[\frac{(\cos\Theta + d/r_0)\cos(k_0 d \cos\Theta) - i\epsilon \sin(k_0 d \cos\Theta)}{\epsilon + \cos\Theta + d/r_0} \right]
$$

$$
\left. - \frac{\epsilon}{r_0^2} \left(\frac{\sin\Theta}{\epsilon + \cos\Theta + d/r_0} \right)^3 \sin\Theta \, e^{ik_0 d \cos\Theta} \right\}.
$$

$$
(8.56)
$$

The two cylindrical components of the electric field can be combined to obtain the spherical components. After extensive algebraic manipulation,

$$
E_{0\Theta}^r(r_0, \Theta) = E_{0\rho}^r(r_0, \Theta) \cos\Theta - E_{0z'}^r(r_0, \Theta) \sin\Theta \tag{8.57}
$$

$$
= -\frac{\omega\mu_0(2h_e I)}{2\pi k_0} e^{ik_0 r_0} \left\{ \frac{ik_0}{r_0} \sin\Theta \right.
$$

$$
\times \left[\frac{\cos\Theta + d/r_0 - \epsilon(d/r_0)\cos\Theta}{\epsilon + \cos\Theta + d/r_0} \cos(k_0 d \cos\Theta) \right.
$$

$$
\left. - i \frac{\epsilon - (d/r_0)\cos\Theta(\cos\Theta + d/r_0)}{\epsilon + \cos\Theta + d/r_0} \sin(k_0 d \cos\Theta) \right]
$$

$$
\left. + \frac{\epsilon}{r_0^2} \left(\frac{\sin\Theta}{\epsilon + \cos\Theta + d/r_0} \right)^3 (\epsilon \cos\Theta - \sin^2\Theta) e^{ik_0 d \cos\Theta} \right\}. \tag{8.58}
$$

Since the terms $\epsilon d/r_0$ and d^2/r_0^2 are negligibly small,

$$
E_{0\Theta}^r(r_0, \Theta) \sim -\frac{\omega\mu_0(2h_e I)}{2\pi k_0} e^{ik_0 r_0} \left\{ \frac{ik_0}{r_0} \sin\Theta \left[\frac{(\cos\Theta + d/r_0)\cos(k_0 d \cos\Theta)}{\epsilon + \cos\Theta + d/r_0} \right. \right.
$$

$$
\left. - \frac{i[\epsilon - (d/r_0)\cos^2\Theta]\sin(k_0 d \cos\Theta)}{\epsilon + \cos\Theta + d/r_0} \right]
$$

$$
\left. + \frac{\epsilon}{r_0^2} \left(\frac{\sin\Theta}{\epsilon + \cos\Theta + d/r_0} \right)^3 (\epsilon \cos\Theta - \sin^2\Theta) e^{ik_0 d \cos\Theta} \right\}. \tag{8.59}
$$

Similarly, with (8.55) and (8.56),

$$E_{0r}^r(r_0, \Theta) = E_{0\rho}^r(r_0, \Theta) \sin \Theta + E_{0z'}^r(r_0, \Theta) \cos \Theta \tag{8.60}$$

$$= \frac{\omega \mu_0 (2h_e I)}{2\pi k_0} e^{ik_0 r_0} \left\{ \frac{ik_0}{r_0} \frac{d}{r_0} \sin^2 \Theta \right.$$

$$\times \left[\frac{\epsilon \cos(k_0 d \cos \Theta) - i(\cos \Theta + d/r_0) \sin(k_0 d \cos \Theta)}{\epsilon + \cos \Theta + d/r_0} \right]$$

$$\left. - \frac{\epsilon}{r_0^2} \left(\frac{\sin \Theta}{\epsilon + \cos \Theta + d/r_0} \right)^3 (\epsilon + \cos \Theta) \sin \Theta \, e^{ik_0 d \cos \Theta} \right\}$$

$$\sim \frac{\omega \mu_0 (2h_e I)}{2\pi k_0} e^{ik_0 r_0} \left\{ \frac{ik_0}{r_0} \frac{d}{r_0} \sin^2 \Theta \right.$$

$$\times \left[\frac{\epsilon \cos(k_0 d \cos \Theta) - i \cos \Theta \sin(k_0 d \cos \Theta)}{\epsilon + \cos \Theta + d/r_0} \right]$$

$$\left. - \frac{\epsilon}{r_0^2} \left(\frac{\sin \Theta}{\epsilon + \cos \Theta + d/r_0} \right)^3 (\epsilon + \cos \Theta) \sin \Theta \, e^{ik_0 d \cos \Theta} \right\}. \tag{8.61}$$

The formulas (8.54), (8.59), and (8.61) give the three components $B_{0\Phi}^r(r_0, \Theta)$, $E_{0\Theta}^r(r_0, \Theta)$, and $E_{0r}^r(r_0, \Theta)$ of the far field of a vertical electric dipole with the electric moment $2h_e I$ located at any height d. The field is defined at any point r_0, Θ; as it is rotationally symmetric, the coordinate Φ does not appear. It contains terms with dependences on Θ in the form $\sin^n \Theta$ with $n = 1, 2, 4$, and 5. Graphs of these functions are in Fig. 8.5. The conditions are: $k_0^2 \ll |k_1^2| \ll |k_2^2|$, $|k_1 l| \leq 0.6$, and $d^2 \ll r_0^2$. When the dielectric layer is absent, $l = 0$ and $\epsilon = (k_0/k_2) - ik_0 l$ becomes $\epsilon = k_0/k_2$.

In many applications the antenna is on the surface of the earth, either as a base-insulated dipole or as a monopole base-driven against a buried ground system. In this case, $d/r_0 \sim 0$, $k_0 d \sim 0$, and

$$B_{0\Phi}^r(r_0, \Theta) = -\frac{\mu_0 (2h_e I)}{2\pi} e^{ik_0 r_0} \left[\frac{ik_0}{r_0} \frac{\sin \Theta \cos \Theta}{\cos \Theta + \epsilon} - \frac{\epsilon}{r_0^2} \left(\frac{\sin \Theta}{\cos \Theta + \epsilon} \right)^3 \right] \tag{8.62}$$

$$E_{0\Theta}^r(r_0, \Theta) = -\frac{\omega \mu_0 (2h_e I)}{2\pi k_0} e^{ik_0 r_0} \left[\frac{ik_0}{r_0} \frac{\sin \Theta \cos \Theta}{\cos \Theta + \epsilon} \right.$$

$$\left. + \frac{\epsilon}{r_0^2} \left(\frac{\sin \Theta}{\cos \Theta + \epsilon} \right)^3 (\epsilon \cos \Theta - \sin^2 \Theta) \right] \tag{8.63}$$

$$E_{0r}^r(r_0, \Theta) = -\frac{\omega \mu_0 (2h_e I)}{2\pi k_0} e^{ik_0 r_0} \frac{\epsilon}{r_0^2} \left[\frac{\sin^4 \Theta}{(\cos \Theta + \epsilon)^2} \right]. \tag{8.64}$$

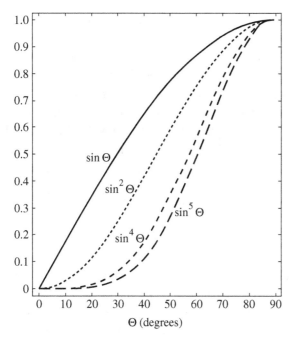

Figure 8.5 The field factors $\sin \Theta$ for the space wave and $\sin^2 \Theta$, $\sin^4 \Theta$, and $\sin^5 \Theta$ for the lateral wave.

The field on the surface of the earth is given by $\Theta \sim \pi/2$, $r_0 = \rho$, so that

$$B_{0\Phi}^r(r_0, \pi/2) = \frac{\mu_0(2h_eI)}{2\pi} \frac{e^{ik_0\rho}}{\epsilon^2\rho^2} \tag{8.65}$$

$$E_{0\Theta}^r(r_0, \pi/2) = \frac{\omega\mu_0(2h_eI)}{2\pi k_0} \frac{e^{ik_0\rho}}{\epsilon^2\rho^2} \tag{8.66}$$

$$E_{0r}^r(r_0, \pi/2) = -\frac{\omega\mu_0(2h_eI)}{2\pi k_0} \frac{e^{ik_0\rho}}{\epsilon\rho^2}. \tag{8.67}$$

Note that these formulas consist entirely of the lateral-wave terms. The direct and image fields cancel.

8.5   Base-driven and grounded monopoles

An important antenna in radio communication is the base-driven and base-grounded monopole. In the AM broadcast band (0.55–1.6 MHz), for example, the monopole is on the earth and driven against a radial ground system of bare conductors. In the maritime radio band (1.5–30 MHz), the monopole may be erected on the upper deck of the ship and be driven against the steel hull which is, of course, grounded in the water – lake or sea.

When the condition $|k_2|^2 \gg k_0^2$ is well satisfied so that the radial distance $\rho = 2|k_2^2|/k_0^3$ to the outer boundary of the intermediate field is at least several hundred meters, the vertical component of the electric field near the monopole is the same as if it were over a perfect conductor. This means that the distribution of current in the base-driven monopole is the same as it would be if it were the upper half of a center-driven dipole isolated in the air. Its impedance is half that of the dipole. If there are several monopoles in a directive array, the properties of the array can be obtained directly from those of the corresponding dipole array. The total radiated power is one-half that radiated by the corresponding dipole array with equal driving-point currents.

The driving-point impedance of the grounded monopole is the sum of the impedance of the monopole over a perfect conductor and the impedance of the ground network. That is,

$$Z_{\text{in}} = Z_0 + Z_g, \tag{8.68}$$

where Z_0 is the impedance of the monopole and Z_g is the impedance of the ground network. This latter is readily evaluated with the theory of the bare dipole in a conducting medium. If the earth is not too dry so that $\sigma_2 > \omega\epsilon_0\epsilon_{2r}$, the current in a monopole in an infinite medium with the wave number $k_2 \sim (i\omega\mu_0\sigma_2)^{1/2}$ is well approximated by

$$I_x(x) \sim \frac{V_0^e}{Z} \frac{\sin k_I(h-x)}{\sin k_I h}, \tag{8.69}$$

where

$$Z = R - iX = \frac{\Psi}{4\pi\sigma_2} k_I \cot k_I h \tag{8.70}$$

$$k_I = \beta_I + i\alpha_I = \left(i\omega\mu_0\sigma_2 - \frac{4\pi z^i \sigma_I}{\Psi}\right)^{1/2} \tag{8.71a}$$

$$\Psi = 2\ln(h/a) - 2. \tag{8.71b}$$

The internal impedance per unit length of the monopole is $z^i = r^i - ix^i$. At frequencies that are sufficiently high so that $(\omega\mu_0)^2 \gg (4\pi r^i/\Psi)^2$, the internal impedance z^i of the copper monopole can be neglected and the monopole treated as perfectly conducting. In this case,

$$k_I = k_2 = (i\omega\mu_0\sigma_2)^{1/2} = (1+i)\left(\frac{\omega\mu_0\sigma_2}{2}\right)^{1/2} = \beta_2 + i\alpha_2. \tag{8.72}$$

In the AM broadcast band, the frequency is high enough to treat the conductor as perfect, so that

$$Z \sim \frac{\Psi}{4\pi\sigma_2}\left(\frac{\omega\mu_0\sigma_2}{2}\right)^{1/2}(1+i)\cot\beta_2 h(1+i)$$

$$= \frac{\Psi}{4\pi}\left(\frac{\omega\mu_0}{2\sigma_2}\right)^{1/2}(1+i)\left(\frac{\cos\beta_2 h\,\cosh\alpha_2 h - i\sin\beta_2 h\,\sinh\alpha_2 h}{\sin\beta_2 h\,\cosh\alpha_2 h + i\cos\beta_2 h\,\sinh\alpha_2 h}\right). \tag{8.73}$$

A study of the behavior of the current in and impedance of a bare antenna in a dissipative medium such as wet or moist earth shows that the current amplitude decreases to negligibly small values at electrical distances from the driving point that exceed $\beta_2 x = \pi/2$. It follows that a monopole length that exceeds $\beta_2 h = \pi/2$ serves no useful purpose. Because the surrounding medium is conducting, a cylindrical conductor serves both as a radiating antenna and as an electrode for transferring the current from the conducting antenna to the conducting earth.

With $\beta_2 h = \pi/2$,

$$Z = \frac{\Psi}{4\pi}\left(\frac{\omega\mu_0}{2\sigma_2}\right)^{1/2}(1+i)(-i\tanh\pi/2)$$

$$= 0.917(1-i)\frac{\Psi}{4\pi}\left(\frac{\omega\mu_0}{2\sigma_2}\right)^{1/2}. \tag{8.74}$$

At a frequency $f = 0.55\,\text{MHz}$ and with $\sigma_2 = 0.04\,\text{S/m}$ for moist earth,

$$k_0 = 0.0115\,\text{m}^{-1}, \qquad k_2 = 0.294(1+i)\,\text{m}^{-1} \tag{8.75}$$

so that

$$h = \frac{\pi}{2\beta_2} = 5.34\,\text{m}. \tag{8.76}$$

If the radius of the monopole is $a = 2.5\,\text{mm}$, $h/a = 2136$ and $\Psi = 13.3$. It follows from (8.74) that

$$Z = 7.1(1-i)\,\text{ohms}. \tag{8.77}$$

The formulas (8.70), (8.73), and (8.74) give half the impedance of a center-driven dipole in the infinite dissipative medium. Because the mutual interaction between the two halves of a dipole is extremely small when the dipole is in such a medium, it contributes negligibly and the impedance of the dipole is well approximated by the sum of the impedances of two independent monopoles, i.e. the impedance of the monopole when driven against any other element is simply half the impedance of the dipole. In the present case, the horizontal monopole in the dissipative earth is driven against a vertical monopole in the air. More specifically, a radial group of N such horizontal monopoles, all connected in parallel, are driven against the vertical monopole in the air.

Actually, the formulas (8.70), (8.73), and (8.74) apply to a monopole in an infinite dissipative medium. In a radial ground network, each element is close to the surface of the earth so that the electric field and the associated currents in the conducting earth are reflected at the air boundary. Since $|k_2|^2 \gg k_0^2$, the reflection coefficient $f_r = (k_2 - k_0)/(k_2 + k_0) \sim 1$. This means that the reflected electric field is in phase with the incident electric field and the reflected magnetic field is in phase opposition with the incident magnetic field. Hence, the complete reflected field is equal to the field of an identical codirectional image monopole at a distance $2d$ from the actual monopole if d is its distance from the air–earth boundary. As the electrical distance $|k_2 d|$ is small, the reflected electric field when combined with the direct field from the monopole is essentially that of two identical monopoles separated by a distance $2d$. In effect, the monopole with its image constitutes a single monopole with the effective radius $a_e = (2da)^{1/2}$, the associated quantity

$$\Psi_e = 2\ln(h/a_e) - 2 \tag{8.78}$$

and a current $I_{1x}(x)$ that is twice that in the actual monopole. That is,

$$I_{1x}(x) = 2I_x(x) = \frac{V_0^e}{Z_e} \frac{\sin k_2(h - x)}{\sin k_2 h} \tag{8.79}$$

$$Z_e = \frac{\Psi_e}{4\pi}\left(\frac{\omega\mu_0}{2\sigma_2}\right)^{1/2}(1 + i)\cot\beta_2 h\,(1 + i). \tag{8.80}$$

The driving-point impedance in the presence of the boundary is

$$Z_1 = \frac{V_0^e}{I_1(0)} = 2Z_e = \frac{2Z\Psi_e}{\Psi}. \tag{8.81}$$

The impedance of N such monopoles in parallel is

$$Z_g = \frac{Z_1}{N} = \frac{2Z\Psi_e}{N\Psi}, \tag{8.82}$$

where Z is the impedance of a single monopole in an infinite medium, as given by (8.70), (8.73) or (8.74). Note that with $N = 10$, the radial conductors are sufficiently far apart in the conducting earth that mutual impedances can be neglected.
   With $d = 0.15\,\text{m}$,

$$a_e = 0.087\,\text{m} \tag{8.83}$$

$$\Psi_e = 6.23 \tag{8.84}$$

so that

$$Z_1 = \frac{2Z\Psi_e}{\Psi} = 0.937Z = 6.66(1 - i)\,\text{ohms}. \tag{8.85}$$

With ten buried radial conductors,

$$Z_g = 0.666(1 - i)\,\text{ohm}. \tag{8.86}$$

It is this value that is substituted in (8.68) to determine the driving-point impedance Z_in of the monopole base-driven against the radial ground network. Note that Z_g is very small compared to the driving-point impedance of a monopole in air over a perfect conductor. If this has $h = \lambda_0/4 = 136.6$ m and the radius $a = 1.24$ cm, it follows that the expansion parameter $\Omega = 2\ln(2h/a) = 20$. Then, with Table 30.1 in [5, p. 168], $Z_0 = 39.3 - i21.8$ ohms and the driving-point impedance of the grounded antenna is

$$Z_{in} = Z_0 + Z_g = 40.0 - i22.5 \text{ ohms.} \tag{8.87}$$

Note that the impedance of the ground network is small enough to be negligible. This means that the properly grounded monopole behaves like the monopole over a perfect conductor for all circuit and field properties within the intermediate-zone range provided $|k_2|^2 \gg k_0^2$. The complete fields in the extended intermediate and far zones are

$$1 \leq k_0\rho \leq \left|\frac{5}{\epsilon^2}\right|: \qquad E_{0\Theta}^i(r_0, \Theta) = -\frac{i\omega\mu_0(2h_eI)}{2\pi} \frac{e^{ik_0r_0}}{r_0} \sin\Theta \tag{8.88}$$

$$E_{0r}^i(r_0, \Theta) = \frac{i\omega\mu_0(2h_eI)}{2\pi} \frac{e^{ik_0r_0}}{r_0} \epsilon \sin^2\Theta \tag{8.89}$$

$$\left|\frac{5}{\epsilon^2}\right| \leq k_0\rho: \qquad E_{0\Theta}^r(r_0, \Theta) = -\frac{\omega\mu_0(2h_eI)}{2\pi k_0} e^{ik_0r_0} \left\{ \frac{ik_0}{r_0}\left(\frac{\sin\Theta\cos\Theta}{\epsilon+\cos\Theta}\right) \right.$$

$$\left. -\frac{\sin^3\Theta\,(\epsilon\sin^2\Theta - \epsilon^2\cos\Theta)}{r_0^2(\epsilon+\cos\Theta)^3} \right\} \tag{8.90}$$

$$E_{0r}^r(r_0, \Theta) = -\frac{\omega\mu_0(2h_eI)}{2\pi k_0} \frac{\epsilon\sin^4\Theta}{r_0^2(\epsilon+\cos\Theta)^2} e^{ik_0r_0} \tag{8.91}$$

with $\epsilon = (k_0/k_2) - ik_0l$. When there is no surface layer of pavement, $l = 0$ and $\epsilon = k_0/k_2$.

The radial components given by (8.89) and (8.91) are due to the lateral wave. In the far field, the transverse component is clearly separated into a space-wave term (that includes the plane-wave reflection coefficient as a factor and decreases with distance as $1/r_0$) and a lateral-wave component (that decreases as $1/r_0^2$). The former vanishes when $\Theta = \pi/2$, the latter has its maximum there. The corresponding formula (8.88) for the intermediate range consists of only one term with its amplitude proportional to $1/r_0$. However, the contribution from the plane-wave reflection coefficient is zero when $\Theta = \pi/2$, whereas in (8.88) the amplitude is at its maximum when $\Theta = \pi/2$. Evidently, (8.88) combines the contributions of the plane-wave reflection coefficient and the lateral wave into one simple term in the intermediate range where the Fresnel-integral term is negligible.

The vertical field patterns for the intermediate range of $|E_{0\Theta}^i(r_0, \Theta)| \sim \sin\Theta$ and of $|E_{0r}^i(r_0, \Theta)| \sim |k_0/k_2|\sin^2\Theta$ are shown in Fig. 8.6. The pattern for $E_{0\Theta}^i(r_0, \Theta)$ applies to a vertical electric dipole on the surface of all types of earth or water. The four patterns for $E_{0r}^i(r_0, \Theta)$ apply to (a) sea water with $\sigma_2 = 4$ S/m, $\epsilon_{2r} = 80$; (b) wet

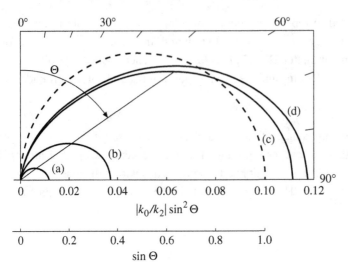

Figure 8.6 The vertical field characteristics of a vertical electric dipole on the boundary between air and earth or water. Broken line: the function $\sin \Theta$ for $|E_{0\Theta}^i(r_0, \Theta)|$ over all types of earth and water. Solid lines: the function $|k_0/k_2| \sin^2 \Theta$ for $|E_{0r}^i(r_0, \Theta)|$ over (a) sea water, (b) wet earth, (c) dry earth, and (d) lake water.

Table 8.1. *Numerical values of $|E_{0\Theta}^r(r_0, \Theta)|_{max}$ and $|E_{0\Theta}^r(r_0, \pi/2)|$ for vertical dipole in air over different media[a]*

Region 2	σ_2 (S/m)	ϵ_{2r}	Θ_{max}	$|E_{0\Theta}^r(r_0, \Theta)|_{max}$ (V/m)	$|E_{0\Theta}^r(r_0, \pi/2)|$ (V/m)
Sea water	4.0	80	78.5°	2.36×10^{-5}	1.73×10^{-6}
Wet earth	0.4	12	73.0°	2.20×10^{-5}	1.73×10^{-7}
Dry earth	0.04	8	66.0°	1.87×10^{-5}	1.74×10^{-8}
Lake water	0.004	80	65.5°	1.80×10^{-5}	1.93×10^{-8}

Frequency $f = 10$ MHz, radial distance $r_0 = 500$ km. See Fig. 8.7.
[a] Taken from King [3, Table 1]. © 1990 American Geophysical Union.

earth with $\sigma_2 = 0.4$ S/m, $\epsilon_{2r} = 12$; (c) dry earth with $\sigma_2 = 0.04$ S/m, $\epsilon_{2r} = 8$; and (d) lake water with $\sigma_2 = 0.004$ S/m, $\epsilon_{2r} = 80$. In all cases, the frequency is $f = 10$ MHz.

The corresponding patterns for the far field $E_{0\Theta}^r(r_0, \Theta)$ as obtained from (8.90) are shown in Figs. 8.7 and 8.8 in conjunction with Table 8.1. Graphs of the far field $E_{0r}^r(r_0, \Theta)$ as obtained from (8.91) are shown in Fig. 8.9 in conjunction with Table 8.2. In all cases, the frequency is $f = 10$ MHz.

8.6 Vertical antennas on the earth for communicating with submarines in the ocean

Shore-based antennas for communicating with submarines are located close to the sea coast at various points around the world. Examples are at Annapolis, MD, and Cutler,

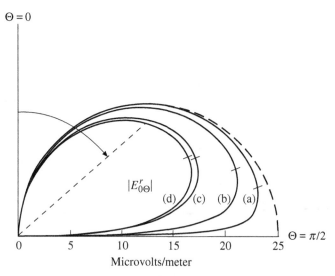

Figure 8.7 Complete field of $|E^r_{0\Theta}(r_0, \Theta)|$ for vertical dipole in air on boundary between air and (a) sea water, (b) wet earth, (c) dry earth, and (d) lake water; the dashed curve is for $\sigma_2 = \infty$. Frequency $f = 10\,\text{MHz}$, radial distance $r_0 = 500\,\text{km}$. Numerical values are in Table 8.1. Taken from King [3, Fig. 3]. © 1990 American Geophysical Union.

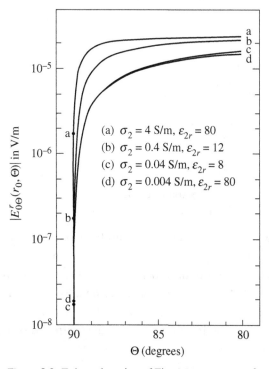

Figure 8.8 Enlarged section of Fig. 8.7 near $\Theta = 90°$. Taken from King [3, Fig. 4]. © 1990 American Geophysical Union.

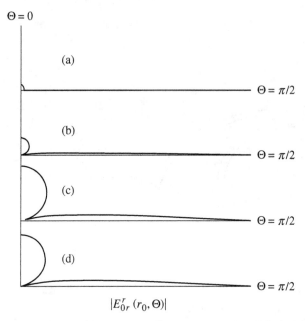

$\Theta = 0$

(a) $\Theta = \pi/2$

(b) $\Theta = \pi/2$

(c) $\Theta = \pi/2$

(d) $\Theta = \pi/2$

$|E_{0r}^r (r_0, \Theta)|$

Figure 8.9 Complete field of $|E_{0r}^r (r_0, \Theta)|$ for vertical dipole in air on boundary between air and (a) sea water, (b) wet earth, (c) dry earth, and (d) lake water. Frequency $f = 10$ MHz, radial distance $r_0 = 500$ km. Numerical values are in Table 8.2. Taken from King [3, Fig. 5]. © 1990 American Geophysical Union.

Table 8.2. *Numerical values of $|E_{0r}^r (r_0, \pi/2)|$ for vertical dipole in air over different media[a]*

Region 2	σ_2 (S/m)	ϵ_{2r}	$|E_{0r}^r (r_0, \pi/2)|$ (V/m)
Sea water	4.0	80	2.04×10^{-8}
Wet earth	0.4	12	6.44×10^{-9}
Dry earth	0.04	8	2.04×10^{-9}
Lake water	0.004	80	2.15×10^{-9}

Frequency $f = 10$ MHz, radial distance $r_0 = 500$ km. See Fig. 8.9.
[a] Taken from King [3, Table 2]. © 1990 American Geophysical Union.

ME, along the Atlantic coast. In order to penetrate to useful depths in the ocean, these antennas must radiate at very low frequencies. Since vertical antennas on the earth are limited for practical reasons to heights of the order of $h = 1200$ ft (366 m), frequencies in the range from 14 kHz to 28.5 kHz involve electrical lengths in the range $k_0 h = 0.107$ to 0.218, where $k_0 = \omega/c$ is the wave number of air. Grounded base-driven monopoles with such small electrical lengths have triangular current distributions and driving-point impedances given by

$$Z = R + jX = 10k_0^2 h^2 - j\frac{30(\Omega - 2 - 2\ln 2)}{k_0 h} \text{ ohms,} \tag{8.92}$$

where $j = -i$ and

$$\Omega = 2\ln(2h/a) \quad \text{and} \quad 2 + 2\ln 2 = 3.39. \tag{8.93}$$

For $f = 14$ to $28.5\,\text{kHz}$ and with $h/a = 75$ and $\Omega = 10$,

$$R = 0.114 \text{ to } 0.475 \text{ ohm}, \qquad X = -1853 \text{ to } -909.6 \text{ ohms}. \tag{8.94}$$

Owing to the very large reactance, the electrically short monopole is not a practical antenna for radiating the very large amounts of power needed to communicate with submerged submarines at large distances in the sea.

In order to reduce the high reactance and increase the very low radiation resistance, it is necessary to increase the length of the antenna. As this cannot be done vertically upward, the only alternative is to do so horizontally. The resulting antenna is known as an inverted L antenna. However, in the case at hand, the vertical section is electrically so short that the horizontal wire with its image in the earth is simply a two-wire transmission line with an open end. If the length of the horizontal wire is a quarter wavelength with an open end, the line is resonant and the impedance seen by the generator in the vertical section is a pure resistance given by $R = 15k_0^2b^2 = 60k_0^2h^2$ where $b = 2h$ is the spacing of the two-wire line. Instead of extending the horizontal wire a full quarter wavelength, it can be much shorter if it is end-loaded with capacitance to ground. This capacitance can consist of large metal panels or areas of wire mesh. Several such radial sections can be arranged in a rotationally symmetric manner and connected in parallel to the feeding transmission line. The resulting umbrella-like structure is known as a top-loaded antenna. Since the top-loading is roughly equivalent to a flat metal disk, it has the general properties of a radial transmission line driven at its center by a generator in series with a vertical conductor that connects the disk to a radial ground network. The total power radiated from under the outer edge of the disk is simply the power radiated by a dipole with the length h and carrying a current with the uniform amplitude $I_z(0)$. The external or radiation resistance of such a dipole is

$$R^e = 40k_0^2h^2, \tag{8.95}$$

where h is the length of the monopole. Since the actual top-loading panels are lower at the edge than at the center and are supported by grounded metal towers, each with its guy wires, the electromagnetic field of the currents in them partly cancel the field of the current in the central monopole. This reduces the radiated field and the effective length which determines its amplitude. The field actually generated by the top-loaded monopole can be measured at suitable distances and the effective length determined for each frequency. It is necessarily smaller than either the length h of the central monopole or the height of the outer edges of the top-loading.

Since the capacitive top-loading usually does not extend out radially far enough to make the antenna resonant, a variable inductance can be connected in series with the monopole at the driving point to tune it to resonance and provide a resistive impedance. This resistance is the sum of the radiation resistance R^e, the resistance R_g of the ground network, and the resistance R_c of the inductance coil. Thus,

$$R_A = R^e + R_g + R_c,\tag{8.96}$$

where

$$R^e = 40k_0^2 h_e^2.\tag{8.97}$$

The ground resistance, if this is due to N radial wires, is approximated by the real part of (8.82) if the wires are not too close together. Since the entire system is tuned to resonance, the total reactance including X_g is zero.

At frequencies in the kilohertz range, the intermediate zone is very extensive. Specifically at $f = 20$ kHz over sea water (wave number k_2), the conditions $|P_0| \leq 1$, $k_0 r_0 \geq 1$ become $|\rho k_0^3/2k_2^2| \leq 1$ or

$$\rho \leq \left|\frac{2k_2^2}{k_0^3}\right| = \frac{2\mu_0\sigma_2 c^3}{\omega^2} = 1.74 \times 10^7 \text{ km}.\tag{8.98}$$

Clearly, the intermediate zone includes the entire useful range so that the formulas (8.36) and (8.37) give the entire electromagnetic field when $2h_e$ for a dipole with the length $2h$ is replaced by h_e for a monopole with the length h. Specifically,

$$E_{0\Theta}^i(r_0, \Theta) = cB_{0\Phi}^i(r_0, \Theta) = -\frac{i\omega\mu_0 h_e I}{2\pi}\frac{e^{ik_0 r_0}}{r_0}\sin\Theta\tag{8.99}$$

$$E_{0r}^i(r_0, \Theta) = \frac{i\omega\mu_0 h_e I}{2\pi}\left(\frac{k_0}{k_2}\right)\frac{e^{ik_0 r_0}}{r_0}\sin^2\Theta.\tag{8.100}$$

The VLF antenna at Cutler, ME, consists of two vertical monopoles, each with the height 298 m and separated by a distance 1870 m. Each monopole is top-loaded by six symmetrically arranged diamond-shaped panels of wire mesh that extend radially outward 935 m. Each panel is supported by the central monopole and three grounded towers. The outermost one is 243 m high. The panels sag significantly between the masts so that the average height is only 201 m. The entire structure is on a peninsula that extends into the Atlantic Ocean. It has a roughly rectangular shape only slightly greater than the area 3740×1870 m under the top-loading of the two monopoles. Each monopole has its own ground system that begins with radial conductors but is interconnected and finally led into the surrounding ocean. Measurements of the vertical electric field indicate that the effective height of the antenna with either one element driven or both driven in parallel is near 150 m. This is half the length of the monopoles and substantially smaller than the outer edge of the panels or their average height. With $h_e = 150$ m and $f = 19.4$ kHz, $k_0 = 4.063 \times 10^{-4}$ m^{-1} and

$$R^e = 40k_0^2 h_e^2 = 0.150 \text{ ohm}.\tag{8.101}$$

The measured resistance of the tuning coil was $R_c = 0.014$ ohm and the total driving-point resistance of one monopole was $R_0 = 0.200$ ohm. It follows that the ground resistance must be

$$R_g = R_0 - R_c - R^e = 0.036 \, \text{ohm}. \tag{8.102}$$

The very extensive ground system ending in salt water makes this quite low resistance reasonable. The radiating efficiency of the antenna is high, namely

$$E = \frac{R^e}{R_0} = 0.75 \quad \text{or} \quad 75\%. \tag{8.103}$$

When operating at full power, each monopole carries a current of 2600 A and radiates a megawatt of power. The array radiates twice that power. The vertical electric field at $\rho = 500 \, \text{km}$ on the surface of the sea is

$$|E_{0\Theta}^i(500 \, \text{km}, \pi/2)| = \frac{\omega\mu_0 h_e I}{2\pi\rho} = 19 \, \text{mV/m}. \tag{8.104}$$

The radial electric field on the surface of the sea is

$$|E_{0r}^i(500 \, \text{km}, \pi/2)| = \left| \frac{k_0}{k_2} E_{0\Theta}^i(500 \, \text{km}, \pi/2) \right|. \tag{8.105}$$

Here,

$$\frac{k_0}{k_2} = \frac{1}{c} \left(\frac{\omega}{\mu_0 \sigma_2} \right)^{1/2} = 0.51 \times 10^{-3}.$$

Hence,

$$|E_{0r}^i(500 \, \text{km}, \pi/2)| = 9.89 \, \mu\text{V/m}. \tag{8.106}$$

The signal received by a submarine at the depth z is

$$|E_{2\rho}(\rho, z)| = |E_{0r}^i(\rho, \pi/2)| \, e^{-\alpha_2 z}, \tag{8.107a}$$

where

$$\alpha_2 = \left(\frac{|k_2|^2}{2} \right)^{1/2} = \left(\frac{\omega\mu_0\sigma_2}{2} \right)^{1/2} = 0.56 \, \text{m}^{-1}. \tag{8.107b}$$

At a depth of $z = 20 \, \text{m}$, the field is

$$|E_{2\rho}(500 \, \text{km}, 20 \, \text{m})| = 1.35 \times 10^{-10} \, \text{V/m} \tag{8.108a}$$

or

$$20 \log_{10} |E_{2\rho}| = -197.4 \, \text{dB}. \tag{8.108b}$$

At $z = 40\,\text{m}$,

$$|E_{2\rho}(500\,\text{km}, 40\,\text{m})| = 1.85 \times 10^{-15}\,\text{V/m} \tag{8.109a}$$

or

$$20\log_{10}|E_{2\rho}| = -294.7\,\text{dB}. \tag{8.109b}$$

This is a large enough signal to be readily detectable by a submarine with a trailing-wire antenna and a sensitive receiver.

The VLF antenna at Annapolis, MD, differs from the Cutler, ME, antenna in that it consists of a single tower that is 1200 ft high and insulated from the ground instead of being connected to a ground network. It is driven by three transmission lines at the 300-ft, 600-ft, and 900-ft heights. These lines lead from the antenna to the transmitter on the ground at some distance from the base of the tower. The top-loading consists of three symmetrical panels that are supported by 600-ft towers. In addition, there is a much longer parallel-wire type of top-loading that extends out beyond one of the three panels. Measurements of the vertical electric field give an effective length of $h_e = 125\,\text{m}$. The radiation efficiency is only 35%.

8.7 High-frequency dipoles over the earth; cellular telephone

Frequencies in the range from 100 to 1800 MHz are used for various types of communication including especially the cellular telephone at frequencies from 900 to 1800 MHz. The antennas involved are dipoles or monopoles on elevated ground planes of finite size. These may be at heights as great as $d = 30\,\text{m}$. Owing to the high frequency, the far-field condition $|P_0| \geq 4$ is satisfied in the practical range of distances. Also, the radial distances ρ are large compared with the height d of the dipole so that the condition $d^2 \ll \rho^2$ is well satisfied. Accordingly, the applicable formula for $E^r_{0\Theta}(r_0, \Theta)$, which is the only component of interest, is (8.59). That is,

$$
\begin{aligned}
E^r_{0\Theta}(r_0, \Theta) = -\frac{\omega\mu_0(2h_e I)}{2\pi k_0} e^{ik_0 r_0} & \left\{ \frac{ik_0}{r_0} \sin\Theta \left[\frac{(\cos\Theta + d/r_0)\cos(k_0 d\cos\Theta)}{\epsilon + \cos\Theta + d/r_0} \right. \right. \\
& \left. - \frac{i[\epsilon - (d/r_0)\cos^2\Theta]\sin(k_0 d\cos\Theta)}{\epsilon + \cos\Theta + d/r_0} \right] \\
& \left. + \frac{\epsilon}{r_0^2} \left(\frac{\sin\Theta}{\epsilon + \cos\Theta + d/r_0}\right)^3 (\epsilon\cos\Theta - \sin^2\Theta) e^{ik_0 d\cos\Theta} \right\}.
\end{aligned} \tag{8.110}
$$

When the dipole is over earth or water, $\epsilon = k_0/k_2$; when there is an electrically thin layer of asphalt or concrete on the earth or a layer of ice on the water, $\epsilon = (k_0/k_2) - ik_0l$. The only restrictions on (8.110) are

$$|k_2| \geq 3k_0, \quad |k_1l| \leq 0.6, \quad d^2 \ll r_0^2. \tag{8.111}$$

The three wave numbers are

$$k_0 = \frac{\omega}{c}, \quad k_1 = k_0\epsilon_{1r}^{1/2}, \quad k_2 = \beta_2 + i\alpha_2 = k_0\epsilon_{2r}^{1/2}\left(1 + \frac{i\sigma_2}{\omega\epsilon_0\epsilon_{2r}}\right)^{1/2}; \tag{8.112}$$

$r_0 = (\rho^2 + z'^2)^{1/2}$ and Θ is measured from the vertical z'-axis. In most cases,

$$\frac{d}{r_0} \ll |\cos\Theta + \epsilon| \tag{8.113}$$

so that (8.110) reduces to

$$
\begin{aligned}
E_{0\Theta}^r(r_0, \Theta) = &-\frac{\omega\mu_0(2h_eI)}{2\pi k_0}e^{ik_0r_0}\Bigg\{\frac{ik_0}{r_0}\left(\frac{\sin\Theta\cos\Theta}{\cos\Theta + \epsilon}\right) \\
&\times \left[\cos(k_0d\cos\Theta) - \frac{i\epsilon\sin(k_0d\cos\Theta)}{\cos\Theta}\right] \\
&+ \frac{\epsilon}{r_0^2}\left(\frac{\sin\Theta}{\cos\Theta + \epsilon}\right)^3(\epsilon\cos\Theta - \sin^2\Theta)e^{ik_0d\cos\Theta}\Bigg\}.
\end{aligned}
\tag{8.114}
$$

In this expression, the term with $1/r_0$ as a factor is the space wave. Its value is zero when $\Theta = \pi/2$ on the boundary surface. The factor containing d takes account of the height of the dipole. It reduces to unity when $d = 0$. The term with $1/r_0^2$ as a factor is the lateral wave. Its maximum occurs when $\Theta = \pi/2$. This is proportional to $(\sin^5\Theta)/\epsilon^2 r_0^2$. Over most of the earth's surface, $\epsilon^2 = k_0^2/|k_2|^2 \ll 1$ so that $1/\epsilon^2 = k_2^2/k_0^2$ is very large. The factor $\sin^5\Theta$ shows that the surface wave is confined to a narrow beam close to the surface where $\Theta = \pi/2$. Only over dry sand is the condition $|k_2| \geq 3k_0$ not satisfied so that $1/\epsilon^2 = k_2^2/k_0^2$ is not large and for all practical purposes there is no surface wave. This means that $|E_{0\Theta}^r(r_0, \pi/2)| \sim 0$.

Graphical representations of the vertical field patterns of a vertical electric dipole at the height $d = 2\,\text{m}$ are shown in a series of figures at the radial distances $r_0 = 100$, 500, and 1000 m, for sea and lake water, two types of earth, and dry sand. Specifically, Figs. 8.10, 8.12, and 8.14 show polar graphs of $|E_{0\Theta}^r(r_0, \Theta)|$ at $f = 100\,\text{MHz}$. The field in the range $80° \leq \Theta \leq 90°$ is shown in logarithmic graphs in Figs. 8.11, 8.13, and 8.15. Similar graphs at $f = 500\,\text{MHz}$ are in Figs. 8.16–8.19, and at $f = 1000\,\text{MHz}$ in Figs. 8.20–8.23. A close study of Figs. 8.10, 8.12, and 8.14 shows that the three sets of diagrams are identical for all values of Θ not too close to $\Theta = \pi/2$ or 90° except for the radial scale which decreases with increasing r_0 exactly as $1/r_0$. Specifically, 1 V/m on Fig. 8.10 with $r_0 = 100\,\text{m}$ appears as

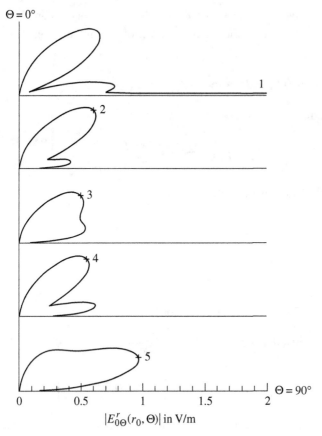

Figure 8.10   Polar graphs of the electric far field (V/m) in air of a vertical electric dipole at the height $d = 2$ m over five different media at $f = 100$ MHz; $r_0 = 100$ m. Taken from King and Sandler [1, Fig. 2]. © 1994 I.E.E.E.

	σ_2 (S/m)	ϵ_{2r}	$|E_{0\Theta}^r(r_0, \Theta)|_{\max}$	at Θ
1. Sea water	4.000	80	4.34×10^0	at 90°
2. Wet earth	0.400	12	8.05×10^{-1}	at 49°
3. Dry earth	0.040	8	6.62×10^{-1}	at 50°
4. Lake water	0.004	80	7.60×10^{-1}	at 47°
5. Dry sand	0.000	2	9.97×10^{-1}	at 74°

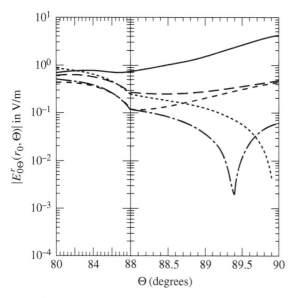

Figure 8.11 Rectangular graphs of the range $80° \leq \Theta \leq 90°$ in Fig. 8.10; $f = 100\,\text{MHz}$, $r_0 = 100\,\text{m}$. Taken from King and Sandler [1, Fig. 5]. © 1994 I.E.E.E.

	σ_2 (S/m)	ϵ_{2r}	$|E^r_{0\Theta}(r_0, 90°)|$
—— Sea water	4.000	80	4.34×10^0
– – – Wet earth	0.400	12	4.37×10^{-1}
– · – Dry earth	0.040	8	6.45×10^{-2}
—— Lake water	0.004	80	4.80×10^{-1}
· · · · · Dry sand	0.000	2	1.20×10^{-2}

0.2 V/m on Fig. 8.12 with $r_0 = 500\,\text{m}$, and as 0.1 V/m on Fig. 8.14 with $r_0 = 1000\,\text{m}$.

In Fig. 8.10, the top curve for sea water with $r_0 = 100\,\text{m}$ has a large peak at $\Theta = 90°$; the peak is much smaller in Fig. 8.12 with $r_0 = 500\,\text{m}$ and does not appear in Fig. 8.14 with $r_0 = 1000\,\text{m}$. This peak is part of the contribution of the lateral wave which decreases with distance as $1/r_0^2$. In order to show its part of the field more clearly, completely separate diagrams are shown in Figs. 8.11, 8.13, and 8.15 for the range $80° \leq \Theta \leq 90°$. Note that the scale between $\Theta = 88°$ and $\Theta = 90°$ is greatly expanded and the amplitude of the field is represented on a logarithmic scale. The large peak at $\Theta = 90°$ for sea water is clearly shown in Figs. 8.11, 8.13, and 8.15 together with the $1/r_0^2$ decrease in amplitude. These figures also show that there is a significant contribution from the lateral wave for all media except dry sand for which the small ratio $|k_2|/k_0 = 2$ makes the field at $\Theta = 90°$ only a little greater than the $1/r_0^2$ near field. Since the space wave decreases to zero at $\Theta = 90°$ over all types of earth or water and the contribution by the lateral wave rises to a maximum there, the transition

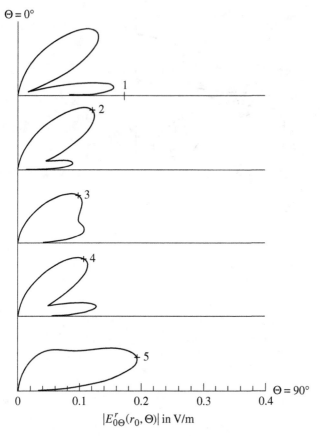

Figure 8.12 Polar graphs of the electric far field (V/m) in air of a vertical electric dipole at the height $d = 2$ m over five different media at $f = 100$ MHz; $r_0 = 500$ m. Taken from King and Sandler [1, Fig. 3]. © 1994 I.E.E.E.

	σ_2 (S/m)	ϵ_{2r}	$|E^r_{0\Theta}(r_0, \Theta)|_{max}$	at Θ
1. Sea water	4.000	80	1.74×10^{-1}	at 90°
2. Wet earth	0.400	12	1.61×10^{-1}	at 49°
3. Dry earth	0.040	8	1.32×10^{-1}	at 50°
4. Lake water	0.004	80	1.52×10^{-1}	at 47°
5. Dry sand	0.000	2	2.00×10^{-1}	at 74°

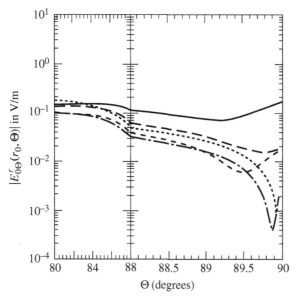

Figure 8.13 Rectangular graphs of the range $80° \leq \Theta \leq 90°$ in Fig. 8.12; $f = 100$ MHz, $r_0 = 500$ m. Taken from King and Sandler [1, Fig. 6]. © 1994 I.E.E.E.

	σ_2 (S/m)	ϵ_{2r}	$\lvert E^r_{0\Theta}(r_0, 90°)\rvert$
—— Sea water	4.000	80	1.74×10^{-1}
– – – Wet earth	0.400	12	1.75×10^{-2}
– · – Dry earth	0.040	8	2.58×10^{-3}
—— Lake water	0.004	80	1.92×10^{-2}
· · · · · Dry sand	0.000	2	4.80×10^{-4}

from the one to the other involves a more or less sharp minimum. This moves closer to $\Theta = 90°$ as r_0 increases because the space wave decreases as $1/r_0$, the surface wave as $1/r_0^2$.

The cellular radiotelephone operates in the 0.9 to 1.8-GHz range of frequencies. Transmitting and receiving antennas in the form of center-driven or loaded dipoles are placed on high towers that are located several miles apart along major highways. The area surrounding each of these antennas is called a cell. Radiotelephone antennas are either attached to hand-held transceivers or are mounted on the metal top or the rear deck of an automobile. The hand-held instruments include a monopole mounted on a typical telephone receiver that now also contains a transmitter. When in use, the monopole extends upward beside and above the head. The metal case serves as the lower part of the antenna. Typically, currents of the order of 0.1 A in the antenna are needed for transmission to the nearest tower. The monopoles on the metal top or rear deck of an automobile are base-driven by a coaxial line with the car top or rear deck serving as the ground plane.

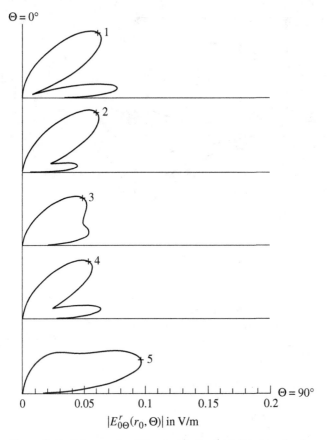

$\Theta = 0°$

$\Theta = 90°$

| 0 | 0.05 | 0.1 | 0.15 | 0.2 |

$|E_{0\Theta}^r(r_0, \Theta)|$ in V/m

Figure 8.14 Polar graphs of the electric far field (V/m) in air of a vertical electric dipole at the height $d = 2$ m over five different media at $f = 100$ MHz; $r_0 = 1000$ m. Taken from King and Sandler [1, Fig. 4]. © 1994 I.E.E.E.

	σ_2 (S/m)	ϵ_{2r}	$|E_{0\Theta}^r(r_0, \Theta)|_{max}$	at Θ
1. Sea water	4.000	80	8.57×10^{-2}	at 48°
2. Wet earth	0.400	12	8.05×10^{-2}	at 49°
3. Dry earth	0.040	8	6.62×10^{-2}	at 50°
4. Lake water	0.004	80	7.60×10^{-2}	at 47°
5. Dry sand	0.000	2	9.98×10^{-2}	at 74°

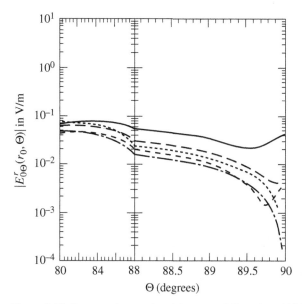

Figure 8.15 Rectangular graphs of the range $80° \leq \Theta \leq 90°$ in Fig. 8.14; $f = 100\,\mathrm{MHz}$, $r_0 = 1000\,\mathrm{m}$. Taken from King and Sandler [1, Fig. 7]. © 1994 I.E.E.E.

	σ_2 (S/m)	ϵ_{2r}	$\|E^r_{0\Theta}(r_0, 90°)\|$
—— Sea water	4.000	80	4.34×10^{-2}
– – – Wet earth	0.400	12	4.37×10^{-3}
– · – Dry earth	0.040	8	6.45×10^{-4}
—— Lake water	0.004	80	4.80×10^{-3}
· · · · · Dry sand	0.000	2	1.20×10^{-4}

The electromagnetic field generated by any of these antennas is simply that of a vertical dipole with an appropriate effective length h_e. The dipole antenna is at a height d over the surface of the earth. The electric field generated by such an antenna is given by (8.110) or (8.114). It is illustrated for a typical hand-held or car-mounted dipole in Figs. 8.20–8.23.

8.8   Vertical dipoles over a two-layered region

All of the formulas in this chapter involve the small parameter $\epsilon = (k_0/k_2) - ik_0l$, where l is the thickness of a dielectric layer (region 1) located between the air (region 0) and the earth or sea (region 2). When there is no such layer, $l = 0$ and $\epsilon = k_0/k_2$ as in the applications in Section 8.7. In this section, applications that involve an electrically thin layer on region 2 are considered.

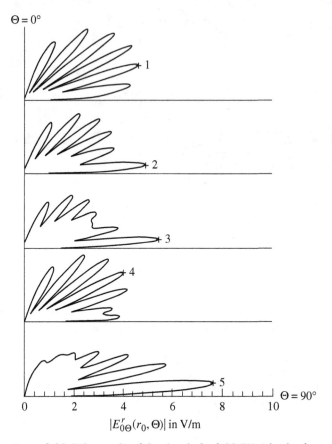

$\Theta = 0°$

$\Theta = 90°$

$|E^r_{0\Theta}(r_0, \Theta)|$ in V/m

Figure 8.16 Polar graphs of the electric far field (V/m) in air of a vertical electric dipole at the height $d = 2$ m over five different media at $f = 500$ MHz; $r_0 = 100$ m. Taken from King and Sandler [1, Fig. 8]. © 1994 I.E.E.E.

	σ_2 (S/m)	ϵ_{2r}	$|E^r_{0\Theta}(r_0, \Theta)|_{max}$	at Θ
1. Sea water	4.000	80	4.91×10^0	at $73°$
2. Wet earth	0.400	12	4.89×10^0	at $86°$
3. Dry earth	0.040	8	5.47×10^0	at $86°$
4. Lake water	0.004	80	4.48×10^0	at $63°$
5. Dry sand	0.000	2	7.72×10^0	at $86°$

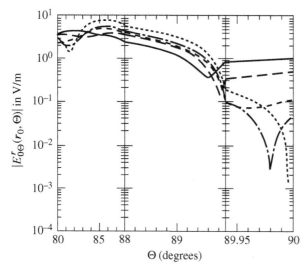

Figure 8.17 Rectangular graphs of the range $80° \leq \Theta \leq 90°$ in Fig. 8.16; $f = 500\,\text{MHz}$, $r_0 = 100\,\text{m}$. Taken from King and Sandler [1, Fig. 9]. © 1994 I.E.E.E.

	σ_2 (S/m)	ϵ_{2r}	$\lvert E^r_{0\Theta}(r_0, 90°)\rvert$
—— Sea water	4.000	80	9.87×10^{-1}
– – – Wet earth	0.400	12	1.12×10^{-1}
– · – Dry earth	0.040	8	4.87×10^{-2}
—— Lake water	0.004	80	4.80×10^{-1}
· · · · · Dry sand	0.000	2	1.20×10^{-2}

Vertical dipoles over asphalt-coated earth

When the earth is asphalt- or cement-coated with a layer that is $l = 0.15\,\text{m}$ thick with $\epsilon_{1r} = 2.65$ over earth with $\sigma_2 = 0.04\,\text{S/m}$ and $\epsilon_{2r} = 8$, the three wave numbers for $f = 100\,\text{MHz}$ are: $k_0 = 2.09\,\text{m}^{-1}$, $k_1 = k_0\epsilon_{1r}^{1/2} = 3.41\,\text{m}^{-1}$, and $k_2 = k_0\epsilon_{2r}^{1/2}(1 + i\sigma_2/\omega\epsilon_0\epsilon_{2r})^{1/2} = 7.43 + i2.85 = 7.96e^{i0.366}\,\text{m}^{-1}$. Also, $k_1 l = 0.51 < 0.6$ and $\epsilon = 0.26e^{-i0.366} - i0.315 = 0.478e^{-i1.03}$. The far field is limited by $\rho \geq \lvert 8/k_0\epsilon^2\rvert = 16.8\,\text{m}$. The far-field patterns of a vertical dipole at the height $d = 2\,\text{m}$ over the asphalt-coated earth are shown in Figs. 8.24, 8.25, and 8.26, respectively, at the radial distances $r_0 = 100$, 500, and 1000 m. The polar graphs in Figs. 8.24a, 8.25a, and 8.26a show the space wave; the logarithmic graphs in Figs. 8.24b, 8.25b, and 8.26b show the complete field in the range $80° \leq \Theta \leq 90°$ where the lateral wave is dominant.

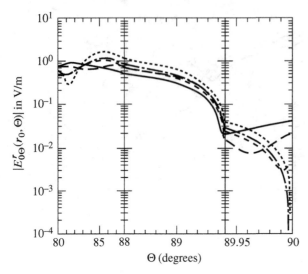

Figure 8.18 Like Fig. 8.17 for $r_0 = 500$ m. Taken from King and Sandler [1, Fig. 10]. © 1994 I.E.E.E.

	σ_2 (S/m)	ϵ_{2r}	$|E_{0\Theta}^r(r_0, 90°)|$
——— Sea water	4.000	80	3.95×10^{-2}
– – – Wet earth	0.400	12	4.49×10^{-3}
– · – Dry earth	0.040	8	1.95×10^{-3}
—— Lake water	0.004	80	1.92×10^{-2}
· · · · · Dry sand	0.000	2	4.80×10^{-4}

Vertical dipoles over the Arctic ice

Communication on the Arctic ice involves a vertical dipole at a height d in the air over a layer of ice with the thickness l on salt water. At $f = 7$ MHz, the relative permittivity of ice is of the order of $\epsilon_{1r} \sim 3.2$. For sea water, $\epsilon_{2r} = 80$ and $\sigma_2 = 4$ S/m. The relevant wave numbers are: for air, $k_0 = 0.147 \, \text{m}^{-1}$; for ice, $k_1 = 0.262 \, \text{m}^{-1}$; and for sea water, $k_2 = 14.87e^{i\pi/4} = 10.52(1+i) \, \text{m}^{-1}$. With $l = 2.5$ m, $k_1 l = 0.658$, which slightly exceeds the condition $k_1 l \leq 0.6$, but is an acceptable value. Note that the small quantity $\epsilon = (k_0/k_2) - ik_0 l = 0.01e^{-i\pi/4} - i0.368$. Evidently, $\epsilon \sim - ik_0 l = -i0.368$ and the sea behaves like a perfect conductor under the ice. The far field occurs when $\rho \geq |8/k_0\epsilon^2| = 8/k_0^3 l^2 = 401$ m. The far-field patterns of a vertical dipole at the height $d = 2$ m over the Arctic ice are shown in Figs. 8.27, 8.28, and 8.29 at the radial distances $r_0 = 500$, 1000, and 5000 m, respectively. As the distance increases, the magnitude of the space wave decreases more slowly than that of the lateral wave, so that the transition from the former to the latter moves nearer to $\Theta = 90°$.

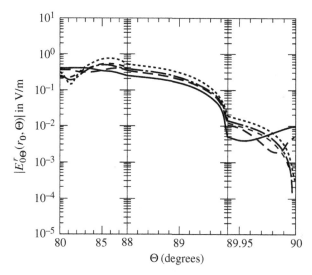

Figure 8.19 Like Fig. 8.17 for $r_0 = 1000$ m. Taken from King and Sandler [1, Fig. 11]. © 1994 I.E.E.E.

	σ_2 (S/m)	ϵ_{2r}	$|E^r_{0\Theta}(r_0, 90°)|$
—— Sea water	4.000	80	9.87×10^{-3}
– – – Wet earth	0.400	12	1.12×10^{-3}
– · – Dry earth	0.040	8	4.87×10^{-4}
— — Lake water	0.004	80	4.80×10^{-3}
· · · · · Dry sand	0.000	2	1.20×10^{-4}

Vertical dipoles on microstrip

Microstrip consists of a thin dielectric layer coating a highly conducting base on which strip transmission lines and antennas are located. Elements of these are horizontal electric dipoles. These are treated in Chapter 9. Vertical connections to the base are vertical electric dipoles. Since the dielectric substrate is thin and k_2 is very large, $\epsilon \sim -ik_0l$. Graphs of both $E_{0z'}(\rho, 0)$ and $E_{0\rho}(\rho, 0)$ as functions of ρ for a vertical dipole in the air on the surface $z' = -z = 0$ of the dielectric layer with the thickness $l = 0.1$ mm are shown in Fig. 8.30 at $f = 10$ GHz. The range $k_0\rho \leq 1$ is the near field. It involves a $1/\rho^2$ decrease with increasing ρ for $E_{0\rho}(\rho, 0)$ and a $1/\rho^3$ decrease for $E_{0z'}(\rho, 0)$. The range $1 \leq k_0\rho \leq 2/k_0^2l^2$ is the intermediate range, where both $E_{0\rho}(\rho, 0)$ and $E_{0z'}(\rho, 0)$ decrease approximately as $1/\rho$. Similar graphs with $f = 5.15$ GHz and $f = 4.21$ GHz and $l = 4.445$ mm are shown in Figs. 8.31 and 8.32. These extend to a much greater range so that a part of the far field where $k_0\rho > 8/k_0^2l^2$ is included. In it, the field has a $1/\rho^2$ dependence. Far-field patterns of $|E_{0\Theta}(r_0, \Theta)|$ for $f = 5.15$ GHz and $l = 4.445$ mm are shown in Figs. 8.33, 8.34, and 8.35 with $\rho = 0.3, 3,$ and 30 m, respectively.

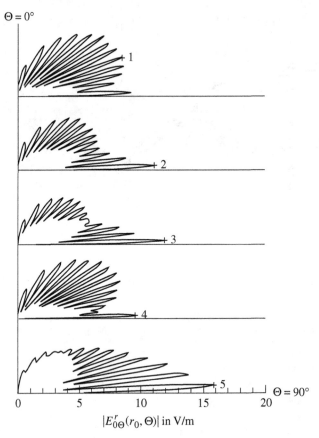

$\Theta = 0°$

$\Theta = 90°$

$|E_{0\Theta}^r(r_0, \Theta)|$ in V/m

Figure 8.20 Polar graphs of the electric far field (V/m) in air of a vertical electric dipole at the height $d = 2\,\text{m}$ over five different media at $f = 1000\,\text{MHz}$; $r_0 = 100\,\text{m}$. Taken from King and Sandler [1, Fig. 12]. © 1994 I.E.E.E.

	σ_2 (S/m)	ϵ_{2r}	$|E_{0\Theta}^r(r_0, \Theta)|_{max}$	at Θ
1. Sea water	4.000	80	9.38×10^0	at $68°$
2. Wet earth	0.400	12	1.14×10^1	at $88°$
3. Dry earth	0.040	8	1.21×10^1	at $88°$
4. Lake water	0.004	80	9.54×10^0	at $88°$
5. Dry sand	0.000	2	1.65×10^1	at $88°$

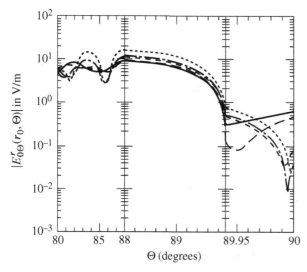

Figure 8.21 Rectangular graphs of the range $80° \leq \Theta \leq 90°$ in Fig. 8.20; $f = 1000\,\text{MHz}$, $r_0 = 100\,\text{m}$. Taken from King and Sandler [1, Fig. 13]. © 1994 I.E.E.E.

	σ_2 (S/m)	ϵ_{2r}	$|E_{0\Theta}^r(r_0, 90°)|$
—— Sea water	4.000	80	6.45×10^{-1}
– – – Wet earth	0.400	12	8.39×10^{-2}
– · – Dry earth	0.040	8	4.82×10^{-2}
—— Lake water	0.004	80	4.80×10^{-1}
· · · · · Dry sand	0.000	2	1.20×10^{-2}

8.9   Propagation over the spherical earth

All of the formulas and applications discussed so far in this chapter apply strictly to a planar earth. Since the radius of the earth is $a = 6378\,\text{km}$, it is to be expected that the planar formulas are a good approximation for a substantial distance along the surface of the earth. For propagation over the sea, much greater distances are involved especially in communicating with submarines and with surface-wave, over-the-horizon radar. These make use of relatively low frequencies and electrically short vertical dipoles on the surface of the earth very close to the sea.

The electric and magnetic fields due to a vertical dipole on the spherical earth are conveniently expressed in the spherical coordinates r, Θ, Φ, where r is the radial distance from the center of the earth, Θ is the angle measured from the radial line through the dipole, and Φ is the circumferential angle about this radial line. The three components of the field are $E_\Theta(r, \Theta)$, $E_r(r, \Theta)$, and $B_\Phi(r, \Theta)$. Note that rotational symmetry obtains with respect to the z'-axis from the center of the earth through the dipole. In the planar limit, $E_\Theta \to E_\rho$, $E_r \to E_{z'}$, and $B_\Phi \to B_{\phi'}$.

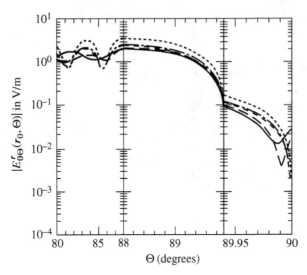

Figure 8.22 Like Fig. 8.21 for $r_0 = 500$ m. Taken from King and Sandler [1, Fig. 14]. © 1994 I.E.E.E.

	σ_2 (S/m)	ϵ_{2r}	$|E_{0\Theta}^r(r_0, 90°)|$
—— Sea water	4.000	80	2.58×10^{-2}
− − − Wet earth	0.400	12	3.36×10^{-3}
− · − Dry earth	0.040	8	1.93×10^{-3}
— — Lake water	0.004	80	1.92×10^{-2}
· · · · · Dry sand	0.000	2	4.80×10^{-4}

The distance along the surface of the earth between the dipole and a point of observation also on the surface of the earth is $\rho_s = a\Theta$. The cylindrical radial distance from the z'-axis to the point of observation is $\rho = a \sin \Theta$. The difference between the two is $\rho_s - \rho = a(\Theta - \sin \Theta) \sim a(\Theta - \Theta + \Theta^3/6 \cdots) \sim a\Theta^3/6$.

A first-order correction for the curvature of the earth is the substitution of $\rho_s = a\Theta$ for $\rho = a \sin \Theta$ in the planar formulas. For propagation over the sea and conducting earth, the direct field in the earth is negligible so that the following formulas for the surface wave give the complete field over the spherical earth:[1]

$$B_{0\Phi}(a, \Theta) \sim -\frac{\mu_0(h_e I)k_0}{4\pi} \frac{e^{ik_0\rho_s + i\pi/4}}{(\pi \rho_s \rho_c)^{1/2}} I_2(\eta_2, g) \tag{8.115}$$

$$E_{0r}(a, \Theta) \sim \frac{k_2}{k_0} E_{0\Theta}(a, \Theta) \sim -\frac{\omega}{k_0} B_{0\Phi}(a, \Theta). \tag{8.116}$$

---

[1] [6] p. II.78.

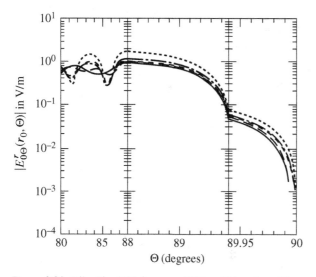

Figure 8.23 Like Fig. 8.21 for $r_0 = 1000$ m. Taken from King and Sandler [1, Fig. 15]. © 1994 I.E.E.E.

	σ_2 (S/m)	ϵ_{2r}	$|E^r_{0\Theta}(r_0, 90°)|$
—— Sea water	4.000	80	6.45×10^{-3}
– – – Wet earth	0.400	12	8.39×10^{-4}
– · – Dry earth	0.040	8	4.82×10^{-4}
—— Lake water	0.004	80	4.80×10^{-3}
· · · · · Dry sand	0.000	2	1.20×10^{-4}

Here,

$$g = \frac{k_0}{k_2}\left(\frac{k_0 a}{2}\right)^{1/3}, \quad \rho_c = a\left(\frac{k_0 a}{2}\right)^{-1/3}, \quad \eta_2 = \frac{\rho_s}{\rho_c} = \Theta\left(\frac{k_0 a}{2}\right)^{1/3}. \tag{8.117}$$

$I_2(\eta_2, g)$ is a complicated integral that is well approximated, when $\eta_2 \gg 1$, by the simple exponential formula

$$I_2(\eta_2, g) = \frac{2\pi i}{\xi_1 + g^2} e^{i\eta_2 \xi_1} = A e^{-\alpha \eta_2} e^{i\beta \eta_2}, \tag{8.118}$$

where $\xi_1 = \beta + i\alpha$ is obtained from the numerically evaluated Table 8.3 for any value of g. Note that

$$A = \frac{2\pi i}{\beta + g^2 + i\alpha} = \frac{2\pi}{[(\beta + g^2)^2 + \alpha^2]^{1/2}} \exp\left[i\left(\frac{\pi}{2} - \tan^{-1}\frac{\alpha}{\beta + g^2}\right)\right].$$

Table 8.3. *Numerically evaluated values of $\xi_1 = \beta + i\alpha$ for $g = 0$ to $g = \infty$. Taken from Houdzoumis [6]*

g	β	α	g	β	α	g	β	α	g	β	α
0	0.509 39	0.882 30	2.30	1.468 87	1.643 56	5.50	1.298 46	1.891 57	8.70	1.250 58	1.942 42
0.01	0.518 87	0.879 80	2.40	1.458 94	1.665 62	5.60	1.296 13	1.894 12	8.80	1.249 65	1.943 38
0.02	0.528 35	0.877 40	2.50	1.449 20	1.685 25	5.70	1.293 88	1.896 57	8.90	1.248 74	1.944 32
0.03	0.537 83	0.875 10	2.60	1.439 77	1.702 81	5.80	1.291 71	1.898 93	9.00	1.247 85	1.945 24
0.04	0.547 31	0.872 89	2.70	1.430 71	1.718 58	5.90	1.289 61	1.901 20	9.10	1.246 98	1.946 14
0.05	0.556 78	0.870 77	2.80	1.422 07	1.732 81	6.00	1.287 58	1.903 39	9.20	1.246 13	1.947 01
0.06	0.566 25	0.868 75	2.90	1.413 86	1.745 72	6.10	1.285 62	1.905 50	9.30	1.245 29	1.947 87
0.07	0.575 71	0.866 82	3.00	1.406 07	1.757 48	6.20	1.283 73	1.907 54	9.40	1.244 48	1.948 71
0.08	0.585 17	0.864 98	3.10	1.398 69	1.768 24	6.30	1.281 89	1.909 50	9.50	1.243 68	1.949 53
0.09	0.594 63	0.863 24	3.20	1.391 71	1.778 11	6.40	1.280 11	1.911 40	9.60	1.242 90	1.950 34
0.10	0.604 07	0.861 58	3.30	1.385 09	1.787 22	6.50	1.278 39	1.913 24	9.70	1.242 13	1.951 12
0.20	0.697 98	0.850 11	3.40	1.378 82	1.795 64	6.60	1.276 72	1.915 02	9.80	1.241 39	1.951 89
0.30	0.790 33	0.847 63	3.50	1.372 89	1.803 45	6.70	1.275 10	1.916 74	9.90	1.240 65	1.952 64
0.40	0.880 35	0.853 87	3.60	1.367 26	1.810 71	6.80	1.273 53	1.918 41	10.00	1.239 93	1.953 38
0.50	0.967 25	0.868 50	3.70	1.361 92	1.817 49	6.90	1.272 00	1.920 02	20.00	1.204 43	1.989 41
0.60	1.050 22	0.891 14	3.80	1.356 84	1.823 83	7.00	1.270 52	1.921 59	30.00	1.192 63	2.001 26

Table 8.3. – *continued*

g	β	α	g	β	α	g	β	α	g	β	α
0.70	1.128 44	0.921 33	3.90	1.352 02	1.829 78	7.10	1.269 08	1.923 10	40.00	1.186 73	2.007 17
0.80	1.201 09	0.958 49	4.00	1.347 43	1.835 36	7.20	1.267 68	1.924 58	50.00	1.183 19	2.010 71
0.90	1.267 35	1.001 91	4.10	1.343 06	1.840 62	7.30	1.266 32	1.926 01	60.00	1.180 83	2.013 07
1.00	1.326 44	1.050 74	4.20	1.338 90	1.845 59	7.40	1.265 00	1.927 40	70.00	1.179 15	2.014 75
1.10	1.377 67	1.103 91	4.30	1.334 92	1.850 28	7.50	1.263 71	1.928 75	80.00	1.177 89	2.016 02
1.20	1.420 48	1.160 19	4.40	1.331 13	1.854 72	7.60	1.262 46	1.930 06	90.00	1.176 91	2.017 00
1.30	1.454 58	1.218 13	4.50	1.327 50	1.858 93	7.70	1.261 24	1.931 34	100.00	1.176 12	2.017 78
1.40	1.479 99	1.276 16	4.60	1.324 03	1.862 92	7.80	1.260 05	1.932 58	200.00	1.172 58	2.021 32
1.50	1.497 16	1.332 75	4.70	1.320 70	1.866 73	7.90	1.258 89	1.933 79	300.00	1.171 41	2.022 50
1.60	1.506 94	1.386 51	4.80	1.317 51	1.870 35	8.00	1.257 76	1.934 97	400.00	1.170 82	2.023 09
1.70	1.510 52	1.436 38	4.90	1.314 46	1.873 80	8.10	1.256 66	1.936 11	500.00	1.170 46	2.023 44
1.80	1.509 23	1.481 76	5.00	1.311 52	1.877 09	8.20	1.255 58	1.937 23	600.00	1.170 23	2.023 68
1.90	1.504 38	1.522 45	5.10	1.308 70	1.880 24	8.30	1.254 53	1.938 32	700.00	1.170 06	2.023 85
2.00	1.497 12	1.558 57	5.20	1.305 99	1.883 26	8.40	1.253 51	1.939 38	800.00	1.169 93	2.023 97
2.10	1.488 36	1.590 50	5.30	1.303 39	1.886 14	8.50	1.252 51	1.940 42	900.00	1.169 83	2.024 07
2.20	1.478 79	1.618 67	5.40	1.300 88	1.888 91	8.60	1.251 53	1.941 43	∞	1.169 05	2.024 86

(a)

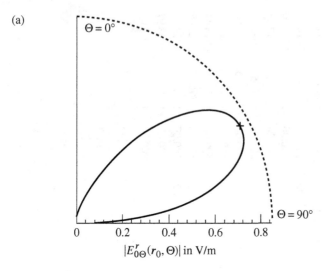

(b)

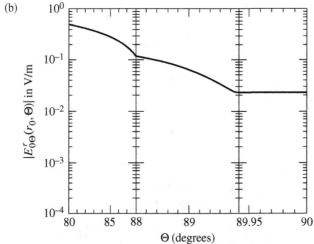

Figure 8.24 Electric far field of vertical dipole at height d over asphalt-coated earth. (a) Polar graph, $0° \leq \Theta \leq 90°$. (b) Rectangular graph, $80° \leq \Theta \leq 90°$. $\epsilon_{1r} = 2.65$, $l = 0.15$ m; $\sigma_2 = 0.04$ S/m, $\epsilon_{2r} = 8$; $f = 100$ MHz, $d = 2$ m, $r_0 = 100$ m. $|E^r_{0\Theta}|_{max} = |E^r_{0\Theta}(100\,\text{m}, 60°)| = 0.813$ V/m; $|E^r_{0\Theta}(100\,\text{m}, 90°)| = 2.3 \times 10^{-2}$ V/m. Taken from King and Sandler [2, Fig. 2]. © 1994 American Geophysical Union.

The amplitude $|A|$ and constants α and β for $|g| = 0.05$, 1, and 20 are given below, together with the associated frequency for sea water:

$$|g| = 0.05: \quad f = 0.265\,\text{MHz}, \quad |A| = 6.0, \quad \alpha = 0.87, \ \beta = 0.56, \quad (8.119a)$$

$$|g| = 1: \quad f = 9.65\,\text{MHz}, \quad |A| = 2.46, \quad \alpha = 1.05, \ \beta = 1.33, \quad (8.119b)$$

$$|g| = 20: \quad f = 351\,\text{MHz}, \quad |A| = 0.0157, \quad \alpha = 1.99, \ \beta = 1.20. \quad (8.119c)$$

(a)

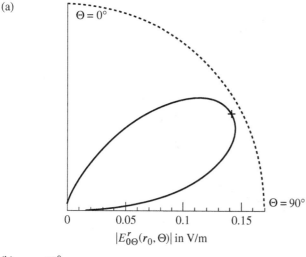

(b)

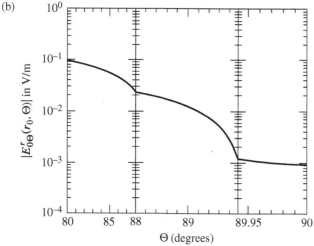

Figure 8.25 Like Fig. 8.24 with $r_0 = 500$ m. $|E_{0\Theta}^r|_{max} = |E_{0\Theta}^r(500\, \text{m}, 60°)| = 0.162$ V/m; $|E_{0\Theta}^r(500\, \text{m}, 90°)| = 9.21 \times 10^{-4}$ V/m. Taken from King and Sandler [2, Fig. 3]. © 1994 American Geophysical Union.

Note that when $\xi_1 \gg g^2$, $I_2 \sim (2\pi i/\xi_1)e^{i\eta_2\xi_1}$ and when $\xi_1 \ll g^2$, $I_2 \sim (2\pi i/g^2)e^{i\eta_2\xi_1}$.

For communicating with submerged submarines at $f = 20\,\text{kHz}$, the quantity of interest is the component of the electric field tangent to the surface of the sea, i.e. $E_{0\Theta}(a, \Theta)$. With (8.115)–(8.119), this is given by the following formula for the field of a base-driven monopole with the effective height h_e:

$$|E_{0\Theta}(a, \Theta)| = \left| \frac{\omega\mu_0(h_e I)k_0}{4\pi k_2} \frac{e^{ik_0\rho_s}}{(\pi\rho_s\rho_c)^{1/2}} A e^{i\xi_1\eta_2} \right|. \tag{8.120}$$

(a)

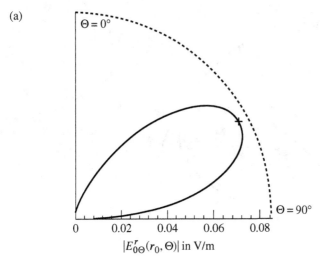

(b)

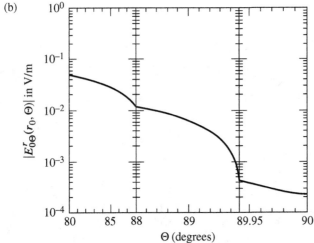

Figure 8.26 Like Fig. 8.24 with $r_0 = 1000$ m. $|E^r_{0\Theta}|_{\max} = |E^r_{0\Theta}(1000$ m, $60°)| = 0.0811$ V/m; $|E^r_{0\Theta}(1000$ m, $90°)| = 2.3 \times 10^{-4}$ V/m. Taken from King and Sandler [2, Fig. 4]. © 1994 American Geophysical Union.

Here,

$$\rho_c = a \left(\frac{k_0 a}{2} \right)^{-1/3} = 579.3 \text{ km.} \tag{8.121}$$

Note that $k_0 = (4\pi/3) \times 10^{-4} \text{ m}^{-1}$ and $k_2 = (\omega\mu_0\sigma_2/2)^{1/2}(1 + i) = (\omega\mu_0\sigma_2)^{1/2}e^{i\pi/4} = 0.795e^{i\pi/4} = 0.56(1 + i) \text{ m}^{-1} = \beta_2 + i\alpha_2$. It follows that $|g| = |(k_0/k_2)(k_0a/2)^{1/3}| = 0.0058$, $\xi_1 = 0.515 + i0.881 = \beta + i\alpha$, and $|A| = |2\pi/(\xi_1 + g^2)| \sim |2\pi/\xi_1| = 6.16$. With $a = 6378$ km and $\rho_s = 5000$ km,

(a)

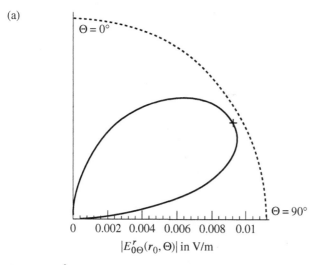

(b)

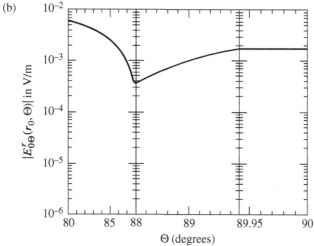

Figure 8.27 Electric far field of vertical dipole at height d over ice-coated sea water. (a) Polar graph, $0° \leq \Theta \leq 90°$. (b) Rectangular graph, $80° \leq \Theta \leq 90°$. $\epsilon_{1r} = 3.2$, $l = 2.5$ m; $\sigma_2 = 4$ S/m, $\epsilon_{2r} = 80$; $f = 7$ MHz, $d = 2$ m, $r_0 = 500$ m. $|E^r_{0\Theta}|_{max} = |E^r_{0\Theta}(500$ m, $60°)| = 1.06 \times 10^{-2}$ V/m; $|E^r_{0\Theta}(500$ m, $90°)| = 1.72 \times 10^{-3}$ V/m. Taken from King and Sandler [2, Fig. 8]. © 1994 American Geophysical Union.

$\eta_2 = \rho_s/\rho_c = 8.63$ and $\Theta = \rho_s/a = 5000/6378 = 0.784$. With these values,

$$|E_{0\Theta}(a, 0.784)| = 6.81 \times 10^{-15}(h_e I). \tag{8.122}$$

It follows that, for a unit dipole,

$$20 \log_{10} |E_{0\Theta}| = -283.3 \text{ dB.} \tag{8.123}$$

(a)

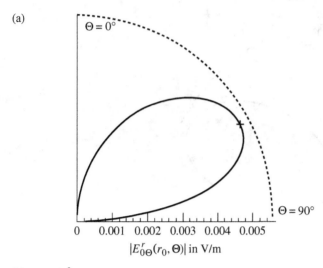

(b)

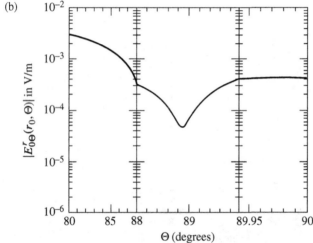

Figure 8.28 Like Fig. 8.27 with $r_0 = 1000$ m.
$|E_{0\Theta}^r|_{max} = |E_{0\Theta}^r(1000 \text{ m}, 60°)| = 5.32 \times 10^{-3}$ V/m; $|E_{0\Theta}^r(1000 \text{ m}, 90°)| = 4.29 \times 10^{-4}$ V/m.
Taken from King and Sandler [2, Fig. 9]. © 1994 American Geophysical Union.

The antenna at Cutler, ME, has an effective length $h_e = 150$ m and carries a maximum current of 2600 A. Hence, $h_e I = 3.9 \times 10^5$ A m and

$$|E_{0\Theta}(a, 0.784)| = 2.65 \times 10^{-9} \text{ V/m} \tag{8.124}$$

$$20 \log_{10} |E_{0\Theta}| = -171.5 \text{ dB.} \tag{8.125}$$

The planar earth values are -225 dB and -112.8 dB, respectively. These are the values at the surface of the sea. At the depth $(a - r)$ in the sea,

$$|E_{2\Theta}(a - r, 0.784)| = |E_{0\Theta}(a, 0.784)|e^{-\alpha_2(a-r)}$$

(a)

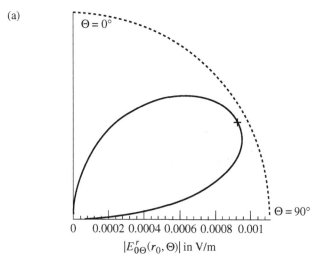

(b)

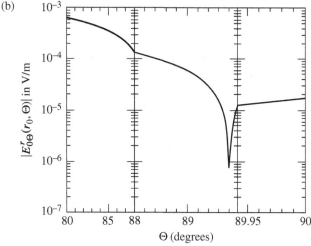

Θ (degrees)

Figure 8.29 Like Fig. 8.27 with $r_0 = 5000$ m.
$|E^r_{0\Theta}|_{max} = |E^r_{0\Theta}(5000\text{ m}, 60°)| = 1.06 \times 10^{-3}$ V/m; $|E^r_{0\Theta}(5000\text{ m}, 90°)| = 1.72 \times 10^{-5}$ V/m.
Taken from King and Sandler [2, Fig. 10]. © 1994 American Geophysical Union.

where $\alpha_2 = 0.56\text{ m}^{-1}$. If a submarine with a trailing-wire antenna can detect a field of the order of $E_{2\Theta} \sim 5.6 \times 10^{-17}$ V/m or $20\log_{10} |E_{2\Theta}| \sim -325$ dB, the submarine can be no deeper than 31.6 m. Thus,

$$20\log_{10} |E_{2\Theta}(31.6, 0.784)e^{-31.6\alpha_2}| = 20\log_{10} 5.47 \times 10^{-17} = -325.2\text{ dB}. \quad (8.126)$$

With the planar earth formula, the 325-dB limit of detectability is reached at $(a - r) = 43.6$ m instead of the actual 31.6 m.

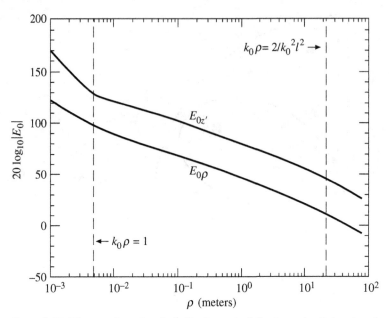

Figure 8.30 The complete electric field of a vertical dipole on the dielectric substrate of microstrip; $f = 10\,\mathrm{GHz}$, $l = 0.1$ mm. Taken from King and Sandler [2, Fig. 11]. © 1994 American Geophysical Union.

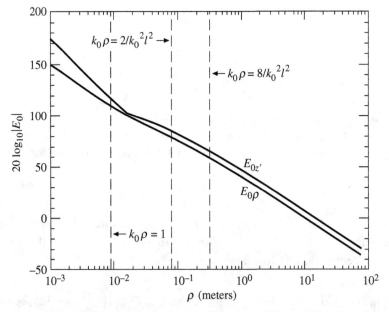

Figure 8.31 Like Fig. 8.30 with $f = 5.15\,\mathrm{GHz}$, $l = 4.445$ mm. Taken from King and Sandler [2, Fig. 12] but with corrected value for frequency. © 1994 American Geophysical Union.

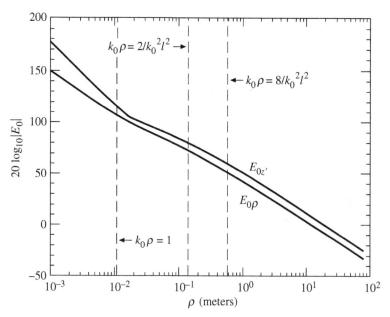

Figure 8.32 Like Fig. 8.30 with $f = 4.21$ GHz, $l = 4.445$ mm. Taken from King and Sandler [2, Fig. 13] but with corrected value for frequency. © 1994 American Geophysical Union.

A second important application of the formulas for the spherical earth is over-the-horizon radar using the surface wave. Because radar using ionospheric reflection is unable to detect low-flying missiles closer than 1000 km, the surface wave must be used for these shorter distances. Since the target is close to the surface of the sea and the shore-based transceiver involves an array of grounded monopoles or base-insulated dipoles, the conditions $z' \sim d \sim 0$ are satisfied.

At a frequency of $f = 9.65$ MHz, $|g| = 1$, $k_0 = 2\pi f/c = 0.202 \, \text{m}^{-1}$, $k_2 = (\omega\mu_0\sigma_2)^{1/2}e^{i\pi/4} = 17.45e^{i\pi/4} = 12.35(1+i) = \beta_2 + i\alpha_2$; $\rho_c = a(k_0a/2)^{-1/3} = 73.88$ km. At $\rho_s = 500$ km, $\eta_2 = \rho_s/\rho_c = 6.76$ and $\Theta = \rho_s/a = 0.0784$. The vertical electric field for a unit dipole is

$$|E_{0r}(a, 0.0784)| = \frac{\omega\mu_0 A e^{-\eta_2\alpha}}{4\pi(\pi\rho_s\rho_c)^{1/2}} = 3.62 \times 10^{-8} \, \text{V/m}. \tag{8.127}$$

The tangential electric field for a unit dipole is

$$|E_{0\Theta}(a, 0.0784)| = \left|\frac{k_0}{k_2} E_{0r}(a, 0.0784)\right| = 4.2 \times 10^{-10} \, \text{V/m}. \tag{8.128}$$

For a grounded vertical monopole with the length h, the electric moment is $h_e I$; for a base-insulated dipole with the length $2h$, the electric moment is $(2h_e I)$.

(a)

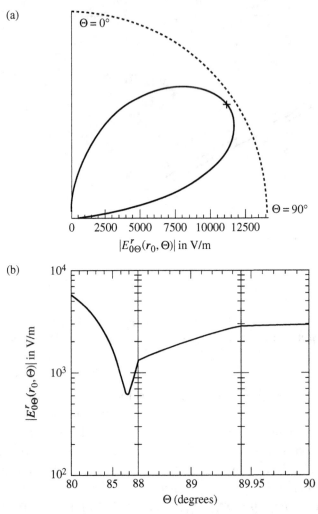

(b)

Figure 8.33 Electric far field $E_{0\Theta}^r(r_0, \Theta)$ of a vertical dipole on the dielectric substrate of microstrip. (a) Polar graph, $0° \leq \Theta \leq 90°$. (b) Rectangular graph, $80° \leq \Theta \leq 90°$. $f = 5.15\,\text{GHz}$, $l = 4.445\,\text{mm}$, $d = 0$, $r_0 = 0.3\,\text{m}$. $|E_{0\Theta}^r|_{\text{max}} = |E_{0\Theta}^r(0.3\,\text{m}, 55°)| = 1.35 \times 10^4$ V/m; $|E_{0\Theta}^r(0.3\,\text{m}, 90°)| = 2.89 \times 10^3$ V/m. Taken from King and Sandler [2, Fig. 14]. © 1994 American Geophysical Union.

The surface-wave, over-the-horizon radar array may consist of a broadside curtain of base-driven, grounded vertical monopoles. Since all elements are in the intermediate zone, their current distributions and impedances are the same as if they were over a perfect conductor. An alternative to the radial ground system is a radial array of traveling-wave horizontal-wire or Beverage antennas. This is discussed in Chapter 9. A novel array of base-insulated vertical dipoles is the resonant circular array described and analyzed in Chapter 11. Both the horizontal-wire antenna and the resonant circular array with two elements driven to produce a pancake-like field pattern are especially

(a)

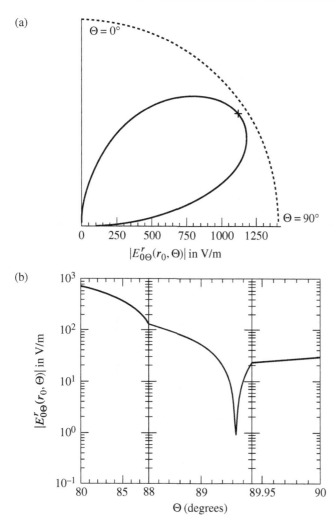

(b)

Figure 8.34 Like Fig. 8.33 with $r_0 = 3$ m. $|E^r_{0\Theta}|_{max} = |E^r_{0\Theta}(3$ m, $55°)| = 1.35 \times 10^3$ V/m; $|E^r_{0\Theta}(3$ m, $90°)| = 28.9$ V/m. Taken from King and Sandler [2, Fig. 15]. © 1994 American Geophysical Union.

suited to excite the lateral wave required for over-the-horizon radar to detect low-flying targets within 1000 km of the coast line.

8.10   Conclusion

In this chapter the properties of vertical electric dipoles in the air over a conducting or dielectric half-space with or without an electrically thin dielectric layer have been described. Analytical formulas for the complete electromagnetic field in the air have

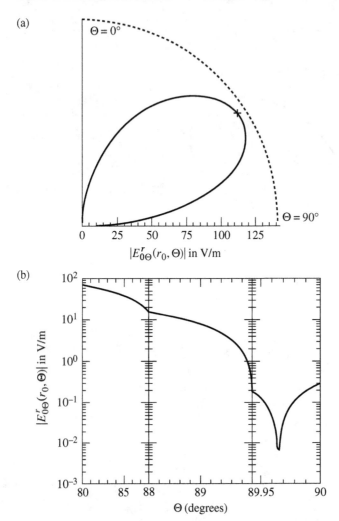

Figure 8.35 Like Fig. 8.33 with $r_0 = 30$ m. $|E^r_{0\Theta}|_{\max} = |E^r_{0\Theta}(30\,\text{m}, 55°)| = 135$ V/m; $|E^r_{0\Theta}(30\,\text{m}, 90°)| = 0.289$ V/m. Taken from King and Sandler [2, Fig. 16]. © 1994 American Geophysical Union.

been discussed in terms of near, intermediate, and far fields. The far field includes a surface-wave term that dominates near the boundary surface. Applications range from communication with submarines in the kilohertz range of frequencies to microstrip circuits at 5–10 GHz.

9 Dipoles parallel to the plane boundaries of layered regions; horizontal dipole over, on, and in the earth or sea

9.1 Introduction

Dipole antennas located parallel to the plane boundaries of a layered region have numerous important applications over a wide range of frequencies. Examples include horizontal-wire (Beverage) antennas in air close to the earth, insulated antennas on or below the surface of the earth or sea, cellular telephone transceivers close to the human head, and patch antennas on microstrip.

The electromagnetic field of a dipole antenna parallel to the surface of a layered region is more complicated than the field of the same dipole when perpendicular to the boundaries. This is a consequence of the fact that all six components of the electromagnetic field are involved. In the cylindrical coordinates ρ, ϕ', z' shown in Fig. 9.1, there are three components of electric type, namely, E_ρ, $E_{z'}$, and $B_{\phi'}$, and three components of magnetic type, namely, B_ρ, $B_{z'}$, and $E_{\phi'}$. The dipole with the length $2h$ and electric moment $2h_e I$ is located at the height d' in the air (region 0, wave number k_0) over the surface of the electrically thin layer (region 1, wave number k_1, thickness l). This coats a dielectric or conducting half-space (region 2, wave number $k_2 = \beta_2 + i\alpha_2$). The vertical $z' = -z$ axis passes through the center of the dipole. The field in the air, $z' \geq 0$, is expressed in terms of the coordinates ρ, ϕ', z'. The fields in the dielectric layer, $0 \leq z \leq l$, and in the conducting or dielectric half-space, $z \geq l$, are expressed in terms of ρ, ϕ, z. Note that $z = -z'$ and $\phi = -\phi'$. The complete field subject only to the conditions

$$9k_0 \leq 3k_1 \leq |k_2|, \qquad k_1 l \leq 0.6 \tag{9.1}$$

is given by

$$E_{0\rho}(\rho, \phi', z') = \frac{\omega\mu_0(2h_e I)}{4\pi k_0} \cos\phi'$$

$$\times \left(e^{ik_0 r_1} \left[\frac{2}{r_1^2} + \frac{2i}{k_0 r_1^3} + \left(\frac{z' - d'}{r_1} \right)^2 \left(\frac{ik_0}{r_1} - \frac{3}{r_1^2} - \frac{3i}{k_0 r_1^3} \right) \right] \right.$$

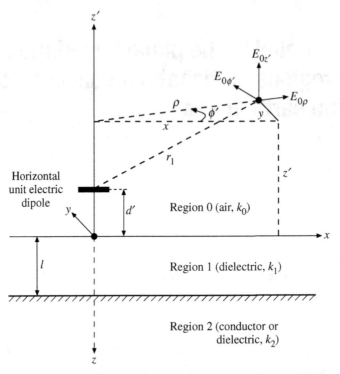

Figure 9.1 Unit horizontal electric dipole at height d' over plane boundary ($z = 0$) between air and a sheet of dielectric with thickness l over a conducting or dielectric half-space.

$$-e^{ik_0 r_2}\left[\frac{2}{r_2^2} + \frac{2i}{k_0 r_2^3} + \left(\frac{z'+d'}{r_2}\right)^2\left(\frac{ik_0}{r_2} - \frac{3}{r_2^2} - \frac{3i}{k_0 r_2^3}\right)\right]$$

$$+ 2e^{ik_0 r_2}\left\{\epsilon\left(\frac{z'+d'}{r_2}\right)\left(\frac{ik_0}{r_2} - \frac{1}{r_2^2}\right) - \epsilon^2\left[\frac{ik_0}{r_2} - \frac{1}{r_2^2} - \frac{i}{k_0 r_2^3}\right.\right.$$

$$\left.\left.\left. - k_0^2\epsilon\left(\frac{r_2}{\rho}\right)\left(\frac{\pi}{k_0 r_2}\right)^{1/2} e^{-iP_2}\mathcal{F}(P_2)\right]\right\}\right)$$

(9.2)

$$E_{0\phi'}(\rho, \phi', z') = \frac{-\omega\mu_0(2h_e I)}{4\pi k_0}\sin\phi'$$

$$\times\left(e^{ik_0 r_1}\left(\frac{ik_0}{r_1} - \frac{1}{r_1^2} - \frac{i}{k_0 r_1^3}\right) - e^{ik_0 r_2}\left(\frac{ik_0}{r_2} - \frac{1}{r_2^2} - \frac{i}{k_0 r_2^3}\right)\right.$$

$$+ 2e^{ik_0 r_2}\left\{\epsilon\left(\frac{z'+d'}{r_2}\right)\left(\frac{ik_0}{r_2} - \frac{1}{r_2^2}\right)\right.$$

$$- \epsilon^2\left[\frac{2}{r_2^2} + \frac{2i}{k_0 r_2^3} + \left(\frac{z'+d'}{r_2}\right)^2\left(\frac{ik_0}{r_2} - \frac{3}{r_2^2} - \frac{3i}{k_0 r_2^3}\right)\right.$$

$$\left.\left.\left. + ik_0\epsilon\left(\frac{r_2^2}{\rho^3}\right)\left(\frac{\pi}{k_0 r_2}\right)^{1/2} e^{-iP_2}\mathcal{F}(P_2)\right]\right\}\right)$$

(9.3)

$$E_{0z'}(\rho, \phi', z') = \frac{-\omega\mu_0(2h_e I)}{4\pi k_0} \cos\phi' \left\{ e^{ik_0 r_1} \left(\frac{\rho}{r_1}\right) \left(\frac{z'-d'}{r_1}\right) \right.$$

$$\times \left(\frac{ik_0}{r_1} - \frac{3}{r_1^2} - \frac{3i}{k_0 r_1^3}\right) - e^{ik_0 r_2} \left(\frac{\rho}{r_2}\right) \left(\frac{z'+d'}{r_2}\right)$$

$$\times \left(\frac{ik_0}{r_2} - \frac{3}{r_2^2} - \frac{3i}{k_0 r_2^3}\right) + 2\epsilon\, e^{ik_0 r_2} \left[\left(\frac{\rho}{r_2}\right)\left(\frac{ik_0}{r_2} - \frac{1}{r_2^2}\right)\right.$$

$$\left.\left. - k_0^2 \epsilon \left(\frac{\pi}{k_0 r_2}\right)^{1/2} e^{-iP_2} \mathcal{F}(P_2) \right] \right\} \tag{9.4}$$

$$B_{0\rho}(\rho, \phi', z') = \frac{\mu_0(2h_e I)}{4\pi} \sin\phi' \left\{ e^{ik_0 r_1} \left(\frac{z'-d'}{r_1}\right)\left(\frac{ik_0}{r_1} - \frac{1}{r_1^2}\right) \right.$$

$$- e^{ik_0 r_2} \left(\frac{z'+d'}{r_2}\right)\left(\frac{ik_0}{r_2} - \frac{1}{r_2^2}\right)$$

$$+ 2\epsilon\, e^{ik_0 r_2} \left[\frac{2}{r_2^2} + \frac{2i}{k_0 r_2^3} + \left(\frac{z'+d'}{r_2}\right)^2 \left(\frac{ik_0}{r_2} - \frac{3}{r_2^2} - \frac{3i}{k_0 r_2^3}\right)\right.$$

$$\left.\left. + ik_0 \epsilon \left(\frac{r_2^2}{\rho^3}\right)\left(\frac{\pi}{k_0 r_2}\right)^{1/2} e^{-iP_2} \mathcal{F}(P_2) \right] \right\} \tag{9.5}$$

$$B_{0\phi'}(\rho, \phi', z') = \frac{\mu_0(2h_e I)}{4\pi} \cos\phi' \left\{ e^{ik_0 r_1} \left(\frac{z'-d'}{r_1}\right)\left(\frac{ik_0}{r_1} - \frac{1}{r_1^2}\right) \right.$$

$$- e^{ik_0 r_2} \left(\frac{z'+d'}{r_2}\right)\left(\frac{ik_0}{r_2} - \frac{1}{r_2^2}\right) + 2\epsilon\, e^{ik_0 r_2} \left[\frac{ik_0}{r_2} - \frac{1}{r_2^2} - \frac{i}{k_0 r_2^3}\right.$$

$$\left.\left. - k_0^2 \epsilon \left(\frac{r_2}{\rho}\right)\left(\frac{\pi}{k_0 r_2}\right)^{1/2} e^{-iP_2} \mathcal{F}(P_2) \right] \right\} \tag{9.6}$$

$$B_{0z'}(\rho, \phi', z') = \frac{-\mu_0(2h_e I)}{4\pi} \sin\phi' \left(e^{ik_0 r_1} \left(\frac{\rho}{r_1}\right)\left(\frac{ik_0}{r_1} - \frac{1}{r_1^2}\right) \right.$$

$$- e^{ik_0 r_2} \left(\frac{\rho}{r_2}\right)\left(\frac{ik_0}{r_2} - \frac{1}{r_2^2}\right) + 2e^{ik_0 r_2} \left\{ \epsilon\left(\frac{\rho}{r_2}\right)\left(\frac{z'+d'}{r_2}\right) \right.$$

$$\times \left(\frac{ik_0}{r_2} - \frac{3}{r_2^2} - \frac{3i}{k_0 r_2^3}\right) - \epsilon^2 \left(\frac{\rho}{r_2}\right) \left[\frac{1}{r_2^2} + \frac{3i}{k_0 r_2^3} - \frac{3}{k_0^2 r_2^4}\right.$$

$$\left.\left.\left. + \left(\frac{z'+d'}{r_2}\right)^2 \left(\frac{ik_0}{r_2} - \frac{6}{r_2^2} - \frac{15i}{k_0 r_2^3}\right) \right] \right\} \right). \tag{9.7}$$

In these formulas,

$$k_0 = \frac{\omega}{c}, \qquad k_1 = k_0\sqrt{\epsilon_{1r}}, \qquad k_2 = k_0\sqrt{\epsilon_{2r}}\left(1 + \frac{i\sigma_2}{\omega\epsilon_0\epsilon_{2r}}\right)^{1/2} \tag{9.8}$$

$$r_1 = \sqrt{\rho^2 + (z' - d')^2}, \qquad r_2 = \sqrt{\rho^2 + (z' + d')^2}, \qquad \epsilon = \frac{k_0}{k_2} - ik_0l \tag{9.9}$$

$$P_2 = \frac{k_0 r_2}{2}\left(\frac{\epsilon r_2 + z' + d'}{\rho}\right)^2, \qquad \mathcal{F}(P_2) = \tfrac{1}{2}(1 + i) - \int_0^{P_2} \frac{e^{it}}{\sqrt{2\pi t}}\, dt. \tag{9.10}$$

Note that ϵ without subscript is the small quantity defined in (9.9); ϵ_{1r} and ϵ_{2r} are the relative permittivities of regions 1 and 2, respectively.

The field in the electrically thin dielectric layer (region 1) is

$$B_{1\phi}(\rho, \phi, z) \sim B_{0\phi}(\rho, \phi, 0) \tag{9.11}$$

$$E_{1\rho}(\rho, \phi, z) \sim E_{0\rho}(\rho, \phi, 0)\left[\frac{k_1(l - z) + i(k_1/k_2)}{k_1 l + i(k_1/k_2)}\right] \tag{9.12}$$

$$E_{1z}(\rho, \phi, z) \sim \frac{k_0^2}{k_1^2} E_{0z}(\rho, \phi, 0) \tag{9.13}$$

$$E_{1\phi}(\rho, \phi, z) \sim E_{0\phi}(\rho, \phi, 0)\left[\frac{k_1(l - z) + i(k_1/k_2)}{k_1 l + i(k_1/k_2)}\right] \tag{9.14}$$

$$B_{1\rho}(\rho, \phi, z) \sim B_{0\rho}(\rho, \phi, 0) \tag{9.15}$$

$$B_{1z}(\rho, \phi, z) \sim B_{0z}(\rho, \phi, 0). \tag{9.16}$$

The field in region 2 is

$$B_{2\phi}(\rho, \phi, z) \sim B_{0\phi}(\rho, \phi, 0)e^{ik_2(z-l)} \tag{9.17}$$

$$E_{2\rho}(\rho, \phi, z) \sim \frac{k_0}{k_2\epsilon} E_{0\rho}(\rho, \phi, 0)e^{ik_2(z-l)} \tag{9.18}$$

$$E_{2z}(\rho, \phi, z) \sim \frac{k_0^2}{k_2^2} E_{0z}(\rho, \phi, 0)e^{ik_2(z-l)} \tag{9.19}$$

$$E_{2\phi}(\rho, \phi, z) \sim \frac{k_0}{k_2\epsilon} E_{0\phi}(\rho, \phi, 0)e^{ik_2(z-l)} \tag{9.20}$$

$$B_{2\rho}(\rho, \phi, z) \sim B_{0\rho}(\rho, \phi, 0)e^{ik_2(z-l)} \tag{9.21}$$

$$B_{2z}(\rho, \phi, z) \sim B_{0z}(\rho, \phi, 0)e^{ik_2(z-l)}. \tag{9.22}$$

The six components of the field consist of three of electric type, namely, E_ρ, E_z, and B_ϕ which are usually dominant, and three of magnetic type, namely, B_ρ, B_z, and E_ϕ.

As for the vertical dipole, the field of the horizontal dipole can be described in terms of the near field, the intermediate field, and the far field. These are defined exactly as for the vertical dipole in Section 8.2.

The practical applications of the horizontal electric dipole in the presence of a layered region are primarily those in which the dipole is close to the surface compared to the radial distance ρ to the point of observation, i.e. $d' \ll \rho$. When this is true, the direct and perfect-image fields virtually cancel. It follows that

$$
E_{0\rho}(\rho, \phi', z') = \frac{\omega\mu_0(2h_eI)\epsilon}{2\pi k_0} \cos\phi' \, e^{ik_0r_2} \left\{ \left(\frac{z'+d'}{r_2}\right)\left(\frac{ik_0}{r_2} - \frac{1}{r_2^2}\right) \right.
$$
$$
\left. - \epsilon\left[\frac{ik_0}{r_2} - \frac{1}{r_2^2} - \frac{i}{k_0r_2^3} - k_0^2\epsilon\left(\frac{r_2}{\rho}\right)\left(\frac{\pi}{k_0r_2}\right)^{1/2} e^{-iP_2}\mathcal{F}(P_2)\right] \right\}
$$
(9.23)

$$
E_{0\phi'}(\rho, \phi', z') = \frac{-\omega\mu_0(2h_eI)\epsilon}{2\pi k_0} \sin\phi' \, e^{ik_0r_2} \left\{ \left(\frac{z'+d'}{r_2}\right)\left(\frac{ik_0}{r_2} - \frac{1}{r_2^2}\right) \right.
$$
$$
- \epsilon\left[\frac{2}{r_2^2} + \frac{2i}{k_0r_2^3} + \left(\frac{z'+d'}{r_2}\right)^2\left(\frac{ik_0}{r_2} - \frac{3}{r_2^2} - \frac{3i}{k_0r_2^3}\right)\right.
$$
$$
\left.\left. + ik_0\epsilon\left(\frac{r_2^2}{\rho^3}\right)\left(\frac{\pi}{k_0r_2}\right)^{1/2} e^{-iP_2}\mathcal{F}(P_2)\right] \right\}
$$
(9.24)

$$
E_{0z'}(\rho, \phi', z') = \frac{-\omega\mu_0(2h_eI)\epsilon}{2\pi k_0} \cos\phi' \, e^{ik_0r_2} \left[\left(\frac{\rho}{r_2}\right)\left(\frac{ik_0}{r_2} - \frac{1}{r_2^2}\right) \right.
$$
$$
\left. - k_0^2\epsilon\left(\frac{\pi}{k_0r_2}\right)^{1/2} e^{-iP_2}\mathcal{F}(P_2) \right]
$$
(9.25)

$$
B_{0\rho}(\rho, \phi', z') = \frac{\mu_0(2h_eI)\epsilon}{2\pi} \sin\phi' \, e^{ik_0r_2}
$$
$$
\times\left[\frac{2}{r_2^2} + \frac{2i}{k_0r_2^3} + \left(\frac{z'+d'}{r_2}\right)^2\left(\frac{ik_0}{r_2} - \frac{3}{r_2^2} - \frac{3i}{k_0r_2^3}\right)\right.
$$
$$
\left. + ik_0\epsilon\left(\frac{r_2^2}{\rho^3}\right)\left(\frac{\pi}{k_0r_2}\right)^{1/2} e^{-iP_2}\mathcal{F}(P_2) \right]
$$
(9.26)

$$
B_{0\phi'}(\rho, \phi', z') = \frac{\mu_0(2h_eI)\epsilon}{2\pi} \cos\phi' \, e^{ik_0r_2} \left[\frac{ik_0}{r_2} - \frac{1}{r_2^2} - \frac{i}{k_0r_2^3}\right.
$$
$$
\left. - k_0^2\epsilon\left(\frac{r_2}{\rho}\right)\left(\frac{\pi}{k_0r_2}\right)^{1/2} e^{-iP_2}\mathcal{F}(P_2) \right]
$$
(9.27)

$$B_{0z'}(\rho, \phi', z') = \frac{-\mu_0(2h_e I)\epsilon}{2\pi} \sin\phi' \, e^{ik_0 r_2} \left(\frac{\rho}{r_2}\right) \left\{ \left(\frac{z'+d'}{r_2}\right) \right.$$

$$\times \left(\frac{ik_0}{r_2} - \frac{3}{r_2^2} - \frac{3i}{k_0 r_2^3}\right) - \epsilon \left[\frac{1}{r_2^2} + \frac{3i}{k_0 r_2^3} - \frac{3}{k_0^2 r_2^4}\right.$$

$$\left. \left. + \left(\frac{z'+d'}{r_2}\right)^2 \left(\frac{ik_0}{r_2} - \frac{6}{r_2^2} - \frac{15i}{k_0 r_2^3}\right)\right]\right\}. \tag{9.28}$$

In these formulas, r_2, ϵ, P_2, and $\mathcal{F}(P_2)$ are the same as defined in (9.9) and (9.10).

When $|P_2| \geq 4$, the far-field formulas in the spherical coordinates r_0, Θ, Φ are useful. Since $d' \ll \rho$, it follows that $r_2 \sim r_0$ and $d/r_2 \sim 0$. In this case, $z'/r_0 = \cos\Theta$ and $\rho/r_0 = \sin\Theta$ so that

$$E_{0\rho}^r(r_0, \Theta, \Phi) = \frac{\omega\mu_0(2h_e I)\epsilon}{2\pi k_0} \cos\Phi \, e^{ik_0 r_0} \left(\frac{ik_0}{r_0}(\cos\Theta - \epsilon) + \frac{\epsilon T^r}{\sin\Theta}\right) \tag{9.29}$$

$$E_{0z'}^r(r_0, \Theta, \Phi) = \frac{-\omega\mu_0(2h_e I)\epsilon}{2\pi k_0} \cos\Phi \, e^{ik_0 r_0} \left(\frac{ik_0}{r_0}\sin\Theta - T^r\right) \tag{9.30}$$

$$B_{0\phi'}^r(r_0, \Theta, \Phi) = \frac{\mu_0(2h_e I)\epsilon}{2\pi} \cos\Phi \, e^{ik_0 r_0} \left(\frac{ik_0}{r_0} - \frac{T^r}{\sin\Theta}\right), \tag{9.31}$$

where

$$T^r = k_0^2 \epsilon \left(\frac{\pi}{k_0 r_0}\right)^{1/2} e^{-iP_0} \mathcal{F}(P_0)$$

$$= \frac{ik_0}{r_0} \frac{\epsilon \sin\Theta}{\epsilon + \cos\Theta} + \frac{\epsilon}{r_0^2} \frac{\sin^3\Theta}{(\epsilon + \cos\Theta)^3}. \tag{9.32}$$

When this value is substituted in the formulas for the three components, these become

$$E_{0r}^r(r_0, \Theta, \Phi) = E_{0\rho}^r(r_0, \Theta, \Phi)\sin\Theta + E_{0z'}^r(r_0, \Theta, \Phi)\cos\Theta$$

$$= \frac{\omega\mu_0(2h_e I)\epsilon}{2\pi k_0} \cos\Phi \, e^{ik_0 r_0} \left[\frac{ik_0}{r_0}\right.$$

$$\times \left(\cos\Theta \sin\Theta - \epsilon \sin\Theta + \frac{\epsilon^2 \sin\Theta}{\epsilon + \cos\Theta}\right.$$

$$\left. - \sin\Theta \cos\Theta + \frac{\epsilon \sin\Theta \cos\Theta}{\epsilon + \cos\Theta}\right)$$

$$\left. + \frac{\epsilon^2}{r_0^2} \frac{\sin^3\Theta}{(\epsilon + \cos\Theta)^3} + \frac{\epsilon}{r_0^2} \frac{\sin^3\Theta \cos\Theta}{(\epsilon + \cos\Theta)^3}\right]$$

$$= \frac{\omega\mu_0(2h_e I)}{2\pi k_0} \cos\Phi \, e^{ik_0 r_0} \frac{\epsilon^2}{r_0^2} \frac{\sin^3\Theta}{(\epsilon + \cos\Theta)^2} \tag{9.33}$$

$$E_{0\Theta}^r(r_0, \Theta, \Phi) = E_{0\rho}^r(r_0, \Theta, \Phi)\cos\Theta - E_{0z'}^r(r_0, \Theta, \Phi)\sin\Theta$$

$$= \frac{\omega\mu_0(2h_eI)\epsilon}{2\pi k_0}\cos\Phi\, e^{ik_0r_0}\left[\frac{ik_0}{r_0}\left(\cos^2\Theta\right.\right.$$

$$- \epsilon\cos\Theta + \frac{\epsilon^2\cos\Theta}{\epsilon+\cos\Theta} + \sin^2\Theta - \frac{\epsilon\sin^2\Theta}{\epsilon+\cos\Theta}\Big)$$

$$\left.+ \frac{\epsilon^2\sin^2\Theta\cos\Theta}{r_0^2(\epsilon+\cos\Theta)^3} - \frac{\epsilon\sin^4\Theta}{r_0^2(\epsilon+\cos\Theta)^3}\right]$$

$$= \frac{\omega\mu_0(2h_eI)\epsilon}{2\pi k_0}\cos\Phi\, e^{ik_0r_0}$$

$$\times \left[\frac{ik_0}{r_0}\left(\frac{\cos\Theta}{\epsilon+\cos\Theta}\right) - \frac{\epsilon}{r_0^2}\frac{\sin^2\Theta(\sin^2\Theta - \epsilon\cos\Theta)}{(\epsilon+\cos\Theta)^3}\right] \tag{9.34}$$

$$B_{0\Phi}^r(r_0, \Theta, \Phi) = \frac{\mu_0(2h_eI)\epsilon}{2\pi}\cos\Phi\, e^{ik_0r_0}$$

$$\times \left[\frac{ik_0}{r_0}\left(1 - \frac{\epsilon}{\epsilon+\cos\Theta}\right) - \frac{\epsilon}{r_0^2}\frac{\sin^2\Theta}{(\epsilon+\cos\Theta)^3}\right]$$

$$= \frac{\mu_0(2h_eI)\epsilon}{2\pi}\cos\Phi\, e^{ik_0r_0}$$

$$\times \left(\frac{ik_0}{r_0}\frac{\cos\Theta}{\epsilon+\cos\Theta} - \frac{\epsilon}{r_0^2}\frac{\sin^2\Theta}{(\epsilon+\cos\Theta)^3}\right). \tag{9.35}$$

9.2 Horizontal traveling-wave antennas over earth or sea; Beverage antenna ($l = 0$, $\epsilon = k_0/k_2$)

Horizontal-wire antennas close to the earth or sea are efficient generators of lateral waves. Although the field of a unit horizontal dipole close to the earth is smaller by the factor $|k_0/k_2|$ than the field of a unit vertical dipole close to the earth, it is possible to make the horizontal-wire antenna very long and terminate it so that the current in it is a traveling wave. This can yield an electric moment that is very much greater than that of a vertical electric dipole. When the traveling-wave antenna is at the small height $d' \ll \rho$, $k_0d' < 1$, over the earth, it can be a bare wire supported on insulating posts. The termination at each end can consist of a suitable resistor in series with a quarter-wave horizontal monopole, or it can consist of a vertical ground connection (Beverage antenna) [1]. These two possibilities are illustrated in Fig. 9.2. Since it

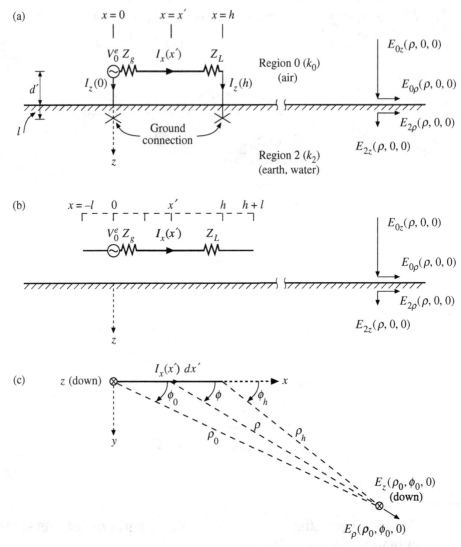

Figure 9.2 Wave antennas. (a) Conventional Beverage antenna; (b) horizontal-wire antenna; (c) coordinates. Taken from King [1, Fig. 1]. © 1983 I.E.E.E.

has been shown[1] that the terminations – whether horizontal or vertical – contribute negligibly to the electromagnetic field since they are very much shorter than the main horizontal wire, only the field of this latter need be determined.

The wave number of the current in the bare x-directed wire at the height d' in the air over the earth is not the wave number k_0 of the current in an isolated antenna in the air. The proximity of the earth greatly modifies the wave number so that it has the value k_L given by

---

[1] [2] Chapter 18.

$$k_L = k_0 \left\{ 1 + \frac{2}{\ln(2d'/a)} \left[\frac{1}{(2k_2d')^2} - \frac{K_1(2k_2d')}{2k_2d'} + \frac{i\pi I_1(2k_2d')}{4k_2d'} \right. \right.$$

$$\left. \left. - i \left(\frac{2k_2d'}{3} + \frac{(2k_2d')^3}{45} + \frac{(2k_2d')^5}{1575} + \cdots \right) \right] \right\}^{1/2}, \tag{9.36}$$

where K_1 and I_1 are modified Bessel functions and the following condition is imposed:

$$k_0 d' \le 0.2\pi. \tag{9.37}$$

When

$$0 \le |k_2 d'| \le 0.8, \tag{9.38}$$

the wave number k_L is well approximated by the following simpler expression:

$$k_L \sim k_0 \left[1 - \frac{\ln(k_2d') + \gamma - \frac{1}{2} - i(\frac{1}{2}\pi - \frac{4}{3}k_2d')}{\ln(2d'/a)} \right]^{1/2}, \tag{9.39}$$

where $\gamma = 0.5772$.

When (9.37) is satisfied, the antenna behaves like a transmission line with the characteristic impedance

$$Z_c = \frac{\zeta_0 k_L}{2\pi k_0} \ln \frac{2d'}{a} = \frac{60 k_L}{k_0} \ln \frac{2d'}{a}. \tag{9.40}$$

In these formulas, a is the radius of the wire. The current in the wire has the general transmission-line form

$$I_x(x) = \frac{-i V_0}{Z_c} \frac{\sin[k_L(h - x) + i\theta_h]}{\cos(k_L h + i\theta_h)}, \tag{9.41}$$

where $\theta_h = \coth^{-1}(Z_h/Z_c)$. When Z_h, the terminating impedance at $x = h$, is equal to the characteristic impedance, i.e. $Z_h = Z_c$, $\theta_h = \infty$ and

$$I_x(x) = \frac{V_0}{Z_{in}} e^{ik_L x} = I_x(0) e^{ik_L x}. \tag{9.42}$$

The impedance of the antenna at the driving point $x = 0$ is

$$Z_{in} = Z + Z_0 + Z_g, \tag{9.43}$$

where $Z = Z_c$ is the impedance of the terminated long wire with the length h_m, Z_0 is the impedance of the terminating sections at $x = 0$ and $x = h$, and Z_g is the impedance of the generator at $x = 0$ (see Fig. 9.2). For traveling-wave operation, $Z_0 + Z_g = Z_0 + Z_L = Z_c$ (where Z_L is the lumped impedance in series with Z_0

at $x = h$) so that $Z_{\text{in}} = 2Z_c$. With the horizontal-wire antenna, Z_0 is the impedance of an open-ended section of the horizontal-wire line with the length l so that $Z_0 = iZ_c \cot k_L l$. With the Beverage antenna, Z_0 is the impedance of each of the grounded vertical conductors at $x = 0$ and $x = h$.

The components of the electric far field of the traveling-wave current in the horizontal-wire antenna with the length h adjusted to maximize the field with $h = h_m$ are

$$E_{0r}^r(r_0, \Theta_0, \Phi_0) = I_x(0)h_e(\Theta_0, \Phi_0)[E_{0r}^r(r_0, \Theta_0, 0)]_h \cos \Phi_0 \tag{9.44}$$

$$E_{0\Theta}^r(r_0, \Theta_0, \Phi_0) = I_x(0)h_e(\Theta_0, \Phi_0)[E_{0\Theta}^r(r_0, \Theta_0, 0)]_h \cos \Phi_0, \tag{9.45}$$

where

$$h_e(\Theta_0, \Phi_0) = \frac{i[1 - e^{i(k_L - k_0 \sin \Theta_0 \cos \Phi_0)h_m}]}{k_L - k_0 \sin \Theta_0 \cos \Phi_0} \tag{9.46a}$$

and

$$(\beta_L - k_0)h_m = \pi - \tan^{-1}\left(\frac{\alpha_L}{\beta_L - k_0}\right). \tag{9.46b}$$

Since α_L is usually small,

$$h_m \sim \frac{\pi}{\beta_L - k_0}. \tag{9.46c}$$

The associated magnetic field is

$$B_{0\Phi}^r(r_0, \Theta_0, \Phi_0) = I_x(0)h_e(\Theta_0, \Phi_0)[B_{0\Phi}^r(r_0, \Theta_0, 0)]_h \cos \Phi_0. \tag{9.47}$$

In these formulas, r_0, Θ_0, and Φ_0 are spherical coordinates referred to the origin at $\rho = 0$, $z' = 0$, on the surface of the earth directly below the generator at $x = 0$ in the horizontal-wire antenna. Also, the subscript h denotes the field of a unit horizontal electric dipole at $x = 0$, $z' = d'$. The three components are given by (9.33), (9.34), and (9.35) with the subscript 0 added to Θ and Φ and with $\epsilon = k_0/k_2$.

In order to display the characteristics of a horizontal-wire antenna, the numerical values of the several parameters can be calculated from the appropriate formulas [3]. For this purpose, consider antennas designed for use at $f = 10\,\text{MHz}$ over earth ($\sigma_2 = 0.04\,\text{S/m}$, $\epsilon_{2r} = 8$) and sea water ($\sigma_2 = 4\,\text{S/m}$, $\epsilon_{2r} = 80$). For operation over the earth, the field for the two heights $d' = 15$ and $45\,\text{cm}$ are studied; for operation over the sea, the height $d' = 4\,\text{cm}$ is used. The several parameters and quantities of interest for these three cases are shown in Table 9.1.

The magnitude of the electric far field $|E_{0\Theta}^r(r_0, \Theta_0, \Phi_0)|$ as obtained from (9.45) is shown in Figs. 9.3, 9.4, and 9.5, respectively, for the three cases A, B, and C

Table 9.1. *Properties of horizontal-wire antenna with maximizing length h_m, radius a, and height d', at $f = 10\,\text{MHz}$*[a]

Case	A (dry earth)	B (dry earth)	C (sea water)
a (m)	0.003	0.003	0.003
k_0 (m^{-1})	0.2094	0.2094	0.2094
σ_2 (S/m)	0.04	0.04	4
ϵ_{2r}	8	8	80
k_2 (m^{-1})	$1.7827e^{i0.73}$	$1.7827e^{i0.73}$	$17.7721e^{i0.78}$
$k_2 = \beta_2 + i\alpha_2$ (m^{-1})	$1.3284 + i1.1888$	$1.3284 + i1.1888$	$12.6365 + i12.4967$
d' (m)	0.15	0.45	0.04
$k_0 d'$	0.03141	0.09423	0.00838
$k_2 d'$	$0.2674e^{i0.73}$	$0.8022e^{i0.73}$	$0.7109e^{i0.78}$
$k_L = \beta_L + i\alpha_L$ (m^{-1})	$0.2407 + i0.0129$	$0.2223 + i0.0079$	$0.2336 + i0.0129$
$\beta_L - k_0$ (m^{-1})	0.0313	0.0129	0.0242
$Z_L = R_L - iX_L$ (ohm)	$317.6 + i17.0$	$363.3 + i12.9$	$219.5 + i12.1$
h_m (m)	$87.85 = 2.93\lambda_0$	$201.96 = 6.73\lambda_0$	$109.57 = 3.65\lambda_0$
$h_e(\pi/2, 0)$ (m)	$38.47e^{i1.10}$	$78.25e^{i0.929}$	$44.47e^{i0.987}$
$h_e(\pi/2, 0)/h_e(\pi/2, \pi)$	$15.33e^{-i0.17}$	$38.93e^{-i0.788}$	$18.55e^{-i0.783}$
$h_e(0, 0)$ (m)	$5.13e^{i1.32}$	$4.00e^{i1.354}$	$3.38e^{i1.38}$
$h_e(0, 0)/h_e(\pi/2, 0)$	$0.13e^{i0.22}$	$0.05e^{i0.425}$	$0.076e^{i0.393}$
$|(k_0/k_2)h_e(\pi/2, 0)|$ (m)	4.52	9.19	0.52
h_{ev} (m) (vertical monopole)	4.78	4.78	4.78

[a] Taken from King [3, Table I]. © 1992 American Institute of Physics.

listed in Table 9.1. Graphs in the xz'-plane are shown for the distances $r_0 = 50$, 100, 500, and 1000 km with $I_x(0) = 10$ A. Graphs in the xy-plane are shown only for $r_0 = 50$ km. The maximum of the field occurs at $\Theta_0 = 76.4$ and $79.5°$ in cases A and B over earth, at $\Theta_0 = 83.4°$ in case C over sea water. (The corresponding maxima for the vertical monopole as obtained in Chapter 8 are $\Theta_0 = 66$ and $78°$.) The horizontal-wire antenna generates a narrow directive beam along the antenna axis that is tilted upward from the earth by only a small angle. The vertical monopole generates a rotationally symmetric pattern that is tilted upward at a much greater angle. The surface wave, with its maximum along the surface $\Theta_0 = \pi/2$, is comparable in magnitude for the horizontal-wire antenna and the vertical monopole when both are over the earth. Over the sea, the small factor k_0/k_2 makes the surface wave due to the horizontal-wire antenna much smaller than that due to the vertical monopole. A complete experimental verification of the current distribution and the field pattern as determined from the above formulas has been reported by Rama Rao [4].

The horizontal field pattern of the terminated horizontal-wire or Beverage antenna is unidirectional and quite directive. The directivity is greatly increased when horizontal-wire antennas are arranged in an array of parallel elements all at the same height above

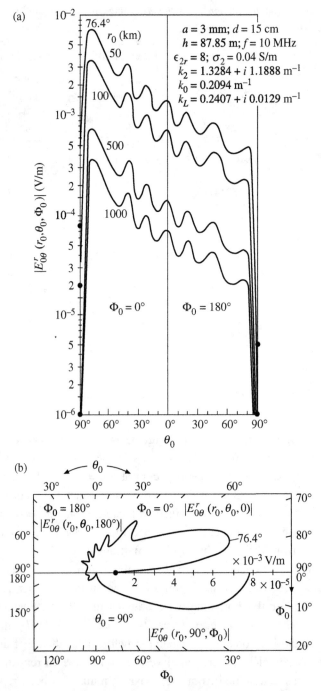

Figure 9.3 Magnitude of electric far field $E^r_{0\Theta}(r_0, \Theta, \Phi)$ of horizontal-wire antenna over the surface of the earth; $d' = 15$ cm, $f = 10$ MHz (case A). (a) Rectangular graphs in vertical xz' plane ($\Phi_0 = 0, \pi$); (b) polar graphs in xy plane, $r_0 = 50$ km. Taken from King [3, Fig. 3]. © 1992 American Institute of Physics.

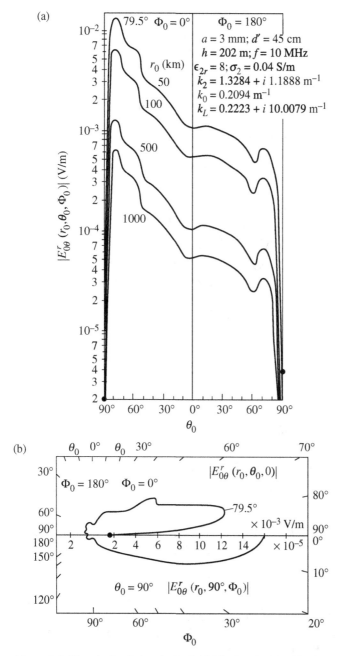

Figure 9.4 Magnitude of electric far field $E_{0\Theta}^r(r_0, \Theta, \Phi)$ of horizontal-wire antenna over the surface of the earth; $d' = 45\,\text{cm}$, $f = 10\,\text{MHz}$ (case B). (a) Rectangular graphs in vertical xz' plane ($\Phi_0 = 0, \pi$); (b) polar graphs in xy plane, $r_0 = 50\,\text{km}$. Taken from King [3, Fig. 4]. © 1992 American Institute of Physics.

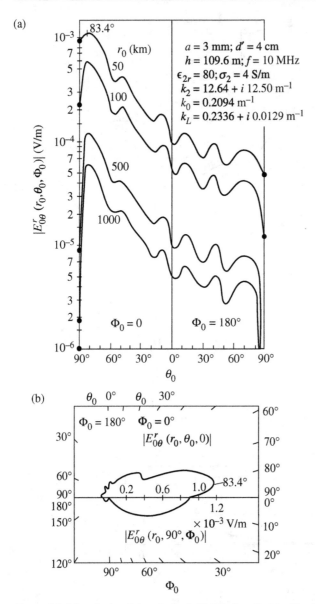

(a)

83.4°

r_0 (km)

50

100

500

1000

$|E_{0\theta}^r (r_0, \theta_0, \Phi_0)|$ (V/m)

$a = 3$ mm; $d' = 4$ cm
$h = 109.6$ m; $f = 10$ MHz
$\epsilon_{2r} = 80$; $\sigma_2 = 4$ S/m
$k_2 = 12.64 + i\ 12.50$ m^{-1}
$k_0 = 0.2094$ m^{-1}
$k_L = 0.2336 + i\ 0.0129$ m^{-1}

$\Phi_0 = 0$ $\Phi_0 = 180°$

θ_0

(b)

θ_0 0° θ_0 30°

$\Phi_0 = 180°$ $\Phi_0 = 0°$

$|E_{0\theta}^r (r_0, \theta_0, 0)|$

0.2   0.6   1.0   -83.4°

1.2

$\times 10^{-3}$ V/m

$|E_{0\theta}^r (r_0, 90°, \Phi_0)|$

Φ_0

Figure 9.5 Magnitude of electric far field $E_{0\theta}^r(r_0, \Theta, \Phi)$ of horizontal-wire antenna over the surface of the sea; $d' = 4$ cm, $f = 10$ MHz (case C). (a) Rectangular graphs in vertical xz' plane ($\Phi_0 = 0, \pi$); (b) polar graphs in xy plane, $r_0 = 50$ km. Taken from King [3, Fig. 5]. © 1992 American Institute of Physics.

the surface of the earth and all driven in phase with currents of the same amplitude. If the distance s between adjacent elements is a half-wavelength or more ($s \geq \lambda_0/2$, $\lambda_0 = 2\pi/k_0$), the mutual interaction is negligible and the current in and driving-point impedance of each element is the same as when isolated. The array is, therefore, uniform and the far field is that of a single element as given by (9.44) or (9.45) multiplied by the array factor of a uniform array, namely,

$$A(\Theta, \Phi) = \frac{\sin(N\pi n \sin \Theta \cos \Phi)}{\sin(\pi n \sin \Theta \cos \Phi)}. \tag{9.48}$$

Here N is the number of elements in the array and n is the distance between adjacent elements in fractions of the wavelength in air.

9.3 The terminated insulated antenna in earth or sea

The amplitude of the current in a bare antenna immersed in the earth or sea is rapidly attenuated due to ohmic losses and radiation. When the copper conductor with the radius a is insulated with a dielectric layer with the radius b or thickness $b - a$, its properties change greatly. If the wave number $k_d = \omega\sqrt{\epsilon_{dr}}/c$ of the insulating dielectric layer is small compared with the magnitude of the wave number $k_2 = \beta_2 + i\alpha_2$ of the ambient medium and the transverse dimension is electrically small, i.e.

$$|k_2| \geq 3k_d, \qquad |k_2a| < |k_2b| < 1 \tag{9.49}$$

the current in the conductor is distributed as in a transmission line with the wave number k_L and characteristic impedance Z_c.

A schematic diagram of the end-driven terminated insulated antenna [5] is shown in Fig. 9.6. The traveling-wave current is given by

$$I_x(x) = I_x(0)e^{ik_Lx}. \tag{9.50}$$

When the antenna is at the depth d below the air–earth or air–sea surface and the conditions

$$|k_2d| < 1, \qquad d^2 \gg b^2 \tag{9.51}$$

are satisfied, the wave number k_L is given by

$$k_L = k_d \left\{ 1 + \frac{1}{\ln(b/a)} \left[i \left(\frac{2\pi r_0}{\omega\mu_0} + \frac{\pi}{2} \right) + \ln \frac{2}{|k_2b|} + \ln \frac{1}{|k_2d|} - 0.90 \right] \right\}^{1/2} \tag{9.52}$$

and the characteristic impedance by

$$Z_c = \frac{\omega\mu_0 k_L}{2\pi k_d^2} \ln \frac{b}{a}. \tag{9.53}$$

In (9.52), r_0 is the internal resistance per unit length of the copper conductor.

The impedance of the antenna at the driving point is

$$Z_{in} = Z + Z_0 + Z_g, \tag{9.54}$$

(a)   Dielectric   Conductor

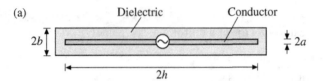

(b)   Bare monopole   Bare monopole

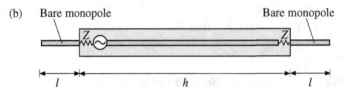

Figure 9.6 Insulated antennas. (a) Completely insulated dipole; (b) terminated insulated antenna. Ambient medium: earth, sea water. Taken from King [5, Fig. 1]. © 1986 I.E.E.E.

where Z is the impedance of the insulated length h of the antenna, Z_0 is the impedance of the bare monopole in series with the impedance Z_g of the generator. At the end $x = h$, the termination consists of a lumped impedance Z_L in series with the bare monopole with impedance Z_0. For matched operation, $Z_0 + Z_g = Z_0 + Z_L = Z_c$ so that $Z = Z_c$ and $Z_{in} = 2Z_c$.

The electromagnetic far field generated in the air by an insulated antenna at the depth d in the earth or sea is the field of a unit electric dipole at $\rho = 0$, multiplied by the quantity

$$h_e = \frac{i[1 - e^{i(k_L - k_0 \sin \Theta_0 \cos \Phi_0)h}]}{k_L - k_0 \sin \Theta_0 \cos \Phi_0}. \qquad (9.55)$$

The maximizing value of h_e is given by (9.55) with

$$h = h_m \sim \frac{\pi}{\beta_L - k_0}, \qquad (9.56)$$

where β_L is the real part of $k_L = \beta_L + i\alpha_L$ and k_0 is the real wave number of air.

The traveling-wave terminated insulated antenna can be arranged in highly directive arrays. N identical elements are located at the same depth d with adjacent elements separated by the distance $s \sim \lambda_2/4$ where $\lambda_2 = 2\pi/\beta_2$. Note that β_2 is the real part of $k_2 = \beta_2 + i\alpha_2$, and k_2 is the wave number of the ambient medium. The far field in the air of the array is the far field of a single element multiplied by the array factor

$$A(N, n) = \frac{\sin(N\pi n \sin \Theta_0 \cos \Phi_0)}{\sin(\pi n \sin \Theta_0 \cos \Phi_0)}. \qquad (9.57)$$

N is the number of parallel elements, n is the distance between elements in fractions of a wavelength in air, i.e. $n = s/\lambda_0$ where $\lambda_0 = 2\pi/k_0$ and k_0 is the wave number of air. Specifically,

$$E_{0\Theta}^r(r_0, \Theta_0, \Phi_0) = [E_{0\Theta}^r(r_0, \Theta_0, \Phi_0)]_h$$

$$\times \frac{i I_x(0)[1 - e^{i(k_L - k_0 \sin \Theta_0 \cos \Phi_0)h}]}{k_L - k_0 \sin \Theta_0 \cos \Phi_0} A(N, n) \qquad (9.58)$$

$$E_{0r}^r(r_0, \Theta_0, \Phi_0) = [E_{0r}^r(r_0, \Theta_0, \Phi_0)]_h$$

$$\times \frac{i I_x(0)[1 - e^{i(k_L - k_0 \sin \Theta_0 \cos \Phi_0)h}]}{k_L - k_0 \sin \Theta_0 \cos \Phi_0} A(N, n). \qquad (9.59)$$

Here,

$$[E_{0\Theta}^r(r_0, \Theta_0, \Phi_0)]_h = \frac{\omega \mu_0}{2\pi k_0} \cos \Phi_0 \, e^{ik_2 d} e^{ik_0 r_0} \left[\frac{ik_0}{r_0} \left(\frac{\cos \Theta_0}{1 + (k_2/k_0) \cos \Theta_0} \right) \right.$$

$$\left. - \frac{k_2}{k_0 r_0^2} \frac{\sin^2 \Theta_0 [\sin^2 \Theta_0 - (k_0/k_2) \cos \Theta_0]}{[1 + (k_2/k_0) \cos \Theta_0]^3} \right] \qquad (9.60)$$

$$[E_{0r}^r(r_0, \Theta_0, \Phi_0)]_h = \frac{\omega \mu_0}{2\pi k_0} \cos \Phi_0 \, e^{ik_2 d} \frac{e^{ik_0 r_0}}{r_0^2} \frac{\sin^3 \Theta_0}{[1 + (k_2/k_0) \cos \Theta_0]^2}. \qquad (9.61)$$

The formulas with the subscript h apply to the unit horizontal dipole at the depth d in the earth or sea. Note that they are the same as for the unit horizontal dipole on the surface of the earth or sea in air except for the factor $e^{ik_2 d}$. When the dipole is below the surface in the earth, the electromagnetic waves travel vertically upward from the dipole to the surface and then propagate as lateral waves parallel to and near the surface in the air.

A comparison of the terminated insulated antenna at a depth d in the earth described in this section with the terminated traveling-wave horizontal-wire antenna described in the preceding section shows great similarity. The reason is quite simple. The horizontal wire is actually an air-insulated antenna lying on the surface of the earth. The conductor is eccentrically located in the infinite air insulation.

9.4 Arrays of horizontal and vertical antennas over the earth

Vertical monopoles on the surface of the earth are usually grounded by means of a network of bare conductors as described in Chapter 8. For most communication purposes at broadcast frequencies, this is entirely satisfactory and the loss of power in the ground system is of no consequence. When the purpose of the antenna or array of antennas is over-the-horizon radar, long-range detection of a target over the sea requires ionospheric reflection and involves all of the complications that go with it. The actual frequency to be used at any given time must be determined by a separate ionosphere-monitoring system. Each of the six frequency bands in the 5 to 28 MHz range has its own 12-element subarray complete with backscreen and ground screen. The separate receiving arrays are quite different. The design and properties of this highly specialized system are not described as the end of the cold war has made it largely inoperative. Furthermore, the ionospheric-reflection method is not useful for the detection of low-flying targets nearer than 500 to 1000 km. For this range, use must be made of the surface wave. The field of the grounded vertical monopole includes a substantial surface wave. This can be increased and all ground losses eliminated if the ground system is replaced by radial traveling-wave antennas either of the horizontal-wire (Beverage) type or in the form of buried insulated traveling-wave antennas.

Consider an omnidirectional array consisting of a vertical monopole base-driven against 10 radial horizontal-wire antennas. The arrangement is shown in Fig. 9.7. Each of the horizontal elements has a length $h_m = 201.96$ m, is at the height d' over the earth, and is terminated at the optimum length in an impedance $Z_h = Z_0 + Z_L = Z_c$. Here Z_0 is the impedance of a quarter-wave section and Z_L is a lumped impedance. With $d' = 45$ cm, the electrical height $k_0 d'$ at $f = 10$ MHz is $k_0 d' = 0.094$, which is sufficiently small so that coupling among the 10 elements is negligible and each can be treated as if isolated. Since the impedance of a single element as given in Table 9.1, case B, is

$$Z_L = Z_c = 363.3 + i12.9 \text{ ohms} \tag{9.62}$$

the combined impedance of the 10 equally spaced elements in parallel is

$$Z_a = \frac{Z_L}{10} = 36.3 + i1.3 \text{ ohms}. \tag{9.63}$$

With a proper choice of the height h_v and radius a_v of the vertical monopole, its impedance Z_v can approximate the complex conjugate of Z_a in (9.63) very closely. For example, with $k_0 h_v = 1.477$ and $a_v/\lambda_0 = 0.007\,022$, $Z_v = 36.6 - i0.13$ ohms. With this choice, the impedance of the monopole in series with the 10-element horizontal array is

$$Z = R - iX = Z_v + Z_a = 72.9 + i1.2 \text{ ohms}. \tag{9.64}$$

The entire array is conveniently driven from a 72-ohm coaxial line that is buried in the earth and extends vertically upward so that the extension of its inner conductor is the vertical monopole and the 10 radial horizontal antennas are connected to the shield. This is illustrated in Fig. 9.7.

The input current to the monopole is

$$I_{z'}(0) = -I_x(0) = \frac{V}{Z}. \tag{9.65}$$

(a)

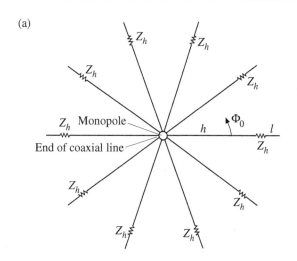

(b)

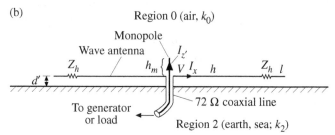

Figure 9.7 Omnidirectional antenna with 10 radial elements. (a) Top view; (b) side view. Taken from King [3, Fig. 6]. © 1992 American Institute of Physics.

The power in the array is

$$P = I_{z'}^2(0)R = I_x^2(0) \times 72.9. \qquad (9.66)$$

The power dissipated in each resistive termination is

$$P_1 = |0.1I_x(0)e^{-\alpha_L h}|^2 R_L = I_x^2(0) \times 0.01 \times e^{-3.2}R_L = 0.0004I_x^2(0)R_L. \qquad (9.67)$$

Since the power in the monopoles with the impedance Z_0 in series with Z_L is radiated, the power dissipated as heat is in

$$R_L = R_h - R_0, \qquad (9.68)$$

where $R_h = R_c$, and R_0 is the resistance of an open-ended section of the horizontal antenna. This has the impedance $Z_0 = Z_c \cot k_L l$, where $k_L = \beta_L + i\alpha_L = 0.2223 + i0.0079$ and $\beta_L l = \pi/2$ so that $l = \pi/0.4446 \sim 7.07$ m. It follows that

$$Z_0 = Z_c \frac{\cos(\frac{1}{2}\pi + i\alpha_L l)}{\sin(\frac{1}{2}\pi + i\alpha_L l)} = Z_c \tanh \alpha_L l \sim Z_c \alpha_L l,$$

$$R_0 \sim \alpha_L l R_c = 0.0079 \times 7.07 \times 363.3 = 20.3 \text{ ohms}.$$

Hence, $R_L = 363.3 - 20.3 = 343.0$ ohms and

$$P_1 = I_x^2(0) \times 0.0004 \times 343.0 = 0.1371 I_x^2(0). \tag{9.69}$$

The power dissipated in the terminations of all 10 elements is

$$P_a = 10 P_1 = 1.371 I_x^2(0). \tag{9.70}$$

The total power radiated is

$$P = (72.9 - 1.37) I_x^2(0) = 71.5 I_x^2(0). \tag{9.71}$$

The radiation efficiency is $71.5/72.9 = 98\%$. Note that this does not mean that 98% of the power is in the space wave. Both the vertical monopole and the horizontal wires generate strong surface waves which transfer power exclusively into the earth or sea as they propagate outward. Since it is entirely by means of the surface wave that low-flying targets can be detected, the power in the surface wave is useful power.

The electric far field of the horizontal-wire antennas in the omnidirectional array is obtained from (9.45) for each of the 10 radial wires. It is

$$[E_{0\Theta}^r(r_0, \Theta_0, \Phi_0)]_{ha} = 0.1 I_x(0) \sum_{n=0}^{9} \frac{i(1 - e^{i[k_L - k_0 \sin \Theta_0 \cos(\Phi_0 + n\pi/5)]h_m})}{k_L - k_0 \sin \Theta_0 \cos(\Phi_0 + n\pi/5)}$$

$$\times \cos(\Phi_0 + n\pi/5)[E_{0\Theta}^r(r_0, \Theta_0, 0)]_h. \tag{9.72}$$

Since the field in the backward direction of any of the 10 wires and the beam width in the horizontal plane are both small, the field of the horizontal array is well approximated by

$$[E_{0\Theta}^r(r_0, \Theta_0, \Phi_0)]_{ha} = 0.1 I_x(0) \left(\frac{i[1 - e^{i(k_L - k_0 \sin \Theta_0 \cos \Phi_0)h_m}]}{k_L - k_0 \sin \Theta_0 \cos \Phi_0} \right) [E_{0\Theta}^r(r_0, \Theta_0, 0)]_h$$

$$\times [\cos \Phi_0 + 2\cos(\Phi_0 + \pi/5) + 2\cos(\Phi_0 + 2\pi/5)]. \tag{9.73}$$

This is approximately circular with a value near that in the direction $\Phi_0 = 0$, so that the final approximation is

$$[E_{0\Theta}^r(r_0, \Theta_0, \Phi_0)]_{ha} \sim [E_{0\Theta}^r(r_0, \Theta_0, 0)]_{ha}$$

$$= 0.324 I_x(0) \left(\frac{i[1 - e^{i(k_L - k_0 \sin \Theta_0)h_m}]}{k_L - k_0 \sin \Theta_0} \right) [E_{0\Theta}^r(r_0, \Theta_0, 0)]_h. \tag{9.74}$$

The electric field of the vertical monopole with $I_{z'}(0) = -I_x(0)$ is

$$[E_{0\Theta}^r(r_0, \Theta_0)]_{vm} = -I_x(0) h_{ev}[E_{0\Theta}^r(r_0, \Theta_0)]_v, \tag{9.75}$$

where h_{ev} is the effective length and $[E^r_{0\Theta}(r_0, \Theta_0)]_v$ is the field of a unit vertical monopole at a height $d' \sim 0$ as given by (8.64).

The complete field of the monopole and horizontal omnidirectional array is

$$[E^r_{0\Theta}(r_0, \Theta_0)]_A = [E^r_{0\Theta}(r_0, \Theta_0)]_{vm} + [E^r_{0\Theta}(r_0, \Theta_0, 0)]_{ha}$$

$$= -I_x(0) h_{ev} [E^r_{0\Theta}(r_0, \Theta_0)]_v + 0.324 I_x(0)$$

$$\times \left(\frac{i[1 - e^{i(k_L - k_0 \sin \Theta_0) h_m}]}{k_L - k_0 \sin \Theta_0} \right) [E^r_{0\Theta}(r_0, \Theta_0, 0)]_h. \tag{9.76}$$

When the explicit formulas for the far fields of unit vertical and horizontal dipoles are substituted in the above expression, it becomes

$$[E^r_{0\Theta}(r_0, \Theta_0)]_A = \frac{\omega \mu_0}{2\pi} I_x(0) e^{ik_0 r_0} \left[\frac{k_2}{k_0} h_{ev} \sin \Theta_0 + 0.324 \right.$$

$$\times \left(\frac{i[1 - e^{i(k_L - k_0 \sin \Theta_0) h_m}]}{k_L - k_0 \sin \Theta_0} \right) \left[\frac{ik_0}{r_0} \left(\frac{\cos \Theta_0}{k_0 + k_2 \cos \Theta_0} \right) \right.$$

$$\left. - \frac{k_2}{k_0^2 r_0^2} \left(\frac{\sin^2 \Theta_0 [\sin^2 \Theta_0 - (k_0/k_2) \cos \Theta_0]}{[1 + (k_2/k_0) \cos \Theta_0]^3} \right) \right]. \tag{9.77}$$

The effective length of a monopole with height $h \sim \lambda_0/4$ is $h_{ev} \sim 2h/\pi \sim \lambda_0/2\pi = k_0^{-1} = 4.78$ m. In the plane $\Theta_0 = \pi/2$, $r_0 = \rho$ and the field reduces to

$$[E^r_{0\Theta}(\rho, \pi/2)]_A = \frac{-\omega \mu_0}{2\pi} I_x(0) e^{ik_0 \rho}$$

$$\times \left(\frac{k_2}{k_0^2} + 0.324 \frac{i[1 - e^{i(k_L - k_0) h_m}]}{k_L - k_0} \right) \left(\frac{k_2}{k_0^2 \rho^2} \right). \tag{9.78}$$

With Table 9.1, case B, $k_2/k_0^2 = 40.65 e^{i0.73} = 30.29 + i27.11$ and

$$h_{eh}(\pi/2, 0) = \frac{i[1 - e^{i(k_L - k_0) h_m}]}{k_L - k_0} = 78.25 e^{i0.929} = 46.84 + i62.68 \tag{9.79}$$

so that

$$[E^r_{0\Theta}(\rho, \pi/2)]_A = -65.81 e^{i0.806} \frac{\omega \mu_0 k_2}{2\pi k_0^2} I_x(0) \frac{e^{ik_0 \rho}}{\rho^2}. \tag{9.80}$$

This is substantially greater than the field of the monopole alone, namely,

$$[E^r_{0\Theta}(\rho, \pi/2)]_{vm} = -40.65 e^{i0.73} \frac{\omega \mu_0 k_2}{2\pi k_0^2} I_x(0) \frac{e^{ik_0 \rho}}{\rho^2}. \tag{9.81}$$

(a)

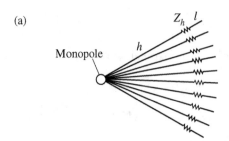

(b)

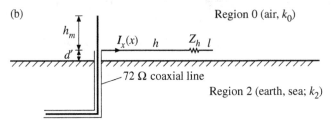

Figure 9.8 Directional fan antenna with 10 traveling-wave elements in a 60° angle and one vertical monopole. (a) Top view; (b) side view. Taken from King [3, Fig. 7]. © 1992 American Institute of Physics.

Evidently, the use of the array of horizontal wave antennas with a vertical monopole in place of the usual grounding network not only increases the space wave but uses the power dissipated in a grounding network to generate a large and useful addition to the surface wave.

An alternative to the omnidirectional array is the directional fan antenna illustrated in Fig. 9.8. It is like the omnidirectional antenna except that the 10 radial horizontal-wire antennas are equally spaced in an arc of 60° or $\pi/3$ radians instead of 360° or 2π radians. Even at the much closer spacing, the coupling among the 10 elements is small because they are close to the earth. Consequently, a good approximation of the complete field of the horizontal elements is the sum of the contributions of the 10 elements with each treated as if isolated. It is

$$[E^r_{0\Theta}(r_0, \Theta_0, \Phi_0)]_{ha} = 0.1i\, I_x(0)[E^r_{0\Theta}(r_0, \Theta_0, 0)]_h$$

$$\times \sum_{n=-9}^{9} \frac{1 - e^{i[k_L - k_0 \sin\Theta_0 \cos(\Phi_0 - n\pi/54)]h_m}}{k_L - k_0 \sin\Theta_0 \cos(\Phi_0 - n\pi/54)}$$

$$\times \cos(\Phi_0 - n\pi/54), \qquad n = \text{odd}. \tag{9.82}$$

The complete field of the array is

$$[E^r_{0\Theta}(r_0, \Theta_0, \Phi_0)]_A = [E^r_{0\Theta}(r_0, \Theta_0)]_{vm} + [E^r_{0\Theta}(r_0, \Theta_0, \Phi_0)]_{ha}, \tag{9.83}$$

where $[E^r_{0\Theta}(r_0, \Theta_0)]_{vm}$ is given in (9.75). This provides a broad unidirectional beam from the horizontal wires and an omnidirectional field from the vertical element. It can serve as a self-contained array or an element in a broadside array directed from the Atlantic seaboard toward the ocean or from the coast of the Gulf of Mexico toward the Gulf. It is an efficient surface-wave antenna for detecting low-flying targets but also radiates an upward-directed space wave for use in ionospheric reflections.

If the use of an array of horizontal-wire antennas in conjunction with a vertical monopole is inconvenient, it can be replaced by a similar array of insulated traveling-wave antennas in the earth or sea a small distance d from the surface. The properties of such elements are discussed in Section 9.3. For the present application, the insulation must be thick and have a wave number k_d close to that of air, so that styrofoam is appropriate with $k_d \sim k_0$. With $f = 10\,\text{MHz}$, $\sigma_2 = 0.04\,\text{S/m}$, $\epsilon_{2r} = 8$, $d = 50\,\text{cm}$, $b = 20\,\text{cm}$, and $a = 1\,\text{mm}$,

$$k_L = 0.229 + i0.0284\,\text{m}^{-1}, \qquad Z_c = 347.7 + i43.1\,\text{ohms}. \tag{9.84}$$

The optimum length for the insulated wire when terminated in Z_c is

$$h_m = \frac{\pi}{\beta_L - k_0} = 160.3\,\text{m}. \tag{9.85}$$

The associated effective length is

$$|h_e| = \left| \frac{1 + e^{-\alpha_L h_m}}{k_L - k_0} \right| = 29.3\,\text{m}. \tag{9.86}$$

The terminating impedances Z_A consist of bare monopoles with the electrical length $\beta_2 l = \pi/2$ in the earth or $l = \pi/2\beta_2 = 1.18\,\text{m}$ in series with lumped impedances Z_L such that $Z_h = Z_0 + Z_L = Z_c = 347.7 + i43.1\,\text{ohms}$. The impedance of the 10 radial elements in parallel is $34.77 + i4.31\,\text{ohms}$. The driving-point impedance of the 10-element array in series with the vertical monopole is $Z_\text{in} = Z_v + Z_c = 36.6 - i0.13 + 34.77 + i4.31 = 71.37 + i4.18\,\text{ohms}$. This value corresponds closely to the impedance with the horizontal-wire array. As the actual and effective lengths of the buried insulated antennas are substantially smaller than those of the horizontal-wire antennas, the contribution to the electromagnetic field by the currents in the radial array is also smaller. But the general properties of the array are much the same whether the radial array consists of horizontal wires in the air or insulated wires in the earth.

9.5   Horizontal antennas over the spherical earth

All of the formulas in this chapter apply to horizontal antennas close to the surface of a planar earth. When the point of observation is also close to the surface with $z' \sim 0$

or $\Theta \sim \pi/2$, $r_0 \sim \rho$, a correction for the spherical earth can be made [6]. Under these conditions, the intermediate and far fields are contained in

$$[E_{0\rho}(\rho, \phi', 0)]_h \sim \frac{-\omega\mu_0 k_0}{2\pi k_2^2} \cos \phi' \, e^{ik_0\rho} \left(\frac{ik_0}{\rho} - T \right) \tag{9.87}$$

$$[E_{0z'}(\rho, \phi', 0)]_h \sim \frac{-\omega\mu_0}{2\pi k_2} \cos \phi' \, e^{ik_0\rho} \left(\frac{ik_0}{\rho} - T \right) \tag{9.88}$$

$$[B_{0\phi'}(\rho, \phi', 0)]_h \sim \frac{\mu_0 k_0}{2\pi k_2} \cos \phi' \, e^{ik_0\rho} \left(\frac{ik_0}{\rho} - T \right), \tag{9.89}$$

where

$$T = \frac{k_0^3}{k_2} \left(\frac{\pi}{k_0\rho} \right)^{1/2} e^{-iP_0} \mathcal{F}(P_0), \qquad P_0 = \frac{k_0^3 \rho}{2k_2^2}. \tag{9.90}$$

The corresponding formulas for the vertical dipole also with $d \sim 0$, $z' \sim 0$, are

$$[E_{0\rho}(\rho, 0)]_v = \frac{\omega\mu_0}{2\pi k_2} e^{ik_0\rho} \left(\frac{ik_0}{\rho} - T \right) = \frac{-k_2}{k_0} [E_{0\rho}(\rho, 0, 0)]_h \tag{9.91}$$

$$[E_{0z'}(\rho, 0)]_v = \frac{\omega\mu_0}{2\pi k_0} e^{ik_0\rho} \left(\frac{ik_0}{\rho} - T \right) = \frac{-k_2}{k_0} [E_{0z'}(\rho, 0, 0)]_h \tag{9.92}$$

$$[B_{0\phi'}(\rho, 0)]_v = \frac{-\mu_0}{2\pi} e^{ik_0\rho} \left(\frac{ik_0}{\rho} - T \right) = \frac{-k_2}{k_0} [B_{0\phi'}(\rho, 0, 0)]_h. \tag{9.93}$$

The above formulas give the intermediate field when $T \sim 0$ and the far field when

$$T \sim \frac{ik_0}{\rho} + \frac{k_2^2}{k_0^2\rho^2}. \tag{9.94}$$

Evidently, any of the three components of the horizontal dipole are given by the corresponding component of the vertical dipole with the simple formula

$$H = \frac{-k_0}{k_2} V \cos \phi'. \tag{9.95}$$

This also applies to the field when expressed in the spherical coordinates r_0, Θ, Φ with $r_0 \to \rho$, $\Theta \to \pi/2$.

In the spherical coordinates r, Θ, Φ with the center of the earth as origin and the angle Θ measured from the radial line through the center of the dipole, the following relations apply:

$$[E_{0\rho}(\rho, 0)]_v = [E_{0\Theta}(a, \Theta)]_v \qquad [E_{0z'}(\rho, 0)]_v = [E_{0r}(a, \Theta)]_v$$
$$[B_{0\phi'}(\rho, 0)]_v = [B_{0\Phi}(a, \Theta)]_v, \tag{9.96}$$

where $r = a$ is the radius of the earth. The corresponding relationships for the horizontal dipole are

$$[E_{0\rho}(\rho, \phi', 0)]_h = [E_{0\Theta}(a, \Theta, \Phi)]_h = \frac{-k_0}{k_2}[E_{0\Theta}(a, \Theta)]_v \cos \Phi \tag{9.97}$$

$$[E_{0z'}(\rho, \phi', 0)]_h = [E_{0r}(a, \Theta, \Phi)]_h = \frac{-k_0}{k_2}[E_{0r}(a, \Theta)]_v \cos \Phi \tag{9.98}$$

$$[B_{0\phi'}(\rho, \phi', 0)]_h = [B_{0\Phi}(a, \Theta, \Phi)]_h = \frac{-k_0}{k_2}[B_{0\Phi}(a, \Theta)]_v \cos \Phi. \tag{9.99}$$

With (8.115) and (8.116) in Chapter 8, it follows that

$$
\begin{aligned}
[E_{0\rho}(\rho, \phi', 0)]_h &= \frac{-k_0}{k_2}\left(-\frac{\omega}{k_2}\right)[B_{0\Phi}(a, \Theta)]_v \cos \Phi \\
&= \frac{-\mu_0\omega k_0^2}{4\pi k_2^2}\frac{e^{ik_0\rho_s + i\pi/4}}{\sqrt{\pi\rho_s\rho_c}} I_2(\eta_2, g) \cos \Phi
\end{aligned}
\tag{9.100}
$$

$$
\begin{aligned}
[E_{0z'}(\rho, \phi', 0)]_h &= \frac{-k_0}{k_2}\left(-\frac{\omega}{k_0}\right)[B_{0\Phi}(a, \Theta)]_v \cos \Phi \\
&= \frac{-\mu_0\omega k_0}{4\pi k_2}\frac{e^{ik_0\rho_s + i\pi/4}}{\sqrt{\pi\rho_s\rho_c}} I_2(\eta_2, g) \cos \Phi
\end{aligned}
\tag{9.101}
$$

$$
\begin{aligned}
[B_{0\phi'}(\rho, \phi', 0)]_h &= \frac{-k_0}{k_2}[B_{0\Phi}(a, \Theta)]_v \cos \Phi \\
&= \frac{\mu_0 k_0^2}{4\pi k_2}\frac{e^{ik_0\rho_s + i\pi/4}}{\sqrt{\pi\rho_s\rho_c}} I_2(\eta_2, g) \cos \Phi.
\end{aligned}
\tag{9.102}
$$

Here,

$$g = \frac{k_0}{k_2}\left(\frac{k_0 a}{2}\right)^{1/3}, \qquad \rho_c = a\left(\frac{k_0 a}{2}\right)^{-1/3}, \qquad \eta_2 = \frac{\rho_s}{\rho_c} = \Theta\left(\frac{k_0 a}{2}\right)^{1/3}. \tag{9.103}$$

The evaluation of $I_2(\eta_2, g)$ is given in (8.118).

As a specific example, consider the array of horizontal traveling-wave antennas and vertical monopole treated in the preceding section, erected on the earth but transmitting over sea water. The component of the field $[E_{0\Theta}^r(r_0, \Theta_0)]_A$ for the array includes the fields of the unit dipoles, namely $[E_{0\Theta}^r(r_0, \Theta_0)]_{vm}$ for the vertical monopole and $[E_{0\Theta}^r(r_0, \Theta_0, 0)]_{ha}$ for each horizontal element. The generalization is valid only for points on the surface, so that it is necessary to set $\Theta_0 \sim \pi/2, r_0 \sim \rho$. In terms of the coordinates (r, Θ, Φ) with the center of the earth as origin and $r = a$, the vertical field of the array is given by (9.76) in the form

$$[E_{0r}^r(a, \Theta, 0)]_A = I_x(0)\{-h_{ev}[E_{0r}^r(a, \Theta)]_v + 0.324h_{eh}[E_{0r}^r(a, \Theta, 0)]_h\}, \tag{9.104}$$

where $\Theta = \rho_s/a$. With (9.98) and (9.101) with $\Phi = 0$,

$$[E_{0r}^r(a, \Theta, 0)]_A = I_x(0) \left(\frac{k_2}{k_0} h_{ev} + 0.324 h_{eh} \right) [E_{0r}^r(a, \Theta, 0)]_h$$

$$= \frac{-\omega\mu_0 k_0 I_x(0)}{4\pi k_2} \left(\frac{k_2}{k_0} h_{ev} + 0.324 h_{eh} \right) \frac{e^{ik_0\rho_s + i\pi/4}}{\sqrt{\pi\rho_s\rho_c}} I_2(\eta_2, g),$$

(9.105)

where, at $f = 10\,\text{MHz}$,

$$h_{ev} = 4.78\,\text{m} \tag{9.106a}$$

$$h_{eh} = \frac{i[1 - e^{i(k_L - k_0)h_m}]}{k_L - k_0} = 46.84 + i62.68\,\text{m}. \tag{9.106b}$$

With these values,

$$[E_{0r}^r(a, \Theta, 0)]_A = -65.81 e^{i0.806} \frac{\omega\mu_0 k_0 I_x(0)}{4\pi k_2} \frac{e^{ik_0\rho_s + i\pi/4}}{\sqrt{\pi\rho_s\rho_c}} I_2(\eta_2, g). \tag{9.107}$$

At $f = 10\,\text{MHz}$,

$$\rho_c = 6378 \times 10^3 \left(\frac{0.2094 \times 6378 \times 10^3}{2} \right)^{-1/3} = 73.0 \times 10^3\,\text{m} = 73\,\text{km}. \tag{9.108}$$

For sea water, $k_2 = 17.77 e^{i0.78}\,\text{m}^{-1}$ and

$$g = \frac{0.2094}{17.77 e^{i0.78}} \left(\frac{0.2094 \times 6378 \times 10^3}{2} \right)^{1/3} = 1.03 e^{-i0.78}. \tag{9.109}$$

For these values,

$$|I_2(\eta_2, g)| \sim 2.46 e^{-1.05\rho_s/\rho_c} \tag{9.110}$$

so that

$$|[E_{0r}^r(a, \Theta, 0)]_A| = 161.8 \left| \frac{\omega\mu_0 k_0 I_x(0)}{4\pi k_2} \right| \frac{e^{-1.05\rho_s/\rho_c}}{\sqrt{\pi\rho_s\rho_c}}. \tag{9.111}$$

This is the vertical electric field at any distance $\rho_s = a\Theta$ along the surface of the sea due to the omnidirectional array consisting of a vertical monopole and a radial 10-element array of horizontal-wire antennas at a height $d' = 45\,\text{cm}$ over the earth close to the sea coast.

At a distance $\rho_s = 500\,\text{km}$, (9.111) gives

$$|[E_{0r}^r(a, \Theta, 0)]_A| = 2.67 \times 10^{-8} I_x(0). \tag{9.112}$$

If the driving-point current in the vertical monopole is $I_z(0) = -I_x(0) = 100\,\text{A}$,

$$\left|[E^r_{0r}(a, \Theta, 0)]_A\right| = 2.67\,\mu\text{V/m}. \tag{9.113}$$

The planar earth formula (9.80) gives

$$\left|[E^r_{0\Theta}(\rho, \pi/2)]_A\right| = 1.34 \times 10^{-4}\,\text{V/m} = 134\,\mu\text{V/m}. \tag{9.114}$$

It is seen that at a distance of 500 km over the sea, the effect of the earth's curvature is to reduce the field by roughly a factor 50 below the value for a plane earth.

9.6 Horizontal electric dipoles for remote sensing on and in the earth, sea, or Arctic ice

Horizontal electric dipoles have a wide range of applications in remote sensing. Antennas for this purpose are located on satellites at altitudes of the order of 800 km, on aircraft flying at various heights, and directly on the surface of the earth, sea, or ice. They are used to map the surface of the earth, monitor the properties and motions of ice sheets, and locate objects buried in the earth or snow and submarines submerged in the ocean or under the Arctic ice. A detailed description of these interesting and in part highly complicated systems is beyond the scope of the book. However, they all involve the electromagnetic field of a dipole antenna parallel to the surface of the earth and the electromagnetic field backscattered from that surface and from objects of arbitrary shape and size at different depths under the surface. A signal from the transmitting antenna induces currents in the earth or sea and in buried or submerged objects. The direction and magnitude of these currents are determined by the polarization of the incident field and the shape, electrical size, orientation, depth, and electrical properties of a buried or submerged object. These currents radiate the scattered field which can be received by the transmitting antenna switched to the receiving mode as in monostatic radar or by a separate receiving antenna as in bistatic radar.

When the transmitting and receiving antennas are *on* the surface of the earth, sea, or ice in a bistatic arrangement, the field transmitted into the earth and that received from a buried scattering object cannot be determined by plane-wave theory. The two components of the electric field in the earth due to an insulated horizontal dipole on the surface of the earth are given by (9.2) and (9.3) with $z' = d' = 0$, $\phi' = -\phi$ and $\epsilon = k_0/k_2$, and multiplied by $e^{ik_2 z}$ where z is directed downward into the earth. They are

$$E_{2\rho}(\rho, \phi, z) = \frac{-\omega\mu_0 k_0 (2h_e I)}{2\pi k_2^2}\left[\frac{ik_0}{\rho} - \frac{1}{\rho^2} - \frac{i}{k_0\rho^3}\right.$$

$$\left. - \frac{k_0^3}{k_2}\left(\frac{\pi}{k_0\rho}\right)^{1/2} e^{-iP_0}\mathcal{F}(P_0)\right] e^{ik_0\rho} e^{ik_2 z}\cos\phi \tag{9.115}$$

$$E_{2\phi}(\rho, \phi, z) = \frac{\omega\mu_0 k_0(2h_e I)}{2\pi k_2^2} \left[\frac{2}{\rho^2} + \frac{2i}{k_0\rho^3} \right.$$

$$\left. + \frac{ik_0^2}{k_2\rho} \left(\frac{\pi}{k_0\rho} \right)^{1/2} e^{-iP_0} \mathcal{F}(P_0) \right] e^{ik_0\rho} e^{ik_2 z} \sin\phi, \tag{9.116}$$

where

$$P_0 = \frac{k_0^3 \rho}{2k_2^2}, \qquad \mathcal{F}(P_0) = \tfrac{1}{2}(1+i) - \int_0^{P_0} \frac{e^{it}}{\sqrt{2\pi t}} \, dt. \tag{9.117}$$

It is this field that propagates downward into the earth and induces currents in any buried objects in the earth. A detailed analysis of the current induced in an insulated metal rod at a depth d below the surface in different locations relative to the transmitting dipole has been made.[2] The field re-radiated by these currents at points on the surface has also been determined. It is this field that is detected by a horizontal receiving antenna moved about on the surface of the earth in the vicinity of the buried rod. The results show a significant change in the electric field over the volume occupied by the rod and that this is sufficiently localized to permit an accurate bounding of the area above the rod in a detail that clearly defines its shape and orientation. It is to be emphasized that the field re-radiated by the buried object and maintained along the air–earth surface is a pure lateral wave. Complete formulas for the field in air due to a horizontal electric dipole at a depth d in the earth or sea are in Chapter 5 of [2].

A similar application with a horizontal electric dipole in the sea close to the surface as the transmitter and a movable crossed dipole also in the sea just below the surface as a receiver has been analyzed [9] and shown to permit the detection of submarines at depths up to 100 m. A generalization of the method to permit the transmitting and receiving antennas to be laid on the surface of the Arctic ice has been made [10].

In the synthetic aperture radar (SAR), the backscattered field is received and recorded successively and over a range of frequencies. It is then processed to form an image. In one system, the radar transmits in a series of narrow bands of about 1 MHz that step up in frequency from 20 to 90 MHz at about 100-μs intervals. The returns are then integrated to produce images. This process of combining the returns from a wide narrow-band series gives the radar the effect of a large band width. Very extensive signal processing is required and this is based on plane-wave theory. Of primary importance is coherence, that is, the phase of the reflected field is defined well enough and contains the information needed to characterize the target. Can the synthetic aperture radar accurately reconstruct the image of a target buried in the earth if the signal processing assumes that the scattered field is a plane wave that obeys the well-known laws of reflection and refraction? Can the essential coherence be achieved with the plane-wave assumption?

[2] See [2], Chapter 7, [7] and [8].

Consider an electrically short insulated metal rod buried at a depth d in the earth. The rod has the length $2h$ and the effective length $2h_e$. If the current induced in the rod by the field incident from the horizontal antenna of the aircraft flying parallel to the x-directed rod at the height z' and radial distance ρ is I, the component of the electric field scattered or re-radiated by the rod parallel to the antenna on the aircraft is

$$E_{0\rho}(\rho, 0, z') = \frac{-\omega\mu_0(2h_e I)}{2\pi k_2} e^{ik_2 d} e^{ik_0 r_0}$$

$$\times \left\{ \frac{k_0}{k_2} \left[\frac{ik_0}{r_0} - \frac{1}{r_0^2} - \frac{i}{k_0 r_0^3} - \frac{r_2}{\rho} T \right] - \left(\frac{z'}{r_0} \right) \left(\frac{ik_0}{r_0} - \frac{1}{r_0^2} \right) \right\}.$$

$$(9.118)$$

The magnetic field is

$$B_{0\phi'}(\rho, 0, z') = \frac{\mu_0 k_0 (2h_e I)}{2\pi k_2} e^{ik_2 d} e^{ik_0 r_0} \left(\frac{ik_0}{r_0} - \frac{1}{r_0^2} - \frac{i}{k_0 r_0^3} - \frac{r_2}{\rho} T \right), \qquad (9.119)$$

where

$$r_0 = \sqrt{\rho^2 + z'^2}, \qquad r_2 = \sqrt{\rho^2 + (z' + d)^2} \qquad (9.120)$$

$$T = \frac{k_0^3}{k_2} \left(\frac{\pi}{k_0 r_2} \right)^{1/2} e^{-iP_2} \mathcal{F}(P_2), \qquad P_2 = \frac{k_0^3 r_2}{2k_2^2} \left(\frac{k_0 r_2 + k_2(z' + d)}{k_0 \rho} \right)^2 \qquad (9.121)$$

$$\mathcal{F}(P_2) = \tfrac{1}{2}(1 + i) - \int_0^{P_2} \frac{e^{it}}{\sqrt{2\pi t}} dt. \qquad (9.122)$$

It is this field that is received by the aircraft at the height z' and radial distance ρ. It is very different from that assumed by plane-wave theory. In its simplest form, this assumes that the induced current in the buried dipole generates a plane wave that propagates directly to the receiving antenna in the air. That is, the signal has the simple form of a free-space plane wave, namely,

$$E_{0\rho} \sim e^{ik_0 r_2} \cos \phi', \qquad (9.123)$$

where r_2 is given in (9.120). This ignores the properties of the earth. A more complete form assumes the emitted plane wave travels to the earth–air boundary which it reaches with an angle of incidence Θ_2. At the surface, it is refracted according to the plane-wave law of refraction and continues in the air at the angle of transmission Θ_0, so that $k_2 \sin \Theta_2 = k_0 \sin \Theta_0$. The transmission coefficient is

$$f_{mt} = \frac{2k_0 \cos \Theta_0}{k_2 \cos \Theta_2 + k_0 \cos \Theta_0}. \qquad (9.124)$$

The field in the air using the refracted plane-wave approximation is

$$E_{0\rho} \sim f_{mt} e^{i(k_2 r_d + k_0 r_0)} \cos \phi', \tag{9.125}$$

where r_d is the distance of travel in the earth from the buried rod at the depth d to the earth–air surface and r_0 is the distance of travel in the air from that surface to the aircraft antenna at the height z'.

A detailed comparison of the signals received and processed by a synthetic aperture radar based on the accurate field in (9.118) and each of the two plane-wave assumptions, (9.123) and (9.125), has been made by Gilbert et al. [11]. Their calculations are for $f = 600\,\text{MHz}$ and $f = 5\,\text{GHz}$ with $d = 0.5\,\text{m}$ and $z' = 3$ and 30 m. They conclude that:

"when the receiving antenna is very close to the ground above a buried source, the use of plane wave approximations to correlate signals will lead to severely degraded images compared to images correlated with accurate analytical solutions to the Maxwell equations [(9.118) and (9.119)]. In contrast, for many other cases of interest, when the receiving antenna is sufficiently high (over 30 m altitude), the quality of images correlated with a plane wave approximation is quite good apart from some loss in image intensity."

For maximum quality of the image, it is clearly desirable to use the accurate formulas rather than plane-wave approximations under all circumstances. This is essential when the transceiver is near the surface of the earth. When $z' \geq \rho$, the condition $P_2 \leq 1$ is generally satisfied so that the Fresnel-integral term is negligible. Thus, when $k_0 \rho \geq 1$ and $P_2 \leq 1$, the intermediate-zone field is

$$E_{0\rho}^i(\rho, \phi', z') = \frac{i\omega\mu_0 k_0 (2h_e I)}{2\pi k_2} \frac{e^{ik_2 d} e^{ik_0 r_0}}{r_0} \left(\frac{z'}{r_0} - \frac{k_0}{k_2}\right) \cos \phi' \tag{9.126}$$

$$B_{0\phi'}^i(\rho, \phi', z') = -\frac{i\mu_0 k_0^2 (2h_e I)}{2\pi k_2} \frac{e^{ik_2 d} e^{ik_0 r_0}}{r_0} \cos \phi'. \tag{9.127}$$

Note that in the spherical coordinates $r_0, \Theta, \Phi, z'/r_0 = \cos \Theta$.

9.7   Horizontal electric dipoles and patch antennas on microstrip

The dipole

As stated in Section 8.8, the horizontal electric dipole is the basic element of strip transmission lines and antennas on microstrip. The complete electromagnetic field in cylindrical coordinates of such a dipole with the electric moment $(2h_e I)$ is given by (9.2)–(9.7) with (9.8)–(9.10). For microstrip with a dielectric layer with the thickness l,

$$\epsilon = \frac{k_0}{k_2} - ik_0 l \sim -ik_0 l. \tag{9.128}$$

The wave number k_2 of the conducting base is generally so large that $|k_0/k_2| \ll k_0 l$ or $|k_2 l| \gg 1$. In effect, the base acts like a perfect conductor. At sufficiently large electrical distances with the dipole on the surface of the dielectric so that $d' = 0$ and

$$k_0 r_0 > 1 \tag{9.129}$$

the point of observation is in the intermediate zone or beyond. The outer limit of the intermediate field is

$$|P_0| \leq 1, \qquad |P_0| = \frac{k_0 r_0}{2} \left| \frac{-ik_0 l r_0 + z'}{\rho} \right|^2 = \frac{k_0 r_0}{2} \left| \frac{-ik_0 l + \cos \Theta}{\sin \Theta} \right|^2, \tag{9.130}$$

where Θ is measured from the z' axis and $\rho/r_0 = \sin \Theta$, $z'/r_0 = \cos \Theta$. It follows that for the intermediate zone,

$$k_0 r_0 \leq 2 \left| \frac{\sin \Theta}{-ik_0 l + \cos \Theta} \right|^2 = \frac{2 \sin^2 \Theta}{k_0^2 l^2 + \cos^2 \Theta}. \tag{9.131}$$

Since $k_0^2 l^2 \ll 1$,

$$k_0 r_0 \leq 2 \tan^2 \Theta, \qquad \Theta \neq \pi/2 \tag{9.132a}$$

$$k_0 r_0 \leq \frac{2}{k_0^2 l^2}, \qquad \Theta = \pi/2. \tag{9.132b}$$

Since $k_0 l$ is approximately constant over the frequency range in which microstrip is useful, a typical value is general. Specifically, with $f = 10\,\text{GHz}$, $l = 0.1\,\text{mm}$, $k_0 l = 2\pi \times 10^{10} \times 10^{-4}/(3 \times 10^8) \sim 2 \times 10^{-2}$ and

$$k_0 r_0 \leq 5 \times 10^3.$$

This is so large that no microstrip circuit can extend beyond the intermediate zone and the far zone in which the field decreases as $1/r^2$ is never reached. The relevant range lies beyond the near field with $k_0 r_0 \geq 1$ entirely in the intermediate range.

The components of the intermediate range include three of electric type, namely,

$$E_{0\rho}^i(\rho, \phi', z') = \frac{-i\omega\mu_0(2h_e I)}{2\pi} (ik_0 l) \frac{e^{ik_0 r_0}}{r_0} \left(\frac{z'}{r_0} + ik_0 l \right) \cos \phi' \tag{9.133}$$

$$E_{0z'}^i(\rho, \phi', z') = \frac{i\omega\mu_0(2h_e I)}{2\pi} (ik_0 l) \frac{e^{ik_0 r_0}}{r_0} \left(\frac{\rho}{r_0} \right) \cos \phi' \tag{9.134}$$

$$B_{0\phi'}^i(\rho, \phi', z') = \frac{-ik_0\mu_0(2h_e I)}{2\pi} (ik_0 l) \frac{e^{ik_0 r_0}}{r_0} \cos \phi' \tag{9.135}$$

and three of magnetic type, namely,

$$B_{0\rho}^i(\rho, \phi', z') = \frac{-ik_0\mu_0(2h_eI)}{2\pi}(ik_0l)\frac{e^{ik_0r_0}}{r_0}\left(\frac{z'^2}{r_0^2}\right)\sin\phi' \tag{9.136}$$

$$B_{0z'}^i(\rho, \phi', z') = \frac{ik_0\mu_0(2h_eI)}{2\pi}(ik_0l)\frac{e^{ik_0r_0}}{r_0}\left(\frac{\rho}{r_0}\right)\left(\frac{z'}{r_0}\right)\sin\phi' \tag{9.137}$$

$$E_{0\phi'}^i(\rho, \phi', z') = \frac{i\omega\mu_0(2h_eI)}{2\pi}(ik_0l)\frac{e^{ik_0r_0}}{r_0}\left(\frac{z'}{r_0}\right)\sin\phi'. \tag{9.138}$$

In the spherical coordinates r_0, Θ, Φ, these are

$$E_{0r}^i(r_0, \Theta, \Phi) = \frac{-i\omega\mu_0(2h_eI)}{2\pi}(ik_0l)\frac{e^{ik_0r_0}}{r_0}\cos\Phi\sin\Theta \tag{9.139}$$

$$E_{0\Theta}^i(r_0, \Theta, \Phi) = \frac{-i\omega\mu_0(2h_eI)}{2\pi}(ik_0l)\frac{e^{ik_0r_0}}{r_0}(1 + ik_0l\cos\Theta)\cos\Phi \tag{9.140}$$

$$B_{0\Phi}^i(r_0, \Theta, \Phi) = \frac{-ik_0\mu_0(2h_eI)}{2\pi}(ik_0l)\frac{e^{ik_0r_0}}{r_0}\cos\Phi \tag{9.141}$$

$$B_{0r}^i(r_0, \Theta, \Phi) = 0 \tag{9.142}$$

$$B_{0\Theta}^i(r_0, \Theta, \Phi) = \frac{-k_0}{\omega}E_{0\Phi}^i(r_0, \Theta, \Phi)$$

$$= \frac{-i\mu_0k_0(2h_eI)}{2\pi}(ik_0l)\frac{e^{ik_0r_0}}{r_0}\sin\Phi\cos\Theta. \tag{9.143}$$

For completeness and possible special cases, the far-field formulas are also given. The far-field condition is

$$|P_0| \geq 4 \quad \text{or} \quad k_0r_0 \geq \left|\frac{8\sin^2\Theta}{\cos^2\Theta + \epsilon\cos\Theta + \epsilon^2}\right|. \tag{9.144}$$

The formulas for the field of electric type are obtained from (9.33)–(9.35) with $\epsilon = -ik_0l$. They are

$$E_{0r}^r(r_0, \Theta, \Phi) = \frac{\omega\mu_0(2h_eI)}{2\pi k_0}\cos\Phi\, e^{ik_0r_0}\frac{\sin^3\Theta}{r_0^2[1 + (i/k_0l)\cos\Theta]^2} \tag{9.145}$$

$$E_{0\Theta}^r(r_0, \Theta, \Phi) = \frac{\omega\mu_0(2h_eI)}{2\pi k_0}\cos\Phi\, e^{ik_0r_0}\left[\frac{ik_0}{r_0}\left(\frac{k_0l\cos\Theta}{k_0l + i\cos\Theta}\right)\right.$$

$$\left. - \frac{\sin^2\Theta\,(ik_0l\sin^2\Theta - k_0^2l^2\cos\Theta)}{k_0^2l^2r_0^2[1 + (i/k_0l)\cos\Theta]^3}\right] \tag{9.146}$$

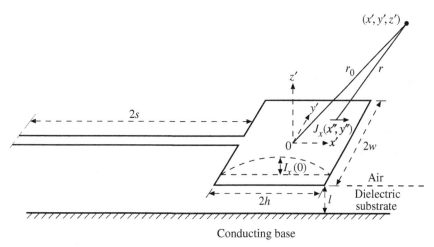

Figure 9.9 Microstrip patch antenna end-driven by microstrip transmission line. Taken from King [12, Fig. 1]. © 1992 American Geophysical Union.

$$B_{0\Phi}^r(r_0, \Theta, \Phi) = \frac{\mu_0(2h_e I)}{2\pi} \cos \Phi\, e^{ik_0 r_0} \left[\frac{ik_0}{r_0} \left(\frac{k_0 l \cos \Theta}{k_0 l + i \cos \Theta} \right) \right.$$

$$\left. - \frac{ik_0 l \sin^2 \Theta}{k_0^2 l^2 r_0^2 [1 + (i/k_0 l) \cos \Theta]^3} \right]. \tag{9.147}$$

As the field of magnetic type includes no lateral wave, the intermediate field for it extends on into the far field.

Patch antennas

In order to increase the electric moment of horizontal electric dipoles on microstrip, they can be enlarged into flat patches that can be rectangular, square, or circular in shape. Such a patch antenna can be driven from a vertical post that pierces the dielectric layer or from a strip transmission line that connects to one end. A schematic diagram of the end-driven rectangular patch antenna [12] is shown in Fig. 9.9. The current on such a patch antenna has not been determined analytically. In practice, it is obtained by numerical methods or simply postulated in terms of a reasonable approximation. Since the radiation field is not sensitive to the details of the current distribution, a usually adequate determination can be made with the latter method.

The current-density distribution appropriate for the rectangular patch antenna shown in Fig. 9.9 has the form

$$J_x(x'', y'') = \frac{I_x(0)}{2w} \cos k_L x''. \tag{9.148}$$

The length of the patch is $2h$ and the width is $2w$ with

$$-h \le x'' \le h, \qquad -w \le y'' \le w. \tag{9.149}$$

In (9.148), $k_L = k_0\sqrt{\epsilon_{1r\,\text{eff}}}$ is the wave number that would characterize the patch as a segment of a microstrip transmission line, $I_x(0)$ is the total current traversing the center line $x'' = 0$ of the patch, and h is so chosen that $k_L h = \pi/2$. The transverse distribution of the x-directed current is assumed to be uniform. Actually, the transverse distribution has large peaks at the edges $|y''| = w$ and a minimum at the center $y'' = 0$. In addition, there is a y''-directed current. Since this is in opposite directions on the top and bottom surfaces of the patch, its contribution to the radiation field is negligible. Since the impedance of the patch as a termination for the strip transmission line is very large – near anti-resonance – the driving-point current is small compared to $I_x(0)$, and the assumed current (9.148) should be a good approximation. Formulas for the characteristic impedance of the microstrip transmission line and the real effective permittivity $\epsilon_{1r\,\text{eff}}$ as given by Hoffman [13] are

$$Z_c = \frac{60}{\sqrt{\epsilon_{1r\,\text{eff}}}} \ln\left(\frac{4l}{w} + \frac{w}{2l}\right) \text{ ohms}, \qquad 0 \le \frac{2w}{l} \le 1 \tag{9.150a}$$

$$Z_c = \frac{120\pi}{\sqrt{\epsilon_{1r\,\text{eff}}}} \left[\frac{2w}{l} + 2.42 - 0.22\frac{l}{w} + \left(1 - \frac{l}{2w}\right)^6\right]^{-1} \text{ ohms}, \qquad \frac{2w}{l} > 1 \tag{9.150b}$$

$$\epsilon_{1r\,\text{eff}} = \frac{\epsilon_{1r} + 1}{2} + \frac{\epsilon_{1r} - 1}{2}\left(1 + \frac{5l}{w}\right)^{-1/2}, \qquad k_L = k_0\sqrt{\epsilon_{1r\,\text{eff}}}. \tag{9.151}$$

These apply to a microstrip transmission line with the width $2w$ on a dielectric substrate with the thickness l and the relative permittivity ϵ_{1r}.

The radiation field of the assumed current distribution in the rectangular patch is obtained by integration using the intermediate-zone formulas (9.140) and (9.143). These have the differential elements

$$dE_{0\Theta}(r, \Theta, \Phi) = \frac{-i\omega\mu_0}{2\pi}(ik_0l)\frac{e^{ik_0r}}{r}(1 + ik_0l\cos\Theta)\cos\Phi$$

$$\times J_x(x'', y'')\,dx''\,dy'' \tag{9.152}$$

$$dE_{0\Phi}(r, \Theta, \Phi) = \frac{i\omega\mu_0}{2\pi}(ik_0l)\frac{e^{ik_0r}}{r}\sin\Phi\cos\Theta\,J_x(x'', y'')\,dx''\,dy''. \tag{9.153}$$

Here,

$$r = \sqrt{(x' - x'')^2 + (y' - y'')^2 + z'^2} \tag{9.154}$$

is the distance from the element $J_x(x'', y'')\,dx''\,dy''$ at $z'' = 0$ to the point of observation at x', y', z'. In the radiation zone, $r \sim r_0 = \sqrt{x'^2 + y'^2 + z'^2}$ is adequate in amplitudes. In phases, the more accurate formula

$$r \sim \sqrt{r_0^2 - 2x'x'' - 2y'y''} \sim r_0 - x'' \frac{x'}{r_0} - y'' \frac{y'}{r_0}$$

$$= r_0 - x'' \cos \Phi \sin \Theta - y'' \sin \Phi \sin \Theta \qquad (9.155)$$

must be used. The integral to be evaluated for all components is

$$J = e^{ik_0 r_0} J(x'') J(y''), \qquad (9.156)$$

where

$$J(x'') = \int_{-h}^{h} \cos k_L x'' \, e^{-ik_0 x'' \cos \Phi \sin \Theta} \, dx'' = \frac{2k_L \cos(k_0 h \cos \Phi \sin \Theta)}{k_L^2 - k_0^2 \cos^2 \Phi \sin^2 \Theta} \qquad (9.157)$$

$$J(y'') = \int_{-w}^{w} e^{-ik_0 y'' \sin \Phi \sin \Theta} \, dy'' = \frac{2 \sin(k_0 w \sin \Phi \sin \Theta)}{k_0 \sin \Phi \sin \Theta}. \qquad (9.158)$$

With these values, the field factor for the patch antenna is

$$P(\Theta, \Phi) = \frac{I_x(0)}{2w} J(x'') J(y'')$$

$$= I_x(0) \left[\frac{2k_L \cos(k_0 h \cos \Phi \sin \Theta)}{k_L^2 - k_0^2 \cos^2 \Phi \sin^2 \Theta} \right] \left[\frac{\sin(k_0 w \sin \Phi \sin \Theta)}{k_0 w \sin \Phi \sin \Theta} \right]. \qquad (9.159)$$

The leading components of the radiation field are

$$E_{0\Theta}^i(r_0, \Theta, \Phi) = \frac{\omega}{k_0} B_{0\Phi}^i(r_0, \Theta, \Phi)$$

$$= \frac{-i\omega\mu_0}{2\pi} (ik_0 l) \frac{e^{ik_0 r_0}}{r_0} \cos \Phi \, P(\Theta, \Phi), \qquad 1 \le k_0 \rho \le \frac{4}{k_0^2 l^2} \qquad (9.160)$$

$$E_{0\Phi}^i(r_0, \Theta, \Phi) = \frac{-\omega}{k_0} B_{0\Theta}^i(r_0, \Theta, \Phi)$$

$$= \frac{i\omega\mu_0}{2\pi} (ik_0 l) \frac{e^{ik_0 r_0}}{r_0} \sin \Phi \cos \Theta \, P(\Theta, \Phi), \qquad 1 \le k_0 \rho. \qquad (9.161)$$

These are valid throughout the intermediate zone which includes the far field for $E_{0\Phi}$ and $B_{0\Theta}$ but excludes the range $|P_0| > 1$ close to the plane $z' = 0$ when this is occupied by microstrip. In practice, the microstrip is finite and the field beyond its edges is modified by edge reflection and diffraction.

Of particular interest is the field in the E-plane when $\Phi = 0$. In this case,

$$P(\Theta, 0) = I_x(0) \left[\frac{2k_L \cos(k_0 h \sin \Theta)}{k_L^2 - k_0^2 \sin^2 \Theta} \right] \qquad (9.162)$$

$$E_{0\Theta}^i(r_0, \Theta, 0) = \frac{\omega}{k_0} B_{0\Phi}^i(r_0, \Theta, 0) = \frac{-i\omega\mu_0}{2\pi} (ik_0 l) \frac{e^{ik_0 r_0}}{r_0} P(\Theta, 0). \qquad (9.163)$$

When $\Phi = \pi/2$, the H-plane pattern is given by

$$P(\Theta, \pi/2) = \frac{I_x(0)}{k_L} \left[\frac{2 \sin(k_0 w \sin \Theta)}{k_0 w \sin \Theta} \right] \tag{9.164}$$

and

$$E_{0\phi}^i(r_0, \Theta, \pi/2) = \frac{-\omega}{k_0} B_{0\Theta}^i(r_0, \Theta, \pi/2)$$

$$= \frac{i \omega \mu_0}{2\pi} (i k_0 l) \frac{e^{i k_0 r_0}}{r_0} \cos \Theta \, P(\Theta, \pi/2). \tag{9.165}$$

10 Application of the two-term theory to general arrays of parallel non-staggered elements

The purpose of this chapter is threefold:

1. To summarize those parts of the two-term theory of Chapters 2–5 which concern arrays with large inter-element spacing. The notation used in this chapter is new and suitable for a general-purpose computer program.
2. To remove the restriction for $N \geq 3$ that the N elements need to be placed on a circle (Chapter 4) or equispaced along a straight line (Chapter 5). The elements must still be parallel, non-staggered, and identical.
3. To prepare for the analysis of the large circular array developed in the next two chapters, for which the new notation of this chapter is particularly well-suited.

Section 10.1 contains a concise derivation of the formulas. Section 10.2 contains the complete formulas in their final form. In Section 10.3, special cases, extensions and computational aspects are discussed. Finally, Section 10.4 derives a special form for the case $kh = \pi/2$.

$[D]$ denotes the $N \times N$ matrix with components D_{nl} $(1 \leq n, l \leq N)$ and $\{t\}$ denotes the vector $(N \times 1$ column matrix) with components $t_1, t_2, \ldots, t_N$. The linear algebra terms used here may be found in any standard linear algebra textbook.

10.1 Brief derivation of the formulas

The two-term formulas for the N coupled integral equations

$$\sum_{n=1}^{N} \int_{-h}^{h} I_n(z') K_{nl}(z - z') \, dz'$$

$$= \frac{-j4\pi}{\zeta_0} \left(C_l \cos kz + \frac{V_l}{2} \sin k|z| \right); \qquad -h < z < h, \qquad l = 1, \ldots, N$$

(10.1)

where the constants C_l are determined from the conditions

$$I_1(\pm h) = I_2(\pm h) = \cdots = I_N(\pm h) = 0$$

(10.2)

are briefly derived here for a general array in free space. Here, "general array" means an array of parallel, identical, non-staggered, center-driven, perfectly conducting cylindrical dipoles satisfying the conditions

$$ka \ll 1, \qquad a \ll h, \qquad \min b_{nl} \gtrsim h, \qquad kh < 3\pi/2. \qquad (10.3)$$

In (10.1)–(10.3), $I_1(z), I_2(z), \ldots, I_N(z)$ are the unknown current distributions on the surface of the N dipoles, h is the half-length of each dipole, and a is its radius. V_l is the voltage driving element l, with $V_l = 0$ if the element is non-driven. The dipoles' centers lie on the plane $z = 0$; $k = 2\pi/\lambda = \omega\sqrt{\mu_0\epsilon_0}$ is the free-space wave number and $\zeta_0 = \sqrt{\mu_0/\epsilon_0} \doteq 376.73$ ohms. The $K_{nl}(z)$ are the kernels of the integral equations. The self-interaction kernels $K_{11}(z), K_{22}(z), \ldots, K_{NN}(z)$ are all equal. They depend on the radius ka of the elements and are given by

$$\frac{K_{ll}(z)}{k} = \frac{K_{11}(z)}{k} = \frac{\cos k\sqrt{z^2 + a^2}}{k\sqrt{z^2 + a^2}} - j\frac{\sin k\sqrt{z^2 + a^2}}{k\sqrt{z^2 + a^2}}, \qquad l = 1, \ldots, N. \quad (10.4)$$

The mutual interaction kernels $K_{nl}(z)$, $n \neq l$, depend on the distance kb_{nl} between the axes of element l and element n. They are given by

$$\frac{K_{nl}(z)}{k} = \frac{K_{ln}(z)}{k} = \frac{\cos k\sqrt{z^2 + b_{nl}^2}}{k\sqrt{z^2 + b_{nl}^2}} - j\frac{\sin k\sqrt{z^2 + b_{nl}^2}}{k\sqrt{z^2 + b_{nl}^2}}, \qquad n \neq l. \quad (10.5)$$

These kernels must be modified for the special case of a large resonant circular array. This modification is discussed in Section 11.8. As discussed in Section 3.2, the third condition in (10.3) can, for many purposes, be replaced by $\min \beta_0 b_{nl} \geq 1$.

First, the integral equations are rearranged in a form suitable for applying the two-term theory approximations: Formula (10.1) for $z = h$ is

$$\frac{j4\pi}{\zeta_0} U_l \equiv \sum_{n=1}^{N} \int_{-h}^{h} I_n(z') K_{nl}(h - z') \, dz' = \frac{-j4\pi}{\zeta_0}\left(C_l \cos kh + \frac{V_l}{2}\sin kh\right). \quad (10.6)$$

If this is subtracted from (10.1), the following equivalent system of integral equations is obtained:

$$\sum_{n=1}^{N} \int_{-h}^{h} I_n(z') K_{dnl}(z, z') \, dz' = \frac{-j4\pi}{\zeta_0}\left(C_l \cos kz + \frac{V_l}{2}\sin k|z| + U_l\right), \quad (10.7)$$

where

$$K_{dnl}(z, z') = K_{nl}(z - z') - K_{nl}(h - z') \quad (10.8)$$

are the difference kernels. In this and the next section, it is assumed that $kh \neq \pi/2$; this restriction is discussed in Section 10.3 and removed in Section 10.4. Formula (10.6) may then be used to write the constants C_l in terms of the constants U_l:

$$C_l = -\frac{\frac{1}{2}V_l \sin kh + U_l}{\cos kh} \tag{10.9}$$

so that the final exact form of the integral equations is found by substituting (10.9) into (10.7) as

$$\sum_{n=1}^{N} \int_{-h}^{h} I_n(z') K_{dnl}(z, z') \, dz'$$

$$= \frac{j4\pi}{\zeta_0 \cos kh} \left[\frac{V_l}{2} \sin k(h - |z|) + U_l(\cos kz - \cos kh) \right], \qquad l = 1, \ldots, N \tag{10.10}$$

which are to be solved together with the conditions

$$\frac{j4\pi}{\zeta_0} U_l = \sum_{n=1}^{N} \int_{-h}^{h} I_n(z') K_{nl}(h - z') \, dz', \qquad l = 1, \ldots, N. \tag{10.11}$$

Both sides of the integral equation (10.10) are proportional to the vector potential difference $A_{zl}(z) - A_{zl}(h)$ and vanish at $z = \pm h$.

Subject to the conditions (10.3), the approximate two-term theory current distributions are of the form

$$I_n(z) = s[V_n \sin k(h - |z|) + t_n(\cos kz - \cos kh)], \qquad n = 1, \ldots, N, \tag{10.12}$$

where the coefficients s and t_n are to be determined.

Thus, one part of the current on each element is the sine term $\sin k(h - |z|)$. The discontinuity in the derivative of this term at the driving point is due to the existence of the idealized delta-function generator there. In the limit $ka \rightarrow 0$, the sine current distribution is the exact current distribution on an isolated element. The coefficient s will turn out to be purely imaginary and to depend on kh and ka only. Thus, the $\sin k(h - |z|)$ part of the current is purely reactive. It is said to be maintained directly by the generator driving element n since its coefficient $s V_n$ is directly proportional to the driving voltage V_n and is independent of the number, location, and driving conditions of the rest of the elements in the array.

The second term in the current (10.12) is the shifted cosine $(\cos kz - \cos kh)$. Its coefficient t_n will turn out to depend linearly on all the driving voltages in the array, namely,

$$t_n = T_{n1} V_1 + T_{n2} V_2 + \cdots + T_{nN} V_N, \qquad n = 1, \ldots, N \tag{10.13}$$

where the T_{nl} are independent of the driving voltages. If element n is non-driven ($V_n = 0$), the distribution of current along its length consists only of the shifted cosine $(\cos kz - \cos kh)$. This is exactly the current distribution on an isolated receiving element in the limit $ka \to 0$.

In terms of the unknowns s and t_n, the constant U_l is found by substituting (10.12) into (10.11),

$$U_l = \frac{\zeta_0 s}{j4\pi} \left[\sum_{n=1}^{N} V_n \Psi_{Vnl} + \sum_{n=1}^{N} t_n \Psi_{Unl} \right], \qquad l = 1, \ldots, N, \tag{10.14}$$

where

$$\Psi_{Vnl} = \int_{-h}^{h} \sin k(h - |z'|) K_{nl}(h - z') \, dz'; \qquad 1 \le n, l \le N \tag{10.15}$$

and

$$\Psi_{Unl} = \int_{-h}^{h} (\cos kz' - \cos kh) K_{nl}(h - z') \, dz'; \qquad 1 \le n, l \le N \tag{10.16}$$

are known constants.

In order to write the left-hand side (LHS) of (10.10) into the form of the right-hand side (RHS), and to reduce the problem of finding the coefficients s and t_n to one of solving a system of linear algebraic equations, the two-term theory approximations are made. These are

$$\int_{-h}^{h} \sin k(h - |z'|) \, \mathrm{Re}\{K_{dll}(z, z')\} \, dz' \doteq \Psi_{dR} \sin k(h - |z|), \qquad l = 1, \ldots, N \tag{10.17}$$

$$\int_{-h}^{h} \sin k(h - |z'|) \, \mathrm{Im}\{K_{dll}(z, z')\} \, dz' \doteq \Psi_{dI}(\cos kz - \cos kh), \qquad l = 1, \ldots, N \tag{10.18}$$

$$\int_{-h}^{h} \sin k(h - |z'|) K_{dnl}(z, z') \, dz' \doteq \Psi_{dVnl}(\cos kz - \cos kh), \qquad n \ne l \tag{10.19}$$

$$\int_{-h}^{h} (\cos kz' - \cos kh) K_{dnl}(z, z') \, dz' \doteq \Psi_{dUnl}(\cos kz - \cos kh), \qquad 1 \le n, l \le N. \tag{10.20}$$

In these approximate formulas, the functions of z on the LHS as well as those on the RHS are even and vanish at $z = \pm h$. In the case $kh < 3\pi/2$, the maxima of the functions of z on the RHS occur at $z = z_{max} = 0$. The exception is when $\pi/2 < kh < 3\pi/2$. In this case, the maxima of the functions on the RHS of (10.17) occur at $z = \pm z_{max} = \pm(h - \lambda/4)$. If the coefficients of proportionality are determined by enforcing the functions on the RHS to coincide with those on the LHS at $\pm z_{max}$, a good approximation results. Thus, the coefficients are given by

$$\Psi_{dR} = \begin{cases} \dfrac{1}{\sin kh} \displaystyle\int_{-h}^{h} \sin k(h - |z'|) \, \mathrm{Re}\{K_{dll}(0, z')\} \, dz', & kh < \pi/2 \quad (10.21) \\[2em] \displaystyle\int_{-h}^{h} \sin k(h - |z'|) \, \mathrm{Re}\{K_{dll}(h - \lambda/4, z')\} \, dz', & \pi/2 < kh < 3\pi/2 \\[0.5em] & \hspace{4em} (10.22) \end{cases}$$

$$\Psi_{dI} = \frac{1}{1 - \cos kh} \int_{-h}^{h} \sin k(h - |z'|) \, \mathrm{Im}\{K_{dll}(0, z')\} \, dz' \qquad (10.23)$$

$$\Psi_{dVnl} = \frac{1}{1 - \cos kh} \int_{-h}^{h} \sin k(h - |z'|) K_{dnl}(0, z') \, dz', \qquad n \neq l \qquad (10.24)$$

$$\Psi_{dUnl} = \frac{1}{1 - \cos kh} \int_{-h}^{h} (\cos kz' - \cos kh) K_{dnl}(0, z') \, dz'. \qquad (10.25)$$

Note from (10.21) and (10.22) that as a function of kh, Ψ_{dR} is continuous at $kh = \pi/2$.

If (10.12) with (10.17)–(10.20) and (10.14) are substituted into the integral equations (10.10) and the coefficients of $\sin k(h - |z|)$ and $(\cos kz - \cos kh)$ are equated, the following equations result. By equating the coefficients of $\sin k(h - |z|)$, it is seen that

$$s = \frac{j2\pi}{\zeta_0 \Psi_{dR} \cos kh} \qquad (10.26)$$

which gives s in terms of the known coefficient Ψ_{dR}. Equating the coefficients of the shifted cosine term, we obtain

$$\sum_{n=1}^{N} D_{ln} t_n = \sum_{n=1}^{N} P_{ln} V_n, \qquad 1 \leq l \leq N, \qquad (10.27)$$

where

$$D_{ln} = D_{nl} = \cos kh \, \Psi_{dUnl} - \Psi_{Unl}, \qquad l = 1, \ldots, N, \qquad n = 1, \ldots, N \qquad (10.28)$$

$$P_{ln} = P_{nl} = -\cos kh \, \Psi_{dVnl} + \Psi_{Vnl}, \qquad 1 \leq n \neq l \leq N \qquad (10.29)$$

$$\mathrm{Re}\{P_{ll}\} = \mathrm{Re}\{P_{11}\} = \mathrm{Re}\{-j\Psi_{dI}\cos kh + \Psi_{V11}\}, \qquad l = 1, \ldots, N \qquad (10.30)$$

$$\mathrm{Im}\{P_{ll}\} = \mathrm{Im}\{P_{11}\} = \mathrm{Im}\{-j\Psi_{dI}\cos kh + \Psi_{V11}\}, \qquad l = 1, \ldots, N. \qquad (10.31)$$

Equation (10.27) is a system of linear equations. The unknowns are the coefficients t_n. It is convenient to substitute (10.15), (10.16), and (10.21)–(10.25) into (10.28)–(10.31) and express the coefficients D_{nl} and P_{nl} as integrals involving the kernels $K_{nl}(z)$. The final complete formulas obtained in this manner are listed in the next section.

10.2   The complete two-term theory formulas

Assuming that $kh \neq \pi/2$ as in the previous section, the currents are

$$I_l(z) = \frac{j2\pi}{\zeta_0 \Psi_{dR} \cos kh} [V_l \sin k(h - |z|) + t_l(\cos kz - \cos kh)], \qquad l = 1, \dots, N.$$
(10.32)

Equation (10.32) can also be written in matrix form as

$$\{I(z)\} = \frac{j2\pi}{\zeta_0 \Psi_{dR} \cos kh} [\{V\} \sin k(h - |z|) + \{t\}(\cos kz - \cos kh)].$$
(10.33)

In (10.32) and (10.33),

$$\Psi_{dR} = \begin{cases} \dfrac{1}{\sin kh} \displaystyle\int_{-h}^{h} \sin k(h - |z'|)[\mathrm{Re}\{K_{11}(z')\} - \mathrm{Re}\{K_{11}(h - z')\}] \, dz', \\ \qquad kh \leq \pi/2 & (10.34) \\[1.5em] \displaystyle\int_{-h}^{h} \sin k(h - |z'|)[\mathrm{Re}\{K_{11}(h - \lambda/4 - z') - \mathrm{Re}\{K_{11}(h - z')\}] \, dz', \\ \qquad \pi/2 \leq kh < 3\pi/2 & (10.35) \end{cases}$$

and the coefficients t_l are determined by solving the $N \times N$ system of linear algebraic equations

$$\sum_{l=1}^{N} D_{nl} t_l = \sum_{l=1}^{N} P_{nl} V_l, \qquad 1 \leq n \leq N.$$
(10.36)

The system (10.36) can be written in matrix form as

$$[D]\{t\} = [P]\{V\}.$$
(10.37)

The matrix components are

$$D_{nl} = \frac{1}{1 - \cos kh} \int_{-h}^{h} (\cos kz - \cos kh)$$
$$\times [\cos kh \, K_{nl}(z) - K_{nl}(h - z)] \, dz, \qquad 1 \leq n, l \leq N.$$
(10.38)

Excluding the principal diagonal, the components of the matrix on the RHS are given by

$$P_{nl} = \frac{-1}{1 - \cos kh} \int_{-h}^{h} \sin k(h - |z|)$$
$$\times [\cos kh \, K_{nl}(z) - K_{nl}(h - z)] \, dz, \qquad n \neq l.$$
(10.39)

The components of the principal diagonal of the matrix on the RHS are

$$P_{ll} = \int_{-h}^{h} \sin k(h - |z|) \, \text{Re}\{K_{11}(h - z)\} \, dz$$

$$- j\frac{1}{1 - \cos kh} \int_{-h}^{h} \sin k(h - |z|)[\cos kh \, \text{Im}\{K_{11}(z)\} - \text{Im}\{K_{11}(h - z)\}] \, dz.$$

$$(10.40)$$

These formulas readily follow from the equations in the previous section. The kernels $K_{nl}(z) = K_{ln}(z)$, $1 \le n, l \le N$, appearing in (10.34), (10.35), and (10.38)–(10.40) are given in (10.4) and (10.5). The relation (10.13) between $[T]$ and $\{t\}$ may be written in matrix notation as

$$\{t\} = [T]\{V\},\tag{10.41}$$

where

$$[T] = [D]^{-1}[P].\tag{10.42}$$

10.3   Remarks and programming considerations

The formulas of the previous section may be easily programmed in the form given above. The inputs of the program are the parameters N, kh, ka, kb_{nl} ($1 \le n, l \le N$), and the driving voltages $V_1, V_2, \ldots, V_N$. Alternatively, the parameters N, kh, ka, kb_{nl}, and the driving-point currents $I_1(0), I_2(0), \ldots, I_N(0)$ may be inputs to the program. Some observations and numerical considerations are given below.

1. The matrix elements D_{nl} and P_{nl} in (10.38)–(10.40) are complex. They depend on N, kh, ka, and kb_{nl}, and they are independent of the driving voltages. Their real (imaginary) part depends only on the real (imaginary) part of the corresponding kernel $K_{nl}(z)$. Thus, for $n \ne l$, each matrix element is completely determined by the distance kb_{nl} and by kh. When $n = l$, the matrix element depends on the self-interaction kernel (10.4) and is completely determined by the radius ka and by kh.

2. It follows that the matrices $[D]$ and $[P]$ are complex and symmetric and that all elements on their principal diagonals are equal.

3. The real and imaginary parts of the matrix elements are easily obtained by numerical integration of the real and imaginary parts of equations (10.38)–(10.40). The change of variables $kz = x$ is convenient. In order to determine all elements D_{nl} and P_{nl} for a general array, a total of $2(N^2 - N + 2)$ different real integrals must be calculated numerically. If the program uses general-purpose adaptive integrators (subroutines), one can specify the desired accuracy for the numerical integrations.

Generally, the elements (integrals) on or close to the principal diagonal are those that must be computed with the highest relative accuracy. The elements far from the diagonal are smaller in absolute value; slight errors in these elements will not appear in the final results.

4. In the case where the base voltages $V_1, V_2, \ldots, V_N$ are specified, the program should first compute the coefficients Ψ_{dR} and t_l. Ψ_{dR} is found from (10.34) or (10.35) by numerical integration. Once the matrix elements D_{nl} and P_{nl} have been found as described above, the program may compute the complex vector $[P]\{V\}$ on the RHS of (10.37) and use a standard routine to solve the complex $N \times N$ linear system (10.37) for the coefficients t_l.

Quantities which may be easily determined once Ψ_{dR} and t_l are found are the following:

(a) The current distributions $I_1(z), I_2(z), \ldots, I_N(z)$ from equation (10.32).
(b) Equation (10.32) and the definition

$$Y_{l,\text{in}} = 1/Z_{l,\text{in}} = I_l(0)/V_l, \qquad l = 1, \ldots, N \tag{10.43}$$

may be used to compute the driving-point admittances[1] $Y_{1,\text{in}}, Y_{2,\text{in}}, \ldots, Y_{N,\text{in}}$ of the elements. $Y_{l,\text{in}}$ is completely determined by the voltage ratios V_1/V_l, $V_2/V_l, \ldots, V_N/V_l$, and by N, kh, ka, and kb_{nl}.

(c) Consider the case where only one element is driven and the rest are (present but) non-driven. Suppose that element n is driven by a voltage V_n ($V_p = 0$ for $p \neq n$). The single driving-point admittance in this case is often called the self-admittance Y_{nn} and the normalized midpoint currents $I_l(0)/V_n, n \neq l$, are often called the mutual admittances Y_{ln}. The matrix $[Y]$ with elements Y_{ln} is completely determined by N, kh, ka, and kb_{nl}. It is easily seen to be related to the matrix $[T]$ of (10.42) by

$$[Y] = \frac{j2\pi}{\zeta_0 \Psi_{dR} \cos kh} \left\{ \sin kh [I^{N \times N}] + (1 - \cos kh)[T] \right\}, \tag{10.44}$$

where $[I^{N \times N}]$ is the $N \times N$ identity matrix. The driving-point admittances in the case when all elements are driven by voltages $V_1, V_2, \ldots, V_N$ are related to Y_{ln} by

$$Y_{l,\text{in}} = \sum_{n=1}^{N} Y_{ln} \frac{V_n}{V_l}, \qquad l = 1, \ldots, N. \tag{10.45}$$

The above equations are useful when one wishes to compute the driving-point admittances for many different sets of driving voltages $\{V\}$. Standard techniques for solving systems with multiple RHS vectors are useful in order to

---

[1] Note that what is referred to in this book as the driving-point admittance (impedance) is often referred to in the literature as the active admittance (impedance).

determine $[T]$ from (10.42). Note that for general arrays with $N \geq 3$, $[T]$ and $[Y]$ are not necessarily symmetric and that the elements on their principal diagonals are not necessarily equal.

(d) Suppose that the center of the axis of dipole l is located at (ρ_l, ϕ_l) in polar coordinates and that the spherical coordinates of a far-field observation point are (r, θ, ϕ). The radiation field is found from Ψ_{dR} and t_l by the equation

$$\mathbf{E}(r, \theta, \phi) = \hat{\boldsymbol{\theta}} E_\theta$$

$$= \hat{\boldsymbol{\theta}} \frac{-1}{\Psi_{dR} \cos kh} \frac{e^{-jkr}}{r} \sum_{l=1}^{N} [V_l F(\theta) + t_l G(\theta)] e^{jk\rho_l \sin\theta \cos(\phi - \phi_l)},$$

(10.46)

where

$$F(\theta) = \frac{\cos(kh \cos\theta) - \cos kh}{\sin\theta}$$

(10.47)

$$G(\theta) = \frac{\sin kh \cos(kh \cos\theta) \cos\theta - \cos kh \sin(kh \cos\theta)}{\sin\theta \cos\theta}$$

(10.48)

are the "element factors" for the sine and shifted-cosine currents, respectively. With the radiation field determined, the gain and directivity of the array may be easily found.

5. The case where the driving-point currents $I_1(0), I_2(0), \ldots, I_N(0)$ are given and the base voltages $V_1, V_2, \ldots, V_N$ are desired may also be treated by a program in a similar manner. The relation between $\{V\}$ and $\{I(0)\}$ is found from (10.33), (10.41), and (10.42) to be

$$[Q]\{V\} = \frac{\zeta_0 \Psi_{dR} \cos kh}{j2\pi} [D]\{I(0)\},$$

(10.49)

where

$$[Q] = \sin kh [D] + (1 - \cos kh)[P].$$

(10.50)

In this case, the program should first compute Ψ_{dR}, $[D]$, and $[P]$ as before, then compute $[Q]$ from (10.50), find the vector on the RHS of (10.49), and then solve the system (10.49) of linear algebraic equations for the unknown vector $\{V\}$. The matrix $[Q]$ has the same form as $[P]$ and $[D]$ (see the second observation of this section).

6. For the special case of a curtain array (Chapter 5), the distances b_{nl} are given by $b_{nl} = |l - n|b$, where b is the distance between adjacent elements. Here, the matrices $[D]$, $[P]$ and $[Q]$, in addition to being symmetric, are Toeplitz (i.e. the matrix elements on all diagonals are equal). The full matrices are determined by the

elements of the first row. In order to determine all elements D_{nl}, P_{nl}, and Q_{nl} for a curtain array, only $4N$ different real integrals need to be calculated numerically. If a routine especially designed for complex Toeplitz systems is used to solve (10.37) or (10.49), further reduction in computer time and storage is accomplished. For $N \geq 3$, the matrices $[T]$ and $[Y]$ are in general neither symmetric nor Toeplitz.

7. For the special case of a circular array, the matrices $[D]$, $[P]$, and $[Q]$ are (symmetric and) circulant so that the solution of the system (10.37) or (10.49) may be written as a superposition over the N phase sequences. A brief, general discussion of symmetric circulant matrices is given in Section 4.5.

All elements in a circular array have the same self-admittance and the mutual admittances satisfy $Y_{nl} = Y_{ln}$. Also, for a given circular array, $Y_{ln} = Y_{pq}$ if $b_{ln} = b_{pq}$. When the array is driven in its mth phase sequence ($V_l = V_1 e^{j2\pi(l-1)m/N}$), the driving-point admittance $Y_l^{(m)}$ of any element in the array does not depend on the element number l; it is denoted by $Y^{(m)}$ in Chapters 4, 11, and 12, and referred to as the mth phase-sequence admittance. For the large circular arrays considered in Chapters 11 and 12, the formulas must be modified as discussed in Sections 11.8 and 12.8. Furthermore, special numerical considerations must be taken into account. These are discussed in Section 13.5.

8. Arrays of vertical monopoles over a ground plane and arrays of horizontal, non-staggered dipoles over a ground plane are discussed here. The ground plane is assumed to be of infinite extent and perfectly conducting in both cases.

(a) The case of N vertical monopoles of length h over a ground plane (located at the plane $z = 0$) reduces by virtue of the theorem of images to that of a general array of N dipoles of length $2h$ in free space: If driving voltages are given and currents are desired, one first solves the latter problem and multiplies the resulting current distributions by a factor of 2. For $z > 0$, these are the desired current distributions on the monopoles.

(b) The theorem of images also applies to the case of N horizontal, non-staggered dipoles over a ground plane. Suppose that the ground plane is located at the plane $x = 0$, and that the N dipoles are in the half-space $x > 0$. Their centers lie on the plane $z = 0$. The b_{nl} ($1 \leq n, l \leq N$) are the axis-to-axis distances. The problem is equivalent to that of $2N$ parallel, non-staggered dipoles in free space. The original dipoles are numbered as $1, 2, \ldots, N$, and the images are numbered as $N+1, N+2, \ldots, 2N$, so that the image of dipole l is the dipole $l+N$ ($l = 1, 2, \ldots, N$). The voltage driving dipole l is the opposite of that driving dipole $l+N$, namely,

$$V_{l+N}^{2N} = -V_l^{2N}, \qquad l = 1, 2, \ldots, N, \tag{10.51}$$

where the superscript $2N$ has been affixed to show that the equivalent array has $2N$ elements.

Denote by $c_{nl} = c_{ln}$ the axis-to-axis distance between element l and the image of element n ($1 \le n, l \le N$). If the original dipoles are sufficiently far from the ground plane, specifically,

$$\min c_{ll} \gtrsim h \tag{10.52}$$

[and if the conditions (10.3) are also satisfied], then the problem may be treated as described above, where there are $2N$ dipoles instead of N. It is possible, however, to reduce the problem to that of solving N linear equations instead of $2N$ as follows.

The matrices $[D]$ and $[P]$ appearing in the $2N \times 2N$ system (10.37) for the $2N$ coefficients t_l^{2N} separate into four $N \times N$ submatrices, namely,

$$[D] = \begin{bmatrix} [D_b] & [D_c] \\ [D_c] & [D_b] \end{bmatrix} \tag{10.53}$$

$$[P] = \begin{bmatrix} [P_b] & [P_c] \\ [P_c] & [P_b] \end{bmatrix}, \tag{10.54}$$

where $[D_b]$ and $[P_b]$ depend only on the distances b_{nl} and are the same as if the N original dipoles (the ones above the plane $x > 0$) were alone in free space. The components of the matrices $[D_c]$ and $[P_c]$ are given by

$$D_{cnl} = \frac{1}{1 - \cos kh} \int_{-h}^{h} (\cos kz - \cos kh)[\cos kh \, K_{cnl}(z) - K_{cnl}(h - z)] \, dz,$$
$$1 \le n, l \le N \tag{10.55}$$

$$P_{cnl} = \frac{-1}{1 - \cos kh} \int_{-h}^{h} \sin k(h - |z|)[\cos kh \, K_{cnl}(z) - K_{cnl}(h - z)] \, dz,$$
$$1 \le n, l \le N, \tag{10.56}$$

where

$$\frac{K_{cnl}(z)}{k} = \frac{K_{cln}(z)}{k} = \frac{\cos k\sqrt{z^2 + c_{nl}^2}}{k\sqrt{z^2 + c_{nl}^2}} - j \frac{\sin k\sqrt{z^2 + c_{nl}^2}}{k\sqrt{z^2 + c_{nl}^2}},$$
$$1 \le n, l \le N \tag{10.57}$$

and the K_{cnl} are similar in form to the kernels K_{nl} but involve the distances c_{nl}. Thus, the matrices $[D_c]$ and $[P_c]$ involve only the distances c_{nl} from dipole to image.

From (10.37), (10.53), and (10.54), it is easily seen that

$$t_{l+N}^{2N} = -t_l^{2N}, \qquad l = 1, 2, \dots, N \tag{10.58}$$

and that

$$\sum_{l=1}^{N} (D_{bnl} - D_{cnl}) t_l^{2N} = \sum_{l=1}^{N} (P_{bnl} - P_{cnl}) V_l^{2N}, \qquad 1 \le n \le N. \qquad (10.59)$$

Formula (10.59) is the desired $N \times N$ system for the N coefficients t_1^{2N}, $t_2^{2N}, \ldots, t_N^{2N}$. With these determined, the currents are found from

$$I_l(z) = \frac{j2\pi}{\Psi_{dR} \zeta_0 \cos kh} [V_l^{2N} \sin k(h - |z|) + t_l^{2N} (\cos kz - \cos kh)],$$

$$l = 1, \ldots, N. \qquad (10.60)$$

The components of the $N \times N$ composite matrices $[D_b] - [D_c]$ and $[P_b] - [P_c]$ are completely determined by the corresponding distances kb_{nl} and kc_{nl} (or, by ka and kc_{ll} in the case of diagonal elements). The matrices are complex and symmetric. Furthermore, they are Toeplitz in the special case of a curtain array above a ground plane, where $b_{nl} = |l - n|b$ and $c_{nl} = \sqrt{c^2 + (l - n)^2 b^2}$; ($c/2$ is the distance from the dipoles' axes to the ground plane).

9. Because of the presence of $\cos kh$ in the denominator of (10.32), the two-term solution apparently is infinite when $kh = \pi/2$. It is shown in the next section that this is not the case. Although each of the terms in (10.32) become infinite when $kh = \pi/2$, their sum remains finite. By rearranging the terms in the two-term theory, an equivalent form will be obtained, none of the terms of which vanish at $kh = \pi/2$. This equivalent form may be used in the case where $kh = \pi/2$. In most applications, however, it is adequate to replace the value $kh = \pi/2$ by another value close to $\pi/2$ and use the formulas listed in Section 10.2.

10.4 Alternative form for the solution and the case $kh = \pi/2$

For the purposes of this section, we will temporarily affix the argument kh to the vector $\{t\}$ and the matrices $[D]$ and $[P]$.

If the vector

$$\{t'(kh)\} = \frac{-1}{\cos kh} (\sin kh \{V\} + \{t(kh)\}) \qquad (10.61)$$

is defined, then in terms of this vector, equation (10.33) takes the equivalent form

$$\{I(z)\} = \frac{j2\pi}{\zeta_0 \Psi_{dR}} [\{V\}(\sin kh - \sin k|z|) - \{t'(kh)\}(\cos kz - \cos kh)]. \qquad (10.62)$$

When $kh = \pi/2$ (but not at any other value of kh), $\{t'(\pi/2)\}$ is simply the derivative of $\{t(kh)\}$ evaluated at $kh = \pi/2$:

$$\{t'(\pi/2)\} = \frac{d}{d(kh)} \{t(\pi/2)\}. \qquad (10.63)$$

This is seen by noting from (10.38)–(10.40) that when $kh = \pi/2$, the relation $[D(\pi/2)] = -[P(\pi/2)]$ holds so that the solution $\{t(\pi/2)\}$ to the system (10.37) when $kh = \pi/2$ may be found by inspection. It is

$$\{t(\pi/2)\} = -\{V\}. \tag{10.64}$$

The definition (10.61) for $\{t'(kh)\}$ is thus indeterminant when $kh = \pi/2$ and its limiting form is (10.63).

For any value of kh, the system for $\{t'(kh)\}$ is readily derived by solving (10.61) for $\{t(kh)\}$ and substituting into the system (10.37). It is seen that

$$[D(kh)]\{t'(kh)\} = [P'(kh)]\{V\}, \tag{10.65}$$

where

$$[P'(kh)] = \frac{-1}{\cos kh}\left(\sin kh\,[D(kh)] + [P(kh)]\right). \tag{10.66}$$

Note that $[P'(kh)]$ is not the derivative of $[P(kh)]$ when $kh = \pi/2$. Using (10.66) and (10.38)–(10.40), it is found that the explicit expressions for the components P'_{nl} of the matrix $[P'] \equiv [P'(kh)]$ are

$$\mathrm{Re}\{P'_{ll}\} = \frac{-\sin kh}{1 - \cos kh} \int_{-h}^{h} (\cos kz - \cos kh)[\mathrm{Re}\{K_{11}(z)\} - \mathrm{Re}\{K_{11}(h-z)\}]\,dz$$

$$-\int_{-h}^{h} (\sin kh - \sin k|z|)\,\mathrm{Re}\{K_{11}(h-z)\}\,dz \tag{10.67}$$

$$\mathrm{Im}\{P'_{ll}\} = \frac{1}{1 - \cos kh} \int_{-h}^{h} (\sin kh - \sin k|z|)$$

$$\times\,[\cos kh\,\mathrm{Im}\{K_{11}(z)\} - \mathrm{Im}\{K_{11}(h-z)\}]\,dz \tag{10.68}$$

and

$$P'_{nl} = \frac{1}{1 - \cos kh} \int_{-h}^{h} (\sin kh - \sin k|z|)$$

$$\times\,[\cos kh\,K_{nl}(z) - K_{nl}(h-z)]\,dz, \qquad 1 \le n \neq l \le N. \tag{10.69}$$

It is seen that there are no apparent infinities when $kh = \pi/2$. The mathematically equivalent form (10.62) and (10.65) with (10.67)–(10.69) of the two-term theory is useful numerically when $kh = \pi/2$.

11 Resonances in large circular arrays of perfectly conducting dipoles

11.1 Introduction

In this and the following chapter, a study of large circular dipole arrays with one or two elements driven is presented. The top view of such an array is shown in Fig. 11.1a, and a realization of a large circular array with one element driven [1] by monopoles over a ground plane is shown in Fig. 11.1b.

The main reason for initiating the study was the belief that such arrays should possess very narrow resonances if the many parameters of the problem are properly chosen. It was further believed that some particular shapes of *non-circular* closed-loop arrays might produce a superdirective field pattern. The large circular array of this chapter is the simplest form of the more general closed-loop array. The latter is a subject of ongoing research.

It was seen in Chapter 4 that circular arrays of a small number of elements possess noticeable resonances; previous studies also supported the idea of the existence of very narrow resonances in large circular arrays. For instance, it was known that the long Yagi array may be thought of as a surface-wave structure [2]. Such a structure does not radiate broadside, and it was observed experimentally that this property is preserved if the array is bent into a semi-circle of sufficiently large radius [3].

Essential initial considerations concerning resonances in large circular arrays came from studies in quantum mechanics. It was found by T. T. Wu and A. Grossmann that an infinite linear array of Fermi pseudopotentials possesses resonances of zero width [4, 5] and that a large circular array of pseudopotentials possesses resonances the width of which is exponentially small in the number N of pseudopotentials in the array [5]. Roughly speaking, a Fermi pseudopotential is a point interaction in the context of the Schrödinger equation, characterized by a single parameter with the dimension of length. These considerations are contained in [6]; this paper also mentions the important possible connection between resonance and superdirectivity in a large closed-loop array.

The belief in the existence of narrow resonances in a large circular array of dipoles with only one element driven led to numerical calculations using the two-term theory in its original form (Chapter 4). Narrow resonances associated with flower-shaped far-field patterns were discovered and illustrated graphically [7] in $N = 60$ and

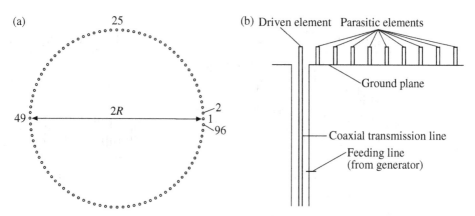

Figure 11.1 (a) Circular array with $N = 96$ elements. (b) Circular array of monopoles over ground plane. Driven element is extension of inner conductor of coaxial transmission line. Part (b) taken from Fikioris *et al.* [1, Fig. 2]. © 1994 EMW Publishing.

$N = 90$ element circular arrays when the dipoles are electrically quite thick and short. If λ is the free-space wavelength, the values of the individual element parameters are $h/\lambda = 0.18$ and $a/\lambda = 0.028$. The quantity d/λ is the varying parameter where d is the spacing between adjacent elements. Resonances were discovered for this rather unusual combination of h/λ and a/λ after unsuccessful attempts were made to discover resonances in arrays with values of the individual element parameters commonly used in other applications.

In [7], each resonance is associated with a particular phase sequence m ($m = N/2$, $N/2 - 1, \ldots$), where resonances with larger m occur at higher frequencies. The complete graphical picture of the basic properties of a $N = 90$ element circular array operating at its last ($m = 45$) resonance is shown in the several parts of Fig. 11.2. These include the conductance $G_{1,l}$, the susceptance $B_{1,l}$, and the magnitude and phase angle of the admittance $Y_{1,l} = G_{1,l} + jB_{1,l}$. Of special interest is the normalized power pattern consisting of 90 sharp peaks separated by 90 sharp nulls.

After the publication of [7], it was found that the two-term theory in its original form gave meaningless results in other cases. This consideration as well as others led D. K. Freeman and T. T. Wu to re-examine the kernels used in the integral equations and to propose a new set of kernels [8, 9]. However, use of these kernels in the two-term theory formulas presents difficulties. The kernels are not of a simple form so that numerical calculations would require a large amount of computer time, and more importantly, such kernels would make an analytical study of the two-term theory formulas difficult. A simpler alternative was found: the "modified" kernels are as simple as the original ones and possess many of the properties of the kernels proposed in [8] and [9]. The modified kernels result very simply from the original ones by setting $a = 0$ in the imaginary part of the self term. They are incorporated into the two-term theory employed in this chapter. Whereas this modification is unimportant in

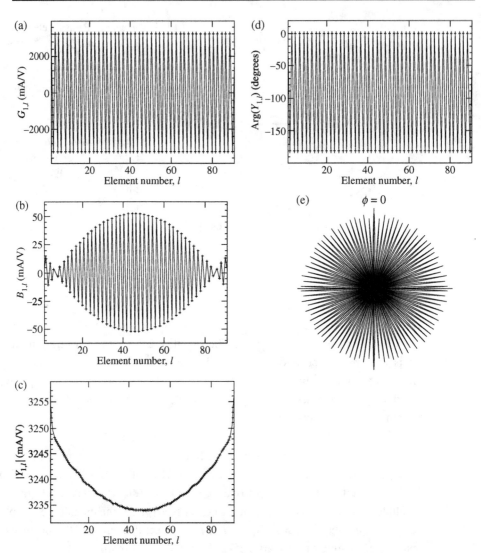

Figure 11.2 Properties of a 90-element circular array, as calculated by the (unmodified) two-term theory of Chapter 4; $N = 90$, $m = 45$, $h/\lambda = 0.18$, $a/\lambda = 0.028$, and $d/\lambda = 0.437\,432\,095$. (a) Self- and mutual conductances $G_{1,l}$ as a function of element number l. (b) Self- and mutual susceptances $B_{1,l}$ as a function of element number l. (c) Magnitudes of self- and mutual admittances $|Y_{1,l}|$ as a function of element number l. (d) Phases of self- and mutual admittances $\arg(Y_{1,l})$ as a function of element number l. (e) Normalized far-field power pattern $|E(\phi)|^2$. Taken from Fikioris *et al.* [7, Fig. 9]. © 1990 American Institute of Physics.

ordinary antenna problems, it is crucial for the accurate description of the phenomenon of resonances in large circular arrays.

This chapter is organized as follows: in Section 11.2, the two-term formulas with the modified kernel are presented in an alternative form. This form is especially suitable for the detailed study to follow and is also suitable for numerical implementation in a computer. This form and the properties of the modified kernel when N is large are the starting points for the analytical study of the large circular array of lossless elements in Sections 11.3–11.6. This study, supplemented by numerical investigations whenever necessary, shows that a large circular array with one element driven may possess very narrow resonances if the parameters are properly chosen. The mth "phase-sequence resonance" occurs when the mth phase-sequence conductance is large. At resonance, all currents are large and their distribution around the circle may be thought of as a standing-wave mode. Design guidelines are derived that help in choosing the number of elements N, the spacing d/λ, and the individual element parameters a/λ and h/λ in order to excite a desired resonance. The behavior of the circular array at or near a narrow resonance is explored. One important conclusion is that the driving-point susceptance always becomes zero near a narrow resonance; in a practical application, the matching of the array to a generator may be accomplished with a transformer. Simple approximate formulas for the radiation field are derived which show that the resonant array is highly directive in the vertical plane with field patterns that involve many sharp, pencil-like beams. The main conclusions in these sections are listed as a series of properties.

In Section 11.7, it is shown that it is possible to excite a resonant *traveling-wave* distribution of current around the array by driving two elements instead of one. The choice of N and of the second driven element must be properly made. The resulting far-field pattern is omnidirectional with a pancake-like shape. After the general conditions needed in order to be able to excite a traveling wave are developed and the meaning of "resonance" when two elements are driven is clarified, the properties of the array over earth are discussed. Applications as a surface-wave generator are proposed. Finally, Section 11.8 is an appendix that discusses the various kernels mentioned throughout this chapter, including, in particular, a detailed discussion about the modified kernels and their relationship to the original kernels.

The analysis in this chapter assumes lossless elements. It is extended to circular arrays of highly conducting dipoles in Chapter 12. In ordinary antenna problems, it is an excellent approximation to assume that highly conducting dipoles are perfectly conducting when admittances and radiation fields are to be calculated. In closed-loop arrays of many elements, such an approximation is no longer valid. It will be seen that the effect of the ohmic losses on both the width of the resonant peaks and the field pattern may be significant.

Interest in this chapter is in circular arrays of perfectly conducting dipoles satisfying the conditions

$$ka \ll 1, \qquad ka \ll kh < \pi/2, \qquad d \gtrsim h, \tag{11.1}$$

where $\lambda = 2\pi/k$ is the free-space wavelength, $k = \omega(\mu_0\epsilon_0)^{1/2}$, a is the radius of the elements, and h is their half-length. Also, $N\ (\gg 1)$ is the number of elements in the array, and $d = 2R \sin(\pi/N)$ is the distance between adjacent elements. It is assumed throughout most of this chapter that N is even, but the formulation for N odd is similar.

11.2   The two-term theory and the modified kernel

Assume that element 1 of the circular array is driven by a voltage V_1 and that the rest of the elements $l = 2, \ldots, N$ are parasitic. The two-term formulas for the N currents $I_l(z), l = 1, 2, \ldots, N$, are readily obtained from equations (10.32)–(10.42) for the general array of Chapter 10 by observing that the matrices $[D]$ and $[P]$ are circulant and applying the method of symmetrical components. The detailed derivation is contained in [10]. The currents are a superposition of the $N/2 + 1$ phase-sequence currents $I^{(m)}(z), m = 0, 1, \ldots, N/2$, where $I^{(m)}(z)$ is the current on element 1 when all elements are driven by voltages $V_l^{(m)} = (V_1/N)e^{j2\pi(l-1)m/N}$. The final formulas are:

$$I_l(z) = \begin{cases} \dfrac{j2\pi V_1}{\zeta_0 \Psi_{dR} \cos kh} [\sin k(h - |z|) + T_1(\cos kz - \cos kh)], & l = 1, \\[2ex] \dfrac{j2\pi V_1}{\zeta_0 \Psi_{dR} \cos kh} T_l(\cos kz - \cos kh), & l = 2, 3, \ldots, N, \end{cases} \tag{11.2}$$

where $\zeta_0 = (\mu_0/\epsilon_0)^{1/2} \doteq 376.73$ ohms. The parameter Ψ_{dR} is real and independent of N and d/λ. It is given by (10.34). The coefficients $T_l = T_{1,l}$ of the shifted-cosine part of the current are complex and depend on all the parameters of the problem. They are obtained by superimposing the phase-sequence coefficients $T^{(m)}$:

$$T_l = \frac{1}{N} \left\{ T^{(0)} - (-1)^l T^{(N/2)} + 2 \sum_{m=1}^{N/2-1} T^{(m)} \cos \left[\frac{2\pi(l-1)m}{N} \right] \right\}, \tag{11.3}$$

where

$$T^{(m)} = \frac{P_R^{(m)} + j P_I^{(m)}}{D_R^{(m)} + j D_I^{(m)}}$$

$$= \frac{P_{1R} + P_{\Sigma R}^{(m)} + j P_I^{(m)}}{D_{1R} + D_{\Sigma R}^{(m)} + j D_I^{(m)}}, \qquad m = 0, 1, \ldots, N/2. \tag{11.4}$$

In (11.4), the parameters in the numerator and the denominator are all real. The subscript I means that the parameter depends only on the imaginary part of the modified phase-sequence kernel. Hence, $P_I^{(m)}$ and $D_I^{(m)}$ are independent of the radius a/λ. The subscript $1R$ means that the parameter depends only on the self-term of the real part of the kernel and is therefore independent of N, m and d/λ. The subscript ΣR

means that the quantity depends only on the mutual terms of the real part of the kernel and is therefore independent of a/λ. The full formulas for the various parameters are

$$P_{1R} = \int_{-h}^{h} \sin k(h - |z|) K_{1R}(h - z)\, dz \tag{11.5}$$

$$P_{\Sigma R}^{(m)} = \frac{-1}{1 - \cos kh} \int_{-h}^{h} \sin k(h - |z|)[\cos kh\, K_{\Sigma R}^{(m)}(z) - K_{\Sigma R}^{(m)}(h - z)]\, dz \tag{11.6}$$

$$P_{I}^{(m)} = \frac{-1}{1 - \cos kh} \int_{-h}^{h} \sin k(h - |z|)[\cos kh\, K_{I}^{(m)}(z) - K_{I}^{(m)}(h - z)]\, dz \tag{11.7}$$

$$D_{1R} = \frac{1}{1 - \cos kh} \int_{-h}^{h} (\cos kz - \cos kh)[\cos kh\, K_{1R}(z) - K_{1R}(h - z)]\, dz \tag{11.8}$$

$$D_{\Sigma R}^{(m)} = \frac{1}{1 - \cos kh} \int_{-h}^{h} (\cos kz - \cos kh)[\cos kh\, K_{\Sigma R}^{(m)}(z) - K_{\Sigma R}^{(m)}(h - z)]\, dz \tag{11.9}$$

$$D_{I}^{(m)} = \frac{1}{1 - \cos kh} \int_{-h}^{h} (\cos kz - \cos kh)[\cos kh\, K_{I}^{(m)}(z) - K_{I}^{(m)}(h - z)]\, dz. \tag{11.10}$$

The various parts of the modified kernel (the use of which is justified in Section 11.8 and in [9] and [10]) are

$$K_{1R}(z) = \frac{\cos k R_1(z)}{R_1(z)} \tag{11.11}$$

$$K_{\Sigma R}^{(m)}(z) = \sum_{l=2}^{N/2+1} \xi_l \cos\left[\frac{2\pi(l-1)m}{N}\right] \frac{\cos k R_l(z)}{R_l(z)} \tag{11.12}$$

$$K_{I}^{(m)}(z) = \frac{-\sin kz}{z} - \sum_{l=2}^{N/2+1} \xi_l \cos\left[\frac{2\pi(l-1)m}{N}\right] \frac{\sin k R_l(z)}{R_l(z)}, \tag{11.13}$$

where

$$\xi_l = \begin{cases} 1, & l = N/2 + 1 \\ 2, & \text{otherwise} \end{cases} \tag{11.14}$$

and

$$R_l(z) = (z^2 + b_{1l}^2)^{1/2}; \qquad b_{1l} = \begin{cases} a, & l = 1 \\ \dfrac{d\,\sin[(l-1)\pi/N]}{\sin(\pi/N)}, & l \neq 1. \end{cases} \tag{11.15}$$

Note that the radius a does not appear in (11.13). Finally, the relation between $T^{(m)}$ and the phase-sequence admittances is

$$
\begin{aligned}
Y^{(m)} &= G^{(m)} + jB^{(m)} \\
&= \frac{j2\pi}{\zeta_0 \Psi_{dR} \cos kh} [\sin kh + T^{(m)}(1 - \cos kh)]
\end{aligned}
\tag{11.16}
$$

and the self- and mutual conductances $G_{1,l}$ (susceptances $B_{1,l}$) are determined only by the phase-sequence conductances $G^{(m)}$ (susceptances $B^{(m)}$) by the relation

$$
\begin{aligned}
Y_{1,l} &= G_{1,l} + jB_{1,l} \\
&= \frac{1}{N}\left\{ Y^{(0)} - (-1)^l Y^{(N/2)} + 2\sum_{m=1}^{N/2-1} Y^{(m)} \cos\left[\frac{2\pi(l-1)m}{N}\right]\right\}.
\end{aligned}
\tag{11.17}
$$

With the original imaginary part of the self term of the kernel, [i.e. $\sin k R_1(z)/R_1(z)$ in place of $\sin kz/z$ in (11.13)], it is easily seen that (11.2)–(11.17) reduce to (4.4)–(4.14).

The modified kernel (11.11)–(11.13) has been evaluated asymptotically for large N and d/λ fixed with $d/\lambda < m/N \le \frac{1}{2}$ (see [8,9], and also Section 11.8). $K_{\Sigma R}^{(m)}(z)$ is well approximated by the kernel of the infinite linear array [replace b_{1l} by ld in (11.12) and let $N \to \infty$ while keeping m/N fixed]; it is therefore roughly independent of N for large N and fixed m/N. The imaginary part $K_I^{(m)}(z)$ must be approximated more carefully because the imaginary part of the kernel of the infinite linear array is exactly zero. The asymptotic formula for $K_I^{(m)}(z)$ is

$$
\begin{aligned}
\frac{K_I^{(m)}(z)}{k} &\sim \frac{-1}{4\pi^{1/2}} \frac{1}{N^{1/2}} \frac{1}{[(m/N)^2 - (d/\lambda)^2]^{3/4}} \exp[-2N(m/N)g(x_m)] \\
&\quad + \{\text{same with } m \to N - m\},
\end{aligned}
\tag{11.18}
$$

where

$$
x_m = \frac{d/\lambda}{m/N}; \qquad g(x) = \cosh^{-1}\left(\frac{1}{x}\right) - (1 - x^2)^{1/2}, \qquad 0 < x < 1.
\tag{11.19}
$$

The approximation is better when z/λ is small and when d/λ is not very close to m/N. These are the cases of interest. When $m \ll N/2$, only the first term needs to be kept; in the extreme case of $m = N/2$, the second term simply contributes a factor of 2. The function $g(x)$ appearing in the exponential is positive and decreasing, with $g(0) = \infty$. The following properties of $K_I^{(m)}(z)$ are noted:

Property 1: $K_I^{(m)}(z)$ is approximately independent of z when $d/\lambda < m/N$.

Property 2: $K_I^{(m)}(z)$ is exponentially small in N for fixed d/λ and fixed m/N with $d/\lambda < m/N$.

Property 3: $K_I^{(m)}(z)$ is a rapidly decreasing function of d/λ when N and m/N are fixed with $d/\lambda < m/N$.

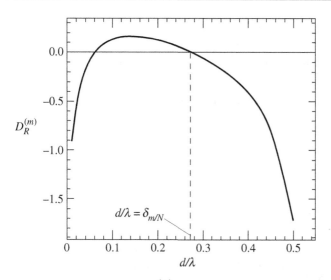

Figure 11.3 Typical plot of $D_R^{(m)}$ as function of d/λ; $N = 90$, $m = 45$, $h/\lambda = 0.2$, and $a/\lambda = 0.05$. Taken from Fikioris *et al.* [1, Fig. 3]. © 1994 EMW Publishing.

With the use of Property 1, approximations for $P_I^{(m)}$ and $D_I^{(m)}$ are obtained when h/λ is not too large. Thus,

$$D_I^{(m)} \sim -2(\sin kh - kh \cos kh)\frac{K_I^{(m)}(0)}{k}; \qquad d/\lambda < m/N \qquad (11.20)$$

$$P_I^{(m)} \sim 2(1 - \cos kh)\frac{K_I^{(m)}(0)}{k}; \qquad\qquad d/\lambda < m/N \qquad (11.21)$$

so that $D_I^{(m)}$ and $P_I^{(m)}$ are slowly varying functions of kh that also possess the Properties 2 and 3 above.

11.3 Phase-sequence resonances

Throughout this section, d/λ is the variable and a/λ, h/λ, and N are fixed. In an experimental study, it is much simpler to vary the frequency and keep the geometrical parameters a, h, and d fixed. The physical picture is very similar in the two cases.

A typical plot of $D_R^{(m)} = D_{1R} + D_{\Sigma R}^{(m)}$ as a function of d/λ is given in Fig. 11.3. It is seen that $D_R^{(m)}$ is a quantity of order 1 that has two zeros in the range $0 < d/\lambda \leq 0.5$. The array is defined to be at its *m*th phase-sequence resonance when $D_R^{(m)}$ is exactly zero. It will be seen that when this occurs, $G^{(m)}$ and $G_{1,1}$ are almost exactly at their maximum and $B_{1,1}$ is very close to its zero. In Fig. 11.3, the smaller root is not in the region of validity of the two-term theory since $d/\lambda < h/\lambda$. Since the position of the resonance is determined only by the real part of the kernel, a particular m/N

phase-sequence resonance will occur at roughly the same value of d/λ for all large N. Here, it is assumed that the resonant spacing d/λ is independent of N so that it is meaningful to examine resonant currents as N becomes larger while keeping m/N and d/λ fixed. Denote by $\delta_{m/N}$ the position of the larger root, so that $D_R^{(m)} = 0$ when $d/\lambda = \delta_{m/N}$. It is seen from (11.4) and (11.16) that, at the mth phase-sequence resonance,

$$T_{\text{res}}^{(m)} = \frac{P_I^{(m)}}{D_I^{(m)}} - j \frac{P_R^{(m)}}{D_I^{(m)}} \tag{11.22}$$

$$G_{\text{res}}^{(m)} = \frac{2\pi}{\zeta_0 \Psi_{dR} \cos kh} \frac{P_R^{(m)}}{D_I^{(m)}} \tag{11.23}$$

$$B_{\text{res}}^{(m)} = \frac{2\pi}{\zeta_0 \Psi_{dR} \cos kh} \left[\sin kh + (1 - \cos kh) \frac{P_I^{(m)}}{D_I^{(m)}} \right], \tag{11.24}$$

where $P_R^{(m)}$, $P_I^{(m)}$, and $D_I^{(m)}$ are evaluated at $d/\lambda = \delta_{m/N}$.

The quantity $P_R^{(m)}$ is of order 1. Because of Properties 2 and 3 and equations (11.20), (11.21), and (11.17), if $\delta_{m/N} < m/N$, it is seen that

Property 4:   At the mth phase-sequence resonance, the phase-sequence conductance $G_{\text{res}}^{(m)}$ is extremely large in N. The self- and mutual conductances $G_{1,l}$ around the array are also extremely large and they vary around the array according to

$$G_{1,l} \propto G_{1,1} \cos\left[\frac{2\pi (l - 1)m}{N} \right], \qquad l = 1, 2, \ldots, N. \tag{11.25}$$

This distribution of current around the array may be recognized as a standing wave.

Property 5: $G_{\text{res}}^{(m)}$ and the $G_{1,l}$'s will be much larger when the resonant spacing $d/\lambda = \delta_{m/N}$ occurs at a smaller value.

It should be pointed out that the conductances are actually predicted by (11.18), (11.20) and (11.23) to be *exponentially* large in N. This is a consequence of the assumption that the resonant spacing $d/\lambda = \delta_{m/N}$ does not depend on N and may or may not be true within the two-term theory.

On the other hand, $B_{\text{res}}^{(m)}$ and the $B_{1,l}$'s are not large when the array is exactly at resonance. In the special case when $\cos[2\pi (l - 1)m/N] = 0$ (this requires N to be a multiple of 4), the current on element l is very small compared to that on all other elements. Figure 11.4 shows the normalized conductances $G_{1,l}$ as a function of the element number l for the $m/N = \frac{3}{8}$ phase-sequence resonance with $N = 72$. With $a/\lambda = 0.05$ and $h/\lambda = 0.2$, this occurs at $d/\lambda \doteq 0.2269$. The data in Fig. 11.4 as well

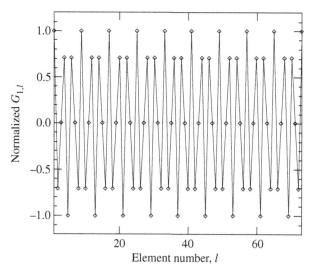

Figure 11.4 Normalized self- and mutual conductances $G_{1,l}$ as function of element number l for $m/N = \frac{3}{8}$ phase-sequence resonance; $N = 72$, $h/\lambda = 0.2$, $a/\lambda = 0.05$, and $d/\lambda = 0.226\,88$. Taken from Fikioris *et al.* [1, Fig. 4]. © 1994 EMW Publishing.

$\phi = 0$

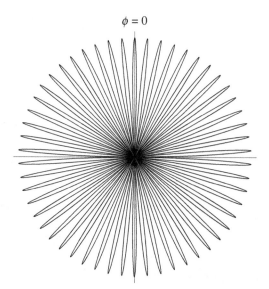

Figure 11.5 Normalized far-field power pattern $|E_\theta(\pi/2, \phi)|^2$ at $\theta = \pi/2$ plane of dipoles' centers for $m = 27$ phase-sequence resonance; $N = 72$, $h/\lambda = 0.2$, $a/\lambda = 0.05$, and $d/\lambda = 0.226\,88$. Taken from Fikioris *et al.* [1, Fig. 5]. © 1994 EMW Publishing.

as those of Figs. 11.5 and 11.6 and of Table 11.1 were obtained with the full two-term theory formulas (11.2)–(11.17) with the imaginary part of the kernel evaluated from (11.13) using quadruple precision and with $T_{\text{res}}^{(m)}$ given by (11.22). In Fig. 11.4, the current distribution of Property 4 is recognized; in this case, the currents divide up into

$\theta = 0$

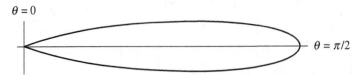

$\theta = \pi/2$

Figure 11.6 Normalized far-field power pattern $|E_\theta(\theta, 0)|^2$ at $\phi = 0$ for $m = 27$ phase-sequence resonance; $N = 72$, $h/\lambda = 0.2$, $a/\lambda = 0.05$, and $d/\lambda = 0.226\,88$.

Table 11.1. *Resonant spacings $d/\lambda = \delta_{m/N}$;[a] values of imaginary part of kernel $K_I^{(m)}(0)/k$ at $z = 0$, $d/\lambda = \delta_{m/N}$; and driving-point conductances $G_{1,1}$ at resonance[b]*

h/λ	$a/\lambda = 0.01$	$a/\lambda = 0.03$	$a/\lambda = 0.05$
0.14	no root	no root	$\delta_{m/N} = 0.479$ $\dfrac{K_I^{(m)}(0)}{k} = -0.25$ $G_{1,1} = 14.7\,\text{mS}$
0.16	no root	$\delta_{m/N} = 0.480$ $\dfrac{K_I^{(m)}(0)}{k} = -0.18$ $G_{1,1} = 10.5\,\text{mS}$	$\delta_{m/N} = 0.439$ $\dfrac{K_I^{(m)}(0)}{k} = -4.8 \times 10^{-3}$ $G_{1,1} = 81.6\,\text{mS}$
0.18	$\delta_{m/N} = 0.494$ $\dfrac{K_I^{(m)}(0)}{k} = -0.47$ $G_{1,1} = 4.8\,\text{mS}$	$\delta_{m/N} = 0.431$ $\dfrac{K_I^{(m)}(0)}{k} = -2.1 \times 10^{-3}$ $G_{1,1} = 109\,\text{mS}$	$\delta_{m/N} = 0.370$ $\dfrac{K_I^{(m)}(0)}{k} = -3.9 \times 10^{-7}$ $G_{1,1} = 4.8 \times 10^5\,\text{mS}$
0.20	$\delta_{m/N} = 0.437$ $\dfrac{K_I^{(m)}(0)}{k} = -4.1 \times 10^{-3}$ $G_{1,1} = 54\,\text{mS}$	$\delta_{m/N} = 0.336$ $\dfrac{K_I^{(m)}(0)}{k} = -7.0 \times 10^{-10}$ $G_{1,1} = 2.1 \times 10^8\,\text{mS}$	$\delta_{m/N} = 0.273$ $\dfrac{K_I^{(m)}(0)}{k} = -2.3 \times 10^{-16}$ $G_{1,1} = 6.3 \times 10^{14}\,\text{mS}$

[a]Roots $\delta_{m/N}$ are sought in interval $h/\lambda < d/\lambda < m/N = \frac{1}{2}$; number of elements $N = 90$ and phase sequence $m = N/2 = 45$.

[b]Taken from Fikioris *et al.* [1, Table 1]. © 1994 EMW Publishing.

five groups. Other combinations are possible. Note that $G_{1,1}$ is repeated at the end of Fig. 11.4 as $G_{1,73}$ for reasons of symmetry.

The parameter $D_R^{(m)} = D_{1R} + D_{\Sigma R}^{(m)}$ depends on a/λ only through D_{1R} and on d/λ only through $D_{\Sigma R}^{(m)}$. D_{1R} is zero when the elements are self-resonant. By plotting D_{1R} for various values of h/λ, it is seen that D_{1R} is a decreasing function of a/λ, at least when $a/\lambda < 0.07$ and $a/\lambda \ll h/\lambda < 0.22$. (It can be shown, in fact, following a procedure similar to that in Appendix II of [11], that the variation with a is linear in the quantity $\Omega = 2\ln(2h/a)$, but the approximation is poor when h/a becomes small.) Hence, making the dipoles electrically thicker will result in decreasing the amplitude of a curve like that in Fig. 11.3, thereby shifting the resonant spacing $\delta_{m/N}$ to a smaller value of d/λ. The resonant currents will therefore become much larger. When the elements are electrically very thin, the array can have no narrow resonances at all, because the resonant root occurs at a value $d/\lambda > m/N$.

The effect of changing the length h/λ is much more involved since both D_{1R} and $D_{\Sigma R}^{(m)}$ depend on h/λ in a complicated way. However, extensive numerical calculations show that the position of the root $\delta_{m/N}$ decreases when h/λ increases, at least when a/λ and h/λ are in the above-mentioned ranges. Table 11.1 shows the resonant spacings $\delta_{m/N}$, the values of $K_I^{(m)}(0)$ evaluated at $d/\lambda = \delta_{m/N}$, and the self-conductance $G_{1,1}$ for 90-element arrays at their $m = N/2$ phase-sequence resonance as a/λ and h/λ vary. The conclusion is that

Property 6: If a specific phase-sequence resonance is desired, making the dipoles electrically longer or thicker will require an electrically smaller circle and will result in much higher resonant currents around the array, at least when a/λ and h/λ are in the ranges

$$a/\lambda < 0.07 \quad \text{and} \quad a/\lambda \ll h/\lambda < 0.22.$$

It is seen from Table 11.1 that the currents are predicted to be extremely large when the perfectly conducting elements are long and thick. Very large currents can be realized in practice only with superconducting elements; in the case of *highly* conducting elements (for example, elements made from brass or aluminum), the currents are severely limited. The effect of ohmic losses is considered in Chapter 12.

It is natural to believe that values of a/λ and h/λ that yield narrow resonances in circular arrays will yield narrow resonances in non-circular closed-loop arrays as long as the minimum radius of curvature is large enough.

11.4 Behavior near a phase-sequence resonance

Consider again that a/λ, h/λ, and N are fixed, and that d/λ is varied but stays very close to a resonant spacing $\delta_{m/N}$ so that the array is very close to its mth phase-sequence resonance. The function $D_R^{(m)}$ is usually a quantity of order 1, but near resonance it is of the order of magnitude of the very small quantity $D_I^{(m)}$; it is the controlling quantity in (11.4). It is a good approximation to assume that $P_R^{(m)}$, $P_I^{(m)}$, and $D_I^{(m)}$ are constant and that $D_R^{(m)}$ varies linearly with d/λ so that the dependence of $T^{(m)}$ on d/λ is explicitly

$$T^{(m)}(d/\lambda) = \frac{P_R^{(m)} + j P_I^{(m)}}{\alpha^{(m)}(d/\lambda - \delta_{m/N}) + j D_I^{(m)}}, \tag{11.26}$$

where $\alpha^{(m)}$ is the slope of $D_R^{(m)}$ near its zero. Using $P_R^{(m)} \gg P_I^{(m)}$, it is seen from (11.26) that $\text{Re}\{T^{(m)}\}$ has extrema when $D_R^{(m)} \doteq \pm D_I^{(m)}$ or $d/\lambda - \delta_{m/N} \doteq \pm D_I^{(m)}/\alpha^{(m)}$ with the corresponding values

$$\text{Re}\{T^{(m)}\} = \pm \frac{P_R^{(m)}}{2D_I^{(m)}} = \mp \frac{1}{2}\,\text{Im}\{T_{\text{res}}^{(m)}\} = \mp \,\text{Im}\{T^{(m)}\}. \tag{11.27}$$

From the relations (11.16) between $T^{(m)}$, $G^{(m)}$, and $B^{(m)}$ and (11.17) between $B_{1,l}$ and $B^{(m)}$, it is seen that

Property 7: $B^{(m)}$ and the $B_{1,l}$'s are very rapidly varying near a narrow resonance. When the spacing d/λ is such that $G^{(m)}$ has decreased to half its maximum value, $B^{(m)}$ is roughly equal to $G^{(m)}$: $B^{(m)} = \pm G^{(m)} = \pm\frac{1}{2}G_{\text{res}}^{(m)}$. Hence, $B^{(m)}$ and the $B_{1,l}$'s have a zero very close to resonance. The $B_{1,l}$'s vary around the array as $\cos[2\pi(l-1)m/N]$.

It is therefore possible to design an array near resonance with a purely resistive driving-point impedance, but this property is extremely sensitive to slight changes in the parameters.

The Q of the resonant array may be estimated from the curve of $G_{1,1}$ as a function of d/λ as

$$Q_r \doteq \frac{\delta_{m/N}}{(d/\lambda)_2 - (d/\lambda)_1}. \tag{11.28}$$

(The actual definition involves the frequency.) In (11.28), the $(d/\lambda)_2$ and $(d/\lambda)_1$ are the spacings at which the power is reduced to one-half the maximum at constant voltage, i.e. when $D_R^{(m)} = \pm D_I^{(m)}$. Using $P_R^{(m)} \gg P_I^{(m)}$ and (11.26), it is seen that

$$Q_r \doteq \frac{\delta_{m/N}|\alpha^{(m)}|}{2|D_I^{(m)}|}. \tag{11.29}$$

Formula (11.29) provides a simple way to estimate the Q. If, in (11.26), $P_I^{(m)}$ is neglected compared to $P_R^{(m)}$, a simpler formula for the behavior of the quantities near a narrow resonance as a function of d/λ (or as a function of frequency) is obtained, which predicts a small, constant self-resistance and a linearly varying self-reactance that becomes zero at resonance. This formula is used in Section 11.7 below.

It is not simple numerically to calculate quantities near resonance from the full formulas (11.2)–(11.17). In addition to the numerical complications created by the smallness of $K_I^{(m)}(z)$ and $D_I^{(m)}$, the calculation of $D_R^{(m)}$ requires high precision. The reason is that $D_R^{(m)}$ is the difference between two nearly equal, complicated integrals of order 1. A more detailed discussion of these and similar numerical difficulties is provided in Section 13.5.

11.5 Radiation field at or near a phase-sequence resonance

With the two-term theory currents, and with the origin of the spherical coordinates (r, θ, ϕ) placed at the center of the array, the radiation field is given by (10.46)–(10.48) with $t_l = V_1 T_{1l} = V_1 T_l$.

$$\mathbf{E}(r, \theta, \phi) = \hat{\boldsymbol{\theta}} E_\theta$$

$$= \hat{\boldsymbol{\theta}} \frac{-V_1}{\Psi_{dR} \cos kh} \frac{e^{-jkr}}{r}$$

$$\times \left\{ F(\theta) e^{jkR \sin\theta \cos(\phi - \phi_1)} + G(\theta) \sum_{l=1}^{N} T_l e^{jkR \sin\theta \cos(\phi - \phi_l)} \right\}, \quad (11.30)$$

where $(R, \pi/2, \phi_l) = (R, \pi/2, 2\pi(l-1)/N)$ is the location of element l, $R = d/(2 \times \sin \pi/N)$ is the radius of the circular array, and

$$F(\theta) = \frac{\cos(kh \cos\theta) - \cos kh}{\sin\theta} \quad (11.31)$$

$$G(\theta) = \frac{\sin kh \cos(kh \cos\theta) \cos\theta - \cos kh \sin(kh \cos\theta)}{\sin\theta \cos\theta}. \quad (11.32)$$

In (11.30), the first term represents radiation from the sine current of the driven element; the second term is radiation from the shifted-cosine currents of all elements; $F(\theta)$ and $G(\theta)$ are the "element factors" for the sine and shifted-cosine currents, respectively. At or near a narrow resonance, we have the standing-wave distribution $T_l \doteq T_1 \cos[2\pi(l-1)m/N]$. It will be seen that the first term in (11.30) may be neglected. Thus, if one defines the array factor for the mth phase-sequence resonance as

$$A^{(m)}(\theta, \phi) = \frac{1}{T_1} \sum_{l=1}^{N} T_l e^{jkR \sin\theta \cos(\phi - \phi_l)} \quad (11.33a)$$

$$= \sum_{l=1}^{N} \cos\left[\frac{2\pi(l-1)m}{N} \right] e^{jkR \sin\theta \cos(\phi - \phi_l)} \quad (11.33b)$$

the radiation field is given approximately by

$$E_\theta = \frac{-V_1}{\Psi_{dR} \cos kh} \frac{e^{-jkr}}{r} G(\theta) T_1 A^{(m)}(\theta, \phi). \quad (11.34)$$

The array factor is the radiation field due to a circular array of isotropic radiators with the mth phase-sequence resonance currents around the array, element 1 having unit current. The sum in (11.33b) may be written exactly as follows (see [12] for a detailed derivation):

$$A^{(m)}(\theta, \phi) = N j^m J_m[N(d/\lambda) \sin\theta] \cos(m\phi)$$
$$+ N \sum_{p=1}^{\infty} j^{Np-m} J_{Np-m}[N(d/\lambda) \sin\theta] \cos[(Np - m)\phi]$$
$$+ N \sum_{p=1}^{\infty} j^{Np+m} J_{Np+m}[N(d/\lambda) \sin\theta] \cos[(Np + m)\phi]. \qquad (11.35)$$

Because of the condition $d/\lambda < m/N \le \frac{1}{2}$, the arguments of the Bessel functions are always smaller than the orders. When N is large, only two terms in (11.35) are significant, namely,

$$A^{(m)}(\theta, \phi) \sim N j^m J_m[N(d/\lambda) \sin\theta] \cos(m\phi)$$
$$+ N j^{N-m} J_{N-m}[N(d/\lambda) \sin\theta] \cos[(N - m)\phi]. \qquad (11.36)$$

As with the imaginary part of the kernel, the first term is adequate when $m \ll N/2$ and the second term is equal to the first one when $m = N/2$. Assuming for simplicity that $m \ll N/2$ and using the asymptotic formula for the Bessel functions, one obtains

$$A^{(m)}(\theta, \phi) \sim j^m \frac{N}{(2\pi m)^{1/2}} \frac{1}{[1 - (x_m \sin\theta)^2]^{1/4}}$$
$$\times \exp[-N(m/N)g(x_m \sin\theta)] \cos(m\phi), \qquad m \ll N/2, \qquad (11.37)$$

where x_m and $g(x)$ are the same as in (11.19). Hence, the array factor is an exponentially small quantity and, in fact, it shares Properties 2 and 3 of the imaginary part of the kernel. Also, it has a zero of order m at $\theta = 0$. From (11.37), the field formula (11.34), and the expressions (11.20)–(11.22) for $T_1 \doteq (2/N) T_{\text{res}}^{(m)}$, it is seen that:

Property 8: The magnitude of the radiation field at any fixed point in space is extremely large in N.

This verifies that radiation from the sine current of element 1 is negligible and justifies the usefulness of the array factor. The largeness of the field should be expected since, for lossless elements, integration of $|E_\theta|^2$ over a large sphere should give the total radiated power $\frac{1}{2}G_{1,1}|V_1|^2$, which is large.

Property 9: The horizontal field pattern ($\theta = \pi/2$) consists of $2m$ spikes.

Property 10: The vertical field pattern is very narrow, with a maximum at $\theta = \pi/2$.

Figures 11.5 and 11.6 show the horizontal and vertical far-field power patterns for the $N = 72$ array of Fig. 11.4 as calculated from (11.30)–(11.32). It is seen that Properties 9 and 10 hold.

The narrowness of the vertical beam can be estimated by neglecting variations of the field in (11.34) due to the slowly varying $G(\theta)$ and defining vertical directivity as the maximum of the array factor divided by its mean value, namely,

$$D_V = \frac{|A^{(m)}(\pi/2, \phi)|}{\frac{1}{2}\int_0^\pi |A^{(m)}(\theta, \phi)| \sin\theta \, d\theta}. \tag{11.38}$$

Subject to the approximation (11.36), the integral can be carried out analytically and the resulting D_V is independent of ϕ when $m \ll N/2$. Thus,

$$D_V = \frac{2 J_m[N(d/\lambda)]}{\pi J_{(m-1)/2}[\frac{1}{2}N(d/\lambda)] J_{(m+1)/2}[\frac{1}{2}N(d/\lambda)]}; \qquad m \ll N/2. \tag{11.39}$$

With the asymptotic expression for the Bessel functions and after some manipulation, it can be shown that

$$D_V \sim (2N/\pi)^{1/2}[(m/N)^2 - (d/\lambda)^2]^{1/4}. \tag{11.40}$$

Hence,

Property 11: For a specific phase-sequence resonance m/N, making N larger will result in a narrower vertical field pattern, and in more spikes in the horizontal plane.

Property 12: For fixed N and for a specific phase-sequence resonance m, making the dipoles thicker or longer will result in a smaller resonant spacing $\delta_{m/N}$, a much narrower resonance, and a slightly more directive vertical field pattern.

The vertical directivity may therefore be made arbitrarily large by making N large (although the increase is slow, roughly as the square root of N). The field strength at any point in space increases very rapidly. The input impedance may be a pure resistance. However, the physical dimensions of the array increase (linearly with N) and the band width decreases very rapidly.

The array factor's smallness has an interesting consequence. For resonant *non-circular* arrays, an array factor $A(\theta, \phi)$ may be defined exactly as in (11.33a). This array factor will depend on the array's geometry and the relative current distribution around the array. It will be a sum of N terms of order 1, each term depending on the location of element l and its relative current (admittance). If a sufficiently large non-circular array with one element driven is thought of as a perturbation of some corresponding circular array, then it is logical to assume that the current distribution around the array will not be significantly affected and will again be of the standing-wave type. Hence, each term in the sum for $A(\theta, \phi)$ will be close to each term in the sum for the circular array. However, any very small quantity that can be written as the sum of terms of order 1 is extremely sensitive to perturbations of these terms. Therefore, the array factor (field pattern) for the non-circular array will not be close to that of the circular array. This means that a wide variety of field patterns may be obtained by resonant non-circular arrays, perhaps even a superdirective pattern. The possibility of using a resonant non-circular array to obtain a highly directive field was proposed in [6].

A far-field pattern consisting of many sharp spikes equally spaced around a circle is unusual and would seem to have no useful purpose. There is, however, one very interesting potential application. Assume that $N = 90$ and $m = 45$ so that there are 90 spikes. If the array of 90 elements is mounted rigidly on a circular disk passing through the center of each element, and if the single driven element is center-driven by a transmission line that extends from the element to the center of the disk and then vertically downward, the entire structure can be rotated about the vertical axis through the center of the disk. When the angular velocity is one revolution in $1\frac{1}{2}$ min or 90 s, the array, operating at a fixed frequency and constant amplitude, emits a sharp pulse once each second in all directions in its horizontal plane. Thus it is a radio beacon that could be used in place of flashing lights along the sea coast. By selecting different angular velocities, each beacon can be made uniquely identifiable. As compared with the conventional flashing lights, which vanish in dense fog, the radio beacon is equally useful in all kinds of weather. Although it sends out short, sharp pulses, it is a structurally and electrically simple, steady-state device.

11.6 Refinements for numerical calculations

Two further improvements to the two-term theory are now presented. These do not change the properties given before but are useful for numerical calculations whenever high precision is necessary. These improvements apply specifically to tubular dipoles (or monopoles over a ground plane) with walls of zero thickness.

Whereas the real part of the self-term of the kernel $K_{1R}(z)$ given in (11.11) assumes interaction from axis to perimeter, the more accurate but more complicated form (see Section 11.8 and Chapter 1)

$$K_{1R}(z) = \frac{1}{2\pi} \int_{-\pi}^{\pi} \frac{\cos\{k[z^2 + 4a^2 \sin^2(\phi/2)]^{1/2}\}}{[z^2 + 4a^2 \sin^2(\phi/2)]^{1/2}} d\phi \tag{11.41}$$

assumes that the interaction is from a point on the perimeter to another point on the perimeter. The use of (11.41) instead of (11.11) provides higher accuracy.

The second improvement comes from the observation that the trigonometric functions $\sin k(h - |z|)$ and $\cos kz - \cos kh$ are not adequate to describe the charge build-up near the ends $z = \pm h$ of the tubular dipole. It is known that $I(z)$ behaves like $\sqrt{h - |z|}$ for $|z|$ close to h for both the driven [13] and the parasitic elements. A simple improvement to the current that takes into account this behavior is achieved [9, 14] by modifying the shifted cosine to

$$
p^S(z) = \begin{cases} \cos kz - \gamma_1, & |z| < z_0 \\ \gamma_2 \sqrt{kh - k|z|}, & z_0 < |z| < h, \end{cases}
\tag{11.42}
$$

where the constants γ_1, γ_2, and z_0 are found by matching $p^S(z)$ and its first two derivatives at $z = z_0$. The resulting equations are

$$
\tan kz_0 = 2(kh - kz_0)
\tag{11.43}
$$

$$
\gamma_1 = \cos kz_0[1 - 4(kh - kz_0)^2]
\tag{11.44}
$$

$$
\gamma_2 = 2 \sin kz_0 \sqrt{kh - kz_0}.
\tag{11.45}
$$

Formula (11.43) is a transcendental equation for kz_0 that has exactly one solution when $kh < \pi/2$. The two-term theory solution then becomes (11.2)–(11.15) with $\cos kz - \cos kh$ in (11.2) and (11.8)–(11.10) replaced by $p^S(z)$ as given by (11.42)–(11.45). This solution gives excellent agreement between theoretically predicted and measured resonant frequencies in the two experimental studies that have been performed [10, 15].

In the extended theory in Chapter 12 that takes into account the effect of a finite but large conductivity of the elements, the current distributions on the elements are assumed to be the same as in the lossless case and the refinements of this section may be included in the extended theory for lossy elements. The values of driving-point admittance obtained from such a theory will be seen to agree very well with those measured.

11.7 Resonant array with two driven elements

In Sections 11.2–11.6, it was seen that properly dimensioned large circular arrays of electrically short, perfectly conducting vertical dipoles possess very narrow resonances when only one element is driven and the rest are parasitic. At each resonance, the currents on all elements are large and are distributed as a standing wave around the circle. The driving-point reactance is zero. The associated field pattern consists of many pencil-like beams.

In recent studies [16, 17], the complete electromagnetic field generated by a vertical electric dipole located in the air above planar earth (salt water, lake water, wet earth, dry earth) has been formulated in simple integrated expressions. Included is the special case when both the vertical dipole and the observation point are on or close to the surface of the earth.

In applications such as broadcast, ground-wave over-the-horizon radar [18], shore-to-ship communication and microwave beacons, it is required to generate a significant electromagnetic field (surface wave) close to the air–earth boundary at $\theta = \pi/2$. In addition to the field close to $\theta = \pi/2$, a typical transmitting antenna generates a significant field at smaller angles θ. This field is unwanted. Furthermore, the upward generated field may reflect off the ionosphere and interfere with the surface wave propagating near $\theta = \pi/2$. In this section, a structurally simple antenna array that is especially suited to generate an omnidirectional surface wave will be described. Instead of directing the outward-traveling electromagnetic field upward toward the ionosphere, the array directs the field along the surface of the earth in a pancake-shaped field pattern. The array is a large circular array as before but in this case two elements are driven instead of one. Each driven element has a driving-point impedance that is purely resistive. A description and analysis of the array are followed by a determination of its complete far field both when the array is in free space and when the array is over planar earth. The generation of a pancake-shaped field pattern by a large circular array with many parasitic elements was first proposed in [19].

In this section, a time dependence $e^{-i\omega t}$ is assumed instead of the $e^{j\omega t}$ of the previous sections. Also, it is assumed that the frequency $f = c/\lambda = \omega/2\pi$ is the varying parameter instead of d/λ of the previous sections.

The basic idea here is to design a resonant circular array with two elements (1 and n) driven that has a *traveling-wave* distribution of current, namely,

$$I_l(0) = I_1(0) \exp\left[-i\,\frac{2\pi(l-1)m}{N}\right], \qquad l = 1, \ldots, N \tag{11.46}$$

instead of the standing wave of (11.25). The resulting array-factor pattern is omnidirectional with a pancake-like shape. The vertical directivity increases as the width of the resonance decreases. The problem will be studied in general; conditions on N, m and n will be developed so that (11.46) is possible; the meaning of "resonance" when two elements are driven will be clarified.

Excitation of a traveling wave with two driven elements

The midpoint currents $I_l(0)$ on the dipoles when elements 1 and n are driven by voltages V_1 and V_n, respectively, are given by (10.43) and (10.45) so that

$$I_l(0) = Y_{l,1}V_1 + Y_{l,n}V_n, \qquad l = 1, \ldots, N, \tag{11.47}$$

where the self- and mutual admittances $Y_{l,j}$ satisfy $Y_{j,l} = Y_{l,j}$, $Y_{j,j} = Y_{1,1}$, and $Y_{j,l} = Y_{1,j-l+1} = Y_{1,j-l+1\pm N}$. The last equality follows from the symmetry of the circular array with one element driven. At or near a narrow resonance (as long as the self-conductance is large), the admittances follow the distribution (11.25) so that

$$I_l(0) = Y_{1,1}\left\{V_1 \cos\left[\frac{2\pi(l-1)m}{N}\right] + V_n \cos\left[\frac{2\pi(l-n)m}{N}\right]\right\}, \qquad l = 1, \ldots, N.$$

$$\tag{11.48}$$

As a function of frequency (see Section 11.4)

$$Y_{1,1}(f) = G_{1,1}(f) - i B_{1,1}(f)$$
$$= \frac{G_{1,1}(f_m)}{1 - i2Q(f - f_m)/f_m}, \qquad \frac{|f - f_m|}{f_m} \ll 1, \tag{11.49}$$

where Q is the quality factor of the resonance curve.

In this section, interest is primarily in cases where $G^{(m)}$ is large enough so that contributions from the rest of the phase sequences are negligible in (11.48). However, the small contributions from the rest of the phase sequences are included in numerical and graphical results.

Defining

$$t_{mn} = \frac{2\pi (n - 1)m}{N} \tag{11.50}$$

it is seen that if the ratio V_n/V_1 is chosen so that

$$V_n = -V_1 e^{i t_{mn}}, \tag{11.51}$$

then the currents satisfy (11.46) with the current on element 1 given by

$$I_1(0) = -V_1 Y_{1,1} i \sin t_{mn} e^{i t_{mn}}. \tag{11.52}$$

It follows that the choice (11.51) is not sufficient for a traveling-wave distribution of current around the array; in addition, the condition

$$\sin t_{mn} \neq 0 \tag{11.53}$$

must be satisfied. This is a restriction on the choice n of the second driven element for given N and m.

The two driving-point admittances are given by

$$Y_{1,\text{in}} = G_{1,\text{in}} - i B_{1,\text{in}} = \frac{I_1(0)}{V_1} = w_{mn} Y_{1,1} \tag{11.54}$$

$$Y_{n,\text{in}} = G_{n,\text{in}} - i B_{n,\text{in}} = \frac{I_n(0)}{V_n} = w_{mn}^* Y_{1,1}, \tag{11.55}$$

where

$$w_{mn} = u_{mn} - i v_{mn} = \sin^2 t_{mn} - i \sin t_{mn} \cos t_{mn} \tag{11.56}$$

and the asterisk denotes the complex conjugate. It follows from $u_{mn} > 0$ that the total power supplied to the array, namely,

$$P_{\text{total,in}} = \tfrac{1}{2} G_{1,\text{in}} |V_1|^2 + \tfrac{1}{2} G_{n,\text{in}} |V_n|^2 = \tfrac{1}{2} (G_{1,\text{in}} + G_{n,\text{in}}) |V_1|^2 \tag{11.57}$$

is positive as one would expect. However, the individual powers supplied may include one that is negative. This means that it is necessary to extract power from one of the elements in order to excite a traveling-wave distribution of current. Although this may be done by center-loading the element (as opposed to driving it by a generator), a case like this is undesirable and can be avoided as will be shown below.

As when one element is driven, the far-field pattern is adequately described by the array factor. With the currents (11.46), this is

$$A^{(m)}(\theta, \phi) = \sum_{l=1}^{N} e^{-i[2\pi(l-1)m/N]} e^{-ikR \sin\theta \cos(\phi-\phi_l)}, \tag{11.58}$$

where $(R, \pi/2, \phi_l)$ is the location of element l in spherical coordinates. This may be evaluated asymptotically for large N as in the case where only one element is driven. The details are in [20]. The final result is

$$A^{(m)}(\theta, \phi) \sim \frac{N}{2\pi} e^{-im\phi} \int_0^{2\pi} e^{-ikR \sin\theta \cos\phi'} e^{-im\phi'} d\phi'$$

$$+ \frac{N}{2\pi} e^{i(N-m)\phi} \int_0^{2\pi} e^{-ikR \sin\theta \cos\phi'} e^{i(N-m)\phi'} d\phi' \tag{11.59a}$$

or

$$A^{(m)}(\theta, \phi) \sim N g_m(\theta) e^{-im\phi} + N g_{N-m}(\theta) e^{i(N-m)\phi}, \tag{11.59b}$$

where

$$g_m(\theta) = (-i)^m J_m[N(d/\lambda) \sin\theta]. \tag{11.60}$$

It is seen from (11.59a) that each term in (11.59a) or (11.59b) is proportional to the radiation field due to a continuous circular traveling wave of current. One is a clockwise traveling wave and the other is a counter-clockwise traveling wave. If $m \ll N/2$, the first term dominates and the resulting radiation field has a pancake-shape, with vertical directivity the same as in (11.40) so that $|A^{(m)}(\theta, \phi)| = A^{(m)}(\theta)$. In the extreme case $m = N/2$ (which is not allowed because (11.53) is not satisfied), the second term would have the same magnitude as the first term and the resulting radiation field would consist of $2m$ pencil-like beams.

Choice of the parameters

If N is chosen to be a multiple of 4, then for certain n there exist values of m such that

$$\cos t_{mn} = 0 \iff v_{mn} = 0 \text{ and } u_{mn} = 1. \tag{11.61}$$

With such a choice of N, n, and m, (11.54)–(11.56) give

$$Y_{1,\text{in}} = Y_{n,\text{in}} = Y_{1,1} \tag{11.62}$$

so that, at or near a narrow resonance, the two driving-point admittances are equal to the self-admittance and all desirable properties of the circular array with one element driven are preserved: The two driving-point susceptances become zero at the same frequency and the driving-point conductances are very large at that frequency. Hence, it is desirable to use as a second driven element one that would have a very small current if only element 1 were driven.

If, furthermore, the second driven element is chosen to be a quarter-way around the circle, i.e. $n = N/4 + 1$, then $u_{mn} = 1$ and $v_{mn} = 0$ for all odd m; with this choice of second driven element, it is possible to excite many phase-sequence resonances; the required voltage ratio is either $e^{i\pi/2}$ or $e^{-i\pi/2}$.

The choice of m depends on various opposing factors. For given large N, if m is chosen to be too large, the following disadvantages apply: (i) the second term in (11.59) might contribute and the field pattern will not have a true pancake-shape, although the value of the ratio $|g_{N-m}(\pi/2)/g_m(\pi/2)|$ decreases very rapidly with decreasing m; and (ii) the antenna might be too frequency-sensitive. On the other hand, the advantages of using a large value of m include a slight increase in directivity as well as smaller contributions from the other phase sequences; these contributions can cause departures from the omnidirectional pancake-like field pattern. The advantage in being able to excite many different phase-sequence resonances with the same construction allows the choice of m to be made to fit a particular application.

Finally, it must be pointed out that when N, m and n are not chosen to satisfy (11.61), a very different frequency dependence may result. Figures 11.7 and 11.8 show the two driving-point admittances as calculated by (11.54)–(11.56) and (11.49) when $Q = 1000$, $N = 90$, $m = 43$, and $n = 24$ (this choice of n corresponds to using as a second driven element one that would have the *largest* possible current if only element 1 were driven at the resonant frequency f_{43}). It is seen that the driving-point conductances both become negative very close to "resonance", that the driving-point susceptances become larger than the conductances, and that all values are much smaller than the resonant self-conductance $G_{1,1}(f_m)$.

A specific example is now given for an operating frequency of about 30 MHz. The number of elements is chosen to be $N = 96$, so that there are many combinations of m and n that satisfy (11.61). The choice $a = 0.28$ m, $h = 1.9$ m, and $d = 3.1$ m is appropriate for 30 MHz. This choice of a, h, and d corresponds to an approximate scaling of the $N = 90$ element experiment[1] over a ground plane with one element driven by a coaxial line. The diameter of the 30-MHz array is $2R \doteq 95$ m. Table 11.2 shows the phase-sequence resonances that may be excited by using elements 1 and 25 ($= N/4 + 1$) as driven elements, the required voltage ratio V_{25}/V_1, the theoretically predicted resonant frequencies f_m, the values of the driving-point conductance at resonance $G_{1,\text{in}}(f_m) = I_1(0)/V_1 = G_{n,\text{in}}(f_m) = G_{1,1}(f_m)$, the ratio of the two

---

[1] [10] Chapter 8.

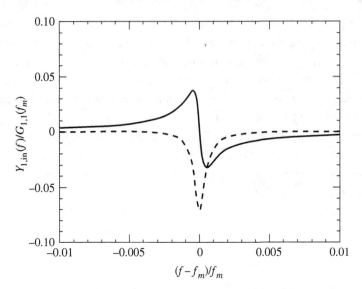

Figure 11.7 Normalized driving-point conductance $G_{1,\,in}(f)/G_{1,1}(f_m)$ (solid line) and driving-point susceptance $B_{1,\,in}(f)/G_{1,1}(f_m)$ (dashed line) of element 1 as function of relative frequency $(f - f_m)/f_m$; $Q = 1000$, $N = 90$, $m = 43$, and $n = 24$. Taken from Fikioris *et al.* [20, Fig. 2]. © 1996 I.E.E.

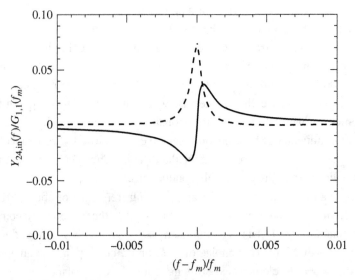

Figure 11.8 Normalized driving-point conductance $G_{24,\,in}(f)/G_{1,1}(f_m)$ (solid line) and driving-point susceptance $B_{24,\,in}(f)/G_{1,1}(f_m)$ (dashed line) of element 24 as function of relative frequency $(f - f_m)/f_m$; $Q = 1000$, $N = 90$, $m = 43$, and $n = 24$.

array-factor terms in (11.59b) at $\theta = \pi/2$ and $\phi = 0$, and the vertical directivity of the array factor D_V as given by (11.40). The values of f_m and $G_{1,1}(f_m)$ were calculated using the full formulas of Section 11.2, and including the two refinements

Table 11.2. *Phase-sequence resonances m;[a] required voltage ratios V_{25}/V_1; predicted resonant frequencies f_m; resonant driving-point conductances $G_{1,\,in}(f_m) = I_1(0)/V_1 = G_{25,\,in}(f_m)$; ratios $|g_{N-m}(\pi/2)/g_m(\pi/2)|$ of two array-factor terms in (11.59) at $\theta = \pi/2$; and vertical directivities D_V[b]*

m	$\dfrac{V_{25}}{V_1}$	f_m (MHz)	$G_{1,\,in}(f_m)$ (mS)	$\left\|\dfrac{g_{N-m}(\pi/2)}{g_m(\pi/2)}\right\|$	D_V
47	i	30.652	2×10^{10}	1×10^{-1}	4.78
45	$-i$	30.589	4×10^{8}	2×10^{-3}	4.60
43	i	30.460	1×10^{7}	3×10^{-5}	4.41
41	$-i$	30.262	5×10^{5}	4×10^{-7}	4.21
39	i	29.987	3×10^{4}	5×10^{-9}	4.01
37	$-i$	29.624	2×10^{3}	6×10^{-11}	3.78
35	i	29.154	3×10^{2}	6×10^{-13}	3.54

[a]Excited when number of elements $N = 96$, radius $a = 0.28$ m, half-length $h = 1.9$ m, element separation $d = 3.1$ m, and number of second driven element $n = 25$.
[b]Taken from Fikioris *et al.* [20, Table 1]. © 1996 I.E.E.

in Section 11.6. It is seen that the second term in (11.59) will have a noticeable effect only in the first case of Table 11.2.

Far field of array in free space

The far field of the omnidirectional array is the product of the field of a single isolated antenna multiplied by the array factor. The far field of a vertical dipole in space that is electrically short and has the effective half-length h_e is

$$E_\theta^r = \frac{-i\omega\mu_0 2h_e I(0)}{4\pi} \frac{e^{ik_2 r_0}}{r_0} \sin\theta, \tag{11.63}$$

where k_2 is the free-space wave number. The far field of the circular array is thus

$$E_\theta^r = \frac{-i\omega\mu_0 2h_e I_1(0)}{4\pi} \frac{e^{ik_2 r_0}}{r_0} A^{(m)}(\theta) \sin\theta, \tag{11.64}$$

where $A^{(m)}(\theta)$ is given by the magnitude of the first term in (11.59b).

Figures 11.9 and 11.10 show the normalized far-field power pattern $|E_\theta^r(\theta,\phi)|^2$ in the plane $\theta = \pi/2$ of the dipoles' centers for the cases $m = 37$ and $m = 45$ of Table 11.2, respectively. Figures 11.9 and 11.10 were obtained using (11.47) and the full formulas in Sections 11.2 and 11.6 so that the effects of the rest of the phase sequences are included. The small oscillatory departure from the smooth omnidirectional field in the $m = 37$ case is due to the contributions from the rest of the phase sequences. The slight ripples in the $m = 45$ case are due to the contribution of the term corresponding to the second term in (11.59). The intermediate cases $m = 41$

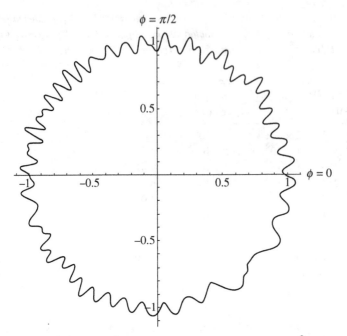

Figure 11.9 Normalized far-field power pattern $|E_\theta^r(\pi/2, \phi)|^2$ as function of polar angle ϕ in plane $\theta = \pi/2$ of dipoles' centers for $m = 37$ case of Table 11.2. Taken from Fikioris *et al.* [20, Fig. 3]. © 1996 I.E.E.

and $m = 43$ appear as smooth circles and are not shown here. Figure 11.11 shows the normalized far-field power patterns in the elevation plane $\phi = 0$ as a function of the polar angle θ. The $m = 45$ case is seen to be slightly more directive.

The array over the earth or sea

Suppose that the resonant array is located in region 2 (air) over region 1 (salt water, wet earth, lake water, dry earth), at a small distance d_0 over the air–earth boundary. Medium 1 is characterized by a complex wave number $k_1 = \omega\sqrt{\mu_0}\sqrt{\epsilon_1 + i(\sigma_1/\omega)}$, so that

$$\frac{|k_1|^2}{k_2^2} = \sqrt{\epsilon_{1r}^2 + \frac{\sigma_1^2}{4\pi^2 f^2 \epsilon_0^2}}, \tag{11.65}$$

where (i) for salt water, $\epsilon_{1r} = 80$ and $\sigma_1 = 4$ S/m, so that $|k_1|^2/k_2^2 = 2400$; (ii) for wet earth, $\epsilon_{1r} = 12$ and $\sigma_1 = 0.4$ S/m, so that $|k_1|^2/k_2^2 = 240$; (iii) for lake water, $\epsilon_{1r} = 80$ and $\sigma_1 = 0.004$ S/m, so that $|k_1|^2/k_2^2 = 80$; and (iv) for dry earth, $\epsilon_{1r} = 8$ and $\sigma_1 = 0.04$ S/m, so that $|k_1|^2/k_2^2 = 25$.

Consider a vertical electric dipole in region 2 (air) over region 1. If the electrical distance $k_2\rho$ from the dipole to the point of observation satisfies $k_2\rho < 2|k_1|^2/k_2^2$,

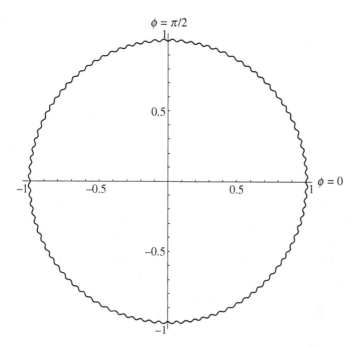

Figure 11.10 Normalized far-field power pattern $|E_\theta^r(\pi/2, \phi)|^2$ as function of polar angle ϕ in plane $\theta = \pi/2$ of dipoles' centers for $m = 45$ case of Table 11.2. Taken from Fikioris *et al.* [20, Fig. 4]. © 1996 I.E.E.

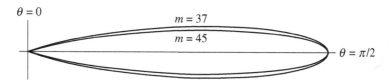

Figure 11.11 Normalized far-field power pattern $|E_\theta^r(\theta, 0)|^2$ as function of polar angle θ in plane $\phi = 0$ of dipoles' centers for $m = 37$ and $m = 45$ cases of Table 11.2. Each pattern is normalized to its maximum at $\theta = \pi/2$. Taken from Fikioris *et al.* [20, Fig. 5]. © 1996 I.E.E.

then the z-component of the electric field E_{2z} in region 2 is the same as when region 1 is a perfect conductor [21]. The range $k_2\rho < 2|k_1|^2/k_2^2$ includes both the intermediate and the near regions, the latter being defined by the stricter condition $k_2\rho < 1$. In the $m = 37$ case of Table 11.2, the smallest element-to-element electrical distance is $k_2\rho = k_2d = 1.95$, and the largest element-to-element electrical distance is $k_2\rho = k_2(2R) = 59.5$. At least in cases (i)–(iii), therefore, all elements in the array are in each other's intermediate region and it is correct to assume that region 2 is perfectly conducting when estimating mutual-coupling effects. Note that these depend entirely on E_{2z}.

In order to provide an estimate of the coupling between the array in region 2 and its perfect image in region 1, the behavior of the z-directed electric field of an array in free space is investigated. The current distribution along the length of all elements in the resonant array is $\cos kz - \cos kh$. Two simplifying assumptions are made: (i) that the elements have infinitesimal thickness; and (ii) that the current distribution along their length is $\sin k(h - |z|)$. Thus, the currents are taken to be

$$\frac{I_l(z)}{I_l(0)} = \frac{\sin k(h - |z|)}{\sin kh}, \tag{11.66}$$

where $I_l(0)$ is given in (11.46). These approximations are adequate for our purposes since the dipoles are electrically short ($kh < \pi/2$) and the two distributions $\sin k(h - |z|)$ and $\cos kz - \cos kh$ are quite similar. Now use is made of the exact formulas (1.38) for the field of an infinitesimally thin dipole with a $\sin k(h - |z|)$ distribution. The origin of the cylindrical coordinates (ρ, ϕ, z) is placed at the center of the array, and the location of the center of dipole l is $(R, \phi_l, 0) = (R, 2\pi(l-1)/N, 0)$ in cylindrical coordinates. From (1.38c), it follows that the z-component e_{lz} of the electric field due to dipole l in the circular array is given everywhere by

$$e_{lz}(\rho, \phi, z) = \frac{i\omega\mu_0 I_l(0)}{4\pi \sin kh} \left(\frac{e^{ik_2 r_{tl}}}{k_2 r_{tl}} + \frac{e^{ik_2 r_{bl}}}{k_2 r_{bl}} - 2\cos kh \frac{e^{ik_2 r_{cl}}}{k_2 r_{cl}} \right), \tag{11.67}$$

where r_{tl}, r_{bl}, and r_{cl} are, respectively, the distances from the observation point to the top ($z = h$), bottom ($z = -h$), and center ($z = 0$) of dipole l in the array.

Upon using (11.46) and setting $\rho = R$, it follows that the total z-directed electric field of the resonant array is given by

$$\frac{4\pi \sin kh}{i\omega\mu_0 I_1(0)} E_z(R, \phi, z) =$$

$$\sum_{l=1}^{N} \exp\left[-i\frac{2\pi(l-1)m}{N} \right] \left(\frac{e^{ik_2 r_{tl}}}{k_2 r_{tl}} + \frac{e^{ik_2 r_{bl}}}{k_2 r_{bl}} - 2\cos kh \frac{e^{ik_2 r_{cl}}}{k_2 r_{cl}} \right). \tag{11.68}$$

The formulas for the distances r_{tl}, r_{bl}, and r_{cl} in the case where $\rho = R$ are

$$r_{tl} = r_{tl}(\phi, z) = \sqrt{4R^2 \sin^2 \frac{\phi - \phi_l}{2} + (h - z)^2} \tag{11.69}$$

$$r_{bl} = r_{bl}(\phi, z) = \sqrt{4R^2 \sin^2 \frac{\phi - \phi_l}{2} + (h + z)^2} \tag{11.70}$$

$$r_{cl} = r_{cl}(\phi, z) = \sqrt{4R^2 \sin^2 \frac{\phi - \phi_l}{2} + z^2}. \tag{11.71}$$

The magnitude of the normalized z-component of the electric field as calculated from (11.68)–(11.71) is plotted in Fig. 11.12 for the $m = 37$ case of Table 11.2 as the distance z varies from -12 to -2 m. It is seen that E_z decays rapidly and monotonically away from the array. The rate of decay is much more rapid than for a single isolated element. This phenomenon is related to the rapid decrease of the field

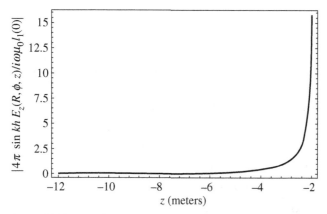

Figure 11.12 Magnitude of normalized z-component $|[(4\pi \sin kh)/i\omega\mu_0 I_1(0)]E_z(R,\phi,z)|$ of electric field as function of z for $\phi = 0$ and for $m = 37$ case of Table 11.2. Taken from Fikioris *et al.* [20, Fig. 6]. © 1996 I.E.E.

(surface wave) observed in infinite linear arrays [2]. This rapid decrease indicates that the coupling between the resonant array and its image is negligible even if the array is placed at a small distance above the surface of the earth.

Field of the array over earth or sea

General formulas for the three cylindrical components of the electromagnetic field of a vertical electric dipole with unit electric moment are given in [16, 17] subject only to the condition that the wave number of air (k_2) be small compared to the magnitude of the complex wave number (k_1) of the earth or sea. That is,

$$|k_1^2| \gg k_2^2 \quad \text{or} \quad |k_1| \geq 3k_2. \tag{11.72}$$

The formulas are valid at all points in the air, $z > 0$, with the dipole at any height d_0. When the three conditions

$$k_2\rho \geq 8|k_1^2|/k_2^2, \qquad d_0^2 \ll r_0^2, \qquad |k_2d_0/k_1r_0| \ll 1 \tag{11.73}$$

are satisfied, the procedure carried out to obtain the formulas for the cylindrical components may be extended to the spherical component $E_\theta^r(r_0, \theta)$. The result is

$$
\begin{aligned}
E_\theta^r(r_0, \theta) = \frac{-\omega\mu_0 2h_e I_0}{2\pi k_2} e^{ik_2r_0} &\left\{ \frac{ik_2}{r_0} \sin\theta \left[\frac{(\cos\theta + d_0/r_0)\cos(k_2d_0\cos\theta)}{\cos\theta + d_0/r_0 + k_2/k_1} \right.\right. \\
&\left. - \frac{i(k_2/k_1 - d_0/r_0\cos^2\theta)\sin(k_2d_0\cos\theta)}{\cos\theta + d_0/r_0 + k_2/k_1} \right] + \frac{k_2}{k_1 r_0^2} \\
&\left. \times \left(\frac{\sin\theta}{\cos\theta + d_0/r_0 + k_2/k_1} \right)^3 \left(\frac{k_2}{k_1}\cos\theta - \sin^2\theta \right) e^{ik_2d_0\cos\theta} \right\}, \tag{11.74}
\end{aligned}
$$

where

$$r_0 = (\rho^2 + z^2)^{1/2}, \tag{11.75}$$

h_e is the effective half-length of the dipole, and I_0 is the current at its center. Of particular interest here is the far field $E_\theta^r(r_0, \theta)$ in spherical coordinates when the dipole is quite close to the surface so that $k_2 d_0 \sim 0$ and $d_0/r_0 \sim 0$. In this case, (11.74) reduces to

$$E_\theta^r(r_0, \theta) = \frac{-\omega\mu_0 2 h_e I_0}{2\pi k_2} e^{ik_2 r_0} \left[\frac{ik_2}{r_0} \left(\frac{k_1 \sin\theta \cos\theta}{k_2 + k_1 \cos\theta} \right) \right.$$
$$\left. - \frac{k_1^2 [\sin^2\theta - (k_2/k_1)\cos\theta]\sin^3\theta}{k_2^2 r_0^2 [1 + (k_1/k_2)\cos\theta]^3} \right]. \tag{11.76}$$

In (11.74) and (11.76), the $1/r_0$ term is that obtained from the plane-wave reflection coefficient. It vanishes along the boundary defined by $\theta = \pi/2$. The $1/r_0^2$ term is the lateral wave that propagates in the air close to the boundary and continuously transfers energy into the earth or sea. Although it decreases with radial distance as $1/r_0^2$, it is multiplied by the very large factor k_1^2/k_2^2. The far field for the surface wave when $(z + d_0)/r_0 \ll 1$ is defined by the first condition in (11.73).

An alternative form is useful when the vertical heights z of the observation point are small compared to the radial distance from the transmitter. Specifically, when

$$|k_1 z| < k_2 r_0, \qquad \sin\theta = \frac{\rho}{r_0}, \qquad \cos\theta = \frac{z}{r_0} \ll 1 \tag{11.77}$$

(11.76) reduces to

$$E_\theta^r(r_0, \theta) = \frac{-\omega\mu_0 2 h_e I_0}{2\pi k_2} \frac{k_1}{k_2} \frac{e^{ik_2 r_0}}{r_0^2} \left(ik_2 z - \frac{k_1}{k_2} \right). \tag{11.78}$$

This formula shows that for observation points at any fixed height $z < |k_2 r_0/k_1|$, the incident electric field is proportional to $1/r_0^2$. This includes both the surface-wave term and the space-wave term. Note that the latter vanishes when $z = 0$ so that the entire field along the surface is due to the lateral wave.

The far field of the resonant circular array of dipoles at a small height $d_0 \sim 0$ over the earth or sea is given by (11.74), (11.76), or (11.78) multiplied by $A^{(m)}(\theta)$ as given by the magnitude of the first term in (11.59). The pancake-like pattern represented by $A^{(m)}(\theta)$ is enhanced by the low-altitude field represented by (11.74), (11.76), or (11.78). In particular, the important field close to the surface of the earth is maximized and virtually no field is maintained at upward angles that are not close to $\theta \sim \pi/2$.

11.8   Appendix: the various kernels for the circular array

In the course of this chapter, four different kernels for the mth phase-sequence integral equation are mentioned. In this appendix, their relationship and applicability are discussed. Consider the coupled integral equations for the current distributions $I_l(z)$ in an array of identical, perfectly conducting, parallel, non-staggered tubular dipoles of radius a and half-length h (the array may or may not be circular, and more than one element may be driven):

$$4\pi\mu_0^{-1} A_{zl}(z) \equiv \sum_n \int_{-h}^h I_n(z') K_{nl}(z - z')\, dz' = -\frac{j4\pi}{\zeta_0}\left(C_l \cos kz + \frac{V_l}{2}\sin k|z|\right).$$

(11.79)

In (11.79), the constants C_l are determined from the conditions $I_l(h) = 0$, $A_{zl}(z)$ is the z-directed vector potential on the surface of dipole l, and V_l is the voltage driving element l, with $V_l = 0$ if the element is parasitic. Each term in the sum on the left-hand side of the integral equation is the vector potential on element l due to the current $I_n(z')$ on element n; the kernel K_{nl} associated with each vector potential is a "self-interaction kernel" if $n = l$ so that $K_{nl}(z) = K_{ll}(z) = K_{11}(z)$ or a "mutual interaction kernel" if $n \neq l$. The various mth phase-sequence kernels referred to in this chapter result from the following four sets of kernels for the general array.

1. The original kernels, employed in Chapters 1–7:

$$K_{ll}(z) = \frac{\exp[-jk(z^2 + a^2)^{1/2}]}{(z^2 + a^2)^{1/2}}$$

(11.80a)

$$K_{nl}(z) = \frac{\exp[-jk(z^2 + b_{nl}^2)^{1/2}]}{(z^2 + b_{nl}^2)^{1/2}}, \qquad n \neq l$$

(11.80b)

where b_{nl} is the distance between the axis of dipole l and the axis of dipole n.

2. The improved kernels of [8] and [9]:

$$K_{ll}(z) = \frac{1}{2\pi}\int_{-\pi}^{\pi} \frac{\exp\{-jk[z^2 + 4a^2\sin^2(\phi'/2)]^{1/2}\}}{[z^2 + 4a^2\sin^2(\phi'/2)]^{1/2}}\, d\phi'$$

(11.81a)

$$K_{nl}(z) = \frac{1}{4\pi^2}\int_{-\pi}^{\pi}\int_{-\pi}^{\pi} \frac{e^{-jkR_{nl}(z,\phi,\phi')}}{R_{nl}(z, \phi, \phi')}\, d\phi'\, d\phi; \qquad n \neq l,$$

(11.81b)

where

$$R_{nl}(z, \phi, \phi') = [z^2 + (a\sin\phi - a\sin\phi')^2 + (a\cos\phi - a\cos\phi' - b_{nl})^2]^{1/2}$$

(11.81c)

is the distance between a point on the surface of dipole l and the surface of dipole n. It is illustrated in Fig. 11.13.

3. The modified kernels introduced in Section 11.2. The mutual interaction kernels $K_{nl}(z)$ are in (11.80b) and the self-interaction kernel is

$$K_{ll}(z) = \frac{\cos[k(z^2 + a^2)^{1/2}]}{(z^2 + a^2)^{1/2}} - j\frac{\sin kz}{z}.$$

(11.82)

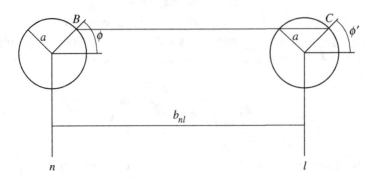

Figure 11.13 The distance BC is the projection of $R_{nl}(z, \phi, \phi')$ onto the $z = 0$ plane.

4. The refined modified kernels discussed in Section 11.6. The mutual interaction kernels are again the same as in (11.80b) and the self-interaction kernel is

$$K_{ll}(z) = \frac{1}{2\pi} \int_{-\pi}^{\pi} \frac{\cos\{k[z^2 + 4a^2 \sin^2(\phi'/2)]^{1/2}\}}{[z^2 + 4a^2 \sin^2(\phi'/2)]^{1/2}} \, d\phi' - j \, \frac{\sin kz}{z}. \qquad (11.83)$$

The improved kernels (11.81a), (11.81b) are, by the nature of their derivation, inherently more accurate than the rest. However, they are significantly more complicated so that adequate simpler alternatives are needed. Note that the imaginary parts of (11.80b), (11.82), and (11.83) are equal to the corresponding imaginary parts of (11.81a), (11.81b) with the radius a set to zero, so that statements concerning the imaginary parts of the improved kernels may be specialized to obtain corresponding statements for the imaginary parts of the modified or refined modified kernels. This is not true, however, for the real parts.

For arrays of a small number of elements N, where narrow resonances do not occur, the original kernels (11.80a), (11.80b) are adequate. A discussion of the relationship between (11.80a) and (11.81a) in the case $N = 1$ (where the latter kernel is exact) is given in Chapter 1 and in [22]. It is believed that the use of any of the sets of kernels in approximate solutions to the integral equations such as the two-term theory would not make a noticeable difference when N is small. Significant differences exist when N is large and narrow resonances occur. The case of a non-driven infinite linear array and that of a large circular array will be examined in turn.

Case 1

It is shown in [8] and [9] that a non-driven infinite linear array of equispaced elements (here, $b_{nl} = |n - l|d$) may possess resonances of zero width where the currents satisfy $I_l^{(\beta)}(z) = I_0^{(\beta)}(z)e^{j\beta l}$. Thus, the integral equation for $I_0^{(\beta)}(z)$ is

$$\int_{-h}^{h} I_0^{(\beta)}(z') K^{(\beta)}(z - z') \, dz' = -\frac{j4\pi}{\zeta_0} C_0 \cos kz, \qquad (11.84)$$

where

$$K^{(\beta)}(z) = \sum_{l=-\infty}^{\infty} K_{0l}(z)e^{j\beta l} = K_{00}(z) + 2\sum_{l=1}^{\infty} K_{0l}(z)\cos\beta l. \tag{11.85}$$

It is shown in [8] and [9] that $K^{(\beta)}(z)$ is real for all z when $d/\lambda < \beta/2\pi < \frac{1}{2}$ and when the improved kernels (11.81a), (11.81b) are used. It follows that this is also true when the modified and refined modified kernels are used. Hence, with these kernels, (11.84) is a real equation and this suggests the possibility of real solutions $I_0^{(\beta)}(z)$ with $I_0^{(\beta)}(h) = 0$ for proper choices of d/λ, h/λ, and a/λ. However, $K^{(\beta)}(z)$ is not real if the original kernels (11.80a), (11.80b) are used, so that (11.80a), (11.80b) are inadequate in this case.

Case 2

Next consider the integral equations for the currents $I_l^{(m)}(z)$ in the mth phase sequence for a large circular array. The driving voltages (and, therefore, the currents) satisfy $V_l^{(m)} = V_1^{(m)} \exp[j2\pi(l-1)m/N]$ and the integral equation for $I_1^{(m)}(z)$ is

$$\int_{-h}^{h} I_1^{(m)}(z')K^{(m)}(z-z')\,dz' = -\frac{j4\pi}{\zeta_0}\left(C_1^{(m)}\cos kz + \frac{V_1^{(m)}}{2}\sin k|z|\right), \tag{11.86}$$

where

$$K^{(m)}(z) = \sum_{l=1}^{N} K_{1l}(z)e^{j2\pi(l-1)m/N} = K_{11}(z) + \sum_{l=2}^{N/2+1} \xi_l K_{1l}(z)\cos\left[\frac{2\pi(l-1)m}{N}\right] \tag{11.87}$$

and ξ_l is defined in (11.14). Here, the distances b_{nl} are $b_{nl} = b_{1,l-n+1} = d\sin(|l - n|\pi/N)/\sin(\pi/N)$. It is shown in [8] and [9] that $\text{Im}\{K^{(m)}(z)\}$ is exponentially small in N for all z when $d/\lambda < m/N < \frac{1}{2}$ and when the improved kernels (11.81a), (11.81b) are used; an asymptotic formula for $\text{Im}\{K^{(m)}(z)\}$ is derived. The asymptotic formula (11.18) for $\text{Im}\{K^{(m)}(z)\}$ when the modified or refined modified kernels are used may then be obtained from the results of [8] and [9]; $\text{Im}\{K^{(m)}(z)\}$ is exponentially small in this case as well – the only difference is a small overall multiplicative factor of $J_0^2(ka)$. If the original kernels (11.80a), (11.80b) were used, $\text{Im}\{K^{(m)}(z)\}$ would not be exponentially small. As seen earlier in this chapter, this property is crucial for an accurate description of the resonances.

The preceding analysis shows: (i) that (11.82) or (11.83) together with (11.80b) are simpler, adequate alternatives to (11.81a), (11.81b); and (ii) that (11.80a), (11.80b) are not adequate for the cases of an infinite linear array or of a large circular array. In fact, one can find cases in which the original theory gives meaningless results. For example, application of the two-term theory for a large circular array with $N = 90$, $h/\lambda = 0.2$, and $a/\lambda = 0.05$ yields the *negative* driving-point conductance $G_{1,1} = -97$ mA/V. It is believed that statements (i) and (ii) above are also valid for large non-circular arrays.

Note that all discussions up to this point concern the imaginary parts of the kernels only; (11.82) differs from (11.83) only in the real part. It seems logical to retain the "exact" real part of the self-interaction kernel for calculations where high precision is needed, especially since the resulting two-term theory formulas are not much more complicated numerically. In any case, the refined modified kernels (11.83) and (11.80b) (together with the square-root end correction of Section 11.6) are the ones that give the best agreement between two-term theory calculations and experiment (see Section 12.7).

12 Resonances in large circular arrays of highly conducting dipoles

In Chapter 11, the phenomenon of resonances in large circular arrays of dipoles is discussed. It is assumed throughout that the dipoles are perfectly conducting. The effect of ohmic losses is considered in this chapter. It is assumed here that one dipole is driven. The case where two dipoles are driven is a very simple extension. In Sections 12.1–12.6, an array of highly conducting dipoles in free space is examined. The two-term theory is extended so that it applies to this theoretical model. The model is mathematically equivalent to the physically unrealizable one of an array consisting of highly conducting monopoles over a ground plane which is *perfectly* conducting: one can determine the currents in the latter model by multiplying those of the former by a factor of 2.

The case of a circular array of highly conducting monopoles over a highly conducting ground plane is examined briefly in Section 12.7. An approximate method is outlined which allows calculation of the admittances in this case by slightly modifying the theory of Sections 12.1–12.6 in which a lossless ground plane is assumed. The model of Section 12.7 closely approximates experimental conditions. The theoretical curve (driving-point admittance as a function of frequency) obtained after the effect of the imperfectly conducting ground plane is taken into account is compared in this section to a corresponding experimental curve and the agreement is very good. Finally, Section 12.8 is an appendix which contains the formulas for the large circular array of highly conducting dipoles in a form convenient for computer implementation.

12.1 Introduction

In the previous chapter, it was seen that the resonances in a lossless array become rapidly narrower and the currents around the array become much larger as the varying parameter f or d/λ (or, as the integer parameter m characterizing the resonance) becomes larger. Also, the resonances become rapidly narrower as N becomes larger; or, to be more precise, a particular m/N resonance becomes rapidly narrower as the number N of elements becomes larger. As a numerical example, the driving-point conductance is predicted by the theory for lossless elements to be of the order of

10^{10} mS for the $m = 45$ resonance of the $N = 90$ element experiment.[1] With the aid of the theory presented here, the two situations described above ("fixed-N array" and "fixed-m/N array") are examined for the case of a lossy array. It is seen that the behavior of the resonant lossy array is quite different from that of the perfectly conducting case.

The interest here is in large arrays where the elements are highly conducting (for example, a $N = 90$ element circular array of brass dipoles at microwave frequencies). After integral equations for the current distributions are developed, an approximate two-term solution is proposed in which the current distributions along the elements are written as a linear combination of $\sin k(h - |z|)$ and $(\cos kz - \cos kh)$, just as in the lossless case. When the conductivity of the dipoles is small, the situation is quite different. The problem of a single isolated dipole of small conductivity has been studied in the past, both theoretically [2–4] and experimentally [5]. It was found that the current distribution along the element changes significantly from the lossless case.

The starting points for the derivation of the integral equations are the well-known concepts of skin effect and internal impedance [6]. Suppose that a current-carrying cylinder has radius a, conductivity σ, and permeability μ_0. The skin depth d_s is defined as

$$d_s = \frac{1}{\sqrt{\pi f \sigma \mu_0}}, \tag{12.1}$$

where $f = \omega/2\pi$ is the operating frequency. Under the condition

$$d_s \ll a \quad \text{or} \quad a\sqrt{\omega\mu_0\sigma} \gg 1 \tag{12.2}$$

the current is principally confined to a thin layer of thickness d_s near the surface of the cylinder. The distribution of the current density $J_z(\rho)$, (and, also, of the vector potential and axial electric field) inside the cylinder as a function of the radial distance ρ is given by

$$J_z(\rho) = J_z(a)\sqrt{\frac{a}{\rho}}\, e^{-(a-\rho)/d_s}\, e^{-j(a-\rho)/d_s}. \tag{12.3}$$

The ratio of the axial electric field $E_z(a)$ at the surface $\rho = a$ at a given cross-section to the total current $I_z = \int_0^a J_z(\rho)2\pi\rho\, d\rho$ across that cross-section is called the internal impedance per unit length z^i. It is given by

$$z^i = r^i + jx^i = \frac{E_z(a)}{I_z} = \frac{1+j}{2\pi a d_s \sigma} = (1+j)\frac{1}{2\pi a}\sqrt{\frac{\pi f \mu_0}{\sigma}}. \tag{12.4}$$

---

[1] [1] Chapter 8.

The same formula for the internal impedance per unit length holds for a tubular conductor of radius a and wall thickness $a - a_1$ provided that the wall thickness is much larger than the skin depth, i.e.

$$d_s \ll a - a_1 \qquad \text{and} \qquad d_s \ll a. \tag{12.5}$$

If, however, the wall thickness is very small compared to the skin depth,

$$a - a_1 \ll d_s \ll a, \tag{12.6}$$

then the internal impedance per unit length is purely resistive and is given by

$$z^i = r^i = \frac{1}{2\pi a(a - a_1)\sigma}. \tag{12.7}$$

12.2 Integral equations

Consider a circular array of N identical, parallel, non-staggered, lossy dipoles of length $2h$ and radius a. Element 1 is center-driven by a voltage V_1 and elements $2, 3, \ldots, N$ are parasitic. Integral equations that take the ohmic losses into account are readily derived from the boundary condition for the tangential electric field $E_{zl}(z)$ on the surface of any dipole l, namely,

$$E_{zl}(z) = -V_1 \delta_{l,1} \delta(z) + z^i I_l(z), \tag{12.8}$$

where z^i is the internal impedance per unit length, $I_l(z)$ is the current on dipole l, and

$$\delta_{l,n} = \begin{cases} 1, & l = n \\ 0, & \text{otherwise.} \end{cases} \tag{12.9}$$

The detailed derivation is contained in [1]. The final form of the integral equations is[2]

$$\sum_{n=1}^{N} \int_{-h}^{h} I_n(z') \left[K_{nl}(z - z') + \delta_{l,n} K_L(z - z') \right] dz'$$

$$= \frac{-j4\pi}{\zeta_0} \left(C_l \cos kz + \delta_{l,1} \frac{V_1}{2} \sin k|z| \right); \quad -h < z < h, \quad l = 1, \ldots, N, \tag{12.10}$$

where

$$\frac{K_L(z)}{k} = \frac{-j2\pi z^i}{k\zeta_0} \sin k|z|. \tag{12.11}$$

[2] [1] equation (6.16).

The only difference between the integral equations (12.10) for lossy elements and the integral equations (10.1) of Chapter 10 for lossless elements is that the self-interaction kernel now includes the additional term $K_L(z)$ proportional to the internal impedance per unit length. This form of the integral equations was derived especially for the case of a large circular array.

The integral equations (12.10) may be decoupled via the method of symmetrical components just as in the lossless case, so that the mth phase-sequence integral equation becomes

$$\int_{-h}^{h} I^{(m)}(z') \left[K^{(m)}(z - z') + K_L(z - z') \right] dz'$$

$$= \frac{-j4\pi}{\zeta_0} \left(C_1^{(m)} \cos kz + \frac{V_1^{(m)}}{2} \sin k|z| \right), \tag{12.12}$$

where $V_1^{(m)} = V_1/N$.

The case of tubular dipoles with walls much thicker than the skin depth is of particular interest. In this case, z^i is given by (12.4) so that $r^i = x^i$ and the real and imaginary parts $K_{RL}(z)$ and $K_{IL}(z)$ of $K_L(z)$ are equal in magnitude. They are given by

$$\frac{K_{RL}(z)}{k} = \frac{-K_{IL}(z)}{k} = \frac{\Phi}{2} \sin k|z|, \tag{12.13}$$

where

$$\frac{\Phi}{2} = \frac{2\pi r^i}{k\zeta_0} = \frac{\lambda r^i}{\zeta_0} = \frac{d_s}{2a} = \frac{1}{\sqrt{2}} \frac{1}{a\sqrt{\omega\mu_0\sigma}} = \frac{1}{2a/\lambda} \sqrt{\frac{\epsilon_0 f}{\pi\sigma}} \tag{12.14}$$

is the dimensionless parameter determining the change in both the real and the imaginary parts of the kernel. The notation Φ is in accordance with the literature [4, 5].

For brass dipoles ($\sigma = 1.4 \times 10^7$ S/m), the skin depth is $d_s = (\pi f \mu_0 \sigma)^{-1/2} = 2.69 \times 10^{-6}$ m at $f = 2.5$ GHz and, for a radius of $a = 3.175 \times 10^{-3}$ m, the parameter $\Phi/2$ has the value

$$\frac{\Phi}{2} = \frac{1}{2a/\lambda} \sqrt{\frac{\epsilon_0 f}{\pi\sigma}} = 4.23 \times 10^{-4}. \tag{12.15}$$

It is seen that $\Phi/2 \ll 1$ so that the real and imaginary parts of $K_{11}(z)/k$ in (12.10) are negligibly affected as functions of kz by the presence of ohmic losses (at least if the dipoles are not many wavelengths long), and the integral equations remain essentially the same. Thus, highly conducting elements may be treated as though they were perfectly conducting in ordinary antenna array problems involving a small number of elements.

If the material is not highly conducting, then a significant change can result. For example, in the experimental study of [5], imperfectly conducting dipoles were constructed by coating a dielectric cylinder with a thin layer of resistive paint. The value of $\Phi/2$ in this experiment ranged from 0.62 to 2.23. In this case, the imaginary part of the kernel and the integral equations change significantly. (The real part of the kernel stays the same here because the thickness of the coating is much smaller than the skin depth so that z^i is given by (12.7) and is purely resistive.)

For a large circular array, the situation is quite different from both of the cases described above. The mth phase-sequence kernel now includes the term $K_L(z) = K_{RL}(z) + jK_{IL}(z)$ of (12.13). It was seen in Section 11.2 that the imaginary part $K_I^{(m)}(z)$ of the lossless kernel is small when N is large and $d/\lambda < m/N$. In this case, the presence of losses in the elements can make a noticeable difference even in the case of highly conducting elements ($\Phi/2 \ll 1$). By contrast, the real part $K_{1R}(z) + K_{\Sigma R}^{(m)}(z)$ of the lossless mth phase-sequence kernel is of order 1 for large N, so that $K_{RL}(z)$ may be neglected when $\Phi/2 \ll 1$.

The effect of the frequency on a large circular array of highly conducting dipoles may be deduced directly from the integral equations. Except for the new term in the integral equations (12.12) involving $\Phi/2$, the integral equations scale (i.e. they do not change if the frequency is changed provided that the electrical parameters h/λ, a/λ, and d/λ remain the same). The case where the electrical parameters are fixed is the one of interest when narrow resonances in circular arrays are desired; resonances are known to occur only if the electrical parameters are chosen from the limited ranges described in Section 11.3. The effect of ohmic losses becomes more pronounced when $\Phi/2$ becomes larger. It is seen from the last expression in (12.14) that $\Phi/2$ is an increasing function of the frequency when h/λ, a/λ, and d/λ are fixed. Thus, if an array of fixed h/λ, a/λ, and d/λ is to be implemented at two different frequencies (physically larger values of h, a, and d are required at the lower frequency), the effect of ohmic losses will be more pronounced at the higher frequency.

12.3   Two-term theory

Assume that the circular array satisfies the conditions (11.1). The theory presented here incorporates the change in the self-part of the kernel directly in the two-term theory in Section 11.2 without making any other changes. Thus, the current distributions on the elements are assumed to remain the same as in the lossless case. The solution is still given by (11.2)–(11.17) but with $K_I^{(m)}(z)$ replaced by $K_I^{(m)}(z) + K_{IL}(z)$. Equivalently, one may replace the parameters $P_I^{(m)}$ and $D_I^{(m)}$ by $P_I^{(m)} + P_{IL}$ and $D_I^{(m)} + D_{IL}$, respectively. The additional parameters P_{IL} and D_{IL} are proportional to $\Phi/2$. They are given by

$$P_{IL} = \frac{-1}{1 - \cos kh} \int_{-h}^{h} \sin k(h - |z|)[\cos kh \, K_{IL}(z) - K_{IL}(h - z)]\,dz \tag{12.16}$$

$$D_{IL} = \frac{1}{1 - \cos kh} \int_{-h}^{h} (\cos kz - \cos kh)[\cos kh \, K_{IL}(z) - K_{IL}(h - z)]\,dz. \tag{12.17}$$

With the expression (12.13) for $K_{IL}(z)$, the integrations in (12.16) and (12.17) may be performed. The resulting formulas for the new parameters are

$$P_{IL} = \frac{\Phi}{2} \frac{1}{1 - \cos kh} \left(-kh + \tfrac{1}{2} \sin 2kh\right) \tag{12.18}$$

$$D_{IL} = \frac{-\Phi}{2} \frac{1}{1 - \cos kh} (2 \cos kh - kh \sin kh - 1 - \cos 2kh). \tag{12.19}$$

Thus, the driving-point admittance is given by

$$Y_{1,1} = \frac{I_1(0)}{V_1}$$

$$= \frac{j2\pi}{\zeta_0 \Psi_{dR} \cos kh} \left[\sin kh + \frac{1}{N} \sum_{m=0}^{N/2} \xi^{(m)} T^{(m)} (1 - \cos kh)\right], \tag{12.20}$$

where

$$T^{(m)} = \frac{P_R^{(m)} + j(P_I^{(m)} + P_{IL})}{D_R^{(m)} + j(D_I^{(m)} + D_{IL})} \tag{12.21}$$

and where, for simplicity, N is assumed to be even.

Similar formulas may be obtained if the square-root end correction of Section 11.6 is taken into account. The complete formulas for both cases are derived in Section 6.5 of [1], and are summarized in Section 12.8 in a form suitable for computer implementation. The qualitative behavior of large resonant circular arrays is of concern in the next section; for this purpose, the simpler version of the two-term theory outlined above is adequate.

12.4   Qualitative behavior

The behavior of a fixed-N array and that of a fixed-m/N array and their differences from the lossless case will now be discussed. The first observation is that the positions of the resonances will remain the same as in the lossless case since the parameter $D_R^{(m)}$ which determines these positions remains unchanged. Although the remaining results of this section may be obtained by studying the lossy two-term theory formulas directly, it is instructive to use the formulas to derive equivalent circuits for the two situations described above and study the equivalent circuits.

If the circular array is at or near its mth phase-sequence resonance, i.e. $D_I^{(m)} + D_{IL}$ is small and $D_R^{(m)}$ is close to its zero, then the formula (12.20) for the self-admittance may be approximated by the methods of Sections 11.3 and 11.4. The details are in Section 6.4 of [1]. If f_m is the resonant frequency, the following approximate formula is obtained:[3]

$$Y_{1,1}(f) = G_{1,1}(f) + jB_{1,1}(f) \doteq \frac{1/(R_{rad}^{(m)} + R_{loss})}{1 + j2Q(f - f_m)/f_m}, \tag{12.22}$$

where

$$R_{rad}^{(m)} = \frac{N}{\xi^{(m)}} \frac{D_I^{(m)}}{H^{(m)}} \qquad \text{(units: ohms)} \tag{12.23}$$

$$R_{loss} = \frac{N}{\xi^{(m)}} \frac{D_{IL}}{H^{(m)}} \qquad \text{(units: ohms)} \tag{12.24}$$

$$Q = \frac{|\alpha^{(m)}| f_m}{2(D_I^{(m)} + D_{IL})} \qquad \text{(dimensionless).} \tag{12.25}$$

In (12.23)–(12.25), $\alpha^{(m)}$ is the slope of $D_R^{(m)}$ near its zero, $\xi^{(m)}$ is given by (12.47) and

$$H^{(m)} = \frac{2\pi(1 - \cos kh)}{\zeta_0 \Psi_{dR} \cos kh} P_R^{(m)} \tag{12.26}$$

is a quantity which depends on the real part of the mth phase-sequence kernel. Equation (12.22) is the same as the formula for the input admittance of a high-Q series RLC circuit when its operating frequency is close to its resonant frequency.[4] Thus, (12.22) shows that a circular array at or near its mth phase-sequence resonance is roughly equivalent to a high-Q RLC circuit. "Equivalence" should be understood in the sense that the driving-point conductance and susceptance in the two cases have the same frequency response. The equivalent circuit for the resonant circular array has two resistances in series and is pictured in Fig. 12.1.

Fixed-N array

This rough equivalence gives a simple picture of the qualitative behavior of an array where the frequency is varied to obtain a series of narrow resonances at frequencies f_m where $m \leq N/2$ (fixed-N array). The array is equivalent to a finite sequence

---

[3] [1] equation (6.40).   [4] [1] equation (6.35).

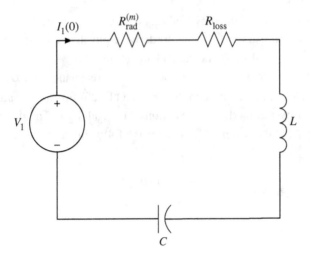

Figure 12.1 Equivalent circuit for circular array at or near its *m*th phase-sequence resonance.

of RLC circuits. The resonant frequency of the *m*th RLC circuit is f_m. The behavior of the array may be determined by studying the dependence of the parameters of the RLC circuits on *m*. It may be shown that $R_{\text{rad}}^{(m)}$ decreases rapidly as *m* increases. R_{loss}, on the other hand, is approximately independent of *m*. The case when $m = N/2$ (when $N =$ even) is an exception: When $m = N/2$, R_{loss} increases by a factor of 2 because of the presence of $\xi^{(m)}$ in (12.23). Hence the finite sequence of high-Q RLC circuits is obtained by rapidly decreasing $R_{\text{rad}}^{(m)}$ as *m* (or the frequency *f*) increases. Assuming that $Q \gg 1$, one can define two regions depending on the relative size between $R_{\text{rad}}^{(m)}$ and R_{loss}:

1. The region of rapid increase, in which $R_{\text{rad}}^{(m)} \gg R_{\text{loss}}$ or $\Phi/2 \ll |K_I^{(m)}| \ll 1$. Here, the ohmic losses do not matter. Since $R_{\text{rad}}^{(m)} + R_{\text{loss}} \doteq R_{\text{rad}}^{(m)}$, the increase of $G_{1,1}$ and Q with *m* is, just as in the lossless case, very rapid.
2. The saturation region, in which $R_{\text{rad}}^{(m)} \ll R_{\text{loss}}$ or $|K_I^{(m)}| \ll \Phi/2 \ll 1$. Here, the ohmic losses are dominant. Since $R_{\text{rad}}^{(m)} + R_{\text{loss}} \doteq R_{\text{loss}}$ which is independent of *m*, $G_{1,1} = G_{1,1 \, \text{sat}}$ stays constant as a function of *m*. In the case when $m = N/2$, the resonant $G_{1,1}$ drops to the value $\frac{1}{2} G_{1,1 \, \text{sat}}$.

Fixed-*m*/*N* array

A similar analysis may be carried out for the case of a fixed-*m*/*N* array. $R_{\text{rad}}^{(m)}$ decreases rapidly as *N* increases. The parameter $H^{(m)}$ is roughly independent of *N* when *m*/*N* is fixed. Thus, the value of the resistance R_{loss} increases linearly as *N* increases. Hence, a "fixed-*m*/*N* array" is equivalent to a series of RLC circuits. For each *N*, the RLC circuit is obtained by rapidly decreasing $R_{\text{rad}}^{(m)}$ and linearly increasing R_{loss}. Assuming again that $Q \gg 1$, two regions are distinguished:

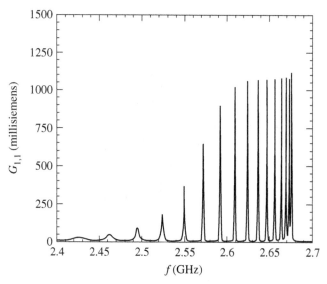

Figure 12.2 Driving-point conductance $G_{1,1}(f)$ as function of frequency for $N = 90$, $h = 0.858$ in, $2a = 1/4$ in, $2R = 40$ in, and $\sigma = 1.4 \times 10^7$ S/m.

1. The region of rapid increase, in which $R_{\text{rad}}^{(m)} \gg R_{\text{loss}}$ or $\Phi/2 \ll |K_I^{(m)}| \ll 1$. Here, the ohmic losses do not matter. Since $R_{\text{rad}}^{(m)} + R_{\text{loss}} \doteq R_{\text{rad}}^{(m)}$, the increase of $G_{1,1}$ and Q with N is, just as in the lossless case, very rapid.
2. The region of decrease as $1/N$, in which $R_{\text{rad}}^{(m)} \ll R_{\text{loss}}$ or $|K_I^{(m)}| \ll \Phi/2 \ll 1$. Here, the ohmic losses are dominant. Since $R_{\text{rad}}^{(m)} + R_{\text{loss}} \doteq R_{\text{loss}}$ which varies linearly with N, $G_{1,1}$ and Q decrease as $1/N$.

The factor of N in the expression (12.23) for R_{loss} comes from the superposition of the phase sequences; the resonant *phase-sequence* conductances $G_{\text{res}}^{(m)}$ eventually become constant as N increases.

12.5 Numerical results

Figures 12.2 and 12.3 show numerical results for the driving-point conductance $G_{1,1}(f)$ and susceptance $B_{1,1}(f)$ for the parameters

$$N = 90, \ h = 0.858 \text{ in}, \ 2a = 1/4 \text{ in}, \ 2R = 40 \text{ in}, \ \sigma = 1.4 \times 10^7 \text{ S/m}, \quad (12.27)$$

of the experimental circular array in Chapter 8 of [1]. The frequency interval is 2.4 GHz $< f <$ 2.7 GHz. The results were obtained using the complete formulas given in Section 12.8, including the "exact" real part of the self term and the square-root end-corrected current for greater accuracy. Dipole admittances were multiplied by a factor of 2, so that the results correspond to the admittances of an array of brass

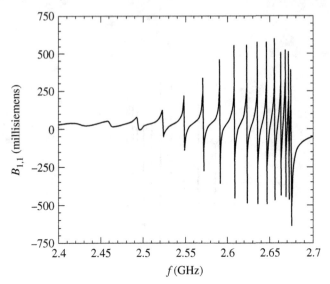

Figure 12.3 Driving-point susceptance $B_{1,1}(f)$ as function of frequency for $N = 90$, $h = 0.858$ in, $2a = 1/4$ in, $2R = 40$ in, and $\sigma = 1.4 \times 10^7$ S/m.

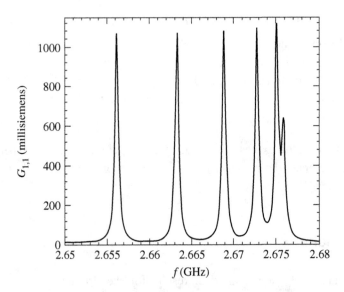

Figure 12.4 Like Fig. 12.2 but in frequency interval 2.65 GHz $< f < 2.68$ GHz.

monopoles over a perfectly conducting ground plane. Table 12.1 shows the resonant frequencies f_m and the values of resonant driving-point conductance $G_{1,1}(f_m)$. For comparison, the values $G'_{1,1}(f_m)$ obtained for the same parameters but assuming lossless elements are in the last column of Table 12.1. It is seen that the effect of the ohmic losses is negligible in the first few resonances and drastic in the later resonances.

Table 12.1. *Resonant frequencies* f_m;[a] *resonant driving-point conductances* $G_{1,1}$ *for* $\sigma = 1.4 \times 10^7$ *S/m; and resonant driving-point conductances* $G'_{1,1}$ *for* $\sigma = \infty$

m	f_m (GHz)	$G_{1,1}$ (mS)	$G'_{1,1}$ (mS)
29	2.4260	28.33	28.62
30	2.4623	47.47	48.77
31	2.4950	89.75	95.92
32	2.5241	183.96	216.44
33	2.5497	373.34	552.37
34	2.5722	653.61	1572.42
35	2.5919	901.02	4946.67
36	2.6090	1026.01	1.711×10^4
37	2.6238	1066.35	6.488×10^4
38	2.6365	1074.19	2.694×10^5
39	2.6473	1073.11	1.224×10^6
40	2.6562	1070.71	6.086×10^6
41	2.6634	1069.38	3.312×10^7
42	2.6689	1070.83	1.974×10^8
43	2.6728	1079.90	1.289×10^9
44	2.6752	1120.60	9.118×10^9
45	2.6759	637.16	1.824×10^{10}

[a] $N = 90$, $h = 0.858$ in, $2a = 1/4$ in, and $2R = 40$ in.

In Fig. 12.4, the results of Fig. 12.2 for $G_{1,1}(f)$ are shown in the frequency interval 2.65 GHz $< f <$ 2.68 GHz only where the resonances $m = 40$–45 occur. It is interesting that the last two resonances ($m = 44$ and $m = 45$) overlap. The same phenomenon would occur for the equivalent composite circuit of the circular array (many high-Q RLC circuits connected in parallel) if the resonant frequencies of two of the RLC circuits were very close.

12.6   Field pattern

The radiation field may be determined from the phase-sequence coefficients $T^{(n)}$ if formula (12.45) for T_l in Section 12.8 is substituted into (11.30). One obtains

$$\mathbf{E}(r, \theta, \phi) = \hat{\boldsymbol{\theta}} E_\theta = \hat{\boldsymbol{\theta}} \frac{-V_1}{\Psi_{dR} \cos kh} \frac{e^{-jkr}}{r} \left\{ F(\theta) e^{jkR \sin\theta \cos(\phi - \phi_1)} \right.$$

$$\left. + G(\theta) \sum_{n=0}^{N/2} \frac{\xi^{(n)}}{N} T^{(n)} A^{(n)}(\theta, \phi) \right\}, \tag{12.28}$$

where $A^{(n)}(\theta, \phi)$ is the array factor for the nth phase sequence (11.33b), and $F(\theta)$ and $G(\theta)$ are defined in (11.31) and (11.32). The field in the lossy case might be different from that of the lossless case for the following reason.

The properties of $A^{(n)}(\theta, \phi)$ were discussed in Section 11.5. In particular, it was seen that for sufficiently large n, the array factor is a small quantity. If a lossless array is at its $n = m$th phase-sequence resonance, $A^{(m)}(\theta, \phi)$ is small, $T^{(m)}$ is large, and the product $T^{(m)} \times A^{(m)}(\theta, \phi)$ remains large. Thus, the term in (12.28) for $n = m$ dominates and the complete radiation field possesses the properties of Section 11.5. If the elements are lossy, it has been seen that $T^{(m)}$ may be much smaller than in the lossless case. If $T^{(m)} \times A^{(m)}(\theta, \phi)$ is sufficiently large, the field of the lossy array will be essentially the same as that of the lossless array. If $T^{(m)} \times A^{(m)}(\theta, \phi)$ is small, contribution from other terms in equation (12.28) will be of importance and $E_\theta(r, \theta, \phi)$ will be different than in the lossless case.

In general, the effect of ohmic losses is more pronounced when m and N are larger. For the omnidirectional array of Section 11.7, numerical calculations based on (12.28) and the formulas in Section 12.8 show that the radiation field is significantly changed in the first few cases of Table 11.2 if the elements are made from copper. However, the omnidirectional radiation field in the $m = 37$ case with copper elements is the same (Figs. 11.9 and 11.11) as if the elements were perfectly conducting. A similar conclusion holds for a 90-element microwave beacon application [7].

12.7 The effect of a highly conducting ground plane

Introduction

In Sections 12.1–12.6, the problem of resonant circular arrays of highly conducting dipoles was addressed and a two-term theory taking the finite conductivity of the dipoles into account was developed. It was remarked that a circular array of highly conducting dipoles of length $2h$ is equivalent to an array of highly conducting monopoles of length h over a perfectly conducting ground plane of infinite extent. In practice, both the monopoles and the ground plane have a large (but finite) conductivity. In the $N = 90$ element experiment in Chapter 8 of [1], for example, the monopoles are made of brass and the ground plane is made of aluminum. The conductivities of the two materials are $\sigma_M = 1.4 \times 10^7$ S/m and $\sigma_G = 3.5 \times 10^7$ S/m, respectively. When the number of elements in the array is large and the array is at or near a narrow resonance, the finite conductivity of the ground plane will make a noticeable difference. In this section, an approximate method is outlined so that the ohmic losses of the ground plane may be taken into account as a perturbation to the theory of Sections 12.1–12.6 which assumes a lossless ground plane. The current distributions on the monopoles are assumed to remain the same.

Consider a circular array consisting of N monopoles of length h, radius a, and conductivity σ_M over an infinite ground plane of conductivity σ_G located at $z = 0$. Element 1 is driven and the rest are parasitic. N is large and the array is at or near its mth phase-sequence resonance so that the currents on the monopoles are the shifted cosine currents, namely,

$$\frac{I_l(z)}{I_1(0)} = \frac{\cos kz - \cos kh}{1 - \cos kh} \cos \frac{2\pi(l-1)m}{N}; \quad l = 1, \ldots, N, \quad 0 < z < h, \qquad (12.29)$$

where, without loss of generality for what follows, the current $I_1(0)$ on the base of element 1 is set to unity.

In the case $\sigma_G = \infty$, a surface current $\mathbf{J}_S(\rho, \phi)$ exists on the ground plane which may be determined from the tangential magnetic field $\mathbf{H}$ by the equation

$$\mathbf{J}_S = \hat{\mathbf{z}} \times \mathbf{H}. \qquad (12.30)$$

In the case of a highly conducting ground plane, there is a quasi-surface current distributed on a thin layer under the surface $z = 0$. In the treatment of problems involving highly conducting materials (for example, when losses in waveguides are calculated), it is usual to assume that the tangential magnetic field is the same as if the surface were perfectly conducting and that the quasi-surface current is a true surface current given by (12.30). Due to the finite conductivity, there is a total time-average power P_G dissipated as heat on the ground plane. If $\mathbf{J}_S$ is known, P_G may be found from the equation

$$P_G = \iint \frac{1}{2} \operatorname{Re}\{Z_G\} |\mathbf{J}_S|^2 \, dS = \iint \frac{1}{2} \sqrt{\frac{\pi f \mu_0}{\sigma_G}} |\mathbf{J}_S|^2 \, dS, \qquad (12.31)$$

where Z_G is the surface impedance of the ground plane. Formula (12.31) is an approximate equation adapted from the problem of a plane wave incident on a highly conducting surface. The integrand is the power dissipated as heat per unit surface area.

Outline of the procedure

A brief description of the method and of the approximations involved is presented here. Detailed calculations and formulas are contained in Chapter 7 of [1].

(A) The first problem is the calculation of the surface currents $\mathbf{J}_S$ on the ground plane for the circular array. This is simple in principle if the field due to a single monopole with a rotationally symmetric current is known.

If the magnetic field due to an isolated monopole placed at the origin is $\mathbf{h} = \hat{\boldsymbol{\phi}} h_\phi(\rho)$, then the surface currents due to an array of monopoles may be determined by superposition. The difficulty is that there are no simple formulas

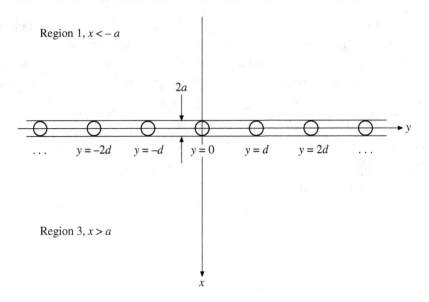

Region 1, $x < -a$

$2a$

$y = -2d$ $y = -d$ $y = 0$ $y = d$ $y = 2d$

Region 3, $x > a$

x

Figure 12.5 Infinite linear array.

for $h_\phi(\rho)$ for an isolated monopole of thickness a with a $\cos kz - \cos kh$ current distribution. A simple formula for $h_\phi(\rho)$ does exist for the case of an infinitely thin monopole with a $\sin k(h - z)$ current and is given in (1.38a).

For the monopole lengths $kh < \pi/2$ of interest, the distributions $\sin k(h - z)$ and $\cos kz - \cos kh$ are quite similar. Thus, the simplifying assumption is made that the monopoles of the array are infinitely thin and that their current distribution is $\sin k(h - z)$ so that the monopole currents (12.29) are replaced by

$$\frac{I_l(z)}{I_1(0)} = \frac{\sin k(h - z)}{\sin kh} \cos \frac{2\pi(l-1)m}{N}, \quad l = 1, \ldots, N. \tag{12.32}$$

(B) Once the surface currents are known, one may attempt to calculate the time-average power P_G dissipated as heat on the ground plane from (12.31). The integration extends to infinity. Consider, however, any array consisting of a finite number of monopoles with $\hat{\mathbf{z}}$-directed currents over a perfectly conducting ground plane. In the far zone, the magnetic field at $z = 0$ is parallel to the ground plane and decreases as $1/\rho$. The resulting surface currents will also decrease as $1/\rho$. Therefore, the integral in (12.31) diverges. The divergence of the integral is a result of the approximation that the currents on the imperfectly conducting ground plane are the same as if it were perfectly conducting. This divergence reveals that the approximation is not valid in the far zone. It is believed, however, that the approximation is valid near the array.

(C) Consider the infinite linear array, taken as the limit of the large circular array as $N \to \infty$. If y is the axis of the array and x is the direction perpendicular to

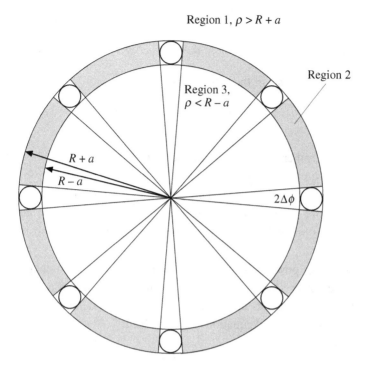

Figure 12.6 Circular array of $N = 8$ elements.

the array (Fig. 12.5), it is seen from the analysis in [1] that the surface currents decrease exponentially when $|x|$ is large and that they are periodic in y with period Nd. Thus, an integration of the form (12.31) is meaningful in the case of the infinite linear array if the integration is carried over a period in y; the integration over a period in y corresponds to the ϕ-integration for the large circular array.

(D) The formulas for $\mathbf{J}_S$ for the infinite linear array derived in [1] are appropriate for numerical evaluation when the observation point is not very close to the axis of the array; they are not suitable when the observation point approaches this axis.

For the reasons outlined above, P_G is calculated as the sum $P_{G2} + P_{G13}$ resulting from separating the plane $z = 0$ into three regions. These regions are illustrated in Fig. 12.6. The original circular array is shown in this figure to consist of $N = 8$ monopoles of radius a. P_{G2} is computed directly for the circular array as the contribution from the shaded region 2 of Fig. 12.6. This calculation is straightforward. Since it is assumed that the monopoles are infinitely thin, a small area near each monopole l is excluded when integrating to determine P_{G2}. Since $R/2a \gg 1$ ($R/2a = 160$ in the experimental study), the particular choice of this small area is not important. For numerical convenience, it is chosen to be

$$R - a < \rho < R + a, \quad \phi_l - \Delta\phi < \phi < \phi_l + \Delta\phi; \qquad \Delta\phi = \sin^{-1}\frac{a}{R} \quad (12.33)$$

so that region 2 consists of the subregions

$$R - a < \rho < R + a, \quad \phi_l + \Delta\phi < \phi < \phi_{l+1} - \Delta\phi; \qquad l = 1, \ldots, N.$$
$$(12.34)$$

P_{G13}, on the other hand, is computed from the infinite linear array of Fig. 12.5 as the contributions for regions 1 $(x < -a, -Nd/2 < y < Nd/2)$ and 3 $(x > a, -Nd/2 < y < Nd/2)$. The properties of the ground plane currents mentioned above (namely, the exponential decrease for large $|x|$ and the periodicity in the y direction) are verified by this calculation, which is somewhat involved.

(E) The time-average power P_M dissipated on the surface of the monopoles and the time-average power $P_G = P_{G2} + P_{G13}$ dissipated on the surface of the ground plane when element 1 has unit current at its base are computed in the manner described above. The final equations have the form

$$P_G = \frac{1}{2}\sqrt{\frac{\pi f \mu_0}{\sigma_G}} \, \psi_G; \qquad P_M = \frac{1}{2}\sqrt{\frac{\pi f \mu_0}{\sigma_M}} \, \psi_M, \qquad (12.35)$$

where the coefficients ψ_G and ψ_M are independent of the conductivities σ_G and σ_M (but depend on m).

An "effective conductivity of the monopoles" $\sigma_{M,\mathrm{eff}}^{(m)}$ is defined for each phase-sequence resonance m by the equation

$$\frac{1}{2}\sqrt{\frac{\pi f \mu_0}{\sigma_{M,\mathrm{eff}}^{(m)}}} \, \psi_M = \frac{1}{2}\sqrt{\frac{\pi f \mu_0}{\sigma_M}} \, \psi_M + \frac{1}{2}\sqrt{\frac{\pi f \mu_0}{\sigma_G}} \, \psi_G = P_M + P_G. \qquad (12.36)$$

The total loss $P_G + P_M$ in the case where the ground plane is imperfectly conducting is equal to the loss in the case where the ground plane is perfectly conducting and the monopoles have the perturbed conductivity $\sigma_{M,\mathrm{eff}}^{(m)}$. Equations (12.35)–(12.36) may be solved for $\sigma_{M,\mathrm{eff}}^{(m)}$ to obtain

$$\sigma_{M,\mathrm{eff}}^{(m)} = \sigma_M \left(\frac{P_M}{P_M + P_G}\right)^2. \qquad (12.37)$$

The perturbation to the theory of the Sections 12.1–12.6 results by replacing σ_M by $\sigma_{M,\mathrm{eff}}^{(m)}$ at each resonance.

Although the method described above involves several simplifying assumptions, the theoretical results obtained (driving-point conductance and susceptance) agree very well with the experimental results.

Table 12.2. *Resonant frequencies f_m; powers P_M, P_{G2}, and P_{G13} dissipated on monopoles and on regions 2 and 1, 3; and effective conductivities $\sigma_{M,\text{eff}}^{(m)}$ for parameters of experiment* [a]

m	f_m (GHz)	P_M (W/A^2)	P_{G2} (W/A^2)	P_{G13} (W/A^2)	$\sigma_{M,\text{eff}}^{(m)}$ (S/m)
29	2.4260	0.2567	0.0276	0.2277	0.3520×10^7
30	2.4623	0.2602	0.0276	0.2171	0.3718×10^7
31	2.4950	0.2634	0.0276	0.2078	0.3903×10^7
32	2.5241	0.2662	0.0277	0.1996	0.4074×10^7
33	2.5497	0.2688	0.0277	0.1923	0.4233×10^7
34	2.5722	0.2711	0.0278	0.1859	0.4378×10^7
35	2.5919	0.2731	0.0278	0.1802	0.4511×10^7
36	2.6090	0.2748	0.0279	0.1752	0.4630×10^7
37	2.6238	0.2763	0.0279	0.1709	0.4736×10^7
38	2.6365	0.2777	0.0279	0.1672	0.4829×10^7
39	2.6473	0.2788	0.0280	0.1640	0.4910×10^7
40	2.6562	0.2797	0.0280	0.1614	0.4977×10^7
41	2.6634	0.2805	0.0280	0.1593	0.5033×10^7
42	2.6689	0.2811	0.0281	0.1577	0.5075×10^7
43	2.6728	0.2815	0.0281	0.1565	0.5106×10^7
44	2.6752	0.2817	0.0281	0.1559	0.5124×10^7
45	2.6759	0.5636	0.0562	0.3113	0.5130×10^7

[a] [1] Chapter 8.

The effective monopole conductivity and numerical results

Table 12.2 shows the values of the resonant frequencies f_m, the time-average powers P_M, P_{G2}, and P_{G13} dissipated on the monopoles and on regions 2 and 1, 3 of the ground plane, and the effective monopole conductivity $\sigma_{M,\text{eff}}^{(m)}$ for the parameters of the experiment in Chapter 8 of [1]. The parameters N, h, a, $d = 2R\sin(\pi/N)$, and $\sigma_M = \sigma$ are given in (12.27). The conductivity σ_G of aluminum is given by $\sigma_G = 3.5 \times 10^7$ S/m.

It is seen from Table 12.2 that P_{G2} is significantly less than P_{G13}, which is the same order of magnitude as P_M. As a result, $\sigma_{M,\text{eff}}^{(m)}$ is significantly smaller than σ_M.

Table 12.3 shows the resulting values of the resonant self-conductance $G_{1,1}$. The corresponding values in the case $\sigma_G = \infty$ from Table 12.1 are repeated here for comparison. It is seen that the effect of the imperfectly conducting ground plane is minimal in the first (broadest) resonances but noticeable in the last resonances.

A continuous frequency-response curve for $G_{1,1}$ or $B_{1,1}$ may be conveniently obtained by using an interpolated value of $\sigma_{M,\text{eff}}^{(m)}$ at each frequency. Figures 12.7 and 12.8 show the frequency-response curves $G_{1,1}(f)$ and $B_{1,1}(f)$ thus obtained.

Table 12.3. *Resonant frequencies f_m; resonant driving-point conductances $G_{1,1}$ for parameters of experiment;[a] and resonant driving-point conductances $G'_{1,1}$ for $\sigma_G = \infty$*

m	f_m (GHz)	$G_{1,1}$ (mS)	$G'_{1,1}$ (mS)
29	2.4260	28.04	28.33
30	2.4623	46.34	47.47
31	2.4950	84.95	89.75
32	2.5241	163.37	183.96
33	2.5497	296.00	373.34
34	2.5722	449.60	653.61
35	2.5919	558.33	901.02
36	2.6090	609.27	1026.01
37	2.6238	628.74	1066.35
38	2.6365	636.60	1074.19
39	2.6473	641.02	1073.11
40	2.6562	644.84	1070.71
41	2.6634	649.49	1069.38
42	2.6689	657.24	1070.83
43	2.6728	675.68	1079.90
44	2.6752	734.49	1120.60
45	2.6759	476.01	637.16

[a] [1] Chapter 8.

In Fig. 12.9, the conductance $G_{1,1}(f)$ is shown in the frequency interval where the $m = 40$–45 phase-sequence resonances occur. The frequency interval is $2.65\,\text{GHz} < f < 2.68\,\text{GHz}$. When compared to the corresponding curve of Fig. 12.4 which assumes $\sigma_G = \infty$, it is seen that the additional losses introduced by the ground plane cause the $m = 45$ phase-sequence resonance to merge with the $m = 44$ resonance. Thus, the last peak of Fig. 12.9 actually corresponds to two phase-sequence resonances namely, $m = 44$ and $m = 45$.

Comparison of theory and experiment

In this section, the theoretical results are compared to experimental results. The experimental study is described in detail in Chapter 8 of [1]. The ground plane is made from aluminum and the monopoles are made from brass. The experimental setup is that of Fig. 11.2. Experimental results are obtained in [1] in four different ways, each corresponding to a different location of the measuring voltage probe in the coaxial line of Fig. 11.2. The resulting frequency-response curves are distinguished by the superscripts AB, AC, BC, and BD.

In Figs. 12.10 and 12.11, the theoretical frequency-response curve of Fig. 12.7 is shown together with the experimentally determined frequency-response curve

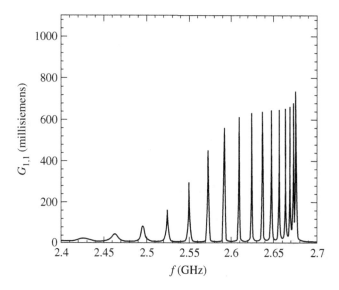

Figure 12.7 Driving-point conductance $G_{1,1}(f)$ for parameters of experiment [1, Chapter 8].

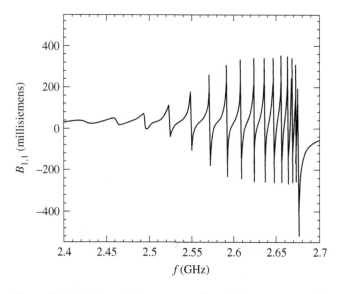

Figure 12.8 Driving-point susceptance $B_{1,1}(f)$ for parameters of experiment [1, Chapter 8].

$G_{1,1}^{AC}(f)$ [7]. Figure 12.10 is in the frequency interval 2.4 GHz $< f <$ 2.6 GHz and Fig. 12.11 is in the frequency interval 2.6 GHz $< f <$ 2.7 GHz. (Note the difference in the horizontal scales in the two figures.) The last peak on the right in Fig. 12.11 belongs to the experimental curve. It is identified with both the $m = 44$ and $m = 45$ resonance since the theory predicts that the $m = 44$ and $m = 45$ resonances merge (Fig. 12.9). The two curves are very much alike except for a small frequency shift. The same is true when the experimentally determined driving-point susceptance $B_{1,1}^{AC}(f)$

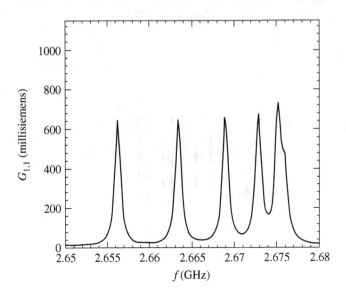

Figure 12.9 Like Fig. 12.7 but in frequency interval 2.65 GHz < f < 2.68 GHz.

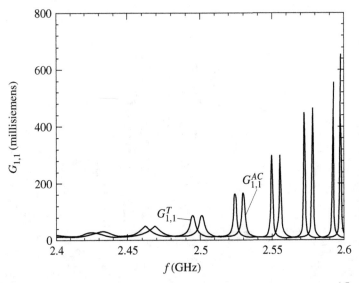

Figure 12.10 Experimentally determined driving-point conductance $G_{1,1}^{AC}$ [1, Chapter 8] together with corresponding theoretical curve $G_{1,1}^{T}$ (from Fig. 12.7) in frequency interval 2.4 GHz < f < 2.6 GHz. Taken from Fikioris [7, Fig. 7a]. © 1998 I.E.E.

and the theoretically computed $B_{1,1}(f)$ (from Fig. 12.8) are compared in Figs. 12.12 and 12.13.

In Table 12.4, the theoretically predicted resonant frequencies f_m of Table 12.3 are compared with the experimentally determined resonant frequencies f_m^{exper} – which were found as the average of the resonant frequencies (frequencies at which

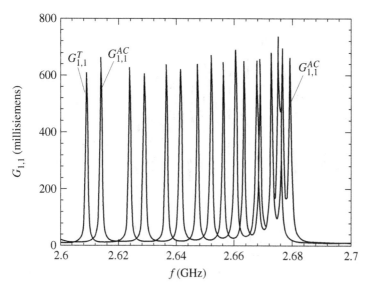

Figure 12.11 Experimentally determined driving-point conductance $G_{1,1}^{AC}$ [1, Chapter 8] together with corresponding theoretical curve $G_{1,1}^{T}$ (from Fig. 12.7) in frequency interval 2.6 GHz $< f <$ 2.7 GHz. Taken from Fikioris [7, Fig. 7b]. © 1998 I.E.E.

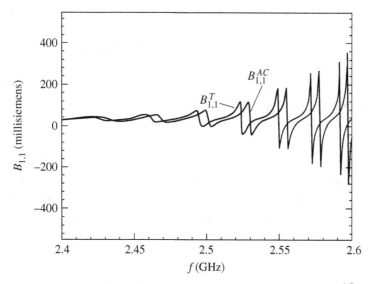

Figure 12.12 Experimentally determined driving-point susceptance $B_{1,1}^{AC}$ [1, Chapter 8] together with corresponding theoretical curve $B_{1,1}^{T}$ (from Fig. 12.8) in frequency interval 2.4 GHz $< f <$ 2.6 GHz.

$G_{1,1}$ attains its maxima) for the four probe locations. The resonant frequencies for the four probe locations differ from their averaged values shown in Table 12.4 at most in the last significant digit. It is seen that the relative difference between

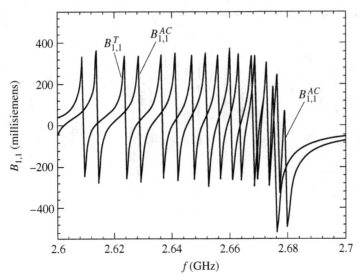

Figure 12.13 Experimentally determined driving-point susceptance $B_{1,1}^{AC}$ [1, Chapter 8] together with corresponding theoretical curve $B_{1,1}^{T}$ (from Fig. 12.8) in frequency interval $2.6\,\text{GHz} < f < 2.7\,\text{GHz}$.

the theoretical and experimental f_m is very small, of the order of a fraction of 1%. Note that the differences become smaller at the last resonances and that the theoretical resonant frequencies are always smaller than the experimentally determined ones.

The theoretically predicted resonant driving-point conductance $G_{1,1}$ of Table 12.3 and the experimentally determined values $G_{1,1}^{\text{exper}}$ (obtained by averaging the results for the four probe locations) are compared in Table 12.5. In general, the agreement is very good. The largest discrepancy (about 17%) occurs at the $m = 44$–45 resonance. Usually, the agreement is within a few percent. The differences observed here are attributed mainly to the experiment; in particular, to imperfections in construction, to sensitivity to mechanical adjustments, and to the difficulty of obtaining accurate, swept-frequency measurements in high-SWR lines. Such difficulties are discussed in detail in Chapter 14.

12.8   Appendix: formulas for the large circular array of highly conducting dipoles

In this appendix, the two-term theory formulas for a circular array of parallel, non-staggered, highly conducting dipoles with one dipole driven (Sections 12.1–12.6) are written in a form convenient for computer implementation. The parameters of the array are assumed to satisfy the conditions (11.1); N may be even or odd.

Table 12.4. *Theoretical (f_m) and experimental (f_m^{exper}) resonant frequencies and percentage error*[a]

m	f_m (GHz)	f_m^{exper} (GHz)	Percentage error
29	2.4260	2.4311	0.214
30	2.4623	2.4681	0.236
31	2.4950	2.5009	0.237
32	2.5241	2.5298	0.227
33	2.5497	2.5554	0.224
34	2.5722	2.5777	0.215
35	2.5919	2.5970	0.199
36	2.6090	2.6137	0.182
37	2.6238	2.6288	0.189
38	2.6365	2.6413	0.183
39	2.6473	2.6519	0.176
40	2.6562	2.6602	0.153
41	2.6634	2.6678	0.165
42	2.6689	2.6728	0.146
43	2.6728	2.6765	0.139
44–45	2.6752	2.6791	0.147

[a] Taken from Fikioris *et al.* [8, Table 2]. © 1994 EMW Publishing.

Table 12.5. *Theoretical ($G_{1,1}$) and experimental ($G_{1,1}^{\text{exper}}$) resonant driving-point conductances*

m	$G_{1,1}$ (mS)	$G_{1,1}^{\text{exper}}$ (mS)
29	28.04	31.2
30	46.34	51.0
31	84.95	91.9
32	163.37	177.
33	296.00	317.
34	449.60	479.
35	558.33	644.
36	609.27	671.
37	628.74	619.
38	636.60	644.
39	641.02	670.
40	644.84	713.
41	649.49	658.
42	657.24	687.
43	675.68	679.
44–45	734.49	612.

The formulas of this section reduce to those of Sections 11.2 and 11.6 when the dipoles are perfectly conducting. When the dipoles are perfectly conducting and extremely narrow resonances occur, special numerical considerations are necessary to implement the formulas; such considerations are discussed in Section 13.5.

There are two choices $p(z)$ of current distributions, namely, the simpler distribution $p^C(z) = \cos kz - \cos kh$ and the square-root end-corrected current $p^S(z)$ of Section 11.6. (The current on the driven element includes the additional term $\sin k(h - |z|)$). Also, there are two choices $K_{1R}(z)$ for the real part of the self-interaction kernel. These are the "approximate" self-term (11.11) and the "exact" self-term (11.41), denoted here by $K_{1R}^A(z)$ and $K_{1R}^E(z)$, respectively. See Section 11.8 with regard to the various kernels.

The simpler version of the two-term theory ("approximate" $K_{1R}(z)$ and shifted-cosine current) is adequate for a qualitative analysis such as that in Sections 11.3–11.5 and 12.4. If high precision is necessary (for example, when comparing theory and experiment in Section 12.7), the refinements of Section 11.6 ("exact" $K_{1R}(z)$ and end-corrected $p(z)$) should be taken into account.

The currents $I_l(z)$ on the elements are given by:

$$
I_l(z) = \begin{cases} \dfrac{j2\pi V_1}{\zeta_0 \Psi_{dR} \cos kh} [\sin k(h - |z|) + T_1 p(z)]; & l = 1 \\[2ex] \dfrac{j2\pi V_1}{\zeta_0 \Psi_{dR} \cos kh} T_l p(z); & l = 2, 3, \ldots, N. \end{cases}
\tag{12.38}
$$

Equation (12.38) and the formulas for the admittances that follow are for dipole admittances. If currents and admittances for monopoles over a perfectly conducting ground plane are desired, the corresponding dipole quantities should be multiplied by a factor of 2. The current $p(z)$ is given by

$$
p(z) = \begin{cases} p^C(z), & \text{shifted cosine} \\ p^S(z), & \text{end-corrected current,} \end{cases}
\tag{12.39}
$$

where

$$
p^C(z) = \cos kz - \cos kh
\tag{12.40}
$$

and the end-corrected current is given by

$$
p^S(z) = \begin{cases} \cos kz - \gamma_1, & k|z| \leq kz_0 \\ \gamma_2 \sqrt{kh - k|z|}, & kz_0 \leq k|z| \leq kh, \end{cases}
\tag{12.41}
$$

where kz_0 is the (unique) solution of the transcendental equation

$$
\tan kz_0 = 2(kh - kz_0)
\tag{12.42}
$$

in the interval $0 < kz_0 < kh$ $(kh < \pi/2)$ and

$$\gamma_1 = \cos kz_0 \left[1 - 4(kh - kz_0)^2 \right] \tag{12.43}$$

$$\gamma_2 = 2 \sin kz_0 \sqrt{kh - kz_0}. \tag{12.44}$$

In (12.38), the parameter Ψ_{dR} is given by (10.34).

The coefficients T_l of the shifted cosine are obtained by superimposing the phase-sequence coefficients $T^{(m)}$:

$$T_l = \frac{1}{N} \sum_{m=0}^{\lfloor N/2 \rfloor} \xi^{(m)} T^{(m)} \cos \left[\frac{2\pi(l-1)m}{N} \right], \tag{12.45}$$

where

$$T^{(m)} = \frac{P^{(m)}}{D^{(m)}} = \frac{P_R^{(m)} + j(P_I^{(m)} + P_{IL})}{D_R^{(m)} + j(D_I^{(m)} + D_{IL})}$$

$$= \frac{P_{1R} + P_{\Sigma R}^{(m)} + j(P_I^{(m)} + P_{IL})}{D_{1R} + D_{\Sigma R}^{(m)} + j(D_I^{(m)} + D_{IL})}, \qquad m = 0, 1, \ldots, \lfloor N/2 \rfloor \tag{12.46}$$

and

$$\xi^{(m)} = \begin{cases} 1, & m = 0 \text{ or } m = N/2 \\ 2, & \text{otherwise} \end{cases} \tag{12.47}$$

$$\lfloor N/2 \rfloor = \begin{cases} N/2, & N = \text{even} \\ (N-1)/2, & N = \text{odd}. \end{cases} \tag{12.48}$$

The various parts of the symmetrical components $P^{(m)}$ and $D^{(m)}$ are given by integrals involving corresponding parts of the mth phase-sequence kernel,

$$P_{1R} = \int_{-h}^{h} \sin k(h - |z|) K_{1R}(h - z) \, dz \tag{12.49}$$

$$P_{\Sigma R}^{(m)} = -\frac{1}{1 - \cos kh} \int_{-h}^{h} \sin k(h - |z|) [\cos kh \, K_{\Sigma R}^{(m)}(z) - K_{\Sigma R}^{(m)}(h - z)] \, dz \tag{12.50}$$

$$P_I^{(m)} = -\frac{1}{1 - \cos kh} \int_{-h}^{h} \sin k(h - |z|) [\cos kh \, K_I^{(m)}(z) - K_I^{(m)}(h - z)] \, dz \tag{12.51}$$

$$D_{1R} = \frac{1}{1 - \cos kh} \int_{-h}^{h} p(z) [\cos kh \, K_{1R}(z) - K_{1R}(h - z)] \, dz \tag{12.52}$$

$$D_{\Sigma R}^{(m)} = \frac{1}{1 - \cos kh} \int_{-h}^{h} p(z) [\cos kh \, K_{\Sigma R}^{(m)}(z) - K_{\Sigma R}^{(m)}(h - z)] \, dz \tag{12.53}$$

$$D_I^{(m)} = \frac{1}{1 - \cos kh} \int_{-h}^{h} p(z)[\cos kh \, K_I^{(m)}(z) - K_I^{(m)}(h - z)] \, dz. \tag{12.54}$$

The integrated formula for the loss parameter P_{IL} is

$$P_{IL} = \frac{\Phi}{2} \frac{1}{1 - \cos kh} (-kh + \tfrac{1}{2} \sin 2kh), \tag{12.55}$$

where $\Phi/2$ is defined in (12.14). An integrated formula for the loss parameter D_{IL} is

$$D_{IL} = \begin{cases} D_{IL}^C, & \text{shifted cosine} \\ D_{IL}^S, & \text{end-corrected current,} \end{cases} \tag{12.56}$$

where

$$D_{IL}^C = -\frac{\Phi}{2} \frac{1}{1 - \cos kh} (2 \cos kh - kh \sin kh - 1 - \cos 2kh) \tag{12.57}$$

$$D_{IL}^S = -\frac{\Phi}{2} \frac{1}{1 - \cos kh} \left\{ 2 \cos(kh - kz_0)[\gamma_1 + \gamma_2 \sqrt{kh - kz_0}] - kz_0 \sin kh \right.$$

$$\left. - \tfrac{1}{2} \cos(kh - 2kz_0) + (\tfrac{1}{2} - 2\gamma_1) \cos kh - \gamma_2 \sqrt{2\pi} \, C\left(\sqrt{\frac{2}{\pi}} \sqrt{kh - kz_0} \right) \right\}. \tag{12.58}$$

In (12.58), $C(x)$ is the Fresnel integral defined as

$$C(x) = \int_0^x \cos\left(\frac{\pi}{2} t^2 \right) dt. \tag{12.59}$$

As noted before, there are two choices for the real part of the self-interaction kernel, namely,

$$K_{1R}(z) = \begin{cases} K_{1R}^A(z), & \text{approximate real part} \\ K_{1R}^E(z), & \text{exact real part,} \end{cases} \tag{12.60}$$

where

$$K_{1R}^A(z) = \frac{\cos k\sqrt{z^2 + a^2}}{\sqrt{z^2 + a^2}} \tag{12.61}$$

and

$$K_{1R}^E(z) = \frac{1}{2\pi} \int_{-\pi}^{\pi} \frac{\cos\left\{ k[z^2 + 4a^2 \sin^2(\phi/2)]^{1/2} \right\}}{[z^2 + 4a^2 \sin^2(\phi/2)]^{1/2}} \, d\phi. \tag{12.62}$$

The mutual interaction part of the mth phase-sequence kernel is

$$K_{\Sigma R}^{(m)}(z) = \sum_{l=2}^{\lfloor N/2 \rfloor + 1} \xi_l \cos\left[\frac{2\pi (l - 1)m}{N} \right] \frac{\cos k R_l(z)}{R_l(z)} \tag{12.63}$$

and the imaginary part of the mth phase-sequence kernel is

$$K_I^{(m)}(z) = -\frac{\sin kz}{z} - \sum_{l=2}^{\lfloor N/2 \rfloor + 1} \xi_l \cos\left[\frac{2\pi(l-1)m}{N}\right]\frac{\sin kR_l(z)}{R_l(z)}, \qquad (12.64)$$

where

$$\xi_l = \begin{cases} 1, & l = 1 \text{ or } l = N/2 + 1 \\ 2, & \text{otherwise} \end{cases} \qquad (12.65)$$

and

$$R_l(z) = (z^2 + b_{1l}^2)^{1/2}, \qquad b_{1l} = \frac{d\,\sin[(l-1)\pi/N]}{\sin(\pi/N)}. \qquad (12.66)$$

Finally, the relation between $T^{(m)}$ and the phase-sequence admittances is

$$Y^{(m)} = G^{(m)} + jB^{(m)} = \frac{j2\pi}{\zeta_0 \Psi_{dR}\cos kh}[\sin kh + T^{(m)}p(0)] \qquad (12.67)$$

and the self- and mutual conductances $G_{1,l}$ (susceptances $B_{1,l}$) are determined by the phase-sequence conductances $G^{(m)}$ (susceptances $B^{(m)}$) by the relation

$$Y_{1,l} = G_{1,l} + jB_{1,l} = \frac{1}{N}\sum_{m=0}^{\lfloor N/2 \rfloor} \xi^{(m)}(G^{(m)} + jB^{(m)})\cos\left[\frac{2\pi(l-1)m}{N}\right]. \qquad (12.68)$$

The necessary modifications for the case of highly conducting monopoles over a highly conducting ground plane are outlined in Section 12.7; the complete formulas for this case are contained in Chapter 7 of [1].

13 Direct numerical methods: a detailed discussion

13.1 Introduction

This chapter discusses numerical methods as applied to the integral equations used in this book. The emphasis is on the integral equations for the current in a single isolated element, although some remarks on the coupled integral equations in the case of arrays and on the special case of resonant circular arrays of lossless elements (see Chapter 11) are contained in Section 13.5.

In Section 1.8, the model of an isolated, center-driven, perfectly conducting tubular dipole was introduced and two integral equations for the current $I(z)$ were derived. Both are often referred to as "Hallén's integral equation", and separately as the "exact integral equation" and the "approximate integral equation". They are

$$\frac{4\pi}{\mu_0} A_z(a, z) \equiv \int_{-h}^{h} K(z - z') I(z') \, dz' = \frac{-j 2\pi V}{\zeta_0} \sin k|z| + C \cos kz, \quad -h < z < h$$

(13.1)

in which $K(z)$ can be either the exact or the approximate kernel, namely,

$K(z) =$

$$
\begin{cases}
K_{\text{ex}}(z) = \dfrac{1}{2\pi} \displaystyle\int_{-\pi}^{\pi} \dfrac{\exp\left[-jk\sqrt{z^2 + 4a^2 \sin^2(\Phi/2)}\,\right]}{\sqrt{z^2 + 4a^2 \sin^2(\Phi/2)}} \, d\Phi, & |z| < 2h \quad \text{or} \quad (13.2\text{a}) \\[4mm]
K_{\text{ap}}(z) = \dfrac{\exp\left(-jk\sqrt{z^2 + a^2}\,\right)}{\sqrt{z^2 + a^2}}, & |z| < 2h. \quad (13.2\text{b})
\end{cases}
$$

In (13.1), $A_z(a, z)$ is the z-directed vector potential on the surface of the tubular dipole, V is the driving voltage located at an infinitesimal gap at $z = 0$, and the constant C is to be determined from the condition that

$$I(h) = 0.$$

(13.3)

The notation in this chapter is identical with that in Sections 1.8 and 1.9, except that k is used instead of β_0 to denote the wave number. Thus, the symbol $I_{\text{ex}}(z)$ will denote the unknown current when the exact kernel $K_{ex}(z)$ is used in (13.1). The corresponding quantity when the approximate kernel $K_{ap}(z)$ is used will be denoted by $I_{ap}(z)$. The symbol $K(z)$ $[I(z)]$ with no subscripts can denote either $K_{ex}(z)$ $[I_{ex}(z)]$ or $K_{ap}(z)$ $[I_{ap}(z)]$.

In Section 1.9, a direct numerical method was applied to the integral equations. This was Galerkin's [1] method with pulse functions, which is a form of the method of moments in [2]. The numerical results of Figs. 1.13 and 1.14 indicated that difficulties are involved with this procedure, at least when the approximate kernel is used. In this chapter, such difficulties are discussed in detail. When possible, improvements to the numerical procedure are proposed. The discussion is based on the mathematical properties of the two integral equations. These are the subject of the following section.

13.2   Properties of the integral equations

When $z = 0$ in (13.2a), the integral diverges. Thus, the exact kernel has a singularity at $z = 0$. This is easily shown to be an integrable, logarithmic singularity. On the other hand, the approximate kernel in (13.2b) is well-defined at $z = 0$ and, in fact, has derivatives of all orders there. Since $ka \ll 1$, $\mathrm{Re}\{K_{ap}(z)\}$ is highly peaked at $z = 0$. As long as $ka \ll k|z| < 2kh$, the two kernels resemble each other as functions of kz, and their difference becomes vanishingly small in the limit $ka \to 0$.

Such a similarity between the two kernels does not guarantee that the solutions of the corresponding integral equations will be close. In fact, the two integral equations have quite different mathematical properties.

Consider the approximate integral equation first. For any function $I(z)$ (satisfying mild admissibility conditions), the left-hand side of (13.1) is differentiable with

$$\frac{\partial}{\partial z} \int_{-h}^{h} K_{ap}(z - z') I(z') \, dz' = \int_{-h}^{h} \frac{\partial K_{ap}(z - z')}{\partial z} I(z') \, dz'. \tag{13.4}$$

The right-hand side of (13.1), however, is not differentiable at $z = 0$. This argument shows that the approximate integral equation can have no solution. The approximate integral equation requires a line current located on the z-axis to maintain a field with a delta-function behavior at $\rho = a$, $z = 0$, and this is not possible.

For the exact integral equation, the argument above does not hold. Note, in particular, that it is not legitimate to interchange the order of differentiation and integration as in (13.4). In fact, one can show [3] that the exact integral equation has a unique solution. Some properties of this solution may be derived without solving the integral equation.

It is instructive to consider first the antenna of infinite length. Although this antenna cannot be realized in practice, it is a useful device for obtaining information about the finite antenna. The current $I^{(\infty)}(z)$ on the antenna of infinite length satisfies an integral equation similar to (13.1), where either the exact or the approximate kernel can be used. When the exact kernel is used, this integral equation can be solved explicitly. Studying an explicit expression for the current $I_{\text{ex}}^{(\infty)}(z)$ is less complicated than studying an integral equation in which the unknown appears implicitly. In this respect, the infinite antenna is much simpler than the finite antenna.

The integral equation for $I_{\text{ex}}^{(\infty)}(z)$ is derived and solved in Section 13.6, which is an appendix to this chapter. (The procedure that gives the solution in the case of the exact kernel yields a contradiction in the case of the approximate kernel, verifying that the approximate integral equation has no solution.) Once $I_{\text{ex}}^{(\infty)}(z)$ is found, its behavior close to the driving point is investigated. It is shown in Section 13.6 that

$$\frac{I_{\text{ex}}^{(\infty)}(z)}{V} = j\,\frac{4ka}{\zeta_0}\,\ln\frac{1}{|z|} + O(1), \quad \text{as } z \to 0. \tag{13.5a}$$

The coefficient in (13.5a) is purely imaginary. This means that there is an infinite driving-point susceptance. This can be understood from a physical point of view: there is a non-zero voltage maintained at the two circular knife edges at $z = 0$. As these are separated by an infinitesimal distance, there is an infinite capacitance.

The finite antenna will exhibit a similar behavior. One does not expect that the singularity at the driving point will be affected by the length of the antenna. This is indeed true, and it can be shown [3] that

$$\frac{I_{\text{ex}}(z)}{V} = j\,\frac{4ka}{\zeta_0}\,\ln\frac{1}{|z|} + O(1), \quad \text{as } z \to 0, \tag{13.5b}$$

i.e. that (13.5a) holds unaltered for the finite antenna of length $2h$.

Whereas $\text{Im}\{I_{\text{ex}}(z)/V\}$ becomes infinite at the driving point, the component of current in phase with the driving voltage is finite there. In Section 13.6 it is shown that for the infinite antenna

$$\text{Re}\left\{\frac{I_{\text{ex}}^{(\infty)}(z)}{V}\right\} = \frac{4k}{\pi\zeta_0}\int_0^k \frac{\cos\zeta z}{(k^2 - \zeta^2)[J_0^2(a\sqrt{k^2 - \zeta^2}) + Y_0^2(a\sqrt{k^2 - \zeta^2})]}\,d\zeta, \tag{13.6}$$

where J_0 (Y_0) is the Bessel function of the first (second) kind of order zero [4].

Although the approximate integral equation has no solution, it is still a useful equation. One should not attempt to find exact solutions but should only seek currents which satisfy the equation approximately. This situation is unusual and constitutes a difficulty associated with the approximate integral equation. The exact integral equation itself is not without difficulties because its solution becomes infinite at the driving point. Since no approximations were made in Section 1.8 when deriving the equation, the infinity is due to the model of the center-driven tubular dipole and, in particular, to the very convenient but overly idealized concept of the delta-function generator. The current in the more realistic model of a base-driven monopole over a ground plane is certainly not infinite at $\rho = a$, $z = 0$.

It has been shown that the approximate integral equation for a center-driven antenna does not have a solution. It can also be shown (following a procedure different from the one used for the center-driven antenna) that the integral equation for the current on a receiving antenna (see Section 2.11) has no solution when the approximate kernel is used.

In light of the mathematical properties of the integral equations just discussed, the next sections examine the behavior of the solutions obtained by the numerical method of Section 1.9.

13.3   On the application of numerical methods

Galerkin's method with pulse functions was applied to (13.1) in Section 1.9. The final numerical solution was obtained in the form of the "staircase"-type approximation

$$I(z) \doteq \sum_{n=-N}^{N} I_n u_n(z) = \sum_{n=-N}^{N} [I_n^{(1)} + C I_n^{(2)}] u_n(z), \tag{13.7}$$

where the $2N + 1$ pulse functions $u_n(z)$ are defined in (1.57). Each pulse function has width z_p where

$$(2N + 1)z_p = 2h. \tag{13.8}$$

In (13.7), $I_n^{(1)}$ and $I_n^{(2)}$, $n = 0, \pm 1, \pm 2, \ldots, \pm N$, are the solutions of the $(2N + 1) \times (2N + 1)$ Toeplitz systems

$$\sum_{n=-N}^{N} A_{l-n} I_n^{(1)} = B_l^{(1)} \quad \text{and} \quad \sum_{n=-N}^{N} A_{l-n} I_n^{(2)} = B_l^{(2)}, \quad l = 0, \pm 1, \ldots, \pm N; \tag{13.9}$$

C is determined from

$$C \doteq -I_N^{(1)}/I_N^{(2)}; \tag{13.10}$$

the matrix coefficients $A_l, l = 0, \pm 1, \pm 2, \ldots, \pm 2N$, are given by

$$A_l = A_{-l} = \int_0^{z_p} (z_p - z)[K(z + l z_p) + K(z - l z_p)] \, dz; \tag{13.11}$$

and integrated expressions for $B_l^{(1)}$ and $B_l^{(2)}$ are given in (1.62a, b). In Section 1.9, the matrix coefficient A_{l-n} is also denoted by A_{ln}. In this chapter, only the former notation will be used.

This section examines the behavior of the numerical solutions obtained by this method. The empirical criterion of making N larger until the solution has converged to a satisfactory final value is also discussed in detail. The symbols $I_{ex,n}/V$ and $I_{ap,n}/V$ are used to distinguish the values of I_n obtained with the exact and the approximate kernels. The simpler case of the exact kernel is discussed first.

The case of the exact integral equation

A valid numerical method must reproduce the true solution of the integral equation. For the case of the exact integral equation, the logarithmic singularity of $\text{Im}\{I_{\text{ex}}(z)/V\}$ must be reproduced. Thus, when applying the criterion of making N larger until the solution has converged, one should ignore the values of $\text{Im}\{I_{\text{ex},n}/V\}$ when n is very small. In particular, the input susceptance converges to ∞ and the input resistance and reactance converge to zero so that the empirical "convergence" criterion should not be used for these quantities. In the plots of Figs. 1.13a, b, the effects mentioned above have not yet appeared. In Fig. 1.14b, however, a slight increase in the value of $B_0 = \text{Im}\{I_{\text{ex},0}/V\}$ can be distinguished.

In the appendix (Section 13.6), the numerical method of Section 1.9 is applied to the infinite antenna. Here, the entire real axis is discretized into segments of length z_p, where kz_p is small. An approximate solution $I_{\text{ex}}^{(\infty)}(z)$ is sought in the form

$$I_{\text{ex}}^{(\infty)}(z) \doteq \sum_{n=-\infty}^{\infty} I_{\text{ex},n}^{(\infty)} u_n(z), \quad -\infty < z < \infty. \tag{13.12}$$

There is an infinite number of unknowns $I_{\text{ex},n}^{(\infty)}$, $n = 0, \pm 1, \ldots$, and an infinite number of equations. It is shown in Section 13.6 that these equations can be solved exactly for non-zero pulse width z_p. Once this is done, and with the distance nz_p from the driving point held fixed, the limit of the numerical solution as $z_p \to 0$ is determined. It is shown that this limit coincides with the exact solution of the integral equation.

The case of the approximate integral equation

The issue of applying the numerical method of Section 1.9 to the approximate integral equation is now examined. The main questions addressed are two:

1. What does one obtain when one applies the numerical method to the integral equation?
2. Under what conditions are numerical solutions obtained with the approximate kernel close to those obtained with the exact kernel, and in what sense?

It was shown in Section 13.2 that the approximate integral equation has no solution. For this reason, the answers to these questions are by no means obvious.

Here, application of the numerical method to the integral equation for the infinite antenna is particularly illuminating. This is carried out in Section 13.6. For the infinite antenna, the following are shown in Section 13.6.

1. The limit of the numerical solution as $z_p \to 0$ does not exist. This is to be expected since the integral equation has no solution.

2. The limit of the real part does exist. Specifically, it is shown that when the distance nz_p from the driving point is held fixed,

$$
\lim_{z_p \to 0} \mathrm{Re} \left\{ \frac{I_{\mathrm{ap},n}^{(\infty)}}{V} \right\}
$$

$$
= \frac{4k}{\pi \zeta_0} \int_0^k \frac{J_0(a\sqrt{k^2 - \zeta^2})\cos(\zeta n z_p)}{(k^2 - \zeta^2)[J_0^2(a\sqrt{k^2 - \zeta^2}) + Y_0^2(a\sqrt{k^2 - \zeta^2})]} \, d\zeta. \tag{13.13}
$$

Comparing (13.13) to the corresponding equation (13.6) for the case of the exact kernel, it is seen that the only difference is the small factor $J_0(a\sqrt{k^2 - \zeta^2})$ in the integrand. Thus, the real part in (13.13) is very close to the corresponding quantity for the exact integral equation, and the two quantities become identical in the limit $ka \to 0$.

3. For small pulse width, and specifically when the conditions

$$
z_p/a \ll 1 \quad \text{and} \quad n z_p/a = O(1) \tag{13.14}
$$

are satisfied, one has asymptotically

$$
\frac{I_{\mathrm{ap},n}^{(\infty)}}{V} \sim j \, \frac{1}{\zeta_0} \frac{\pi^3}{32\sqrt{2}} k z_p \sqrt{\frac{z_p}{a}} (-1)^n \exp(a\pi/z_p) \frac{1}{\cosh[(nz_p/a)\pi/2]}
$$

$$
\times \left[1 - \frac{5}{2\pi} \frac{z_p}{a} + \frac{5}{4} n \left(\frac{z_p}{a} \right)^2 \tanh[(nz_p/a)\pi/2] \right], \tag{13.15}
$$

where the quantity in brackets is simply a correction factor. Equation (13.15) reveals that, when the pulse width is small, the numerical method yields an exponentially large, purely imaginary "driving-point admittance", and a large, purely imaginary, rapidly oscillating "current", at least for points on the antenna not too far from the driving point.

With the analytical results for the infinite antenna at hand, assume that $z_p \ll a$ and consider the case of the *finite* antenna. The condition $z_p \ll a$ here is the same as $N \gg h/a$. It is true for the finite antenna that:

1. The limit of the numerical solution as $N \to \infty$ does not exist.
2. Near the driving point, the real part $\mathrm{Re}\{I_{\mathrm{ap},n}/V\}$ obtained with the approximate kernel is close to $\mathrm{Re}\{I_{\mathrm{ex},n}/V\}$ obtained with the exact kernel.
3. Near the driving point, the values of $\mathrm{Im}\{I_{\mathrm{ap},n}/V\}$ are large and oscillate rapidly. In fact, these values are very closely approximated by the corresponding values of $\mathrm{Im}\{I_{\mathrm{ap},n}^{(\infty)}/V\}$ for the case of the infinite antenna and the asymptotic formula (13.15).

It seems difficult to prove the last two assertions, or even to state them in a more precise manner. They have been verified, however, by extensive numerical

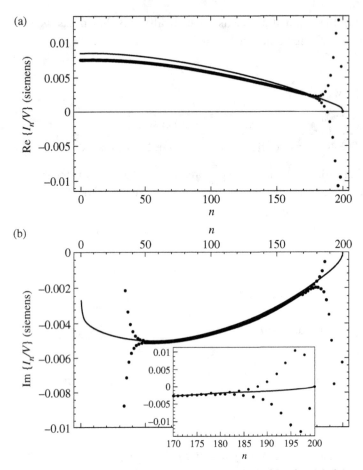

Figure 13.1 Numerical results obtained with method of Section 1.9; $h/\lambda = 0.25$, $a/\lambda = 0.006$, and $N = 200$. (a) $\mathrm{Re}\{I_{\mathrm{ex},n}/V\}$ (solid line) and $\mathrm{Re}\{I_{\mathrm{ap},n}/V\}$ (dots). (b) $\mathrm{Im}\{I_{\mathrm{ex},n}/V\}$ (solid line) and $\mathrm{Im}\{I_{\mathrm{ap},n}/V\}$ (dots). For $\mathrm{Im}\{I_{\mathrm{ap},n}/V\}$, only values with $n \geq 32$ are shown; insert shows all values with $n \geq 170$.

calculations. Figures 13.1a, b and Table 13.1 show typical numerical results, obtained for $N = 200$, $h/\lambda = 0.25$, and $a/\lambda = 0.006$ so that $a/z_p = 4.8$, which is moderately large. In Fig. 13.1a, $\mathrm{Re}\{I_n/V\}$ is shown. The values $\mathrm{Re}\{I_{\mathrm{ex},n}/V\}$, $n = 0, 1, \ldots, N$, have been joined by straight lines, whereas the values $\mathrm{Re}\{I_{\mathrm{ap},n}/V\}$ are shown as points. Except in the vicinity of $z = h$ (or $n = 200$), the two solutions are seen to be quite close. In Fig. 13.1b, the imaginary parts are compared. A behavior resembling a logarithmic singularity is apparent in the solution for the case of the exact kernel. For the case of the approximate kernel, only the values of $\mathrm{Im}\{I_{\mathrm{ap},n}/V\}$ for n larger than 32 are shown. The remaining values (near the driving point) are larger and out of scale. They are listed in Table 13.1, where they are compared to results obtained from the asymptotic formula (13.15) for the infinite antenna. The agreement is very good up to about $n = 20$, so that the two quantities are close even when n is larger than a/z_p. For

Table 13.1. *Comparison of* $\mathrm{Im}\{I_{\mathrm{ap},n}/V\}$ *for finite antenna as calculated by numerical method of Section 1.9 and* $I_{\mathrm{ap},n}^{(\infty)}/jV$ *for infinite antenna as calculated by asymptotic formula (13.15);* $h/\lambda = 0.25$, $a/\lambda = 0.006$, *and* $N = 200$

n	$\mathrm{Im}\{I_{\mathrm{ap},n}/V\}$ (S)	$I_{\mathrm{ap},n}^{(\infty)}/jV$ (S)	n	$\mathrm{Im}\{I_{\mathrm{ap},n}/V\}$ (S)	$I_{\mathrm{ap},n}^{(\infty)}/jV$ (S)
0	2.18×10^1	1.99×10^1	16	4.39×10^{-1}	4.37×10^{-1}
1	-2.08×10^1	-1.93×10^1	17	-3.40×10^{-1}	-3.26×10^{-1}
2	1.82×10^1	1.75×10^1	18	2.49×10^{-1}	2.42×10^{-1}
3	-1.50×10^1	-1.50×10^1	19	-1.96×10^{-1}	-1.80×10^{-1}
4	1.20×10^1	1.23×10^1	20	1.40×10^{-1}	1.34×10^{-1}
5	-9.32×10^0	-9.75×10^0	21	-1.14×10^{-1}	-9.91×10^{-2}
6	7.16×10^0	7.57×10^0	22	7.78×10^{-2}	7.34×10^{-2}
7	-5.47×10^0	-5.80×10^0	23	-6.72×10^{-2}	-5.44×10^{-2}
8	4.15×10^0	4.40×10^0	24	4.23×10^{-2}	4.03×10^{-2}
9	-3.15×10^0	-3.32×10^0	25	-4.05×10^{-2}	-2.98×10^{-2}
10	2.38×10^0	2.50×10^0	26	2.20×10^{-2}	2.20×10^{-2}
11	-1.81×10^0	-1.88×10^0	27	-2.52×10^{-2}	-1.63×10^{-2}
12	1.36×10^0	1.41×10^0	28	1.04×10^{-2}	1.20×10^{-2}
13	-1.03×10^0	-1.05×10^0	29	-1.65×10^{-2}	-8.87×10^{-3}
14	7.73×10^{-1}	7.86×10^{-1}	30	3.80×10^{-3}	6.54×10^{-3}
15	-5.92×10^{-1}	-5.87×10^{-1}	31	-1.16×10^{-2}	-4.83×10^{-3}

Table 13.2. *Driving-point admittances* $I_{\mathrm{ex},0}/V$ *and* $I_{\mathrm{ap},0}/V$ *obtained by numerical method of Section 1.9 for exact and approximate integral equations;* $z_p/\lambda \sim 0.006$ *and* $a/\lambda = 0.02$

h/λ	N	$I_{\mathrm{ex},0}/V$ (S)	$I_{\mathrm{ap},0}/V$ (S)
0.25	40	$9.15 \times 10^{-3} + j9.17 \times 10^{-4}$	$7.44 \times 10^{-3} + j0.933$
0.75	121	$7.34 \times 10^{-3} + j2.49 \times 10^{-3}$	$6.32 \times 10^{-3} + j0.935$

choices of parameters leading to larger values of a/z_p, the agreement becomes better. (See, however, the discussion on the effect of roundoff errors in the next section.)

For the case of the approximate kernel, an oscillatory behavior is also observed in the curves of both Figs. 13.1a, b near the endpoint $z = h$. The oscillations here are smaller than those of the imaginary part near the driving point. Naturally, this effect does not occur in the case of the center-driven infinite antenna. It was found, however, that similar oscillations do occur when the numerical method is applied to the integral equation for the finite, unloaded receiving antenna (see Section 2.11). Therefore, this effect cannot be attributed to the delta-function generator.

When $z_p/\lambda \ll a/\lambda$, it follows from the third assertion that the initial values of $\mathrm{Im}\{I_{\mathrm{ap},n}/V\}$ are independent of the length h/λ of the antenna and depend only on a/λ and z_p/λ. This point is further illustrated in Table 13.2. For both integral equations, the driving-point admittances obtained for $h/\lambda = 0.25$ and $h/\lambda = 0.75$ are shown. The dipole radius is $a/\lambda = 0.02$. For the shorter length, $N = 40$; for the longer length, N has been chosen to be 121 so that z_p/λ is the same for both cases (roughly 0.006). Here a/z_p is roughly 3.3. It is seen that $\text{Im}\{I_{ap,0}/V\}$ is nearly identical for the two antenna lengths.

The similarity of the driving-point conductances $\text{Re}\{I_{ex,0}/V\}$ and $\text{Re}\{I_{ap,0}/V\}$ is now discussed in more detail, assuming that z_p/λ is small and the same for both. When h/λ is fixed, numerical calculations verify that, as in the case of the infinite antenna, the difference between the two real parts becomes smaller when a/λ becomes smaller. When a/λ is fixed, both real parts exhibit a strong dependence on h/λ, and so does their difference. Numerical calculations with $h/\lambda \leq 0.5$ show that, with the exception of a region near resonance, the two solutions are closer when h/λ becomes larger. The reason for this is that the difference between the two kernels becomes less pronounced when h/λ is large. In the region near resonance, the value of h/λ is more critical (see Chapter 2). Near resonance, or when h/λ is small and a/λ is large, the difference between $\text{Re}\{I_{ex,0}/V\}$ and $\text{Re}\{I_{ap,0}/V\}$ may be significant. In the first case of Table 13.2, for example, the difference is about 19%. If N is increased to 60, this difference slightly increases.

In practice, both the real and the imaginary parts are of equal importance. Having examined the behavior of the solution when N is large, the problem of choosing N is addressed. Here, the criterion of making N larger until the solution has converged cannot be used. Since oscillations appear near the driving point when $z_p/\lambda \ll a/\lambda$ or $N \gg h/a$, it is necessary to choose a pulse width $z_p/\lambda \gtrsim a/\lambda$ or $N \lesssim h/a$ when $\text{Im}\{I(z)/V\}$ is desired. With h/λ and a/λ as in the first case of Table 13.2, h/a is only 12.5, and oscillations are noticeable for N as low as 15. The choice of N should be based on the important parameter h/a rather than on the commonly used criterion of the number of points per wavelength. The optimal value of N should be considered to be the one that gives results closest to results obtained with the exact kernel. *A priori* determination of this value is not possible. Indeed, the criterion of closeness of the solutions may itself be formulated in a number of different ways. For example, the solutions may be considered close when the driving-point conductances are close, or when $\text{Im}\{I_{ex,n}/V\}$ is close to $\text{Im}\{I_{ap,n}/V\}$ for $1 \ll n \ll N$.

13.4   Additional remarks

The behavior of the numerical solutions was discussed in detail in the previous section. Here, some additional comments for both the exact and the approximate integral equations are made. For the former case, two improvements of the direct numerical method are proposed.

Roundoff errors

For the case of the approximate integral equation, the unavoidable appearance of oscillations when $N \gg h/a$ is a consequence of the properties of the approximate kernel and the right-hand side of the integral equation. Such oscillations would appear in a hypothetical computer with an infinite wordlength and no roundoff error. Roundoff error can also be an important effect. It is discussed here briefly, for both the exact and the approximate integral equations. The focus is on the properties of the matrix in (13.9) with elements A_{l-n}; the fact that the approximate integral equation has no solution is not pertinent to these properties.

The matrix in (13.9) results from the discretization of a Fredholm integral equation of the first kind. It is typical of such matrices to be ill-conditioned [5] so that difficulties may be present when solving the systems of algebraic equations in (13.9). The usual rule of thumb that more ill-conditioning occurs when the kernel is smoother [6–8] is applicable in this case: $K_{ex}(z)$ is logarithmically singular at $z = 0$ for any value of ka, and therefore serious ill-conditioning does not occur. In other words, a typical modern computer can easily solve the systems of equations in (13.9) for the case of the exact kernel, even if N is quite large.

For the case of the approximate kernel, the situation is different. Although $K_{ap}(z)$ is an analytic function of z, its real part is highly peaked at $z = 0$ when $ka \ll 1$. Numerical investigations (specifically, estimations of the L_1 condition number [9]) show that the parameter a/z_p or Na/h may be taken as a rough measure of ill-conditioning. When this parameter is small, ill-conditioning is not prohibitive. When this parameter becomes larger, the matrix becomes rapidly ill-conditioned. For given h/λ and a/λ, as N is increased, roundoff error will quickly become the dominant factor. When roundoff error is dominant, the results obtained depend on the particular software and hardware used.

Exact integral equation: improvements

For the exact integral equation, the quantities $I^{(1)}(z)$ and $I^{(2)}(z)$ in (1.54a, b) behave like $(h^2 - z^2)^{-1/2}$ as $z \to \pm h$, whereas the total current vanishes like $(h^2 - z^2)^{1/2}$ (see, for example, [10, 11]). Thus, the second equation in (1.55) should be written more precisely as $C = -\lim_{z \to h} I^{(1)}(z)/I^{(2)}(z)$. It is noteworthy that the intermediate solutions $I^{(1)}(z)$ and $I^{(2)}(z)$ in (1.54) are not square-integrable.

The aforementioned behavior creates a minor difficulty in the application of the numerical method to the exact integral equation if C is determined from (13.10). Equation (13.10) involves large values so that, as N is made larger, C (and, with it, the entire numerical solution) converges rather slowly. The convergence may be slightly accelerated by noting that the square-root behavior of $I(z)$ near $z = h$ implies

$\lim_{z_p \to 0} [I(h - z_p/2)]/[I(h - 3z_p/2)] = 1/\sqrt{3}$. Thus, instead of (13.10), C may be calculated from

$$C \doteq -\frac{\sqrt{3}\, I_N^{(1)} - I_{N-1}^{(1)}}{\sqrt{3}\, I_N^{(2)} - I_{N-1}^{(2)}}. \tag{13.16}$$

Such a refinement is meaningful only in the case of the exact kernel. For the parameters of Table 13.2, the driving-point conductances are changed to 8.46×10^{-3} siemens and 6.94×10^{-3} siemens for $h/\lambda = 0.25$ and $h/\lambda = 0.75$, respectively. When z_p/λ is small, use of (13.16) instead of (13.10) effectively amounts to an increase in h/λ. In the examples just given, the antenna lengths are slightly larger than the resonant lengths for the radius $a/\lambda = 0.02$, and therefore the $I_{\text{ex},0}/V$ values decrease.

A significantly better method for finding the solution to the exact integral equation is to solve (13.1) with (13.2a) numerically for the quantity $I_{\text{ex}}(z) + j4kaV\zeta_0^{-1}\ln k|z|$, which, as discussed in Section 13.2, is bounded at $z = 0$. The relation between such quantities and the apparent admittance of a base-driven monopole over a ground plane is discussed in [12–17].

Approximate integral equation: application of other numerical methods

For the case of the approximate kernel, other forms of the method of moments are discussed here briefly. A large number of subsectional basis functions is assumed. One cannot expect to obtain the same answers by applying different numerical methods, so that quantitative results such as the approximate formula (13.15) apply only when the particular numerical procedure of Section 1.9 is used. In a qualitative sense, however, many of the results of Section 13.3 hold for other numerical methods. For instance, one can use Galerkin's method with an even number $2N$ of pulse functions. This is the same as writing (1.54a) in the form

$$\int_0^h [K(z - z') + K(z + z')] I^{(1)}(z')\, dz' = \frac{i}{2\zeta_0} V \sin k|z|, \quad 0 < z < h \tag{13.17}$$

[and similarly for (1.54b)], and using N pulse functions in $(0, h)$. When $N \gg h/a$, large, rapidly oscillating solutions are obtained. Such solutions are also obtained when the point-matching technique [2] is used to determine the coefficients of the N pulse functions, or when Galerkin's method is employed with overlapping triangles [2] (with half-triangles at the endpoints $z = 0$ and $z = h$) as basis functions.

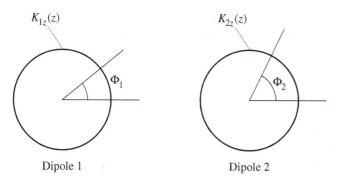

Figure 13.2 Top view of two-element array.

13.5 Notes on arrays of cylindrical dipoles

In the two previous sections, the application of numerical methods to the integral equations for the case of a single isolated antenna was discussed. With the exact kernel, the main difficulty is the fact that $\mathrm{Im}\{I(0)/V\} = \infty$, a feature that will eventually show up in a direct numerical solution. For the case of the approximate kernel, the situation is more complicated and was discussed in detail.

A main subject of study in this book is the array of parallel, identical, non-staggered, center-driven cylindrical dipoles. In this case, there is no exact integral equation similar to (13.1) since the model of an array of such dipoles with only z-directed, azimuthally independent currents is inconsistent. This is discussed in [18, 19], where a method to overcome this difficulty is also proposed. The difficulty is easily understood from Fig. 13.2, where the top view of a two-element array is shown. If the separation between dipoles is large, one assumes that the currents have no circulating component and that each current is independent of the position Φ_1 or Φ_2 around the respective element. The z-component E_{1z} of the electric field on dipole 1 is here a sum of two terms, one due to the current on the dipole itself, the other due to the current on dipole 2. By the initial assumption, the first term is independent of Φ_1. For $z \neq 0$, the boundary condition requires $E_{1z} = 0$ and, therefore, the second term is also independent of Φ_1. It is obvious from Fig. 13.2, however, that the electric field on dipole 1 due to the Φ_2-independent current on dipole 2 must depend on Φ_1. This contradiction shows that the theoretical model is inconsistent.

Since the coupled integral equations of Chapters 3–5 have no exact counterpart, comparisons like those performed in Figs. 13.1a, b are not easy to implement. However, if a numerical method similar to the one in Section 1.9 is applied to these coupled integral equations and the pulse width is small, one can expect effects such as those discussed previously. The problem of choosing N is more severe than the much simpler case of the isolated antenna studied here. It is also more difficult to distinguish between effects due to roundoff and effects inherent in the process of discretizing a system of integral equations with no solution.

Resonant circular arrays of perfectly conducting dipoles

The problem of determining the currents on a large resonant circular array of lossless dipoles presents special difficulties. It was shown in Chapter 11 (see, in particular, Table 11.1) that proper choices of the parameters lead to very large currents and extremely narrow resonances. Inherently, such phenomena are very sensitive to small changes in the parameters. Therefore, unless special considerations are taken into account, methods that are sufficient for conventional arrays will be highly susceptible to truncation and roundoff errors. As an indication of the complications that might arise if other methods are applied, the considerations required to implement the two-term theory formulas (11.2)–(11.17) in a computer [20] are discussed here. In order to make the following discussion meaningful, it is necessary that the modified kernel be used, as discussed in Section 11.8.

The underlying reason for the occurrence of extremely narrow resonances is the exponential smallness of the imaginary part of the modified phase-sequence kernel $K_I^{(m)}(z)/k$. Because this quantity is small, the quantity $D_I^{(m)}$ in (11.10) is also small. As defined in (11.13), $K_I^{(m)}(z)/k$ is the sum of $N/2$ terms of order 1. Therefore, if calculated directly, the sum must be evaluated with high precision. Alternatively, or in cases where extended computer precision is inadequate, the asymptotic formula (11.18) can be used. The calculation of $D_I^{(m)}$ by numerical integration then becomes a simple task, and there is no need to require high accuracy when evaluating (11.10) numerically. Note that high accuracy in the numerical integrations would be necessary if the order of integration in (11.10) and summation in (11.13) were interchanged.

In (11.2)–(11.17), an important quantity that is especially susceptible to errors is $D_R^{(m)} = D_{1R} + D_{\Sigma R}^{(m)}$. By the discussion in Section 11.3, this quantity becomes exactly zero at the mth phase-sequence resonance. If one programs the formulas (11.2)–(11.17) without taking this into account and attempts to find resonances by varying the frequency, there is no hope that the value zero can be obtained in a computer. Instead, a small non-zero value for $D_R^{(m)}$ will be obtained. As a consequence, the value of $G^{(m)}$ (or $G_{1,1}$) obtained by the computer will be smaller than the very large value actually predicted by the theory. The remedy here is simple: one first determines the resonant frequency with a standard root-finding method. Subsequently, one calculates the maximum value of $G^{(m)}$ (or $G_{1,1}$) by *setting* $D_R^{(m)}$ to zero at the determined frequency. The numerical tradeoff for doing this is relatively unimportant: the maximum value predicted by the two-term theory is obtained, but at a slightly different value of the frequency.

These considerations apply to the case of extremely narrow resonances. When the resonant currents are smaller, the numerical difficulties are less pronounced; away from the mth phase-sequence resonance, a highly accurate calculation of $K_I^{(m)}(z)/k$ and $D_I^{(m)}$ is not at all important. Interestingly, it is much simpler to program the formulas of Section 12.8 for resonant circular arrays of highly conducting dipoles; the smallness of the various quantities is in this case less important.

It was possible to identify and subsequently remove the source of the numerical difficulties because of the relative simplicity of the two-term theory solution in (11.2)–(11.17). For approaches such as the direct numerical solution of integral equations for the current distributions, identifying and removing the numerical difficulties may not be a simple task.

13.6 Appendix: the infinite antenna

Many questions concerning the center-driven, finite antenna of length $2h$ are conveniently addressed by first resolving analogous questions for the center-driven infinite antenna, for which $h = \infty$. The basic reason is that the integral equation for the current on the infinite antenna can be solved exactly. Questions of this type include the behavior of the solutions obtained by the numerical method of Section 1.9.

In this appendix, the infinite antenna is studied in some detail, starting from first principles. A time dependence $e^{-i\omega t}$ instead of $e^{j\omega t}$ is assumed. Additional information about the infinite antenna, as well as many discussions on the relation between the finite and infinite antennas can be found in [3]. Note that in [3], a factor of $1/4\pi$ has been absorbed into the kernel and the right-hand side of the integral equation.

Integral equation and its solution

The integral equation for the current $I^{(\infty)}(z)$ on the infinite tubular antenna can be derived by following the steps of Section 1.8. The left-hand side of the integral equation is the expression for $4\pi A_z(a, z)/\mu_0$ resulting from integrating over the length of the antenna, where either the exact or the approximate kernel can be used. To determine the right-hand side, note that the most general even solution (1.46) to the differential equation (1.45) can be written as

$$A_z(a, z) = \left(\frac{C_1}{2} + \frac{V}{4c}\right)e^{ik|z|} + \left(\frac{C_1}{2} - \frac{V}{4c}\right)e^{-ik|z|}.$$

In this case, C_1 is determined by setting the coefficient of $e^{-ik|z|}$ to zero. One way to understand this condition is to note that the vector potential must represent an outgoing wave from the origin. Alternatively, note that the expression for $A_z(a, z)$ remains valid when the antenna is surrounded by a lossy medium for which $\text{Im}\{k\} > 0$. Here, the second term is exponentially increasing for large $|z|$, so that its coefficient must vanish. Thus, the desired integral equation for the current $I^{(\infty)}(z)$ on the infinite antenna is

$$\frac{4\pi}{\mu_0} A_z(a, z) \equiv \int_{-\infty}^{\infty} K(z - z') I^{(\infty)}(z') \, dz' = \frac{2\pi}{\zeta_0} V e^{ik|z|}, \quad -\infty < z < \infty, \quad (13.18)$$

where $K(z)$ stands for either $K_{ex}(z)$ or $K_{ap}(z)$ [eqs. (13.2a) or (13.2b), with $-i$ in place of j].

In this appendix, the use of Fourier transforms is made. The symbols $\bar{K}(\zeta)$, $\bar{K}_{ex}(\zeta)$, and $\bar{K}_{ap}(\zeta)$ stand for the respective Fourier transforms

$$\bar{K}(\zeta) = \begin{cases} \bar{K}_{ex}(\zeta) = \displaystyle\int_{-\infty}^{\infty} K_{ex}(z) e^{i\zeta z} \, dz \quad \text{or} & (13.19a) \\[3mm] \bar{K}_{ap}(\zeta) = \displaystyle\int_{-\infty}^{\infty} K_{ap}(z) e^{i\zeta z} \, dz. & (13.19b) \end{cases}$$

When ζ and k are real, these are easily determined as follows.

The Fourier transform of the approximate kernel can be found from integrals tabulated in [21, p. 472] to be

$$\bar{K}_{ap}(\zeta) = \begin{cases} i\pi H_0^{(1)}(a\sqrt{k^2 - \zeta^2}), & \text{if } |\zeta| < k \\[3mm] 2K_0(a\sqrt{\zeta^2 - k^2}), & \text{if } |\zeta| > k, \end{cases} \quad (13.20)$$

where $H_0^{(1)} = J_0 + iY_0$ is the Hankel function of order zero and K_0 is the modified Bessel function of order zero [4]. To determine the Fourier transform of the exact kernel, substitute (13.2a) into (13.19a), interchange the order of integration, and use (13.20) to obtain

$$\bar{K}_{ex}(\zeta) = \begin{cases} \dfrac{i}{2} \displaystyle\int_{-\pi}^{\pi} H_0^{(1)}\left(2a \sin(|\Phi|/2) \sqrt{k^2 - \zeta^2}\right) d\Phi, & \text{if } |\zeta| < k \\[4mm] \dfrac{1}{\pi} \displaystyle\int_{-\pi}^{\pi} K_0\left(2a \sin(|\Phi|/2) \sqrt{\zeta^2 - k^2}\right) d\Phi, & \text{if } |\zeta| > k. \end{cases} \quad (13.21)$$

These can both be evaluated with the help of integrals tabulated in [21, pp. 738–739], and one obtains

$$\bar{K}_{ex}(\zeta) = \begin{cases} i\pi J_0(a\sqrt{k^2 - \zeta^2}) H_0^{(1)}(a\sqrt{k^2 - \zeta^2}), & \text{if } |\zeta| < k \\[3mm] 2I_0(a\sqrt{\zeta^2 - k^2}) K_0(a\sqrt{\zeta^2 - k^2}), & \text{if } |\zeta| > k, \end{cases} \quad (13.22)$$

where I_0 is the modified Bessel function of order zero.

The properties of the special functions in (13.20) and (13.22) that will be used throughout this appendix can be found in [4]. Note that both $\bar{K}_{ex}(\zeta)$ and $\bar{K}_{ap}(\zeta)$ are real when ζ is real and $|\zeta| > k$. Note also that $\bar{K}_{ap}(\zeta)$ is exponentially small when ζ is real and large, whereas $\bar{K}_{ex}(\zeta)$ decreases only as $1/\zeta$ for large ζ:

$$\bar{K}_{ex}(\zeta) = \frac{1}{a\zeta} + O\left(\frac{1}{\zeta^3}\right), \quad \text{as } \zeta \to +\infty. \tag{13.23}$$

Equations (13.20) and (13.22) can be analytically continued to complex ζ. In the complex ζ-plane, the branch cut originating from $\zeta = k$ lies in the upper-half ζ-plane, and the one originating from $\zeta = -k$ lies in the lower-half ζ-plane [3].

First assume that $K = K_{ex}$ in (13.18). This integral equation can be solved by first determining the Fourier transform $\bar{I}_{ex}^{(\infty)}(\zeta)$ and then using the Fourier inversion formula

$$I_{ex}^{(\infty)}(z) = \frac{1}{2\pi} \int_{-\infty}^{\infty} \bar{I}_{ex}^{(\infty)}(\zeta) e^{-i\zeta z} \, dz. \tag{13.24}$$

$\bar{I}_{ex}^{(\infty)}(\zeta)$ can be found by taking the Fourier transform of (13.18) with respect to z and noting that the convolution integral on the left-hand side transforms to the product $\bar{K}_{ex}(\zeta)\bar{I}_{ex}^{(\infty)}(\zeta)$. The only difficulty here is that the transform of the right-hand side does not exist when k is real. To overcome this, it is initially assumed that k has a small positive imaginary part. In other words, the integral equation is first solved for the case of a lossy surrounding medium, and the solution is then extended to the case where the antenna is in free space.

With $\text{Im}\{k\} > 0$, the Fourier transform of the right-hand side of (13.18) is

$$\int_{-\infty}^{\infty} \frac{2\pi V}{\zeta_0} e^{ik|z|} e^{i\zeta z} \, dz = \frac{4\pi}{\zeta_0} \frac{iVk}{k^2 - \zeta^2} \tag{13.25}$$

so that the current on the antenna is

$$\begin{aligned} I_{ex}^{(\infty)}(z) &= \frac{i2kV}{\zeta_0} \int_{-\infty}^{\infty} \frac{e^{-i\zeta z}}{(k^2 - \zeta^2)\bar{K}_{ex}(\zeta)} \, d\zeta \\ &= \frac{i4kV}{\zeta_0} \int_{0}^{\infty} \frac{\cos \zeta z}{(k^2 - \zeta^2)\bar{K}_{ex}(\zeta)} \, d\zeta, \quad -\infty < z < \infty. \end{aligned} \tag{13.26}$$

The restriction $\text{Im}\{k\} > 0$ can now be removed, and the solution can be extended to real k if the path of integration in (13.26) passes below the point $\zeta = k$ (and above the point $\zeta = -k$). This path is shown in Fig. 13.3a.

Suppose now that $K = K_{ap}$ in (13.18), and that one attempts to solve the integral equation by Fourier transformation. The integral corresponding to (13.26) diverges in this case because the exponentially small quantity $\bar{K}_{ap}(\zeta)$ appears in the denominator. Thus one is led to a contradiction. This shows that when the approximate kernel is used, the integral equation cannot have a solution which possesses a Fourier transform.

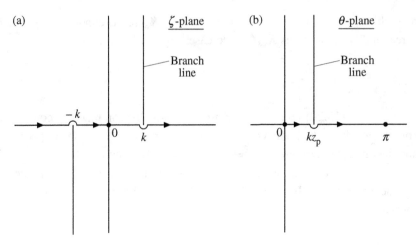

Figure 13.3 (a) Path of integration in (13.26) when k is real. (b) Path of integration in (13.43) when k is real.

Although the solution (13.26) for the case of the exact kernel has a complicated form, it reveals some basic properties of the current $I_{ex}^{(\infty)}(z)$ on the infinite antenna. It is easily seen that the current is finite for all z except $z = 0$. When $z = 0$, the integral diverges logarithmically and, in fact,

$$\frac{I_{ex}^{(\infty)}(z)}{V} = -i \frac{4ka}{\zeta_0} \ln \frac{1}{|z|} + O(1), \quad \text{as } z \to 0. \tag{13.27}$$

Equation (13.27) is a direct consequence of (13.23). Note that the coefficient in (13.27) is purely imaginary.

Although $\text{Im}\{I_{ex}^{(\infty)}(z)/V\}$ is infinite at the driving point, the component of current in phase with V is finite for all z, including $z = 0$. It is given by

$$\text{Re}\left\{\frac{I_{ex}^{(\infty)}(z)}{V}\right\} = \frac{4k}{\pi \zeta_0} \int_0^k \frac{\cos \zeta z}{(k^2 - \zeta^2)[J_0^2(a\sqrt{k^2 - \zeta^2}) + Y_0^2(a\sqrt{k^2 - \zeta^2})]} \, d\zeta. \tag{13.28}$$

Equation (13.28) is obtained from (13.26) as follows. Assume initially that $z \neq 0$ and write (13.26) as

$$\frac{I_{ex}^{(\infty)}(z)}{V} = h_1(z, \epsilon) + h_2(z, \epsilon) + h_3(z, \epsilon)$$

$$= \left(\int_0^{k-\epsilon} + \int_{L_\epsilon} + \int_{k+\epsilon}^\infty\right) \frac{i4k}{\zeta_0} \frac{\cos \zeta z}{(k^2 - \zeta^2)\bar{K}_{ex}(\zeta)} \, d\zeta, \tag{13.29}$$

where $\epsilon > 0$ and where L_ϵ is a path that starts at $k - \epsilon$, ends at $k + \epsilon$, and lies in the lower-half ζ-plane. Then use (13.22) to separate the integral $h_1(z, \epsilon)$ into its real and imaginary parts. The real part is

$$\mathrm{Re}\{h_1(z, \epsilon)\} = \frac{4k}{\pi \zeta_0} \int_0^{k-\epsilon} \frac{\cos \zeta z}{(k^2 - \zeta^2)\left[J_0^2(a\sqrt{k^2 - \zeta^2}) + Y_0^2(a\sqrt{k^2 - \zeta^2})\right]} \, d\zeta.$$

(13.30)

An examination of the behavior of the integrand near $\zeta = k$ shows that the limit of $\mathrm{Re}\{h_1(z, \epsilon)\}$ exists as $\epsilon \to 0$. Furthermore, from (13.22), it is seen that the integrand of $h_2(z, \epsilon)$ behaves according to

$$\frac{\cos \zeta z}{(k^2 - \zeta^2)\bar{K}_{\mathrm{ex}}(\zeta)} = \frac{\cos kz}{2k} \frac{1}{(\zeta - k)\ln(\zeta - k)} + O\left(\frac{1}{(\zeta - k)[\ln(\zeta - k)]^2}\right)$$

(13.31)

as $\zeta \to k$ in the lower-half ζ-plane. As a direct consequence of (13.31) and of the symmetric limits of integration, $\lim_{\epsilon \to 0} h_2(z, \epsilon)$ exists and equals zero. One can thus write

$$\frac{I_{\mathrm{ex}}^{(\infty)}(z)}{V} = \mathrm{Re}\{h_1(z, 0)\} + \lim_{\epsilon \to 0} [i \, \mathrm{Im}\{h_1(z, \epsilon)\} + h_3(z, \epsilon)].$$

(13.32)

As $h_3(z, \epsilon)$ is purely imaginary, $\mathrm{Re}\{h_1(z, 0)\}$ can be identified with $\mathrm{Re}\{I_{\mathrm{ex}}^{(\infty)}(z)/V\}$, and, with (13.30), (13.28) has been derived. The restriction $z \neq 0$ may be removed in the real part of $I_{\mathrm{ex}}^{(\infty)}(z)/V$ only.

Numerical method

In this section, the numerical method of Section 1.9 is applied to the infinite antenna. More precisely, an approximate solution of (13.18) is sought in the form

$$I^{(\infty)}(z) \doteq \sum_{n=-\infty}^{\infty} I_n^{(\infty)} u_n(z), \quad -\infty < z < \infty,$$

(13.33)

where the pulse functions $u_n(z)$ are given by (1.57). Here, the entire real axis is divided into segments of length z_p, where kz_p is small. The usual convention for the symbols $I_{\mathrm{ex},n}^{(\infty)}$, $I_{\mathrm{ap},n}^{(\infty)}$, and $I_n^{(\infty)}$ is employed.

As in the previous section, it is initially assumed that $\mathrm{Im}\{k\} > 0$. The procedure of Section 1.9 (i.e. the substitution of (13.33) into (13.18), multiplication by $u_l(z)$, and integration from $z = -\infty$ to $z = \infty$) yields in this case an infinite system of equations, namely,

$$\sum_{n=-\infty}^{\infty} A_{l-n} I_n^{(\infty)} = B_l^{(\infty)}, \quad l = 0, \pm 1, \pm 2, \ldots,$$

(13.34)

where the matrix elements A_l are given by (13.11) and where

$$B_l^{(\infty)} = \frac{2\pi V}{\zeta_0} \int_{(l-1/2)z_p}^{(l+1/2)z_p} e^{ik|z|} dz$$

$$= \begin{cases} \dfrac{i8\pi V}{\zeta_0 k} \sin^2(kz_p/4) + \dfrac{4\pi V}{\zeta_0 k} \sin(kz_p/2), & \text{if } l = 0 \\[4mm] \dfrac{4\pi V}{\zeta_0 k} \sin(kz_p/2)\, e^{i|l|kz_p}, & \text{if } l = \pm 1, \pm 2, \dots . \end{cases} \tag{13.35}$$

The solution of the infinite Toeplitz system (13.34) can be found in closed form in the following manner: multiply each side by $e^{il\theta}$, where $-\pi < \theta \le \pi$, and sum with respect to l. Then, interchange the order of summation, and introduce the Fourier series

$$\bar{A}(\theta) = \sum_{l=-\infty}^{\infty} A_l e^{il\theta}, \quad -\pi < \theta \le \pi \tag{13.36}$$

$$\bar{B}(\theta) = \sum_{l=-\infty}^{\infty} B_l^{(\infty)} e^{il\theta}, \quad -\pi < \theta \le \pi \tag{13.37}$$

and

$$\bar{I}(\theta) = \sum_{l=-\infty}^{\infty} I_l^{(\infty)} e^{il\theta}, \quad -\pi < \theta \le \pi. \tag{13.38}$$

One obtains $\bar{A}(\theta)\bar{I}(\theta) = \bar{B}(\theta)$ so that the Fourier coefficients $I_n^{(\infty)}$ are given by

$$I_n^{(\infty)} = \frac{1}{2\pi} \int_{-\pi}^{\pi} \bar{I}(\theta) e^{-in\theta}\, d\theta = \frac{1}{2\pi} \int_{-\pi}^{\pi} \frac{\bar{B}(\theta)}{\bar{A}(\theta)} e^{-in\theta}\, d\theta \quad (\text{Im}\{k\} > 0). \tag{13.39}$$

Since it has been assumed that $\text{Im}\{k\} > 0$, the sum in (13.37) converges and, with (13.35), it is straightforward to show that

$$\bar{B}(\theta) = \frac{-i V}{\zeta_0} \frac{8\pi}{k} \sin^2(kz_p/4) \frac{\cos(kz_p/2) + \cos^2(\theta/2)}{\sin[(\theta + kz_p)/2]\sin[(\theta - kz_p)/2]}, \quad -\pi < \theta \le \pi. \tag{13.40}$$

The substitution of (13.11) into (13.36) and the application of the Poisson summation formula [22] lead to

$$\bar{A}(\theta) = \sum_{m=-\infty}^{\infty} \int_{-\infty}^{\infty} \int_0^{z_p} (z_p - z)[K(z - xz_p) + K(z + xz_p)] dz\, e^{ix(\theta - 2m\pi)}\, dx \tag{13.41}$$

from which it is seen that

$$\bar{A}(\theta) = z_p \sum_{m=-\infty}^{\infty} \bar{K}\left(\frac{2m\pi - \theta}{z_p}\right) \frac{\sin^2(\theta/2)}{[m\pi - (\theta/2)]^2}, \tag{13.42}$$

where $\bar{K}$ – the Fourier transform of the kernel – is given by (13.20) or (13.22).

The expression (13.40) is also meaningful for real k except for $k = \pm\theta/z_p$. Thus, (13.39) can be analytically continued to real k if the path of integration in (13.39) passes below the point $\theta = kz_p$ and above the point $\theta = -kz_p$. The final exact expression for the coefficients $I_n^{(\infty)}$ can therefore be written as

$$I_n^{(\infty)} = \frac{1}{\pi} \int_0^\pi \frac{\bar{B}(\theta)}{\bar{A}(\theta)} \cos n\theta \, d\theta, \quad -\infty < n < \infty \quad (k \text{ real}), \tag{13.43}$$

where the path of integration passes below the branch point at $\theta = kz_p$. This path is shown in Fig. 13.3b. Thus far, the results hold for both the exact and the approximate kernels.

First, interpret $K(z)$ as $K_{\text{ex}}(z)$. From (13.42) and (13.40),

$$\bar{A}(\theta) \sim z_p \bar{K}_{\text{ex}}(\theta/z_p)[\sin(\theta/2)/(\theta/2)]^2 \tag{13.44}$$

and

$$\bar{B}(\theta) \sim -\frac{i2\pi V}{k\zeta_0} \frac{(kz_p)^2}{\theta^2 - (kz_p)^2} \frac{1 + \cos^2(\theta/2)}{[\sin(\theta/2)/(\theta/2)]^2} \tag{13.45}$$

as $z_p \to 0$, uniformly for all θ. Substituting into (13.43), changing the variable $\zeta = \theta/z_p$, and taking the limit as $z_p \to 0$ with nz_p fixed give

$$I_{\text{ex},n}^{(\infty)} \sim \frac{i4kV}{\zeta_0} \int_0^\infty \frac{\cos \zeta nz_p}{(k^2 - \zeta^2)\bar{K}_{\text{ex}}(\zeta)} \, d\zeta, \tag{13.46}$$

where the path passes below $\zeta = k$. Therefore, as one might expect, the limit of the numerical solution for zero pulse width is precisely the exact solution (13.26) of the integral equation. In particular, the numerical method reproduces the logarithmic singularity at the driving point.

Numerical method: the case of the approximate integral equation

The more interesting case of the approximate kernel is now examined. If $\bar{K} = \bar{K}_{\text{ap}}$ in (13.42), then, as expected, (13.43) has no limit as $z_p \to 0$: the integral corresponding to (13.46) diverges because it is seen from (13.20) that

$$\bar{K}_{\text{ap}}(\zeta) \sim \sqrt{\frac{2\pi}{a|\zeta|}} \, e^{-a|\zeta|} \tag{13.47}$$

for $a|\zeta| \gg 1$. The nature of the divergence will now be examined. Specifically, the asymptotic behavior of (13.43) is investigated subject to the conditions

$$z_p/a \ll 1 \quad \text{and} \quad nz_p/a = O(1). \tag{13.48}$$

Note that the first condition implies $kz_p \ll 1$. It is observed from (13.42), (13.20), and (13.47) that

$$\bar{A}(\theta) \sim 4z_p \sin^2(\theta/2) \left[\bar{K}_{ap}(\theta/z_p) \frac{1}{\theta^2} + \bar{K}_{ap}[(2\pi - \theta)/z_p] \frac{1}{(2\pi - \theta)^2} \right] \tag{13.49}$$

as $z_p/a \to 0$, uniformly for $0 \le \theta \le \pi$. The right-hand side of (13.49) consists of the terms $m = 0$ and $m = 1$ in the summation (13.42); the neglected terms are exponentially smaller than those retained.

Substituting (13.49) into (13.43) and setting $\phi = \pi - \theta$ give

$$I_{ap,n}^{(\infty)} \sim \frac{(-1)^n}{4\pi z_p} \int_0^\pi \frac{\bar{B}(\pi - \phi)/\cos^2(\phi/2)}{\bar{K}_{ap}[(\pi - \phi)/z_p]/(\pi - \phi)^2 + \bar{K}_{ap}[(\pi + \phi)/z_p]/(\pi + \phi)^2}$$

$$\times \cos n\phi \, d\phi, \tag{13.50}$$

where the path of integration passes above the point $\phi = \pi - kz_p$.

Because of the conditions (13.48), the main contribution to this integral comes from a narrow region near $\phi = 0$. It is therefore legitimate to neglect the contribution $\int_1^\pi$ and replace the upper limit π in (13.50) by 1. Once this is done, (13.47) can be applied for both $\bar{K}_{ap}[(\pi - \phi)/z_p]$ and $\bar{K}_{ap}[(\pi + \phi)/z_p]$. Note that this approximation was not possible in the whole original interval. Note also that the choice of 1 for an upper limit makes no difference in the sense that any other choice of $O(1)$ will lead to the same final result (13.57). The aforementioned approximations and the change of variable $x = \phi(a/z_p)$ in the resulting integral lead to

$$I_{ap,n}^{(\infty)} \sim \frac{1}{\sqrt{2\pi}} kz_p \sqrt{\frac{z_p}{a}} (-1)^n e^{\pi a/z_p} f(z_p/a, ka, nz_p/a), \tag{13.51}$$

where f is the integral

$$f(z_p/a, ka, nz_p/a) = \int_0^{a/z_p} g(x; z_p/a, ka) \cos[(nz_p/a)x] \, dx \tag{13.52}$$

in which

$$g(x; z_p/a, ka) = \frac{1}{4\pi} \frac{\bar{B}[\pi - (z_p/a)x]/(kz_p^2 \cos^2[(z_p/a)x/2])}{e^x[\pi - (z_p/a)x]^{-5/2} + e^{-x}[\pi + (z_p/a)x]^{-5/2}}. \tag{13.53}$$

The function f depends on its argument z_p/a because a/z_p is the upper limit of integration in (13.52), and also because g depends on z_p/a. It can be verified from (13.40) that, apart from the factor V/ζ_0, g is indeed a function of x, z_p/a, and ka.

The next step is to approximate f by the first two terms in its Taylor-series expansion about the point $z_p/a = 0$, while keeping ka and nz_p/a fixed. It is readily checked that this can be carried out by expanding g in powers of z_p/a and integrating from $x = 0$ to $x = \infty$ in (13.52). With (13.40), it is found that the expansion of g is

$$g(x; z_p/a, ka) = \frac{-iV}{\zeta_0} \frac{\pi^{5/2}}{16} \frac{1}{\cosh x} \left(1 - \frac{5}{2\pi} \frac{z_p}{a} x \tanh x \right) + O\left((z_p/a)^2\right). \quad (13.54)$$

When (13.54) is substituted into (13.52) and the upper limit of integration is set to ∞, two integrals occur. With $y = nz_p/a$, these are [21, pp. 503–505]

$$\int_0^\infty \frac{\cos yx}{\cosh x} dx = \frac{\pi}{2\cosh(\pi y/2)} \quad (13.55)$$

and

$$\begin{aligned}
\int_0^\infty \frac{x \tanh x}{\cosh x} \cos yx \, dx &= \frac{d}{dy} \int_0^\infty \frac{\tanh x}{\cosh x} \sin yx \, dx \\
&= \frac{d}{dy} \frac{\pi y}{2\cosh(\pi y/2)} \\
&= \frac{\pi}{2\cosh(\pi y/2)} \left[1 - \frac{\pi}{2} y \tanh(\pi y/2) \right].
\end{aligned} \quad (13.56)$$

The substitution of the resulting formula for $f(z_p/a, ka, nz_p/a)$ into (13.51) yields the final result:

$$\begin{aligned}
I_{ap,n}^{(\infty)} &\sim \frac{-iV}{\zeta_0} \frac{\pi^3}{32\sqrt{2}} kz_p \sqrt{\frac{z_p}{a}} (-1)^n \exp(a\pi/z_p) \frac{1}{\cosh[(nz_p/a)\pi/2]} \\
&\times \left[1 - \frac{5}{2\pi} \frac{z_p}{a} + \frac{5}{4} n \left(\frac{z_p}{a}\right)^2 \tanh[(nz_p/a)\pi/2] \right],
\end{aligned} \quad (13.57)$$

where the quantity in brackets is simply a correction factor.

Thus, when the pulse width is small, the numerical method yields an exponentially large, purely imaginary "driving-point admittance", and a large, purely imaginary, rapidly oscillating "current", at least for points on the antenna not too far from the driving point.

On the other hand, it can be shown that as $z_p \to 0$, $\mathrm{Re}\{I_{ap,n}^{(\infty)}/V\}$ exists and is finite for all n. Specifically, with nz_p fixed,

$$\lim_{z_p \to 0} \mathrm{Re}\left\{\frac{I_{ap,n}^{(\infty)}}{V}\right\} = \frac{4k}{\pi\zeta_0} \int_0^k \frac{J_0(a\sqrt{k^2 - \zeta^2}) \cos(\zeta nz_p)}{(k^2 - \zeta^2)[J_0^2(a\sqrt{k^2 - \zeta^2}) + Y_0^2(a\sqrt{k^2 - \zeta^2})]} d\zeta, \quad (13.58)$$

so that the numerical method yields a finite real part of the current. It is seen from (13.28) that this real part is very close to the corresponding quantity for the exact integral equation, and that the two quantities become identical in the limit $ka \to 0$.

The derivation of (13.58) from (13.43) is similar to the derivation of (13.28) from (13.26) The intermediate formulas are lengthy, and only an outline of the procedure is given here.

Assume initially that both z_p and n are non-zero, and divide both sides of (13.43) by V. Change the variable $\zeta = \theta/z_p$, and split the resulting integral as in (13.29). The parameter $\epsilon > 0$ plays the same role, and the upper limit of integration in the third integral is π/z_p instead of ∞. Determine the real part of the integral $\int_0^{k-\epsilon}$, and note that this real part has a limit as $\epsilon \to 0$. As $\zeta \to k$ in the lower-half ζ-plane, only the $m = 0$ term in the sum for $\bar{A}(\zeta z_p)$ is important, so that the integrand in $\int_{k-\epsilon}^{k+\epsilon}$ again behaves according to (13.31). In this case, the coefficient of the leading term depends on z_p.

In this manner, a relation of the form

$$\frac{I_{ap,n}^{(\infty)}}{V} = \mathrm{Re}\{h_1(n, z_p, \epsilon = 0)\} + \lim_{\epsilon \to 0} [i\, \mathrm{Im}\{h_1(n, z_p, \epsilon)\} + h_3(n, z_p, \epsilon)] \tag{13.59}$$

with $h_3(n, z_p, \epsilon)$ purely imaginary is obtained. Thus, the quantity $\mathrm{Re}\{h_1(n, z_p, 0)\}$ is the real part of $I_{ap,n}^{(\infty)}/V$ for non-zero z_p. The remaining task – to show that the limit $\lim_{z_p \to 0} \mathrm{Re}\{h_1(n, z_p, \epsilon = 0)\}$ exists and equals the right-hand side of (13.58) – is easily carried out, because the terms with $m \neq 0$ in the sum for $\bar{A}(\zeta z_p)$ vanish when $z_p = 0$.

The infinite antenna: summary

The integral equation for the current on an antenna of infinite length, center-driven by a delta-function generator has been derived. This integral equation has two forms depending on the choice of kernel. For the case of the exact kernel, the integral equation was solved explicitly. Having done this, it was easy to deduce that the current is logarithmically singular at the driving point. The coefficient of the logarithm was determined, and was found to be purely imaginary when the driving voltage is real. On the other hand, the component of current in phase with the driving voltage is finite along the entire length of the antenna, including the driving point.

For the case of the approximate kernel, the integral equation has no solution. Nevertheless, the question "What does one obtain when one applies a numerical method to this equation?" is meaningful and, thus, Galerkin's method with pulse functions was applied to the integral equation. An infinite system of equations resulted; this system was solved exactly for non-zero pulse width, and the solution was developed asymptotically for the case in which the pulse width is much smaller than the radius of the infinite antenna. The asymptotic study predicts that one obtains a large, purely imaginary "driving-point admittance", and rapid oscillations in the imaginary part of the current near the driving point. It also predicts that the real part is close to the corresponding real part obtained with the exact kernel, and that the two quantities become identical in the limit of zero antenna radius.

14 Techniques and theory of measurements

Theory and experiment have run hand-in-hand leading the way toward new concepts and new radiating and receiving structures throughout the history of antennas. Sometimes new antennas were developed through experiment with supporting analysis coming later and perhaps pointing the way to optimization. In other instances, the analysis pointed the way with experiment showing that the analytical model could be achieved in the physically real world. Regardless of the origin of the idea, the most effective results have occurred when the theoretical analysis and the experimental measurements come together. For the experimental and theoretical results to agree, both must describe a model that has exactly the same electromagnetic boundary conditions. Since it is rarely possible to have identical models in the theoretical and the real world, the models must be critically examined, their differences identified, and at least approximately taken into account. Theoretical concepts such as "less than" and "thin" must be reconciled in each experimental model. Similarly, the measurement process must be critically examined to understand exactly what is being measured. Are sampling probes extracting enough power to distort the interference pattern on the transmission line? Are the fields being sampled too close to a discontinuity? Are loop probes being excited only in their basic mode? Finally, there is always the basic question in antenna experiments: "To what in the world are the elements really coupling?"

Measurement procedures on uniform sections of transmission lines have been discussed many times [1–7]. Measurements on the uniform sections of lines include the effects of discontinuities at the end or driving point of the line, such as those that occur when the transmission lines are attached to dipoles or monopoles over ground planes. The theoretical models usually assume the antenna elements to be isolated in free space and do not account for the effects of the attached transmission lines. Hence, measurements of such quantities as driving-point impedances that are made on the uniform sections of lines will not agree with the results calculated from theoretical analyses unless these differences in the experimental and theoretical models are taken into account [8–10].

Some of the basic considerations in measurements of the impedances or admittances and the current distributions of dipole arrays and arrays of monopoles on ground planes as well as the general properties of probes are discussed in the following sections.

Equations relating impedances and admittances with the measured quantities are given in Section 14.4. Considerable emphasis is placed on the measurement of driving-point impedances or admittances. These are quantities that can be especially difficult to reconcile with theoretical results because their values depend on the exact ratio of the fields at a single point on the transmission line, and this is a point where the differences between the experimental and analytical models are greatest. In addition, the elements in the experimental model will be coupled not only to each other but to the feeding structures and generally to anything else in their vicinity, which may seriously modify the currents and voltages at the driving point. Although the discussions are not restricted to any particular frequency range, applications have been primarily in the 100 MHz to 3 GHz range.

14.1   Transmission lines with coupled loads

If the termination of a transmission line is to be described as a lumped circuit element and to be characterized in a useful manner as an impedance or admittance, it should ideally be independent of the circuit to which it is attached. However, this ideal condition is rarely met for real dipoles and monopoles because they are coupled to their feeding lines over at least small regions near the driving points. Except for the model of a monopole over a ground plane, for which a formal solution is available [11], the complete analysis of antennas coupled to their feeding lines has proven too complicated for theoretical analysis. In principle, moment methods [12] or other computer modeling methods can be used to account for the difficult regions of coupling or discontinuity, but the physical phenomena in the junction region must be well understood in order to construct an accurate model and the limits of the model should be explored and explicitly delineated. Therefore, the measurement line or feeding line on which measurements are to be made is usually treated as a uniform line outside of a small region over which the line–load coupling takes place. This small region near the line–load junction is called the terminal zone and the coupling effects are called terminal-zone effects or end effects.

Properties of the terminal zones can be determined from the differential equations for the voltage and current along the transmission line. In the usual method of deriving these equations, the line is divided into identical infinitesimal sections, each section is represented by a lumped capacitance, an inductance, and a resistance, and Kirchhoff's laws of ordinary circuits are assumed to apply to each section. The results obtained by this method apply only to infinite, unloaded lines. They can contain no information about the terminal zone which requires a derivation based on a more complete theoretical model. Detailed steps of the more exact derivation are given in Chapters 2 of [9] and [10]. Regardless of the particular transmission-line model and termination used in the derivation, the following generalized equations are obtained:

$$\frac{\partial^2 V(w)}{\partial w^2} - \gamma^2(w)V(w) = 0 \tag{14.1a}$$

$$I_{1L}(w) = \frac{1}{z(w)}\frac{\partial V(w)}{\partial w}. \tag{14.1b}$$

The distance, w, is measured along the transmission line from the line–load junction. $V(w)$ is the scalar potential difference or voltage between conductors of the transmission line at w, $I_{1L}(w)$ is the total current in one of the conductors at w, and the line is assumed to be perfectly balanced so that $I_{2L}(w) = -I_{1L}(w)$.

In the following definitions, $W(w)$ is the vector potential difference between the conductors at w, and a subscript p indicates the component of the vector potential parallel to the transmission line. A subscript L denotes that part of a quantity which is determined only from currents and charges in the line, and a subscript T denotes that part which is determined only from currents and charges in the termination or load. The various quantities in (14.1a, b) are

$$\gamma(w) = \sqrt{z(w)y(w)a_1(w)\phi_1(w)}; \qquad \text{propagation "constant"} \tag{14.2}$$

$$a_1(w) = \frac{W_{pL}(w) + W_{pT}(w)}{W_{pL}(w)}; \qquad \text{coefficient of inductive coupling} \tag{14.3}$$

$$\phi_1(w) = \frac{V_L(w)}{V_L(w) + V_T(w)}; \qquad \text{coefficient of capacitive coupling} \tag{14.4}$$

$$z(w) = r(w) + j\omega l^e(w) \qquad \text{impedance per unit length} \tag{14.5}$$

$$\doteq j\omega l^e(w)$$
$$= j\omega[l_L^e(w) + l_T^e(w)];$$

$$y(w) = g(w) + j\omega c(w) \qquad \text{admittance per unit length} \tag{14.6}$$

$$\doteq j\omega c(w);$$

$$y^{-1}(w) = y_L^{-1}(w) + y_T^{-1}(w)$$

$$\beta^2 = \omega^2 \mu\epsilon; \tag{14.7}$$

$$Z_c(w) = \sqrt{\frac{z(w)}{y(w)}} \doteq \sqrt{\frac{l^e(w)}{c(w)}}; \qquad \text{"characteristic" impedance} \tag{14.8}$$

μ and ϵ are the permeability and permittivity of the material in which the transmission-line conductors are embedded; $\mu = \mu_r \mu_0$, $\epsilon = \epsilon_r \epsilon_0$, where μ_0, ϵ_0 are the values for free space.

Vector and scalar potentials are calculated from the Helmholtz integrals of currents and charges over all conductors. These integrals are defined in (1.12a, b). A detailed discussion of the transmission-line parameters is in [9] and [10], where it is shown that differences in the potential of equipotential rings located just outside the conductors of the transmission line at a distance w from the line–load junction are given approximately by

$$W_p(w) = W_{pL}(w) + W_{pT}(w) \doteq l^e(w) I_{1L}(w) \tag{14.9}$$

$$V(w) = V_L(w) + V_T(w) \doteq \frac{q_L(w)}{c(w)}, \tag{14.10}$$

where $q_L(w)$ is the charge per unit length along one of the transmission-line conductors. Those parts due only to currents and charges in the line are

$$W_{pL}(w) \doteq l_L^e(w) I_{1L}(w) \tag{14.11}$$

$$V_L(w) \doteq \frac{q_L(w)}{c_L(w)}. \tag{14.12}$$

With these approximations, (14.3) and (14.4) become

$$a_1(w) \doteq l^e(w)/l_L^e(w) \tag{14.13}$$

$$\phi_1(w) \doteq c(w)/c_L(w). \tag{14.14}$$

When coupling between the line and its termination is expressed in terms of the vector potential, it is inductive; if there is no inductive coupling, $a_1(w) = 1$. Note that if all conductors of a termination are perpendicular to all conductors of the transmission line, there is no inductive coupling between them. When coupling between the line and its termination is expressed in terms of the scalar potential, it is capacitive; if there is no capacitive coupling, $\phi_1(w) = 1$.

Because the form of $\gamma(w)$ differs for each line and termination, (14.1a, b) have no general solution. For most specific lines and loads, they are too complicated to yield useful analytical solutions, although numerical computer techniques could be used to obtain results for each specific case. The development of an approximate but more general procedure that provides some physical insight and is especially useful for experimental work is outlined in the following discussion.

A detailed examination of the parameters in (14.2)–(14.14) reveals that the non-uniformities decrease rapidly with distance from the line–load junction. Along the transmission line, $a_1(w)$ and $\phi_1(w)$ usually differ negligibly from one and $z(w)$ and $y(w)$ are sensibly constant at distances from the line–load junction that exceed ten times the center-to-center spacing between the conductors of a two-wire line, or ten times the difference in radii between outer and inner conductors of a coaxial line. This is a rough measure of the extent of the terminal zone. For most transmission lines it is less than 0.1λ. At greater distances from the line–load junction, all parameters are constant and (14.1a, b) reduce to the usual linear form:

$$\frac{d^2V(w)}{dw^2} - \gamma^2 V(w) = 0 \tag{14.15a}$$

$$I_{1L}(w) = \frac{1}{z_0}\frac{dV(w)}{dw} \tag{14.15b}$$

since

$$z(w) = z_0 = r_0 + j\omega l_0^e \tag{14.16a}$$

$$y(w) = y_0 = g_0 + j\omega c_0 \tag{14.16b}$$

$$\gamma^2(w) = \gamma^2 = z_0 y_0 \tag{14.16c}$$

$$\gamma = \alpha + j\beta. \tag{14.16d}$$

Thus, except within a small terminal zone, conventional transmission-line theory applies and the usual measuring techniques are valid. Changes that occur in the line parameters over short distances near the line–load junction appear as lumped inductances and capacitances in series and in parallel with the actual terminating impedance. When a load impedance is determined in the usual manner from measurements on the uniform part of the line, the quantity determined is always a combination of the actual load impedance with the inductances and capacitances caused by changes of the line parameters within the terminal zone. Approximate account can be taken of such changes if it is assumed that the uniform line parameters of an infinite unloaded line apply everywhere including the terminal zone, and the differences that occur within the terminal zone between the actual parameters and the assumed ones are represented by a balanced network of equivalent lumped series inductances and shunting capacitances, as shown in Figs. 14.1a or 14.1b. The lumped elements are defined as follows:

$$L_T = \int_0^d [l^e(w) - l_0^e]\,dw \tag{14.17a}$$

$$C_T = \int_0^d [c(w) - c_0]\,dw, \tag{14.17b}$$

where $l^e(w) = l_L^e(w) + l_T^e(w)$ is the true inductance per unit length, $c^{-1}(w) = c_L^{-1}(w) + c_T^{-1}(w)$ is the true reciprocal capacitance or elastance per unit length, and l_0^e and c_0^{-1} are the corresponding quantities for an infinite line. Everywhere along an infinite line or outside of the terminal zone of a terminated line, the ratio of

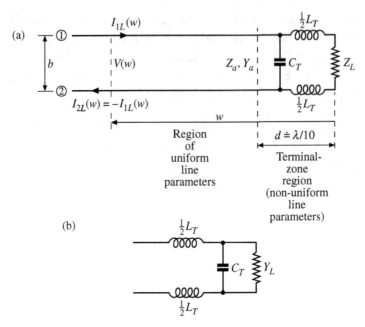

Figure 14.1 (a) Terminal-zone region and network. (b) Alternative representation for terminal-zone network.

the tangential component of the vector potential difference to the current and the ratio of the scalar potential difference to the charge per unit length are constant and given by

$$W_p(w)/I_{1L}(w) = l_0^e \tag{14.18a}$$

$$V(w)/q_{1L}(w) = 1/c_0. \tag{14.18b}$$

With (14.13) and (14.14), the integrals for L_T and C_T are

$$L_T = \int_0^d [l_L^e(w)a_1(w) - l_0^e]\,dw \tag{14.19a}$$

$$C_T = \int_0^d [c_L(w)\phi_1(w) - c_0]\,dw. \tag{14.19b}$$

One advantage of the procedure of assuming line parameters to be uniform throughout the terminal zone and representing terminal-zone non-uniformities by a network of lumped elements is that useful approximate expressions for L_T and C_T can frequently be derived from considerations of the static and induction fields or quite easily be calculated for specific configurations using computer modeling techniques. Also, C_T and L_T can be obtained from measurements. If the load is characterized by its impedance and the terminal-zone network is like Fig. 14.1a, the actual load impedance Z_L, and the apparent load impedance Z_a (or the apparent load admittance Y_a), are related by

$$Y_a = \frac{1}{Z_a} = \frac{1}{Z_L + j\omega L_T} + j\omega C_T \qquad (14.20)$$

$$Z_L = \frac{Z_a}{1 - j\omega C_T Z_a} - j\omega L_T = \frac{1}{Y_a - j\omega C_T} - j\omega L_T. \qquad (14.21)$$

If the load is characterized by its admittance and the network of Fig. 14.1b is used,

$$Z_a = Y_a^{-1} = \frac{1}{Y_L + j\omega C_T} + j\omega L_T \qquad (14.22)$$

$$Y_L = \frac{1}{Z_a - j\omega L_T} - j\omega C_T = \frac{Y_a}{1 - j\omega L_T Y_a} - j\omega C_T. \qquad (14.23)$$

The terminology used here is that of the lower-frequency transmission lines. Problems involving waveguides and some of those involving coaxial lines are conveniently solved in terms of propagating and evanescent modes [13, 14]. The latter decay rapidly with distance from a discontinuity and it is this distance which defines the extent of the terminal zone.

14.2   Equivalent lumped elements for terminal-zone networks

Whenever an antenna that is ideally approximated by an independent impedance Z_L is attached to a transmission line, the impedance that appears to be loading the transmission line is Z_a, not Z_L. The apparent impedance, Z_a, is a combination of Z_L and a terminal-zone network consisting of a series reactance $X_T = \omega L_T$, and a shunting susceptance, $B_T = \omega C_T$. This network takes account of changes in the parameters of the line as its end is approached and of coupling between the transmission line and the antenna.

L_T and C_T can be evaluated theoretically, or determined from the measured values of Z_a and Z_L. For many transmission lines, Z_L can be obtained by repeating the measurement of Z_a as the distance between the conductors of the transmission line is decreased successively, and then extrapolating the results to a fictitious "zero" spacing. The extrapolated value of Z_a is Z_L.

One common use of a terminal-zone network is to transform driving-point impedances which have been calculated from an established theory into those which can be measured on a particular transmission line. For this purpose, a single model of the desired antenna and its attached transmission line can be constructed,

$Y_a = G_a + jB_a$ measured, and $Z_L = R_L + jX_L$ computed from the theory. Then, from (14.20),

$$\omega C_T = B_a \pm \sqrt{(G_a/R_L) - G_a^2} \tag{14.24a}$$

$$\omega L_T = -X_L \pm \sqrt{(R_L/G_a) - R_L^2}. \tag{14.24b}$$

For some models, Y_L may be more convenient to compute than Z_L. From (14.22),

$$\omega C_T = -B_L \pm \sqrt{(G_L/R_a) - G_L^2} \tag{14.25a}$$

$$\omega L_T = X_a \pm \sqrt{(R_a/G_L) - R_a^2}. \tag{14.25b}$$

The resulting values of ωL_T and ωC_T can then be used for all other elements of the same kind in the array, as long as the element spacing is not so small that the terminal zones are directly coupled to one another.

The sign to be used in (14.24) and (14.25) is usually the one which makes the magnitudes of ωL_T and ωC_T smallest; in any case, their correct values are the ones that satisfy the imaginary parts of (14.20) and (14.22). Equations (14.24) and (14.25) may involve small differences between quite large numbers so that high accuracy is required in Y_a or Z_a. Therefore account must be taken of adapters, bends, or connectors which are between the antenna and the point where the measurements are made.

Theoretical determinations of L_T and C_T can be based on (14.17) or (14.19). Expressions for the inductances and capacitances in the integrands of (14.17) are themselves integrals of the static or induction fields for the particular load and transmission line that is being analyzed. An evaluation of these integrals is readily carried out by computer with numerical methods no more complicated than Simpson's rule. Some examples of approximate formulas that are applicable to dipoles as end-loads on two-wire lines and to monopoles as end-loads on coaxial lines are summarized in the following paragraphs. Many additional examples will be found in [13–18].

Symmetrical dipole as a load on two-wire lines

This model is shown in Fig. 14.2. Approximate expressions for L_T and C_T are

$$L_T \doteq \frac{\mu}{2\pi}(b - a), \qquad -C_T/c_0 b \doteq 1.5/\ln(b/a), \tag{14.26}$$

where a is the radius of the conductors and b is the distance between the conductors of the transmission line. These expressions are derived under the assumption that the conductors are thin compared to their separation, $a \ll b$. They are accurate to within about 20% for $b/a = 3$ and improve in accuracy as b/a increases. Expressions with higher accuracy have been derived.[1]

---

[1] [10], p. 50.

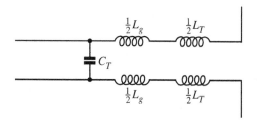

Figure 14.2 Network for terminal zone of dipole as end-load of two-wire line.

The inductance given by (14.26) accounts for non-uniformities near the end of the transmission line. Most theoretical models assume the antenna to extend from $z = 0$ to $z = \pm h$, whereas in some experimental models a section of the antenna may be missing between $z = \pm \frac{1}{2}b$. Account can be taken of this missing section by subtracting from the measured input reactance the difference in zero-order input reactance between an antenna of length $(h - b/2)$ and one of length h [19]. That is,

$$X_g = \omega L_g = \frac{-\zeta_0 \psi}{2\pi} [\cot \beta_0 (h - b/2) - \cot \beta_0 h] \qquad (14.27)$$

$$\psi = \begin{cases} \dfrac{|C_a(h, 0) \sin \beta_0 h - S_a(h, 0) \cos \beta_0 h|}{\sin \beta_0 h}, & \beta_0 h < \pi/2 \quad (14.28a) \\[2mm] |C_a(h, h - \lambda/4) \sin \beta_0 h - S_a(h, h - \lambda/4) \cos \beta_0 h|, & \beta_0 h > \pi/2. \quad (14.28b) \end{cases}$$

The integral functions $C_a(h, z)$ and $S_a(h, z)$ are defined as

$$C_a(h, z) = \int_{-h}^{h} \cos \beta_0 z' \frac{e^{-j\beta_0 R_1}}{R_1} dz'$$

$$= \int_{0}^{h} \cos \beta_0 z' \left| \frac{e^{-j\beta_0 R_1}}{R_1} + \frac{e^{-j\beta_0 R_2}}{R_2} \right| dz' \qquad (14.28c)$$

$$S_a(h, z) = \int_{-h}^{h} \sin \beta_0 z' \frac{e^{-j\beta_0 R_1}}{R_1} dz'$$

$$= \int_{0}^{h} \sin \beta_0 z' \left| \frac{e^{-j\beta_0 R_1}}{R_1} + \frac{e^{-j\beta_0 R_2}}{R_2} \right| dz' \qquad (14.28d)$$

with

$$R_1 = \sqrt{(z - z')^2 + a^2}, \qquad R_2 = \sqrt{(z + z')^2 + a^2}. \qquad (14.28e)$$

The total correction for end-effect is a shunt susceptance $B_T = \omega C_T$ with C_T given by (14.26) and a reactance $X_E/2$ in series with each conductor where

$$X_E = X_T + X_g = \omega(L_T + L_g) \qquad (14.29)$$

with L_T given by (14.26) and L_g by (14.27). The location of C_T in the network shown in Fig. 14.2 is arbitrary. It may be connected across the terminals of the dipole or across the line between L_g and L_T if more convenient.

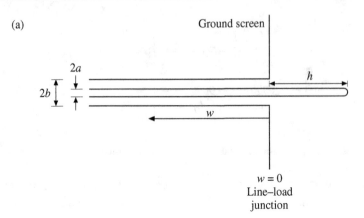

(a)

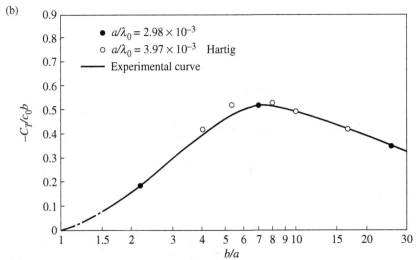

(b)

Figure 14.3 (a) Monopole over a ground plane. (b) Experimentally determined capacitive end correction for monopole over a ground screen driven by a coaxial line.

Monopole over a ground plane fed by a coaxial line[2]

In this model the outer conductor of the coaxial line ends at the surface of the ground plane, and the inner conductor continues through the ground plane to form the monopole shown in Fig. 14.3a. When the coaxial line and monopole are perpendicular to the ground plane, currents on the ground plane are not coupled inductively to the antenna or to the transmission line; hence, $L_T = 0$. As the current on the inner conductor is continuous at the line–load junction, $L_g = 0$. Hence, the terminal network consists only of a shunting capacitance that is given in Fig. 14.3b. When considered on

---

[2] [9], p. 430; also *Trans. I.R.E.*, **AP-3**, 66, April 1955.

an admittance basis, this model is especially simple since the terminal-zone correction applies only to the susceptance

$$G_L + j(B_L + \omega C_T) = G_a + jB_a \tag{14.30}$$

so that $G_a = G_L$ and B_a and B_L differ only by an additive constant for any particular combination of a coaxial transmission line and monopole antenna.

Change in conductor radius or spacing of two-wire lines[3] or coaxial lines[4]

Applications frequently occur in which the conductors of the dipole or monopole must have diameters different from those of the attached feeding lines, or the base separation at the driving point must be different from the spacing between the conductors of the feeding lines. End-effects associated with these various changes have been analyzed separately in the literature and the results can be applied directly, provided the conductors of the dipole are extended to form a short section of two-wire line, or the monopole and the associated hole in the ground plane are extended to form a short section of coaxial line. This section of line should be at least twice as long as the terminal zone to prevent coupling between the different terminal regions.

Consider first a two-wire line (Fig. 14.4a) with constant spacing between the axes of its conductors but with conductors of radius a_l for $y \le s$ and a_r for $y \ge s$. The current in each conductor is continuous and in only one direction near the junction at $y = s$; therefore, $L_T = 0$ and the terminal-zone network consists only of a shunting susceptance. The value of $B_T = \omega C_T$ for the two-wire line is one-half of that obtained for a coaxial line (Fig. 14.4b) which has a corresponding change in radius of the inner conductor, as long as the ratios b/a_l and b/a_r are the same for the two-wire line and the coaxial line.[5]

Consider next a two-wire line (Fig. 14.4c) with conductors of constant radius a but with a distance b_l between their axes $y \le s$ and b_r for $y \ge s$. Short sections of conductors normal to the axis of the transmission line join the two parts at $y = s$. The junction and its terminal-zone network are shown in Fig. 14.4c. Approximate formulas for calculating L_{Tl}, L_{Ty}, C_{Tl}, C_{Tr} and C_{Tc} are straightforward but quite long.[6] Terminal-zone networks for many other combinations and junctions are given in [9, 13–18].

[3]   [9], p. 368 and p. 411.

[4]   [9], p. 377; [13], p. 380; [17], pp. 111, 112.

[5]   Formulas and graphs for determining B_T due to changes in the radius of both inner and outer conductors of a coaxial line are in [9], pp. 368–377 and [17], pp. 111, 112.

[6]   They are given respectively by (8), (13), (29), (38), and (39) of [9], pp. 411–418.

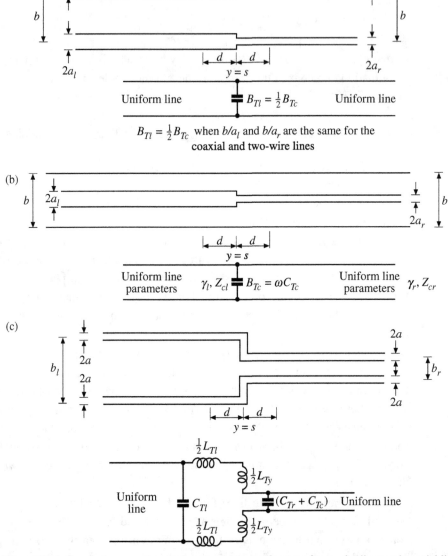

Figure 14.4 Terminal-zone networks for changes in conductors of two-wire lines and coaxial lines. Within terminal zone of length $d \doteq \lambda/10$, line parameters are non-uniform. (a) Change in radius of two-wire line conductors; (b) change in radius of inner conductor of coaxial line; (c) change in spacing of two-wire line.

14.3   Voltages, currents, and impedances of uniform sections of lines

Whenever an array is driven from a single generator, the various non-parasitic elements are connected to the generator through a network of transmission lines that supply at the several terminals currents and voltages which have the necessary amplitudes and phases to produce the desired radiation pattern. In addition, they must give correct impedance matches for a maximum transfer of power. A detailed consideration of the design of power-dividers, phasing and matching networks is beyond the scope of this book.[7] Experimental procedures for evaluating an array and for measuring the impedances and admittances of the elements are based on the solutions of the linearized transmission-line equations. A short review of relevant forms of the solution and their properties is given in this section.

Near line–load junctions and the ends of a transmission line the propagation "constant" γ is usually a function of position along the line. A practical procedure for taking account of terminal-zone effects with lumped networks and uniform sections of line has already been discussed. Outside the terminal zones the line is essentially uniform and the simple wave equations with constant coefficients, (14.15a) and (14.15b), apply. Solutions of these equations may have many forms. One of the most useful is

$$V(w) = Ae^{\gamma w} + Be^{-\gamma w},\tag{14.31}$$

where $\gamma = \alpha + j\beta$ and α is the attenuation constant in nepers per unit length, β the phase constant in radians per unit length. This solution is fitted to a particular line and load when the terminal conditions at the ends of the line are used to determine A and B. Note that these conditions must be specified within the terminal zones at the line–load junction or the line–generator junction. Hence, the apparent terminal impedance must be used in determining A and B. If the line is terminated at $w = 0$ by an apparent impedance Z_a with current $I(0)$ and voltage drop $V(0)$, it follows that

$$V(0) = I(0)Z_a.\tag{14.32}$$

With (14.32), (14.31) and (14.15b) A and B are readily evaluated and the following expressions obtained:

$$V(w) = \frac{I(0)}{2}[(Z_a + Z_c)e^{\gamma w} + (Z_a - Z_c)e^{-\gamma w}]\tag{14.33a}$$

$$I(w) = \frac{I(0)}{2Z_c}[(Z_a + Z_c)e^{\gamma w} - (Z_a - Z_c)e^{-\gamma w}],\tag{14.33b}$$

where Z_c is the characteristic impedance of the transmission line.

---

[7] See, for example, [8–10], [14], and [18].

These solutions suggest that an incident wave, traveling in the $-w$ direction from the generator at $w = s$ toward the load at $w = 0$, strikes the load and is partially or completely reflected back toward the generator. The incident and reflected parts are

$$V^+(w) = \frac{I(0)}{2}(Z_a + Z_c)e^{\gamma w}; \qquad \text{incident wave} \tag{14.34a}$$

$$V^-(w) = \frac{I(0)}{2}(Z_a - Z_c)e^{-\gamma w}; \qquad \text{reflected wave.} \tag{14.34b}$$

The ratio of the reflected wave to the incident wave is called the reflection coefficient and designated by Γ. At a distance w from the line–load junction,

$$\Gamma(w) = \frac{V^-(w)}{V^+(w)} = \frac{I^-(w)}{I^+(w)} = \frac{(Z_a - Z_c)}{(Z_a + Z_c)}e^{-2\gamma w}. \tag{14.35a}$$

At $w = 0$,

$$\Gamma_a = \frac{Z_a - Z_c}{Z_a + Z_c} = \frac{Y_c - Y_a}{Y_c + Y_a} = |\Gamma_a|e^{j\psi_a} \tag{14.35b}$$

is the reflection coefficient of the load. $Y_c = 1/Z_c$ is the characteristic admittance of the line. Therefore,

$$\Gamma(w) = \Gamma_a e^{-2\gamma w}. \tag{14.35c}$$

It follows that

$$V(w) = \frac{V(0)}{(1 + \Gamma_a)}[e^{\gamma w} + \Gamma_a e^{-\gamma w}] = \frac{I(0)Z_c}{(1 - \Gamma_a)}[e^{\gamma w} + \Gamma_a e^{-\gamma w}] \tag{14.36}$$

$$I(w) = \frac{I(0)}{(1 - \Gamma_a)}[e^{\gamma w} - \Gamma_a e^{-\gamma w}] = \frac{V(0)Y_c}{(1 + \Gamma_a)}[e^{\gamma w} - \Gamma_a e^{-\gamma w}]. \tag{14.37}$$

The superposition of the incident and reflected waves yields an interference pattern called a standing wave along the transmission line. When $Z_a = Z_c$, $\Gamma_a = 0$, $V^-(0) = 0$, the line is matched. For pure traveling waves outside the terminal zones the line appears to be infinite in length. When $|Z_a| \ll |Z_c|$, as when the load is a short circuit, $\Gamma_a \to -1$, and the incident wave is reflected with a $180°$ shift in phase. The voltage and current distributions are pure standing waves given by

$$V(w) = I(0)Z_c \sinh \gamma w, \qquad I(w) = I(0) \cosh \gamma w. \tag{14.38}$$

When $Z_a \gg Z_c$, as with an open circuit, the entire incident wave is again reflected but with no change in phase so that $\Gamma_a = 1$. The distributions of current and voltage are given by (14.38) with the sinh and cosh interchanged.

The impedance looking toward the load at any point w is given by

$$Z(w) = V(w)/I(w) = Z_c \left| \frac{1 + \Gamma_a e^{-2\gamma w}}{1 - \Gamma_a e^{-2\gamma w}} \right| \tag{14.39}$$

and the admittance is $Y(w) = 1/Z(w)$.

Alternative expressions in terms of the hyperbolic functions are

$$V(w)/V(0) = \cosh \gamma w + (Y_a/Y_c) \sinh \gamma w \tag{14.40a}$$

$$I(w)/I(0) = \cosh \gamma w + (Z_a/Z_c) \sinh \gamma w \tag{14.40b}$$

$$V(w)/[I(0)Z_c] = \sinh \gamma w + (Z_a/Z_c) \cosh \gamma w \tag{14.41a}$$

$$I(w)/[V(0)Y_c] = \sinh \gamma w + (Y_a/Y_c) \cosh \gamma w \tag{14.41b}$$

$$Z(w)/Z_c = \frac{Z_a/Z_c + \tanh \gamma w}{1 + (Z_a/Z_c) \tanh \gamma w}. \tag{14.42}$$

The preceding equations for current, voltage and impedance express $V(w)$ and $I(w)$ in terms of $V(0)$ and $I(0)$ at the load. They do not involve the actual driving voltage of the generator. A complete solution is obtained by imposing boundary conditions at both ends of the line[8] in terms of a generator with apparent internal impedance Z_g and voltage V^e at $w = s$ or $y = s - w = 0$ and a load with an apparent impedance Z_a at $w = 0$ or $y = s$. Specifically

$$y = 0: \qquad V_0 = V^e - I_0 Z_g$$

$$y = s: \qquad V_s = I_s Z_a.$$

The elimination of A and B in (14.31) yields

$$V(y) = \frac{V^e Z_c}{Z_c + Z_g} \frac{e^{-\gamma y} + \Gamma_a e^{-\gamma(2s-y)}}{1 - \Gamma_g \Gamma_a e^{-2\gamma s}} \tag{14.43a}$$

$$I(y) = \frac{V^e}{Z_c + Z_g} \frac{e^{-\gamma y} - \Gamma_a e^{-\gamma(2s-y)}}{1 - \Gamma_g \Gamma_a e^{-2\gamma s}}, \tag{14.43b}$$

where Γ_g is the reflection coefficient corresponding to Z_g.

---

[8]   [9], p. 75.

The introduction of functions to describe separately the attenuation and the phase characteristics of the terminations makes it possible to express currents and voltages in a completely hyperbolic form. These terminal functions are defined as follows:

$$\rho + j\phi = \coth^{-1} \frac{Z}{Z_c} = \tanh^{-1} \frac{Y}{Y_c} \tag{14.44a}$$

or

$$\rho + j\phi' = \coth^{-1} \frac{Y}{Y_c} = \tanh^{-1} \frac{Z}{Z_c}. \tag{14.44b}$$

The corresponding expressions for the currents and voltages are:

$$\frac{V(w)}{V^e} = \frac{\sinh(\rho_g + j\phi_g)\cosh[(\alpha w + \rho_a) + j(\beta w + \phi_a)]}{\sinh[(\alpha s + \rho_g + \rho_a) + j(\beta s + \phi_g + \phi_a)]} \tag{14.45a}$$

$$\frac{I(w)}{V^e} = \frac{\sinh(\rho_g + j\phi_g)\sinh[(\alpha w + \rho_a) + j(\beta w + \phi_a)]}{\sinh[(\alpha s + \rho_g + \rho_a) + j(\beta s + \phi_g + \phi_a)]}. \tag{14.45b}$$

The effects of a termination are now shown to be equivalent to those of a section of transmission line with a total loss specified by ρ and a total phase shift specified by ϕ. Note that the denominator of (14.45a, b) includes the total loss and total phase shift of the line plus its terminations at both ends. Impedance and admittance are given by

$$Z(w) = \coth[(\alpha w + \rho_a) + j(\beta w + \phi_a)] \tag{14.46a}$$

$$Y(w) = \tanh[(\alpha w + \rho_a) + j(\beta w + \phi_a)]. \tag{14.46b}$$

The terminal functions and the reflection coefficient are related as follows:

$$\Gamma = |\Gamma|\, e^{j\psi} = e^{-2(\rho + j\phi)}$$
$$|\Gamma| = e^{-2\rho} = \frac{\coth \rho - 1}{\coth \rho + 1}, \qquad \psi = -2\phi \tag{14.47}$$

$$\rho = \tfrac{1}{2}\ln 1/|\Gamma| = \coth^{-1}\frac{1 + |\Gamma|}{1 - |\Gamma|}. \tag{14.48}$$

For most transmission lines that are useful as feeders for an array or for measuring sections, the line losses are very small and can often be neglected. Under these conditions

$$\gamma \doteq j\beta, \qquad Z_c \doteq R_c = \sqrt{l^e/c}$$

and (14.36) and (14.37) give

$$V(w) \doteq \frac{V(0)e^{j\beta w}}{1 + \Gamma_a}\,[1 + |\Gamma_a|e^{-j(2\beta w - \psi_a)}] \tag{14.49a}$$

$$I(w) \doteq \frac{I(0)e^{j\beta w}}{1 - \Gamma_a}\,[1 - |\Gamma_a|e^{-j(2\beta w - \psi_a)}] \tag{14.49b}$$

and

$$\Gamma(w) = |\Gamma_a| e^{-j(2\beta w - \psi_a)}. \tag{14.49c}$$

These have the following maxima and minima:

$$|V_{\max}(w)| = \left| \frac{V(0)}{1 + \Gamma_a} \right| [1 + |\Gamma_a|], \qquad |I_{\min}(w)| = \left| \frac{I(0)}{1 - \Gamma_a} \right| [1 - |\Gamma_a|],$$

$$2\beta w - \psi_a = 0, \ 2\pi, \ \cdots = 2n\pi \tag{14.50a}$$

$$|V_{\min}(w)| = \left| \frac{V(0)}{1 + \Gamma_a} \right| [1 - |\Gamma_a|], \qquad |I_{\max}(w)| = \left| \frac{I(0)}{1 - \Gamma_a} \right| [1 + |\Gamma_a|],$$

$$2\beta w - \psi_a = \pi, \ 3\pi, \ \cdots = (2n+1)\pi, \ n = 0, \ 1, \ 2\dots. \tag{14.50b}$$

The ratio of the maximum-to-minimum of either the current or the voltage on a lossless line is called the standing wave ratio and abbreviated SWR.

$$\text{SWR} = |V_{\max}(w)/V_{\min}(w)| = |I_{\max}(w)/I_{\min}(w)| = (1 + |\Gamma_a|)/(1 - |\Gamma_a|) \tag{14.51a}$$

$$= \frac{|Z_a + Z_c| + |Z_a - Z_c|}{|Z_a + Z_c| - |Z_a - Z_c|} = \frac{|Y_c + Y_a| + |Y_c - Y_a|}{|Y_c + Y_a| - |Y_c - Y_a|} = \coth \rho. \tag{14.51b}$$

The distributions of voltage and current are periodic and repeat every half wavelength; the adjacent maxima and minima of voltage or current are separated by a quarter wavelength; the current maxima occur at voltage minima and vice versa. The impedance and admittance looking toward the load also repeat every half wavelength. At maxima and minima of the current or voltage the impedance and admittance are real with the following values [from (14.39)]:

Voltage maxima or current minima

$$2\beta w - \psi_a = 0, 2\pi, 4\pi, \dots$$

$$Z(w) = R_c \left[\frac{1 + |\Gamma_a|}{1 - |\Gamma_a|} \right] \qquad Y(w) = G_c \left[\frac{1 - |\Gamma_a|}{1 + |\Gamma_a|} \right]$$

$$= R_c[\text{SWR}]; \qquad\qquad = G_c/[\text{SWR}]. \tag{14.52a}$$

Voltage minima or current maxima

$$2\beta w - \psi_a = \pi, \ 3\pi, \ 5\pi, \dots$$

$$Z(w) = R_c \left[\frac{1 - |\Gamma_a|}{1 + |\Gamma_a|} \right] \qquad Y(w) = G_c \left[\frac{1 + |\Gamma_a|}{1 - |\Gamma_a|} \right]$$

$$= R_c/[\text{SWR}]; \qquad\qquad = G_c[\text{SWR}]. \tag{14.52b}$$

The relative distributions of current and voltage for lossless lines are:

$$V(w)/V(0) = \cos \beta w + j(Y_a/G_c) \sin \beta w \tag{14.53a}$$

$$I(w)/I(0) = \cos \beta w + j(Z_a/R_c) \sin \beta w \tag{14.53b}$$

$$V(w)/[I(0)R_c] = (Z_a/R_c) \cos \beta w + j \sin \beta w \tag{14.53c}$$

$$I(w)/[V(0)G_c] = (Y_a/G_c) \cos \beta w + j \sin \beta w. \tag{14.53d}$$

The corresponding impedance is

$$Z(w) = R_c \left[\frac{(Z_a/R_c) + j \tan \beta w}{1 + j(Z_a/R_c) \tan \beta w} \right]. \tag{14.54a}$$

The admittance is given by an identical expression with Z_a and R_c replaced by Y_a and G_c.

The input power to a section of line is $P = VI^* = |V|^2/Z$ with the asterisk denoting the complex conjugate. At a voltage or current maximum (14.52a, b) give

$$P = |V_{max}|^2/R_c[\text{SWR}] = |I_{max}|^2 R_c/[\text{SWR}]. \tag{14.54b}$$

Equation (14.54b) is sometimes of help in measuring the relative power in the branches of a feeding network.

Useful properties of a quarter-wave section of transmission line follow from (14.53). If $I(0)$ and $V(0)$ are the required driving-point current and voltage and if $\beta w = \pi/2$,

$$V\left(\frac{\pi}{2}\right) = jR_c I(0), \qquad I\left(\frac{\pi}{2}\right) = jV(0)/R_c, \qquad Z\left(\frac{\pi}{2}\right) = R_c^2 Y_a. \tag{14.54c}$$

The preceding expressions have traditionally been considered to be primarily functions of the distance from the line–load junction w, but since

$$\beta w = \frac{2\pi w}{\lambda} = \frac{2\pi f w}{c}, \tag{14.54d}$$

where c is the velocity of propagation, they are equally functions of the frequency.

14.4   Theoretical basis of impedance measurements

Over the past 15 to 20 years, developments in electronic instrumentation have produced a revolution in the design and construction of instruments for measuring impedances. For many decades, phase-stable RF signals were possible only at a single frequency and mechanical components of the experiment were varied to determine the behavior of a system. However, the development of synthesized RF sources that are very stable in frequency, phase, and output, and that can be readily changed in precise, small or large frequency steps over wide frequency bands has made the frequency the easiest and most cost effective parameter to vary. The change to frequency as the independent variable of choice has been further supported by the development of high-quality, wide-band directional couplers and network analyzers that combine many individual instruments into integrated packages, and the use of computers to control the instruments, aid in their calibration, and process the data. Because of their compatibility with the integrated electronics and inherently higher data rates, directional couplers and various other impedance bridges have largely replaced slotted lines as the preferred instruments for measuring impedances. In effect, the complicated electronics is replacing the artful machining work required to build high-quality slotted lines.

For the investigation of monopoles over a ground plane, a special type of coaxial transmission line has proven very useful.[9] Measurements along these lines can be made either by multiple-probe techniques[10] or by converting the line into a slotted line. In the latter technique, a loop current probe or voltage probe protrudes from a slot in the inner conductor of the coaxial line and the RF information is carried in a second coaxial line inside the inner conductor that forms the antenna above the ground plane, thus removing the connecting cables and equipment from the fields being measured. The slotted conductor is extended above the ground plane, permitting the probe to be used both for measuring the current or charge distribution along the antenna as well as the driving-point impedance.

Variations of this approach can be used with dipoles driven by two-wire lines and for the probing of surface fields of more complex shapes. Because of the utility of this technique, two different slotted-line impedance-measurement procedures – the distribution-curve and the resonance-curve methods – will be discussed. These procedures can also be easily automated. The motion of the probe can be controlled by a computer through the addition of a D/A converter and a stepping motor, and the RF output from the probe can be sent directly to a network analyzer and back to the computer for processing. If only the impedance is of interest, a multiple-probe method will be simpler and offer much higher data rates. General details of the measurement techniques, the instruments, and the errors and calibration of network analyzers are discussed in the literature [1–4]. The objective here is to review the theoretical basis for the techniques, and to summarize pertinent equations for calculating impedances from the measured quantities.

The directional-coupler and bridge methods determine the magnitude of Γ_a by measuring the change in amplitude and phase of the reflected voltage at a fixed point on the line when a short circuit is replaced by the unknown load. The multiple-probe method determines the load admittance by comparing the ratio of voltages at two or more fixed points on the line. No matched load or short circuit is required at

---

[9] See [20], and [10], pp. 129 and 228. [10] [2] p. 109, and [4] p. 122.

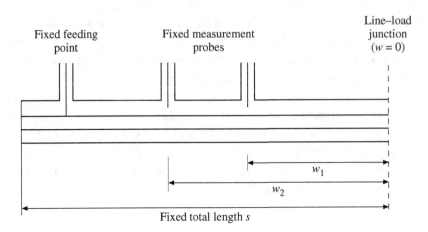

Figure 14.5 Dual probe method of measurement.

the line–load junction, but the probes must have identical electrical properties, or be calibrated relative to one of them. Alternatively, with modern phase-stable RF sources that can be reset in frequency to a few cycles or better, the need for calibration can be avoided by using a single probe in different fixed positions, as discussed in Section 14.8. In the distribution-curve method, the total length of the transmission line and its excitation point remain fixed while a loosely coupled probe is moved along the line to locate a current or voltage minimum and the SWR. In the resonance-curve or Chipman method, a movable short circuit is used to tune the line with its terminations to resonance by adjusting the total length of the line. A small loop probe projecting from the short circuit is used to locate a resonant maximum and measure the SWR. A particular advantage of the resonance-curve method is that it can be used with a receiving antenna as well as with a transmitting antenna. These measurement methods are schematically illustrated in Figs. 14.5, 14.6, and 14.7.

Multiple-probe method

Let the voltage measured at position w_1 be V_1 and that measured at w_2 be V_2. Both the amplitude and the phase of the voltages must be measured. The probes are assumed to be identical, or it is assumed that a single probe is moved from w_1 to w_2. Equation (14.53c) written for V_2 and again for V_1 and solved for Z_a and Y_a yields

$$Z_a = jR_c \left[\frac{\sin \beta w_2 - V_R \sin \beta w_1}{V_R \cos \beta w_1 - \cos \beta w_2} \right] \tag{14.55a}$$

$$Y_a = jY_c \left[\frac{\cos \beta w_2 - V_R \cos \beta w_1}{\sin \beta w_2 - V_R \sin \beta w_1} \right], \tag{14.55b}$$

where $V_R = V_2/V_1$ is the ratio of two measured complex numbers.

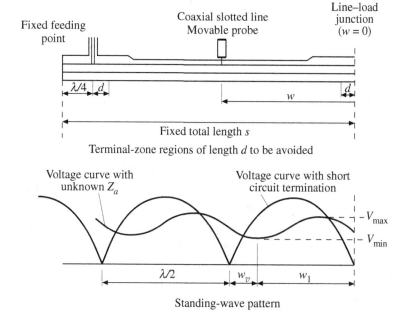

Figure 14.6 Distribution-curve method of measurement.

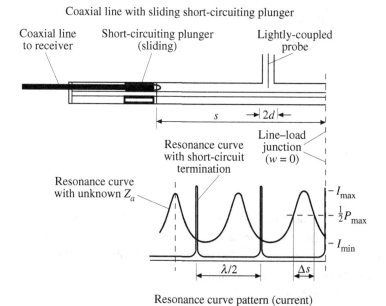

Figure 14.7 Resonance-curve method of measurement.

Similarly, the reflection coefficient Γ_a follows from (14.36),

$$\Gamma_a = e^{j2\beta w_2} \left[\frac{1 - |V_2/V_1| e^{j[\psi_V - \beta(w_2 - w_1)]}}{|V_2/V_1| e^{j[\psi_V + \beta(w_2 - w_1)]} - 1} \right], \tag{14.55c}$$

where $\psi_V = \psi_2 - \psi_1$, and ψ_2, ψ_1 are the measured phases of V_2 and V_1.

The unknown load admittance follows from (14.35b) and (14.35c):

$$\frac{Y_a}{Y_c} = \frac{1 - \Gamma_a}{1 + \Gamma_a}. \tag{14.55d}$$

The probe separation, $w_2 - w_1$, should not be a multiple of a half-wavelength within the frequency range of interest. Additional probe positions can be used to avoid this situation. If a single probe is moved between the positions, care must be exercised to avoid the introduction of errors from flexing cables. The probe holder must be designed to provide good and repeatable contact with the outer conductor. Also, covers with good repeatable contact must generally be provided for the empty positions. The application of a variation of this technique to the measurement of the sharp resonances in circular arrays is discussed in Section 14.8.

Directional-coupler method

From (14.35a) and (14.49c) for lossless transmission lines, the reflection coefficient at any point w along the line is

$$\Gamma(w) = |\Gamma(w)| e^{j\psi_V} = \left| \frac{V^-(w)}{V^+(w)} \right| e^{j\psi_{Va}} = |\Gamma_a| e^{j(\psi_a - 2\beta w)}. \tag{14.56a}$$

Thus,

$$|\Gamma_a| = \left| \frac{V^-(w)}{V^+(w)} \right| \tag{14.56b}$$

and

$$\psi_a = \psi_{Va} + 2\beta w. \tag{14.56c}$$

With a properly calibrated vector voltmeter that is based on a dual directional coupler, the phase and amplitude of $V^-(w)$ and $V^+(w)$ can be measured directly. However, the calibration requires a matched load to be placed at the line–load junction. Generally, this is not practical with the experimental models that are used to investigate monopole antennas over ground planes.

Let the reflected voltage only be measured, first with a short circuit at the ground-plane surface, and then with the unknown load. For the short circuit,

$$\Gamma_s(w) = \left| \frac{V_s^-(w)}{V^+(w)} \right| e^{j(\psi_s - 2\beta w)}. \tag{14.57a}$$

From the unknown load,

$$\Gamma_a(w) = \left| \frac{V_a^-(w)}{V^+(w)} \right| e^{j(\psi_a - 2\beta w)}. \tag{14.57b}$$

Using the short-circuited properties of Γ,

$$\Gamma_s = -1 = e^{j(2n+1)\pi} \tag{14.57c}$$

it follows that

$$\left| \frac{\Gamma_a(w)}{\Gamma_s(w)} \right| = |\Gamma_a(w)| = \left| \frac{V_a^-(w)}{V_s^-(w)} \right| \tag{14.58a}$$

$$\psi_a = \psi_d + (2n+1)\pi, \tag{14.58b}$$

where

$$\psi_d = \psi_a - \psi_s = \psi_a - (2n+1)\pi. \tag{14.58c}$$

The magnitude of the reflection coefficient is commonly expressed in dB,

$$|\Gamma_a|_{dB} = 20 \log_{10} \left| \frac{V_a^-(w)}{V_s^-(w)} \right|. \tag{14.58d}$$

The ultimate accuracy of this technique rests on the directivity of the directional coupler. Couplers with directivities of 30–40 dB or better over very wide frequency bands are available. Calibration procedures have been devised to reduce the effects of internal reflections and increase measurement accuracy when the couplers are used as part of vector network analyzers.[11,12]

Distribution-curve method

Assume that a voltage probe is used to measure the SWR and the location of a voltage minimum. Let w_n be the distance from the line–load junction to the nth voltage minimum; in Fig. 14.6, $n = 1$. If transmission-line losses are negligible over the section of line used in the measurements, the impedance at a voltage minimum is $R_c/[\text{SWR}]$. From (14.52b) and (14.54a),

$$\frac{1}{[\text{SWR}]} = \left[\frac{Z_a/R_c + j \tan \beta w_n}{1 + j(Z_a/R_c) \tan \beta w_n} \right].$$

[11] [1] Chapters 10 and 11, and [3] Chapter 6.
[12] A continuing discussion of procedures related to vector network analyzers may be found in HP 8510/8720 News, Hewlett-Packard Co., 1400 Fountaingrove Pkwy., Santa Rosa, CA 95403-1799.

When this equation is solved for $Z_a = R_a + jX_a$, the result is

$$Z_a = R_a + jX_a = \begin{cases} R_c\left[\dfrac{1 - j[\text{SWR}]\tan\beta w_n}{[\text{SWR}] - j\tan\beta w_n}\right] & \text{Voltage probe} & (14.59a) \\[2em] \dfrac{R_c}{[\text{SWR}]^2 + \tan^2\beta w_n}\{[\text{SWR}](1 + \tan^2\beta w_n) \\[1em] \qquad + j(1 - [\text{SWR}]^2)\tan\beta w_n\}. & & (14.59b) \end{cases}$$

$\beta = 2\pi/\lambda$, n is the number of the minimum corresponding to w_n, λ is the wavelength along the transmission line.

If a current probe is used and w_n is the distance from the line–load junction to a current minimum,

$$Z_a = R_a + jX_a = \begin{cases} R_c\left[\dfrac{[\text{SWR}] - j\tan\beta w_n}{1 - j[\text{SWR}]\tan\beta w_n}\right] & \text{Current probe} & (14.59c) \\[2em] \dfrac{R_c}{1 + [\text{SWR}]^2\tan^2\beta w_n}\{[\text{SWR}](1 + \tan^2\beta w_n) \\[1em] \qquad + j([\text{SWR}]^2 - 1)\tan\beta w_n\}. & & (14.59d) \end{cases}$$

The reflection coefficient is defined by (14.35b) with magnitude and phase given by

$$|\Gamma_a| = \frac{[\text{SWR}] - 1}{[\text{SWR}] + 1} \tag{14.60a}$$

$$\psi_a = \begin{cases} 2[\beta w_n - (n + \tfrac{1}{2})\pi] & \text{Voltage probe} & (14.60b) \\ 2[\beta w_n - n\pi] & \text{Current probe.} & (14.60c) \end{cases}$$

If the admittance is to be determined and a voltage probe is used, (14.59c) and (14.59d) apply with $G_a + jB_a$ substituted for $R_a + jX_a$ and with G_c substituted for R_c. If a current probe is used for admittance measurements, (14.59a) and (14.59b) apply with the indicated substitutions.

High SWR's are difficult to measure because the SWR minimum may be in the noise level of the measuring system or its maximum-to-minimum range may exceed the linear range of the detecting system. For these cases the curve width method may be more convenient.[13] In this method the width of the distribution curve is measured at half power points and the terminal functions can be obtained from (14.47) and (14.48). The relations are

$$\rho_a = \coth^{-1}\frac{1 + |\Gamma_a|}{1 - |\Gamma_a|} = \coth^{-1}\text{SWR} = \tfrac{1}{2}\ln\frac{\text{SWR} + 1}{\text{SWR} - 1} \tag{14.61a}$$

$$\doteq \pi\,\Delta w/\lambda, \qquad \Delta w = \text{curve width} \tag{14.61b}$$

---

13 [2], pp. 102–104 and [9], pp. 266–269.

$$\phi_a = -\psi_a/2 = \begin{cases} (n + \tfrac{1}{2})\pi - \beta w_n & \text{Voltage probe} & (14.62\text{a}) \\ n\pi - \beta w_n & \text{Current probe.} & (14.62\text{b}) \end{cases}$$

Real and imaginary parts of impedances or admittances in terms of the terminal functions can be found from (14.44a, b). The results are

$$Z_a = R_a + jX_a = Z_c \left\{ \frac{\sinh 2\rho_a}{\cosh 2\rho_a - \cos 2\phi_a} - j \frac{\sin 2\phi_a}{\cosh 2\rho_a - \cos 2\phi_a} \right\} \qquad (14.63\text{a})$$

$$Y_a = G_a + jB_a = Y_c \left\{ \frac{\sinh 2\rho_a}{\cosh 2\rho_a + \cos 2\phi_a} + j \frac{\sin 2\phi_a}{\cosh 2\rho_a + \cos 2\phi_a} \right\}. \qquad (14.63\text{b})$$

Frequently the total distance w_n from the line–load junction to a convenient minimum is difficult to measure accurately. Since on a lossless line the impedance is repeated at intervals of $\lambda/2$ or $\beta w = \pi$ radians, it is necessary only to determine the location of a minimum with respect to an integral number of half wavelengths from the junction. If the load being investigated is removed and a short circuit is placed at the junction, a voltage null will appear at the junction and along the line at each half wavelength from the junction. Let w_v be the distance from the nth voltage null with the short circuit as a load to the nearest voltage minimum with the antenna as a load. Distances toward the generator are positive, those toward the load are negative. Then,

$$\beta w_n = n\pi \pm \beta w_v \qquad (14.64)$$

and

$$\tan(n\pi \pm \beta w_v) = \frac{\tan n\pi \pm \tan \beta w_v}{1 \pm \tan n\pi \tan \beta w_v} = \pm \tan \beta w_v$$

so that w_v, the shift in the location of a minimum when the unknown impedance is substituted for the short circuit, can be used directly in (14.59a) and (14.59b) if due regard is given to the sign. Similar results hold for a current minimum and (14.59c) and (14.59d). In terms of the minimum shift, the phase of the terminal functions and of the reflection coefficient is

$$\phi_a = -\psi_a/2 = \begin{cases} \pi/2 \mp \beta w_v & \text{Voltage probe} & (14.65\text{a}) \\ \mp \beta w_v & \text{Current probe.} & (14.65\text{b}) \end{cases}$$

The various distances are illustrated in Fig. 14.6.

Resonance-curve method

Let the solution of (14.45) be written for a balanced generator located at an arbitrary point $y = y_g$ along the transmission line instead of at $y = 0$. Also let the hyperbolic functions be separated into their real and imaginary parts.[14] Then the magnitudes of currents and voltages along the line are:

$$
|I| = \frac{V^e}{R_c}
\frac{\sqrt{\begin{aligned}&\sinh^2(\alpha y_g + \rho_g) + \sin^2(\beta y_g + \phi_g)\\&\times\sqrt{\sinh^2(\alpha w + \rho_a) + \sin^2(\beta w + \phi_a)}\end{aligned}}}{\sqrt{\sinh^2(\alpha s + \rho_g + \rho_a) + \sin^2(\beta s + \phi_g + \phi_a)}}
$$

$$
|V| = V^e
\frac{\sqrt{\begin{aligned}&\sinh^2(\alpha y_g + \rho_g) + \sin^2(\beta y_g + \phi_g)\\&\times\sqrt{\sinh^2(\alpha w + \rho_a) + \cos^2(\beta w + \phi_a)}\end{aligned}}}{\sqrt{\sinh^2(\alpha s + \rho_g + \rho_a) + \sin^2(\beta s + \phi_g + \phi_a)}}
$$

$$\left. \right\} \quad y_g \leq y \leq s$$

$$
|I| = \frac{V^e}{R_c}
\frac{\sqrt{\begin{aligned}&\sinh^2(\alpha w_g + \rho_a) + \sin^2(\beta w_g + \phi_a)\\&\times\sqrt{\sinh^2(\alpha y + \rho_g) + \sin^2(\beta y + \phi_g)}\end{aligned}}}{\sqrt{\sinh^2(\alpha s + \rho_g + \rho_a) + \sin^2(\beta s + \phi_g + \phi_a)}}
$$

$$
|V| = V^e
\frac{\sqrt{\begin{aligned}&\sinh^2(\alpha w_g + \rho_a) + \sin^2(\beta w_g + \phi_a)\\&\times\sqrt{\sinh^2(\alpha y + \rho_g) + \cos^2(\beta y + \phi_g)}\end{aligned}}}{\sqrt{\sinh^2(\alpha s + \rho_g + \rho_a) + \sin^2(\beta s + \phi_g + \phi_a)}}
$$

$$\left. \right\} \quad 0 \leq y \leq y_g$$

where $w_g = s - y_g$ is the distance from the load to the generator. The first two equations give current and voltage distributions between the load at $y = s$ and the generator as illustrated in Fig. 14.7. The last two equations give the distributions between the generator and the end of the line at $y = 0$. If the distance y_g or w_g between the appropriate ends of the line and the generator, and the distance w or y between the appropriate ends of the line and the probe are held constant, while the total length s is changed, the current and voltage vary in the same manner,

$$I(w) \sim V(w) \sim [\sinh^2(\alpha s + \rho_g + \rho_a) + \sin^2(\beta s + \phi_g + \phi_a)]^{-1/2}. \tag{14.66}$$

The line is resonant when (14.66) has its maximum value. Maxima and minima of (14.66) are defined by

$$
(\beta s + \phi_g + \phi_a) =
\begin{cases}
n\pi - \frac{1}{2}\sin^{-1}\left[\dfrac{\alpha}{\beta}\sinh 2(\alpha s + \rho_g + \rho_a)\right] & \text{Maxima}\\[2ex]
(n + \frac{1}{2})\pi + \frac{1}{2}\sin^{-1}\left[\dfrac{\alpha}{\beta}\sinh 2(\alpha s + \rho_g + \rho_a)\right] & \text{Minima.}
\end{cases}
$$

---

[14] See Chapter 4 of [9].

For lossless lines, $\alpha/\beta = 0$, and

$$(\beta s_{\max} + \phi_g + \phi_a) = n\pi \qquad \text{Maxima} \qquad (14.67a)$$

$$(\beta s_{\min} + \phi_g + \phi_a) = (n + \tfrac{1}{2})\pi \qquad \text{Minima.} \qquad (14.67b)$$

With a short circuit at $w = 0$, $\phi_a = \pi/2$; this value is commonly used as a reference. Let s_s be a convenient resonant length with a short circuit as the termination at $w = 0$ and let s_1 be the corresponding resonant length with the unknown impedance at $w = 0$. When (14.67a) is written successively for both loads and the one is subtracted from the other, the result is

$$\phi_a = \pi/2 + \beta(s_s - s_1). \qquad (14.68)$$

For lines with very small losses, $\alpha s \ll 1$, the ratio of maximum-to-minimum in (14.66) is

$$\text{SWR} \doteq \coth(\rho_g + \rho_a). \qquad (14.69)$$

This equation illustrates an important additional requirement in the resonance-curve method that does not occur in the distribution-curve method. In the distribution-curve method, only the parameters that characterize the generator are involved in the measurements; in the resonance-curve method, ρ_g must be known or the generator must be lightly coupled so that $\rho_g \ll \rho_a$ for the loads under investigation. When these conditions are satisfied, ρ_a is given by (14.69), and impedances or admittances can be computed directly from (14.63a) or (14.63b). The magnitude of the reflection coefficient can be calculated from (14.60a) and its phase, ψ_a, from

$$\psi_a = -2\phi_a = \beta(s_1 - s_s) - \pi. \qquad (14.70)$$

With the resonance-curve method, it is frequently more convenient to measure the curve width than the SWR, and sometimes it is simpler to measure the curve widths at power levels other than 1/2. For low-loss transmission lines and symmetrical resonance curves, it can be shown that

$$\rho_a \doteq \frac{\pi}{\sqrt{p^2 - 1}} \frac{\Delta s}{\lambda},$$

where Δs is the width of the resonance curve at a level $1/p$ of the maximum.

The bridge methods, directional-coupler methods, slotted-line methods, and resonant-line methods all use a short circuit at the line–load junction as a reference. If the feeding line is coaxial, this may be simply a conducting plug that makes good and repeatable contacts with the inner and outer conductors at a known position. If the feeding line is an unshielded two-wire line, the short circuit may be a conducting disk that makes good contact with both conductors and that has a radius of at least five times the center-to-center spacing of the conductors.

When measurements are made on open two-wire lines, the principal difficulty that is encountered is keeping the lines balanced so that $I_2(w) = -I_1(w)$. This requires that perfect symmetry be maintained everywhere in the vicinity of the lines. A small probe placed midway between the conductors and connected to a sensitive detector is usually necessary to monitor the condition of balance. When the lines are perfectly balanced, nothing is received on the monitor probe.

A probe is required in several of the measurement procedures and additional probes may be convenient for monitoring at various points. The probes must be loosely coupled in order to avoid distorting the interference pattern along the transmission line.[15] Generally the probe should be tuned to provide maximum sensitivity with minimum intrusion into the line. Any distortion or loading introduced by the probe is most pronounced at a current maximum with a current probe and at a voltage maximum with a voltage probe. When a probe is too tightly coupled to the transmission line, the measured SWR is less than the true one and maxima are shifted from points midway between adjacent minima. A simple test for excessive probe coupling is to measure a moderate SWR with probes of different sizes. If there is no change in the measured SWR, the probes are not introducing significant errors. Another useful test is to measure the location of a maximum, the adjacent minima, and the curve width at half maximum power with a short circuit as the termination. If the probe is introducing no errors, power variations about the maximum should be like $\cos^2 \beta w$ [from (14.53)] and the maximum should fall exactly midway between the minima. This test is particularly severe because only the probe is absorbing power from the line. A movable short circuit that maintains good electrical contact during its motion as required in the resonance-curve method is very difficult to construct. If the load being investigated has only a small loss, the resonance-curve maximum is especially sensitive to erratic electrical contacts. When there is sufficient room between the inner and outer conductors, a "non-contacting" short circuit can be used [21].

14.5   The measurement of self- and mutual impedance or admittance

At the driving points of the several elements in an array, currents and voltages are related by the usual coupled circuit equations. Let V_k be the driving voltage across the terminals of element k in an array of N elements; let $I_k(0)$ be the current in the same

---

[15] [2] p. 93.

terminals. Then, if a Kirchhoff equation is written for each element, the following set is obtained:

$$V_1 = I_1(0)Z_{11} + I_2(0)Z_{12} + \cdots I_k(0)Z_{1k} + \cdots I_p(0)Z_{1p} + \cdots I_N(0)Z_{1N}$$

$$\vdots$$

$$V_k = I_1(0)Z_{k1} + I_2(0)Z_{k2} + \cdots I_k(0)Z_{kk} + \cdots I_p(0)Z_{kp} + \cdots I_N(0)Z_{kN} \qquad (14.71)$$

$$\vdots$$

$$V_N = I_1(0)Z_{N1} + I_2(0)Z_{N2} + \cdots I_k(0)Z_{Nk} + \cdots I_p(0)Z_{Np} + \cdots I_N(0)Z_{NN}.$$

The coefficient Z_{kp}, $p \neq k$, is the mutual impedance between element k and element p. As long as the array is in an isotropic medium such as air, $Z_{kp} = Z_{pk}$. Z_{kk} is the self-impedance of element k. The input or driving-point impedance of element k is

$$Z_{kin} = \frac{V_k}{I_k(0)} = \frac{I_1(0)}{I_k(0)} Z_{k1} + \cdots Z_{kk} + \cdots \frac{I_p(0)}{I_k(0)} Z_{kp} + \cdots \frac{I_N(0)}{I_k(0)} Z_{kN}. \qquad (14.72)$$

If the elements are fed by transmission lines, Z_{kin} is the apparent load impedance of the transmission line. The driving terminals of an antenna coincide with the line–load junction between it and its feeding transmission line. The self-impedance of an element is the input impedance at the terminals of that element when the driving-point currents of all other elements in the array are zero – that is, when all other elements are open-circuited at their driving points. The mutual impedance between element k and element p is the open-circuit voltage at the driving point of element p per unit current at the driving terminals of element k, with the driving points of all elements but k open-circuited.

The relations that involve the admittances are the duals of (14.71). They are

$$I_1(0) = V_1 Y_{11} + V_2 Y_{12} + \cdots V_k Y_{1k} + \cdots V_p Y_{1p} + \cdots V_N Y_{1N}$$

$$\vdots$$

$$I_k(0) = V_1 Y_{k1} + V_2 Y_{k2} + \cdots V_k Y_{kk} + \cdots V_p Y_{kp} + \cdots V_N Y_{kN} \qquad (14.73)$$

$$\vdots$$

$$I_N(0) = V_1 Y_{N1} + V_2 Y_{N2} + \cdots V_k Y_{Nk} + \cdots V_p Y_{Np} + \cdots V_N Y_{NN}$$

$$Y_{kin} = \frac{I_k(0)}{V_k} = \frac{V_1}{V_k} Y_{k1} + \frac{V_2}{V_k} Y_{k2} + \cdots Y_{kk} + \cdots \frac{V_p}{V_k} Y_{kp} + \cdots \frac{V_N}{V_k} Y_{kN}. \qquad (14.74)$$

The self-admittance of an element is its input admittance when the driving-point voltages of all other elements in the array are zero – that is, when all other elements are short-circuited at their driving points. The mutual admittance between element k and element p is the short-circuit current at the driving point of element p per unit voltage at the terminals of element k, when the terminals of all elements except k are short-circuited.

Self- and mutual impedances or admittances depend upon the geometrical config- uration of each element, the relative orientation and location of the elements in the array, and the total number of elements. Once the self- and mutual impedances or admittances have been determined for an array, they can be used in equations like (14.72) and (14.74) to predict the driving-point impedances or admittances for any set of driving voltages or currents that may be applied to the array.

In principle, there is no difficulty in determining self- and mutual impedances. If known sets of currents or voltages are maintained at the terminals of the several elements and a sufficient number of input impedances or admittances are mea- sured, (14.72) or (14.74) can be inverted and the self- and mutual impedances evaluated. There are, however, two practical difficulties. The first is that the only set of excitation coefficients useful in measuring self- and mutual impedances is that which can be adjusted with high accuracy independently of the driving-point impedances. The second difficulty is that so many quantities must be determined. In a linear array of N identical, equally spaced elements there are $N/2$ different self-impedances and $N^2/4$ different mutual impedances if N is even; $(N + 1)/2$ different self-impedances and $(N^2 - 1)/4$ different mutual impedances if N is odd. For $N = 100$ there are 2550 different quantities to be determined. Fortunately, many of the mutual impedances are sufficiently small so that they can be neglected and many of the self-impedances are alike within tolerable limits. The measuring procedure should provide a rapid indication of such possible simplifications as well as relatively simple steps for determining the significant self- and mutual impedances.

An important especially simple array consists of identical elements uniformly spaced about the circumference of a circle. Since all elements have the same self- impedance, symmetry reduces the total number of unknowns to $N/2$ if N is even or $(N + 1)/2$ if N is odd. Self- and mutual admittances can be measured as follows. Let element k be driven, all others short-circuited at their driving points. Measure the apparent input admittance of element k. From (14.74)

$$Y_{kin1} = Y_{kk}. \tag{14.75}$$

Next, let elements k and p be driven with the voltages $V_p = -V_k$; let the terminals of all other elements be short-circuited. Again measure the apparent input admittance of element k. Then

$$Y_{kin2} = Y_{kk} - Y_{kp}, \qquad Y_{kp} = Y_{kin1} - Y_{kin2}. \tag{14.76}$$

Thus, for the circular array, all mutual admittances can be found by successively driving element k and one other element with equal voltages in opposite phases, and each time measuring the input admittance of only element k. The driving voltages $V_k = -V_p$ produce a null in the electromagnetic field along the perpendicular bisector of a chord joining elements k and p. This property can be used to obtain the correct voltages by locating a probe at the center of the array and adjusting the phase and amplitude of element p or k until no signal is observed in the probe. A 30–40 dB range of receiving sensitivity provides an accurate adjustment of the voltages. The short circuits in the elements other than p and k can be placed at the driving terminals or at the ends of lossless sections of transmission lines which are electrically one-half wavelength long. This length is critical and must take account of phase shifts in connectors, terminal zones, etc. For monopoles driven over a ground plane by coaxial lines in the manner shown in Fig. 14.3a, the short circuits can consist of very thin plugs in the ends of the coaxial lines. End-effects are quite simple for this model since the terminal-zone network consists of a shunting capacitance. If the measured apparent input admittances given in (14.75) and (14.76) are Y_{ka1} and Y_{ka2}, then

$$Y_{kk} = Y_{ka1} - j\omega C_T \tag{14.77a}$$

$$Y_{kp} = Y_{ka1} - Y_{ka2}. \tag{14.77b}$$

Note that the end-effect contributes only to the self-susceptance.

The self- and mutual impedances of a circular array can be measured with analogous procedures. To determine the self-impedance, drive one element, open-circuit all others so that all the driving-point currents are zero, and measure the input impedance of the driven element. To determine the mutual admittances, drive the elements successively in pairs with a receiving probe at the center of the circle to aid in setting $I_p(0) = -I_k(0)$, and measure the input impedance of a driven element. Equations (14.75) and (14.76) apply with the corresponding impedances substituted for admittances. On an impedance basis, the experimental model of monopoles driven by coaxial lines offers little simplification in terminal-zone effects, and the actual load or input impedances are obtained from the measured apparent ones with (14.21).

In a linear, planar or more general array, the self-admittances or impedances can be measured by the method used for circular arrays, i.e. by driving the element of interest, loading the other elements with short circuits for admittances or open circuits for impedances, and measuring the input admittance or impedance of the driven element. Difficulties arise in measuring mutual admittances by the method of driving the elements in pairs so that $I_p(0) = -I_k(0)$ or $V_p = -V_k$, since, in general, there is no simple null in the field at which a receiving probe can be located to aid in the adjustment of the current or voltage. For example, in a seven-element curtain array of identical equispaced elements, there are four self- and eight mutual impedances which must be determined. Among the eight mutual impedances only Z_{17}, Z_{26} and Z_{35} correspond to pairs of elements symmetrically placed about the center of the array. For these elements, a null at the center ensures that each pair is driven by voltages with equal amplitudes and opposite phases.

The open circuit–short circuit method is a traditional procedure for measuring self- and mutual impedances. The self-impedances are measured as already discussed by driving the element of interest and open circuiting the driving points of all other elements. If element k is the driven element, (14.72) becomes

$$Z_{kin1} = V_k/I_k(0) = Z_{kk}. \tag{14.78}$$

To determine a given mutual impedance a short circuit is substituted for the open circuit in the appropriate element, and the input impedance of the driven element is again measured. If element k is driven and element p is short-circuited, the applicable pair of equations is

$$V_k = I_k(0)Z_{kk} + I_p(0)Z_{kp} \tag{14.79a}$$

$$0 = I_k(0)Z_{kp} + I_p(0)Z_{pp}. \tag{14.79b}$$

From (14.79a)

$$Z_{kin2} = \frac{V_k}{I_k(0)} = Z_{kk} + \frac{I_p(0)}{I_k(0)} Z_{kp}. \tag{14.79c}$$

The use of (14.79b) to eliminate the current ratio in (14.79c), the subsequent solution for Z_{kp}, and the expression of Z_{kk} and Z_{pp} in terms of their measured values, Z_{kin1} and Z_{pin1}, yield

$$Z_{kp} = \pm\sqrt{Z_{pin1}(Z_{kin1} - Z_{kin2})}. \tag{14.80}$$

Mutual admittances may be determined in the same manner by an interchange of open and short circuits and the substitution of the appropriate admittances in (14.80). A satisfactory method of providing the required open circuits is to short-circuit the feeding line at an electrical quarter wavelength from the line–load junction.

An alternative procedure[16] for determining mutual impedances is based on the measurement of both the relative amplitude and phase of the driving-point currents or

---

16 See [10], p. 349.

voltages. Let element k in an array be driven and all other elements be open-circuited. The complete set (14.71) then becomes

$$V_1 = I_k(0)Z_{1k}$$

$$\vdots$$

$$V_k = I_k(0)Z_{kk}$$

$$\vdots \qquad\qquad\qquad\qquad (14.81)$$

$$V_p = I_k(0)Z_{pk}$$

$$\vdots$$

$$V_N = I_k(0)Z_{Nk}.$$

It follows that

$$Z_{kin} = \frac{V_k}{I_k(0)} = Z_{kk} \qquad\qquad (14.82a)$$

$$\frac{V_p}{V_k} = \frac{Z_{pk}}{Z_{kk}}. \qquad\qquad (14.82b)$$

The relative amplitudes of the voltages immediately indicate which mutual impedances are large enough to be important, and the relative phases need be measured only for these.

Let the open circuits in the elements other than k be provided by identical sections of transmission line that are terminated in short circuits. It follows from (14.54a) that the transmission line may be either $\lambda/4$ or $3\lambda/4$ in length, and (14.53) suggests that the apparent voltages across the loads can be measured by a probe placed at a distance $w = \lambda/2$ from the line–load junction. Thus, if a section of transmission line is assembled with a short circuit at $w = 3\lambda/4$ and a probe at $w = \lambda/2$, the apparent driving-point voltages can be measured by interchanging this measuring section with the other loads. When the short circuit is removed, the measuring section can be incorporated in the line feeding the driven element and used to measure V_k. Note that the probe must be loosely coupled to the line and the electrical distances carefully adjusted. A coaxial measuring section for use with this procedure is shown in Fig. 14.8.

For the measurement of admittances, one element is driven while the others are short-circuited. From (14.73) it is seen that

$$Y_{kin} = I_k(0)/V_k = Y_{kk} \qquad\qquad (14.83a)$$

$$I_p(0)/I_k(0) = Y_{pk}/Y_{kk}. \qquad\qquad (14.83b)$$

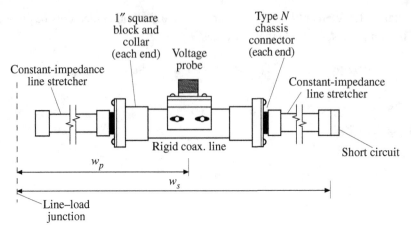

Figure 14.8 Coaxial measuring section. For apparent open-circuited load voltages: $w_p = \lambda/2$, $w_s = 3\lambda/4$. For apparent open-circuit load currents: $w_p = \lambda/4$, $w_s = \lambda/2$.

If a current probe is used in the measuring section, it must be placed at $w = \lambda/2$, with the short circuit at $w = \lambda$. However, from (14.53c) with $\beta w = \pi/2$, it is seen that

$$V(\lambda/4) = jI(0)R_c \tag{14.84}$$

so that a voltage probe may be used at $w = \lambda/4$ and the short circuit placed at $w = \lambda/2$.

14.6   Theory and properties of probes

Successful techniques for sampling fields, currents and charges must be based on the responses of physically real probes, not ideal infinitesimal electric and magnetic doublets. Electric-field or charge probes are usually one-dimensional, short thin dipoles or monopoles that have a simple behavior without serious errors. The usual magnetic-field or current probes, on the other hand, are small loops that have complicated behavior because they are two-dimensional and can be excited in more than one mode. For electrically small loops only the first two modes are important. Because of the manner in which current is distributed around the loop, they are called the circulating or transmission-line mode and the dipole mode, respectively. In the transmission-line mode, there is a continuous current circulating around the loop; currents on opposite sides are equal but in opposite directions in space. In the dipole mode, currents on opposite sides are equal but in opposite directions around the loop, hence in the same direction in space; there is no net circulating current and the probe resembles a small folded dipole. As is shown below, currents in the transmission-line mode are related to the amplitude of the magnetic field at the center of the loop; currents in the dipole mode are related to the amplitude of the electric field at the center of the loop. Generally, currents in both modes can maintain a potential difference across a load. Hence, when the objective is to measure magnetic fields, the presence of dipole-mode currents in the loop may introduce an error that must be eliminated or corrected.

Charge or electric-field probes

To examine the properties of a small charge or electric field probe[17] consider the short, thin, center-loaded dipole in a linearly polarized field of $\mathbf{E}^i$ volts per meter shown in Fig. 14.9a (see also Fig. 14.10a). In the figure $\mathbf{E}^i$ and $\mathbf{E}_p^i$ are in the plane wave front perpendicular to the propagation vector $\mathbf{k}$; $\mathbf{E}_p^i$ and $\mathbf{k}$ are in the plane containing the axis of the antenna. The equivalent circuit, Fig. 14.9b (see also Fig. 14.10b), consists of a Thévenin generator of voltage $V_g(Z_L = \infty)$ in a series combination with the load impedance Z_L and the input impedance of the antenna Z_0. $V_g(Z_L = \infty)$ is the open-circuit voltage at the terminals. It is given by

$$V_g(Z_L = \infty) = -2h_e(\Theta)E^i \cos \psi, \tag{14.85}$$

where $h_e(\Theta)$ is the complex effective half-length of a short dipole and $E^i \cos \psi = E_p^i$ is the projection of E^i onto the plane containing the axis of the antenna and the direction of advance of the incident plane wave through the center of the antenna. The load current is

$$I_L = \frac{V_g(Z_L = \infty)}{Z_0 + Z_L} = \frac{-2h_e(\Theta)E^i \cos \psi}{Z_0 + Z_L} = S_c E_p^i, \tag{14.86}$$

where

$$S_c = \frac{-2h_e(\Theta)}{Z_0 + Z_L}$$

is a sensitivity constant. As indicated in (14.86), the load current is proportional to the average tangential electric field along the dipole. Directions of the field can be determined by rotating the probe until I_L is maximum. If the incident electric field is elliptically polarized, it can be resolved into two linearly polarized components along the major and minor axes of the ellipse and an open-circuit voltage defined for each. The total current is the algebraic sum of the currents due to each generator.

For many applications a short monopole over a conducting surface is an effective electric field probe. Such a probe is easily made by extending the inner conductor of a coaxial line. Equations (14.85)–(14.86) still apply if the appropriate value of Z_0 is used. For a monopole of length h over a ground plane, the input admittance is twice that of a dipole of the same thickness and length $2h$. With either dipole or monopole

---

[17] See [10], p. 184 and p. 475, [22] and [23].

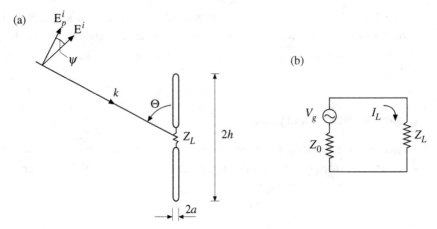

Figure 14.9 Center-loaded receiving dipole for electric field probe. (a) Idealized with no feeding lines; (b) idealized equivalent circuit.

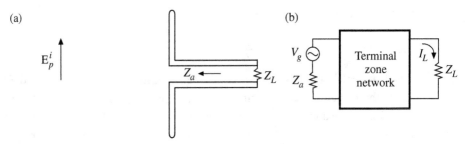

Figure 14.10 Center-loaded receiving dipole for electric field probe. (a) Actual with feeding lines; (b) actual equivalent circuit.

probes, most errors of measurement are introduced because the probe is too long or bent or both.

Computed and measured sensitivities S_c of some monopole probes are given in Fig. 14.11. Usually, the probe is loaded by a section of transmission line terminated in a matched detector so that Z_L can be calculated from (14.39), (14.42) or (14.46a). The complex electrical effective half-length of a short dipole is

$$\beta_0 h_e(\Theta) = \tfrac{1}{2}\beta_0 h \sin \Theta. \tag{14.87a}$$

The input impedance of a short dipole, $\beta_0 h \leq 1$, with $\Omega = 2\ln(2h/a) = 10$ is [10]

$$Z_0 = 18.3\beta_0^2 h^2(1 + 0.086\beta_0^2 h^2) - j(396.0/\beta_0 h)(1 - 0.383\beta_0^2 h^2). \tag{14.87b}$$

When $\beta_0 h \leq 0.5$ and $a \ll h$, the reactance is quite accurately given by

$$X_0 \doteq \frac{-60(\Omega - 3.39)}{\beta_0 h} \tag{14.87c}$$

and the resistance by

$$R_0 \doteq 20\beta_0^2 h^2(1 + 0.133\beta_0^2 h^2). \tag{14.87d}$$

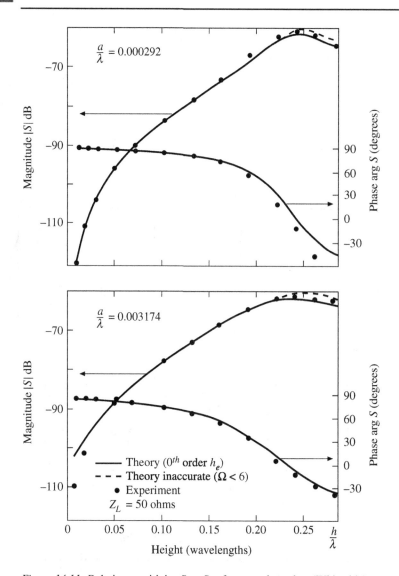

Figure 14.11 Relative sensitivity $S = S_c$ of monopole probes (Whiteside).

If terminal-zone effects are significant, account must be taken of them in determining the apparent resistance and reactance.

Current or magnetic-field probes

To illustrate the important features of small loops as probes,[18] consider an unloaded square loop of side w, and perimeter l immersed in a linearly polarized electromagnetic

---

[18] [22] p. 270, [23], [24], [25] and [26].

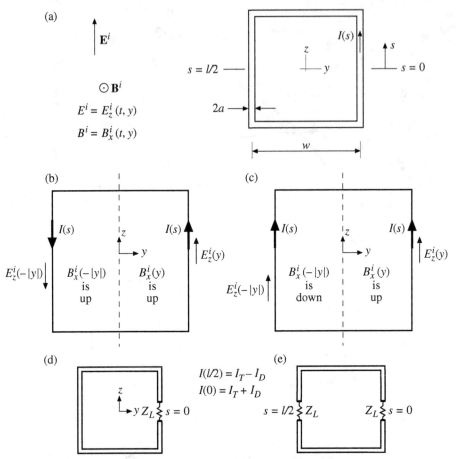

Figure 14.12 Loop probe in linearly polarized field. (a) Coordinates; (b) even magnetic field; (c) odd magnetic field; (d) singly-loaded loop; (e) doubly-loaded loop.

field as shown in Fig. 14.12a. A convenient starting point in the analysis is the integral form of the Maxwell equation, $\nabla \times \mathbf{E} = -j\omega \mathbf{B}$ together with $\mathbf{B} = \nabla \times \mathbf{A}$. That is

$$\oint_s \mathbf{E} \cdot \mathbf{ds} = j\omega \iint_S \hat{\mathbf{n}} \cdot \mathbf{B} \, dS = j\omega \oint \mathbf{A} \cdot \mathbf{ds}, \qquad (14.88)$$

where s is measured along the contour of the loop, S is the plane area bounded by the contour, $\hat{\mathbf{n}}$ is a unit normal to this area in the right-hand screw sense with respect to integration around s, $\mathbf{E}$ is the complex amplitude of the total electric field and $\mathbf{B}$ of the total magnetic field at any point on the surface S of the loop. The radius of the wire, a, is assumed small, so that a quasi-one-dimensional analysis is adequate. The analysis is no more complicated for a rectangular than for a square loop, but the latter can be shown to have the optimum shape for minimizing averaging errors in a general incident field. $\mathbf{E}$ on the surface of the wire can be related to the total axial current by $\mathbf{E} \cdot \mathbf{ds} = z^i I \, ds$ where z^i is the internal impedance per unit length of the wire. In general, it is convenient to treat the total field in two parts, the incident field and the reradiated field maintained by the currents induced in the loop, i.e. $\mathbf{E} = \mathbf{E}^i + \mathbf{E}^r$. With these relations and (1.23a), (14.88) becomes

$$-j\omega \int \int_S \hat{\mathbf{n}} \cdot \mathbf{B}^i \, dS = \oint_s z^i I(s) \, ds + \frac{j\omega\mu_0}{4\pi} \oint_s \oint_s I(s') \frac{e^{-j\beta_0 R}}{R} \, d\mathbf{s}' \cdot d\mathbf{s}, \qquad (14.89)$$

where $\beta_0 = 2\pi/\lambda$ and R is the distance from the element $d\mathbf{s}'$ at s' along the axis of the wire to the element $d\mathbf{s}$ on its surface. At this point, the assumption is usually made that the loop is sufficiently small to replace $\mathbf{B}^i$ by its value at the center of the loop and $I(s)$ by a constant I. Actually a more careful treatment is often required.

Suppose that $\mathbf{B}^i$ can be resolved into an even and an odd part with respect to an axis through the center of the loop. For example, if the loop lies in the yz plane and $\mathbf{B}^i$ is a function of y only and is directed parallel to the x-axis as in Fig. 14.12, the even and odd parts of $\mathbf{B}^i$ with respect to y are

$$\mathbf{B}^i = \mathbf{B}^i_T + \mathbf{B}^i_D \qquad (14.90)$$

even: $\qquad \mathbf{B}^i_T(y) = \frac{1}{2}[\mathbf{B}^i(y) + \mathbf{B}^i(-y)] \qquad (14.91)$

odd: $\qquad \mathbf{B}^i_D(y) = \frac{1}{2}[\mathbf{B}^i(y) - \mathbf{B}^i(-y)] \qquad (14.92)$

with the following symmetry conditions:

$$\mathbf{B}^i_T(-y) = \mathbf{B}^i_T(y); \qquad \mathbf{B}^i_D(-y) = -\mathbf{B}^i_D(y). \qquad (14.93)$$

The subscripts T and D denote the transmission-line and dipole modes of the induced currents.

The electric field is related to the magnetic field by the Maxwell equation

$$\mathbf{E}^i = \frac{j}{\omega\mu_0\epsilon_0} \nabla \times \mathbf{B}^i. \qquad (14.94)$$

Since $\mathbf{E}^i$ is obtained from the first spatial derivative of $\mathbf{B}^i$, $\mathbf{E}^i$ is odd when $\mathbf{B}^i$ is even, and vice versa. That is,

$$E^i_T(-y) = -E^i_T(y); \qquad E^i_D(-y) = E^i_D(y). \qquad (14.95)$$

Note that $\mathbf{B}^i_T \doteq \hat{\mathbf{x}} B^i_0$ over the area bounded by the loop when $\beta_0 w \ll 1$; the current, $I_T(s)$, is, then, essentially constant around the loop as indicated in Fig. 14.12b. With the symmetry conditions (14.91) and (14.94), (14.89) becomes

$$-j\omega B^i_0 S = I_T \left\{ \oint_s z^i \, ds + \frac{j\omega\mu}{4\pi} \oint_s \oint_s \frac{e^{-jkR}}{R} \, d\mathbf{s}' \cdot d\mathbf{s} \right\} = I_T Z_0. \qquad (14.96)$$

The quantity in braces in (14.96) is the impedance $Z_0 = Y_0^{-1}$ of the loop with constant current.[19] Therefore

$$I_T = I_T(0) = -j\omega B_0^i Y_0 S = \lambda S_B(c B_0^i), \tag{14.97}$$

where the sensitivity constant S_B for the unloaded loop has been introduced. It is defined by

$$S_B = -jkSY_0/\lambda \tag{14.98}$$

and depends only on the geometry of the probe. The magnetic field is conveniently multiplied by $c = 3 \times 10^8$ m/s to give it the same dimensions as $\mathbf{E}^i$. Note that the current I_T is directly proportional to and, hence, a measure of the incident magnetic field $B_0^i = B_T^i(0)$ at the center of the loop.

$\mathbf{B}_D^i(y)$ is odd in y and therefore zero at the center of the loop; it makes no contribution to the surface integral in (14.89). The associated electric field $\mathbf{E}_D^i$ in (14.95) has a non-zero value at the center of the loop and is approximately constant over the space occupied by the small loop. It maintains equal and co-directional currents in the two sides of the loop which are parallel to the z-axis and hence parallel to $\mathbf{E}_D^i$ as shown in Fig. 14.12c. These dipole-mode currents are zero at $s = \pm l/4$ ($y = 0$, $z = \pm w/2$). In so far as they are concerned, the loop could be cut at these points and treated as an array of two bent receiving antennas. The current at the center of each side is

$$I_D(0) = h_{eD} Y_D E_0^i = \lambda S_E E_0^i, \tag{14.99}$$

where h_{eD} is the effective length of each half of the array for the dipole mode, Y_D is the input admittance at the center of each antenna when both are driven with equal and co-directional currents, and $E_0^i = E_0^i(0)$ is the electric field at the center of the loop. The electric sensitivity constant S_E for the unloaded loop has been introduced in (14.99). It is defined as follows:

$$S_E = h_{eD} Y_D/\lambda. \tag{14.100}$$

Note that $I_D(0)$ is proportional to and, therefore, a measure of the incident electric field E_0^i at the center of the loop.

Transmission-line and dipole-mode currents have the following symmetries with respect to the anti-clockwise direction:

$$I_T\left(s + \frac{l}{2}\right) = I_T(s); \qquad I_D\left(s + \frac{l}{2}\right) = -I_D(s). \tag{14.101}$$

---

[19] See Chapter 6 of [27].

Hence, $I_T(s)$ corresponds to the zero-sequence current $I^{(0)}(s)$, and $I_D(s)$ to the first-sequence current $I^{(1)}(s)$. Higher-order sequence currents are assumed to be negligible in a sufficiently small loop. The total currents at points s and $s + l/2$ in the loop are

$$I(s) = I_T(s) + I_D(s) \tag{14.102}$$

$$I\left(s + \frac{l}{2}\right) = I_T(s) - I_D(s). \tag{14.103}$$

A very important fact is now evident. The magnetic field that contributes to the surface integral is related only to loop currents in the transmission-line (zero-sequence) mode and not to those in the dipole (first-sequence) mode. Conversely, if the magnetic field at the center of the loop is to be determined from the current induced in the loop, measurements must involve currents in the transmission-line mode only.

When a small loop is used as a probe, the quantity of primary concern is the load current. Let a load Z_L be located at $s = 0$, as shown in Fig. 14.12d. The loaded loop can be analyzed by replacing the load by a Thévenin generator of voltage $V = -I_L(0)Z_L$, where $I_L(0)$ is the current in the load. This generator maintains a current $VY(s) = -I(0)Z_L Y(s)$ where $Y(s)$ is the input admittance when the loop is driven at the point s. The total current at $s = 0$ is then

$$I_L(0) = I_T(0) + I_D(0) - I_L(0)Z_L Y(0). \tag{14.104}$$

With (14.97) and (14.99), it follows that

$$I_L(0) = \lambda S_B^{(1)} c B_0^i + \lambda S_E^{(1)} E_0^i, \tag{14.105}$$

where the sensitivity constants for the singly-loaded loop are defined as follows:

$$S_B^{(1)} = \frac{Y_L}{Y_L + Y(0)} S_B \tag{14.106}$$

$$S_E^{(1)} = \frac{Y_L}{Y_L + Y(0)} S_E. \tag{14.107}$$

The importance of the location of the load with respect to the incident fields is now evident. When the load is located at $s = l/2$ instead of $s = 0$ (a change that is equivalent to rotating the loop through $180°$ about the x-axis), the input admittance of the loop is still the same but the current is given by (14.103) so that

$$I_L\left(\frac{l}{2}\right) = \lambda S_B^{(1)} c B_0^i - \lambda S_E^{(1)} E_0^i. \tag{14.108}$$

This equation and (14.105) are useful for determining the relative importance of I_D. If the load current remains constant when the probe is rotated $180°$ about the x-axis,

I_D is negligible and $I_L = I_T$. If the load current does not remain constant, readings of amplitude and phase may be taken in both positions. Then

$$I_T = \lambda S_B^{(1)} c B_0^i = \frac{1}{2}\left[I_L(0) + I_L\left(\frac{l}{2}\right)\right] \qquad (14.109\mathrm{a})$$

$$I_D = \lambda S_E^{(1)} E_0^i = \frac{1}{2}\left[I_L(0) - I_L\left(\frac{l}{2}\right)\right]. \qquad (14.109\mathrm{b})$$

Instead of rotating the probe and taking two readings of amplitude and phase, one may use two loads with a hybrid junction to evaluate the sum and the difference.

In the simple example of a linearly polarized electric field indicated in Fig. 14.12, I_D is easily eliminated by a simple rotation of the loop until the side containing the load is perpendicular to the electric field. That is, the load is located at $s = \pm l/4$ in Fig. 14.12. However, when the electric field is elliptically polarized this expedient is unavailable and a doubly-loaded probe is probably the simplest solution in spite of the increased constructional difficulties.

The analysis of a doubly-loaded loop with identical loads Z_L at $s = 0$ and $s = l/2$, as shown in Fig. 14.12e, parallels that of the singly-loaded loop. The two load currents are

$$I_{L1} = I(0) = I_T(0) + I_D(0) - \left[I(0)Y(0) + I\left(\frac{l}{2}\right)Y\left(\frac{l}{2}\right)\right]/Y_L \qquad (14.110\mathrm{a})$$

$$I_{L2} = I\left(\frac{l}{2}\right) = I_T(0) - I_D(0) - \left[I(0)Y\left(\frac{l}{2}\right) + I\left(\frac{l}{2}\right)Y(0)\right]/Y_L. \qquad (14.110\mathrm{b})$$

The driving-point admittances $Y(0)$ and $Y(l/2)$ may be resolved into the zero- and first-sequence admittances $Y^{(0)}$ and $Y^{(1)}$. These can be introduced as follows:

$$Y(0) = Y^{(0)} + Y^{(1)}; \qquad Y\left(\frac{l}{2}\right) = Y^{(0)} - Y^{(1)}.$$

Let

$$I_\Sigma = I_{L1} + I_{L2} = \frac{2Y_L}{Y_L + 2Y^{(0)}} I_T(0) = \lambda S_B^{(2)} c B_0^i \qquad (14.111\mathrm{a})$$

$$I_\Delta = I_{L1} - I_{L2} = \frac{2Y_L}{Y_L + 2Y^{(1)}} I_D(0) = \lambda S_E^{(2)} E_0^i. \qquad (14.111\mathrm{b})$$

The sensitivity constants for the doubly-loaded probe are defined as follows:

$$S_B^{(2)} = \frac{2Y_L}{Y_L + 2Y^{(0)}} S_B \qquad (14.111\mathrm{c})$$

$$S_E^{(2)} = \frac{2Y_L}{Y_L + 2Y^{(1)}} S_E. \qquad (14.111\mathrm{d})$$

In actual practice, the hybrid junctions used to perform the summing and differencing operations will have good but not infinite isolation, so that the actual measurable currents are

$$I_B = I_\Sigma + \gamma I_\Delta = \lambda S_B^{(2)} c B_0^i + \gamma \lambda S_E^{(2)} E_0^i \tag{14.112a}$$

$$I_E = I_\Delta + \gamma' I_\Sigma = \lambda S_E^{(2)} E_0^i + \gamma' \lambda S_B^{(2)} c B_0^i, \tag{14.112b}$$

where γ and γ' are the coefficients of cross-coupling between the adding and subtracting circuits. It is assumed that they are small.

In the measurement of the magnetic field (especially near the end of a dipole antenna where the polarization of the electric field is highly elliptical) it is particularly important that those parts of the current in the load that are excited in the dipole mode, namely I_D, be negligible. To provide a measure of the ability of a probe and loading system to discriminate against such currents, a system error ratio $\epsilon^{(n)}$ can be defined as the ratio of the output current due to unit parallel electric field ($E_0^i = 1$ volt/meter) to the output current due to unit normal magnetic field ($c B_0^i = 1$ volt/meter),

$$\epsilon^{(1)} = S_E^{(1)}/S_B^{(1)} \tag{14.113a}$$

$$\epsilon^{(2)} = S_E^{(2)}/S_B^{(2)}, \tag{14.113b}$$

where the superscript indicates the number of loads in the probe. Note that (14.113b) applies to the combination of the probe and its summing and differencing circuits. The actual ratio of the two currents depends on the ratio of the fields $E_0^i/c B_0^i$ and generally equals $\epsilon^{(n)}$ only in a plane-wave field. For a system to be capable of measuring the magnetic field with an error of no more than 10%, it is necessary that $\epsilon^{(n)} \leq -20$ dB, where $\epsilon^{(n)}$ in dB $= 20 \log_{10} \epsilon^{(n)}$.

So far, the discussion has been concerned with square loops, although circular loops are often more desirable. Actually, a comparable analysis of circular loops follows precisely the steps outlined for the square loop including the definition of sensitivity constants, error ratios, etc; differences between the two shapes arise in the theoretical expressions for evaluating the sensitivity constants.

The final step in the practical analysis of loops as probes is to obtain expressions for the sensitivity constants. Consider first the square loop. Y_0, required in the definition of S_B in (14.98), may be found from (14.96) by an expansion of the exponential in a power series. The result for a loop of side w and wire radius a is [26]

$$Y_0 = \frac{-j\pi}{\zeta_0 \beta_0 w (\Omega - 4.32 + 0.37 \beta_0^2 w^2)}, \tag{14.114}$$

where $\zeta_0 = \sqrt{\mu_0/\epsilon_0} \doteq 120\pi$ ohms and $\Omega = 2 \ln(4w/a)$. Hence

$$S_B = \frac{-\pi w}{\lambda \zeta_0 (\Omega - 4.32 + 14.6 w^2/\lambda^2)}. \tag{14.115}$$

The unloaded electric sensitivity S_E is defined by (14.100) in terms of h_{eD} and Y_D. The effective length for the dipole mode is found by cutting the loop at $s = \pm l/4$, treating the two halves as a transmitting array, and applying the Rayleigh–Carson reciprocal theorem.[20] The result is

$$h_{eD} = \frac{2}{I_0} \int_0^{l/4} I(s)\,ds,$$ (14.116)

where $I(s)$ is the transmitting current when the array is driven with co-directional currents, and I_0 is its value at the driving point. To zero order, this current is

$$I(z) \approx \frac{j2\pi V}{\zeta_0(\Omega - 3.17)} \frac{\sin \beta_0(w - |z|)}{\cos \beta_0 w}.$$ (14.117)

With (14.117), (14.116) becomes

$$h_{eD} = \frac{\cos \frac{1}{2}\beta_0 w - \cos \beta_0 w}{\beta_0 \sin \frac{1}{2}\beta_0 w}.$$ (14.118)

The input admittance is[21]

$$Y_D = \frac{j2\pi \tan \beta_0 w}{\zeta_0(\Omega - 3.17)}.$$ (14.119)

It follows that for the square loop,

$$S_E = \frac{j}{\zeta_0(\Omega - 3.17)} \frac{\tan \beta_0 w (\cos \frac{1}{2}\beta_0 w - \cos \beta_0 w)}{\sin \frac{1}{2}\beta_0 w}.$$ (14.120)

In order to calculate the sensitivity constants S_E, S_B for loaded loops, note that $Y^{(1)} = Y_D/2$ and $Y^{(0)} \approx Y_0$ so that the values from (14.119) and (14.114) can be directly substituted into the following equations for $S_B^{(2)}$ and $S_E^{(2)}$: (14.106), (14.107), (14.111c) and (14.111d).

For a circular loop of diameter w and wire radius a,

$$Y_0 = \frac{-j4}{\zeta_0 \beta_0 w (\Omega - 3.52 + 0.33\beta_0^2 w^2)}$$ (14.121)

and

$$S_B = \frac{-\pi w}{\lambda \zeta_0 (\Omega - 3.52 + 13.0 w^2/\lambda^2)},$$ (14.122)

where $\Omega = 2 \ln(\pi w/a)$.

20 [10], p. 568.   21 [10], p. 568.

The unloaded electric sensitivity is found from the response of the loop to an electric field that is uniform in the plane of the loop and pointing in the z-direction. The result is[22]

$$I(\phi) = E_{z0}^i \frac{w}{j\zeta_0 a_1} \cos \phi, \qquad (14.123a)$$

where ϕ is the angular coordinate measured from the y-axis and a_1 is an expansion parameter calculated by Storer [29]. For $w \leq 0.1\lambda$,

$$a_1 \doteq -(\Omega - 3.52)(1 - \beta_0^2 w^2/4)/\pi \beta_0 w \qquad (14.123b)$$

and

$$S_E = \frac{j2\pi^2 w^2/\lambda^2}{\zeta_0(\Omega - 3.52)(1 - 9.8w^2/\lambda^2)} \qquad (14.123c)$$

for the circular loop. Zero- and first-phase-sequence admittances are again needed for evaluating the sensitivity constant for the loaded loop. They can be found from Storer's results [29]. The expressions are complicated but subject to the conditions $w \leq 0.03\lambda$ and $Y_L > 10Y^{(1)}$, $Y^{(0)}$ and $Y^{(1)}$ are

$$Y^{(0)} \doteq -j2\lambda[\pi w\zeta_0(\Omega - 3.52)]^{-1} \qquad (14.124a)$$

$$Y^{(1)} \doteq j4\pi w[\lambda\zeta_0(\Omega - 3.52)]^{-1}. \qquad (14.124b)$$

Generally (14.114)–(14.124b) provide quite accurate results for loop diameters or sides $w \leq 0.03\lambda$ and serve as a useful guide for $w \leq 0.1\lambda$. When $w \leq 0.03\lambda$, most of the expressions can be simplified. Note that $Y^{(0)} \gg Y^{(1)}$. They are summarized in Table 14.1.

The simplified relations of Table 14.1 reveal that the error ratio $\epsilon^{(1)}$ of singly-loaded probes is independent of the load and approximately a linear function of the length of the side for square probes and of the diameter for circular probes. The magnetic fields measured with a circular loop will have an error less than 10% provided $w \leq 0.016\lambda$. At 600 MHz this corresponds to $w \sim 0.5$ cm; at 3000 MHz to $w \sim 1$ mm. It is obviously advantageous to make such measurements at frequencies below 1000 MHz.

Sensitivities and error ratios as functions of loop size are shown in Fig. 14.13 for typical square loops, in Fig. 14.14 for circular loops. The sensitivities are in dB referred to 1 siemens, the error ratios are in dB referred to 1, and magnitudes are absolute, phases are relative. Important characteristics of the graphs are the relatively slow increase in sensitivity as w/λ increases beyond about 0.03 or 0.04, and the minimum of $\epsilon^{(2)}$ at about $w/\lambda = 0.04$, indicating that this size may be a good compromise for probes. From the curves of Figs. 14.13c and 14.14c, a singly-loaded loop with dimensions as large as $w/\lambda = 0.1$ is seen to respond nearly as well to the electric field in the dipole mode as to the normal magnetic field in the circulating mode.

[22] [28], Chapter 10.

Table 14.1. *Probe characteristics of electrically small loops*

Square loop, side w, $w \leq 0.03\lambda$

$$S_B^{(1)} = \frac{Y_L}{Y_L + Y^{(0)}} \frac{-\pi w}{\lambda \zeta_0 (\Omega - 4.32)} \qquad S_E^{(1)} = \frac{Y_L}{Y_L + Y^{(0)}} \frac{j3\pi^2 w^2}{\lambda^2 \zeta_0 (\Omega - 3.17)}$$

$$\epsilon^{(1)} = -j3\pi \, \frac{w}{\lambda} \frac{\Omega - 4.32}{\Omega - 3.17}$$

$$Y^{(0)} = -j\lambda [2w\zeta_0(\Omega - 4.32)]^{-1} \qquad Y^{(1)} = j2\pi^2 w[\lambda \zeta_0(\Omega - 3.17)]^{-1}$$

$$\Omega = 2\ln(4w/a)$$

Circular loop, diameter w, $w \leq 0.03\lambda$

$$S_B^{(1)} = \frac{Y_L}{Y_L + Y^{(0)}} \frac{-\pi w}{\lambda \zeta_0 (\Omega - 3.52)} \qquad S_E^{(1)} = \frac{Y_L}{Y_L + Y^{(0)}} \frac{j2\pi^2 w^2}{\lambda^2 \zeta_0 (\Omega - 3.52)}$$

$$\epsilon^{(1)} = -j2\pi \, \frac{w}{\lambda}$$

$$Y^{(0)} = -j2\lambda [\pi w \zeta_0(\Omega - 3.52)]^{-1} \qquad Y^{(1)} = j4\pi w[\lambda \zeta_0(\Omega - 3.52)]^{-1}$$

$$\Omega = 2\ln(\pi w/a)$$

Square and circular loops, $w \leq 0.03\lambda$

$$S_B^{(2)} = 2\frac{Y_L + Y^{(0)}}{Y_L + 2Y^{(0)}} S_B^{(1)} \qquad S_E^{(2)} = 2\frac{Y_L + Y^{(0)}}{Y_L + 2Y^{(1)}} S_E^{(1)}$$

$$\epsilon^{(2)} = \frac{Y_L + 2Y^{(0)}}{Y_L + 2Y^{(1)}} \epsilon^{(1)}$$

A comparison of the theoretical and experimental results in Figs. 14.13–14.14 shows good agreement and suggests that the theoretical results are adequate guides for the design of probes. Graphs of the limiting loop sizes and wire thicknesses required to keep error ratios below given limits are in Fig. 14.15 for singly-loaded loops, in Fig. 14.16 for doubly-loaded loops. In Fig. 14.16, γ, the cross-coupling coefficient between the adding and subtracting circuits, is assumed to be 1; in an actual system it will be at least $-20\,\mathrm{dB}$, reducing the indicated error ratio by this amount. Since the effects of changes in wire thickness and load resistance are similar for circular and square loops, the curves of Figs. 14.15 and 14.16 provide a useful qualitative guide for the former.

14.7   Construction and use of field probes

In many applications probes are used either in a free-standing arrangement or in conjunction with an image plane.[23] In the former, shown in Fig. 14.17, the probe

---

[23] See [10], p. 127 and [30].

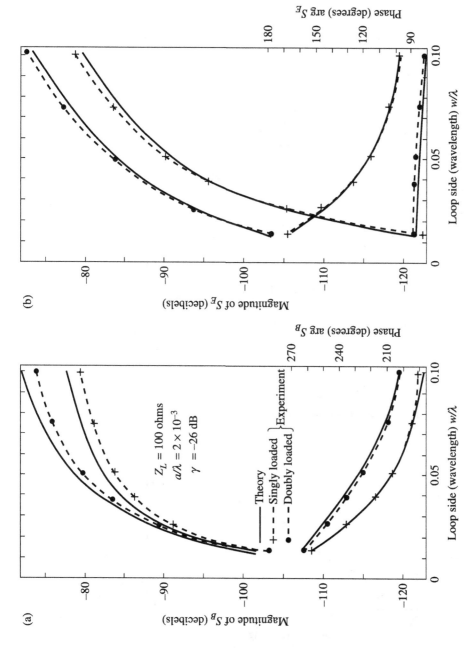

Figure 14.13 (a) Typical magnetic sensitivity of square loops. (b) Typical electric sensitivity of square loops.

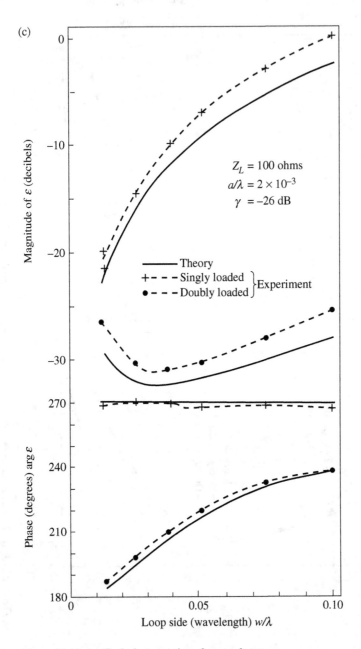

Figure 14.13 (c) Typical error ratios of square loops.

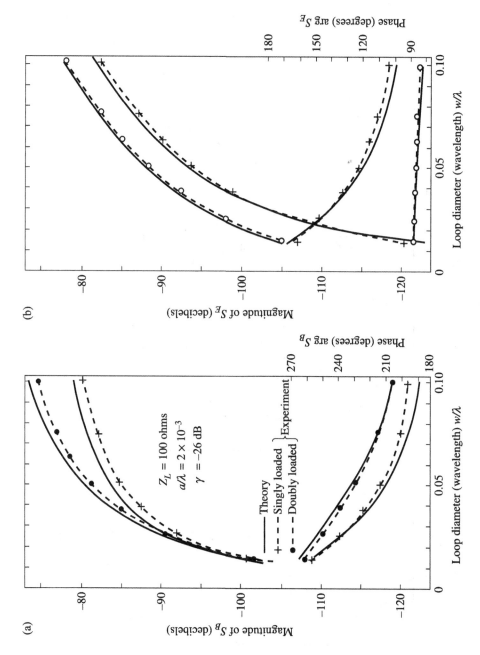

Figure 14.14 (a) Typical magnetic sensitivity of circular loops. (b) Typical electric sensitivity of circular loops.

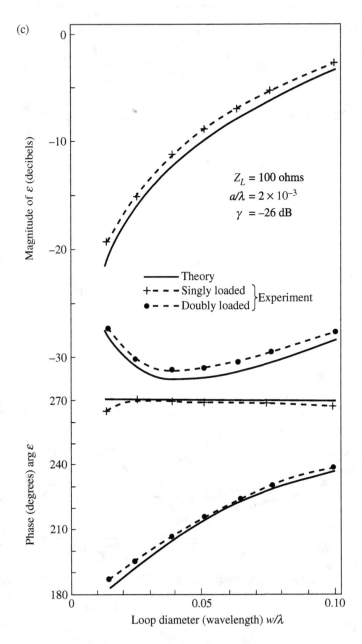

Figure 14.14 (c) Typical error ratios of circular loops.

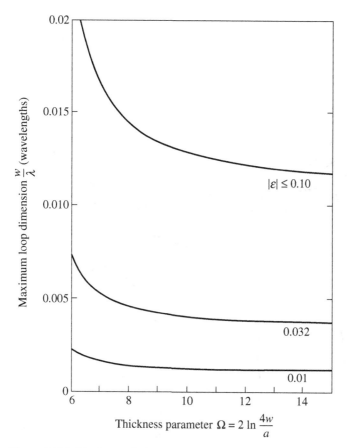

Figure 14.15 Maximum dimension of singly-loaded square loop for given error ratio $|\epsilon^{(1)}|$. Theoretical curves independent of Z_L (Whiteside).

is supported at the end of a long rigid tube which contains the feeder lines to the receiving equipment. The supporting tube is attached by a movable carriage to a track on a pivoting arm that permits the accurate placement and orientation of the probe. A loop is usually mounted with its plane perpendicular to the axis of the supporting tube. The free-standing arrangement is versatile and useful for measuring near-zone fields and surface currents on three-dimensional models. A principal disadvantage is that the supporting tube is always present in the field. Its disturbing effects can be reduced with quarter-wave sleeves and absorbing material. An alternative procedure incorporates a rectifying crystal directly in the probe and makes use of resistive wire to measure the d.c. voltage.

The image-plane arrangement, Fig. 14.18, is well-suited for measuring the surface currents on symmetrical models that can themselves be mounted on an image plane. It has the advantage that all cables and supports are contained within the metal walls of the object under investigation or behind the image plane. The probe is mounted at the

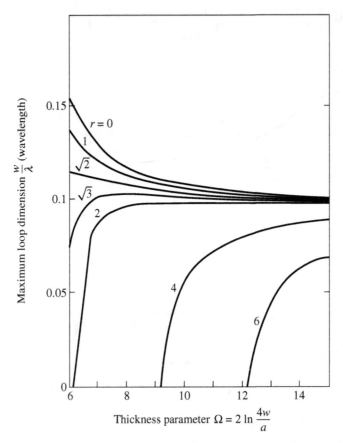

Figure 14.16 Maximum dimension of doubly-loaded square loop for given error ratio $|\epsilon^{(2)}| \leq 1$. Theoretical curves, load impedance $Z_L = 60\,r$ ohms (Whiteside).

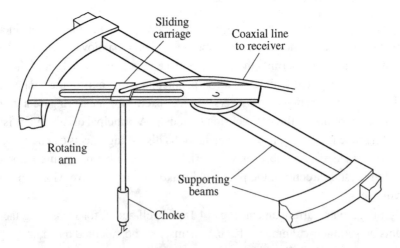

Figure 14.17 Arrangement for free-space probe measurements.

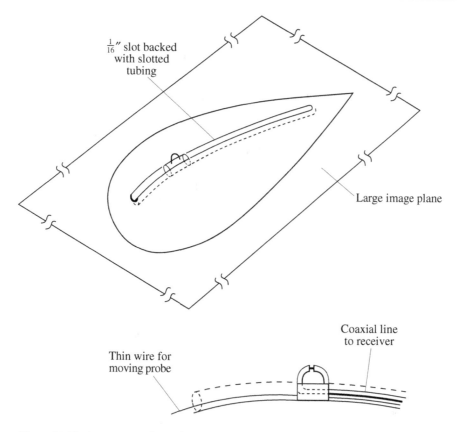

Figure 14.18 Arrangement for probe measurements on an image plane.

end of a tube that serves both to move the probe and as the outer conductor of a coaxial line that connects the probe to its receiving system. The entire assembly is contained within a second slotted tube that serves to guide the probe at a constant height along the slot. In an alternative arrangement that permits the probe to move more easily along curved surfaces, the probe is mounted in a short cylindrical block, and a flexible feed line is used in conjunction with only the outer slotted tube, which can be bent to guide the probe along the surface. If the slot is parallel to the direction of the current, it has no significant effect. If it cuts the lines of flow, it must be covered with conducting tape except in the immediate vicinity of the probe.

Examples of probes that may be used in a free-standing arrangement are shown in Fig. 14.19. A balanced charge probe consists of an electrically short dipole formed by 90° bends in the conductors of a two-wire transmission line, or by bends in the inner conductors of a shielded-pair line or in a pair of adjacent miniature coaxial lines (Fig. 14.19a). Loop probes (Figs. 14.19b and 14.19c) are made one-half from a solid brass rod and one-half from a miniature rigid coaxial line. At the junction, which is the location of the load, a small gap is left in the outer conductor and insulator of

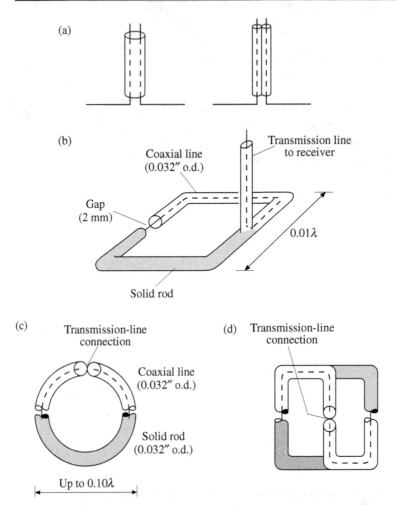

Figure 14.19 Probes for free-space system. (a) Balanced charge probes; (b) singly-loaded square loop; (c) doubly-loaded circular loop; (d) bridged loop.

the coaxial line. Typical gap widths are 1–2 mm. The load Z_L is the impedance seen looking into the coaxial line at the gap. For singly-loaded loops, the gap is located symmetrically with respect to the axis of the supporting tube. Typical dimensions of the coaxial line are: outer diameter 0.032 inch, wall thickness 0.004 inch, and an inner conductor of 34 gauge wire. The characteristic impedance is 50 ohms. With doubly-loaded loops, the vertical supporting tube causes some degradation of the error ratio and the bridged loop shown in Fig. 14.19d may be more satisfactory. The mechanical construction must preserve a high degree of symmetry.

Probes for use with the image-plane technique (Fig. 14.20) have the same basic construction as those just described, except that the charge probes are usually monopoles and the current probes are half-loops. The half-loops may be either singly- or doubly-loaded.

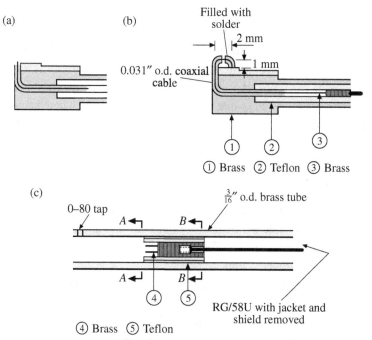

(a)

(b)

Filled with
solder

2 mm

0.031" o.d. coaxial
cable

1 mm

① Brass ② Teflon ③ Brass

(c)

0–80 tap

$\frac{3}{16}$" o.d. brass tube

A   B

A   B

④ Brass ⑤ Teflon

RG/58U with jacket and
shield removed

Note: i.d. of tubing stepped 0.002" at *AA*, and 0.001" at *BB*.

Figure 14.20 Probes for image-plane equipment. (a) Charge probe; (b) current probe; (c) probe socket.

The probes just described are shielded loops. Open-wire loops could be used instead of shielded loops,[24] but they are less convenient in the elimination of dipole-mode currents. With shielded loops the currents induced on the outside of the shield maintain a potential difference across the gap which is small and located at the point where dipole-mode currents vanish. If the gap is not at this point a dipole-mode voltage is also developed across the load.

Several simple tests can be used to reveal the sensitivity of loop probes to dipole-mode currents.[25] If the current in the load circuit of the probe is constant under a 180° rotation of the probe, dipole-mode currents are negligible. If the load current is not constant, readings must be taken with the probe in each position and averaged. With the image-plane technique, the probe cannot be rotated but can be tested in a short-circuited coaxial line. Let the probe current be measured with the probe successively at $w = w_0 = \lambda/4$ and $w = w_1 = \lambda/2$. Then

$$\epsilon_{dB}^{(1)} = 20 \log |I_L(w_0)/I_L(w_1)|.$$

When doubly-loaded probes are used with summing and differencing circuits, the output currents must be balanced because neither the probe nor the attached lines

[24] See [9], p. 209.   [25] [23], Chapter VII, p. 5.

and loads are perfectly symmetrical. For this purpose, variable attenuators and phase shifters (or line stretchers) are used in the feed lines. The differencing circuit may be balanced by placing the probe so that it is symmetrically excited and adjusting the attenuators and phase shifters until the difference current is constant under 180° rotations of the probe. An alternative procedure is to measure individually the difference currents $I_{\Delta 1}$ and $I_{\Delta 2}$ due to the probe loads 1 and 2 when the probe is placed so that it is symmetrically excited. The probe is rotated 180° and the new currents $I'_{\Delta 1}$ and $I'_{\Delta 2}$ are measured, and the attenuators and phase shifters adjusted until

$$I_{\Delta 2}/I_{\Delta 1} = I'_{\Delta 1}/I'_{\Delta 2}. \tag{14.125}$$

Similar procedures can be used to balance the summing circuits. If a single hybrid junction is used to provide the outputs for both the sum and difference, the balancing adjustment cannot be optimized for both arms simultaneously, but a satisfactory compromise can usually be found.

The measurement of surface distributions of current and charge on good conductors actually involves the measurement of magnetic and electric fields near the surface. Most of the current in a good conductor at high frequencies is concentrated within a very small distance of the surface, d_s, called the skin depth and given by[26]

$$d_s = (2/\omega\mu\sigma)^{1/2}. \tag{14.126}$$

Such a thin layer of current is well approximated by the surface density $\mathbf{K}$ on a perfect conductor and is related to the total magnetic field at the surface by the boundary condition

$$\hat{\mathbf{n}} \times \mathbf{B} = \hat{\mathbf{t}}B_t = -\mathbf{K}\mu_0. \tag{14.127}$$

Similarly, the surface charge η is related to the total electric field by

$$\hat{\mathbf{n}} \cdot \mathbf{E} = E_n = -\eta/\epsilon_0, \tag{14.128}$$

where $\hat{\mathbf{n}}$ is an outward unit normal from the surface. On thin cylinders, $\mathbf{K}$ and η have no angular variation around the cylinder, so that the total axial current and charge per unit length are $\mathbf{I}(z) = 2\pi a \mathbf{K}(z)$ and $q(z) = 2\pi a \eta(z)$. Except near the ends or edges of conductors distributions of B_t and E_n are often unchanged at very small fractions of a wavelength from the surface, so that probes placed sufficiently near the surface and moved parallel to it sample fields which are proportional to $\mathbf{K}$ and η. In the image-plane method, the effective center of the probe is usually quite near the surface; in the free-standing method, it is at least a probe radius away. At distances from a surface that are less than a few probe diameters, a probe is tightly coupled to its

---

[26] [23], Chapter VII, p. 5.

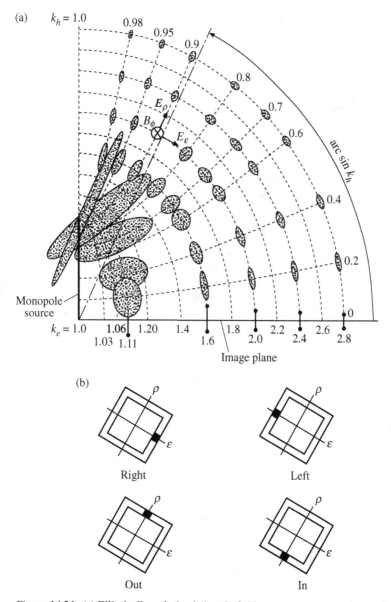

Figure 14.21 (a) Elliptically polarized electric field near quarter-wave monopole over an image plane; (b) probe orientations for measurements of Figs. 14.22 and 14.23.

image so that its distance from the surface must be kept constant. Since the coupling between a probe and an ideal image does not exist near edges and corners, meaningful measurements cannot be made.

The most difficult fields to measure are linearly polarized magnetic fields associated with elliptically polarized electric fields. Figures 14.21–14.23 show the effects of size and orientation of probes in such fields and illustrate the effective use of singly-

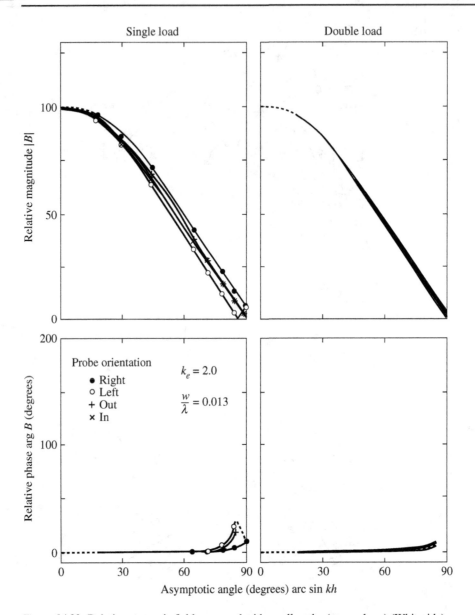

Figure 14.22 Relative magnetic field measured with small probe (square loop) (Whiteside).

and doubly-loaded loops to make measurements. The near-zone elliptically polarized electric field of a quarter-wave monopole over an image plane is shown in confocal coordinates in Fig. 14.21a.[27] In Fig. 14.22 are graphs of measurements made along the coordinate $k_e = 2$ with a singly-loaded square loop with $w/\lambda = 0.013$, oriented in the four positions indicated in Fig. 14.21b. Owing to its small size, this loop was relatively insensitive to dipole-mode fields and its orientation with respect to

[27] See Chapter 5 in [10].

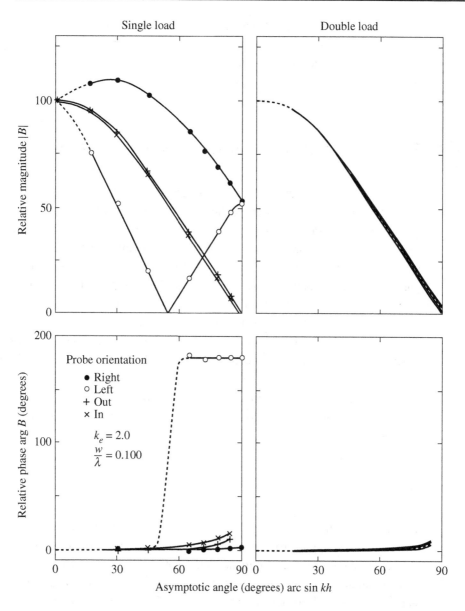

Figure 14.23 Relative magnetic field measured with large probe (square loop) (Whiteside).

the electric field was not critical. In contrast, for a loop with $w/\lambda = 0.1$, the orientation is seen in Fig. 14.23 to be very important. When the probe is oriented in the positions marked "In" and "Out" the dipole-mode current excited by the component E_ε maintains a voltage across the load, that excited by E_ρ does not. In the positions marked "Right" and "Left", the dipole-mode currents due to E_ρ maintain a voltage across the load, those due to E_ε do not. Since E_ε is nearly proportional to B_Φ, whereas E_ρ is not, no significant error is introduced in a relative measurement with the probe in the "In" and "Out" positions, a very large error in the "Right" and "Left" positions. The doubly-loaded loop with its summing and differencing circuits is seen to provide accurate results regardless of orientation even for sizes as large as $w/\lambda = 0.1$.

14.8   The measurement of sharp resonances in circular arrays

The objective of the experiment to be described was to determine if the very sharp resonances that had earlier been predicted theoretically could be found in a physically real array. The array consisted of equally spaced monopoles over a ground plane. Only one of the elements was driven. The experimental investigation presented special difficulties because of the very high values of conductance at the resonant points of the array and its rapid change with frequency due to the sharpness of the resonance. Theoretical investigations predicted maximum conductance values of 700–800 millisiemens with adjacent resonant maxima separated by as little as 2.4 MHz. The theoretical investigations also indicated that the number, length, thickness, and spacing of the array elements would be very critical. Conductive losses in the elements and the transmission line as well as contact resistance between the elements and the ground plane and end effects of the transmission line had to be considered to obtain good agreement between results from the experimental and theoretical models.

The measurement procedure

The measurement procedure adopted [31,32] was a variation of the multiple-probe method with a single probe used at four different probe stations. A complete set of measurements of the admittance terminating the transmission line was taken in very small frequency steps with the probe at one station, and then at another station. This traced out the resonance curve of the array, permitting the maximum of the conductance curve – and hence the resonant frequency – to be accurately located. The probe was then moved to a different station and the measurements repeated at the same set of frequencies. This procedure minimized possible errors due to moving the probe to its different stations and gave a relatively high data rate. The RF source and receiver were part of an integrated vector network analyzer that used a synthesized source providing a resolution of 1 Hz over the range of 300 kHz to 3 GHz. The measurement procedure was possible because of the high stability and repeatability of the synthesized signal generator and receiver.

Ohmic losses on the transmission line were accounted for by using the hyperbolic form of (14.55b):

$$Y_a = Y_c \left[\frac{\cosh \gamma w_2 - V_R \cosh \gamma w_1}{- \sinh \gamma w_2 + V_R \sinh \gamma w_1} \right], \tag{14.129a}$$

where γ is the complex propagation constant of the line, defined by

$$\gamma = \sqrt{(R + j2\pi f L) j 2\pi f C}. \tag{14.129b}$$

R is the resistance per unit length due to ohmic losses on the walls of the brass transmission line that was used,

$$R = \frac{1}{\pi d_s \sigma_B} \left(\frac{1}{2a} + \frac{1}{2b} \right). \tag{14.129c}$$

Here, d_s is the skin depth, σ_B is the conductivity of brass, and a and b are, respectively, the outer radius of the inner conductor and the inner radius of the outer conductor. L and C are, respectively, the inductance and conductance per unit length of the line, given by

$$L = \frac{\mu_0}{2\pi} \ln \frac{b}{a}, \qquad C = \frac{2\pi \epsilon_0}{\ln(b/a)}. \tag{14.129d}$$

When ohmic losses are included, the characteristic admittance Y_c of the line has a small imaginary component given by

$$Y_c = \sqrt{\frac{j 2\pi f C}{R + j 2\pi f L}}. \tag{14.129e}$$

The experimental model [32]

The theoretical calculations indicated that a large number of elements were required to obtain the very sharp resonances, and 90 elements were chosen of which 89 were parasitic and one driven. The choice of element parameters was based on a theoretical model but tempered by practical considerations of available ground plane sizes, the frequency range of the available network analyzer, and reasonable matching tolerances. The final element parameters were height $h = 0.858$ inch and diameter $2a = 0.25$ inch; the array diameter was $2R = 40$ inches, resulting in a spacing between the elements of approximately $d/\lambda = 0.3$ at an average wavelength. With these parameters, 13 resonances of interest were predicted with the lowest $f_{33} = 2.55$ GHz and the highest $f_{45} = 2.68$ GHz.

The parameters of the parasitic elements are shown in Fig. 14.24. The elements were hollow brass tubes with an inner diameter of 7/32 inch, resulting in a wall thickness of 1/64 inch that provided an acceptable approximation to the zero thickness of the

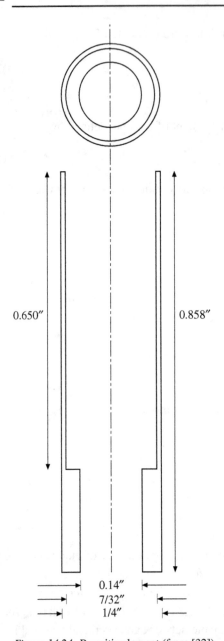

0.650"   0.858"

0.14"
7/32"
1/4"

Figure 14.24 Parasitic element (from [32]).

theoretical model. The elements were fastened to the ground plane with NC 6-32 screws to insure good contact with the ground plane. They were machined to a length tolerance of ±0.001 inch. The evanescent fields inside the rods were estimated from waveguide theory and found to decay rapidly.

    The ground plane was a 60 inch by 60 inch piece of 1/4 inch aluminum jigplate. Although losses in the ground plane were a concern and copper has a higher

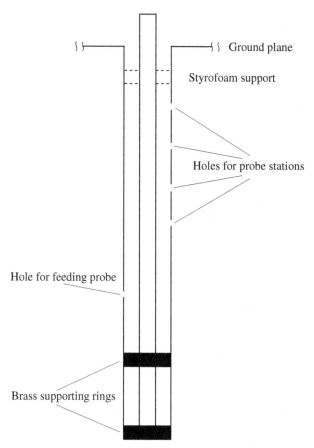

Figure 14.25 The driven element (adapted from [32]). Note that the six possible distances between the probe-station holes are all different.

conductivity than aluminum, the aluminum jigplate is more rigid, easier to machine, much lighter, and its oxidation is easier to remove if necessary. The ground plane containing the array rested on a larger aluminum table of 144 by 93 inches. With the above array parameters, the smallest distance between the edge of the ground plane and the array was 10 inches, or about 2.2 times the average wavelength of interest. The NC 6-32 tapped holes for the parasitic elements in the ground plane had radial tolerances of ± 0.001 inch and angular tolerances of $\pm 0°0'1''$. An additional hole of 0.75 inch equal to the outer diameter of the transmission-line feeder was provided for the driven element. A very precisely made jig that fitted over three adjacent elements was used for the final location of each element. The driving-point conductance at resonance was found to be very sensitive to the exact placement of the driven element.

The driven element was the 0.858-inch extension above the ground plane of the 0.25-inch diameter inner conductor of a coaxial line. The inner diameter of the outer conductor of the line was $2b = 0.687$ inch, resulting in a characteristic impedance of

Figure 14.26 The probe (from [32]).

$Z_c = 60 \ln(2b/2a) = 60$ ohms or $Y_c = 1/Z_c = 17$ millisiemens. For the transverse electromagnetic (TEM) propagation mode, the ohmic losses in the brass walls of the line were on the order of 2 ohms. The inner conductor was supported by two brass shorting plugs located before the feeding stub near the lower end of the line and a styrofoam ring approximately 1 inch from the ground-plane end of the line, near a voltage minimum. The cutoff frequency for the next higher propagating mode was 8 GHz, well above the operating frequency. Four 1/4-inch holes for probe stations were provided in the outer conductor. Their distances from the end of the line were chosen so that the distances between any pair would not be $n\lambda/2$ within the frequency interval of interest, and so that the distance between each of the six possible measurement pairs would be different. The line is shown schematically in Fig. 14.25.

The probe, shown in Fig. 14.26, extended through a block that had a curved inner surface to be flush with the inner surface of the outer conductor of the line. Covers for the three probe stations not in use at any given time were provided with similar curved inner surfaces.

The measured results are presented in Section 12.7, where they are also compared to theoretical results. Generally, there was very good agreement between theory and experiment.

Appendix I

Tables of Ψ_{dR}, $T^{(m)}$ or $T'^{(m)}$ and self- and mutual admittances for single elements and circular arrays

Notation

The self- and mutual admittances are written in the form

$$Y_{1(m+1)} = G_{1(m+1)} + jB_{1(m+1)}$$

with the self-admittance Y_{11} given by the row corresponding to $m = 0$, the first mutual admittance Y_{12} by the row corresponding to $m = 1$, etc. A factor of 10^{-3} has been suppressed in the admittances; hence, tabulated values are in millisiemens. The characteristic impedance of free space, ζ_0, was taken to be $\zeta_0 = 376.730$ ohms.

Table 1. Ψ_{dR}, $T(h)$ or $T'(h)$ and admittances for isolated antenna; $N = 1$, $a/\lambda = 7.022 \times 10^{-3}$

$\beta_0 h$	h/λ	Ψ_{dR}	$T(h)$ Real	or	$T'(h)$ Imag.	G_0	B_0	$B_0 + 0.72$
1.200	0.1910	5.31670	0.27602		−0.74791	4.12999	9.59504	10.315
1.350	0.2150	5.69058	−0.37945		−1.16933	12.28333	9.12585	9.846
1.432	0.2280	5.88844	−0.98304		−0.87529	15.51296	2.93684	3.657
1.501	0.2390	6.05385	−1.13164		−0.36538	13.56922	−2.22898	−1.509
1.570	0.2500	6.21771	2.65166		3.79157	10.17040	−4.43037	−3.710
1.652	0.2630	6.37511	−0.78390		0.20461	7.09591	−4.77136	−4.051
1.796	0.2860	6.54009	−0.51411		0.30471	4.24183	−3.92426	−3.204
1.997	0.3180	6.61380	−0.32658		0.30590	2.63298	−2.72756	−2.008
2.355	0.3750	6.44947	−0.19988		0.26239	1.63816	−1.33809	−0.618
2.751	0.4380	6.05835	−0.16459		0.21601	1.23747	−0.18730	0.533
3.141	0.5000	5.73687	−0.17204		0.17559	1.02096	1.00032	1.720
3.398	0.5410	5.66947	−0.19117		0.15329	0.91729	1.91899	2.639
3.649	0.5810	5.74850	−0.22145		0.14008	0.87181	2.99706	3.717
3.800	0.6050	5.86161	−0.24791		0.14155	0.91248	3.80520	4.525
3.926	0.6250	5.98717	−0.27769		0.15443	1.03854	4.65317	5.373

Table 2. Ψ_{dR}, $T'^{(m)}$ and admittances of circular array[1] of N elements; $a/\lambda = 7.022 \times 10^{-3}$, $h/\lambda = 0.25$, $\beta_0 h = \pi/2$, $\Omega = 8.54$

Sequence				Sequence admittance		Self- and mutual admittances			
m	Ψ_{dR}	Re $T'^{(m)}$	Im $T'^{(m)}$	$G^{(m)}$	$B^{(m)}$	$G_{1(m+1)}$	$B_{1(m+1)}$	d/λ	N
0	6.21771	1.08745	2.72968	7.32201	−0.23458	6.29694	−7.25537	0.1875	2
1	6.21771	6.32221	1.96537	5.27186	−14.27616	1.02507	7.02079	0.1875	2
0	6.21771	0.89928	3.12073	8.37095	0.27016	6.92620	−5.08924	0.2500	2
1	6.21771	4.89530	2.04351	5.48146	−10.44863	1.44475	5.35939	0.2500	2
0	6.21771	0.74401	3.60819	9.67851	0.68667	7.71191	−3.72010	0.3125	2
1	6.21771	4.02973	2.14188	5.74531	−8.12687	1.96660	4.40677	0.3125	2
0	6.21771	0.68981	4.23444	11.35834	0.83205	8.70338	−2.84976	0.3750	2
1	6.21771	3.43500	2.25488	6.04843	−6.53158	2.65496	3.68181	0.3750	2
0	6.21771	0.88064	5.00919	13.43651	0.32017	9.91292	−2.50992	0.4375	2
1	6.21771	2.99078	2.38197	6.38933	−5.34001	3.52359	2.83009	0.4375	2
0	6.21771	1.53690	5.78029	15.50489	−1.44016	11.13910	−2.91835	0.5000	2
1	6.21771	2.63905	2.52512	6.77331	−4.39654	4.36579	1.47819	0.5000	2
0	6.21771	2.69527	6.08303	16.31694	−4.54733	11.76374	−4.08367	0.5625	2
1	6.21771	2.34955	2.68812	7.21053	−3.62001	4.55321	−0.46366	0.5625	2
0	6.21771	3.78922	5.55813	14.90897	−7.48172	11.31236	−5.22543	0.6250	2
1	6.21771	2.10691	2.87646	7.71574	−2.96914	3.59661	−2.25629	0.6250	2
0	6.21771	4.22845	4.62546	12.40720	−8.65991	10.35762	−5.54486	0.6875	2
1	6.21771	1.90584	3.09727	8.30803	−2.42981	2.04958	−3.11505	0.6875	2
0	6.21771	4.12243	3.84925	10.32511	−8.37552	9.66718	−5.19435	0.7500	2
1	6.21771	1.75053	3.35869	9.00926	−2.01319	0.65793	−3.18117	0.7500	2
0	6.21771	0.65962	1.97164	5.28868	0.91303	5.27747	−9.21309	0.1875	3
1	6.21771	6.32221	1.96537	5.27186	−14.27616	0.00561	5.06306	0.1875	3
0	6.21771	0.33433	2.29021	6.14320	1.78558	5.70204	−6.37056	0.2500	3
1	6.21771	4.89530	2.04351	5.48146	−10.44863	0.22058	4.07807	0.2500	3
0	6.21771	−0.08089	2.69241	7.22203	2.89935	6.23755	−4.45146	0.3125	3
1	6.21771	4.02973	2.14188	5.74531	−8.12687	0.49224	3.67541	0.3125	3
0	6.21771	−0.62328	3.27474	8.78406	4.35423	6.96030	−2.90297	0.3750	3
1	6.21771	3.43500	2.25488	6.04843	−6.53158	0.91188	3.62860	0.3750	3
0	6.21771	−1.33434	4.27609	11.47006	6.26158	8.08291	−1.47282	0.4375	3
1	6.21771	2.99078	2.38197	6.38933	−5.34001	1.69358	3.86720	0.4375	3
0	6.21771	−2.06303	6.38546	17.12817	8.21617	10.22493	−0.19230	0.5000	3
1	6.21771	2.63905	2.52512	6.77331	−4.39654	3.45162	4.20424	0.5000	3
0	6.21771	−0.52523	10.97641	29.44281	4.09124	14.62129	−1.04959	0.5625	3
1	6.21771	2.34955	2.68812	7.21053	−3.62001	7.41076	2.57042	0.5625	3
0	6.21771	7.51808	10.33528	27.72306	−17.48391	14.38485	−7.80740	0.6250	3
1	6.21771	2.10691	2.87646	7.71574	−2.96914	6.66911	−4.83826	0.6250	3

[1] Note that $T'^{(m)} = T'^{(m)}(\lambda/4)$.

Table 2. – *continued*

m	Ψ_{dR}	Re $T'^{(m)}$	Im $T'^{(m)}$	$G^{(m)}$	$B^{(m)}$	$G_{1(m+1)}$	$B_{1(m+1)}$	d/λ	N
0	6.21771	7.73778	4.56984	12.25800	−18.07322	9.62469	−7.64428	0.6875	3
1	6.21771	1.90584	3.09727	8.30803	−2.42981	1.31666	−5.21447	0.6875	3
0	6.21771	5.86472	2.79150	7.48783	−13.04899	8.50211	−5.69179	0.7500	3
1	6.21771	1.75053	3.35869	9.00926	−2.01319	−0.50714	−3.67860	0.7500	3
0	6.21771	0.40910	1.61526	4.33274	1.58500	4.23392	−9.94442	0.1875	4
1	6.21771	4.64950	2.06586	5.54141	−9.78933	0.70316	5.84226		
2	6.21771	9.12118	0.56670	1.52011	−21.78403	−1.30749	−0.15509		
0	6.21771	−0.01844	1.89994	5.09635	2.73183	4.80273	−6.56582	0.2500	4
1	6.21771	3.61790	2.21455	5.94025	−7.02217	0.71557	4.42065		
2	6.21771	6.57371	0.83287	2.23406	−14.95077	−1.13752	0.45635		
0	6.21771	−0.63097	2.25790	6.05652	4.37485	5.44649	−4.28694	0.3125	4
1	6.21771	2.96320	2.39158	6.41511	−5.26603	0.78932	3.84135		
2	6.21771	5.09732	1.08085	2.89924	−10.99055	−0.96862	0.97909		
0	6.21771	−1.59127	2.81668	7.55539	6.95074	6.25385	−2.36381	0.3750	4
1	6.21771	2.49201	2.60144	6.97803	−4.00212	1.01286	3.83811		
2	6.21771	4.13220	1.30629	3.50394	−8.40172	−0.72418	1.63832		
0	6.21771	−3.34395	4.05396	10.87422	11.65209	7.56365	−0.23979	0.4375	4
1	6.21771	2.12940	2.85620	7.66138	−3.02948	1.70415	4.55110		
2	6.21771	3.44273	1.51270	4.05763	−6.55232	−0.09773	2.78968		
0	6.21771	−6.93658	9.19894	24.67498	21.28884	11.57104	2.89748	0.5000	4
1	6.21771	1.85172	3.17452	8.51526	−2.28462	5.02408	6.60463		
2	6.21771	2.91237	1.70695	4.57867	−5.12969	3.05578	5.18210		
0	6.21771	12.59562	19.02475	51.03145	−31.10376	18.82977	−9.67092	0.5625	4
1	6.21771	1.67463	3.57861	9.59917	−1.80960	11.48554	−6.78576		
2	6.21771	2.47658	1.89731	5.08929	−3.96072	9.23060	−7.86132		
0	6.21771	9.49153	3.80603	10.20918	−22.77744	9.42201	−7.31992	0.6250	4
1	6.21771	1.66405	4.07540	10.93173	−1.78124	1.14845	−4.95942		
2	6.21771	2.09596	2.09344	5.61538	−2.93977	−1.50973	−5.53868		
0	6.21771	6.13225	2.43284	6.52578	−13.76660	9.33672	−5.19927	0.6875	4
1	6.21771	1.93857	4.59131	12.31559	−2.51759	0.08397	−2.94282		
2	6.21771	1.74386	2.30763	6.18991	−1.99530	−2.97887	−2.68168		
0	6.21771	5.74651	2.36377	6.34051	−12.73189	9.43757	−5.02606	0.7000	4
1	6.21771	2.03672	4.67782	12.54766	−2.78086	0.00652	−2.73032		
2	6.21771	1.67501	2.35405	6.31443	−1.81063	−3.11010	−2.24520		
0	6.21771	4.63831	2.26944	6.08749	−9.75930	9.77373	−4.78684	0.7500	4
1	6.21771	2.54993	4.87384	13.07345	−4.15749	−0.19326	−2.17155		
2	6.21771	1.40005	2.55764	6.86054	−1.07309	−3.29972	−0.62935		
0	6.21771	0.21514	1.40228	3.76143	2.10529	3.54866	−10.26034	0.1875	5
1	6.21771	3.38329	1.96748	5.27752	−6.39288	0.85015	6.20351		
2	6.21771	8.57189	0.63877	1.71342	−20.31061	−0.74377	−0.02069		

Table 2. – *continued*

Sequence				Sequence admittance		Self- and mutual admittances			
m	Ψ_{dR}	Re $T'^{(m)}$	Im $T'^{(m)}$	$G^{(m)}$	$B^{(m)}$	$G_{1(m+1)}$	$B_{1(m+1)}$	d/λ	N
0	6.21771	−0.33422	1.66115	4.45583	3.57888	4.20839	−6.53350	0.2500	5
1	6.21771	2.57552	2.16720	5.81324	−4.22614	0.80724	4.69057		
2	6.21771	6.18088	0.92448	2.47981	−13.89704	−0.68352	0.36562		
0	6.21771	−1.21851	1.99594	5.35384	5.95086	4.92644	−3.95320	0.3125	5
1	6.21771	1.99727	2.40790	6.45887	−2.67505	0.83996	4.15493		
2	6.21771	4.79641	1.18564	3.18031	−10.18338	−0.62626	0.79710		
0	6.21771	−2.92127	2.65004	7.10840	10.51830	5.86009	−1.54167	0.3750	5
1	6.21771	1.50745	2.71553	7.28405	−1.36117	1.08846	4.44406		
2	6.21771	3.89003	1.42112	3.81196	−7.75215	−0.46430	1.58593		
0	6.21771	−4.15869	3.23605	8.68029	13.83752	6.43347	−0.37358	0.4000	5
1	6.21771	1.32071	2.86853	7.69445	−0.86027	1.37683	4.92397		
2	6.21771	3.60681	1.50951	4.04908	−6.99243	−0.25343	2.18158		
0	6.21771	−7.54050	5.70219	15.29540	22.90879	8.19364	2.13276	0.4375	5
1	6.21771	1.04301	3.14776	8.44346	−0.11536	2.68116	6.51145		
2	6.21771	3.23949	1.63771	4.39293	−6.00714	0.86972	3.87657		
0	6.21771	−9.81216	8.71608	23.37976	29.00223	10.01690	3.62199	0.45313	5
1	6.21771	0.92770	3.28782	8.81917	0.19394	4.29908	7.64959		
2	6.21771	3.10265	1.69000	4.53321	−5.64009	2.38234	5.04053		
0	6.21771	−3.91456	26.67988	71.56534	13.18266	20.12878	0.97151	0.48438	5
1	6.21771	0.70033	3.62681	9.72844	0.80383	13.95875	4.34305		
2	6.21771	2.85149	1.79350	4.81084	−4.96639	11.75953	1.76253		
0	6.21771	11.71804	22.64512	60.74263	−28.74975	18.24252	−7.17352	0.5000	5
1	6.21771	0.59141	3.83461	10.28586	1.09600	11.81836	−4.10811		
2	6.21771	2.73538	1.84506	4.94912	−4.65492	9.43170	−6.68000		
0	6.21771	13.94954	12.57851	33.74024	−34.73548	13.15613	−8.14286	0.51563	5
1	6.21771	0.48991	4.07569	10.93251	1.36826	6.45297	−5.36780		
2	6.21771	2.62455	1.89671	5.08769	−4.35766	3.83909	−7.92851		
0	6.21771	8.42858	4.54611	12.19435	−19.92622	10.09068	−4.65915	0.5625	5
1	6.21771	0.31035	5.07731	13.61924	1.84991	2.33913	−2.61274		
2	6.21771	2.31774	2.05426	5.51030	−3.53468	−1.28730	−5.02080		
0	6.21771	5.44546	3.30165	8.85624	−11.92436	11.93591	−3.68524	0.6250	5
1	6.21771	1.26107	7.19498	19.29960	−0.70028	2.17891	−1.64603		
2	6.21771	1.95089	2.27860	6.11205	−2.55063	−3.71874	−2.47353		
0	6.21771	4.22835	3.18268	8.53713	−8.65964	12.14514	−6.63446	0.6875	5
1	6.21771	4.95667	7.19221	19.29217	−10.61325	1.89086	−2.51209		
2	6.21771	1.61255	2.53586	6.80212	−1.64308	−3.69486	1.49950		
0	6.21771	3.96685	3.20896	8.60762	−7.95818	10.98986	−7.20681	0.7100	5
1	6.21771	5.73836	5.99691	16.08594	−12.71004	1.41714	−2.73297		
2	6.21771	1.49504	2.64128	7.08489	−1.32788	−2.60826	2.35729		

Table 2. – *continued*

m	Ψ_{dR}	Re $T'^{(m)}$	Im $T'^{(m)}$	$G^{(m)}$	$B^{(m)}$	$G_{1(m+1)}$	$B_{1(m+1)}$	d/λ	N
				Sequence admittance		Self- and mutual admittances			
0	6.21771	3.64504	3.29276	8.83240	−7.09499	9.21245	−6.73060	0.7500	5
1	6.21771	5.65993	4.08688	10.96254	−12.49968	0.64516	−2.71185		
2	6.21771	1.29053	2.85284	7.65238	−0.77932	−0.83519	2.52965		
0	14.83942	0.76543	1.60211	1.80064	0.26364	1.12981	−44.14067	0.0625	8
1	8.04324	5.28009	1.57354	3.26285	−8.87507	0.87168	24.88961		
2	3.73120	9.62409	0.15395	0.68816	−38.54906	0.01520	−2.53078		
3	4.39219	22.19073	−0.04765	−0.18092	−80.46642	−0.34588	−0.42171		
4	2.56902	16.03502	−0.04660	−0.30252	−97.60802	−0.41117	0.53006		
0	8.48656	0.21045	1.19329	2.34512	1.55166	1.77822	−18.48555	0.1250	8
1	7.82203	2.78655	1.76586	3.76519	−3.80931	0.93309	10.62944		
2	6.21771	7.19363	0.75322	2.02043	−16.61362	−0.21774	−0.81075		
3	4.61340	10.31149	0.04917	0.17777	−33.66265	−0.33526	0.07470		
4	3.94887	10.77021	−0.01094	−0.04619	−41.26493	−0.19327	0.25043		
0	6.21771	−0.30497	1.04952	2.81521	3.50041	2.43903	−10.55638	0.1875	8
1	6.21771	1.51389	1.60947	4.31719	−1.37844	0.97192	6.99193		
2	6.21771	4.52861	1.20601	3.23497	−9.46505	−0.44869	−0.40912		
3	6.21771	8.56067	0.28472	0.76372	−20.28053	−0.28443	0.30903		
4	6.21771	10.58235	0.02433	0.06525	−25.70343	−0.10143	0.27310		
0	6.21771	−1.61604	1.26374	3.38982	7.01720	3.18540	−6.05024	0.2500	8
1	6.21771	0.86391	1.85441	4.97421	0.36504	0.97095	5.48956		
2	6.21771	3.22073	1.58968	4.26410	−5.95683	−0.59988	0.18818		
3	6.21771	5.99746	0.61091	1.63868	−13.40505	−0.20834	0.62110		
4	6.21771	7.49628	0.12651	0.33936	−17.42544	−0.12105	0.46977		
0	6.21771	−3.07274	1.59436	4.27665	10.92459	3.66117	−4.11294	0.28125	8
1	6.21771	0.51587	1.99731	5.35752	1.29862	1.03474	5.39567		
2	6.21771	2.76714	1.76689	4.73947	−4.74013	−0.57855	0.72395		
3	6.21771	5.16270	0.79124	2.12240	−11.16592	−0.10905	0.98879		
4	6.21771	6.44788	0.21396	0.57391	−14.61324	−0.07879	0.82071		
0	6.21771	−6.26287	3.48135	9.33828	19.48172	4.68497	−1.79227	0.3125	8
1	6.21771	0.11521	2.16787	5.81502	2.37333	1.62530	6.06000		
2	6.21771	2.38182	1.94322	5.21243	−3.70656	−0.02668	1.81987		
3	6.21771	4.50764	0.97178	2.60668	−9.40879	0.49098	1.89439		
4	6.21771	5.59887	0.32552	0.87318	−12.33586	0.47411	1.72546		
0	6.21771	−3.84144	14.56160	39.05962	12.98655	8.83417	−1.45440	0.34375	8
1	6.21771	−0.37948	2.39001	6.41089	3.70027	5.31836	4.99896		
2	6.21771	2.03733	2.12667	5.70452	−2.78250	3.60935	1.01128		
3	6.21771	3.98168	1.14798	3.07932	−7.99797	4.14048	0.86301		
4	6.21771	4.90003	0.45641	1.22425	−10.46134	4.08906	0.69445		

Table 2. – *continued*

Sequence				Sequence admittance		Self- and mutual admittances			
m	Ψ_{dR}	$\mathrm{Re}\,T'^{(m)}$	$\mathrm{Im}\,T'^{(m)}$	$G^{(m)}$	$B^{(m)}$	$G_{1(m+1)}$	$B_{1(m+1)}$	d/λ	N
0	6.21771	4.74469	8.23353	22.08539	−10.04466	7.22937	−3.18958	0.3750	8
1	6.21771	−1.03911	2.71812	7.29101	5.46964	3.22335	2.03307		
2	6.21771	1.71403	2.32753	6.24331	−1.91530	1.40099	−1.88938		
3	6.21771	3.54993	1.31820	3.53589	−6.83985	1.89571	−2.31900		
4	6.21771	4.31831	0.59990	1.60915	−8.90094	1.81592	−2.50447		
0	6.21771	3.60511	5.20813	13.97015	−6.98788	6.92141	−1.54358	0.40625	8
1	6.21771	−1.99948	3.29915	8.84954	8.04573	2.35624	2.53477		
2	6.21771	1.39645	2.56053	6.86828	−1.06342	0.28046	−1.55617		
3	6.21771	3.18782	1.48290	3.97768	−5.86853	0.63578	−2.38466		
4	6.21771	3.82897	0.74937	2.01009	−7.58834	0.50779	−2.63218		
0	6.21771	2.61065	4.38490	11.76194	−4.32035	7.89420	0.37131	0.4375	8
1	6.21771	−3.51453	4.63664	12.43719	12.10963	2.58768	3.29992		
2	6.21771	1.07151	2.84911	7.64236	−0.19182	−0.13872	−1.30088		
3	6.21771	2.87791	1.64397	4.40975	−5.03724	−0.25045	−2.76242		
4	6.21771	3.41217	0.89960	2.41305	−6.47035	−0.52927	−3.16489		
0	6.21771	0.97143	2.43161	6.52248	0.07664	1.43500	−12.49961	0.1700	20
1	6.21771	−0.58939	0.96375	2.58513	4.26334	1.04357	8.55105		
2	6.21771	0.64311	1.12504	3.01778	0.95730	0.30332	−0.63750		
3	6.21771	1.72937	1.05916	2.84107	−1.95645	−0.02529	−0.20626		
4	6.21771	3.18240	0.70144	1.88152	−5.85400	0.06437	−0.05895		
5	6.21771	5.16224	0.26348	0.70676	−11.16468	0.19925	−0.12010		
6	6.21771	7.38030	0.04658	0.12494	−17.11434	0.22424	−0.22323		
7	6.21771	9.39761	−0.00104	−0.00279	−22.52551	0.20842	−0.27115		
8	6.21771	10.99347	−0.00840	−0.02253	−26.80619	0.20638	−0.28878		
9	6.21771	12.02388	−0.01052	−0.02823	−29.57014	0.21217	−0.30248		
10	6.21771	12.38084	−0.01120	−0.03004	−30.52763	0.21460	−0.30893		

Table 2. *– continued*

m	Ψ_{dR}	Re $T'^{(m)}$	Im $T'^{(m)}$	$G^{(m)}$	$B^{(m)}$	$G_{1(m+1)}$	$B_{1(m+1)}$	d/λ	N
				Sequence admittance		Self- and mutual admittances			
0	6.21771	−1.02714	1.71496	4.60017	5.43753	2.85913	−5.81890	0.2500	20
1	6.21771	1.00569	3.01550	8.08869	−0.01526	1.73291	5.81862		
2	6.21771	−1.67840	2.02008	5.41861	7.18446	0.20562	−0.80449		
3	6.21771	0.34365	1.59237	4.27133	1.76059	−0.01856	−0.36063		
4	6.21771	1.51934	1.42374	3.81899	−1.39307	0.09549	−0.41945		
5	6.21771	2.71476	1.03755	2.78309	−4.59963	−0.06486	−0.39799		
6	6.21771	4.11322	0.53003	1.42174	−8.35082	−0.20028	−0.11090		
7	6.21771	5.62312	0.16006	0.42935	−12.40092	−0.23897	0.22478		
8	6.21771	6.89325	0.02510	0.06733	−15.80788	−0.25558	0.52470		
9	6.21771	7.69158	−0.00100	−0.00270	−17.94931	−0.25782	0.74258		
10	6.21771	7.96098	−0.00395	−0.01059	−18.67192	−0.25483	0.82200		
0	6.21771	1.05103	3.85805	10.34872	−0.13687	6.02583	−1.62248	0.3400	20
1	6.21771	−2.34603	3.18249	8.53661	8.97530	2.94009	4.10157		
2	6.21771	0.65817	3.07359	8.24451	0.91691	−0.50792	−0.95743		
3	6.21771	−0.87100	7.04184	18.88884	5.01871	−1.01616	−0.19390		
4	6.21771	−0.43591	2.47257	6.63236	3.85164	−0.83124	0.06052		
5	6.21771	1.11157	1.93507	5.19058	−0.29926	0.06403	0.32155		
6	6.21771	2.22382	1.45216	3.89524	−3.28274	1.02681	0.06635		
7	6.21771	3.25400	0.88492	2.37368	−6.04608	1.15323	−0.41290		
8	6.21771	4.24964	0.37783	1.01349	−8.71674	0.41960	−0.78701		
9	6.21771	5.02288	0.10007	0.26843	−10.79085	−0.57409	−0.95892		
10	6.21771	5.31204	0.02995	0.08034	−11.56648	−1.02580	−0.99404		

Table 3. Ψ_{dR}, $T^{(m)}$ and admittances of circular array[2] of N elements; $a/\lambda = 7.022 \times 10^{-3}$, $h/\lambda = 3/8$, $\beta_0 h = 3\pi/4$, $\Omega = 9.34$

	Sequence			Sequence admittance		Self- and mutual admittances			
m	Ψ_{dR}	Re $T^{(m)}$	Im $T^{(m)}$	$G^{(m)}$	$B^{(m)}$	$G_{1(m+1)}$	$B_{1(m+1)}$	d/λ	N
0	6.44947	−0.34647	0.34841	2.17514	−0.42294	1.32680	−1.12131	0.1875	2
1	6.44947	−0.12274	0.07664	0.47845	−1.81968	0.84834	0.69837		
0	6.44947	−0.30504	0.39203	2.44750	−0.68161	1.55035	−1.05823	0.2500	2
1	6.44947	−0.18438	0.10463	0.65319	−1.43485	0.89716	0.37662		
0	6.44947	−0.24661	0.42127	2.63003	−1.04640	1.72902	−1.11269	0.3125	2
1	6.44947	−0.22537	0.13263	0.82801	−1.17899	0.90101	0.06630		
0	6.44947	−0.17613	0.42372	2.64534	−1.48639	1.82429	−1.24458	0.3750	2
1	6.44947	−0.25359	0.16070	1.00324	−1.00277	0.82105	−0.24181		
0	6.44947	−0.11086	0.39101	2.44112	−1.89389	1.81048	−1.38879	0.4375	2
1	6.44947	−0.27267	0.18898	1.17984	−0.88369	0.63064	−0.50510		
0	6.44947	−0.07291	0.33159	2.07016	−2.13080	1.71430	−1.47116	0.5000	2
1	6.44947	−0.28423	0.21759	1.35844	−0.81153	0.35586	−0.65964		
0	6.44947	−0.07068	0.26849	1.67624	−2.14471	1.60735	−1.46392	0.5625	2
1	6.44947	−0.28878	0.24643	1.53846	−0.78313	0.06889	−0.68079		
0	6.44947	−0.09430	0.22053	1.37682	−1.99728	1.54683	−1.39861	0.6250	2
1	6.44947	−0.28608	0.27500	1.71683	−0.79995	−0.17001	−0.59867		
0	6.44947	−0.12872	0.19294	1.20452	−1.78236	1.54531	−1.32429	0.6875	2
1	6.44947	−0.27546	0.30211	1.88610	−0.86622	−0.34079	−0.45807		
0	6.44947	−0.16365	0.18248	1.13923	−1.56431	1.58566	−1.27515	0.7500	2
1	6.44947	−0.25628	0.32549	2.03209	−0.98599	−0.44643	−0.28916		
0	6.44947	−0.47138	0.35091	2.19075	0.35693	1.04922	−1.09414	0.1875	3
1	6.44947	−0.12274	0.07664	0.47845	−1.81968	0.57076	0.72553		
0	6.44947	−0.44657	0.43938	2.74308	0.20198	1.34982	−0.88924	0.2500	3
1	6.44947	−0.18438	0.10463	0.65319	−1.43485	0.69663	0.54561		
0	6.44947	−0.38446	0.54228	3.38553	−0.18574	1.68051	−0.84791	0.3125	3
1	6.44947	−0.22537	0.13263	0.82801	−1.17899	0.85251	0.33108		
0	6.44947	−0.24901	0.63644	3.97336	−1.03141	1.99328	−1.01232	0.3750	3
1	6.44947	−0.25359	0.16070	1.00324	−1.00277	0.99004	−0.00954		
0	6.44947	−0.02896	0.63299	3.95182	−2.40516	2.10384	−1.39084	0.4375	3
1	6.44947	−0.27267	0.18898	1.17984	−0.88369	0.92399	−0.50716		
0	6.44947	0.13984	0.45236	2.82412	−3.45905	1.84700	−1.69403	0.5000	3
1	6.44947	−0.28423	0.21759	1.35844	−0.81153	0.48856	−0.88251		
0	6.44947	0.12450	0.23956	1.49557	−3.36327	1.52417	−1.64317	0.5625	3
1	6.44947	−0.28878	0.24643	1.53846	−0.78313	−0.01430	−0.86005		
0	6.44947	0.02068	0.12744	0.79560	−2.71512	1.40975	−1.43834	0.6250	3
1	6.44947	−0.28608	0.27500	1.71683	−0.79995	−0.30708	−0.63839		

[2] Note that $T^{(m)} = T^{(m)}(3\lambda/8)$.

Table 3. – *continued*

m	Ψ_{dR}	Re $T^{(m)}$	Im $T^{(m)}$	$G^{(m)}$	$B^{(m)}$	$G_{1(m+1)}$	$B_{1(m+1)}$	d/λ	N
				Sequence admittance		Self- and mutual admittances			
0	6.44947	−0.07859	0.09350	0.58373	−2.09536	1.45198	−1.27594	0.6875	3
1	6.44947	−0.27546	0.30211	1.88610	−0.86622	−0.43412	−0.40971		
0	6.44947	−0.15342	0.09624	0.60083	−1.62816	1.55500	−1.20004	0.7500	3
1	6.44947	−0.25628	0.32549	2.03209	−0.98599	−0.47709	−0.21406		
0	6.44947	−0.55329	0.35643	2.22523	0.86829	0.92702	−1.05015	0.1875	4
1	6.44947	−0.19581	0.11142	0.69561	−1.36349	0.53340	0.80255		
2	6.44947	−0.03909	0.01468	0.09163	−2.34192	0.23141	0.31334		
0	6.44947	−0.54954	0.48678	3.03902	0.84487	1.28146	−0.76256	0.2500	4
1	6.44947	−0.24507	0.15105	0.94301	−1.05600	0.70955	0.65699		
2	6.44947	−0.12860	0.03216	0.20081	−1.78309	0.33845	0.29345		
0	6.44947	−0.48519	0.68375	4.26873	0.44314	1.75057	−0.67078	0.3125	4
1	6.44947	−0.27373	0.19101	1.19247	−0.87708	0.98003	0.45381		
2	6.44947	−0.19443	0.05584	0.34860	−1.37211	0.55810	0.20630		
0	6.44947	−0.19693	0.91255	5.69715	−1.35650	2.27841	−0.99994	0.3750	4
1	6.44947	−0.28731	0.23157	1.44574	−0.79228	1.29303	−0.07445		
2	6.44947	−0.24463	0.08409	0.52501	−1.05872	0.83267	−0.20767		
0	6.44947	0.32272	0.70498	4.40129	−4.60077	2.13046	−1.75241	0.4375	4
1	6.44947	−0.28670	0.27216	1.69915	−0.79610	0.91976	−0.94602		
2	6.44947	−0.28340	0.11569	0.72225	−0.81668	0.43131	−0.95631		
0	6.44947	0.29755	0.22178	1.38458	−4.44362	1.54804	−1.71731	0.5000	4
1	6.44947	−0.27039	0.30999	1.93530	−0.89794	0.11189	−0.95347		
2	6.44947	−0.31334	0.15009	0.93700	−0.62975	−0.38725	−0.81938		
0	6.44947	0.07700	0.06927	0.43245	−3.06671	1.45636	−1.44191	0.5625	4
1	6.44947	−0.23702	0.33817	2.11123	−1.10623	−0.18452	−0.64456		
2	6.44947	−0.33597	0.18749	1.17052	−0.48847	−0.65487	−0.33568		
0	6.44947	−0.06893	0.05805	0.36244	−2.15564	1.52608	−1.33380	0.6250	4
1	6.44947	−0.19080	0.34549	2.15694	−1.39480	−0.26640	−0.44142		
2	6.44947	−0.35175	0.22874	1.42802	−0.38996	−0.63085	0.06100		
0	6.44947	−0.15616	0.08162	0.50959	−1.61108	1.56757	−1.32067	0.6875	4
1	6.44947	−0.14737	0.32377	2.02131	−1.66597	−0.30213	−0.31785		
2	6.44947	−0.35980	0.27520	1.71810	−0.33968	−0.45373	0.34529		
0	6.44947	−0.16904	0.08762	0.54702	−1.53067	1.56935	−1.31981	0.7000	4
1	6.44947	−0.14096	0.31630	1.97472	−1.70593	−0.30848	−0.29848		
2	6.44947	−0.36028	0.28527	1.78094	−0.33673	−0.40537	0.38611		
0	6.44947	−0.21007	0.11293	0.70504	−1.27449	1.56698	−1.30361	0.7500	4
1	6.44947	−0.12720	0.28125	1.75587	−1.79184	−0.33652	−0.22955		
2	6.44947	−0.35715	0.32854	2.05113	−0.35628	−0.18889	0.48822		
0	6.44947	−0.61970	0.36751	2.29438	1.28290	0.86350	−0.99648	0.1875	5
1	6.44947	−0.27081	0.14413	0.89982	−0.89527	0.53394	0.86995		
2	6.44947	−0.05584	0.01790	0.11174	−2.23737	0.18150	0.26974		

Table 3. – *continued*

Sequence				Sequence admittance		Self- and mutual admittances			
m	Ψ_{dR}	$\mathrm{Re}\,T^{(m)}$	$\mathrm{Im}\,T^{(m)}$	$G^{(m)}$	$B^{(m)}$	$G_{1(m+1)}$	$B_{1(m+1)}$	d/λ	N
0	6.44947	−0.64486	0.55417	3.45972	1.43995	1.28095	−0.64140	0.2500	5
1	6.44947	−0.31193	0.19746	1.23277	−0.63857	0.76674	0.75430		
2	6.44947	−0.14433	0.03840	0.23975	−1.68489	0.32265	0.28637		
0	6.44947	−0.54344	0.93294	5.82443	0.80675	1.96639	−0.55634	0.3125	5
1	6.44947	−0.33241	0.25558	1.59564	−0.51070	1.23005	0.51358		
2	6.44947	−0.20862	0.06537	0.40812	−1.28353	0.69897	0.16796		
0	6.44947	0.32595	1.11700	6.97358	−4.62090	2.43923	−1.52249	0.3750	5
1	6.44947	−0.33217	0.32142	2.00669	−0.51224	1.44711	−0.66922		
2	6.44947	−0.25667	0.09684	0.60459	−0.98354	0.82007	−0.87999		
0	6.44947	0.57666	0.70317	4.39000	−6.18612	2.02805	−1.81601	0.4000	5
1	6.44947	−0.32429	0.35016	2.18607	−0.56143	0.92523	−1.02005		
2	6.44947	−0.27237	0.11037	0.68904	−0.88555	0.25574	−1.16500		
0	6.44947	0.42187	0.23114	1.44306	−5.21980	1.60258	−1.63129	0.4375	5
1	6.44947	−0.30048	0.39465	2.46383	−0.71005	0.32744	−0.88634		
2	6.44947	−0.29276	0.13152	0.82111	−0.75828	−0.40720	−0.90791		
0	6.44947	0.31785	0.14604	0.91174	−4.57038	1.56449	−1.52095	0.45313	5
1	6.44947	−0.28518	0.41283	2.57737	−0.80556	0.21681	−0.78337		
2	6.44947	−0.30023	0.14063	0.87798	−0.71161	−0.54318	−0.74135		
0	6.44947	0.14751	0.07588	0.47375	−3.50693	1.60478	−1.38079	0.48438	5
1	6.44947	−0.24284	0.44530	2.78003	−1.06989	0.11638	−0.63020		
2	6.44947	−0.31352	0.15938	0.99503	−0.62863	−0.68189	−0.43286		
0	6.44947	0.08327	0.06638	0.41440	−3.10585	1.64758	−1.35482	0.5000	5
1	6.44947	−0.21527	0.45754	2.85645	−1.24202	0.09445	−0.58309		
2	6.44947	−0.31937	0.16903	1.05530	−0.59209	−0.71104	−0.29243		
0	6.44947	0.03044	0.06516	0.40683	−2.77603	1.69002	−1.35474	0.51563	5
1	6.44947	−0.18355	0.46529	2.90484	−1.44006	0.07902	−0.55239		
2	6.44947	−0.32471	0.17888	1.11679	−0.55877	−0.72062	−0.15826		
0	6.44947	−0.07925	0.08411	0.52513	−2.09120	1.74275	−1.45800	0.5625	5
1	6.44947	−0.07441	0.44607	2.78486	−2.12141	0.02551	−0.52578		
2	6.44947	−0.33765	0.20974	1.30944	−0.47798	−0.63432	0.20918		
0	6.44947	−0.15781	0.12414	0.77502	−1.60074	1.56994	−1.56058	0.6250	5
1	6.44947	0.01552	0.31211	1.94857	−2.68287	−0.11828	−0.51643		
2	6.44947	−0.34722	0.25448	1.58877	−0.41822	−0.27918	0.49635		
0	6.44947	−0.19469	0.16288	1.01690	−1.37050	1.40669	−1.40661	0.6875	5
1	6.44947	−0.02918	0.17778	1.10991	−2.40384	−0.27375	−0.43291		
2	6.44947	−0.34575	0.30407	1.89837	−0.42743	0.07886	0.45097		
0	6.44947	−0.20126	0.17513	1.09336	−1.32949	1.40201	−1.32741	0.7100	5
1	6.44947	−0.06173	0.15075	0.94115	−2.20057	−0.31778	−0.39124		
2	6.44947	−0.34162	0.32311	2.01720	−0.45322	0.16345	0.39020		

Table 3. *– continued*

m	Ψ_{dR}	$\mathrm{Re}\,T^{(m)}$	$\mathrm{Im}\,T^{(m)}$	$G^{(m)}$	$B^{(m)}$	$G_{1(m+1)}$	$B_{1(m+1)}$	d/λ	N
				Sequence admittance		Self- and mutual admittances			
0	6.44947	−0.20643	0.19347	1.20785	−1.29723	1.45463	−1.20877	0.7500	5
1	6.44947	−0.12027	0.12776	0.79763	−1.83511	−0.38310	−0.31211		
2	6.44947	−0.32801	0.35800	2.23501	−0.53821	0.25971	0.26788		
0	8.34383	−0.02515	0.40052	1.93278	−1.87751	0.51502	−1.13221	0.1500	20
1	8.25111	−0.83304	0.41564	2.02828	2.04386	0.43561	0.84404		
2	7.98204	−0.53929	0.24478	1.23478	0.63094	0.25643	0.19158		
3	7.56295	−0.37900	0.12488	0.66487	−0.18747	0.09682	−0.09386		
4	7.03486	−0.24489	0.03860	0.22095	−0.96914	0.01625	−0.14187		
5	6.44947	−0.15155	0.00584	0.03649	−1.63983	−0.00505	−0.16493		
6	5.86408	−0.11432	0.00036	0.00248	−2.05918	−0.00889	−0.19133		
7	5.33600	−0.11294	−0.00012	−0.00089	−2.27337	−0.01514	−0.21710		
8	4.91690	−0.12607	−0.00015	−0.00122	−2.35962	−0.02320	−0.23417		
9	4.64783	−0.13954	−0.00015	−0.00131	−2.37952	−0.02871	−0.24255		
10	4.55512	−0.14497	−0.00015	−0.00133	−2.37994	−0.03047	−0.24494		
0	6.44947	0.35836	0.88997	5.55618	−4.82325	1.53249	−0.96196	0.2800	20
1	6.44947	−0.44839	0.45566	2.84473	0.21335	1.00305	0.09476		
2	6.44947	0.09217	0.37118	2.31731	−3.16144	0.17140	−0.59347		
3	6.44947	−0.62997	0.63638	3.97298	1.34698	−0.10521	−0.31997		
4	6.44947	−0.50165	0.29433	1.83756	0.54586	0.01334	−0.02415		
5	6.44947	−0.40107	0.15795	0.98612	−0.08206	0.18889	0.13000		
6	6.44947	−0.31037	0.06941	0.43331	−0.64832	0.30382	0.11621		
7	6.44947	−0.22595	0.02062	0.12870	−1.17535	0.29188	−0.07330		
8	6.44947	−0.16422	0.00380	0.02375	−1.56073	0.16145	−0.34748		
9	6.44947	−0.13077	0.00038	0.00240	−1.76954	0.01048	−0.57893		
10	6.44947	−0.12053	−0.00003	−0.00016	−1.83351	−0.05450	−0.66864		

Table 4. Ψ_{dR}, $T^{(m)}$ and admittances of circular array[3] of N elements; $a/\lambda = 7.022 \times 10^{-3}$, $h/\lambda = 0.5$, $\beta_0 h = \pi$, $\Omega = 9.92$

	Sequence			Sequence admittance		Self- and mutual admittances			
m	Ψ_{dR}	Re $T^{(m)}$	Im $T^{(m)}$	$G^{(m)}$	$B^{(m)}$	$G_{1(m+1)}$	$B_{1(m+1)}$	d/λ	N
0	5.73687	−0.24563	0.26072	1.51595	1.42818	0.90483	1.15817	0.1875	2
1	5.73687	−0.15275	0.05051	0.29370	0.88815	0.61112	0.27001		
0	5.73687	−0.21332	0.28152	1.63686	1.24030	1.02739	1.15671	0.2500	2
1	5.73687	−0.18456	0.07187	0.41791	1.07312	0.60948	0.08359		
0	5.73687	−0.17004	0.28854	1.67767	0.98868	1.11071	1.09623	0.3125	2
1	5.73687	−0.20704	0.09352	0.54375	1.20379	0.56696	−0.10755		
0	5.73687	−0.12443	0.27467	1.59702	0.72347	1.13360	1.00855	0.3750	2
1	5.73687	−0.22249	0.11526	0.67017	1.29364	0.46342	−0.28509		
0	5.73687	−0.09046	0.23989	1.39480	0.52596	1.09581	0.93779	0.4375	2
1	5.73687	−0.23212	0.13704	0.79681	1.34962	0.29900	−0.41183		
0	5.73687	−0.07897	0.19532	1.13564	0.45915	1.02932	0.91689	0.5000	2
1	5.73687	−0.23642	0.15874	0.92300	1.37462	0.10632	−0.45774		
0	5.73687	−0.08921	0.15582	0.90602	0.51869	0.97645	0.94372	0.5625	2
1	5.73687	−0.23541	0.18005	1.04688	1.36875	−0.07043	−0.42503		
0	5.73687	−0.11205	0.12974	0.75438	0.65153	0.95933	0.99088	0.6250	2
1	5.73687	−0.22878	0.20024	1.16429	1.33024	−0.20495	−0.33936		
0	5.73687	−0.13840	0.11752	0.68333	0.80471	0.97538	1.03084	0.6875	2
1	5.73687	−0.21618	0.21798	1.26743	1.25697	−0.29205	−0.22613		
0	5.73687	−0.16285	0.11602	0.67459	0.94690	1.00918	1.04805	0.7500	2
1	5.73687	−0.19765	0.23111	1.34377	1.14920	−0.33459	−0.10115		
0	5.73687	−0.32803	0.29439	1.71168	1.90727	0.76636	1.22786	0.1875	3
1	5.73687	−0.15275	0.05051	0.29370	0.88815	0.47266	0.33971		
0	5.73687	−0.29727	0.35169	2.04486	1.72846	0.96023	1.29157	0.2500	3
1	5.73687	−0.18456	0.07187	0.41791	1.07312	0.54232	0.21845		
0	5.73687	−0.22826	0.40826	2.37378	1.32720	1.15376	1.24493	0.3125	3
1	5.73687	−0.20704	0.09352	0.54375	1.20379	0.61001	0.04114		
0	5.73687	−0.10932	0.42957	2.49770	0.63563	1.27935	1.07430	0.3750	3
1	5.73687	−0.22249	0.11526	0.67017	1.29364	0.60918	−0.21934		
0	5.73687	0.02165	0.35891	2.08686	−0.12586	1.22683	0.85779	0.4375	3
1	5.73687	−0.23212	0.13704	0.79681	1.34962	1.43002	−0.49183		
0	5.73687	0.06652	0.21555	1.25327	−0.38679	1.03309	0.78749	0.5000	3
1	5.73687	−0.23642	0.15874	0.92300	1.37462	0.11009	−0.58714		
0	5.73687	0.01613	0.10388	0.60400	−0.09378	0.89925	0.88124	0.5625	3
1	5.73687	−0.23541	0.18005	1.04688	1.36875	−0.14763	−0.48751		
0	5.73687	−0.05914	0.05632	0.32744	0.34386	0.88534	1.00145	0.6250	3
1	5.73687	−0.22878	0.20024	1.16429	1.33024	−0.27895	−0.32879		
0	5.73687	−0.12272	0.04855	0.28230	0.71352	0.93905	1.07582	0.6875	3
1	5.73687	−0.21618	0.21798	1.26743	1.25697	−0.32838	−0.18115		

[3] Note that $T^{(m)} = T^{(m)}(\lambda/2)$.

Table 4. – *continued*

m	Ψ_{dR}	Re $T^{(m)}$	Im $T^{(m)}$	$G^{(m)}$	$B^{(m)}$	$G_{1(m+1)}$	$B_{1(m+1)}$	d/λ	N
				Sequence admittance		Self- and mutual admittances			
0	5.73687	−0.16955	0.05915	0.34394	0.98581	1.01050	1.09474	0.7500	3
1	5.73687	−0.19765	0.23111	1.34377	1.14920	−0.33328	−0.05446		
0	5.73687	−0.38958	0.32298	1.87795	2.26516	0.70749	1.28318	0.1875	4
1	5.73687	−0.19076	0.07711	0.44835	1.10916	0.45566	0.40398		
2	5.73687	−0.11166	0.00951	0.05532	0.64925	0.25914	0.17403		
0	5.73687	−0.36068	0.42208	2.45413	2.09712	0.95895	1.38535	0.2500	4
1	5.73687	−0.21789	0.10779	0.62675	1.26688	0.58149	0.29665		
2	5.73687	−0.15660	0.02205	0.12818	0.91051	0.33220	0.11847		
0	5.73687	−0.24528	0.54869	3.19030	1.42614	1.25828	1.31478	0.3125	4
1	5.73687	−0.23260	0.13859	0.80581	1.35240	0.73978	0.07449		
2	5.73687	−0.19403	0.03976	0.23119	1.12817	0.45247	−0.03762		
0	5.73687	0.03603	0.57029	3.31589	−0.20949	1.41026	0.96170	0.3750	4
1	5.73687	−0.23660	0.16917	0.98359	1.37571	0.73948	−0.37859		
2	5.73687	−0.22442	0.06156	0.35796	1.30489	0.42667	−0.41400		
0	5.73687	0.21558	0.28252	1.64266	−1.25348	1.11283	0.71545	0.4375	4
1	5.73687	−0.22971	0.19830	1.15297	1.33565	0.28499	−0.67437		
2	5.73687	−0.24835	0.08646	0.50271	1.44399	−0.04014	−0.62020		
0	5.73687	0.09549	0.06553	0.38103	−0.55524	0.90838	0.86177	0.5000	4
1	5.73687	−0.21098	0.22272	1.29499	1.22672	−0.07037	−0.52603		
2	5.73687	−0.26639	0.11394	0.66250	1.54890	−0.38661	−0.36495		
0	5.73687	−0.03885	0.01820	0.10581	0.22588	0.92235	0.98819	0.5625	4
1	5.73687	−0.18107	0.23615	1.37308	1.05281	−0.18291	−0.34885		
2	5.73687	−0.27884	0.14403	0.83744	1.62126	−0.45073	−0.06462		
0	5.73687	−0.12339	0.02822	0.16409	0.71746	0.96987	1.02010	0.6250	4
1	5.73687	−0.14652	0.23091	1.34259	0.85195	−0.21653	−0.23539		
2	5.73687	−0.28533	0.17718	1.03020	1.65902	−0.37272	0.16814		
0	5.73687	−0.17437	0.05243	0.30486	1.01388	0.98500	1.01961	0.6875	4
1	5.73687	−0.12132	0.20555	1.19513	0.70539	−0.23500	−0.15998		
2	5.73687	−0.28443	0.21410	1.24487	1.65379	−0.21013	0.31422		
0	5.73687	−0.18188	0.05774	0.33572	1.05751	0.98483	1.02087	0.7000	4
1	5.73687	−0.11866	0.19889	1.15643	0.68991	−0.23876	−0.14716		
2	5.73687	−0.28312	0.22199	1.29075	1.64615	−0.17160	0.33096		
0	5.73687	−0.20546	0.07919	0.46045	1.19464	0.98512	1.03626	0.7500	4
1	5.73687	−0.11730	0.17165	0.99803	0.68205	−0.25588	−0.09792		
2	5.73687	−0.27283	0.25522	1.48397	1.58631	−0.01291	0.35421		
0	5.73687	−0.44337	0.35405	2.05857	2.57796	0.68373	1.33527	0.1875	5
1	5.73687	−0.23258	0.10524	0.61191	1.35233	0.46530	0.45724		
2	5.73687	−0.11985	0.01172	0.06813	0.69687	0.22212	0.16411		
0	5.73687	−0.41387	0.51736	3.00811	2.40640	1.00601	1.46290	0.2500	5
1	5.73687	−0.25661	0.14733	0.85663	1.49204	0.65756	0.35440		
2	5.73687	−0.16545	0.02654	0.15434	0.96200	0.34349	0.11736		

Table 4. – *continued*

	Sequence			Sequence admittance		Self- and mutual admittances			
m	Ψ_{dR}	$\text{Re}\,T^{(m)}$	$\text{Im}\,T^{(m)}$	$G^{(m)}$	$B^{(m)}$	$G_{1(m+1)}$	$B_{1(m+1)}$	d/λ	N
0	5.73687	−0.14964	0.75703	4.40167	0.87007	1.43714	1.26009	0.3125	5
1	5.73687	−0.26445	0.19253	1.11942	1.53760	0.93049	−0.01700		
2	5.73687	−0.20253	0.04688	0.27258	1.17759	0.55178	−0.17801		
0	5.73687	0.35041	0.42661	2.48049	−2.03744	1.22362	0.72145	0.3750	5
1	5.73687	−0.25376	0.24152	1.40429	1.47545	0.53554	−0.66098		
2	5.73687	−0.23165	0.07129	0.41452	1.34690	0.09290	−0.71847		
0	5.73687	0.30621	0.20277	1.17897	−1.78041	1.03469	0.76885	0.4000	5
1	5.73687	−0.24246	0.26157	1.52085	1.40974	0.26962	−0.63572		
2	5.73687	−0.24123	0.08193	0.47638	1.40260	−0.19748	−0.63891		
0	5.73687	0.14471	0.04886	0.28409	−0.84143	0.96034	0.92211	0.4375	5
1	5.73687	−0.21531	0.28980	1.68502	1.25190	0.07942	−0.49057		
2	5.73687	−0.25353	0.09868	0.57378	1.47410	−0.41754	−0.39120		
0	5.73687	0.08545	0.02679	0.15579	−0.49686	0.97479	0.96515	0.45313	5
1	5.73687	−0.19977	0.29980	1.74317	1.16155	0.04731	−0.44113		
2	5.73687	−0.25794	0.10593	0.61591	1.49975	−0.45681	−0.28988		
0	5.73687	−0.00645	0.01543	0.08970	0.03748	1.02753	0.99888	0.48438	5
1	5.73687	−0.16073	0.31321	1.82113	0.93455	0.01560	−0.37660		
2	5.73687	−0.26553	0.12088	0.70284	1.54390	−0.48451	−0.10409		
0	5.73687	−0.04103	0.01824	0.10608	0.23859	1.05256	0.99281	0.5000	5
1	5.73687	−0.13765	0.31486	1.83072	0.80033	0.00556	−0.35896		
2	5.73687	−0.26871	0.12859	0.74764	1.56240	−0.47880	−0.01815		
0	5.73687	−0.06970	0.02385	0.13868	0.40525	1.07040	0.97532	0.51563	5
1	5.73687	−0.11303	0.31186	1.81328	0.65720	−0.00487	−0.34852		
2	5.73687	−0.27148	0.13645	0.79337	1.57848	−0.46099	0.06349		
0	5.73687	−0.12986	0.04795	0.27882	0.75507	1.05588	0.89736	0.5625	5
1	5.73687	−0.04373	0.26898	1.56397	0.25424	−0.05392	−0.33909		
2	5.73687	−0.27718	0.16104	0.93633	1.61162	−0.33461	0.26795		
0	5.73687	−0.17222	0.08301	0.48267	1.00134	0.92501	0.89242	0.6250	5
1	5.73687	−0.01972	0.15994	0.92995	0.11463	−0.15783	−0.30843		
2	5.73687	−0.27789	0.19628	1.14124	1.61574	−0.06334	0.36289		
0	5.73687	−0.18928	0.11386	0.66204	1.10052	0.87378	1.02151	0.6875	5
1	5.73687	−0.07632	0.08466	0.49228	0.44377	−0.24722	−0.22979		
2	5.73687	−0.26826	0.23410	1.36115	1.55976	0.14135	0.26930		
0	5.73687	−0.19114	0.12315	0.71602	1.11135	0.89069	1.06776	0.7100	5
1	5.73687	−0.10210	0.07336	0.42653	0.59366	−0.27077	−0.19626		
2	5.73687	−0.26143	0.24804	1.44218	1.52007	0.18344	0.21805		
0	5.73687	−0.19034	0.13638	0.79297	1.10669	0.95087	1.12366	0.7500	5
1	5.73687	−0.14431	0.06840	0.39769	0.83910	−0.30452	−0.13340		
2	5.73687	−0.24366	0.27226	1.58301	1.41671	0.22557	0.12492		

Appendix II
Tables of matrix elements Φ_u and Φ_v for curtain arrays

This appendix presents a set of five tables of (1) Ψ_{dR}, Φ_u and Φ_v for single elements as functions of Ω and h/λ, and (2) Φ_u and Φ_v for off-diagonal elements as functions of $(k - i)(b/\lambda)$ and h/λ. The left column, i.e. $(k - i)(b/\lambda)$, is the electrical distance between elements k and i. The two columns under each Φ_{kiu} and Φ_{kiv} are the real and imaginary parts.

Table 1. *Elements of Φ_u and Φ_v matrices for $h/\lambda = 0.125$*

	Values for single element (Φ_{kku}, Φ_{kkv} and $\Psi_{kkdR} = \Psi_{dR}$)				
Ω	Ψ_{dR}	Φ_{kku}		Φ_{kkv}	
7.00	4.05418	2.64345	0.19772	0.78608	−0.38223
7.50	4.51933	2.97163	0.19776	0.79628	−0.38230
8.00	4.99202	3.30540	0.19778	0.80428	−0.38234
8.50	5.47065	3.64353	0.19779	0.81054	−0.38237
9.00	5.95394	3.98508	0.19780	0.81544	−0.38238
9.50	6.44089	4.32928	0.19781	0.81926	−0.38239
10.00	6.93071	4.67555	0.19781	0.82224	−0.38240
10.50	7.42276	5.02343	0.19781	0.82457	−0.38240
11.00	7.91657	5.37257	0.19781	0.82638	−0.38240
11.50	8.41174	5.72268	0.19781	0.82780	−0.38240
12.00	8.90797	6.07356	0.19781	0.82890	−0.38240
12.50	9.40504	6.42503	0.19781	0.82976	−0.38240
13.00	9.90275	6.77696	0.19781	0.83043	−0.38241
15.00	11.89766	8.18757	0.19781	0.83192	−0.38241

	Off diagonal values of Φ_u and Φ_v			
$(k-i)(b/\lambda)$	Φ_{kiu}		Φ_{kiv}	
0.250	0.09815	0.11183	−0.19198	−0.21626
0.500	0.08361	−0.03104	−0.16178	0.05987
0.750	−0.01455	−0.06033	0.02796	0.11660
1.000	−0.04655	0.00834	0.08992	−0.01602
1.250	0.00539	0.03774	−0.01034	−0.07287
1.500	0.03167	−0.00376	−0.06116	0.00722
1.750	−0.00277	−0.02727	0.00532	0.05265
2.000	−0.02393	0.00213	0.04619	−0.00408
2.250	0.00168	0.02131	−0.00323	−0.04114
2.500	0.01921	−0.00136	−0.03708	0.00262
2.750	−0.00113	−0.01748	0.00216	0.03374
3.000	−0.01603	0.00095	0.03095	−0.00182
3.250	0.00081	0.01481	−0.00155	−0.02859
3.500	0.01376	−0.00070	−0.02656	0.00134
3.750	−0.00061	−0.01285	0.00117	0.02480
4.000	−0.01205	0.00053	0.02325	−0.00102
4.250	0.00047	0.01134	−0.00091	−0.02189
4.500	0.01071	−0.00042	−0.02068	0.00081
4.750	−0.00038	−0.01015	0.00073	0.01960
5.000	−0.00965	0.00034	0.01862	−0.00066
5.250	0.00031	0.00919	−0.00059	−0.01773
5.500	0.00877	−0.00028	−0.01693	0.00054
5.750	−0.00026	−0.00839	0.00050	0.01620
6.000	−0.00804	0.00024	0.01552	−0.00046
6.250	0.00022	0.00772	−0.00042	−0.01490

Table 1. – *continued*

$(k-i)(b/\lambda)$	Φ_{kiu}		Φ_{kiv}	
6.500	0.00742	−0.00020	−0.01433	0.00039
6.750	−0.00019	−0.00715	0.00036	0.01380
7.000	−0.00690	0.00017	0.01331	−0.00033
7.250	0.00016	0.00666	−0.00031	−0.01285
7.500	0.00644	−0.00015	−0.01242	0.00029
7.750	−0.00014	−0.00623	0.00027	0.01202
8.000	−0.00603	0.00013	0.01165	−0.00026
8.250	0.00013	0.00585	−0.00024	−0.01129
8.500	0.00568	−0.00012	−0.01096	0.00023
8.750	−0.00011	−0.00552	0.00021	0.01065
9.000	−0.00536	0.00011	0.01035	−0.00020
9.250	0.00010	0.00522	−0.00019	−0.01007
9.500	0.00508	−0.00009	−0.00981	0.00018
9.750	−0.00009	−0.00495	0.00017	0.00956
10.000	−0.00483	0.00009	0.00932	−0.00016
10.250	0.00008	0.00471	−0.00016	−0.00909
10.500	0.00460	−0.00008	−0.00888	0.00015
10.750	−0.00007	−0.00449	0.00014	0.00867
11.000	−0.00439	0.00007	0.00847	−0.00014
11.250	0.00007	0.00429	−0.00013	−0.00828
11.500	0.00420	−0.00006	−0.00810	0.00012
11.750	−0.00006	−0.00411	0.00012	0.00793
12.000	−0.00402	0.00006	0.00777	−0.00011
12.250	0.00006	0.00394	−0.00011	−0.00761
12.500	0.00386	−0.00005	−0.00746	0.00011
12.750	−0.00005	−0.00379	0.00010	0.00731
13.000	−0.00371	0.00005	0.00717	−0.00010
13.250	0.00005	0.00364	−0.00009	−0.00703
13.500	0.00358	−0.00005	−0.00690	0.00009
13.750	−0.00005	−0.00351	0.00009	0.00678
14.000	−0.00345	0.00004	0.00606	−0.00008
14.250	0.00004	0.00339	−0.00008	−0.00654
14.500	0.00333	−0.00004	−0.00643	0.00008
14.750	−0.00004	−0.00327	0.00008	0.00632
15.000	−0.00322	0.00004	0.00621	−0.00007
15.250	0.00004	0.00317	−0.00007	−0.00611
15.500	0.00312	−0.00004	−0.00601	0.00007
15.750	−0.00003	−0.00307	0.00007	0.00592
16.000	−0.00302	0.00003	0.00583	−0.00006
16.250	0.00003	0.00297	−0.00006	−0.00574
16.500	0.00293	−0.00003	−0.00565	0.00006
16.750	−0.00003	−0.00288	0.00006	0.00557
17.000	−0.00284	0.00003	0.00548	−0.00006

Table 1. – *continued*

$(k-i)(b/\lambda)$	Φ_{kiu}		Φ_{kiv}	
17.250	0.00003	0.00280	−0.00006	−0.00540
17.500	0.00276	−0.00003	−0.00533	0.00005
17.750	−0.00003	−0.00272	0.00005	0.00525
18.000	−0.00268	0.00003	0.00518	−0.00005
18.250	0.00003	0.00265	−0.00005	−0.00511
18.500	0.00261	−0.00003	−0.00504	0.00005
18.750	−0.00002	−0.00258	0.00005	0.00497
19.000	−0.00254	0.00002	0.00491	−0.00005
19.250	0.00002	0.00251	−0.00004	−0.00484
19.500	0.00248	−0.00002	−0.00478	0.00004
19.750	−0.00002	−0.00245	0.00004	0.00472
20.000	−0.00241	0.00002	0.00466	−0.00004
20.250	0.00002	0.00239	−0.00004	−0.00460
20.500	0.00236	−0.00002	−0.00455	0.00004
20.750	−0.00002	−0.00233	0.00004	0.00449
21.000	−0.00230	0.00002	0.00444	−0.00004
21.250	0.00002	0.00227	−0.00004	−0.00439
21.500	0.00225	−0.00002	−0.00434	0.00004
21.750	−0.00002	−0.00222	0.00003	0.00429
22.000	−0.00220	0.00002	0.00424	−0.00003
22.250	0.00002	0.00217	−0.00003	−0.00419
22.500	0.00215	−0.00002	−0.00414	0.00003
22.750	−0.00002	−0.00212	0.00003	0.00410
23.000	−0.00210	0.00002	0.00405	−0.00003
23.250	0.00002	0.00208	−0.00003	−0.00401
23.500	0.00206	−0.00002	−0.00397	0.00003
23.750	−0.00002	−0.00203	0.00003	0.00393
24.000	−0.00201	0.00001	0.00388	−0.00003
24.250	0.00001	0.00199	−0.00003	−0.00384
24.500	0.00197	−0.00001	−0.00381	0.00003
24.750	−0.00001	−0.00195	0.00003	0.00377
25.000	−0.00193	0.00001	0.00373	−0.00003
25.250	0.00001	0.00191	−0.00003	−0.00369
25.500	0.00189	−0.00001	−0.00366	0.00003
25.750	−0.00001	−0.00188	0.00002	0.00362
26.000	−0.00186	0.00001	0.00359	−0.00002
26.250	0.00001	0.00184	−0.00002	−0.00355
26.500	0.00182	−0.00001	−0.00352	0.00002
26.750	−0.00001	−0.00181	0.00002	0.00349
27.000	−0.00179	0.00001	0.00345	−0.00002
27.250	0.00001	0.00177	−0.00002	−0.00342
27.500	0.00176	−0.00001	−0.00339	0.00002
27.750	−0.00001	−0.00174	0.00002	0.00336

Table 1. – *continued*

$(k-i)(b/\lambda)$	Φ_{kiu}		Φ_{kiv}	
28.000	−0.00173	0.00001	0.00333	−0.00002
28.250	0.00001	0.00171	−0.00002	−0.00330
28.500	0.00169	−0.00001	−0.00327	0.00002
28.750	−0.00001	−0.00168	0.00002	0.00324
29.000	−0.00167	0.00001	0.00321	−0.00002
29.250	0.00001	0.00165	−0.00002	−0.00319
29.500	0.00164	−0.00001	−0.00316	0.00002
29.750	−0.00001	−0.00162	0.00002	0.00313
30.000	−0.00161	0.00001	0.00311	−0.00002
30.250	0.00001	0.00160	−0.00002	−0.00308
30.500	0.00158	−0.00001	−0.00306	0.00002
30.750	−0.00001	−0.00157	0.00002	0.00303
31.000	−0.00156	0.00001	0.00301	−0.00002
31.250	0.00001	0.00155	−0.00002	−0.00298
31.500	0.00153	−0.00001	−0.00296	0.00002
31.750	−0.00001	−0.00152	0.00002	0.00294
32.000	−0.00151	0.00001	0.00291	−0.00002
32.250	0.00001	0.00150	−0.00002	−0.00289
32.500	0.00149	−0.00001	−0.00287	0.00002
32.750	−0.00001	−0.00147	0.00002	0.00285
33.000	−0.00146	0.00001	0.00283	−0.00002
33.250	0.00001	0.00145	−0.00001	−0.00280
33.500	0.00144	−0.00001	−0.00278	0.00001
33.750	−0.00001	−0.00143	0.00001	0.00276
34.000	−0.00142	0.00001	0.00274	−0.00001
34.250	0.00001	0.00141	−0.00001	−0.00272
34.500	0.00140	−0.00001	−0.00270	0.00001
34.750	−0.00001	−0.00139	0.00001	0.00268
35.000	−0.00138	0.00001	0.00266	−0.00001
35.250	0.00001	0.00137	−0.00001	−0.00264
35.500	0.00136	−0.00001	−0.00263	0.00001
35.750	−0.00001	−0.00135	0.00001	0.00261
36.000	−0.00134	0.00001	0.00259	−0.00001
36.250	0.00001	0.00133	−0.00001	−0.00257
36.500	0.00132	−0.00001	−0.00255	0.00001
36.750	−0.00001	−0.00131	0.00001	0.00254
37.000	−0.00131	0.00001	0.00252	−0.00001
37.250	0.00001	0.00130	−0.00001	−0.00250
37.500	0.00129	−0.00001	−0.00249	0.00001
37.750	−0.00001	−0.00128	0.00001	0.00247
38.000	−0.00127	0.00001	0.00245	−0.00001
38.250	0.00001	0.00126	−0.00001	−0.00244
38.500	0.00125	−0.00001	−0.00242	0.00001
38.750	−0.00001	−0.00125	0.00001	0.00241
39.000	−0.00124	0.00001	0.00239	−0.00001

Table 1. – *continued*

$(k-i)(b/\lambda)$	Φ_{kiu}		Φ_{kiv}	
39.250	0.00001	0.00123	−0.00001	−0.00238
39.500	0.00122	−0.00001	−0.00236	0.00001
39.750	−0.00001	−0.00122	0.00001	0.00235
40.000	−0.00121	0.00001	0.00233	−0.00001
40.250	0.00001	0.00120	−0.00001	−0.00232
40.500	0.00119	−0.00001	−0.00230	0.00001
40.750	−0.00001	−0.00119	0.00001	0.00229
41.000	−0.00118	0.00001	0.00227	−0.00001
41.250	0.00001	0.00117	−0.00001	−0.00226
41.500	0.00116	−0.00000	−0.00225	0.00001
41.750	−0.00000	−0.00116	0.00001	0.00223
42.000	−0.00115	0.00000	0.00222	−0.00001
42.250	0.00000	0.00114	−0.00001	−0.00221
42.500	0.00114	−0.00000	−0.00219	0.00001
42.750	−0.00000	−0.00113	0.00001	0.00218
43.000	−0.00112	0.00000	0.00217	−0.00001
43.250	0.00000	0.00112	−0.00001	−0.00216
43.500	0.00111	−0.00000	−0.00214	0.00001
43.750	−0.00000	−0.00110	0.00001	0.00213
44.000	−0.00110	0.00000	0.00212	−0.00001
44.250	0.00000	0.00109	−0.00001	−0.00211
44.500	0.00109	−0.00000	−0.00210	0.00001
44.750	−0.00000	−0.00108	0.00001	0.00208
45.000	−0.00107	0.00000	0.00207	−0.00001
45.250	0.00000	0.00107	−0.00001	−0.00206
45.500	0.00106	−0.00000	−0.00205	0.00001
45.750	−0.00000	−0.00106	0.00001	0.00204
46.000	−0.00105	0.00000	0.00203	−0.00001
46.250	0.00000	0.00104	−0.00001	−0.00202
46.500	0.00104	−0.00000	−0.00200	0.00001
46.750	−0.00000	−0.00103	0.00001	0.00199
47.000	−0.00103	0.00000	0.00198	−0.00001
47.250	0.00000	0.00102	−0.00001	−0.00197
47.500	0.00102	−0.00000	−0.00196	0.00001
47.750	−0.00000	−0.00101	0.00001	0.00195
48.000	−0.00101	0.00000	0.00194	−0.00001
48.250	0.00000	0.00100	−0.00001	−0.00193
48.500	0.00100	−0.00000	−0.00192	0.00001
48.750	−0.00000	−0.00099	0.00001	0.00191
49.000	−0.00099	0.00000	0.00190	−0.00001
49.250	0.00000	0.00098	−0.00001	−0.00189
49.500	0.00098	−0.00000	−0.00188	0.00001
49.750	−0.00000	−0.00097	0.00001	0.00187
50.000	−0.00097	0.00000	0.00186	−0.00001

Table 2. *Elements of* Φ_u *and* Φ_v *matrices for* $h/\lambda = 0.250$

Ω	Ψ_{dR}	Φ_{kku}		Φ_{kkv}	
		Values for single element (Φ_{kku}, Φ_{kkv} and $\Psi_{kkdR} = \Psi_{dR}$)			
7.00	4.73675	0.61497	−1.21658	4.73675	−0.00000
7.50	5.21607	0.63565	−1.21746	5.21607	−0.00000
8.00	5.69991	0.65181	−1.21800	5.69991	−0.00000
8.50	6.18729	0.66443	−1.21832	6.18729	−0.00000
9.00	6.67744	0.67427	−1.21852	6.67744	−0.00000
9.50	7.16976	0.68196	−1.21864	7.16976	−0.00000
10.00	7.66377	0.68794	−1.21871	7.66377	−0.00000
10.50	8.15911	0.69261	−1.21876	8.15911	−0.00000
11.00	8.65547	0.69625	−1.21878	8.65547	−0.00000
11.50	9.15263	0.69908	−1.21880	9.15263	−0.00000
12.00	9.65042	0.70129	−1.21881	9.65042	−0.00000
12.50	10.14870	0.70301	−1.21882	10.14870	−0.00000
13.00	10.64736	0.70435	−1.21882	10.64736	−0.00000
15.00	12.64438	0.70734	−1.21883	12.64438	−0.00000

$(k-i)(b/\lambda)$	Φ_{kiu}		Φ_{kiv}	
	Off diagonal values of Φ_u and Φ_v			
0.250	−0.47248	−0.67976	0.00000	−0.00000
0.500	−0.49881	0.20887	−0.00000	−0.00000
0.750	0.11054	0.37495	−0.00000	0.00000
1.000	0.29570	−0.06686	0.00000	−0.00000
1.250	−0.04437	−0.24260	0.00000	−0.00000
1.500	−0.20507	0.03146	−0.00000	0.00000
1.750	0.02341	0.17734	−0.00000	0.00000
2.000	0.15607	−0.01807	0.00000	−0.00000
2.250	−0.01436	−0.13929	0.00000	−0.00000
2.500	−0.12573	0.01168	−0.00000	−0.00000
2.750	0.00968	0.11455	−0.00000	0.00000
3.000	0.10517	−0.00816	0.00000	−0.00000
3.250	−0.00696	−0.09721	0.00000	−0.00000
3.500	−0.09036	0.00601	−0.00000	−0.00000
3.750	0.00524	0.08440	−0.00000	0.00000
4.000	0.07918	−0.00461	0.00000	−0.00000
4.250	−0.00409	−0.07457	−0.00000	−0.00000
4.500	−0.07046	0.00365	−0.00000	−0.00000
4.750	0.00328	0.06678	−0.00000	0.00000
5.000	0.06346	−0.00296	0.00000	0.00000
5.250	−0.00269	−0.06046	0.00000	−0.00000
5.500	−0.05772	0.00245	−0.00000	−0.00000
5.750	0.00224	0.05522	0.00000	0.00000
6.000	0.05293	−0.00206	0.00000	−0.00000
6.250	−0.00190	−0.05083	0.00000	−0.00000

Table 2. *– continued*

$(k-i)(b/\lambda)$	Φ_{kiu}		Φ_{kiv}	
6.500	−0.04888	0.00175	−0.00000	−0 00000
6.750	0.00163	0.04707	−0.00000	0.00000
7.000	0.04540	−0.00151	0.00000	0.00000
7.250	−0.00141	−0.04384	0.00000	−0.00000
7.500	−0.04238	0.00132	−0.00000	−0.00000
7.750	0.00124	0.04102	−0.00000	0.00000
8.000	0.03974	−0.00116	0.00000	−0.00000
8.250	−0.00109	−0.03854	−0.00000	−0.00000
8.500	−0.03741	0.00103	−0.00000	0.00000
8.750	0.00097	0.03634	−0.00000	0.00000
9.000	0.03533	−0.00092	0.00000	0.00000
9.250	−0.00087	−0.03438	0.00000	−0.00000
9.500	−0.03348	0.00082	−0.00000	−0.00000
9.750	0.00078	0.03262	−0.00000	0.00000
10.000	0.03181	−0.00074	0.00000	0.00000
10.250	−0.00071	−0.03103	0.00000	−0.00000
10.500	−0.03029	0.00067	−0.00000	−0.00000
10.750	0.00064	0.02959	−0.00000	0.00000
11.000	0.02892	−0.00061	0.00000	0.00000
11.250	−0.00059	−0.02828	0.00000	−0.00000
11.500	−0.02766	0.00056	−0.00000	0.00000
11.750	0.00054	0.02707	0.00000	0.00000
12.000	0.02651	−0.00052	0.00000	−0.00000
12.250	−0.00049	−0.02597	−0.00000	−0.00000
12.500	−0.02545	0.00048	−0.00000	−0.00000
12.750	0.00046	0.02495	−0.00000	0.00000
13.000	0.02447	−0.00044	0.00000	0.00000
13.250	−0.00042	−0.02401	0.00000	−0.00000
13.500	−0.02357	0.00041	−0.00000	−0.00000
13.750	0.00039	0.02314	−0.00000	0.00000
14.000	0.02273	−0.00038	0.00000	0.00000
14.250	−0.00037	−0.02233	0.00000	−0.00000
14.500	−0.02194	0.00035	−0.00000	−0.00000
14.750	0.00034	0.02157	−0.00000	0.00000
15.000	0.02121	−0.00033	0.00000	0.00000
15.250	−0.00032	−0.02087	0.00000	−0.00000
15.500	−0.02053	0.00031	−0.00000	−0.00000
15.750	0.00030	0.02020	−0.00000	0.00000
16.000	0.01989	−0.00029	0.00000	−0.00000
16.250	−0.00028	−0.01958	−0.00000	−0.00000
16.500	−0.01929	0.00027	−0.00000	0.00000
16.750	0.00026	0.01900	0.00000	0.00000
17.000	0.01872	−0.00026	0.00000	−0.00000
17.250	−0.00025	−0.01845	−0.00000	−0.00000

Table 2. *−continued*

$(k-i)(b/\lambda)$	Φ_{kiu}		Φ_{kiv}	
17.500	−0.01818	0.00024	−0.00000	−0.00000
17.750	0.00024	0.01793	−0.00000	0.00000
18.000	0.01768	−0.00023	0.00000	0.00000
18.250	−0.00022	−0.01744	0.00000	−0.00000
18.500	−0.01720	0.00022	−0.00000	−0.00000
18.750	0.00021	0.01697	−0.00000	0.00000
19.000	0.01675	−0.00021	0.00000	0.00000
19.250	−0.00020	−0.01653	0.00000	−0.00000
19.500	−0.01632	0.00020	−0.00000	−0.00000
19.750	0.00019	0.01611	−0.00000	0.00000
20.000	0.01591	−0.00019	0.00000	0.00000
20.250	−0.00018	−0.01572	0.00000	−0.00000
20.500	−0.01552	0.00018	−0.00000	−0.00000
20.750	0.00017	0.01534	−0.00000	0.00000
21.000	0.01515	−0.00017	0.00000	0.00000
21.250	−0.00016	−0.01498	0.00000	−0.00000
21.500	−0.01480	0.00016	−0.00000	−0.00000
21.750	0.00016	0.01463	−0.00000	0.00000
22.000	0.01447	−0.00015	0.00000	0.00000
22.250	−0.00015	−0.01430	0.00000	−0.00000
22.500	−0.01414	0.00015	−0.00000	−0.00000
22.750	0.00014	0.01399	0.00000	0.00000
23.000	0.01384	−0.00014	0.00000	−0.00000
23.250	−0.00014	−0.01369	−0.00000	−0.00000
23.500	−0.01354	0.00013	−0.00000	0.00000
23.750	0.00013	0.01340	0.00000	0.00000
24.000	0.01326	−0.00013	0.00000	−0.00000
24.250	−0.00013	−0.01312	−0.00000	−0.00000
24.500	−0.01299	0.00012	−0.00000	0.00000
24.750	0.00012	0.01286	0.00000	0.00000
25.000	0.01273	−0.00012	0.00000	0.00000
25.250	−0.00012	−0.01260	0.00000	−0.00000
25.500	−0.01248	0.00011	−0.00000	−0.00000
25.750	0.00011	0.01236	−0.00000	0.00000
26.000	0.01224	−0.00011	0.00000	0.00000
26.250	−0.00011	−0.01212	0.00000	−0.00000
26.500	−0.01201	0.00011	−0.00000	−0.00000
26.750	0.00010	0.01190	−0.00000	0.00000
27.000	0.01179	−0.00010	0.00000	0.00000
27.250	−0.00010	−0.01168	0.00000	−0.00000
27.500	−0.01157	0.00010	−0.00000	−0.00000
27.750	0.00010	0.01147	−0.00000	0.00000
28.000	0.01137	−0.00009	0.00000	0.00000
28.250	−0.00009	−0.01127	0.00000	−0.00000

Table 2. – *continued*   •

$(k-i)(b/\lambda)$	Φ_{kiu}		Φ_{kiv}	
28.500	−0.01117	0.00009	−0.00000	−0.00000
28.750	0.00009	0.01107	−0.00000	0.00000
29.000	0.01098	−0.00009	0.00000	0.00000
29.250	−0.00009	−0.01088	0.00000	−0.00000
29.500	−0.01079	0.00009	−0.00000	−0.00000
29.750	0.00008	0.01070	−0.00000	0.00000
30.000	0.01061	−0.00008	0.00000	0.00000
30.250	−0.00008	−0.01052	0.00000	−0.00000
30.500	−0.01044	0.00008	−0.00000	−0.00000
30.750	0.00008	0.01035	−0.00000	0.00000
31.000	0.01027	−0.00008	0.00000	0.00000
31.250	−0.00008	−0.01019	0.00000	−0.00000
31.500	−0.01010	0.00007	−0.00000	−0.00000
31.750	0.00007	0.01002	−0.00000	0.00000
32.000	0.00995	−0.00007	0.00000	−0.00000
32.250	−0.00007	−0.00987	−0.00000	−0.00000
32.500	−0.00979	0.00007	−0.00000	0.00000
32.750	0.00007	0.00972	0.00000	0.00000
33.000	0.00965	−0.00007	0.00000	−0.00000
33.250	−0.00007	−0.00957	−0.00000	−0.00000
33.500	−0.00950	0.00007	−0.00000	0.00000
33.750	0.00007	0.00943	0.00000	0.00000
34.000	0.00936	−0.00006	0.00000	−0.00000
34.250	−0.00006	−0.00929	−0.00000	−0.00000
34.500	−0.00923	0.00006	−0.00000	0.00000
34.750	0.00006	0.00916	0.00000	0.00000
35.000	0.00909	−0.00006	0.00000	0.00000
35.250	−0.00006	−0.00903	0.00000	−0.00000
35.500	−0.00897	0.00006	−0.00000	−0.00000
35.750	0.00006	0.00890	−0.00000	0.00000
36.000	0.00884	−0.00006	0.00000	0.00000
36.250	−0.00006	−0.00878	0.00000	−0.00000
36.500	−0.00872	0.00006	−0.00000	−0.00000
36.750	0.00006	0.00866	−0.00000	0.00000
37.000	0.00860	−0.00005	0.00000	0.00000
37.250	−0.00005	−0.00854	0.00000	−0.00000
37.500	−0.00849	0.00005	−0.00000	−0.00000
37.750	0.00005	0.00843	−0.00000	0.00000
38.000	0.00838	−0.00005	0.00000	0.00000
38.250	−0.00005	−0.00832	0.00000	−0.00000
38.500	−0.00827	0.00005	−0.00000	−0.00000
38.750	0.00005	0.00821	−0.00000	0.00000
39.000	0.00816	−0.00005	0.00000	0.00000
39.250	−0.00005	−0.00811	0.00000	−0.00000

Table 2. *– continued*

$(k-i)(b/\lambda)$	Φ_{kiu}		Φ_{kiv}	
39.500	−0.00806	0.00005	−0.00000	−0.00000
39.750	0.00005	0.00801	−0.00000	0.00000
40.000	0.00796	−0.00005	0.00000	0.00000
40.250	−0.00005	−0.00791	0.00000	−0.00000
40.500	−0.00786	0.00005	−0.00000	−0.00000
40.750	0.00004	0.00781	−0.00000	0.00000
41.000	0.00776	−0.00004	0.00000	0.00000
41.250	−0.00004	−0.00772	0.00000	−0.00000
41.500	−0.00767	0.00004	−0.00000	−0.00000
41.750	0.00004	0.00762	−0.00000	0.00000
42.000	0.00758	−0.00004	0.00000	0.00000
42.250	−0.00004	−0.00753	0.00000	−0.00000
42.500	−0.00749	0.00004	−0.00000	−0.00000
42.750	0.00004	0.00745	−0.00000	0.00000
43.000	0.00740	−0.00004	0.00000	0.00000
43.250	−0.00004	−0.00736	0.00000	−0.00000
43.500	−0.00732	0.00004	−0.00000	−0.00000
43.750	0.00004	0.00728	−0.00000	0.00000
44.000	0.00723	−0.00004	0.00000	0.00000
44.250	−0.00004	−0.00719	0.00000	−0.00000
44.500	−0.00715	0.00004	−0.00000	−0.00000
44.750	0.00004	0.00711	−0.00000	0.00000
45.000	0.00707	−0.00004	0.00000	0.00000
45.250	−0.00004	−0.00703	0.00000	−0.00000
45.500	−0.00700	0.00004	−0.00000	0.00000
45.750	0.00004	0.00696	0.00000	0.00000
46.000	0.00692	−0.00004	0.00000	−0.00000
46.250	−0.00003	−0.00688	−0.00000	−0.00000
46.500	−0.00685	0.00003	−0.00000	0.00000
46.750	0.00003	0.00681	0.00000	0.00000
47.000	0.00677	−0.00003	0.00000	−0.00000
47.250	−0.00003	−0.00674	−0.00000	−0.00000
47.500	−0.00670	0.00003	−0.00000	0.00000
47.750	0.00003	0.00667	0.00000	0.00000
48.000	0.00663	−0.00003	0.00000	−0.00000
48.250	−0.00003	−0.00660	−0.00000	−0.00000
48.500	−0.00656	0.00003	−0.00000	0.00000
48.750	0.00003	0.00653	0.00000	0.00000
49.000	0.00650	−0.00003	0.00000	−0.00000
49.250	−0.00003	−0.00646	−0.00000	−0.00000
49.500	−0.00643	0.00003	−0.00000	−0.00000
49.750	0.00003	0.00640	−0.00000	0.00000
50.000	0.00637	−0.00003	0.00000	0.00000

Table 3. *Elements of* Φ_u *and* Φ_v *matrices for* $h/\lambda = 0.375$

		Values for single element (Φ_{kku}, Φ_{kkv} and $\Psi_{kkdR} = \Psi_{dR}$)			
Ω	Ψ_{dR}	Φ_{kku}		Φ_{kkv}	
7.00	4.19919	−3.23123	2.57558	0.21311	−1.80626
7.50	4.67051	−3.60298	2.57993	0.24365	−1.80933
8.00	5.14769	−3.97013	2.58257	0.26761	−1.81119
8.50	5.62962	−4.33386	2.58417	0.28638	−1.81232
9.00	6.11538	−4.69506	2.58514	0.30105	−1.81301
9.50	6.60417	−5.05439	2.58573	0.31251	−1.81343
10.00	7.09538	−5.41231	2.58609	0.32146	−1.81368
10.50	7.58849	−5.76918	2.58631	0.32844	−1.81383
11.00	8.08311	−6.12526	2.58644	0.33388	−1.81393
11.50	8.57890	−6.48075	2.58652	0.33812	−1.81398
12.00	9.07561	−6.83578	2.58657	0.34143	−1.81402
12.50	9.57305	−7.19047	2.58660	0.34401	−1.81404
13.00	10.07105	−7.54490	2.58661	0.34602	−1.81405
15.00	12.06658	−8.96103	2.58664	0.35049	−1.81407

	Off diagonal values of Φ_u and Φ_v			
$(k-i)(b/\lambda)$	Φ_{kiu}		Φ_{kiv}	
0.250	0.91495	1.41028	−0.64760	−0.98392
0.500	1.02366	−0.50347	−0.71398	0.36254
0.750	−0.30409	−0.80374	0.22402	0.56460
1.000	−0.65434	0.19794	0.46271	−0.14798
1.250	0.13707	0.54781	−0.10339	−0.38918
1.500	0.46915	−0.09974	−0.33434	0.07566
1.750	−0.07549	−0.40923	0.05748	0.29226
2.000	−0.36234	0.05896	0.25917	−0.04501
2.250	0.04724	0.32477	−0.03613	−0.23255
2.500	0.29408	−0.03866	−0.21074	0.02960
2.750	−0.03220	−0.26857	0.02468	0.19258
3.000	−0.24705	0.02721	0.17724	−0.02088
3.250	0.02329	0.22868	−0.01788	−0.16412
3.500	0.21281	−0.02016	−0.15278	0.01548
3.750	−0.01761	−0.19898	0.01353	0.14288
4.000	−0.18682	0.01552	0.13418	−0.01193
4.250	0.01377	0.17604	−0.01059	−0.12646
4.500	0.16643	−0.01231	−0.11957	0.00946
4.750	−0.01106	−0.15781	0.00851	0.11339
5.000	−0.15003	0.01000	0.10782	−0.00769
5.250	0.00908	0.14298	−0.00698	−0.10276
5.500	0.13656	−0.00828	−0.09815	0.00637
5.750	−0.00758	−0.13068	0.00583	0.09393
6.000	−0.12529	0.00697	0.09006	−0.00536
6.250	0.00642	0.12033	−0.00494	−0.08650

Table 3. – *continued*

$(k - i)(b/\lambda)$	Φ_{kiu}		Φ_{kiv}	
6.500	0.11574	−0.00594	−0.08320	0.00457
6.750	−0.00551	−0.11148	0.00424	0.08015
7.000	−0.10753	0.00513	0.07731	−0.00395
7.250	0.00478	0.10385	−0.00368	−0.07467
7.500	0.10041	−0.00447	−0.07219	0.00344
7.750	−0.00419	−0.09719	0.00322	0.06988
8.000	−0.09417	0.00393	0.06771	−0.00303
8.250	0.00370	0.09133	−0.00285	−0.06567
8.500	0.08866	−0.00348	−0.06375	0.00268
8.750	−0.00329	−0.08613	0.00253	0.06194
9.000	−0.08375	0.00311	0.06023	−0.00239
9.250	0.00294	0.08150	−0.00227	−0.05861
9.500	0.07936	−0.00279	−0.05707	0.00215
9.750	−0.00265	−0.07733	0.00204	0.05561
10.000	−0.07541	0.00252	0.05423	−0.00194
10.250	0.00240	0.07358	−0.00185	−0.05291
10.500	0.07183	−0.00229	−0.05166	0.00176
10.750	−0.00218	−0.07016	0.00168	0.05046
11.000	−0.06857	0.00208	0.04932	−0.00161
11.250	0.00199	0.06705	−0.00153	−0.04822
11.500	0.06560	−0.00191	−0.04718	0.00147
11.750	−0.00183	−0.06421	0.00141	0.04618
12.000	−0.06287	0.00175	0.04522	−0.00135
12.250	0.00168	0.06159	−0.00130	−0.04430
12.500	0.06036	−0.00161	−0.04342	0.00124
12.750	−0.00155	−0.05918	0.00120	0.04257
13.000	−0.05805	0.00149	0.04175	−0.00115
13.250	0.00144	0.05695	−0.00111	−0.04096
13.500	0.05590	−0.00138	−0.04021	0.00107
13.750	−0.00134	−0.05489	0.00103	0.03948
14.000	−0.05391	0.00129	0.03877	−0.00099
14.250	0.00124	0.05297	−0.00096	−0.03809
14.500	0.05205	−0.00120	−0.03744	0.00092
14.750	−0.00116	−0.05117	0.00089	0.03681
15.000	−0.05032	0.00112	0.03619	−0.00086
15.250	0.00109	0.04950	−0.00084	−0.03560
15.500	0.04870	−0.00105	−0.03503	0.00081
15.750	−0.00102	−0.04793	0.00078	0.03447
16.000	−0.04718	0.00099	0.03394	−0.00076
16.250	0.00096	0.04646	−0.00074	−0.03341
16.500	0.04575	−0.00093	−0.03291	0.00071
16.750	−0.00090	−0.04507	0.00069	0.03242
17.000	−0.04441	0.00087	0.03194	−0.00067
17.250	0.00085	0.04377	−0.00065	−0.03148

Table 3. – *continued*

$(k-i)(b/\lambda)$	Φ_{kiu}		Φ_{kiv}	
17.500	0.04314	−0.00082	−0.03103	0.00064
17.750	−0.00080	−0.04253	0.00062	0.03059
18.000	−0.04194	0.00078	0.03017	−0.00060
18.250	0.00076	0.04137	−0.00058	−0.02976
18.500	0.04081	−0.00074	−0.02936	0.00057
18.750	−0.00072	−0.04027	0.00055	0.02896
19.000	−0.03974	0.00070	0.02858	−0.00054
19.250	0.00068	0.03922	−0.00053	−0.02821
19.500	0.03872	−0.00066	−0.02785	0.00051
19.750	−0.00065	−0.03823	0.00050	0.02750
20.000	−0.03775	0.00063	0.02716	−0.00049
20.250	0.00062	0.03729	−0.00047	−0.02682
20.500	0.03683	−0.00060	−0.02649	0.00046
20.750	−0.00059	−0.03639	0.00045	0.02618
21.000	−0.03596	0.00057	0.02586	−0.00044
21.250	0.00056	0.03553	−0.00043	−0.02556
21.500	0.03512	−0.00055	−0.02526	0.00042
21.750	−0.00053	−0.03472	0.00041	0.02497
22.000	−0.03432	0.00052	0.02469	−0.00040
22.250	0.00051	0.03394	−0.00039	−0.02441
22.500	0.03356	−0.00050	−0.02414	0.00038
22.750	−0.00049	−0.03319	0.00038	0.02388
23.000	−0.03283	0.00048	0.02362	−0.00037
23.250	0.00047	0.03248	−0.00036	−0.02336
23.500	0.03123	−0.00046	−0.02311	0.00035
23.750	−0.00045	−0.03180	0.00035	0.02287
24.000	−0.03147	0.00044	0.02263	−0.00034
24.250	0.00043	0.03114	−0.00033	−0.02240
24.500	0.03082	−0.00042	−0.02217	0.00032
24.750	−0.00041	−0.03051	0.00032	0.02195
25.000	−0.03021	0.00040	0.02173	−0.00031
25.250	0.00040	0.02991	−0.00031	−0.02151
25.500	0.02962	−0.00039	−0.02130	0.00030
25.750	−0.00038	−0.02933	0.00029	0.02110
26.000	−0.02905	0.00037	0.02089	−0.00029
26.250	0.00037	0.02877	−0.00028	−0.02069
26.500	0.02850	−0.00036	−0.02050	0.00028
26.750	−0.00035	−0.02823	0.00027	0.02031
27.000	−0.02797	0.00035	0.02012	−0.00027
27.250	0.00034	0.02771	−0.00026	−0.01994
27.500	0.02746	−0.00033	−0.01975	0.00026
27.750	−0.00033	−0.02722	0.00025	0.01958
28.000	−0.02697	0.00032	0.01940	−0.00025
28.250	0.00032	0.02673	−0.00024	−0.01923

Table 3. – *continued*

$(k-i)(b/\lambda)$	Φ_{kiu}		Φ_{kiv}	
28.500	0.02650	−0.00031	−0.01906	0.00024
28.750	−0.00031	−0.02627	0.00024	0.01890
29.000	−0.02604	0.00030	0.01873	−0.00023
29.250	0.00030	0.02582	−0.00023	−0.01857
29.500	0.02560	−0.00029	−0.01842	0.00022
29.750	−0.00029	−0.02539	0.00022	0.01826
30.000	−0.02518	0.00028	0.01811	−0.00022
30.250	0.00028	0.02497	−0.00021	−0.01796
30.500	0.02476	−0.00027	−0.01781	0.00021
30.750	−0.00027	−0.02456	0.00021	0.01767
31.000	−0.02436	0.00026	0.01753	−0.00020
31.250	0.00026	0.02417	−0.00020	−0.01738
31.500	0.02398	−0.00025	−0.01725	0.00020
31.750	−0.00025	−0.02379	0.00019	0.01711
32.000	−0.02360	0.00025	0.01698	−0.00019
32.250	0.00024	0.02342	−0.00019	−0.01685
32.500	0.02324	−0.00024	−0.01672	0.00018
32.750	−0.00024	−0.02306	0.00018	0.01659
33.000	−0.02289	0.00023	0.01646	−0.00018
33.250	0.00023	0.02272	−0.00018	−0.01634
33.500	0.02255	−0.00023	−0.01622	0.00017
33.750	−0.00022	−0.02238	0.00017	0.01610
34.000	−0.02221	0.00022	0.01598	−0.00017
34.250	0.00022	0.02205	−0.00017	−0.01586
34.500	0.02189	−0.00021	−0.01575	0.00016
34.750	−0.00021	−0.02174	0.00016	0.01563
35.000	−0.02158	0.00021	0.01552	−0.00016
35.250	0.00020	0.02143	−0.00016	−0.01541
35.500	0.02128	−0.00020	−0.01530	0.00015
35.750	−0.00020	−0.02113	0.00015	0.01520
36.000	−0.02098	0.00019	0.01509	−0.00015
36.250	0.00019	0.02084	−0.00015	−0.01499
36.500	0.02069	−0.00019	−0.01489	0.00015
36.750	−0.00019	−0.02055	0.00014	0.01478
37.000	−0.02041	0.00018	0.01468	−0.00014
37.250	0.00018	0.02028	−0.00014	−0.01459
37.500	0.02014	−0.00018	−0.01449	0.00014
37.750	−0.00018	−0.02001	0.00014	0.01439
38.000	−0.01988	0.00018	0.01430	−0.00013
38.250	0.00017	0.01975	−0.00013	−0.01420
38.500	0.01962	−0.00017	−0.01411	0.00013
38.750	−0.00017	−0.01949	0.00013	0.01402
39.000	−0.01937	0.00017	0.01393	−0.00013
39.250	0.00016	0.01924	−0.00013	−0.01384

Table 3. – *continued*

$(k-i)(b/\lambda)$	Φ_{kiu}		Φ_{kiv}	
39.500	0.01912	−0.00016	−0.01375	0.00012
39.750	−0.00016	−0.01900	0.00012	0.01367
40.000	−0.01888	0.00016	0.01358	−0.00012
40.250	0.00016	0.01877	−0.00012	−0.01350
40.500	0.01865	−0.00015	−0.01342	0.00012
40.750	−0.00015	−0.01854	0.00012	0.01333
41.000	−0.01842	0.00015	0.01325	−0.00012
41.250	0.00015	0.01831	−0.00011	−0.01317
41.500	0.01820	−0.00015	−0.01309	0.00011
41.750	−0.00014	−0.01809	0.00011	0.01301
42.000	−0.01798	0.00014	0.01294	−0.00011
42.250	0.00014	0.01788	−0.00011	−0.01286
42.500	0.01777	−0.00014	−0.01278	0.00011
42.750	−0.00014	−0.01767	0.00011	0.01271
43.000	−0.01757	0.00014	0.01264	−0.00011
43.250	0.00014	0.01746	−0.00010	−0.01256
43.500	0.01736	−0.00013	−0.01249	0.00010
43.750	−0.00013	−0.01726	0.00010	0.01242
44.000	−0.01717	0.00013	0.01235	−0.00010
44.250	0.00013	0.01707	−0.00010	−0.01228
44.500	0.01697	−0.00013	−0.01221	0.00010
44.750	−0.00013	−0.01688	0.00010	0.01214
45.000	−0.01679	0.00012	0.01207	−0.00010
45.250	0.00012	0.01669	−0.00010	−0.01201
45.500	0.01660	−0.00012	−0.01194	0.00009
45.750	−0.00012	−0.01651	0.00009	0.01188
46.000	−0.01642	0.00012	0.01181	−0.00009
46.250	0.00012	0.01633	−0.00009	−0.01175
46.500	0.01624	−0.00012	−0.01168	0.00009
46.750	−0.00012	−0.01616	0.00009	0.01162
47.000	−0.01607	0.00011	0.01156	−0.00009
47.250	0.00011	0.01599	−0.00009	−0.01150
47.500	0.01590	−0.00011	−0.01144	0.00009
47.750	−0.00011	−0.01582	0.00009	0.01138
48.000	−0.01574	0.00011	0.01132	−0.00008
48.250	0.00011	0.01565	−0.00008	−0.01126
48.500	0.01557	−0.00011	−0.01120	0.00008
48.750	−0.00011	−0.01549	0.00008	0.01115
49.000	−0.01542	0.00011	0.01109	−0.00008
49.250	0.00010	0.01534	−0.00008	−0.01103
49.500	0.01526	−0.00010	−0.01098	0.00008
49.750	−0.00010	−0.01518	0.00008	0.01092
50.000	−0.01511	0.00010	0.01087	−0.00008

Table 4. *Elements of Φ_u and Φ_v matrices for $h/\lambda = 0.50$*

		Values for single element (Φ_{kku}, Φ_{kkv} and $\Psi_{kkdR} = \Psi_{dR}$)		
Ω	Ψ_{dR}	Φ_{kku}		Φ_{kkv}
7.00	2.96068	−3.65469	2.87405	0.48701	−1.64559
7.50	3.42057	−4.16756	2.88308	0.52709	−1.65089
8.00	3.88899	−4.67618	2.88857	0.55866	−1.65410
8.50	4.36420	−5.18191	2.89191	0.58344	−1.65605
9.00	4.84477	−5.68568	2.89393	0.60287	−1.65724
9.50	5.32957	−6.18815	2.89516	0.61807	−1.65796
10.00	5.81769	−6.68975	2.89590	0.62995	−1.65839
10.50	6.30841	−7.19079	2.89635	0.63923	−1.65866
11.00	6.80117	−7.69146	2.89662	0.64647	−1.65882
11.50	7.29552	−8.19189	2.89679	0.65212	−1.65891
12.00	7.79112	−8.69217	2.89689	0.65653	−1.65897
12.50	8.28768	−9.19235	2.89695	0.65996	−1.65901
13.00	8.78501	−9.69246	2.89699	0.66264	−1.65903
15.00	10.77905	−11.69262	2.89704	0.66860	−1.65906

	Off diagonal values of Φ_u and Φ_v			
$(k-i)(b/\lambda)$	Φ_{kiu}		Φ_{kiv}	
0.250	1.02173	1.52473	−0.68124	−0.85698
0.500	1.10298	−0.66347	−0.62956	0.40704
0.750	−0.45462	−0.90710	0.29109	0.51517
1.000	−0.76996	0.32113	0.44407	−0.21742
1.250	0.23434	0.66433	−0.16567	−0.39044
1.500	0.58100	−0.17648	−0.34702	0.12866
1.750	−0.13673	−0.51431	0.10187	0.31107
2.000	−0.46021	0.10856	0.28103	−0.08216
2.250	0.08803	0.41571	−0.06739	−0.25573
2.500	0.37862	−0.07267	−0.23423	0.05612
2.750	−0.06093	−0.34732	0.04737	0.21582
3.000	−0.32062	0.05177	0.19993	−0.04046
3.250	0.04450	0.29760	−0.03492	−0.18610
3.500	0.27757	−0.03864	−0.17397	0.03042
3.750	−0.03385	−0.26000	0.02673	0.16327
4.000	−0.24448	0.02989	0.15376	−0.02365
4.250	0.02659	0.23067	−0.02107	−0.14527
4.500	0.21832	−0.02379	−0.13764	0.01889
4.750	−0.02141	−0.20720	0.01702	0.13076
5.000	−0.19714	0.01937	0.12451	−0.01542
5.250	0.01761	0.18800	−0.01403	−0.11882
5.500	0.17966	−0.01607	−0.11362	0.01282
5.750	−0.01473	−0.17202	0.01175	0.10885
6.000	−0.16500	0.01354	0.10446	−0.01082
6.250	0.01250	0.15852	−0.00999	−0.10040
6.500	0.15253	−0.01157	−0.09664	0.00925

Table 4. – *continued*

$(k-i)(b/\lambda)$	Φ_{kiu}		Φ_{kiv}	
6.750	−0.01074	−0.14697	0.00859	0.09315
7.000	−0.14180	0.00999	0.08990	−0.00800
7.250	0.00932	0.13698	−0.00746	−0.08687
7.500	0.13247	−0.00872	−0.08403	0.00698
7.750	−0.00817	−0.12825	0.00654	0.08137
8.000	−0.12429	0.00767	0.07888	−0.00615
8.250	0.00722	0.12056	−0.00578	−0.07653
8.500	0.11706	−0.00680	−0.07431	0.00545
8.750	−0.00642	−0.11374	0.00515	0.07222
9.000	−0.11061	0.00607	0.07024	−0.00487
9.250	0.00575	0.10765	−0.00461	−0.06837
9.500	0.10484	−0.00545	−0.06659	0.00438
9.750	−0.00518	−0.10217	0.00416	0.06491
10.000	−0.09964	0.00493	0.06330	−0.00395
10.250	0.00469	0.09722	−0.00376	−0.06177
10.500	0.09492	−0.00447	−0.06032	0.00359
10.750	−0.00427	−0.09273	0.00342	0.05893
11.000	−0.09063	0.00407	0.05760	−0.00327
11.250	0.00390	0.08863	−0.00313	−0.05633
11.500	0.08672	−0.00373	−0.05512	0.00300
11.750	−0.00357	−0.08488	0.00287	0.05396
12.000	−0.08312	0.00343	0.05284	−0.00275
12.250	0.00329	0.08143	−0.00264	−0.05177
12.500	0.07981	−0.00316	−0.05074	0.00254
12.750	−0.00304	−0.07826	0.00244	0.04976
13.000	−0.07676	0.00292	0.04881	−0.00235
13.250	0.00281	0.07531	−0.00226	−0.04789
13.500	0.07393	−0.00271	−0.04701	0.00218
13.750	−0.00261	−0.07259	0.00210	0.04616
14.000	−0.07130	0.00252	0.04534	−0.00203
14.250	0.00243	0.07005	−0.00196	−0.04455
14.500	0.06885	−0.00235	−0.04379	0.00189
14.750	−0.00227	−0.06768	0.00183	0.04305
15.000	−0.06656	0.00220	0.04233	−0.00177
15.250	0.00212	0.06547	−0.00171	−0.04164
15.500	0.06442	−0.00206	−0.04097	0.00165
15.750	−0.00199	−0.06340	0.00160	0.04033
16.000	−0.06241	0.00193	0.03970	−0.00155
16.250	0.00187	0.06145	−0.00150	−0.03909
16.500	0.06052	−0.00182	−0.03850	0.00146
16.750	−0.00176	−0.05962	0.00142	0.03793
17.000	−0.05875	0.00171	0.03737	−0.00138
17.250	0.00166	0.05790	−0.00134	−0.03683
17.500	0.05707	−0.00161	−0.03631	0.00130

Table 4. – *continued*

$(k-i)(b/\lambda)$	Φ_{kiu}		Φ_{kiv}	
17.750	−0.00157	−0.05627	0.00126	0.03580
18.000	−0.05549	0.00153	0.03531	−0.00123
18.250	0.00148	0.05473	−0.00119	−0.03482
18.500	0.05400	−0.00144	−0.03435	0.00116
18.750	−0.00141	−0.05328	0.00113	0.03390
19.000	−0.05258	0.00137	0.03345	−0.00110
19.250	0.00133	0.05190	−0.00107	−0.03302
19.500	0.05123	−0.00130	−0.03260	0.00105
19.750	−0.00127	−0.05059	0.00102	0.03219
20.000	−0.04995	0.00124	0.03179	−0.00099
20.250	0.00121	0.04934	−0.00097	−0.03139
20.500	0.04874	−0.00118	−0.03101	0.00095
20.750	−0.00115	−0.04815	0.00092	0.03064
21.000	−0.04758	0.00112	0.03028	−0.00090
21.250	0.00110	0.04702	−0.00088	−0.02992
21.500	0.04647	−0.00107	−0.02957	0.00086
21.750	−0.00105	−0.04594	0.00084	0.02923
22.000	−0.04542	0.00102	0.02890	−0.00082
22.250	0.00100	0.04491	−0.00080	−0.02858
22.500	0.04441	−0.00098	−0.02826	0.00079
22.750	−0.00096	−0.04392	0.00077	0.02795
23.000	−0.04345	0.00094	0.02765	−0.00075
23.250	0.00092	0.04298	−0.00074	−0.02735
23.500	0.04252	−0.00090	−0.02706	0.00072
23.750	−0.00088	−0.04208	0.00071	0.02678
24.000	−0.04164	0.00086	0.02650	−0.00069
24.250	0.00084	0.04121	−0.00068	−0.02623
24.500	0.04079	−0.00082	−0.02596	0.00066
24.750	−0.00081	−0.04038	0.00065	0.02570
25.000	−0.03998	0.00079	0.02544	−0.00064
25.250	0.00078	0.03958	−0.00062	−0.02519
25.500	0.03919	−0.00076	−0.02494	0.00061
25.750	−0.00075	−0.03881	0.00060	0.02470
26.000	−0.03844	0.00073	0.02446	−0.00059
26.250	0.00072	0.03807	−0.00058	−0.02423
26.500	0.03772	−0.00070	−0.02400	0.00057
26.750	−0.00069	−0.03736	0.00056	0.02378
27.000	−0.03702	0.00068	0.02356	−0.00055
27.250	0.00067	0.03668	−0.00054	−0.02334
27.500	0.03635	−0.00065	−0.02313	0.00053
27.750	−0.00064	−0.03602	0.00052	0.02292
28.000	−0.03570	0.00063	0.02272	−0.00051
28.250	0.00062	0.03538	−0.00050	−0.02252
28.500	0.03507	−0.00061	−0.02232	0.00049

Table 4. – *continued*

$(k-i)(b/\lambda)$	Φ_{kiu}		Φ_{kiv}	
28.750	−0.00060	−0.03477	0.00048	0.02213
29.000	−0.03447	0.00059	0.02194	−0.00047
29.250	0.00058	0.03417	−0.00047	−0.02175
29.500	0.03388	−0.00057	−0.02157	0.00046
29.750	−0.00056	−0.03360	0.00045	0.02139
30.000	−0.03332	0.00055	0.02121	−0.00044
30.250	0.00054	0.03304	−0.00044	−0.02103
30.500	0.03277	−0.00053	−0.02086	0.00043
30.750	−0.00052	−0.03251	0.00042	0.02069
31.000	−0.03225	0.00052	0.02052	−0.00041
31.250	0.00051	0.03199	−0.00041	−0.02036
31.500	0.03173	−0.00050	−0.02020	0.00040
31.750	−0.00049	−0.03148	0.00040	0.02004
32.000	−0.03124	0.00048	0.01988	−0.00039
32.250	0.00048	0.03100	−0.00038	−0.01973
32.500	0.03076	−0.00047	−0.01958	0.00038
32.750	−0.00046	−0.03052	0.00037	0.01943
33.000	−0.03029	0.00045	0.01928	−0.00037
33.250	0.00045	0.03007	−0.00036	−0.01914
33.500	0.02984	−0.00044	−0.01899	0.00035
33.750	−0.00043	−0.02962	0.00035	0.01885
34.000	−0.02940	0.00043	0.01871	−0.00034
34.250	0.00042	0.02919	−0.00034	−0.01858
34.500	0.02898	−0.00042	−0.01844	0.00033
34.750	−0.00041	−0.02877	0.00033	0.01831
35.000	−0.02856	0.00040	0.01818	−0.00033
35.250	0.00040	0.02836	−0.00032	−0.01805
35.500	0.02816	−0.00039	−0.01792	0.00032
35.750	−0.00039	−0.02796	0.00031	0.01780
36.000	−0.02777	0.00038	0.01768	−0.00031
36.250	0.00038	0.02758	−0.00030	−0.01755
36.500	0.02739	−0.00037	−0.01743	0.00030
36.750	−0.00037	−0.02720	0.00029	0.01732
37.000	−0.02702	0.00036	0.01720	−0.00029
37.250	0.00036	0.02684	−0.00029	−0.01708
37.500	0.02666	−0.00035	−0.01697	0.00028
37.750	−0.00035	−0.02648	0.00028	0.01686
38.000	−0.02631	0.00034	0.01675	−0.00028
38.250	0.00034	0.02614	−0.00027	−0.01664
38.500	0.02597	−0.00033	−0.01653	0.00027
38.750	−0.00033	−0.02580	0.00027	0.01642
39.000	−0.02563	0.00033	0.01632	−0.00026
39.250	0.00032	0.02547	−0.00026	−0.01621
39.500	0.02531	−0.00032	−0.01611	0.00026

Table 4. – *continued*

$(k-i)(b/\lambda)$	Φ_{kiu}		Φ_{kiv}	
39.750	−0.00031	−0.02515	0.00025	0.01601
40.000	−0.02499	0.00031	0.01591	−0.00025
40.250	0.00031	0.02484	−0.00025	−0.01581
40.500	0.02469	−0.00030	−0.01571	0.00024
40.750	−0.00030	−0.02453	0.00024	0.01562
41.000	−0.02438	0.00029	0.01552	−0.00024
41.250	0.00029	0.02424	−0.00023	−0.01543
41.500	0.02409	−0.00029	−0.01534	0.00023
41.750	−0.00028	−0.02395	0.00023	0.01524
42.000	−0.02380	0.00028	0.01515	−0.00023
42.250	0.00028	0.02366	−0.00022	−0.01506
42.500	0.02352	−0.00027	−0.01497	0.00022
42.750	−0.00027	−0.02339	0.00022	0.01489
43.000	−0.02325	0.00027	0.01480	−0.00022
43.250	0.00026	0.02312	−0.00021	−0.01472
43.500	0.02298	−0.00026	−0.01463	0.00021
43.750	−0.00026	−0.02285	0.00021	0.01455
44.000	−0.02272	0.00026	0.01446	−0.00021
44.250	0.00025	0.02259	−0.00020	−0.01438
44.500	0.02247	−0.00025	−0.01430	0.00020
44.750	−0.00025	−0.02234	0.00020	0.01422
45.000	−0.02222	0.00024	0.01414	−0.00020
45.250	0.00024	0.02210	−0.00019	−0.01407
45.500	0.02197	−0.00024	−0.01399	0.00019
45.750	−0.00024	−0.02185	0.00019	0.01391
46.000	−0.02174	0.00023	0.01384	−0.00019
46.250	0.00023	0.02162	−0.00019	−0.01376
46.500	0.02150	−0.00023	−0.01369	0.00018
46.750	−0.00023	−0.02139	0.00018	0.01361
47.000	−0.02127	0.00022	0.01354	−0.00018
47.250	0.00022	0.02116	−0.00018	−0.01347
47.500	0.02105	−0.00022	−0.01340	0.00018
47.750	−0.00022	−0.02094	0.00017	0.01333
48.000	−0.02083	0.00021	0.01326	−0.00017
48.250	0.00021	0.02072	−0.00017	−0.01319
48.500	0.02062	−0.00021	−0.01312	0.00017
48.750	−0.00021	−0.02051	0.00017	0.01306
49.000	−0.02041	0.00021	0.01299	−0.00017
49.250	0.00020	0.02030	−0.00016	−0.01292
49.500	0.02020	−0.00020	−0.01286	0.00016
49.750	−0.00020	−0.02010	0.00016	0.01279
50.000	−0.02000	0.00020	0.01273	−0.00016

Table 5. *Elements of Φ_u and Φ_v matrices for $h/\lambda = 0.625$*

	Values for single element (Φ_{kku}, Φ_{kkv} and $\Psi_{kkdR} = \Psi_{dR}$)				
Ω	Ψ_{dR}	Φ_{kku}		Φ_{kkv}	
7.00	2.81599	−2.06379	1.28068	0.83173	−1.02685
7.50	3.26383	−2.40814	1.28795	0.88239	−1.03206
8.00	3.72310	−2.75320	1.29236	0.92221	−1.03523
8.50	4.19134	−3.09919	1.29505	0.95343	−1.03716
9.00	4.66659	−3.44620	1.29668	0.97785	−1.03833
9.50	5.14730	−3.79420	1.29766	0.99694	−1.03904
10.00	5.63227	−4.14312	1.29826	1.01185	−1.03947
10.50	6.12057	−4.49286	1.29863	1.02348	−1.03973
11.00	6.61145	−4.84330	1.29885	1.03255	−1.03989
11.50	7.10435	−5.19435	1.29898	1.03963	−1.03998
12.00	7.59882	−5.54589	1.29906	1.04514	−1.04004
12.50	8.09452	−5.89783	1.29911	1.04944	−1.04008
13.00	8.59116	−6.25011	1.29914	1.05278	−1.04010
15.00	10.58370	−7.66140	1.29918	1.06024	−1.04013

	Off diagonal values of Φ_u and Φ_v			
$(k-i)(b/\lambda)$	Φ_{kiu}		Φ_{kiv}	
0.250	0.45164	0.59842	−0.56525	−0.53538
0.500	0.42851	−0.45552	−0.40583	0.24351
0.750	−0.38272	−0.42507	0.17624	0.28315
1.000	−0.41223	0.29489	0.23084	−0.16312
1.250	0.22288	0.38447	−0.15177	−0.21024
1.500	0.35176	−0.16988	−0.19944	0.13674
1.750	−0.13186	−0.31989	0.12048	0.19073
2.000	−0.29107	0.10445	0.18209	−0.10500
2.250	0.08436	0.26579	−0.09121	−0.17329
2.500	0.24383	−0.06935	−0.16453	0.07932
2.750	−0.05790	−0.22480	0.06922	0.15603
3.000	−0.20826	0.04901	0.14794	−0.06068
3.250	0.04198	0.19382	−0.05348	−0.14033
3.500	0.18113	−0.03635	−0.13324	0.04738
3.750	−0.03176	−0.16993	0.04221	0.12666
4.000	−0.15997	0.02798	0.12058	−0.03778
4.250	0.02484	0.15108	−0.03399	−0.11496
4.500	0.14310	−0.02219	−0.10977	0.03071
4.750	−0.01994	−0.13589	0.02787	0.10497
5.000	−0.12936	0.01801	0.10053	−0.02539
5.250	0.01635	0.12342	−0.02322	−0.09641
5.500	0.11799	−0.01491	−0.09259	0.02131
5.750	−0.01365	−0.11300	0.01962	0.08904
6.000	−0.10842	0.01255	0.08574	−0.01812
6.250	0.01157	0.10419	−0.01678	−0.08265

Table 5. – *continued*

$(k-i)(b/\lambda)$	Φ_{kiu}		Φ_{kiv}	
6.500	0.10027	−0.01070	−0.07977	0.01558
6.750	−0.00993	−0.09663	0.01450	0.07707
7.000	−0.09325	0.00923	0.07454	−0.01353
7.250	0.00861	0.09009	−0.01266	−0.07217
7.500	0.08714	−0.00805	−0.06993	0.01186
7.750	−0.00754	−0.08437	0.01114	0.06783
8.000	−0.08177	0.00707	0.06584	−0.01047
8.250	0.00665	0.07933	−0.00987	−0.06396
8.500	0.07702	−0.00627	−0.06218	0.00932
8.750	−0.00592	−0.07485	0.00881	0.06050
9.000	−0.07279	0.00559	0.05890	−0.00834
9.250	0.00530	0.07085	−0.00791	−0.05739
9.500	0.06900	−0.00502	−0.05594	0.00751
9.750	−0.00477	−0.06725	0.00713	0.05457
10.000	−0.06558	0.00453	0.05326	−0.00679
10.250	0.00431	0.06400	−0.00647	−0.05201
10.500	0.06249	−0.00411	−0.05082	0.00617
10.750	−0.00392	−0.06105	0.00589	0.04968
11.000	−0.05967	0.00375	0.04859	−0.00563
11.250	0.00358	0.05835	−0.00539	−0.04754
11.500	0.05709	−0.00343	−0.04654	0.00516
11.750	−0.00328	−0.05589	0.00495	0.04558
12.000	−0.05473	0.00315	0.04465	−0.00475
12.250	0.00302	0.05362	−0.00456	−0.04377
12.500	0.05255	−0.00290	−0.04291	0.00438
12.750	−0.00279	−0.05153	0.00421	0.04209
13.000	−0.05054	0.00268	0.04130	−0.00406
13.250	0.00258	0.04959	−0.00391	−0.04054
13.500	0.04868	−0.00249	−0.03981	0.00376
13.750	−0.00240	−0.04780	0.00363	0.03910
14.000	−0.04695	0.00231	0.03842	−0.00350
14.250	0.00223	0.04613	−0.00338	−0.03776
14.500	0.04534	−0.00216	−0.03712	0.00327
14.750	−0.00209	−0.04457	0.00316	0.03650
15.000	−0.04383	0.00202	0.03590	−0.00306
15.250	0.00195	0.04312	−0.00296	−0.03532
15.500	0.04242	−0.00189	−0.03476	0.00286
15.750	−0.00183	−0.04175	0.00277	0.03422
16.000	−0.04110	0.00177	0.03370	−0.00269
16.250	0.00172	0.04047	−0.00261	−0.03318
16.500	0.03986	−0.00167	−0.03269	0.00253
16.750	−0.00162	−0.03927	0.00246	0.03221
17.000	−0.03869	0.00157	0.03174	−0.00238
17.250	0.00153	0.03813	−0.00232	−0.03129

Table 5. – *continued*

$(k-i)(b/\lambda)$	Φ_{kiu}		Φ_{kiv}	
17.500	0.03759	−0.00148	−0.03085	0.00225
17.750	−0.00144	−0.03706	0.00219	0.03042
18.000	−0.03655	0.00140	0.03000	−0.00213
18.250	0.00136	0.03605	−0.00207	−0.02959
18.500	0.03556	−0.00133	−0.02920	0.00202
18.750	−0.00129	−0.03509	0.00196	0.02881
19.000	−0.03463	0.00126	0.02844	−0.00191
19.250	0.00122	0.03418	−0.00186	−0.02807
19.500	0.03375	−0.00119	−0.02772	0.00182
19.750	−0.00116	−0.03332	0.00177	0.02737
20.000	−0.03290	0.00113	0.02703	−0.00173
20.250	0.00111	0.03250	−0.00168	−0.02670
20.500	0.03210	−0.00108	−0.02638	0.00164
20.750	−0.00105	−0.03172	0.00160	0.02606
21.000	−0.03134	0.00103	0.02576	−0.00157
21.250	0.00101	0.03097	−0.00153	−0.02546
21.500	0.03061	−0.00098	−0.02516	0.00150
21.750	−0.00096	−0.03026	0.00146	0.02488
22.000	−0.02992	0.00094	0.02460	−0.00143
22.250	0.00092	0.02958	−0.00140	−0.02432
22.500	0.02925	−0.00090	−0.02405	0.00137
22.750	−0.00088	−0.02893	0.00134	0.02379
23.000	−0.02862	0.00086	0.02354	−0.00131
23.250	0.00084	0.02831	−0.00128	−0.02328
23.500	0.02801	−0.00082	−0.02304	0.00125
23.750	−0.00080	−0.02772	0.00123	0.02280
24.000	−0.02743	0.00079	0.02256	−0.00120
24.250	0.00077	0.02715	−0.00118	−0.02233
24.500	0.02687	−0.00076	−0.02210	0.00115
24.750	−0.00074	−0.02660	0.00113	0.02188
25.000	−0.02633	0.00073	0.02167	−0.00111
25.250	0.00071	0.02607	−0.00109	−0.02145
25.500	0.02582	−0.00070	−0.02124	0.00106
25.750	−0.00068	−0.02557	0.00104	0.02104
26.000	−0.02532	0.00067	0.02084	−0.00102
26.250	0.00066	0.02508	−0.00100	−0.02064
26.500	0.02484	−0.00065	−0.02045	0.00099
26.750	−0.00063	−0.02461	0.00097	0.02026
27.000	−0.02439	0.00062	0.02007	−0.00095
27.250	0.00061	0.02416	−0.00093	−0.01989
27.500	0.02394	−0.00060	−0.01971	0.00092
27.750	−0.00059	−0.02373	0.00090	0.01953
28.000	−0.02352	0.00058	0.01936	−0.00088
28.250	0.00057	0.02331	−0.00087	−0.01919

Table 5. – *continued*

$(k-i)(b/\lambda)$	Φ_{kiu}		Φ_{kiv}	
28.500	0.02310	−0.00056	−0.01902	0.00085
28.750	−0.00055	−0.02290	0.00084	0.01885
29.000	−0.02271	0.00054	0.01869	−0.00082
29.250	0.00053	0.02251	−0.00081	−0.01853
29.500	0.02232	−0.00052	−0.01838	0.00080
29.750	−0.00051	−0.02213	0.00078	0.01822
30.000	−0.02195	0.00050	0.01807	−0.00077
30.250	0.00050	0.02177	−0.00076	−0.01792
30.500	0.02159	−0.00049	−0.01778	0.00074
30.750	−0.00048	−0.02141	0.00073	0.01763
31.000	−0.02124	0.00047	0.01749	−0.00072
31.250	0.00046	0.02107	−0.00071	−0.01735
31.500	0.02091	−0.00046	−0.01722	0.00070
31.750	−0.00045	−0.02074	0.00069	0.01708
32.000	−0.02058	0.00044	0.01695	−0.00068
32.250	0.00044	0.02042	−0.00067	−0.01682
32.500	0.02026	−0.00043	−0.01669	0.00066
32.750	−0.00042	−0.02011	0.00065	0.01656
33.000	−0.01996	0.00042	0.01644	−0.00064
33.250	0.00041	0.01981	−0.00063	−0.01631
33.500	0.01966	−0.00040	−0.01619	0.00062
33.750	−0.00040	−0.01951	0.00061	0.01607
34.000	−0.01937	0.00039	0.01595	−0.00060
34.250	0.00039	0.01923	−0.00059	−0.01584
34.500	0.01909	−0.00038	−0.01572	0.00058
34.750	−0.00038	−0.01895	0.00057	0.01561
35.000	−0.01882	0.00037	0.01550	−0.00057
35.250	0.00037	0.01868	−0.00056	−0.01539
35.500	0.01855	−0.00036	−0.01528	0.00055
35.750	−0.00036	−0.01842	0.00054	0.01518
36.000	−0.01829	0.00035	0.01507	−0.00054
36.250	0.00035	0.01817	−0.00053	−0.01497
36.500	0.01804	−0.00034	−0.01486	0.00052
36.750	−0.00034	−0.01792	0.00051	0.01476
37.000	−0.01780	0.00033	0.01466	−0.00051
37.250	0.00033	0.01768	−0.00050	−0.01457
37.500	0.01756	−0.00032	−0.01447	0.00049
37.750	−0.00032	−0.01745	0.00049	0.01437
38.000	−0.01733	0.00031	0.01428	−0.00048
38.250	0.00031	0.01722	−0.00047	−0.01419
38.500	0.01711	−0.00031	−0.01409	0.00047
38.750	−0.00030	−0.01700	0.00046	0.01400
39.000	−0.01689	0.00030	0.01391	−0.00046
39.250	0.00029	0.01678	−0.00045	−0.01383

Table 5. – *continued*

$(k-i)(b/\lambda)$	Φ_{kiu}		Φ_{kiv}	
39.500	0.01667	−0.00029	−0.01374	0.00044
39.750	−0.00029	−0.01657	0.00044	0.01365
40.000	−0.01647	0.00028	0.01357	−0.00043
40.250	0.00028	0.01636	−0.00043	−0.01348
40.500	0.01626	−0.00028	−0.01340	0.00042
40.750	−0.00027	−0.01616	0.00042	0.01332
41.000	−0.01606	0.00027	0.01324	−0.00041
41.250	0.00027	0.01597	−0.00041	−0.01316
41.500	0.01587	−0.00026	−0.01308	0.00040
41.750	−0.00026	−0.01578	0.00040	0.01300
42.000	−0.01568	0.00026	0.01292	−0.00039
42.250	0.00025	0.01559	−0.00039	−0.01285
42.500	0.01550	−0.00025	−0.01277	0.00038
42.750	−0.00025	−0.01541	0.00038	0.01270
43.000	−0.01532	0.00025	0.01262	−0.00038
43.250	0.00024	0.01523	−0.00037	−0.01255
43.500	0.01514	−0.00024	−0.01248	0.00037
43.750	−0.00024	−0.01505	0.00036	0.01241
44.000	−0.01497	0.00023	0.01234	−0.00036
44.250	0.00023	0.01488	−0.00035	−0.01227
44.500	0.01480	−0.00023	−0.01220	0.00035
44.750	−0.00023	−0.01472	0.00035	0.01213
45.000	−0.01464	0.00022	0.01206	−0.00034
45.250	0.00022	0.01456	−0.00034	−0.01200
45.500	0.01448	−0.00022	−0.01193	0.00034
45.750	−0.00022	−0.01440	0.00033	0.01187
46.000	−0.01432	0.00021	0.01180	−0.00033
46.250	0.00021	0.01424	−0.00032	−0.01174
46.500	0.01416	−0.00021	−0.01167	0.00032
46.750	−0.00021	−0.01409	0.00032	0.01161
47.000	−0.01401	0.00021	0.01155	−0.00031
47.250	0.00020	0.01394	−0.00031	−0.01149
47.500	0.01387	−0.00020	−0.01143	0.00031
47.750	−0.00020	−0.01379	0.00030	0.01137
48.000	−0.01372	0.00020	0.01131	−0.00030
48.250	0.00019	0.01365	−0.00030	−0.01125
48.500	0.01358	−0.00019	−0.01119	0.00030
48.750	−0.00019	−0.01351	0.00029	0.01114
49.000	−0.01344	0.00019	0.01108	−0.00029
49.250	0.00019	0.01337	−0.00029	−0.01102
49.500	0.01331	−0.00019	−0.01097	0.00028
49.750	−0.00018	−0.01324	0.00028	0.01091
50.000	−0.01317	0.00018	0.01086	−0.00028

Appendix III

Tables of admittance and impedance for curtain arrays

This set of tables is abridged from the report "Tables for Curtain Arrays" by Ronold W. P. King, Barbara H. Sandler and Sheldon S. Sandler, Cruft Laboratory Scientific Report No. 4 (Series 3), Harvard University, May 1964.

The calculations for the individual elements are given for $\Omega = 2 \ln 2h/a = 8.6138$ for $\beta_0 h = \pi/4$ and for $\Omega = 10$ for all other electrical lengths. The admittances are given in millisiemens and the impedances in ohms. The vertical listings begin with the first element at the top. The unilateral endfire patterns are prescribed to point in the direction away from the first element toward the last element.

Table 1. *Broadside array (driving-point currents specified);* $\beta_0 h = 0.78539$

Admittance	Impedance	Admittance	Impedance
$\beta_0 b = 1.57080$		$\beta_0 b = 3.14159$	
N = 1		*N* = 1	
0.122 + j3.250	11.564 − j307.244	0.122 + j3.250	11.564 − j307.244
N = 4		*N* = 4	
0.123 + j3.135	12.532 − j318.521	0.100 + j3.204	9.690 − j311.824
0.223 + j3.047	23.894 − j326.465	0.078 + j3.170	7.786 − j315.273
0.223 + j3.047	23.894 − j326.465	0.078 + j3.170	7.786 − j315.273
0.123 + j3.135	12.532 − j318.521	0.100 + j3.204	9.690 − j311.824
N = 10		*N* = 10	
0.152 + j3.154	15.277 − j316.276	0.101 + j3.210	9.786 − j311.261
0.202 + j3.084	21.113 − j322.911	0.076 + j3.161	7.615 − j316.127
0.170 + j3.020	18.620 − j330.103	0.083 + j3.185	8.207 − j313.805
0.151 + j3.030	16.446 − j329.235	0.080 + j3.172	7.963 − j315.015
0.166 + j3.078	17.493 − j323.927	0.081 + j3.178	8.058 − j314.481
0.166 + j3.078	17.493 − j323.927	0.081 + j3.178	8.058 − j314.481
0.151 + j3.030	16.446 − j329.235	0.080 + j3.172	7.963 − j315.015
0.170 + j3.020	18.620 − j330.103	0.083 + j3.185	8.207 − j313.805
0.202 + j3.084	21.113 − j322.911	0.076 + j3.161	7.615 − j316.127
0.152 + j3.154	15.277 − j316.276	0.101 + j3.210	9.786 − j311.261
N = 20		*N* = 20	
0.140 + j3.146	14.158 − j317.252	0.101 + j3.211	9.804 − j311.079
0.210 + j3.071	22.183 − j324.119	0.076 + j3.159	7.599 − j316.331
0.184 + j3.029	19.973 − j328.914	0.084 + j3.187	8.233 − j313.549
0.141 + j3.045	15.134 − j327.710	0.080 + j3.169	7.934 − j315.316
0.148 + j3.066	15.754 − j325.408	0.082 + j3.182	8.106 − j314.096
0.180 + j3.057	19.217 − j325.963	0.081 + j3.173	7.998 − j314.970
0.177 + j3.046	18.957 − j327.166	0.082 + j3.179	8.071 − j314.345
0.150 + j3.053	16.056 − j326.795	0.081 + j3.175	8.024 − j314.780
0.151 + j3.058	16.158 − j326.263	0.081 + j3.178	8.052 − j314.506
0.176 + j3.052	18.842 − j326.573	0.081 + j3.176	8.039 − j314.639
0.176 + j3.052	18.842 − j326.573	0.081 + j3.176	8.039 − j314.639
0.151 + j3.058	16.158 − j326.263	0.081 + j3.178	8.052 − j314.506
0.150 + j3.053	16.056 − j326.795	0.081 + j3.175	8.024 − j314.780
0.177 + j3.046	18.957 − j327.166	0.082 + j3.179	8.071 − j314.345
0.180 + j3.057	19.217 − j325.963	0.081 + j3.173	7.998 − j314.970
0.148 + j3.066	15.754 − j325.408	0.082 + j3.182	8.106 − j314.096
0.141 + j3.045	15.134 − j327.710	0.080 + j3.169	7.934 − j315.316
0.184 + j3.029	19.973 − j328.914	0.084 + j3.187	8.233 − j313.549
0.210 + j3.071	22.183 − j324.119	0.076 + j3.159	7.599 − j316.331
0.140 + j3.146	14.158 − j317.252	0.101 + j3.211	9.804 − j311.079

Table 2. *Broadside array (base voltages specified);* $\beta_0 h = 0.78539$

Admittance	Impedance	Admittance	Impedance
$\beta_0 b = 1.57080$		$\beta_0 b = 3.14159$	
$N = 1$		$N = 1$	
$0.122 + j3.250$	$11.564 - j307.244$	$0.122 + j3.250$	$11.564 - j307.244$
$N = 4$		$N = 4$	
$0.118 + j3.137$	$11.937 - j318.359$	$0.100 + j3.204$	$9.724 - j311.807$
$0.229 + j3.045$	$24.512 - j326.608$	$0.078 + j3.170$	$7.753 - j315.290$
$0.229 + j3.045$	$24.512 - j326.608$	$0.078 + j3.170$	$7.753 - j315.290$
$0.118 + j3.137$	$11.937 - j318.359$	$0.100 + j3.204$	$9.724 - j311.807$
$N = 10$		$N = 10$	
$0.153 + j3.157$	$15.298 - j316.064$	$0.102 + j3.210$	$9.848 - j311.209$
$0.204 + j3.087$	$21.338 - j322.502$	$0.075 + j3.161$	$7.517 - j316.212$
$0.168 + j3.014$	$18.457 - j330.739$	$0.084 + j3.185$	$8.263 - j313.749$
$0.148 + j3.026$	$16.125 - j329.718$	$0.080 + j3.172$	$7.931 - j315.051$
$0.169 + j3.083$	$17.702 - j323.419$	$0.082 + j3.178$	$8.068 - j314.470$
$0.169 + j3.083$	$17.702 - j323.419$	$0.082 + j3.178$	$8.068 - j314.470$
$0.148 + j3.026$	$16.125 - j329.718$	$0.080 + j3.172$	$7.931 - j315.051$
$0.168 + j3.014$	$18.457 - j330.739$	$0.084 + j3.185$	$8.263 - j313.749$
$0.204 + j3.087$	$21.338 - j322.502$	$0.075 + j3.161$	$7.517 - j316.212$
$0.153 + j3.157$	$15.298 - j316.064$	$0.102 + j3.210$	$9.848 - j311.209$
$N = 20$		$N = 20$	
$0.137 + j3.147$	$13.778 - j317.153$	$0.102 + j3.211$	$9.844 - j311.084$
$0.215 + j3.071$	$22.654 - j324.092$	$0.074 + j3.158$	$7.449 - j316.530$
$0.187 + j3.025$	$20.346 - j329.273$	$0.084 + j3.187$	$8.278 - j313.535$
$0.136 + j3.046$	$14.591 - j327.645$	$0.079 + j3.168$	$7.842 - j315.477$
$0.146 + j3.069$	$15.423 - j325.119$	$0.082 + j3.181$	$8.116 - j314.110$
$0.184 + j3.056$	$19.673 - j326.009$	$0.080 + j3.172$	$7.932 - j315.103$
$0.180 + j3.044$	$19.327 - j327.375$	$0.082 + j3.179$	$8.064 - j314.379$
$0.146 + j3.054$	$15.636 - j326.695$	$0.080 + j3.174$	$7.966 - j314.901$
$0.148 + j3.059$	$15.791 - j326.090$	$0.081 + j3.178$	$8.062 - j314.490$
$0.180 + j3.050$	$19.235 - j326.688$	$0.081 + j3.176$	$8.036 - j314.645$
$0.180 + j3.050$	$19.235 - j326.688$	$0.081 + j3.176$	$8.036 - j314.645$
$0.148 + j3.059$	$15.791 - j326.090$	$0.081 + j3.178$	$8.062 - j314.490$
$0.146 + j3.054$	$15.636 - j326.695$	$0.080 + j3.174$	$7.966 - j314.901$
$0.180 + j3.044$	$19.327 - j327.375$	$0.082 + j3.179$	$8.064 - j314.379$
$0.184 + j3.056$	$19.673 - j326.009$	$0.080 + j3.172$	$7.932 - j315.103$
$0.146 + j3.069$	$15.423 - j325.119$	$0.082 + j3.181$	$8.116 - j314.110$
$0.136 + j3.046$	$14.591 - j327.645$	$0.079 + j3.168$	$7.842 - j315.477$
$0.187 + j3.025$	$20.346 - j329.273$	$0.084 + j3.187$	$8.278 - j313.535$
$0.215 + j3.071$	$22.654 - j324.092$	$0.074 + j3.158$	$7.449 - j316.530$
$0.137 + j3.147$	$13.778 - j317.153$	$0.102 + j3.211$	$9.844 - j311.084$

Table 3. *Endfire array (driving-point currents specified);* $\beta_0 h = 0.78539$

Admittance	Impedance	Admittance	Impedance
$\beta_0 b = 1.57080$		$\beta_0 b = 3.14159$	
$N = 1$		$N = 1$	
$0.122 + j3.250$	$11.564 - j307.244$	$0.122 + j3.250$	$11.564 - j307.244$
$N = 4$		$N = 4$	
$0.059 + j3.193$	$5.778 - j313.123$	$0.168 + j3.367$	$14.796 - j296.304$
$0.151 + j3.309$	$13.761 - j301.623$	$0.195 + j3.404$	$16.735 - j292.822$
$0.150 + j3.308$	$13.659 - j301.691$	$0.195 + j3.404$	$16.735 - j292.822$
$0.257 + j3.427$	$21.769 - j290.196$	$0.168 + j3.367$	$14.796 - j296.304$
$N = 10$		$N = 10$	
$0.063 + j3.203$	$6.120 - j312.060$	$0.183 + j3.442$	$15.431 - j289.678$
$0.142 + j3.292$	$13.121 - j303.166$	$0.217 + j3.498$	$17.657 - j284.743$
$0.169 + j3.333$	$15.135 - j299.277$	$0.228 + j3.526$	$18.267 - j282.387$
$0.192 + j3.399$	$16.544 - j293.286$	$0.233 + j3.541$	$18.520 - j281.159$
$0.196 + j3.409$	$16.847 - j292.350$	$0.235 + j3.548$	$18.619 - j280.616$
$0.215 + j3.469$	$17.814 - j287.133$	$0.235 + j3.548$	$18.619 - j280.616$
$0.209 + j3.457$	$17.416 - j288.192$	$0.233 + j3.541$	$18.520 - j281.159$
$0.236 + j3.528$	$18.875 - j282.157$	$0.228 + j3.526$	$18.267 - j282.387$
$0.203 + j3.482$	$16.706 - j286.198$	$0.217 + j3.498$	$17.657 - j284.743$
$0.309 + j3.580$	$23.931 - j277.234$	$0.183 + j3.442$	$15.431 - j289.678$
$N = 20$		$N = 20$	
$0.064 + j3.207$	$6.188 - j311.699$	$0.193 + j3.500$	$15.723 - j284.807$
$0.141 + j3.288$	$13.044 - j303.581$	$0.229 + j3.564$	$17.992 - j279.463$
$0.170 + j3.338$	$15.223 - j298.777$	$0.243 + j3.599$	$18.651 - j276.631$
$0.189 + j3.392$	$16.416 - j293.886$	$0.250 + j3.621$	$18.964 - j274.831$
$0.199 + j3.418$	$17.012 - j291.597$	$0.255 + j3.637$	$19.145 - j273.590$
$0.211 + j3.458$	$17.564 - j288.089$	$0.258 + j3.649$	$19.258 - j272.700$
$0.215 + j3.472$	$17.802 - j286.910$	$0.260 + j3.657$	$19.335 - j272.063$
$0.224 + j3.507$	$18.174 - j283.963$	$0.261 + j3.663$	$19.385 - j271.619$
$0.226 + j3.513$	$18.255 - j283.463$	$0.262 + j3.667$	$19.415 - j271.339$
$0.235 + j3.547$	$18.571 - j280.729$	$0.263 + j3.668$	$19.428 - j271.203$
$0.234 + j3.546$	$18.548 - j280.770$	$0.263 + j3.668$	$19.428 - j271.203$
$0.243 + j3.580$	$18.869 - j278.028$	$0.262 + j3.667$	$19.415 - j271.339$
$0.240 + j3.573$	$18.740 - j278.613$	$0.261 + j3.663$	$19.385 - j271.619$
$0.251 + j3.611$	$19.130 - j275.641$	$0.260 + j3.657$	$19.335 - j272.063$
$0.245 + j3.595$	$18.836 - j276.916$	$0.258 + j3.649$	$19.258 - j272.700$
$0.259 + j3.640$	$19.423 - j273.359$	$0.255 + j3.637$	$19.145 - j273.590$
$0.245 + j3.610$	$18.749 - j275.767$	$0.250 + j3.621$	$18.964 - j274.831$
$0.272 + j3.673$	$20.021 - j270.784$	$0.243 + j3.599$	$18.651 - j276.631$
$0.232 + j3.611$	$17.682 - j275.761$	$0.229 + j3.564$	$17.992 - j279.463$
$0.344 + j3.706$	$24.834 - j267.513$	$0.193 + j3.500$	$15.723 - j284.807$

Table 4. *Endfire array (base voltages specified);* $\beta_0 h = 0.78539$

Admittance	Impedance	Admittance	Impedance
$\beta_0 b = 1.57080$		$\beta_0 b = 3.14159$	
N = 1		*N = 1*	
$0.122 + j3.250$	$11.564 - j307.244$	$0.122 + j3.250$	$11.564 - j307.244$
N = 4		*N = 4*	
$0.053 + j3.191$	$5.215 - j313.290$	$0.169 + j3.367$	$14.886 - j296.227$
$0.149 + j3.309$	$13.563 - j301.579$	$0.193 + j3.403$	$16.646 - j292.899$
$0.142 + j3.306$	$12.988 - j301.907$	$0.193 + j3.403$	$16.646 - j292.899$
$0.245 + j3.423$	$20.806 - j290.646$	$0.169 + j3.367$	$14.886 - j296.226$
N = 10		*N = 10*	
$0.059 + j3.202$	$5.753 - j312.223$	$0.187 + j3.446$	$15.675 - j289.334$
$0.140 + j3.294$	$12.839 - j303.024$	$0.217 + j3.499$	$17.654 - j284.681$
$0.161 + j3.329$	$14.500 - j299.689$	$0.227 + j3.526$	$18.207 - j282.464$
$0.187 + j3.396$	$16.167 - j293.555$	$0.232 + j3.540$	$18.439 - j281.305$
$0.189 + j3.403$	$16.230 - j292.965$	$0.234 + j3.546$	$18.531 - j280.791$
$0.210 + j3.464$	$17.472 - j287.643$	$0.234 + j3.546$	$18.531 - j280.791$
$0.201 + j3.450$	$16.837 - j288.855$	$0.232 + j3.540$	$18.439 - j281.305$
$0.230 + j3.520$	$18.527 - j282.918$	$0.227 + j3.526$	$18.207 - j282.464$
$0.198 + j3.477$	$16.294 - j286.701$	$0.217 + j3.499$	$17.654 - j284.681$
$0.293 + j3.571$	$22.792 - j278.154$	$0.187 + j3.446$	$15.675 - j289.334$
N = 20		*N = 20*	
$0.061 + j3.206$	$5.895 - j311.808$	$0.199 + j3.508$	$16.092 - j284.146$
$0.138 + j3.289$	$12.707 - j303.472$	$0.231 + j3.567$	$18.068 - j279.167$
$0.163 + j3.335$	$14.649 - j299.156$	$0.243 + j3.600$	$18.651 - j276.528$
$0.184 + j3.389$	$15.985 - j294.162$	$0.249 + j3.621$	$18.933 - j274.843$
$0.192 + j3.411$	$16.458 - j292.213$	$0.254 + j3.636$	$19.096 - j273.675$
$0.205 + j3.453$	$17.166 - j288.556$	$0.257 + j3.647$	$19.200 - j272.836$
$0.208 + j3.464$	$17.287 - j287.644$	$0.259 + j3.655$	$19.270 - j272.234$
$0.219 + j3.501$	$17.813 - j284.541$	$0.260 + j3.660$	$19.316 - j271.815$
$0.219 + j3.504$	$17.772 - j284.267$	$0.261 + j3.664$	$19.343 - j271.550$
$0.229 + j3.539$	$18.242 - j281.386$	$0.261 + j3.666$	$19.356 - j271.421$
$0.227 + j3.536$	$18.087 - j281.618$	$0.261 + j3.666$	$19.356 - j271.421$
$0.238 + j3.572$	$18.568 - j278.749$	$0.261 + j3.664$	$19.343 - j271.550$
$0.233 + j3.563$	$18.293 - j279.483$	$0.260 + j3.660$	$19.316 - j271.815$
$0.246 + j3.601$	$18.853 - j276.425$	$0.259 + j3.655$	$19.270 - j272.234$
$0.237 + j3.584$	$18.398 - j277.783$	$0.257 + j3.647$	$19.200 - j272.836$
$0.254 + j3.629$	$19.162 - j274.232$	$0.254 + j3.636$	$19.096 - j273.675$
$0.239 + j3.600$	$18.329 - j276.574$	$0.249 + j3.621$	$18.933 - j274.843$
$0.265 + j3.659$	$19.715 - j271.845$	$0.243 + j3.600$	$18.651 - j276.528$
$0.228 + j3.605$	$17.447 - j276.327$	$0.231 + j3.567$	$18.068 - j279.167$
$0.324 + j3.693$	$23.582 - j268.702$	$0.199 + j3.508$	$16.092 - j284.146$

Table 5. *Broadside array (driving-point currents specified); $\beta_0 h = 1.57079$*

Admittance	Impedance	Admittance	Impedance
$\beta_0 b = 1.57080$		$\beta_0 b = 3.14159$	
$N = 1$		$N = 1$	
$10.449 - j3.889$	$84.059 + j31.286$	$10.499 - j3.889$	$84.059 + j31.286$
$N = 4$		$N = 4$	
$12.709 + j3.576$	$72.913 - j20.518$	$14.944 - j0.824$	$66.715 + j\ 3.678$
$5.807 + j3.319$	$129.803 - j74.198$	$18.493 + j1.493$	$53.723 - j\ 4.336$
$5.807 + j3.319$	$129.803 - j74.198$	$18.493 + j1.493$	$53.723 - j\ 4.336$
$12.709 + j3.576$	$72.913 - j20.518$	$14.944 - j0.824$	$66.715 + j\ 3.678$
$N = 10$		$N = 10$	
$10.113 + j1.428$	$96.946 - j13.689$	$14.720 - j0.850$	$67.711 + j\ 3.909$
$7.202 + j2.746$	$121.234 - j46.219$	$18.832 + j1.491$	$52.771 - j\ 4.177$
$5.923 + j5.391$	$92.343 - j84.042$	$17.703 + j1.442$	$56.116 - j\ 4.572$
$6.572 + j6.018$	$82.765 - j75.783$	$18.161 + j1.434$	$54.721 - j\ 4.321$
$8.176 + j3.532$	$103.068 - j44.525$	$17.978 + j1.440$	$55.268 - j\ 4.428$
$8.176 + j3.532$	$103.068 - j44.525$	$17.978 + j1.440$	$55.268 - j\ 4.428$
$6.572 + j6.018$	$82.765 - j75.783$	$18.161 + j1.434$	$54.721 - j\ 4.321$
$5.923 + j5.391$	$92.343 - j84.042$	$17.703 + j1.442$	$56.116 - j\ 4.572$
$7.202 + j2.746$	$121.234 - j46.219$	$18.832 + j1.491$	$52.771 - j\ 4.177$
$10.113 + j1.428$	$96.946 - j13.689$	$14.720 - j0.850$	$67.711 + j\ 3.909$
$N = 20$		$N = 20$	
$11.322 + j2.266$	$84.924 - j16.998$	$14.668 - j0.855$	$67.946 + j\ 3.961$
$6.558 + j3.070$	$125.075 - j58.542$	$18.882 + j1.487$	$52.634 - j\ 4.145$
$6.008 + j4.532$	$106.088 - j80.025$	$17.645 + j1.450$	$56.294 - j\ 4.625$
$7.939 + j6.197$	$78.268 - j61.095$	$18.234 + j1.427$	$54.509 - j\ 4.267$
$8.713 + j4.980$	$86.512 - j49.440$	$17.883 + j1.452$	$55.553 - j\ 4.509$
$6.889 + j3.994$	$108.645 - j62.989$	$18.112 + j1.431$	$54.870 - j\ 4.336$
$6.640 + j4.415$	$104.430 - j69.432$	$17.957 + j1.447$	$55.328 - j\ 4.459$
$8.040 + j5.486$	$84.866 - j57.908$	$18.060 + j1.436$	$55.023 - j\ 4.374$
$8.225 + j5.215$	$86.717 - j54.984$	$17.997 + j1.443$	$55.210 - j\ 4.427$
$6.835 + j4.270$	$105.235 - j65.740$	$18.027 + j1.440$	$55.120 - j\ 4.402$
$6.835 + j4.270$	$105.235 - j65.740$	$18.027 + j1.440$	$55.120 - j\ 4.402$
$8.225 + j5.215$	$86.717 - j54.984$	$17.997 + j1.443$	$55.210 - j\ 4.427$
$8.040 + j5.486$	$84.866 - j57.908$	$18.060 + j1.436$	$55.023 - j\ 4.374$
$6.640 + j4.415$	$104.430 - j69.432$	$17.957 + j1.447$	$55.328 - j\ 4.459$
$6.889 + j3.994$	$108.645 - j62.989$	$18.112 + j1.431$	$54.870 - j\ 4.336$
$8.713 + j4.980$	$86.512 - j49.440$	$17.883 + j1.452$	$55.553 - j\ 4.509$
$7.939 + j6.197$	$78.268 - j61.095$	$18.234 + j1.427$	$54.509 - j\ 4.267$
$6.008 + j4.532$	$106.088 - j80.025$	$17.645 + j1.450$	$56.294 - j\ 4.625$
$6.558 + j3.070$	$125.075 - j58.542$	$18.882 + j1.487$	$52.634 - j\ 4.145$
$11.322 + j2.266$	$84.924 - j16.998$	$14.668 - j0.855$	$67.946 + j\ 3.961$

Table 6. *Endfire array (driving-point currents specified);* $\beta_0 h = 1.57079$

Admittance	Impedance	Admittance	Impedance
$\beta_0 b = 1.57080$		$\beta_0 b = 3.14159$	
$N = 1$		$N = 1$	
$10.449 - j3.889$	$84.059 + j\ 31.286$	$10.449 - j3.889$	$84.059 + j\ 31.286$
$N = 4$		$N = 4$	
$21.021 - j0.091$	$47.571 + j\ \ 0.205$	$5.122 - j3.827$	$125.295 + j\ 93.607$
$7.128 - j4.119$	$105.169 + j\ 60.777$	$4.399 - j3.396$	$142.437 + j109.966$
$7.400 - j4.432$	$99.455 + j\ 59.563$	$4.399 - j3.396$	$142.437 + j109.966$
$3.999 - j2.972$	$161.068 + j119.715$	$5.122 - j3.827$	$125.295 + j\ 93.607$
$N = 10$		$N = 10$	
$18.995 - j1.976$	$52.082 + j\ \ 5.419$	$3.533 - j3.361$	$148.595 + j141.338$
$7.821 - j4.265$	$98.548 + j\ 53.747$	$2.986 - j2.896$	$172.544 + j167.362$
$6.462 - j4.040$	$111.269 + j\ 69.563$	$2.754 - j2.736$	$182.750 + j181.551$
$4.552 - j3.546$	$136.734 + j106.504$	$2.640 - j2.661$	$187.898 + j189.374$
$4.458 - j3.576$	$136.500 + j109.499$	$2.591 - j2.629$	$190.138 + j192.923$
$3.461 - j3.112$	$159.762 + j143.670$	$2.591 - j2.629$	$190.138 + j192.923$
$3.638 - j3.296$	$150.965 + j136.780$	$2.640 - j2.661$	$187.898 + j189.374$
$2.863 - j2.783$	$179.606 + j174.555$	$2.754 - j2.736$	$182.750 + j181.551$
$3.232 - j3.222$	$155.171 + j154.712$	$2.986 - j2.896$	$172.544 + j167.362$
$2.450 - j2.360$	$211.705 + j203.950$	$3.533 - j3.361$	$148.595 + j141.338$
$N = 20$		$N = 20$	
$18.250 - j2.610$	$53.695 + j\ \ 7.678$	$2.755 - j2.976$	$167.509 + j180.951$
$8.076 - j4.268$	$96.793 + j\ 51.152$	$2.366 - j2.566$	$194.207 + j210.621$
$6.252 - j4.000$	$113.486 + j\ 72.618$	$2.175 - j2.404$	$206.910 + j228.769$
$4.693 - j3.602$	$134.084 + j102.908$	$2.060 - j2.313$	$214.737 + j241.090$
$4.300 - j3.500$	$139.895 + j113.871$	$1.984 - j2.254$	$220.090 + j249.939$
$3.591 - j3.196$	$155.387 + j138.288$	$1.932 - j2.213$	$223.909 + j256.432$
$3.467 - j3.172$	$156.992 + j143.647$	$1.895 - j2.184$	$226.647 + j261.167$
$3.019 - j2.926$	$170.781 + j165.519$	$1.870 - j2.165$	$228.549 + j264.491$
$2.993 - j2.945$	$169.756 + j167.026$	$1.855 - j2.152$	$229.753 + j266.612$
$2.659 - j2.729$	$183.170 + j187.995$	$1.847 - j2.146$	$230.338 + j267.646$
$2.683 - j2.778$	$179.903 + j186.225$	$1.847 - j2.146$	$230.338 + j267.646$
$2.405 - j2.573$	$193.846 + j207.450$	$1.855 - j2.152$	$229.753 + j266.612$
$2.465 - j2.651$	$188.140 + j202.324$	$1.870 - j2.165$	$228.549 + j264.491$
$2.211 - j2.443$	$203.633 + j225.004$	$1.895 - j2.184$	$226.647 + j261.167$
$2.305 - j2.556$	$194.603 + j215.766$	$1.932 - j2.213$	$223.909 + j256.432$
$2.053 - j2.325$	$213.437 + j241.665$	$1.984 - j2.254$	$220.090 + j249.939$
$2.190 - j2.497$	$198.569 + j226.323$	$2.060 - j2.313$	$214.737 + j241.090$
$1.912 - j2.196$	$225.537 + j258.968$	$2.175 - j2.404$	$206.910 + j228.769$
$2.119 - j2.521$	$195.412 + j232.494$	$2.366 - j2.566$	$194.207 + j210.621$
$1.787 - j1.965$	$253.380 + j278.536$	$2.755 - j2.976$	$167.509 + j180.951$

Table 7. *Broadside array (driving-point currents specified);* $\beta_0 h = 2.35620$

Admittance	Impedance	Admittance	Impedance
$\beta_0 b = 1.57080$		$\beta_0 b = 3.14159$	
N = 1		*N = 1*	
$1.416 - j1.335$	$373.800 + j352.429$	$1.416 - j1.335$	$373.800 + j352.429$
N = 4		*N = 4*	
$1.772 - j2.166$	$226.222 + j276.570$	$1.688 - j1.860$	$267.545 + j294.791$
$2.885 + j0.558$	$334.125 - j\ 64.674$	$1.761 - j2.489$	$189.461 + j267.725$
$2.885 + j0.558$	$334.125 - j\ 64.674$	$1.761 - j2.489$	$189.461 + j267.725$
$1.772 - j2.166$	$226.222 + j276.570$	$1.688 - j1.860$	$267.545 + j294.791$
N = 10		*N = 10*	
$2.546 - j1.081$	$332.848 + j141.276$	$1.668 - j1.738$	$287.394 + j299.522$
$2.273 - j0.367$	$428.840 + j\ 69.202$	$1.706 - j2.798$	$158.797 + j260.557$
$4.977 - j1.391$	$186.363 + j\ 52.072$	$1.767 - j2.111$	$233.130 + j278.516$
$5.319 - j2.786$	$147.544 + j\ 77.279$	$1.743 - j2.477$	$189.977 + j270.003$
$2.547 - j0.541$	$375.674 + j\ 79.816$	$1.766 - j2.302$	$209.859 + j273.425$
$2.547 - j0.541$	$375.674 + j\ 79.816$	$1.766 - j2.302$	$209.859 + j273.425$
$5.319 - j2.786$	$147.544 + j\ 77.279$	$1.743 - j2.477$	$189.977 + j270.003$
$4.977 - j1.391$	$186.363 + j\ 52.072$	$1.767 - j2.111$	$233.130 + j278.516$
$2.273 - j0.367$	$428.840 + j\ 69.202$	$1.706 - j2.798$	$158.797 + j260.557$
$2.546 - j1.081$	$332.848 + j141.276$	$1.668 - j1.738$	$287.394 + j299.522$
N = 20		*N = 20*	
$1.768 - j1.569$	$316.473 + j280.864$	$1.674 - j1.677$	$298.066 + j298.738$
$3.253 + j0.018$	$307.387 - j\ \ \ 1.730$	$1.622 - j2.887$	$147.942 + j263.273$
$3.757 + j1.121$	$244.412 - j\ 72.928$	$1.797 - j2.006$	$247.732 + j276.631$
$2.666 - j1.466$	$288.002 + j158.418$	$1.660 - j2.600$	$174.426 + j273.219$
$2.367 - j1.387$	$314.522 + j184.290$	$1.814 - j2.142$	$230.212 + j271.886$
$3.422 - j0.000$	$292.212 + j\ \ \ 0.020$	$1.690 - j2.481$	$187.563 + j275.284$
$3.590 + j0.494$	$273.339 - j\ 37.627$	$1.802 - j2.227$	$219.579 + j271.342$
$2.689 - j1.268$	$304.189 + j143.435$	$1.721 - j2.406$	$196.649 + j274.916$
$2.397 - j1.327$	$319.318 + j176.796$	$1.779 - j2.291$	$211.463 + j272.272$
$3.683 + j0.277$	$269.994 - j\ 20.304$	$1.751 - j2.347$	$204.192 + j273.668$
$3.683 + j1.277$	$269.994 - j\ 20.304$	$1.751 - j2.347$	$204.192 + j273.668$
$2.397 - j1.327$	$319.318 + j176.796$	$1.779 - j2.291$	$211.463 + j272.272$
$2.689 - j1.268$	$304.189 + j143.435$	$1.721 - j2.406$	$196.649 + j274.916$
$3.590 + j0.494$	$273.339 - j\ 37.627$	$1.802 - j2.227$	$219.579 + j271.342$
$3.422 - j0.000$	$292.212 + j\ \ \ 0.020$	$1.690 - j2.481$	$187.563 + j275.284$
$2.367 - j1.387$	$314.522 + j184.290$	$1.814 - j2.142$	$230.212 + j271.886$
$2.666 - j1.466$	$288.002 + j158.418$	$1.660 - j2.600$	$174.426 + j273.219$
$3.757 + j1.121$	$244.412 - j\ 72.928$	$1.797 - j2.006$	$247.732 + j276.631$
$3.253 + j0.018$	$307.387 - j\ \ \ 1.730$	$1.622 - j2.887$	$147.942 + j263.273$
$1.768 - j1.569$	$316.473 + j280.864$	$1.674 - j1.677$	$298.066 + j298.738$

Table 8. *Broadside array (base voltages specified);* $\beta_0 h = 2.35620$

Admittance	Impedance	Admittance	Impedance
$\beta_0 b = 1.57080$		$\beta_0 b = 3.14159$	
N = 1		*N* = 1	
$1.416 - j1.335$	$373.800 + j352.429$	$1.416 - j1.335$	$373.800 + j352.429$
N = 4		*N* = 4	
$2.378 - j1.235$	$331.264 + j172.017$	$1.667 - j1.929$	$256.499 + j296.734$
$3.440 - j0.465$	$285.452 + j\ 38.575$	$1.796 - j2.409$	$198.923 + j266.764$
$3.440 - j0.465$	$285.452 + j\ 38.575$	$1.796 - j2.409$	$198.923 + j266.764$
$2.378 - j1.235$	$331.264 + j172.017$	$1.667 - j1.929$	$256.499 + j296.734$
N = 10		*N* = 10	
$2.326 - j1.136$	$347.120 + j169.481$	$1.638 - j1.901$	$260.172 + j301.832$
$3.232 - j0.566$	$300.210 + j\ 52.564$	$1.845 - j2.474$	$193.668 + j259.699$
$3.597 - j0.879$	$262.334 + j\ 64.099$	$1.719 - j2.288$	$209.960 + j279.353$
$3.509 - j1.021$	$262.762 + j\ 76.455$	$1.782 - j2.366$	$203.171 + j269.662$
$3.341 - j0.911$	$278.568 + j\ 75.945$	$1.756 - j2.334$	$205.799 + j273.575$
$3.341 - j0.911$	$278.568 + j\ 75.945$	$1.756 - j2.334$	$205.799 + j273.575$
$3.509 - j1.021$	$262.762 + j\ 76.455$	$1.782 - j2.366$	$203.171 + j269.662$
$3.597 - j0.879$	$262.334 + j\ 64.099$	$1.719 - j2.288$	$209.960 + j279.353$
$3.232 - j0.566$	$300.210 + j\ 52.564$	$1.845 - j2.474$	$193.668 + j259.699$
$2.326 - j1.136$	$347.120 + j169.481$	$1.638 - j1.901$	$260.172 + j301.832$
N = 20		*N* = 20	
$2.266 - j1.086$	$358.924 + j172.045$	$1.630 - j1.884$	$262.611 + j303.586$
$3.186 - j0.506$	$306.136 + j\ 48.569$	$1.850 - j2.475$	$193.750 + j259.242$
$3.508 - j0.761$	$272.249 + j\ 59.083$	$1.707 - j2.269$	$211.723 + j281.430$
$3.359 - j0.926$	$276.676 + j\ 76.317$	$1.790 - j2.370$	$202.871 + j268.669$
$3.197 - j0.897$	$289.962 + j\ 81.384$	$1.738 - j2.309$	$208.073 + j276.419$
$3.189 - j0.676$	$300.102 + j\ 63.641$	$1.773 - j2.350$	$204.593 + j271.231$
$3.205 - j0.709$	$297.486 + j\ 65.803$	$1.749 - j2.322$	$206.964 + j274.743$
$3.184 - j0.826$	$294.272 + j\ 76.330$	$1.765 - j2.341$	$205.359 + j272.381$
$3.206 - j0.798$	$293.755 + j\ 73.121$	$1.755 - j2.329$	$206.355 + j273.839$
$3.267 - j0.771$	$289.930 + j\ 68.405$	$1.760 - j2.335$	$205.877 + j273.139$
$3.267 - j0.771$	$289.930 + j\ 68.405$	$1.760 - j2.335$	$205.877 + j273.139$
$3.206 - j0.798$	$293.755 + j\ 73.121$	$1.755 - j2.329$	$206.355 + j273.839$
$3.184 - j0.826$	$294.272 + j\ 76.330$	$1.765 - j2.341$	$205.359 + j272.381$
$3.205 - j0.709$	$297.486 + j\ 65.803$	$1.749 - j2.322$	$206.964 + j274.743$
$3.189 - j0.676$	$300.102 + j\ 63.641$	$1.773 - j2.350$	$204.593 + j271.231$
$3.197 - j0.897$	$289.962 + j\ 81.384$	$1.738 - j2.309$	$208.073 + j276.419$
$3.359 - j0.926$	$276.676 + j\ 76.317$	$1.790 - j2.370$	$202.871 + j268.669$
$3.508 - j0.761$	$272.249 + j\ 59.083$	$1.707 - j2.269$	$211.723 + j281.430$
$3.186 - j0.506$	$306.136 + j\ 48.569$	$1.850 - j2.475$	$193.750 + j259.242$
$2.266 - j1.086$	$358.924 + j172.045$	$1.630 - j1.884$	$262.611 + j303.586$

Table 9. *Endfire array (driving-point currents specified);* $\beta_0 h = 2.35620$

Admittance	Impedance	Admittance	Impedance
	$\beta_0 b = 1.57080$		$\beta_0 b = 3.14159$
	$N = 1$		$N = 1$
$1.416 - j1.335$	$373.800 + j352.429$	$1.416 - j1.335$	$373.800 + j352.429$
	$N = 4$		$N = 4$
$0.768 - j2.562$	$107.348 + j358.072$	$1.009 - j0.493$	$800.083 + j391.021$
$1.181 - j0.897$	$536.798 + j407.909$	$0.932 - j0.393$	$910.638 + j384.363$
$0.995 - j1.165$	$423.857 + j496.201$	$0.932 - j0.393$	$910.638 + j384.363$
$0.811 - j0.485$	$908.772 + j543.270$	$1.009 - j0.493$	$800.083 + j391.021$
	$N = 10$		$N = 10$
$0.719 - j1.989$	$160.692 + j444.648$	$0.816 - j0.100$	$1207.569 + j147.854$
$1.399 - j0.929$	$496.162 + j329.370$	$0.718 - j0.035$	$1390.182 + j\ 68.698$
$0.823 - j0.999$	$491.292 + j596.033$	$0.671 - j0.010$	$1489.227 + j\ 22.603$
$0.942 - j0.508$	$822.512 + j443.818$	$0.646 - j0.004$	$1547.295 - j\ \ 8.699$
$0.751 - j0.657$	$754.637 + j659.352$	$0.635 - j0.010$	$1573.637 - j\ 23.788$
$0.750 - j0.337$	$1108.900 + j498.327$	$0.635 - j0.010$	$1573.637 - j\ 23.788$
$0.694 - j0.473$	$984.268 + j670.369$	$0.646 - j0.004$	$1547.295 - j\ \ 8.699$
$0.633 - j0.227$	$1399.044 + j502.042$	$0.671 - j0.010$	$1489.227 + j\ 22.603$
$0.656 - j0.359$	$1173.727 + j642.474$	$0.718 - j0.035$	$1390.182 + j\ 68.698$
$0.537 - j0.150$	$1727.200 + j481.071$	$0.816 - j0.100$	$1207.569 + j147.854$
	$N = 20$		$N = 20$
$1.317 - j1.946$	$238.516 + j352.369$	$0.757 + j0.160$	$1263.562 - j267.384$
$2.159 - j0.457$	$443.325 + j\ 93.749$	$0.652 + j0.191$	$1412.576 - j413.497$
$1.140 - j0.949$	$518.065 + j431.501$	$0.595 + j0.206$	$1500.532 - j519.963$
$1.353 - j0.279$	$709.164 + j146.166$	$0.558 + j0.216$	$1558.218 - j603.151$
$1.018 - j0.603$	$727.155 + j430.732$	$0.532 + j0.223$	$1597.740 - j669.409$
$1.079 - j0.213$	$892.289 + j176.135$	$0.514 + j0.228$	$1625.266 - j721.834$
$0.886 - j0.440$	$905.806 + j449.935$	$0.501 + j0.232$	$1644.340 - j762.222$
$0.899 - j0.164$	$1076.254 + j196.561$	$0.491 + j0.235$	$1657.144 - j791.705$
$0.781 - j0.332$	$1084.904 + j460.938$	$0.485 + j0.236$	$1665.037 - j811.004$
$0.754 - j0.102$	$1302.915 + j175.574$	$0.483 + j0.237$	$1668.794 - j820.548$
$0.719 - j0.235$	$1255.957 + j410.737$	$0.483 + j0.237$	$1668.794 - j820.548$
$0.652 - j0.051$	$1523.627 + j118.167$	$0.485 + j0.236$	$1665.037 - j811.004$
$0.684 - j0.151$	$1394.547 + j308.085$	$0.491 + j0.235$	$1657.144 - j791.705$
$0.588 - j0.055$	$1686.099 + j156.772$	$0.501 + j0.232$	$1644.340 - j762.222$
$0.603 - j0.142$	$1571.217 + j368.572$	$0.514 + j0.228$	$1625.266 - j721.834$
$0.518 - j0.081$	$1884.893 + j293.189$	$0.532 + j0.223$	$1597.740 - j669.409$
$0.494 - j0.164$	$1824.240 + j606.325$	$0.558 + j0.216$	$1558.218 - j603.151$
$0.445 - j0.052$	$2218.386 + j260.922$	$0.595 + j0.206$	$1500.532 - j519.963$
$0.462 - j0.131$	$2003.505 + j567.693$	$0.652 + j0.191$	$1412.576 - j413.497$
$0.383 - j0.026$	$2597.795 + j175.311$	$0.757 + j0.160$	$1263.562 - j267.384$

Table 10. *Endfire array (base voltages specified);* $\beta_0 h = 2.35620$

Admittance	Impedance	Admittance	Impedance
$\beta_0 b = 1.57080$		$\beta_0 b = 3.14159$	
$N = 1$		$N = 1$	
$1.416 - j1.335$	$373.800 + j\ 352.429$	$1.416 - j1.335$	$373.800 + j352.429$
$N = 4$		$N = 4$	
$1.587 - j1.820$	$272.125 + j\ 312.111$	$1.004 - j0.597$	$735.858 + j437.483$
$1.187 - j0.653$	$646.777 + j\ 355.615$	$0.931 - j0.300$	$972.956 + j313.400$
$1.038 - j0.426$	$824.490 + j\ 338.479$	$0.931 - j0.300$	$972.956 + j313.400$
$0.739 + j0.188$	$1271.479 - j\ 323.466$	$1.004 - j0.597$	$735.858 + j437.483$
$N = 10$		$N = 10$	
$1.568 - j1.797$	$275.752 + j\ 316.025$	$0.833 - j0.377$	$996.584 + j451.392$
$1.220 - j0.689$	$621.279 + j\ 350.921$	$0.734 - j0.054$	$1355.209 + j\ 99.818$
$0.979 - j0.326$	$919.565 + j\ 305.944$	$0.670 + j0.055$	$1482.051 - j121.931$
$0.814 - j0.133$	$1195.813 + j\ 196.023$	$0.637 + j0.101$	$1529.885 - j243.575$
$0.736 - j0.056$	$1351.188 + j\ 103.261$	$0.624 + j0.120$	$1545.965 - j297.753$
$0.658 + j0.033$	$1516.022 - j\ 76.287$	$0.624 + j0.120$	$1545.965 - j297.753$
$0.641 + j0.053$	$1550.594 - j\ 128.146$	$0.637 + j0.101$	$1529.885 - j243.575$
$0.560 + j0.162$	$1646.217 - j\ 476.624$	$0.670 + j0.055$	$1482.051 - j121.931$
$0.605 + j0.072$	$1631.125 - j\ 193.964$	$0.734 - j0.054$	$1355.209 + j\ 99.819$
$0.487 + j0.468$	$1066.800 - j1025.906$	$0.833 - j0.377$	$996.584 + j451.392$
$N = 20$		$N = 20$	
$1.881 - j1.656$	$299.529 + j\ 263.639$	$0.742 - j0.258$	$1201.877 + j418.102$
$1.445 - j0.580$	$595.950 + j\ 239.114$	$0.650 + j0.057$	$1526.390 - j132.914$
$1.149 - j0.242$	$833.526 + j\ 175.946$	$0.584 + j0.171$	$1576.159 - j462.011$
$1.004 - j0.058$	$993.159 + j\ 57.727$	$0.545 + j0.229$	$1560.743 - j655.797$
$0.837 + j0.001$	$1194.570 - j\ 1.535$	$0.520 + j0.264$	$1530.486 - j776.945$
$0.631 + j0.003$	$1583.581 - j\ 6.286$	$0.503 + j0.287$	$1500.443 - j856.095$
$0.607 + j0.054$	$1634.380 - j\ 145.843$	$0.491 + j0.303$	$1475.286 - j908.707$
$0.507 + j0.076$	$1929.067 - j\ 290.680$	$0.484 + j0.313$	$1456.283 - j943.151$
$0.516 + j0.121$	$1837.440 - j\ 429.232$	$0.479 + j0.320$	$1443.647 - j964.120$
$0.509 + j0.179$	$1748.511 - j\ 615.388$	$0.477 + j0.323$	$1437.399 - j974.022$
$0.491 + j0.195$	$1757.837 - j\ 698.636$	$0.477 + j0.323$	$1437.399 - j974.022$
$0.489 + j0.250$	$1621.102 - j\ 827.013$	$0.479 + j0.320$	$1443.647 - j964.120$
$0.442 + j0.240$	$1748.191 - j\ 947.672$	$0.484 + j0.313$	$1456.283 - j943.151$
$0.361 + j0.248$	$1880.501 - j1295.187$	$0.491 + j0.303$	$1475.286 - j908.707$
$0.335 + j0.238$	$1986.679 - j1409.322$	$0.503 + j0.287$	$1500.443 - j856.095$
$0.174 + j0.210$	$2335.692 - j2818.250$	$0.520 + j0.264$	$1530.486 - j776.945$
$0.249 + j0.214$	$2307.799 - j1986.459$	$0.545 + j0.229$	$1560.743 - j655.797$
$0.234 + j0.310$	$1552.071 - j2055.252$	$0.584 + j0.171$	$1576.159 - j462.011$
$0.282 + j0.211$	$2274.708 - j1702.714$	$0.650 + j0.057$	$1526.390 - j132.914$
$0.253 + j0.559$	$671.650 - j1484.457$	$0.742 - j0.258$	$1201.877 + j418.102$

Table 11. *Broadside array (driving-point currents specified);* $\beta_0 h = 3.14159$

Admittance	Impedance	Admittance	Impedance
$\beta_0 b = 1.57080$		$\beta_0 b = 3.14159$	
N = 1		*N = 1*	
$0.985 + j1.000$	$499.710 - j507.494$	$0.985 + j1.000$	$499.710 - j507.494$
N = 4		*N = 4*	
$1.300 + j1.158$	$428.935 - j382.052$	$1.122 + j0.538$	$724.851 - j347.681$
$2.605 + j0.524$	$368.900 - j\ 74.220$	$1.051 + j0.284$	$886.543 - j239.267$
$2.605 + j0.524$	$368.900 - j\ 74.220$	$1.051 + j0.284$	$886.543 - j239.267$
$1.300 + j1.158$	$428.935 - j382.052$	$1.122 + j0.538$	$724.851 - j347.681$
N = 10		*N = 10*	
$1.417 + j1.159$	$422.724 - j345.951$	$1.133 + j0.556$	$711.388 - j349.311$
$2.785 + j0.718$	$336.688 - j\ 86.845$	$1.031 + j0.271$	$907.602 - j238.628$
$2.318 + j0.788$	$386.673 - j131.498$	$1.101 + j0.324$	$835.998 - j246.013$
$2.325 + j0.865$	$377.747 - j140.553$	$1.071 + j0.304$	$864.154 - j245.716$
$2.438 + j0.876$	$363.270 - j130.614$	$1.082 + j0.311$	$853.393 - j245.529$
$2.438 + j0.876$	$363.270 - j130.614$	$1.082 + j0.311$	$853.393 - j245.529$
$2.325 + j0.865$	$377.747 - j140.553$	$1.071 + j0.304$	$864.154 - j245.716$
$2.318 + j0.788$	$386.673 - j131.498$	$1.101 + j0.324$	$835.998 - j246.013$
$2.785 + j0.718$	$336.688 - j\ 86.845$	$1.031 + j0.271$	$907.602 - j238.628$
$1.417 + j1.159$	$422.724 - j345.951$	$1.133 + j0.556$	$711.388 - j349.311$
N = 20		*N = 20*	
$1.410 + j1.161$	$422.735 - j348.012$	$1.135 + j0.561$	$708.268 - j349.951$
$2.761 + j0.700$	$340.320 - j\ 86.216$	$1.028 + j0.270$	$910.264 - j239.139$
$2.337 + j0.773$	$385.711 - j127.522$	$1.105 + j0.326$	$832.672 - j245.894$
$2.343 + j0.890$	$372.969 - j141.735$	$1.066 + j0.302$	$868.043 - j246.271$
$2.405 + j0.899$	$364.852 - j136.406$	$1.088 + j0.315$	$848.027 - j245.190$
$2.409 + j0.834$	$370.623 - j128.334$	$1.074 + j0.307$	$860.805 - j246.146$
$2.381 + j0.833$	$374.226 - j130.931$	$1.084 + j0.312$	$852.290 - j245.320$
$2.378 + j0.878$	$370.031 - j136.654$	$1.077 + j0.309$	$857.942 - j245.875$
$2.390 + j0.879$	$368.540 - j135.585$	$1.081 + j0.311$	$854.451 - j245.464$
$2.392 + j0.840$	$372.168 - j130.632$	$1.079 + j0.310$	$856.102 - j245.671$
$2.392 + j0.840$	$372.168 - j130.632$	$1.079 + j0.310$	$856.102 - j245.671$
$2.390 + j0.879$	$368.540 - j135.585$	$1.081 + j0.311$	$854.451 - j245.464$
$2.378 + j0.878$	$370.031 - j136.654$	$1.077 + j0.309$	$857.942 - j245.875$
$2.381 + j0.833$	$374.226 - j130.931$	$1.084 + j0.312$	$852.290 - j245.320$
$2.409 + j0.834$	$370.623 - j128.334$	$1.074 + j0.307$	$860.805 - j246.146$
$2.405 + j0.899$	$364.852 - j136.406$	$1.088 + j0.315$	$848.027 - j245.190$
$2.343 + j0.890$	$372.969 - j141.735$	$1.066 + j0.302$	$868.043 - j246.271$
$2.337 + j0.773$	$385.711 - j127.522$	$1.105 + j0.326$	$832.672 - j245.894$
$2.761 + j0.700$	$340.320 - j\ 86.216$	$1.028 + j0.270$	$910.264 - j239.139$
$1.410 + j1.161$	$422.735 - j348.012$	$1.135 + j0.561$	$708.268 - j349.951$

Table 12. *Broadside array (base voltages specified);* $\beta_0 h = 3.14159$

Admittance	Impedance	Admittance	Impedance
$\beta_0 b = 1.57080$		$\beta_0 b = 3.14159$	
N = 1		*N* = 1	
$0.985 + j1.000$	$499.710 - j507.494$	$0.985 + j1.000$	$499.710 - j507.494$
N = 4		*N* = 4	
$1.568 + j0.815$	$502.274 - j260.992$	$1.078 + j0.563$	$728.585 - j380.868$
$2.530 + j1.186$	$324.034 - j151.902$	$1.085 + j0.255$	$873.469 - j205.141$
$2.530 + j1.186$	$324.034 - j151.902$	$1.085 + j0.255$	$873.469 - j205.141$
$1.568 + j0.815$	$502.274 - j260.992$	$1.078 + j0.563$	$728.585 - j380.868$
N = 10		*N* = 10	
$1.624 + j0.957$	$457.034 - j269.359$	$1.059 + j0.600$	$714.585 - j405.086$
$2.329 + j1.180$	$341.631 - j173.109$	$1.109 + j0.198$	$873.705 - j156.039$
$2.562 + j0.859$	$350.885 - j117.687$	$1.058 + j0.363$	$845.679 - j290.370$
$2.452 + j0.762$	$371.908 - j115.630$	$1.092 + j0.285$	$857.310 - j223.673$
$2.310 + j0.894$	$376.473 - j145.732$	$1.076 + j0.317$	$855.336 - j252.234$
$2.310 + j0.894$	$376.473 - j145.732$	$1.076 + j0.317$	$855.336 - j252.234$
$2.452 + j0.762$	$371.908 - j115.630$	$1.092 + j0.285$	$857.310 - j223.673$
$2.562 + j0.859$	$350.885 - j117.687$	$1.058 + j0.363$	$845.679 - j290.370$
$2.329 + j1.180$	$341.631 - j173.109$	$1.109 + j0.198$	$873.705 - j156.039$
$1.624 + j0.957$	$457.034 - j269.359$	$1.059 + j0.600$	$714.585 - j405.086$
N = 20		*N* = 20	
$1.626 + j0.929$	$463.503 - j264.895$	$1.053 + j0.609$	$711.428 - j411.334$
$2.368 + j1.183$	$337.980 - j168.790$	$1.115 + j0.190$	$871.458 - j148.401$
$2.558 + j0.901$	$347.775 - j122.478$	$1.051 + j0.374$	$844.861 - j300.449$
$2.400 + j0.756$	$379.037 - j119.387$	$1.101 + j0.272$	$856.088 - j211.894$
$2.319 + j0.832$	$382.028 - j136.976$	$1.065 + j0.334$	$855.035 - j268.393$
$2.393 + j0.910$	$365.151 - j138.824$	$1.090 + j0.294$	$855.212 - j230.435$
$2.428 + j0.879$	$364.115 - j131.830$	$1.072 + j0.321$	$855.872 - j256.175$
$2.376 + j0.826$	$375.550 - j130.485$	$1.084 + j0.303$	$855.462 - j238.823$
$2.362 + j0.838$	$376.066 - j133.411$	$1.077 + j0.314$	$855.814 - j249.498$
$2.408 + j0.883$	$366.082 - j134.306$	$1.081 + j0.309$	$855.664 - j244.371$
$2.408 + j0.883$	$366.082 - j134.306$	$1.081 + j0.309$	$855.664 - j244.371$
$2.362 + j0.838$	$376.066 - j133.411$	$1.077 + j0.314$	$855.814 - j249.498$
$2.376 + j0.826$	$375.550 - j130.485$	$1.084 + j0.303$	$855.462 - j238.823$
$2.428 + j0.879$	$364.115 - j131.830$	$1.072 + j0.321$	$855.872 - j256.175$
$2.393 + j0.910$	$365.151 - j138.824$	$1.090 + j0.294$	$855.212 - j230.435$
$2.319 + j0.832$	$382.028 - j136.976$	$1.065 + j0.334$	$855.035 - j268.393$
$2.400 + j0.756$	$379.037 - j119.387$	$1.101 + j0.272$	$856.088 - j211.894$
$2.558 + j0.901$	$347.775 - j122.478$	$1.051 + j0.374$	$844.861 - j300.449$
$2.368 + j1.183$	$337.980 - j168.790$	$1.115 + j0.190$	$871.458 - j148.401$
$1.626 + j0.929$	$463.503 - j264.895$	$1.053 + j0.609$	$711.428 - j411.334$

Table 13. *Endfire array (driving-point currents specified);* $\beta_0 h = 3.14159$

Admittance	Impedance	Admittance	Impedance
	$\beta_0 b = 1.57080$		$\beta_0 b = 3.14159$
	$N = 1$		$N = 1$
$0.985 + j1.000$	$499.710 - j507.494$	$0.985 + j1.000$	$499.710 - j507.494$
	$N = 4$		$N = 4$
$0.823 + j1.798$	$626.331 - j607.101$	$0.866 + j1.595$	$262.842 - j484.190$
$0.438 + j1.615$	$156.481 - j576.885$	$0.665 + j1.803$	$180.000 - j488.263$
$0.336 + j1.896$	$90.012 - j511.616$	$0.665 + j1.803$	$180.000 - j488.263$
$0.058 + j2.375$	$10.328 - j420.763$	$0.866 + j1.595$	$262.842 - j484.190$
	$N = 10$		$N = 10$
$0.842 + j0.819$	$610.374 - j593.897$	$0.822 + j1.788$	$212.309 - j461.649$
$0.453 + j1.600$	$163.769 - j578.767$	$0.585 + j2.028$	$131.295 - j455.109$
$0.470 + j1.926$	$119.479 - j489.972$	$0.528 + j2.138$	$108.895 - j440.821$
$0.464 + j2.050$	$104.931 - j464.035$	$0.504 + j2.187$	$100.117 - j434.080$
$0.439 + j2.119$	$93.669 - j452.559$	$0.495 + j2.207$	$96.704 - j431.406$
$0.424 + j2.191$	$85.172 - j439.884$	$0.495 + j2.207$	$96.704 - j431.406$
$0.403 + j2.220$	$79.130 - j436.040$	$0.504 + j2.187$	$100.117 - j434.080$
$0.402 + j2.279$	$75.038 - j425.585$	$0.528 + j2.138$	$108.895 - j440.821$
$0.363 + j2.280$	$68.208 - j427.764$	$0.585 + j2.028$	$131.295 - j455.109$
$0.241 + j2.621$	$34.754 - j378.347$	$0.822 + j1.788$	$212.309 - j461.649$
	$N = 20$		$N = 20$
$0.842 + j0.819$	$610.364 - j593.637$	$0.787 + j1.896$	$186.738 - j449.810$
$0.452 + j1.599$	$163.874 - j579.215$	$0.544 + j2.129$	$112.526 - j440.877$
$0.470 + j1.928$	$119.354 - j489.435$	$0.481 + j2.246$	$91.227 - j425.781$
$0.463 + j2.047$	$105.074 - j464.774$	$0.450 + j2.307$	$81.435 - j417.597$
$0.440 + j2.123$	$93.566 - j451.597$	$0.431 + j2.344$	$75.825 - j412.756$
$0.422 + j2.185$	$85.189 - j441.258$	$0.417 + j2.368$	$72.217 - j409.642$
$0.407 + j2.229$	$79.317 - j434.217$	$0.409 + j2.384$	$69.844 - j407.524$
$0.394 + j2.270$	$74.339 - j427.698$	$0.402 + j2.395$	$68.253 - j406.119$
$0.384 + j2.298$	$70.687 - j423.376$	$0.399 + j2.402$	$67.288 - j405.231$
$0.374 + j2.329$	$67.160 - j418.519$	$0.397 + j2.405$	$66.835 - j404.819$
$0.365 + j2.349$	$64.664 - j415.738$	$0.397 + j2.405$	$66.835 - j404.819$
$0.357 + j2.376$	$61.925 - j411.645$	$0.399 + j2.402$	$67.288 - j405.231$
$0.350 + j2.387$	$60.187 - j410.030$	$0.402 + j2.395$	$68.253 - j406.119$
$0.344 + j2.413$	$57.913 - j406.109$	$0.409 + j2.384$	$69.844 - j407.524$
$0.337 + j2.417$	$56.628 - j405.766$	$0.417 + j2.368$	$72.217 - j409.642$
$0.334 + j2.446$	$54.847 - j401.269$	$0.431 + j2.344$	$75.825 - j412.756$
$0.324 + j2.441$	$53.464 - j402.653$	$0.450 + j2.307$	$81.435 - j417.597$
$0.329 + j2.474$	$52.816 - j397.207$	$0.481 + j2.246$	$91.227 - j425.781$
$0.304 + j2.461$	$49.407 - j400.177$	$0.544 + j2.129$	$112.526 - j440.877$
$0.205 + j2.754$	$26.829 - j361.165$	$0.787 + j1.896$	$186.738 - j449.810$

Table 14. *Endfire array (base voltages specified);* $\beta_0 h = 3.14159$

Admittance	Impedance	Admittance	Impedance
$\beta_0 b = 1.57080$		$\beta_0 b = 3.14159$	
$N = 1$		$N = 1$	
$0.985 + j1.000$	$499.710 - j507.494$	$0.985 + j1.000$	$499.710 - j507.494$
$N = 4$		$N = 4$	
$1.062 + j0.567$	$732.748 - j391.436$	$0.776 + j1.612$	$242.440 - j503.755$
$0.912 + j1.492$	$298.230 - j487.943$	$0.761 + j1.800$	$199.335 - j471.228$
$0.860 + j1.706$	$235.711 - j467.372$	$0.761 + j1.800$	$199.335 - j471.228$
$0.680 + j2.257$	$122.374 - j406.108$	$0.776 + j1.612$	$242.440 - j503.755$
$N = 10$		$N = 10$	
$1.015 + j0.656$	$694.918 - j449.322$	$0.636 + j1.827$	$169.953 - j488.127$
$0.940 + j1.444$	$316.569 - j486.420$	$0.608 + j2.045$	$133.598 - j449.294$
$0.826 + j1.754$	$219.728 - j466.639$	$0.585 + j2.135$	$119.452 - j435.688$
$0.719 + j1.948$	$166.737 - j451.867$	$0.570 + j2.180$	$112.239 - j429.418$
$0.663 + j2.032$	$145.060 - j444.764$	$0.562 + j2.199$	$109.080 - j426.883$
$0.595 + j2.136$	$120.953 - j434.437$	$0.562 + j2.199$	$109.080 - j426.883$
$0.577 + j2.155$	$116.025 - j433.058$	$0.570 + j2.180$	$112.239 - j429.418$
$0.512 + j2.251$	$96.063 - j422.314$	$0.585 + j2.135$	$119.452 - j435.688$
$0.547 + j2.186$	$107.662 - j430.552$	$0.608 + j2.045$	$133.598 - j449.294$
$0.506 + j2.472$	$79.476 - j388.190$	$0.636 + j1.827$	$169.953 - j488.127$
$N = 20$		$N = 20$	
$1.012 + j0.660$	$693.386 - j451.865$	$0.556 + j1.946$	$135.869 - j475.068$
$0.943 + j1.439$	$318.711 - j486.074$	$0.530 + j2.155$	$107.519 - j437.577$
$0.822 + j1.760$	$217.821 - j466.334$	$0.505 + j2.249$	$95.051 - j423.302$
$0.723 + j1.940$	$168.775 - j452.604$	$0.484 + j2.305$	$87.325 - j415.556$
$0.657 + j2.042$	$142.835 - j443.665$	$0.468 + j2.340$	$82.177 - j410.927$
$0.602 + j2.123$	$123.550 - j436.002$	$0.456 + j2.363$	$78.648 - j407.944$
$0.567 + j2.172$	$112.536 - j431.044$	$0.447 + j2.379$	$76.234 - j405.963$
$0.531 + j2.223$	$101.632 - j425.476$	$0.441 + j2.390$	$74.611 - j404.642$
$0.512 + j2.251$	$96.033 - j422.381$	$0.437 + j2.397$	$73.614 - j403.834$
$0.485 + j2.291$	$88.376 - j417.712$	$0.435 + j2.400$	$73.133 - j403.451$
$0.474 + j2.307$	$85.487 - j415.897$	$0.435 + j2.400$	$73.133 - j403.451$
$0.451 + j2.343$	$79.227 - j411.531$	$0.437 + j2.397$	$73.614 - j403.834$
$0.446 + j2.349$	$78.112 - j410.911$	$0.441 + j2.390$	$74.611 - j404.642$
$0.425 + j2.386$	$72.358 - j406.303$	$0.447 + j2.379$	$76.234 - j405.963$
$0.425 + j2.380$	$72.675 - j407.110$	$0.456 + j2.363$	$78.648 - j407.944$
$0.403 + j2.423$	$66.816 - j401.672$	$0.468 + j2.340$	$82.177 - j410.927$
$0.410 + j2.404$	$68.970 - j404.249$	$0.484 + j2.305$	$87.325 - j415.556$
$0.376 + j2.462$	$60.644 - j396.977$	$0.505 + j2.249$	$95.051 - j423.302$
$0.417 + j2.388$	$70.880 - j406.343$	$0.530 + j2.155$	$107.519 - j437.577$
$0.408 + j2.615$	$58.233 - j373.316$	$0.556 - j1.946$	$135.869 - j475.068$

References

Preface

1 Some relevant discussions are contained in Chapter 13 of this book, and in: T. T. Wu (1969), Introduction to linear antennas, *Antenna Theory, Part I*, R. E. Collin and F. J. Zucker (eds.), Chapter 8, McGraw-Hill, New York; and D. K. Freeman and T. T. Wu (1991), An improved kernel for arrays of cylindrical dipoles, 1991 Digest of the I.E.E.E./AP-S International Symposium, London, Ontario, June 24–28, I.E.E.E. Press, Piscataway, NJ.

Preface to first edition

1 M. A. Bontsch-Bruewitsch (1926), Die Strahlung der Komplizierten Recht-winkeligen Antennen mit Gleichbeschaffenen Vibratoren, *Ann. Physik.* **81**, 425.
2 G. C. Southworth (1930), Certain factors affecting the gain of directive antennas, *Proc. I.R.E.* **18**, 1502.
3 E. J. Sterba (1931), Theoretical and practical aspects of directional transmitting systems, *Proc. I.R.E.* **19**, 1184.
4 P. S. Carter, C. W. Hansell, and N. E. Lindenblad (1931), Development of directive transmitting antennas by R.C.A. Communications, Inc., *Proc. I.R.E.* **19**, 1773.
5 S. A. Schelkunoff (1943), A mathematical theory of linear arrays, *Bell System Tech. J.* **22**, 80.
6 C. L. Dolph (1946), A current distribution for broadside arrays which optimizes the relationship between beam width and side-lobe level, *Proc. I.R.E.* **34**, 335.
7 T. T. Taylor and J. R. Whinnery (1951), Application of potential theory to the design of linear arrays, *J. Appl. Phys.* **22**, 19.
8 P. S. Carter (1932), Circuit relations in radiating systems and applications to antenna problems, *Proc. I.R.E.* **20**, 1004.
9 G. H. Brown (1937), Directional antennas, *Proc. I.R.E.* **25**, 78.
10 W. Walkinshaw (1946), Theoretical treatment of short Yagi aerials, *J. Inst. Elec. Engrs. (London)* **93**, Part III(A), 598.
11 C. R. Cox (1947), Mutual impedance between vertical antennas of unequal heights, *Proc. I.R.E.* **35**, 1367.
12 G. Barzilai (1948), Mutual impedance of parallel aerials, *Wireless Engr.* **25**, 343 and (1949) **26**, 73.

13   B. Starnecki and E. Fitch (1948), Mutual impedance of two center-driven parallel aerials, *Wireless Engr.* **25**, 385.

14   H. Brückmann (1939), *Antennen*, S. Hirzel, Leipzig.

15   C. T. Tai (1948), Coupled antennas, *Proc. I.R.E.* **36**, 487.

16   R. King (1950), Theory of *N*-coupled parallel antennas, *J. Appl. Phys.* **21**, 94.

17   R. W. P. King (1956), *Theory of Linear Antennas*, Chapter 3, Harvard University Press, Cambridge, MA.

18   J. Aharoni (1946), *Antennae*, Clarendon Press, Oxford.

19   J. Stratton (1941), *Electromagnetic Theory*, McGraw-Hill, New York.

20   R. C. Hansen (1966), *Microwave Scanning Antennas*, Vol. 2, Academic Press, New York.

21   R. King (1959), Linear arrays: currents, impedances and fields, *Trans. I.R.E.* **AP-7**, S440.

22   R. B. Mack (1963), *A Study of Circular Arrays*, Technical Reports, Nos. 381–386, Cruft Laboratory, Harvard University.

23   R. W. P. King and S. S. Sandler (1963), Linear arrays: currents, impedances and fields, II, *Electromagnetic Theory and Antennas*, E. C. Jordan (ed.), 1307, Macmillan Co., New York.

24   R. W. P. King and S. S. Sandler (1964), Theory of broadside arrays; Theory of endfire arrays, *Trans. I.E.E.E.* **AP-12**, 269, 276.

25   R. J. Mailloux (1966), The long Yagi–Uda array, *Trans. I.E.E.E.* **AP-14**, 128.

26   I. L. Morris (1965), *Optimization of the Yagi Array*, Scientific Reports, Nos. 6 and 10 (Series 3), Cruft Laboratory, Harvard University.

27   W.-M. Cheong (1967), *Arrays of Unequal and Unequally-Spaced Dipoles*, Ph.D. Thesis, Harvard University.

28   R. W. P. King and T. T. Wu (1965), Currents, charges and near fields of cylindrical antennas, *Radio Science, Jour. Res. NBS* **69D**, 429.

Chapter 1

1   R. W. P. King and S. Prasad (1986), *Fundamental Electromagnetic Theory and Applications*, Prentice-Hall, Englewood Cliffs, NJ.

2   R. W. P. King and C. W. Harrison, Jr. (1969), *Antennas and Waves: A Modern Approach*, The M.I.T. Press, Cambridge, MA.

3   R. W. P. King (1956), *Theory of Linear Antennas*, Harvard University Press, Cambridge, MA.

4   The method originated from the paper by L. Brillouin (1922), Sur l'origine de la resistance de rayonnement, *Radioélectricité* **3**, 147. Some related early papers are: A. A. Pistolkors (1929), The radiation resistance of beam antennas, *Proc. I.R.E.* **17**, 562; R. Bechmann (1931), On the calculation of radiation resistance of antennas and antenna combinations, *Proc. I.R.E.* **19**, 1471; P. S. Carter (1932),

Circuit relations in radiating systems and applications to antenna problems, *Proc. I.R.E.* **20**, 1004.

5　E. C. Jordan and K. G. Balmain (1968), *Electromagnetic Waves and Radiating Systems*, 2nd edition, pp. 540–547, Prentice-Hall, Englewood Cliffs, NJ.

6　M. Abramowitz and I. A. Stegun (eds.) (1972), *Handbook of Mathematical Functions*, Dover Publications, New York.

7　T. Morita (1950), Current distributions on transmitting and receiving antennas, *Proc. I.R.E.* **38**, 898.

8　R. W. P. King and T. T. Wu (1965), Currents, charges, and near fields of cylindrical antennas, *Radio Science* **69D**, 429.

9　R. W. P. King (1967), The linear antenna – eighty years of progress, *Proc. I.E.E.E.* **55**, 2.

10　T. T. Wu (1969), Introduction to linear antennas, *Antenna Theory, Part I*, R. E. Collin and F. J. Zucker (eds.), Chapter 8, McGraw-Hill, New York.

11　Many such methods are described or referred to in [3] and [9].

12　R. F. Harrington (1968), *Field Computation by Moment Methods*, Macmillan Co., New York; reprinted, I.E.E.E. Press, Piscataway, NJ, 1993.

13　The name originated from the paper by B. G. Galerkin (1915), Series solutions of some problems of elastic equilibrium of rods and plates (in Russian), *Vestnik Inzh. i Tekh.* **19**, 897.

14　Finite Toeplitz matrices are discussed concisely in Chapter 2 of W. H. Press, B. P. Flannery, S. A. Teukolsky, and W. T. Vetterling (1992), *Numerical Recipes in FORTRAN. The Art of Scientific Computing*, 2nd edition, Cambridge University Press, New York.

15　One such technique is described in Chapter 9 of G. Fikioris (1993), *Resonant Arrays of Cylindrical Dipoles: Theory and Experiment*, Ph.D. Thesis, Harvard University.

16　R. Mittra and C. A. Klein (1975), Stability and convergence of moment method solutions, Chapter 5 in *Numerical and Asymptotic Techniques in Electromagnetics*, R. Mittra (ed.), Springer-Verlag, New York.

17　R. S. Elliott (1981), *Antenna Theory and Design*, Prentice Hall, Englewood Cliffs, NJ.

18　T. K. Sarkar (1983), A study of various methods for computing electromagnetic field utilizing thin wire integral equations, *Radio Science* **18**, 29.

19　R. E. Collin (1984), Equivalent line current for cylindrical dipole antennas and its asymptotic behavior, *Trans. I.E.E.E.* **AP-32**, 200.

20　R. E. Collin (1985), *Antennas and Radiowave Propagation*, McGraw-Hill, New York.

21　A. G. Tijhuis, P. Zhongqiu, and A. R. Bretones (1992), Transient excitation of a straight thin-wire segment: a new look at an old problem, *Trans. I.E.E.E.* **AP-40**, 1132.

Chapter 2

1 R. W. P. King (1956), *Theory of Linear Antennas*, Harvard University Press, Cambridge, MA.

Chapter 3

1 R. W. P. King (1990), Electric fields and vector potentials of thin cylindrical antennas, *Trans. I.E.E.E.* **AP-38**, 1456.

Chapter 4

1 R. B. Mack (1963), *A Study of Circular Arrays: Part 4, Tables of Quasi-Zeroth-Order $\Psi_{dR}^{(m)}(h)$, $T^{(m)}(h)$, $T'^{(m)}(\lambda/4)$, Admittances, and Quasi-First-Order Susceptances*, Cruft Laboratory Technical Report 384, Harvard University.

2 R. B. Mack (1963), *A Study of Circular Arrays: Part 2, Self and Mutual Admittances*, Cruft Laboratory Technical Report 382, Harvard University.

Chapter 5

1 R. W. P. King, B. H. Sandler, and S. S. Sandler (1964), *Tables for Curtain Arrays*, Cruft Laboratory Scientific Report 4 (Series 3), Harvard University.

2 R. W. P. King (1956), *Theory of Linear Antennas*, Harvard University Press, Cambridge, MA.

3 R. B. Mack and E. W. Mack (1960), *Tables of $E(h, z)$, $C(h, z)$, $S(h, z)$*, Cruft Laboratory Technical Report 331, Harvard University.

4 E. C. Jordan and K. G. Balmain (1968), *Electromagnetic Waves and Radiating Systems*, 2nd edition, Prentice-Hall, Englewood Cliffs, NJ.

5 S. S. Sandler (1962), *A General Theory of Curtain Arrays*, Ph.D. Thesis, Harvard University.

Chapter 6

1 R. J. Mailloux (1966), The long Yagi–Uda array, *Trans. I.E.E.E.* **AP-14**, 128–137.

2 S. S. Sandler (1962), *A General Theory of Curtain Arrays*, Ph.D. Thesis, Harvard University.

3 I. L. Morris (1965), *Optimization of the Yagi Array*, Ph.D. Thesis, Harvard University.

4 R. W. P. King (1956), *Theory of Linear Antennas*, Harvard University Press, Cambridge, MA.

5 R. W. P. King, A. H. Wing, and H. R. Mimno (1965), *Transmission Lines, Antennas and Wave Guides*, Dover Publications, New York.

6   W.-M. Cheong (1967), *Arrays of Unequal and Unequally-Spaced Dipoles*, Ph.D. Thesis, Harvard University.

7   D. E. Isbell (1960), Log-periodic dipole arrays, *Trans. I.R.E.* **AP-8**, 260–267.

8   E. C. Jordan, G. A. Deschamps, J. D. Dyson, and R. E. Mayes (1964), Developments in broadband antennas, *I.E.E.E. Spectrum* **1**, 58–71.

9   R. L. Carrel (1961), *Analysis and Design of the Log-Period Dipole Antenna*, Technical Report 52, Electrical Engineering Research Laboratory, University of Illinois, Urbana, IL.

10 V. H. Rumsey (1966), *Frequency-Independent Antennas*, Academic Press, New York.

11 R. W. P. King (1965), *Transmission-Line Theory*, Dover Publications, New York.

12 W.-M. Cheong and R. W. P. King (1967), Log-periodic dipole antenna, *Radio Science*, **2**, 1315–1325.

13 W.-M. Cheong and R. W. P. King (1967), Arrays of unequal and unequally spaced dipoles, *Radio Science*, **2**, 1303–1314.

Chapter 7

1   R. W. P. King and T. T. Wu (1965), The cylindrical antenna with arbitrary driving point, *Trans. I.E.E.E.* **AP-13**, 710–718.

2   R. W. P. King (1956), *Theory of Linear Antennas*, Harvard University Press, Cambridge, MA.

3   V. W. H. Chang and R. W. P. King (1968), Theoretical study of dipole array of N parallel elements, *Radio Science*, **3**, 411–424.

Chapter 8

1   R. W. P. King and S. S. Sandler (1994), The electromagnetic field of a vertical electric dipole over the earth or sea, *Trans. I.E.E.E.*, **42**, 382.

2   R. W. P. King and S. S. Sandler (1994), The electromagnetic field of a vertical electric dipole in the presence of a three-layered region, *Radio Science* **29**, 97.

3   R. W. P. King (1990), Electromagnetic field of a vertical dipole over an imperfectly conducting half-space, *Radio Science* **25**, 149.

4   R. W. P. King, M. Owens, and T. T. Wu (1992), *Lateral Electromagnetic Waves*, Springer-Verlag, New York.

5   R. W. P. King (1956), *Theory of Linear Antennas*, Harvard University Press, Cambridge, MA.

6   V. Houdzoumis (1994), *Part I: Scattering of Electromagnetic Missiles. Part II: Vertical Electric Dipole Radiation over Spherical Earth*, Ph.D. Thesis, Harvard University.

Chapter 9

1 R. W. P. King (1983), The wave antenna for transmission and reception, *Trans. I.E.E.E.*, **AP-31**, 956.
2 R. W. P. King, M. Owens, and T. T. Wu (1992), *Lateral Electromagnetic Waves*, Springer-Verlag, New York.
3 R. W. P. King (1992), The circuit properties and complete fields of horizontal-wire antennas and arrays over earth and sea, *J. Appl. Phys.*, **71**, 1499.
4 B. Rama Rao (1991), Paper No. 68.6, 1991 Digest of the I.E.E.E./AP-S International Symposium, London, Ontario, June 24–28, I.E.E.E. Press, Piscataway, NJ.
5 R. W. P. King (1986), Antennas in material media near boundaries with application to communication and geophysical exploration, Part II: The terminated insulated antenna, *Trans. I.E.E.E.*, **AP-34**, 490.
6 V. Houdzoumis (1994), *Part I: Scattering of Electromagnetic Missiles. Part II: Vertical Electric Dipole Radiation over Spherical Earth*, Ph.D. Thesis, Harvard University.
7 R. W. P. King (1985), Scattering of lateral waves by buried or submerged objects, I. The incident lateral-wave field, *J. Appl. Phys.* **57**, 1453.
8 R. W. P. King (1985), Scattering of lateral waves by buried or submerged objects, II. The electric field on the surface above a buried insulated wire, *J. Appl. Phys.* **57**, 1460.
9 R. W. P. King (1989), Lateral electromagnetic waves from a horizontal antenna for remote sensing in the ocean, *Trans. I.E.E.E.* **AP-37**, 1250.
10 R. W. P. King (1991), The electromagnetic field of a horizontal electric dipole in the presence of a three-layered region, *J. Appl. Phys.* **69**, 7987.
11 G. Gilbert, M. Braunstein, and J. Ralston (1994), Characterizing and correcting media-induced focus errors in synthetic aperture radar imagery, *Proc. SPIE* **2230**, 87.
12 R. W. P. King (1992), Electromagnetic field of dipoles and patch antennas on microstrip, *Radio Science* **27**, 71.
13 R. K. Hoffman (1987), *Handbook of Microwave Integrated Circuits*, Sec. 3.2, Artech House, Norwood, MA.

Chapter 11

1 G. Fikioris, R. W. P. King, and T. T. Wu (1994), A novel resonant circular array: improved analysis, *Progress in Electromagnetic Research*, Vol. 8, J. A. Kong (ed.), pp. 1–30, EMW Publishing, Cambridge, MA.
2 R. J. Mailloux (1965), Antenna and wave theories of infinite Yagi–Uda arrays, *Trans. I.E.E.E.* **AP-13**, 499.
3 J. Shefer (1963), Periodic cylinder arrays as transmission lines, *Trans. I.E.E.E.* **MTT-11**, 55.

4   T. T. Wu (1984), Fermi pseudopotentials and resonances in arrays, *Resonances–Models and Phenomena: Proceedings, Bielefeld 1984*, S.Albeverio, L. S. Ferreira, and L. Streit (eds.), pp. 293–306, Springer-Verlag, Berlin.

5   A. Grossmann and T. T. Wu (1987), A class of potentials with extremely narrow resonances, *Chin. J. Phys.* **25**, 129.

6   R. W. P. King (1989), Supergain antennas and the Yagi and circular arrays, *Trans. I.E.E.E.* **AP-37**, 178.

7   G. Fikioris, R. W. P. King, and T. T. Wu (1990), The resonant circular array of electrically short elements, *J. Appl. Phys.* **68**, 431.

8   D. K. Freeman and T. T. Wu (1991), An improved kernel for arrays of cylindrical dipoles, 1991 Digest of the I.E.E.E./AP-S International Symposium, London, Ontario, June 24–28, I.E.E.E. Press, Piscataway, NJ.

9   D. K. Freeman (1992), *Extremely Narrow Resonances in Closed-Loop Arrays of Quantum Mechanical and Electromagnetic Interactions*, Ph.D. Thesis, Harvard University.

10 G. Fikioris (1993), *Resonant Arrays of Cylindrical Dipoles: Theory and Experiment*, Ph.D. Thesis, Harvard University.

11 R. W. P. King (1990), The large circular array with one element driven, *Trans. I.E.E.E.* **AP-38**, 1462.

12 J. D. Tillman, Jr. (1966), *The Theory and Design of Circular Antenna Arrays*, University of Tennessee Experiment Station, Knoxville, TN.

13 H.-M. Shen and T. T. Wu (1989), The universal current distribution near the end of a tubular antenna, *J. Math. Phys.* **30**, 2721.

14 D. K. Freeman and T. T. Wu (1995), Variational-principle formulation of the two-term theory for arrays of cylindrical dipoles, *Trans. I.E.E.E.* **AP-43**, 340.

15 H.-M. Shen (1991), Experimental study of the resonance of a circular array, *Proc. SPIE* **1407**, 306.

16 R. W. P. King (1990), Electromagnetic field of a vertical dipole over an imperfectly conducting half-space, *Radio Science* **25**, 149.

17 R. W. P. King (1990), On the radiation efficiency and the electromagnetic field of a vertical electric dipole in the air above a dielectric or conducting half-space, *Progress in Electromagnetic Research*, Vol. 4, J. A. Kong (ed.), pp. 1–43, Elsevier, New York.

18 M. I. Skolnik (1980), *Introduction to Radar Systems*, McGraw-Hill, New York.

19 A. Grossman, J. Schwartz, and T. T. Wu (1988), Antenne électromagnétique émettrice ou réceptrice très directive, French Patent No. 2 605 148.

20 G. Fikioris, R. W. P. King, and T. T. Wu (1996), Novel surface-wave antenna, *I.E.E. Proc. Microw. Antennas Propagat.* **143**, 1.

21 R. W. P. King and S. S. Sandler (1994), The electromagnetic field of a vertical electric dipole in the presence of a three-layered region, *Radio Science* **29**, 97.

22 T. T. Wu (1969), Introduction to linear antennas, *Antenna Theory, Part I*, R. E. Collin and F. J. Zucker (eds.), Chapter 8, McGraw-Hill, New York.

Chapter 12

1   G. Fikioris (1993), *Resonant Arrays of Cylindrical Dipoles: Theory and Experiment*, Ph.D. Thesis, Harvard University.
2   R. W. P. King (1969), Cylindrical antennas and arrays, *Antenna Theory, Part I*, R. E. Collin and F. J. Zucker (eds.), Chapter 9, McGraw–Hill, New York.
3   R. W. P. King and T. T. Wu (1966), The imperfectly conducting cylindrical transmitting antenna, *Trans. I.E.E.E.* **AP-14**, 524.
4   R. W. P. King, C. W. Harrison, Jr., and E. A. Aronson (1966), The imperfectly conducting cylindrical transmitting antenna: numerical results, *Trans. I.E.E.E.* **AP-14**, 535.
5   L. C. Shen (1967), An experimental study of imperfectly conducting dipoles, *Trans. I.E.E.E.* **AP-15**, 782.
6   R. W. P. King (1963), *Fundamental Electromagnetic Theory*, Chapter V, Dover Publications, New York.
7   G. Fikioris (1998), Experimental study of novel resonant circular arrays, *I.E.E. Proc. Microw. Antennas Propagat.* **145**, 92.
8   G. Fikioris, R. W. P. King, and T. T. Wu (1994), A novel resonant circular array: improved analysis, *Progress in Electromagnetic Research*, Vol. 8, J. A. Kong (ed.), pp. 1–30, EMW Publishing, Cambridge, MA.

Chapter 13

1   B. G. Galerkin (1915), Series solutions of some problems of elastic equilibrium of rods and plates (in Russian), *Vestnik Inzh. i Tekh.* **19**, 897.
2   R. F. Harrington (1968), *Field Computation by Moment Methods*, Macmillan Co., New York; reprinted, I.E.E.E. Press, Piscataway, NJ, 1993.
3   T. T. Wu (1969), Introduction to linear antennas, *Antenna Theory, Part I*, R. E. Collin and F. J. Zucker (eds.), Chapter 8, McGraw-Hill, New York.
4   M. Abramowitz and I. A. Stegun (eds.) (1972), *Handbook of Mathematical Functions*, Dover Publications, New York.
5   Matrix condition numbers and matrix ill-conditioning are discussed in many linear algebra and numerical analysis textbooks. A good discussion is given in B. Noble and J. W. Daniel (1988), *Applied Linear Algebra*, pp. 271–274, Prentice-Hall, Englewood Cliffs, NJ.
6   F. R. de Hoog (1980), Review of Fredholm equations of the first kind, *The Application and Numerical Solution of Integral Equations*, R. S. Anderssen, F. R. de Hoog, and M. A. Lukas (eds.), Sijthoff and Noordjoff, Alphen aan den Rijn, The Netherlands.
7   B. Noble (1977), The numerical solution of integral equations, *The State of the Art in Numerical Analysis*, D. Jacobs (ed.), Chapter VII.3, Academic Press, New York.

8   C. T. H. Baker (1978), *The Numerical Treatment of Integral Equations*, Chapter 5, Clarendon Press, Oxford, England.

9   IMSL User's Manual (1991), *MATH/LIBRARY FORTRAN Subroutines for Mathematical Applications*, Version 2.0, IMSL, Houston, TX.

10 D. S. Jones (1981), Note on the integral equation for a straight wire antenna, *I.E.E. Proc.* **128**, pt. H, 114.

11 H.-M. Shen and T. T. Wu (1989), The universal current distribution near the end of a tubular antenna, *J. Math. Phys.* **30**, 2721.

12 R. W. P. King (1971), *Tables of Antenna Characteristics*, IFI/Plenum, New York.

13 C.-L. Chen and T. T. Wu (1969), Theory of the long dipole antenna, *Antenna Theory, Part I*, R. E. Collin and F. J. Zucker (eds.), Chapter 10, McGraw-Hill, New York.

14 T. T. Wu (1962), Input admittance of infinitely long dipole antennas driven from coaxial lines, *J. Math. Phys.* **3**, 1298.

15 R. H. Duncan (1962), Theory of the infinite cylindrical antenna including the feedpoint singularity in antenna current, *J. Res. Natl. Bur. Std.* **66D**, 181.

16 D. C. Chang and T. T. Wu (1968), A note on the theory of end-corrections for thick monopoles, *Radio Science* **3**, 639.

17 D. C. Chang (1967), On the electrically thick cylindrical antenna, *Radio Science* **2**, 1043.

18 D. K. Freeman and T. T. Wu (1991), An improved kernel for arrays of cylindrical dipoles, 1991 Digest of the I.E.E.E./AP-S International Symposium, London, Ontario, June 24–28, I.E.E.E. Press, Piscataway, NJ.

19 D. K. Freeman (1992), *Extremely Narrow Resonances in Closed-Loop Arrays of Quantum Mechanical and Electromagnetic Interactions*, Chapter 5, Ph.D. Thesis, Harvard University.

20 G. Fikioris (1993), *Resonant Arrays of Cylindrical Dipoles: Theory and Experiment*, Chapter 9, Ph.D. Thesis, Harvard University.

21 I. S. Gradshteyn and I. M. Ryzhik (1980), *Tables of Integrals, Series, and Products*, Academic Press, New York.

22 P. M. Morse and H. Feshbach (1953), *Methods of Theoretical Physics, Part I*, pp. 466–467, McGraw-Hill, New York.

Chapter 14

1   A. E. Bailey (1989), *Microwave Measurements*, Peter Peregrinus Ltd., London.

2   P. I. Somlo and J. D. Hunter (1985), *Microwave Impedance Measurements*, Peter Peregrinus Ltd., London.

3   T. S. Laverghetta (1978), *Microwave Measurements and Techniques*, Artech House, Norwood, MA.

4   M. Sucher and J. Fox (1963), *Handbook of Microwave Measurements*, Polytechnic Press, Polytechnic Institute of Brooklyn, New York.

5   R. Rubin (1961), Antenna measurements, *Antenna Engineering Handbook*, H. Jasik (ed.), Chapter 34, McGraw-Hill, New York.

6   E. L. Ginzton (1957), *Microwave Measurements*, McGraw-Hill, New York.

7   G. C. Montgomery (1947), *Technique of Microwave Measurements*, Vol. 11, Radiation Laboratory Series, McGraw-Hill, New York.

8   R. S. Elliot (1981), *Antenna Theory and Design*, Prentice-Hall, Englewood Cliffs, NJ.

9   R. W. P. King (1965), *Transmission Line Theory*, Dover Publications, New York.

10   R. W. P. King (1956), *Theory of Linear Antennas*, Harvard University Press, Cambridge, MA.

11   T. T. Wu (1963), Input admittance of linear antennas driven from a coaxial line, *J. Res. NBS* **67D**, 83.

12   R. F. Harrington (1968), *Field Computation by Moment Methods*, Macmillan Co., New York; reprinted, I.E.E.E. Press, Piscataway, NJ, 1993.

13   N. Marcuvitz (1964), *Wave-Guide Handbook*, Vol. 10, Radiation Laboratory Series, Boston Technical Publishers, Lexington, MA.

14   F. Sporleder and H. G. Unger (1979), *Waveguide Tapers, Transitions and Couplers*, Peter Peregrinus Ltd., London.

15   J. S. Izadian and S. M. Izadian (1985), *Microwave Transition Design*, Artech House, Norwood, MA.

16   B. C. Wadell (1991), *Transmission Line Design Handbook*, Artech House, Norwood, MA.

17   T. Saad (1971), *Microwave Engineers' Handbook*, Vols. 1 and 2, Artech House, Norwood, MA.

18   P. L. D. Abrie (1985), *The Design of Impedance-Matching Networks for Radio-Frequency and Microwave Amplifiers*, Artech House, Norwood, MA.

19   K. Iizuka and R. W. P. King (1962), *Terminal-Zone Corrections for a Dipole Driven by a Two-Wire Line*, Cruft Laboratory Technical Report 352, Harvard University.

20   R. B. Mack (1963), *A Study of Circular Arrays: Part I, Experimental Equipment*, Cruft Laboratory Technical Report 381, Harvard University.

21   W. H. Huggins (1947), Broadband noncontacting short circuit for coaxial lines, *Proc. I.R.E.* **35**, 966.

22   R. W. P. King (1958), Quasi-stationary and nonstationary currents in electric circuits, *Handbuch der Physik*, Vol. 16, Springer-Verlag, Berlin.

23   H. Whiteside (1962), *Electromagnetic Field Probes*, Cruft Laboratory Technical Report 377, Harvard University.

24   H. Whiteside and R. W. P. King (1964), The loop antenna as a probe, *Trans. I.E.E.E.* **AP-12**, 291.

25   R. W. P. King (1957), *The Loop Antenna as a Probe in Arbitrary Electromagnetic Fields*, Cruft Laboratory Technical Report 262, Harvard University.

26   R. W. P. King (1959), The rectangular loop as a dipole, *Trans. I.R.E.* **AP-7**, 53.

27   R. W. P. King (1963), *Fundamental Electromagnetic Theory*, Dover Publications, New York.

28   R. W. P. King and C. W. Harrison, Jr. (1969), *Antennas and Waves: A Modern Approach*, The M.I.T. Press, Cambridge, MA.

29   J. E. Storer (1956), Impedance of thin wire loop antennas, *Trans. A.I.E.E.* **75**, Part 1 (Communications and Electronics), 606.

30   E. F. Knott (1965), A surface field measurement facility, *Proc. I.E.E.E.* **53**, 1105.

31   H.-M. Shen (1991), Experimental study of the resonance of a circular array, *Proc. SPIE* **1407**, 306.

32   G. Fikioris (1993), *Resonant Arrays of Cylindrical Dipoles: Theory and Experiment*, Ph.D. Thesis, Harvard University.

List of symbols

$$\mathbf{A} \qquad \text{vector potential, } 4, 5$$

$$A \qquad \text{complex factor, } 329$$

$$A_i, A_k, A'_k \qquad \text{complex coefficients, } 113, 120, 162, 164, 248$$

$$A_{zk}^{\text{even}}(z_k), A_{zk}^{\text{odd}}(z_k) \qquad \text{even and odd parts of vector potential, } 243, 244$$

$$A(N, n) \qquad \text{array factor, } 358$$

$$A(\Theta, \Phi) \qquad \text{array factor of uniform array, } 357$$

$$A^{(m)}(\theta, \phi) \qquad \text{array factor for } m\text{th phase-sequence current, } 405$$

$$a \qquad \text{radius of antenna, } 2$$

$$a \qquad \text{radius of spherical earth, } 367$$

$$a_e \qquad \text{effective radius of monopole with reflected image, } 306$$

$$a_1(w) \qquad \text{coefficient of inductive coupling of line and load, } 477$$

$$\mathbf{B} \qquad \text{magnetic vector, } 3$$

$$\mathbf{B}^r \qquad \text{magnetic vector in radiation zone, } 10$$

$$\mathbf{B}_D^i(y) \qquad \text{incident magnetic field in dipole mode of loop, } 513$$

$$\mathbf{B}_T^i(y) \qquad \text{incident magnetic field in transmission-line mode of loop, } 513$$

$$B^i, B^n, B^r \qquad \text{magnetic field in intermediate zone, near field, and far field, } 294\text{–}6$$

$$B^{(m)} \qquad m\text{th phase-sequence susceptance, } 398$$

$$B_c \qquad \text{lumped corrective susceptance, } 42$$

$$B_i, B_{iR}, B_{iI} \qquad \text{complex coefficient and real and imaginary parts, } 113, 120, 121, 162, 163$$

$$B_k, B'_k \qquad \text{complex coefficients, } 248$$

$$B_{1,l} \qquad \text{self- and mutual susceptances for circular array with one driven element, } 398$$

$$B_T \qquad \text{lumped susceptance for terminal zone, } 42$$

$$B_{\text{res}}^{(m)} \qquad \text{resonant } m\text{th phase-sequence susceptance, } 400$$

$$B_\Phi \qquad \text{cylindrical or spherical component of magnetic field, } 9$$

$$B_\Phi^r \qquad \text{cylindrical or spherical component of magnetic field in radiation zone, } 9$$

$$B_0 \qquad \text{driving-point susceptance, } 2$$

$$b \qquad \text{radius of insulating dielectric layer, } 357$$

$$b_{ki}, b_{nl} \qquad \text{axis-to-axis distance between antennas } i \text{ and } k, \text{ or } n \text{ and } l, \ 80, 380$$

G_0	driving-point conductance, 2
g	parameter for spherical earth, 329
$g_m(\theta)$	parameter, 412
$g(\Theta, \beta_0 h), g'(\Theta, \beta_0 h)$	field functions of array, 69
$H^{(m)}$	parameter, 431
H_{0z}	component in three-term and five-term current, 109
$H(z)$	step function, 3
$H_m(\Theta, \beta_0 h)$	field factor, 45
h	half-length of antenna, 2
h	half-length of rectangular patch antenna, 376
h_e	effective half-length of vertical dipole, 292
h_{eD}	effective length of each half of loop for dipole mode, 514
h_m	maximizing length of terminated wire, 352
$h_e(\pi/2)$	effective half-length of receiving antenna, 53
$h_e(\Theta_0, \Phi_0)$	effective length of horizontal-wire antenna, 352
$h_{eN}(\Theta, \Phi)$	effective half-length of N-element Yagi array, 208
I_m	maximum of sinusoidal current, 13
I_V, I_V'	complex coefficients, 37, 39
$I_V^{(m)}$	complex coefficient for phase sequences, 59, 62
$I(z)$	total current along antenna, 7
$I^{(m)}(z)$	mth phase-sequence current, 62, 84
$I^{(\infty)}(z)$	current on infinitely long antenna, 453
$I_{ap}(z), I_{ex}(z)$	currents obtained using approximate and exact kernels, 452, 453
$I_D(s), I_T(s)$	current induced in loop in dipole and transmission-line modes, 515
$I_L(s)$	load current in loop, 515
$I_{1L}(w)$	current along transmission line, 477
$I_x(x)$	current in monopole in infinite conducting medium, 304
$I_z(z)$	total axial current, 31, 37
$I_z(0)$	current in load of receiving antenna, 52
$I_{zi}^{even}(z_i), I_{zi}^{odd}(z_i)$	even and odd parts of current, 244
$I_D(\Theta, \Phi)$	current in receiving antenna, 207
$I_2(\eta_2, g)$	integral function for spherical earth, 329
$\mathbf{J}$	volume density of current, 3
$\mathbf{J}_S$	surface current on ground plane, 437

w	half-width of rectangular patch antenna,	376
w_{mn}	complex coefficient,	411
$w(z)$	normalized current distribution,	64
X_i, Y_i, Z_i	Cartesian coordinates for center of element i,	255
X_0	driving-point reactance,	2
$Y^{(m)}$	mth phase-sequence admittance,	62, 398
$Y^{(0)}, Y^{(1)}$	zero- and first-phase-sequence admittances of loop,	516
Y_a	apparent admittance of load terminating line,	481
Y_c	characteristic admittance of line,	488
Y_{1in}	driving-point (input) admittance,	66
$Y_{1,\text{in}}, Y_{n,\text{in}}$	driving-point admittances of two driven antennas in resonant array,	411
Y_{kin}	driving-point (input) admittance of element k,	88
Y_{kk}, Y_{kp}	self- and mutual admittances,	504
Y_L	admittance of load terminating line,	481
Y_L	admittance loading loop,	515
$Y_{l,\text{in}}$	driving-point admittance of element l,	386
$Y_{1,l}$	self- and mutual admittances for circular array with one driven element,	398
Y_{s1}, Y_{s2}	self-admittances,	65
Y_T	terminating admittance,	220
$Y(s)$	driving-point admittance of loop driven at point s,	515
Y_0	admittance of circular loop with constant current,	518
Y_0	admittance of square loop with constant current,	514, 517
Y_0	driving-point admittance of antenna,	2
Y_1	driving-point admittance of log-periodic dipole array,	221
Y_{12}, Y_{21}	mutual admittances,	65
y_T	normalized terminating admittance,	222
$y(w)$	admittance per unit length of line,	477
y_0	admittance per unit length of uniform line,	479
Z	impedance of monopole,	304
Z	impedance of terminated long wire,	351
$Z^{(m)}$	mth phase-sequence impedance,	95
Z_a	apparent impedance of load terminating line,	481
Z_c	characteristic impedance of line,	2, 216
Z_e	impedance of monopole with reflected image,	306
Z_g	impedance of generator,	206, 351
Z_g	impedance of ground network,	304, 306

Z_G	surface impedance of ground plane, 437
Z_{in}	driving-point impedance of grounded monopole, 304
Z_{1in}	driving-point (input) impedance, 66
Z_{kk}, Z_{kp}	self- and mutual impedances, 503
Z_L	impedance of load terminating line, 481
Z_L	load impedance for loop, 515
Z_L	load impedance of receiving antenna, 52
Z_L	lumped impedance, 358
Z_{s1}, Z_{s2}	self-impedances, 65
Z_T	terminating impedance, 216
Z_0	driving-point impedance of antenna, 2
Z_0	impedance of loop with constant current, 514
Z_0	impedance of terminating sections, 351
Z_1, Z_2	series impedances for antennas, 66
Z_{11}, Z_{22}	impedances of primary and secondary circuits, 66
Z_{12}, Z_{21}	mutual impedances, 65
z	cylindrical coordinate, 7
z'	height above source dipole, 293
z^i	internal impedance per unit length, 304, 512
z_p	width of pulse function, 24, 455
$z(w)$	impedance per unit length of line, 477
z_0	impedance per unit length of uniform line, 479
α	attenuation constant of uniform line, 479
α	imaginary part of ξ_i, 329
α_{ik}	cofactor divided by determinant, 214
α_L	attenuation constant for current in bare horizontal wire over earth or sea, 361
α_2	attenuation constant of earth or sea, 290
β	phase constant of uniform line, 479
β	real part of ξ_i, 329
β_{ik}	cofactor divided by determinant, 214
β_L	phase constant for current in bare horizontal wire over earth or sea, 361
β_0	wave number in air, 4
β_2	wave number of earth or sea, 290
Γ_a	complex apparent reflection coefficient of load terminating line, 488
$\Gamma(w)$	reflection coefficient along transmission line, 488
γ	Euler's constant, 351

γ propagation constant of uniform line, 479

γ, γ' coefficients of cross coupling for loop, 517

γ_{ik} cofactor divided by determinant, 214

$\gamma(w)$ propagation constant of line, 477

Δ determinant, 211

Δ_1, Δ_2 denominators, 169

Δs width of resonance curve, 501

δ_{ik} Kronecker delta, 118

$\delta_{m/N}$ resonant spacing for mth phase-sequence resonance, 400

$\delta(z)$ Dirac delta function, 2

ϵ small parameter, 292

$\epsilon^{(1)}, \epsilon^{(2)}$ error ratios for singly and doubly loaded loops, 517

ϵ_{dr} relative permittivity of dielectric layer, 357

$\epsilon_{1r\,\text{eff}}$ real effective permittivity of microstrip transmission line, 376

ϵ_0 permittivity of free space, 3

ζ_0 characteristic impedance of free space, 3, 10

η surface density of charge, 4

η_2 parameter for spherical earth, 329

Θ spherical coordinate, 7

Θ^i angle of incidence on earth surface in far field, 299

Θ_{hp} half-power beam width, 13

Θ_0 angle of transmission, 371

Θ_2 angle of incidence, 371

λ wavelength in free space, 4

μ_0 permeability of free space, 3

ξ_i ratio of voltages, 135

ξ_i parameter for spherical earth, 329

ρ radial distance from source dipole, 293

ρ volume density of charge, 3

ρ, ρ_a, ρ_g apparent terminal attenuation function in general, for load, and for generator, 490

ρ_c critical distance for spherical earth, 329

ρ_s distance along surface of spherical earth, 328

ρ, Φ, z cylindrical coordinates, 7

Index